Handbook of Mechanical Alloy Design

MECHANICAL ENGINEERING

A Series of Textbooks and Reference Books

Founding Editor

L. L. Faulkner

Columbus Division, Battelle Memorial Institute
and Department of Mechanical Engineering
The Ohio State University
Columbus, Ohio

1. *Spring Designer's Handbook*, Harold Carlson
2. *Computer-Aided Graphics and Design*, Daniel L. Ryan
3. *Lubrication Fundamentals*, J. George Wills
4. *Solar Engineering for Domestic Buildings*, William A. Himmelman
5. *Applied Engineering Mechanics: Statics and Dynamics*, G. Boothroyd and C. Poli
6. *Centrifugal Pump Clinic*, Igor J. Karassik
7. *Computer-Aided Kinetics for Machine Design*, Daniel L. Ryan
8. *Plastics Products Design Handbook, Part A: Materials and Components; Part B: Processes and Design for Processes*, edited by Edward Miller
9. *Turbomachinery: Basic Theory and Applications,* Earl Logan, Jr.
10. *Vibrations of Shells and Plates*, Werner Soedel
11. *Flat and Corrugated Diaphragm Design Handbook,* Mario Di Giovanni
12. *Practical Stress Analysis in Engineering Design,* Alexander Blake
13. *An Introduction to the Design and Behavior of Bolted Joints,* John H. Bickford
14. *Optimal Engineering Design: Principles and Applications,* James N. Siddall
15. *Spring Manufacturing Handbook,* Harold Carlson
16. *Industrial Noise Control: Fundamentals and Applications,* edited by Lewis H. Bell
17. *Gears and Their Vibration: A Basic Approach to Understanding Gear Noise,* J. Derek Smith
18. *Chains for Power Transmission and Material Handling: Design and Applications Handbook,* American Chain Association
19. *Corrosion and Corrosion Protection Handbook,* edited by Philip A. Schweitzer
20. *Gear Drive Systems: Design and Application,* Peter Lynwander
21. *Controlling In-Plant Airborne Contaminants: Systems Design and Calculations,* John D. Constance
22. *CAD/CAM Systems Planning and Implementation,* Charles S. Knox
23. *Probabilistic Engineering Design: Principles and Applications,* James N. Siddall
24. *Traction Drives: Selection and Application,* Frederick W. Heilich III and Eugene E. Shube
25. *Finite Element Methods: An Introduction,* Ronald L. Huston and Chris E. Passerello
26. *Mechanical Fastening of Plastics: An Engineering Handbook,* Brayton Lincoln, Kenneth J. Gomes, and James F. Braden
27. *Lubrication in Practice: Second Edition,* edited by W. S. Robertson
28. *Principles of Automated Drafting,* Daniel L. Ryan
29. *Practical Seal Design,* edited by Leonard J. Martini
30. *Engineering Documentation for CAD/CAM Applications,* Charles S. Knox
31. *Design Dimensioning with Computer Graphics Applications,* Jerome C. Lange
32. *Mechanism Analysis: Simplified Graphical and Analytical Techniques,* Lyndon O. Barton
33. *CAD/CAM Systems: Justification, Implementation, Productivity Measurement,* Edward J. Preston, George W. Crawford, and Mark E. Coticchia
34. *Steam Plant Calculations Manual,* V. Ganapathy
35. *Design Assurance for Engineers and Managers,* John A. Burgess

36. *Heat Transfer Fluids and Systems for Process and Energy Applications,* Jasbir Singh
37. *Potential Flows: Computer Graphic Solutions,* Robert H. Kirchhoff
38. *Computer-Aided Graphics and Design: Second Edition,* Daniel L. Ryan
39. *Electronically Controlled Proportional Valves: Selection and Application,* Michael J. Tonyan, edited by Tobi Goldoftas
40. *Pressure Gauge Handbook,* AMETEK, U.S. Gauge Division, edited by Philip W. Harland
41. *Fabric Filtration for Combustion Sources: Fundamentals and Basic Technology,* R. P. Donovan
42. *Design of Mechanical Joints,* Alexander Blake
43. *CAD/CAM Dictionary,* Edward J. Preston, George W. Crawford, and Mark E. Coticchia
44. *Machinery Adhesives for Locking, Retaining, and Sealing,* Girard S. Haviland
45. *Couplings and Joints: Design, Selection, and Application,* Jon R. Mancuso
46. *Shaft Alignment Handbook,* John Piotrowski
47. *BASIC Programs for Steam Plant Engineers: Boilers, Combustion, Fluid Flow, and Heat Transfer,* V. Ganapathy
48. *Solving Mechanical Design Problems with Computer Graphics,* Jerome C. Lange
49. *Plastics Gearing: Selection and Application,* Clifford E. Adams
50. *Clutches and Brakes: Design and Selection,* William C. Orthwein
51. *Transducers in Mechanical and Electronic Design,* Harry L. Trietley
52. *Metallurgical Applications of Shock-Wave and High-Strain-Rate Phenomena,* edited by Lawrence E. Murr, Karl P. Staudhammer, and Marc A. Meyers
53. *Magnesium Products Design,* Robert S. Busk
54. *How to Integrate CAD/CAM Systems: Management and Technology,* William D. Engelke
55. *Cam Design and Manufacture: Second Edition; with cam design software for the IBM PC and compatibles, disk included*, Preben W. Jensen
56. *Solid-State AC Motor Controls: Selection and Application,* Sylvester Campbell
57. *Fundamentals of Robotics,* David D. Ardayfio
58. *Belt Selection and Application for Engineers,* edited by Wallace D. Erickson
59. *Developing Three-Dimensional CAD Software with the IBM PC,* C. Stan Wei
60. *Organizing Data for CIM Applications,* Charles S. Knox, with contributions by Thomas C. Boos, Ross S. Culverhouse, and Paul F. Muchnicki
61. *Computer-Aided Simulation in Railway Dynamics,* by Rao V. Dukkipati and Joseph R. Amyot
62. *Fiber-Reinforced Composites: Materials, Manufacturing, and Design,* P. K. Mallick
63. *Photoelectric Sensors and Controls: Selection and Application,* Scott M. Juds
64. *Finite Element Analysis with Personal Computers,* Edward R. Champion, Jr., and J. Michael Ensminger
65. *Ultrasonics: Fundamentals, Technology, Applications: Second Edition, Revised and Expanded,* Dale Ensminger
66. *Applied Finite Element Modeling: Practical Problem Solving for Engineers,* Jeffrey M. Steele
67. *Measurement and Instrumentation in Engineering: Principles and Basic Laboratory Experiments,* Francis S. Tse and Ivan E. Morse
68. *Centrifugal Pump Clinic: Second Edition, Revised and Expanded,* Igor J. Karassik
69. *Practical Stress Analysis in Engineering Design: Second Edition, Revised and Expanded,* Alexander Blake
70. *An Introduction to the Design and Behavior of Bolted Joints: Second Edition, Revised and Expanded,* John H. Bickford
71. *High Vacuum Technology: A Practical Guide,* Marsbed H. Hablanian
72. *Pressure Sensors: Selection and Application,* Duane Tandeske
73. *Zinc Handbook: Properties, Processing, and Use in Design,* Frank Porter
74. *Thermal Fatigue of Metals,* Andrzej Weronski and Tadeusz Hejwowski
75. *Classical and Modern Mechanisms for Engineers and Inventors,* Preben W. Jensen
76. *Handbook of Electronic Package Design,* edited by Michael Pecht
77. *Shock-Wave and High-Strain-Rate Phenomena in Materials*, edited by Marc A. Meyers, Lawrence E. Murr, and Karl P. Staudhammer
78. *Industrial Refrigeration: Principles, Design and Applications*, P. C. Koelet

79. *Applied Combustion*, Eugene L. Keating
80. *Engine Oils and Automotive Lubrication*, edited by Wilfried J. Bartz
81. *Mechanism Analysis: Simplified and Graphical Techniques, Second Edition, Revised and Expanded*, Lyndon O. Barton
82. *Fundamental Fluid Mechanics for the Practicing Engineer*, James W. Murdock
83. *Fiber-Reinforced Composites: Materials, Manufacturing, and Design, Second Edition, Revised and Expanded*, P. K. Mallick
84. *Numerical Methods for Engineering Applications*, Edward R. Champion, Jr.
85. *Turbomachinery: Basic Theory and Applications, Second Edition, Revised and Expanded*, Earl Logan, Jr.
86. *Vibrations of Shells and Plates: Second Edition, Revised and Expanded*, Werner Soedel
87. *Steam Plant Calculations Manual: Second Edition, Revised and Expanded*, V. Ganapathy
88. *Industrial Noise Control: Fundamentals and Applications, Second Edition, Revised and Expanded*, Lewis H. Bell and Douglas H. Bell
89. *Finite Elements: Their Design and Performance*, Richard H. MacNeal
90. *Mechanical Properties of Polymers and Composites: Second Edition, Revised and Expanded*, Lawrence E. Nielsen and Robert F. Landel
91. *Mechanical Wear Prediction and Prevention*, Raymond G. Bayer
92. *Mechanical Power Transmission Components*, edited by David W. South and Jon R. Mancuso
93. *Handbook of Turbomachinery*, edited by Earl Logan, Jr.
94. *Engineering Documentation Control Practices and Procedures*, Ray E. Monahan
95. *Refractory Linings: Thermomechanical Design and Applications*, Charles A. Schacht
96. *Geometric Dimensioning and Tolerancing: Applications and Techniques for Use in Design, Manufacturing, and Inspection*, James D. Meadows
97. *An Introduction to the Design and Behavior of Bolted Joints: Third Edition, Revised and Expanded*, John H. Bickford
98. *Shaft Alignment Handbook: Second Edition, Revised and Expanded*, John Piotrowski
99. *Computer-Aided Design of Polymer-Matrix Composite Structures*, edited by S. V. Hoa
100. *Friction Science and Technology*, Peter J. Blau
101. *Introduction to Plastics and Composites: Mechanical Properties and Engineering Applications*, Edward Miller
102. *Practical Fracture Mechanics in Design*, Alexander Blake
103. *Pump Characteristics and Applications*, Michael W. Volk
104. *Optical Principles and Technology for Engineers*, James E. Stewart
105. *Optimizing the Shape of Mechanical Elements and Structures*, A. A. Seireg and Jorge Rodriguez
106. *Kinematics and Dynamics of Machinery*, Vladimír Stejskal and Michael Valášek
107. *Shaft Seals for Dynamic Applications*, Les Horve
108. *Reliability-Based Mechanical Design*, edited by Thomas A. Cruse
109. *Mechanical Fastening, Joining, and Assembly*, James A. Speck
110. *Turbomachinery Fluid Dynamics and Heat Transfer*, edited by Chunill Hah
111. *High-Vacuum Technology: A Practical Guide, Second Edition, Revised and Expanded*, Marsbed H. Hablanian
112. *Geometric Dimensioning and Tolerancing: Workbook and Answerbook*, James D. Meadows
113. *Handbook of Materials Selection for Engineering Applications*, edited by G. T. Murray
114. *Handbook of Thermoplastic Piping System Design*, Thomas Sixsmith and Reinhard Hanselka
115. *Practical Guide to Finite Elements: A Solid Mechanics Approach*, Steven M. Lepi
116. *Applied Computational Fluid Dynamics*, edited by Vijay K. Garg
117. *Fluid Sealing Technology*, Heinz K. Muller and Bernard S. Nau
118. *Friction and Lubrication in Mechanical Design*, A. A. Seireg
119. *Influence Functions and Matrices*, Yuri A. Melnikov
120. *Mechanical Analysis of Electronic Packaging Systems*, Stephen A. McKeown

121. *Couplings and Joints: Design, Selection, and Application, Second Edition, Revised and Expanded,* Jon R. Mancuso
122. *Thermodynamics: Processes and Applications*, Earl Logan, Jr.
123. *Gear Noise and Vibration*, J. Derek Smith
124. *Practical Fluid Mechanics for Engineering Applications,* John J. Bloomer
125. *Handbook of Hydraulic Fluid Technology*, edited by George E. Totten
126. *Heat Exchanger Design Handbook*, T. Kuppan
127. *Designing for Product Sound Quality*, Richard H. Lyon
128. *Probability Applications in Mechanical Design*, Franklin E. Fisher and Joy R. Fisher
129. *Nickel Alloys,* edited by Ulrich Heubner
130. *Rotating Machinery Vibration: Problem Analysis and Troubleshooting*, Maurice L. Adams, Jr.
131. *Formulas for Dynamic Analysis*, Ronald Huston and C. Q. Liu
132. *Handbook of Machinery Dynamics*, Lynn L. Faulkner and Earl Logan, Jr.
133. *Rapid Prototyping Technology: Selection and Application*, Ken Cooper
134. *Reciprocating Machinery Dynamics: Design and Analysis*, Abdulla S. Rangwala
135. *Maintenance Excellence: Optimizing Equipment Life-Cycle Decisions,* edited by John D. Campbell and Andrew K. S. Jardine
136. *Practical Guide to Industrial Boiler Systems*, Ralph L. Vandagriff
137. *Lubrication Fundamentals: Second Edition, Revised and Expanded,* D. M. Pirro and A. A. Wessol
138. *Mechanical Life Cycle Handbook: Good Environmental Design and Manufacturing,* edited by Mahendra S. Hundal
139. *Micromachining of Engineering Materials,* edited by Joseph McGeough
140. *Control Strategies for Dynamic Systems: Design and Implementation,* John H. Lumkes, Jr.
141. *Practical Guide to Pressure Vessel Manufacturing,* Sunil Pullarcot
142. *Nondestructive Evaluation: Theory, Techniques, and Applications,* edited by Peter J. Shull
143. *Diesel Engine Engineering: Dynamics, Design, and Control,* Andrei Makartchouk
144. *Handbook of Machine Tool Analysis,* Ioan D. Marinescu, Constantin Ispas, and Dan Boboc
145. *Implementing Concurrent Engineering in Small Companies,* Susan Carlson Skalak
146. *Practical Guide to the Packaging of Electronics: Thermal and Mechanical Design and Analysis*, Ali Jamnia
147. *Bearing Design in Machinery: Engineering Tribology and Lubrication,* Avraham Harnoy
148. *Mechanical Reliability Improvement: Probability and Statistics for Experi-mental Testing,* R. E. Little
149. *Industrial Boilers and Heat Recovery Steam Generators: Design, Applications, and Calculations*, V. Ganapathy
150. *The CAD Guidebook: A Basic Manual for Understanding and Improving Computer-Aided Design,* Stephen J. Schoonmaker
151. *Industrial Noise Control and Acoustics,* Randall F. Barron
152. *Mechanical Properties of Engineering Materials,* Wolé Soboyejo
153. *Reliability Verification, Testing, and Analysis in Engineering Design*, Gary S. Wasserman
154. *Fundamental Mechanics of Fluids: Third Edition*, I. G. Currie
155. *Intermediate Heat Transfer*, Kau-Fui Vincent Wong
156. *HVAC Water Chillers and Cooling Towers: Fundamentals, Application, and Operations*, Herbert W. Stanford III
157. *Gear Noise and Vibration: Second Edition, Revised and Expanded,* J. Derek Smith
158. *Handbook of Turbomachinery: Second Edition, Revised and Expanded*, Earl Logan, Jr., and Ramendra Roy
159. *Piping and Pipeline Engineering: Design, Construction, Maintenance, Integrity, and Repair,* George A. Antaki
160. *Turbomachinery: Design and Theory*, Rama S. R. Gorla and Aijaz Ahmed Khan

161. *Target Costing: Market-Driven Product Design,* M. Bradford Clifton, Henry M. B. Bird, Robert E. Albano, and Wesley P. Townsend
162. *Fluidized Bed Combustion,* Simeon N. Oka
163. *Theory of Dimensioning: An Introduction to Parameterizing Geometric Models,* Vijay Srinivasan
164. *Handbook of Mechanical Alloy Design*, George E. Totten, Lin Xie, and Kiyoshi Funatani
165. *Structural Analysis of Polymeric Composite Materials,* Mark E. Tuttle

Additional Volumes in Preparation

Handbook of Pneumatic Conveying Engineering, David Mills, Mark G. Jones, and Vijay K. Agarwal

Mechanical Wear Fundamentals and Testing: Second Edition, Revised and Expanded, Raymond G. Bayer

Engineering Design for Wear: Second Edition, Revised and Expanded, Raymond G. Bayer

Clutches and Brakes: Design and Selection, Second Edition, William C. Orthwein

Progressing Cavity Pumps, Downhole Pumps, and Mudmotors, Lev Nelik

Mechanical Engineering Software

Spring Design with an IBM PC, Al Dietrich

Mechanical Design Failure Analysis: With Failure Analysis System Software for the IBM PC, David G. Ullman

Handbook of Mechanical Alloy Design

edited by

George E. Totten
G.E. Totten & Associates, LLC
Seattle, Washington, U.S.A.

Lin Xie
SolidWorks Corporation
Concord, Massachusettes, U.S.A.

Kiyoshi Funatani
IMST Institute
Nagoya, Japan

MARCEL DEKKER

MARCEL DEKKER, INC. NEW YORK • BASEL

Library of Congress Cataloging-in-Publication Data
A catalog record for this book is available from the Library of Congress.

ISBN: 0-8247-4308-3

This book is printed on acid-free paper.

Headquarters
Marcel Dekker, Inc., 270 Madison Avenue, New York, NY 10016, U.S.A.
tel: 212-696-9000; fax: 212-685-4540

Distribution and Customer Service
Marcel Dekker, Inc., Cimarron Road, Monticello, New York 12701, U.S.A.
tel: 800-228-1160; fax: 845-796-1772

Eastern Hemisphere Distribution
Marcel Dekker AG, Hutgasse 4, Postfach 812, CH-4001 Basel, Switzerland
tel: 41-61-260-6300; fax: 41-61-260-6333

World Wide Web
http://www.dekker.com

The publisher offers discounts on this book when ordered in bulk quantities. For more information, write to Special Sales/Professional Marketing at the headquarters address above.

Current printing (last digit):

10 9 8 7 6 5 4 3 2 1

PRINTED IN THE UNITED STATES OF AMERICA

Preface

Selection of appropriate materials for mechanical design is an ever-increasing challenge as performance demands continue to push the limits of the mechanical properties that can be achieved with available materials. Also, the range of materials being considered for design applications continues to increase beyond a single material type. For example, in automobile production, the weight/strength properties for steel, cast iron, aluminum, magnesium, metal-matrix composites, and many other materials are vitally important considerations during the initial design process for the various production components that are utilized in manufacturing and assembly.

This book focuses on alloy composition versus microstructural and mechanical properties of a wide range of metallic materials that are typically involved in the mechanical engineering process. The text begins with a fundamental overview of the elements of the design process and a discussion of the particularly important risk-based design process. The metal alloys discussed include: carbon, low- and medium-alloy, tool, stainless, high-strength low-alloy (HSLA), microalloyed and other steels, cast iron, aluminum, titanium, magnesium, nickel-based alloys, copper, metal-matrix composites, and powder metal alloys.

The objective of this text is to provide the mechanical, metallurgical and materials engineer, and the designer with an in-depth reference on the microstructure and mechanical property design features as a function of alloy composition. This book is suitable as an advanced undergraduate and graduate level textbook. The breadth and depth of treatment of the subject makes this text an invaluable long-term reference book on the topic of mechanical alloy design. This is one of the most thorough and integrated treatments of the effect of alloy material composition on mechanical properties available anywhere.

To produce a text of this scope and magnitude is an enormous task. My coeditors and I are indebted to the persistence and thorough work of all the contributors to this volume. We are especially grateful for the invaluable assistance provided by the staff at Marcel Dekker, Inc. throughout the preparation of this book.

George E. Totten
Lin Xie
Kiyoshi Funatani

Contents

Contributors

Trevor B. Abbott, Ph.D. Monash University, Clayton, Victoria, Australia

Carlos H. Cáceres, Ph.D. The University of Queensland, Brisbane, Queensland, Australia

David N. Collins, M.Sc., Dip. Met., C.Eng., M.I.M. University College Dublin, Dublin, Ireland

Mark A. Easton, Ph.D. Monash University, Clayton, Victoria, Australia

David V. Edmonds, Ph.D., F.R.Eng. University of Leeds, Leeds, England

John Fanning, M.S. TIMET Henderson Technical Laboratory, Henderson, Nevada, U.S.A.

Morris E. Fine, B.Met.E., M.S., Ph.D. Northwestern University, Evanston, Illinois, U.S.A.

Gerhard E. Fuchs, Ph.D. University of Florida, Gainesville, Florida, U.S.A.

David U. Furrer, Dr.Eng. Ladish Company, Inc., Cudahy, Wisconsin, U.S.A.

J. T. Hellman, M.Sc. (Tech.) Helsinki University of Technology, Espoo, Finland

Nack J. Kim, Ph.D. Pohang University of Science and Technology, Pohang, Korea

Ronald A. Kohser, Ph.D. University of Missouri-Rolla, Rolla, Missouri, U.S.A.

Joseph Ki Leuk Lai, M.A. (Oxon.), Ph.D. City University of Hong Kong, Kowloon, Hong Kong, China

Lin Li, Ph.D. Shanghai University, Shanghai, China

Veikko K. Lindroos, D.Sc. (Tech.) Helsinki University of Technology, Espoo, Finland

Guoxun Liu, B.S., M.S. University of Science and Technology, Beijing, China

X. W. Liu, M.Sc. (Tech.) Helsinki University of Technology, Espoo, Finland

D. Lou, Ph.D. Helsinki University of Technology, Espoo, Finland

Michelle L. McCann, B.S., M.E. TIMET Henderson Technical Laboratory, Henderson, Nevada, U.S.A.

Junji Miyake, B.S., M.E., Ph.D. Nippon Mining and Metals Company, Ltd., Koza, Kanagawa, Japan

Joseph W. Newkirk, Ph.D. University of Missouri-Rolla, Rolla, Missouri, U.S.A.

R. Nowak, Ph.D. Helsinki University of Technology, Espoo, Finland

E. Pagounis, D.Sc. (Tech.) Helsinki University of Technology, Espoo, Finland

I. M. Penttinen, Lic. Sc. (Tech.) Helsinki University of Technology, Espoo, Finland

Mario Solari, Dr.Ing. Consejo Nacional de Investigaciones Científicas y Técnicas, Buenos Aires, Argentina

Henry W. Stoll, Ph.D. Northwestern University, Evanston, Illinois, U.S.A.

Ingvar L. Svensson, Prof. Dr.Ing. Jöngköping University, Jöngköping, Sweden

Magnus Wessén, Ph.D. Jöngköping University, Jöngköping, Sweden

Luoping Xu, B.S. Shanghai University, Shanghai, China

Lin-Jiang Yang, Ph.D. Nanyang Technological University, Singapore, Republic of Singapore

1

Basic Principles

Henry W. Stoll
Northwestern University, Evanston, Illinois, U.S.A.

I. INTRODUCTION

The purpose of this chapter is to provide a context for metallurgical design by presenting basic design principles and concepts that apply, in general, to the creation and improvement of products and equipment. At this level, material properties are one of many aspects of the product or equipment design. Similarly, material selection is one of many steps involved in the design process. Later chapters of this book focus more particularly on specific aspects of metallurgical design.

Although the principles discussed in this chapter are generally applicable to the full spectrum of design including the design of systems, products, software, and services, the chapter focus is on engineered designs. Engineered designs are considered to be finished products, equipment, devices, and hardware that have been designed to meet specific end-user needs and are sold by an enterprise to customers. In this context, computers, electrical circuit breakers, cell phones, automobiles, machine tools, construction equipment, consumer products, and aircraft are all considered to be engineered designs.

We begin our discussion by briefly reviewing a variety of considerations that pertain to design in general, and that provide insight for much of the discussion that follows.

II. GENERAL CONSIDERATIONS

A. Definition

When used as a noun, the term "design" refers to the *plan* or arrangement of elements or details in a physical entity, such as a material, a component, a finished product, or a machine. When used as a verb, the term "design" refers to the *process* by which the plan for the physical entity is created or improved. Hence, design is both the plan and the process for making the plan.

B. Clients and Customers

Designs are created by clients to meet customer needs. The *client* is typically a manufacturing enterprise seeking to market and sell products, a developer or business seeking to provide a service or facility, or a government authority responsible for providing physical infrastructure or military hardware. Clients pay for the design either by providing money or by providing resources needed to create the design. To be acceptable to the client, the design must generate a reasonable profit or offer a satisfactory cost/benefit ratio.

Customers use the design and benefit from its functionality, appearance, ease of use, reliability, long life, safety, and so forth. To be acceptable to the customer, the design must meet well-defined and clearly understood customer needs in a way that delights the customer and insures long-term, sustainable customer satisfaction. A particular design can have many customers. For example, the customers for a simple consumer product not only include the end user, but also all those concerned with producing, distributing, marketing, selling, and ultimately disposing or recycling the product. This includes the retail store that displays and sells the product, the shipping company that transports the product from the manufacturing plant to the store, and the workers who manufacture the parts and assemble them into the finished product. A well-designed product will consider the needs of all customers involved in the production–consumption cycle for the product. We call this *customer-focused design*.

C. Viewpoints

The design of modern products and industrial equipment generally involves several different points of view that must all be considered from the beginning of the design. For example, *industrial design* is concerned with the external appearance of the design and with how it interacts with the user. The primary goal of industrial design is to create demand. This is done in various ways, such as by creating an emotional feeling for the design, making ownership and use a pleasing experience, establishing and managing brand identification, and so forth.

Engineering design, on the other hand, is concerned with the "internals" of the design including how it works, the form, fit, and function of individual components and assemblies, material selection and properties, and cost. The primary goal of engineering design is to develop and optimize the working principle and functionality of the design while also insuring that goals related to quality, reliability, safety, cost, and so forth are met.

Another point of view is that of the manufacturing engineer who is concerned with how the individual parts are to be made and assembled. This viewpoint, which has become widely known as design for manufacture (DFM), focuses on geometric details of individual parts and assemblies to insure ease of manufacture and assembly. Often, DFM is the key for harmonizing client goals and customer needs.

To insure proper consideration of all points of view, all stakeholders in the design must work together from the beginning. This means that all aspects of the design, including market analysis, process planning, facilities planning, manufacturing and automation equipment design and procurement, tooling design and procurement, and procurement of supplied parts, must be performed in a concurrent or overlapping manner as an integral part of the design process. This concept, which involves early and constant interaction between all stakeholders in the project, has become widely known as *concurrent engineering* or *simultaneous engineering*.

D. Standard and Designed Components

Most designs are combinations of standard and designed components. A standard component is usually supplied by a vendor and is used interchangeably in a variety of products or design applications. Examples include electric motors, light bulbs, electrical connectors, and mechanical fasteners. A designed component is a unique part or subassembly that must be specifically designed and fabricated as part of the design being developed. Choosing between designed and standard components can be an important consideration. For example, the choice between designing a special purpose component, optimized for performance and weight, or purchasing a standard, "off-the-shelf" component from a supplier, can have far-reaching performance, cost, quality, and timing consequences.

E. Design Situations

Problems of design are not all the same and each must be approached according to the unique circumstances that define it. This can be illustrated by considering four situations in mechanical engineering design that seem to be the corners of a field in which most mechanical design problems are located [1]. They can be called: (1) the in-house tool, (2) the mass product, (3) the megabuck machine, and (4) the code design. An engineer given the job of designing an "in-house" tool will probably think about it, sketch several possible solutions, select one by intuition or with the help of some quick analysis, and proceed to sketch the idea and have it built by the shop. In this case, the engineer does little in the way of formal analysis, optimization, or documentation and relies on his ability to quickly fix any problem that may arise.

The design team in charge of a mass-produced consumer product, on the other hand, will probably expend significant efforts to understand customer needs, build several prototypes, test them extensively with customers and for functionality, and document the solution in great detail using a CAD system. The designer of a large-scale oil drilling platform cannot afford functional prototypes and must learn as much as possible from existing field data and from the use of analysis even if it is complex and expensive. A *code* is a set of specifications for the analysis, design, manufacture, and construction of something. Hence, the designer of a boiler must pay more attention to codes and may be more concerned with legal interpretation of the code and calculation of stresses to three decimal places than he is with customer needs or industrial design.

Most design problems fall between these extremes. An automobile is a mass product, but subject to various codes. The design of a personal computer, a satellite, or of an oil refinery each proceeds in their own way. All are within the general principles discussed in this chapter, but each is very different from the others.

F. Cycles in Design

There may be multiple design–build–test cycles involved in the creation of a design (Fig. 1). In the case of one of the products, such as oil drilling platforms and chemical processing plants, generally only one cycle occurs. It is typically too risky to develop batch and mass-produced products in one cycle. In these cases, a first round of planning, design, manufacture, and assembly is undertaken to produce a model, which when tested may indicate ways of improving the design. This information, along with possibly extensive computer-aided analysis and optimization, is incorporated in a new cycle either to manufacture an improved model for further tests (dashed line in Fig. 1), or to manufacture a prototype. It may then be worthwhile for still further refinements to be made, especially with respect to manufactureability of the design, before finally putting the finished product into full-scale production.

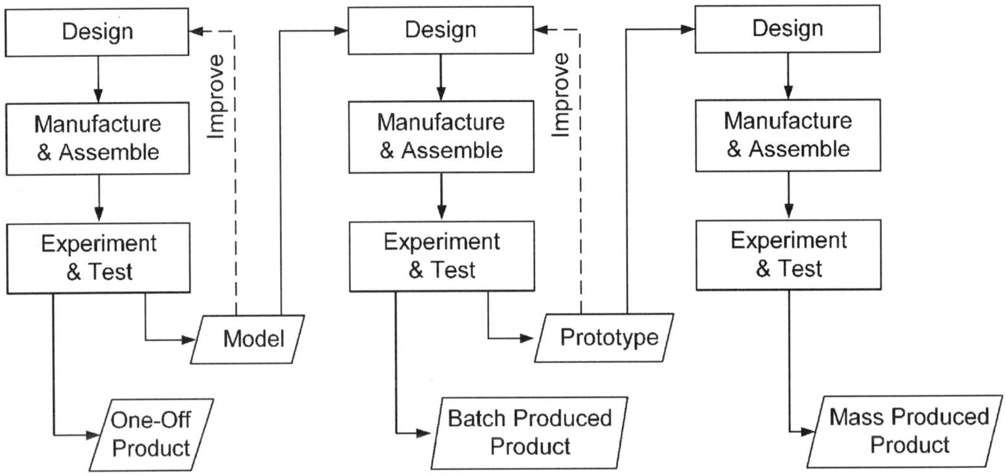

Figure 1 Cycles in product creation. (From Ref. 2.)

G. Design Project Anchors

In most design situations, many alternative design directions are possible. The design direction that is selected often depends on the over-riding concerns that constrain or anchor the project. Major concerns or *design project anchors* include manufacturing cost, product performance, development time, and development cost. One or more of these concerns anchors most design situations. For example, many high volume mass-produced products, especially commodity-type products, are anchored by manufacturing cost. Hence, design decisions are made to minimize manufacturing cost while still meeting threshold performance and quality standards and expectations.

In other products, such as aircraft and some military hardware, performance is the key design consideration and design decisions are made to insure acceptable performance regardless of cost. In still other cases, such as computers and some consumer products, time to market is critical. Although reduced part count is always important, it is not unusual when time to market is critical for the designer to make design decisions that lead to numerous easy-to-tool parts rather than fewer parts that involve long tooling lead times. Development cost can also anchor a project, especially for start-up companies that must develop and introduce new products on a very limited budget.

In most situations, the design is anchored by more than one over-riding concern. For example, a sport motorcycle must be lightweight and meet high performance goals, and at the same time, it must also be mass produced for low cost and must be introduced into the market in a timely manner. It is interesting to note that experience has shown that it is possible to make design decisions that are constrained by three of the anchors, but not all four. That is, a company can probably design a successful sport motorcycle that meets performance goals and cost targets in a timely manner, but to do so, design decisions cannot also be constrained to minimize development cost. At least one of the four anchors has to give.

III. OVERVIEW OF THE DESIGN PROCESS

Design is a process. That is, design is a series of steps that often must be repeated iteratively and that progress from the abstract to the concrete. The activities involved in this progression can often be divided into a time sequence of phases (Fig. 2). As part of each phase of the design process, many questions of design must be resolved and technical and economic decisions made. These decisions generally require a great deal of information and the quality of such decisions often depends directly on the completeness, correctness, and availability of needed information. If the required information is not available, the designer or design team makes the best decision possible and then re-examine the decision at a later date as more complete information becomes available. This process of re-examination is the iterative nature of design.

A. Engineering Design

Engineering design typically involves the following design activities:

1. Clarify and define the requirements of the design problem to be solved.
2. Develop a working principle or physical concept for solving the design problem.
3. Determine the geometric arrangement (layout) of components and establish dimensional relationships between components.
4. Decide which components are standard and which need to be designed. A *designed* component is a unique part or assembly that must be defined as part of the design. A *standard* component, on the other hand, is a supplied or "off-the-shelf" component, such as an electric motor, light bulb, electrical connector, or mechanical fastener.
5. Select the general type of material (e.g., polymer, metal) and basic manufacturing process (e.g., casting, machining) to be used for each designed component, if not already determined.
6. Determine the configuration (i.e., size, shape, external and internal geometric features, etc.) of each designed component.
7. Select a specific material and manufacturing process for each designed component.
8. Establish dimensions and tolerances for each designed component.
9. Supply additional dimensions, tolerances, and detailed information required for manufacture and assembly of the components.

These activities proceed from the general to the specific and are typically performed in the order listed when developing a new design.

The engineering design process begins by conceiving a physical concept for the design and then creating a preliminary layout of the design that embodies the physical concept (Fig. 3). This initial phase is often referred to as the conceptual design stage and typically involves activities 1–5 listed above. The preliminary layout represents a conceptual arrangement of parts that implement the physical concept and working principle of the design. It is preliminary because, at most, only key dimensions and relationships between parts have been specified; the actual size, shape, and detail features of the parts are, as yet, either undefined or only partially defined.

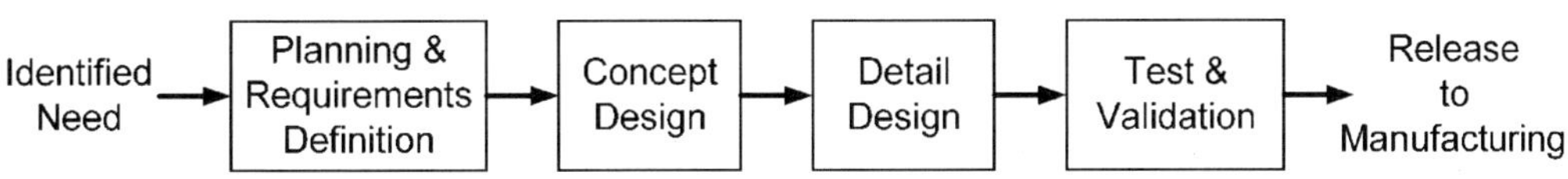

Figure 2 Typical phases of design.

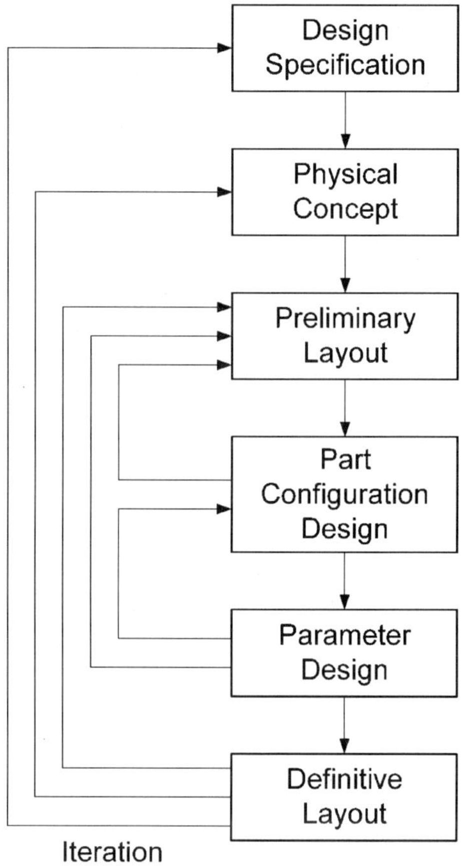

Figure 3 The engineering design process.

The preliminary layout is then developed into a completed design by developing a configuration design and parametric design for each designed component. Configuration design involves determining the size, shape, and detail features of the designed components (activities 5 and 6), while parametric design involves assigning specific material properties, dimensions, and tolerances (activities 7–9). Collectively, these design activities are often referred to as the detail design stage. During detail design, the preliminary layout changes and evolves as questions are answered and uncertainties resolved. The end result is the definitive layout or final design. The definitive layout contains the design information required to fabricate and assemble the parts.

B. Iterative Nature of Design

Design iteration can be modeled as a design, analyze, redesign process in which design solutions are proposed, evaluated and modified until an acceptable solution is achieved (Fig. 4). The process begins with a *design problem*, such as a specification for a design that needs to be created or a particular problem that needs to be solved. Using both general information and specific information about the design problem, together with design knowledge gained through past experience, the designer or design team first analyzes the problem to find the best way to approach the design, and then, based on this analysis and the insights gained, generates an initial design. The team then evaluates this candidate solution using the best available engineering practices and methods. Based on this evaluation, the team makes a judgment regarding the acceptability of the design. If the candidate design is unacceptable in one or more ways, the design solution is modified in an attempt to correct the identified shortcomings. This is the redesign step. The redesign is then evaluated and the process repeated until either an acceptable design solution is found, or the team concludes that the design problem cannot be solved as formulated.

Design iteration occurs on all levels of design and during all phases of design. For example, consider the detail design of an engineered component, such as a casting or plastic injection molded part. As shown in Fig. 5, the design engineer goes through an iterative design process to select the material and specify the component geometry. This geometry is then passed on to the manufacturing or tooling engineer who repeats the iterative design process to specify the tooling and process design. Problems discovered during this stage generate additional iterations if component geometry changes are required. Additional iterations to the component geometry and tool design may also be required during tool fabrication and preparation for production of the first article. Finally, iterative changes

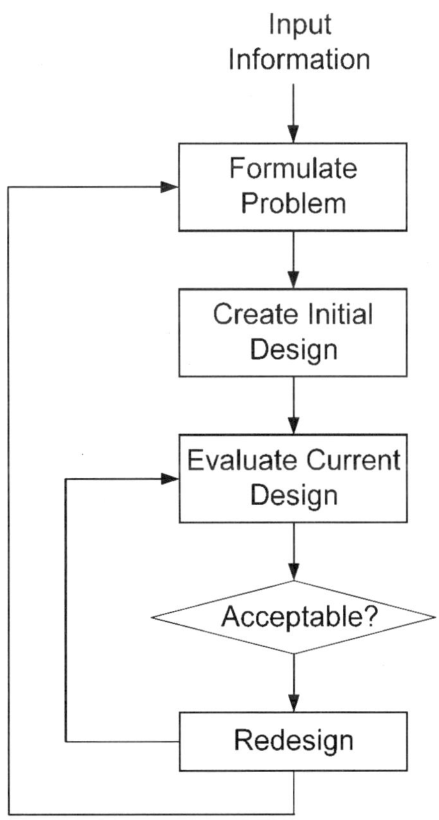

Figure 4 Iterative design process.

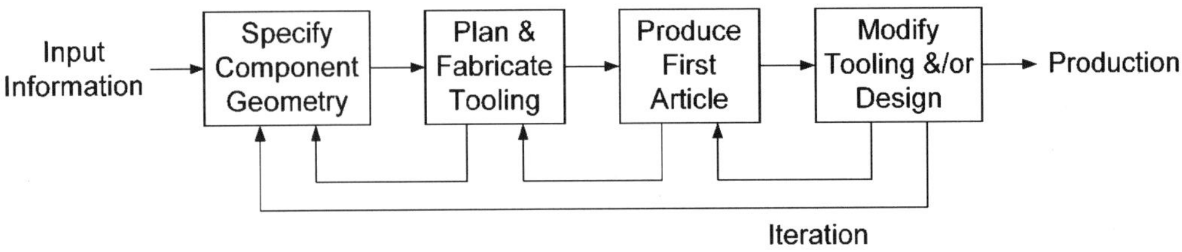

Figure 5 Iterations in component design.

to the tooling and perhaps the component geometry may be necessary to "tweak" the component design to meet production requirements.

C. Design Process Best Practice Recommendations

Design iteration, which is an inherent characteristic of the design process, is best performed in the early stages of design when conceptual maneuverability is wide and hardware is still remote. Excessive design iteration, especially if performed late in the design process, can significantly increase design cost and time. Most importantly, design iterations performed late in the process can lead to suboptimal design. The following best practices are recommended as a means for avoiding excessive design iteration in the later stages of the design process [3]:

1. Design component geometry, tooling, and manufacturing process as a coordinated system in one concurrent process. Consider geometry, material, and process interactions and design-related cost drivers early in the engineering design process.
2. Develop a thorough understanding of all customer needs including downstream processing constraints before beginning component design.
3. Focus on creating an acceptable initial design. By spending the time "up front" to create the best possible initial design, a large number of lengthy analyze–redesign iterations are avoided. The evaluation phase should confirm the design rather than create it.
4. Use manufacturing process simulation software and other modern computer-aided analysis and inspection methods to quickly optimize the design.
5. Develop a consistent, well-defined "science base" for component design by developing design guidelines and structured methodologies for each core manufacturing process or method used by the firm.

These recommendations are, in fact, the essence of a sound DFM approach to design. Design for manufacture is discussed in more detail later in this chapter.

IV. CUSTOMER-FOCUSED DESIGN

Customer-focused design is the recognition that, to be successful, the design must delight and satisfy the customers of the design in a way that creates demand for the design and sustains it economically for the long term. Most designs have several different customers. These include not only the primary and secondary end users of the design, but also a multitude of others that come in contact with the design, such as regulatory agencies, manufacturing representatives and distributors, installation, service and maintenance personnel, and so forth.

The key to customer-focused design is the systematic identification of customers and customer needs. Some proven practices for doing this include the following [3]:

- Each activity in the production–consumption cycle of the design (Fig. 6) involves different customers of the design. Study these activities to identify customers and their needs.
- Observe the use of environment. Look for contextual behavior and coping mechanisms. "Feel the pain" associated with using current products.
- Watch more than ask. Interview real customers rather than relying on focus groups or surveys. Bring "props" to user interviews. Props help to stimulate discussion about needs and can often reveal opportunities. Useful props include existing products, a prototype or mockup of an idea, or any one of many other items related to the use environment.
- Never make assumptions about customer needs or their importance. Test all ideas and concepts with real customers.

As revealed by the above suggestions, the goal of customer-focused design is the development of a comprehensive understanding of customers and their needs. This understanding is gained by observing customers, experimenting with new ideas through the

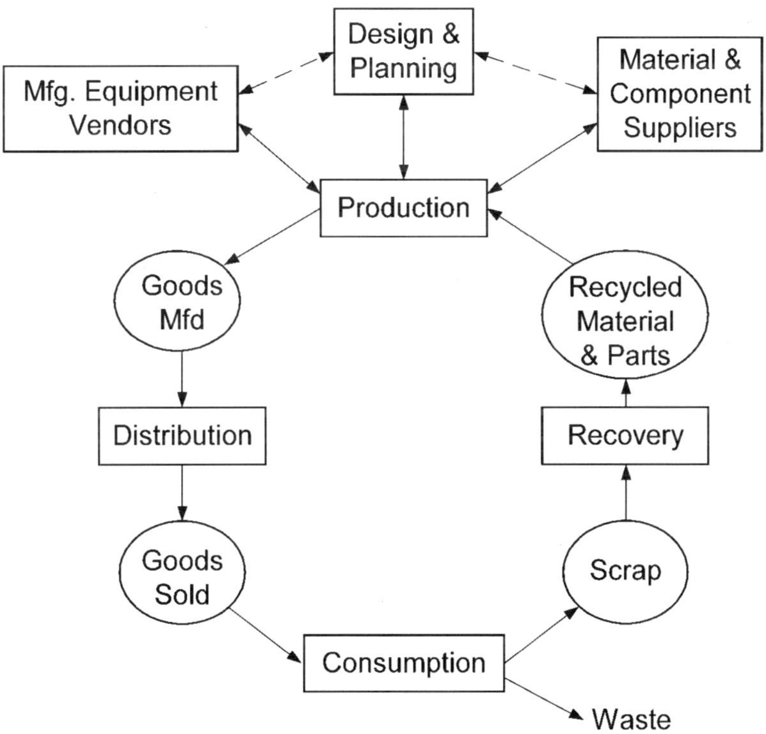

Figure 6 Each activity in the production–consumption cycle involves different customers. (From Ref. 3.)

use of simple models and prototypes, and refining the design based on customer likes and dislikes.

V. CREATIVE DESIGN

The heart of design is the development of creative ideas and concepts for solving the design problem. New ideas and concepts for designs appear to be generated in three different ways:

1. Adapting or rearranging existing design solutions.
2. Using imagination and ingenuity alone.
3. Using idea-stimulating techniques.

Adapting and rearranging existing solutions is one of the richest sources of new ideas. Existing solutions can be identified in a variety of ways, such as interviewing customers, benchmarking competitor products, examining patents, visiting suppliers, studying design solutions in other fields and applications, and networking with "idea brokers" and universities. The generation of creative design solutions by imagination and ingenuity requires the ability to get out of a "mental rut" and the ability to look at things in new and different ways. Using methods, such as "brainstorming", often stimulates creativity. The objective of brainstorming is to generate numerous ideas by focusing on a specific problem during a brainstorming session.

Systematic methods can also be used to generate design ideas. One successful technique is to decompose the design problem into subproblems or problem fragments. A variety of solutions for each problem fragment are then generated by brainstorming and other techniques. Different design concepts are created by visualizing various combinations of subsolutions using a matrix (Fig. 7).

Great designs are creative designs. Creativity requires focus and discipline. The best practices for creative design, such as the following, can help [3]:

- It is critical that the team searches both internally and externally for creative design solutions. Brainstorming and other idea-stimulating techniques help the team search internally. To search externally, the team should consult lead users, patents, literature, competitive products, similar or related products or equipment, manufacturing personnel, and so forth to be sure the range of possible approaches has been considered.
- Both individuals and groups generate good ideas. Therefore, to maximize the number of creative ideas, it is wise to seek a balance between individual and group approaches.
- Set stretch goals for brainstorming, such as "we want 400 ideas."
- The best ideas are not necessarily the first to be thought of. Avoid becoming overly enamored with the first good idea that is generated.

VI. MATERIAL AND PROCESS SELECTION

Material selection is the process of selecting the best or most appropriate material for each designed part. Similarly, *process selection* is the process of deciding on the method of

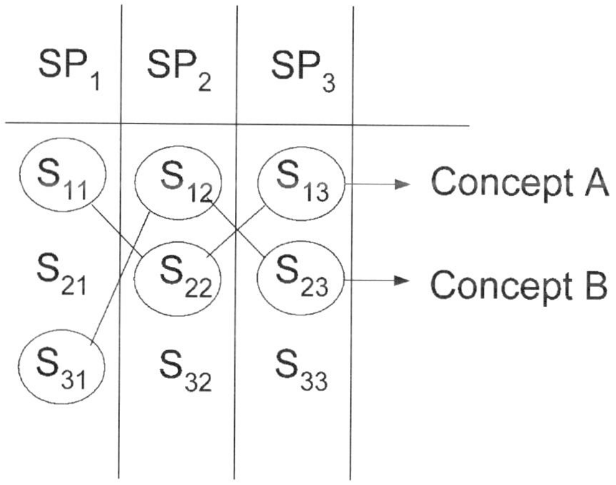

Figure 7 Creative design concepts are created by visualizing different combinations of subsolutions.

manufacture to be used to produce each designed part. Depending on the design situation, material and process selections can be a key consideration in achieving the goals of the design. For example, part count, assembly complexity, and secondary processing cost can be greatly reduced by integrating parts together using near net shape manufacturing processes, such as plastic injection molding, investment casting, or powder metallurgy. However, the selection process can also be extremely complex because the material, process, and part geometry are closely coupled. For example, steel cannot be injection molded. Similarly, a powder metal part cannot be designed with an undercut that prevents proper compaction or removal from the mold. Ultimately, material and process selection come down to trade-offs made between manufacturing cost, investment cost, performance and quality requirements, and tooling lead time.

In general, material and process selection is a process of progressively narrowing from a large range of possibilities to one highly specified choice. Depending on the type of product or design situation, many of the possible choices may be predetermined or implied. For example, the exterior components of a typical passenger car are usually formed sheet steel. However, depending on how unconventional the auto company wishes to be, formed aluminum or molded plastic may also be viable alternatives. In other cases, the choice of material and process may be wide open, in which case, a variety of factors, such as production volume, functionality, and in-house manufacturing capability and expertise must be considered in reaching a final choice.

Dixon and Poli [4] suggest two alternative approaches for narrowing the field of possibilities (Fig. 8). In the "material-first" approach, the material class (e.g., metal, plastic, ceramic) is selected first by considering the material property requirements of the application. Manufacturing processes that are consistent with the selected material class are then evaluated and selected based on considerations of production volume and on component size, shape, and complexity. In the "process-first" approach, the process class (e.g., machining, metal forming, casting) is selected first by considering the production volume and size, shape, and complexity of the component. Materials that are compatible with the process class are then evaluated and narrowed based on the material properties

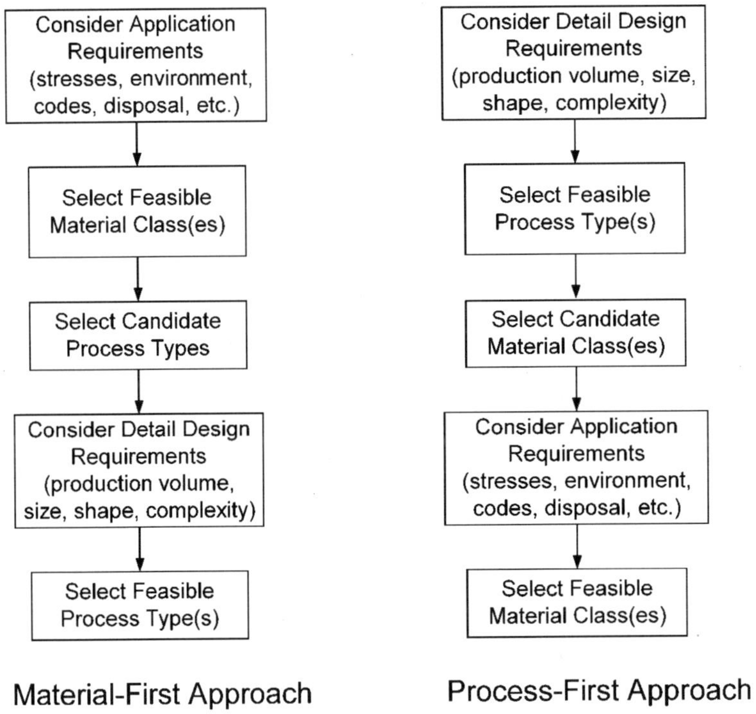

Figure 8 Alternative approaches for selecting material and process classes. (From Ref. 4.)

required and the target cost. In both cases, final detail specifications are developed by simultaneously considering both the required material properties and the process requirements and constraints.

Because the range of possible material choices can be large, experience has shown that use of a formal material selection process is highly advisable. A formal material selection process involves three main elements: [1] a detailed and comprehensive requirement definition, [2] two or more candidate materials from which to choose, and [3] a formal written set of recommendations together with a clearly defined selection rational. One way to identify candidate materials and develop selection recommendations is to use a "filtering" process in which the range of possible choices is first narrowed by considering "must haves" and "key factors" and then developing selection recommendations based on producibility issues (Fig. 9). *Must haves* are primary constraints, i.e., they are properties or characteristics that the material must possess to be acceptable. *Key factors*, on the other hand, are material considerations that make a particular material more or less desirable. For example, a "must have" may be a certain minimum tensile strength while a "key factor" may be density with lighter weight being preferable. *Producibility* issues include a range of practical considerations, such as availability, cost, compatibility with available manufacturing processes, ease of joining, esthetics, and so forth.

VII. FAILURE PREVENTION

The primary focus of most designs is on minimizing the probability of failure. A variety of approaches and practices are commonly used.

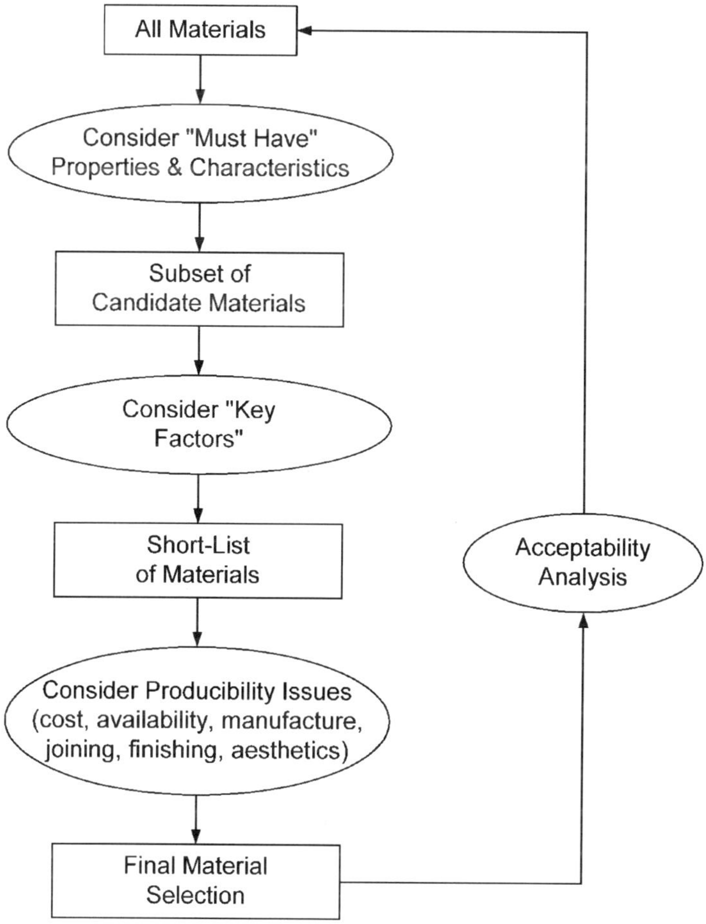

Figure 9 A systematic material selection procedure.

A. Factor of Safety

It is often necessary to guard against uncertainties associated with material properties, magnitude of external loading, part-to-part dimensional variation, and so forth. Let P_f designate the critical failure value and P_w the safe working value associated with a particular design factor or failure mode. Then, the *factor of safety* (FS) guarding against failure with respect to the failure mode or failure condition is defined as

$$FS = \frac{P_f}{P_w} \tag{1}$$

The value selected for the factor of safety in a particular design situation typically depends on the failure mode, the uncertainties involved, and on experience with similar situations. This value is often referred to as the *design factor of safety*. The design factor of safety is a design criterion that must be satisfied for the design to be acceptable. Often, because of a variety of considerations, such as the use of stock sizes, the actual calculated factor of safety is different from the specified value. For this reason, the *realized factor of safety* is defined as the ratio of critical failure value to actual calculated value [5]. To be acceptable, the realized factor of safety must be greater than or equal to the design factor of safety.

B. Design Allowable

In some cases, the critical failure value (P_f) for a particular design consideration is reduced to provide added safety. The term *design allowable* refers to this reduced value. For example, the *Manual of Steel Construction* published by AISC (American Institute of Steel Construction) specifies the allowable yield strength for tension to be 45–60% of the minimum value for yield strength published in the manual for a particular steel [5].

Uncertainties due to manufacturing process can also form the basis for a design allowable. In the design of safety–critical castings, such as those used in automotive suspension components and in aerospace applications, for example, uncertainties regarding casting strength are accounted for by using a casting factor. The *casting factor* is used to essentially increase the design factor of safety by reducing the design allowable even further. That is

$$S_D = S/CF \tag{2}$$

where S is the design allowable for static strength of the casting alloy based on rigorous statistical analysis of strength data, CF is the casting factor, and S_D is the allowable design value for strength to be used in selecting the dimensions of the casting.

C. Design Codes

In certain design situations, such as the design of a bridge or boiler, design codes are used to achieve a specified degree of safety, efficiency, and performance or quality. A *code* is a set of specifications for the analysis, design, manufacture, and construction of a particular machine, structure, or piece of equipment intended for a specified purpose. Consider the casting factor discussed above. The code for aerospace castings specifies the methods and procedures for determining the value of static strength (S) that is to be used in Eq. 2. In addition, the code specifies the amount and type of casting inspection and testing that must be performed for the particular value of the casting factor selected. For example, if a casting factor of 1.25 is used, the code requires 100% visual, magnetic/penetrant and x-ray inspection of all castings plus destructive testing of three castings. A comprehensive discussion of casting factors is given in Ref. [6].

D. Fail-Safe Design

This design philosophy seeks to design in such a way that, if failure occurs, the failure will be economically acceptable and will not lead to accidents. The use of fuses, shear-pins, and other replaceable elements that protect the design by planned failure is an example of fail-safe design that protects users and equipment from overloads. A helical compression spring that, when compressed solid alters the load path through a mechanism, or a clutch that slips under overload conditions, or a pressure relief valve, are examples of self-protecting designs, which is another form of fail-safe design.

Fail-safe design is also used by aircraft designers who cannot tolerate the added weight required by large safety factors nor the danger inherent in small safety factors. Multiple load paths and crack stoppers built into the structure are some of the techniques that are commonly employed.

Interlocks, which guard against human error, are another form of fail-safe design. In this approach, safety switches and other elements are installed to insure that the design cannot be operated improperly. For example, in some cars with manual transmissions, the engine cannot be started unless the clutch pedal is depressed. Similarly, a safety switch is often used to automatically disconnect electrical power from an electric room heater in the event that the heater is accidentally tipped over.

E. Infinite-Life Design, Safe-Life Design, and Damage Tolerant Design

A variety of design criteria are used to guard against fatigue failure. The oldest criterion is unlimited safety or *infinite-life design*, which requires that design stresses be safely below the appropriate fatigue limit. *Safe-life* design recognizes that worst case loading conditions may rarely or never occur during the life of the design. Hence, in this approach, it is assumed that designing for finite-life under maximum or worst case loading conditions is satisfactory. The factor of safety in safe-life design may be taken in terms of life (e.g., calculated life $= FS \times$ desired life), in terms of load (e.g., assumed load $= FS \times$ expected load), or by specifying that both margins must be satisfied, as in the ASME Boiler Code [7]. *Damage tolerant design* is a refinement of fail-safe design in which it is assumed that cracks caused either by processing or by fatigue will exist. To avoid failure, the structure is designed using fracture mechanics techniques to insure that the cracks can be detected during periodic inspection before they grow large enough to produce failures.

F. Derating

This design practice is an extension of the factor of safety concept to electrical, electronic, and mechanical components and subsystems. The basic assumption is that when the load factor of equipment is reduced, so is the probability of failure. Hence, in a *derated design*, the operating conditions (e.g., power, voltage, and temperature) for purchased components, such as electronic devices and electric motors are selected to insure that these components operate below their rated or nameplate values.

G. Redundancy

This design practice is aimed at increasing the reliability of the overall design. In parallel, redundant designs, multiple duplicate systems are used to insure continued operation even when one or more systems fail. Multiengine aircraft are a good example. When the cost or weight penalty for parallel redundant design is prohibitive, other strategies, such as the use of "back-up redundancy", can be employed. An automobile spare tire is an example.

H. Simplicity

Simple components and assemblies often have fewer failure modes and less opportunity for error or malfunction. Therefore, identifying the minimum number of simply shaped components that are easily analyzed and manufactured is an important design goal. The idea of simplification extends across all aspects of design. A truly simple design will be easy to manufacture, operate, service, maintain, recycle, and will have no or little impact on the environment.

I. Failure Mode and Effects Analysis

Failure mode and effects analysis (FMEA) is the name given to a group of activities that are performed to identify possible failure modes of a design, assess the likelihood that a failure might actually occur, and insure that appropriate corrective measures are taken to prevent failure. The primary goal of FMEA is to avoid surprises and unnecessary quality risk from reaching the customer. To identify potential failure modes and their effects, the design team seeks answers to the following key questions [3]:

1. What is the intended function of the component, subsystem, or system?
2. What are the possible failure modes? How could the design conceivably fail to perform its intended function?
3. What would be the effect if the failure did occur?
4. What mechanisms or causes might produce these failure modes?
5. What current controls or countermeasures are provided to prevent the failure or to compensate for it?

Once these questions and others like them have been answered, the risk associated with each potential failure mode is assessed. Typically, three risk factors are commonly considered: severity, occurrence, and detection. *Severity* (S) is an assessment of the seriousness of the effect produced by the potential failure mode if it occurs. *Occurrence* (O) is an estimation of the likelihood that a specific mechanism or cause will occur. *Detection* (D) is an assessment of the ability to detect a potential mechanism or cause or an actual failure before it reaches the customer. Evaluation criterion for each risk factor is usually based on a point scale, such as that shown in Table 1. Using the agreed upon rating criteria, the overall risk is computed as

$$RPN = (S) \times (O) \times (D) \tag{3}$$

where RPN stands for "risk priority number." If the five-point scale suggested in Table 1 is used, the RPN values will be between "1" and "125". Obviously, an RPN value of "125" would be of great concern while a "1" could probably be completely ignored. RPN values between these extremes require an interpretation criterion, such as that given in Table 2. As a general rule, failure modes that have a high severity rating should be given special attention regardless of the resultant RPN value.

The intent of corrective action is to reduce any one or all the risk factors. Typically, possible corrective actions are identified by "brainstorming" or by investigating the actions that were taken for a previous or similar design. When possible, the best corrective action is usually a design change. A design change can reduce any or all the three risk factors. Also, a design change is the only way that the severity rating and occurrence rating can be reduced. The detection rating, on the other hand, can be reduced by making a design change, or in some cases, by increasing preventive measures or by the addition of validation and/or verification testing.

Table 1 Typical Design FMEA Rating System

Rating	Severity (S)	Occurrence (O)	Detection (D)
1	No effect to minor; defect may be noticed by customer	Rare	Will almost certainly be detected
2	Customer is inconvenienced	Infrequent	Reasonably detectable by current controls
3	Item is operable, but at a reduced performance level; customer is dissatisfied	Frequent	Detectable before reaching the customer
4	Item is inoperable; loss of function	Very frequent to high	Detectable only by the customer andor during service
5	Safety-related catastrophic failure; regulatory noncompliance involved	High to very high	Undetectable until catastrophe occurs

VIII. EFFICIENT DESIGN

In general, there are many different ways of solving a design problem. Finding the right and/or best solution, i.e., the most *efficient design*, for a given design situation and set of functional requirements and constraints is a central consideration in design. To illustrate this important concept, we briefly overview efficient design as it applies to the design of members that transmit force.

A. Force Flow

Many machine designs must transmit force as part of their function. In these applications, it is often useful for the designer to visualize the force flow through the machine or component as an aid to identifying the most efficient design [8]. Consider the axially loaded stepped shaft shown in Fig. 10a. The force flow is the approximate paths taken by the force as it passes through the member and is determined by inspection as shown in Fig. 10b. By following the lines of force through the various parts, the designer is able to note possible critical sections and failure modes. For example, bearing pressure exerted on the tensile bar at sections "A" (and "J") could cause unacceptable deformation or wear. Continuing along the force flow, shear tear out could occur at section "B" (and "H"). Similarly, tensile failure must be checked at sections "C", "D", "E", and "F".

Force-flow analysis can also provide insight and stimulate creativity. Because the force must flow into a smaller section of the shaft at section "E", the lines of force deflect

Table 2 Typical RPN Interpretation Criterion

RPN value	InterpretationAction
$1 \leq RPN \leq 17$	*Minor risk*: little or no action required
$18 \leq RPN \leq 63$	*Moderate risk*: this requires selective design validation andor redesign to reduce the risk priority number
$64 \leq RPN \leq 125$	*Major risk*: high priority. Extensive design revision should be taken to reduce the risk priority number

and bunch together. As shown in Fig. 11, sudden changes in the direction of force flow and bunching of the flow lines cause stress concentration. Figure 12 illustrates various ways the designer can reduce the severity of stress concentration by visualizing various geometry modifications that ease and smooth the flow of force through the step.

The concept of force flow is also the basis of guiding design principles, such as the principle of direct and short force transmission path [9]. In general, a change in the direction of the force flow generally tends to weaken the part. Therefore, direct and short force transmission paths generally help insure minimum volume, weight, and deformation, while at the same time, allowing the use of simple component shapes that can be easily analyzed and manufactured. The application of this principle suggests that it is best, if possible, to solve a problem using tensile or compressive stresses alone, because these stresses, unlike bending and torsional stresses, produce smaller and more uniform deformations. This principle is particularly applicable if a rigid component or support structure is required.

B. Balanced Design

In a *balanced design*, all the load carrying members of the machine or product are sized so that each member is equally able to survive an overload condition. A member is consid-

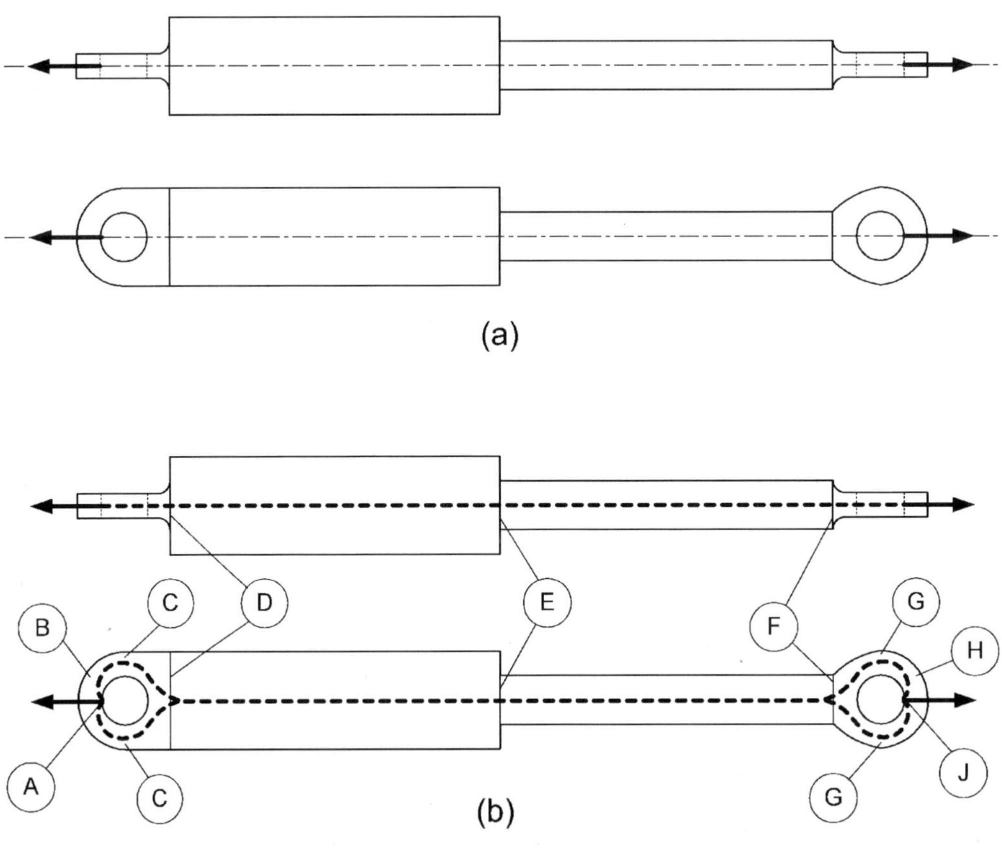

Figure 10 Pin-loaded tensile bar: (a) top and front views; (b) force-flow lines and critical sections.

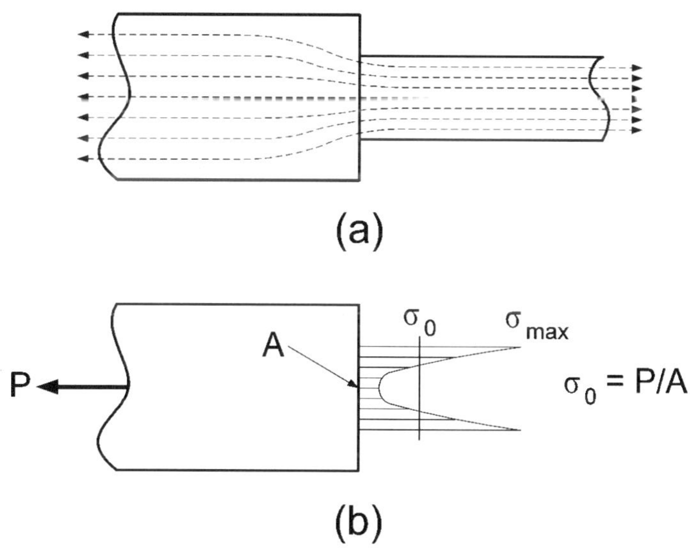

Figure 11 Force flow at section "E" of the tensile bar shown in Fig. 10: (a) force flow lines; (b) stress distribution on the smaller cross-section at section "E". Note that the stress concentration factor at section "E" is the ratio of maximum stress (σ_{max}) to nominal stress (σ_0).

ered to be overdesigned if it is able to carry significantly greater load than other members without failing. Similarly, a member is underdesigned if it fails at a load significantly below that of other members. The most underdesigned member is the "weak link" in the design. Balanced design seeks to avoid overdesign and weak links. Ideally, in a perfectly balanced design, each member would fail simultaneously when the design overload condition is reached.

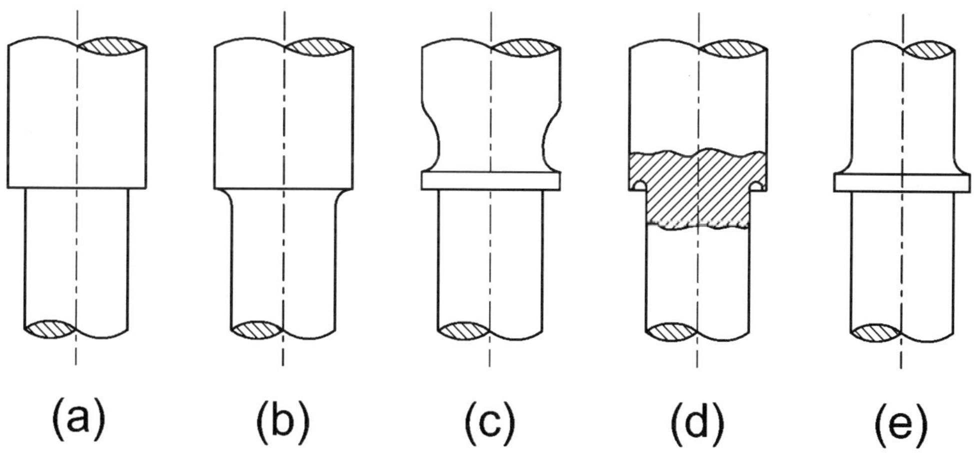

Figure 12 Reducing stress concentration in a stepped shaft: (a) severe stress concentration; (b) use large fillet radius if possible; (c) add groove if large radius is not possible; (d) undercut shoulder helps if (b) or (c) cannot be used; (e) a short step might also help.

The concept of balanced design also applies to individual components. For example, the design of the tensile bar, shown in Fig. 10, would be balanced if the individual features of the bar were sized so that the safety factor at each critical section "A" through section "J" was about the same. This idea can be used as the basis for creating efficient designs that have high strength-to-weight ratios. In this approach, the designer visualizes the flow of force through the member and cuts holes or thins down sections where there is little force flow. This practice, which is often referred to as *shape optimization* or *carve-out design* [10], has been greatly facilitated by the availability of powerful engineering workstations and solid modeling software that significantly enhances the engineer's ability to visualize complex three-dimensional geometry and to analyze stress levels and deflections of complex three-dimensional shapes. By using finite element analysis (FEA), the designer is able to experiment with different geometry until the best or most efficient design is identified. Ideally, each section in the most efficient design is loaded to its full capacity. In critical applications, balanced designs should be verified experimentally. The design is correctly balanced when a large number of the "balanced design" are tested to failure and the number of failures at each section is about the same.

Reinforcement design is another way for achieving balanced design of a machine or structural member [10]. In this approach, weak areas are strengthened by reinforcing them with additional material or by using composite materials. For example, steel, glass, or carbon fibers may be used to strengthen materials, such as concrete or plastic that are weak in tension. Since the stiffer material will carry more of the load, the reinforcing material should be stiffer than the reinforced material (the matrix). Also, to have an efficient composite material system, force flow must be such that each material in the system is subjected to stresses in its strongest direction. The concept of reinforcement design is illustrated by the steel reinforced concrete beam shown in Fig. 13. Since concrete is weak in tension, the beam is reinforced by steel bars embedded in the concrete below the neutral axis of bending, which is a region of high tensile stress for the loading shown. It should be noted that the steel bars have "ridged surfaces" that facilitate force flow from the concrete into the steel bars by loading the ridges in shear.

C. Design Optimization

Design optimization is the systematic search for the best parameter values for a given design configuration. The need for design optimization arises out of the fact that, in most design situations, there is more than one possible set of parameter values. Hence, for a given set of design requirements and criteria, there likely exists one particular set of parameter values that is best or optimal. The importance of identifying the optimal design obviously depends on the particular design situation or application involved. In some cases, all that is needed is an acceptable solution, that is one that satisfies all the functional requirements and constraints of the design. In these cases, the cost of identifying the optimum design may not be justified. In other cases, where certain criteria, such as weight or natural frequency are of critical importance, finding the optimal design may mean the difference between success and failure.

Design optimization problems are typically formulated in terms of design parameters, performance measures, and constraints. *Design parameters*, which are the set of independent design variables that specify the design, can either be continuous or discrete. *Continuous* design parameters can take any value over a predetermined range, while *discrete* parameters are restricted to a finite set of values. A hole diameter or its location in a sheet metal part is a continuous design parameter because it can take

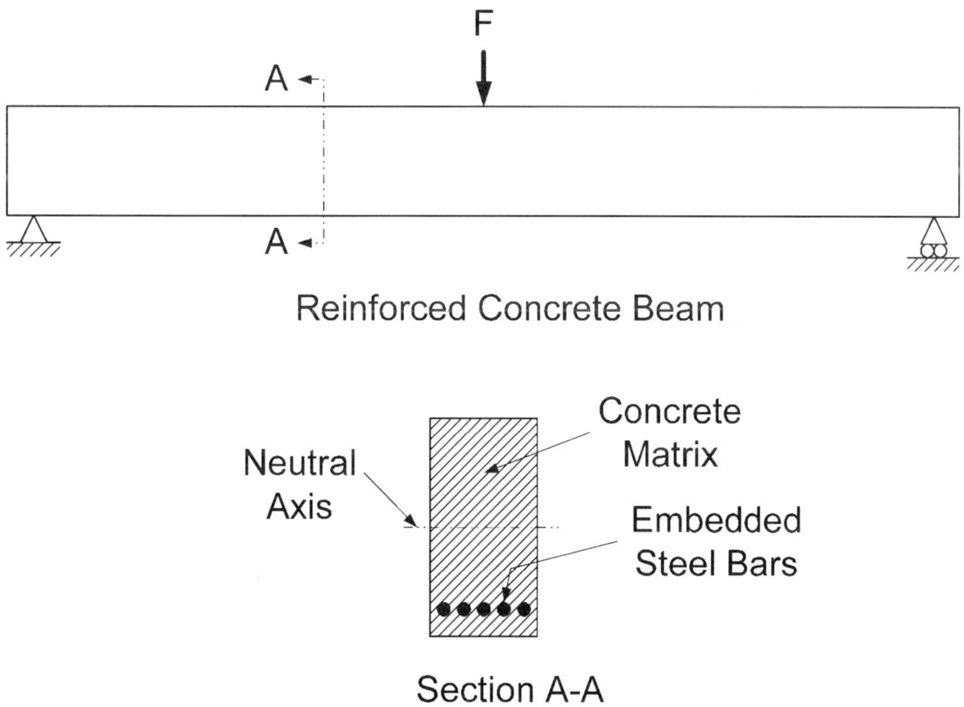

Figure 13 Steel reinforced concrete beam.

any value within a permissible range. The bolt that passes through the hole and the gage thickness of the sheet metal part, on the other hand, are discrete design parameters because they are limited to the available sizes. Modulus of elasticity and other properties of a material are also discrete design parameters because they have specific values for each material considered.

To illustrate the concept of design optimization, consider the tensioned belt shown in Fig. 14. Suppose that the designer desires to select dimensional values for the width, b, and thickness, t, of the belt cross-section and for the pulley radius, R, such that the normal stress carried by the belt for a given tensile preload, P, is equal to or less than the allowable value, σ_{all}, and that the weight of the belt is as low as possible. Suppose also that the belt thickness, t, is only available in increments of Δt inches, that R must be between R_{min} and R_{max}, that b must be between b_{min} and b_{max}, and that P must be between P_{min} and P_{max} lbs. This optimization problem may be formulated mathematically as follows:

$$\text{Weight} = 2\rho bt(\pi R + h) \rightarrow \min \tag{4}$$

$$\sigma = \frac{P}{2bt} + \frac{Et}{2R} \leq \sigma_{all} \tag{5}$$

$$R_{min} \leq R \leq R_{max} \tag{6}$$

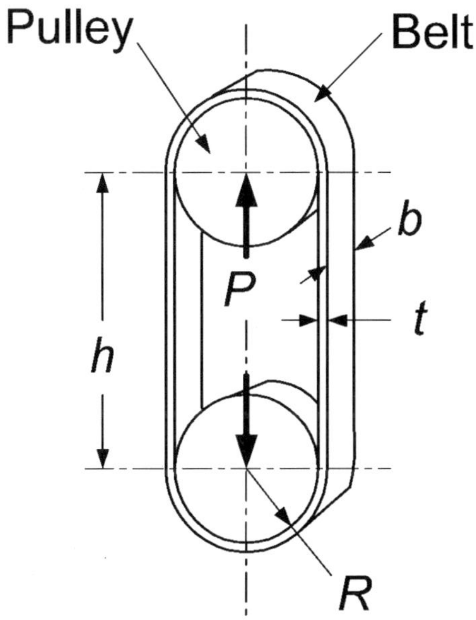

Figure 14 Belt design for minimum weight.

$$b_{\min} \leq b \leq b_{\max} \tag{7}$$

$$P_{\min} \leq P \leq P_{\max} \tag{8}$$

where ρ is the weight density of the material, h the center distance between the belt pulleys, and E is the modulus of elasticity of the belt material. This problem can be solved using a variety of different computer-based optimization and/or search algorithms. In general, the process would consist of using the algorithm to calculate the *optimum* values of b, t, R, and P that minimize the belt weight for a given belt material and center distance, h. The *optimal* design would then be determined by calculating the belt weight for the discrete values of t that bracket the optimum value and specifying the discrete value that results in the lowest belt weight.

Alternatively, for relatively simple optimization problems, such as this, the designer can use reason and intuition to identify the optimal design. First, it might be noted that the belt tensile stress given by Eq. 5 contains two terms. The first term, which is due to the axial preload imposed on the belt, has the thickness, t, in the denominator. The second term, which accounts for bending stress caused by the belt wrap around the pulley, has the belt thickness, t, in the numerator. Hence, stress due to the preload decreases as t increases, while stress due to bending increases. This implies that there is an optimum value for t that minimizes tensile stress in the belt for given values of b, R, P, and E. This optimum value can be found by differentiating Eq.5 with respect to t, setting the result equal to 0, and solving for t to give

$$t_{\text{opt}} = \sqrt{\frac{PR}{bE}} \tag{9}$$

When Eq. 9 is substituted into Eq. 4, the designer can see that the belt weight will be minimized by letting $P = P_{min}$, $R = R_{min}$, $b = b_{min}$ and using Eq. 9 and these values to calculate t, rounded to the nearest acceptable discrete value. Also, the stiffest material (i.e., large E) possible that has an acceptable allowable strength should be considered. If this design results in $\sigma > \sigma_{all}$, then R may need to be increased and/or a material having a lower value of E or a higher strength may need to be considered.

D. Value Engineering

Value engineering is a systematic approach for design improvement and cost reduction. The goal is to maximize the value of a design where value is defined as

$$\text{value} = \frac{\text{function}}{\text{cost}} \tag{10}$$

The value engineering approach is predicated on the recognition that every component contributes both to the cost and the function of the design and that design efficiency can be improved and cost reduced by providing each function for the lowest possible cost. Value engineering seeks to understand the relationship between function and cost. In analyzing a design, the focus is always on function, cost, and the value formula given by Eq. 10. How can value be added? Can the function be increased? The cost lowered? Can we do both?

For any expenditure or cost, two kinds or value are received: use (functional) value and esteem (prestige) value. *Use value* reflects the properties or qualities of the design that accomplish the intended work or service. To achieve maximum use value is to achieve the lowest possible cost in providing the performance function of the design. *Esteem value* is composed of properties, features, or attractiveness that make ownership of the design desirable. To achieve maximum esteem value is to achieve the lowest possible cost in providing the necessary appearance, attractiveness, and features the customers want. Examples of prestige items include surface finish, streamlining, packaging, decorative trim, ornamentation, attachments, special features, adjustments, and so forth. In addition to the two kinds of value received, additional costs are incurred due to unnecessary aspects of the design. Termed waste, these are features or properties of the design that provide neither use value or esteem value.

Value engineering is generally performed in two phases, the analytical phase and the creative phase. In the analytical phase, the use value and esteem value offered by the design is systematically investigated by a multidisciplinary team. Findings generated in the analytical phase are then used by the team in the creative phase to define innovative design solutions which maintain the desired balance between use and esteem value, maximize these values by providing required function for the lowest cost, and eliminate waste or unneeded features. Some steps in a typical value engineering design improvement exercise are as follows [3]:

1. Identify the basic and secondary functions that the design is intended to perform. Basic functions relate to the specific work that is to be performed. Secondary functions are subordinate functions that are performed in providing the basic functions.
2. Determine the value of all functions performed using Eq. 10. This step is usually carried out in tabular form with each function forming a separate row. Percent of total performance contributed by each function is estimated and its cost as a percent of design cost is determined. Value is calculated as the ratio of

percent performance-to-percent cost. Insight into what aspects of the design constitute waste and should be eliminated and/or where improvements are needed is obtained by comparing the value calculated for each function.

3. Search for creative ways of reducing cost without reducing value, or of adding value without adding cost. Do this by questioning the need for stated design specifications, analyzing material selection, questioning high cost features or components, and so forth.

Value engineering is unique in its focus on function and on the cost of providing function. Because of its reliance on the availability of accurate cost data, value engineering can be difficult to use in the very early stages of design. On the other hand, it is an excellent starting point for the redesign or cost reduction of an existing product.

IX. DESIGN FOR QUALITY

Traditionally, the term *quality* in the context of design has implied precision and accuracy, especially as it relates to dimensional tolerances and other tolerance specifications. *Design for quality*, on the other hand, involves the much broader context of *total quality*, which is the totality of features and characteristics of a product or device including design, manufacture, distribution, sale, service, use, and disposal, that bear on its ability to satisfy stated or implied needs.

Each design decision, both large and small, contributes to total quality. To design for quality, it is necessary to systematically consider all aspects of the design. One approach, suggested in Ref. [3], is to decompose total quality into "external" and "internal" components as shown in Fig. 15. *External quality* is a reflection of how well the design satisfies customer requirements. This quality is "external" because it is of primary importance to the customers who specify, buy, distribute, use, service, and dispose of the design. *Internal quality*, on the other hand, relates to design characteristics, such as ease of manufacture and assembly, consistency from build to build, sensitivity to hard-to-control disturbances, scrap rate, need for rework, and warrantee claims. This quality is "internal" because the business enterprise or client that creates and profits from the design reaps the primary benefits of this quality. Ultimately, external quality leads to customer satisfaction, internal quality leads to lower cost, and design for quality leads to higher profits that sustain over the long term.

A. External Quality

As shown in Fig. 15, external quality is composed of three components: quality of concept, quality of ownership, and operational robustness. Each of these are briefly discussed as follows.

1. Quality of Concept

This quality is what makes the design desirable to the end user. For most designs, this quality is a composite of performance, features, esthetics, and ergonomics. Performance relates to the basic functionality of the design. Features supplement the basic functionality. Esthetics and ergonomics involve how the design looks, feels, and interacts with the user. Maximizing this quality is a primary goal of customer-focused design discussed previously.

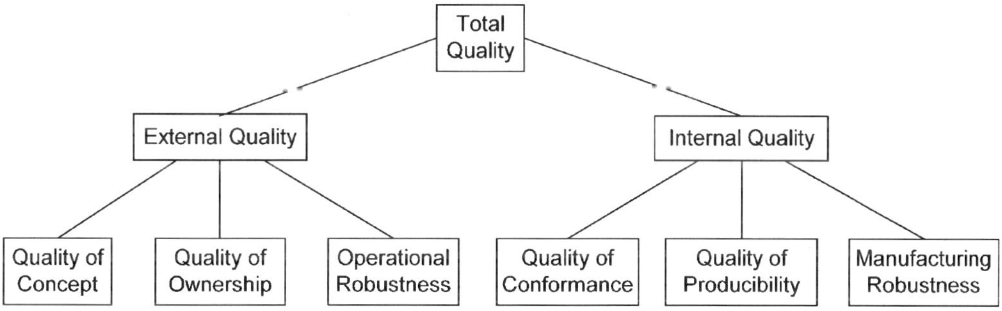

Figure 15 Components of total quality. (From Ref. 3.)

2. Quality of Ownership

This quality relates to the experience the customer has as a result of owning and operating the design. Considerations, such as ease of use, operating cost, reliability, durability, serviceability, maintainability, condition of the design when delivered or accepted, and customer service, influence this quality. High quality of ownership is important because it is what causes a customer to become a repeat buyer and to become an advocate of the design. This quality, more than any other, contributes to the firm's reputation for manufacturing and selling high quality designs.

3. Operational Robustness

A *robust design* is one that is insensitive to and/or tolerant of change and variation in hard-to-control variables that disturb or degrade function. *Operational robustness* refers to the design's ability to tolerate variation in hard-to-control variables that influence the design function. Typically, these variables [1] are associated with the environment in which the design operates, [2] arise as the result of changes and degradations that occur over time and with use, or [3] occur due to variation from build to build manufactured under the same specifications. Examples of hard-to-control environmental variables that influence the design during use include temperature, humidity, external load, time rate of load application, type of use, and so forth. Loss of strength due to corrosion, wear of mating parts, shifts in calibration or adjustment exemplify the types of hard-to-control variation that occur during use. Variation in dimensions and calibration values that result from manufacturing processes are representative of hard-to-control build to build variation. In robust design methods, such as the Taguchi method, the design team seeks to identify robust values for key design parameters by taking advantage of nonlinearities between desired functional characteristics and the design variables (see Fig. 16).

B. Internal Quality

Internal quality depends on the degree to which the design and operating characteristics meet established standards, how easy the product is to manufacture and assemble, and how sensitive the design is to changes in demand, business needs, and product and manufacturing technology. In other words, it depends on the qualities of conformance, producibility, and manufacturing robustness. These qualities are discussed as follows.

1. Quality of Conformance

This quality refers to how well the actual manufactured design conforms to design intent, where *design intent* is the intended value or target value of the design variable or functional

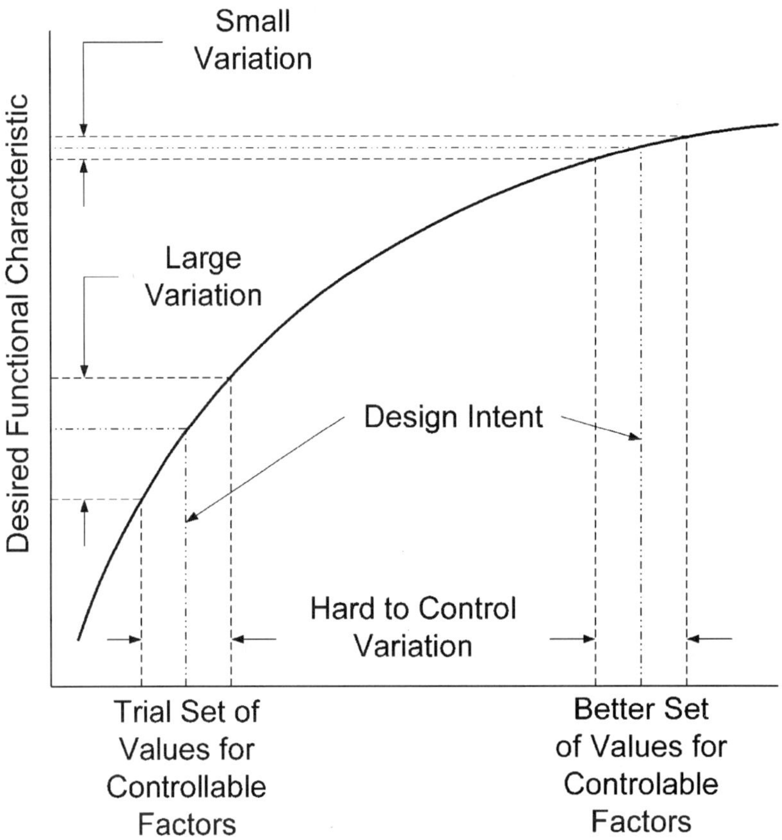

Figure 16 Parameter design exploits nonlinearities between desired functional characteristics and design variables. (From Ref. 11.)

characteristic. Conformance is acceptable when the actual value is within the design specification. Quality of conformance improves as the actual value approaches design intent and is a maximum when the actual and target values are the same. Hence, the quality of conformance depends on both the amount of variation and on the mean of the actual value. The Capability Index (C_p) is commonly used to quantify the quality of conformance:

$$C_p = \frac{\text{design specification width}}{\text{process width}} = \frac{\text{upper limit} - \text{lower limit}}{6\hat{\sigma}}$$

where $\hat{\sigma}$ represents the standard deviation of the manufacturing process. It is interesting to note that the popular concept of "six-sigma quality" requires $C_p > 2$, which is something that is often very difficult to achieve in practice, especially for designs that are essentially mechanical in nature. To maximize C_p, the design should have wide specification widths (i.e., loose tolerances) and should be manufactureable using highly repeatable processes (i.e., low standard deviations).

2. Producibility

This quality relates to the ease with which the design can be manufactured, assembled, inspected, and tested using readily available materials and components. A high degree of producibility reduces quality risks, manufacturing complexity, hard-to-control randomness, production uncertainty, and life cycle support complexity. Producibility is maximized by (1) identifying design concepts that are inherently easy to manufacture for the best balance of cost and time, (2) focusing on component design for ease of manufacture and assembly, and (3) integrating the design of the product and manufacturing process to insure the best matching of needs and requirements.

3. Manufacturing Robustness

This quality reflects the ability of the manufacturing system to tolerate design changes and production volume changes that occur due to changing market needs, business needs, and technology innovation. Platform products, that utilize a large number of standard components and, that are easy to redesign to meet evolving needs because only a few subsystems and components need to be changed, are examples of designs that exhibit a high quality of manufacturing robustness.

X. DESIGN FOR MANUFACTURE

Design for manufacture is a philosophy of design that seeks to identify the best possible design that meets the functional requirements of the design and that is also easy to manufacture and assemble. Design for manufacture considers the limitations of people, materials, and manufacturing processes. It is most effective when performed early in the design process when conceptual maneuverability is still wide and, few or no hard, irreversible decisions have been made. Design for manufacture is often implemented by using a design–analyze–redesign process in which the design team first proposes a design concept and then analyzes it to identify opportunities for improving manufacturability (Fig. 17). Often, this process begins with an existing design that is then improved iteratively by applying proven DFM and assembly design guidelines.

A. Design for Manufacture Guidelines

The following guidelines or design suggestions, which are taken from Ref. [3], are relatively independent of any particular method of manufacture or assembly. When correctly applied and adhered to, they will result in a design that is inherently easier to manufacture, assemble, test, service, and maintain regardless of the extent or degree of manufacturing mechanization employed. The guidelines are based on three basic producibility principles: eliminate, simplify, and standardize where practical.

1. Optimize the Assembly Sequence

Develop a design that clearly delineates the assembly sequence of components. Consider a "building block" approach, or a layered or stacked construction, or the use of a base component that locates, orients, and relates the various components of the product to each other. Select a starting point for assembly that facilitates a minimum number of assembly operations. Visualize the order of incorporation of parts into the assembly envelope as

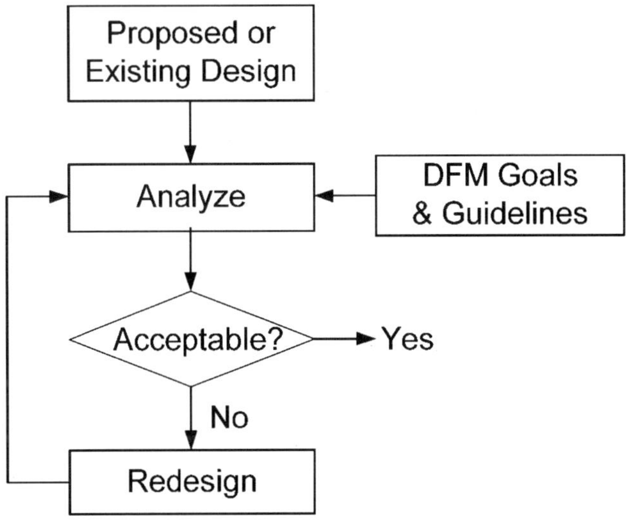

Figure 17 The DFM design–analyze–redesign process.

well as the spatial relationships involved and seek to simplify the assembly as much as possible. Anticipate where interference between previously assembled parts and/or subsequent operations might occur and design to eliminate potential trouble spots.

2. Design for a Minimum Number of Parts

Always seek the minimum number of simply shaped components. Fewer parts mean less of everything that is needed to manufacture the design. Consolidate parts by combining mating or contacting parts that do not move independent of each other into an integral part. Consider alternative materials, net shape manufacturing processes and so forth.

3. Eliminate Separate Fasteners

Eliminate fasteners by integrating their function into mating parts. Fasteners themselves may be relatively inexpensive but their installation is costly and prone to quality risk. If all the fasteners cannot be eliminated, design to reduce the fastener count. Consider "snap together" or interlocking designs. Rationalize standard fasteners to reduce fastener types, head styles, drive styles, sizes, and so forth.

4. Eliminate Adjustments

Identify critical dimensions that, if not confined to a single part, require slots and other features to permit adjustment between parts. If possible, incorporate such dimensions into a single part. Avoid the need for adjustment between mating parts.

5. Eliminate Randomness

Use rigid parts wherever possible. Flexible parts are more difficult to handle and orient. Plan the layout of electrical connections, flexible tubing, and control cables early in the design. Provide features that orient, locate, and restrain wire harnesses and other components that would otherwise occupy random orientations or positions. Avoid dangling connectors, lead wires, unrestrained parts, ambiguities between mating parts. Avoid part-to-part or part-to-process dependencies, such as "fit at assembly".

6. Design Parts for Easy Handling

Design symmetrical parts. Part symmetry means fewer orientation steps, resulting in faster assembly time and lower automated handling system costs. If symmetry is difficult or asymmetry is important, then design easily recognized orientation and positioning features or provide clear identifying marks. When possible, avoid very large or very small sizes. Try to use rigid parts and avoid flexible components. Provide adequate, easily recognized and accessed surfaces to allow mechanical gripping for transfer and location retention. Design to eliminate part tangling, nesting, and interlocking.

7. Reduce Processing Surfaces

Simplify components so that the number of surfaces being processed is minimized. Try to design parts for orthogonal and/or parallel fabrication since most fabrication equipment is designed using orthogonal and/or parallel constructs. Try to complete all processing on one surface before moving to the next. This is especially important in assembly. Do not design the assembly sequence so that the operator must jump back and forth from surface to surface.

8. Minimize Insertion Motion

Avoid complex, multimotion insertion. Develop a "top–down *Z*-axis" assembly approach. *Z*-axis assembly means simple robot or insertion tooling with gravity serving to assist in the assembly process. Tooling is usually less expensive; part clamping becomes either not necessary or less costly. Try to design so that parts are assembled using a "sandwich" approach with unidirectional *Z*-axis insertion. Avoid multimotion insertion. Not using *Z*-axis assembly usually means special tooling.

9. Design Parts for Easy Insertion

Design for easy part insertion by providing generous tapers, leads, chamfers, and radii on mating parts. Such guiding features help align and provide centering forces to assist and facilitate the insertion process. They also offset the detrimental effect of part inconsistencies, misalignments, and tolerance stack-up.

10. Process in the Open

Design out restricted access and obstructed vision. Provide adequate clearance for *standard* tooling. Provide a clear, unobstructed view for assembly operations. This is important for eye–hand coordination for manual assembly and, in most cases, essential for automatic assembly.

11. Retain Parts

Once parts are in place, keep them in place. Design location features that eliminate repositioning or holding down a component after it has been "mated" to another part. Provide these features either in the part being assembled or in its mating part.

12. Design for Error-Free Assembly

Design parts so that they are impossible to assemble incorrectly. Provide tabs, "nests", and location features that prevent incorrect mating of parts. Poor quality is no longer an issue since it is impossible to achieve.

B. Design for Assembly

Design for assembly (DFA) is at the heart of DFM since assembly is where most manufacturability problems and issues arise. The goal of design for assembly is to eliminate parts and to design the parts that remain to be easy to handle and assemble. Essentially, this is done by systematically applying the DFM guidelines given above. Unfortunately, it is not uncommon for assembly considerations to be overlooked, especially under the schedule and budget pressures that drive most design projects. As a result, a variety of design for assembly methods have been developed to help assist designers to analyze their designs and improve them with respect to assembly. Perhaps the best known and most widely used method is the design for assembly method developed by Boothroyd and Dewhurst. Details of the methodology are presented in Ref. [12]. Based largely on industrial engineering time study methods, the design for assembly method developed by Boothroyd and Dewhurst seeks to minimize cost of assembly within constraints imposed by other design requirements. Essentially, the method is a systematic computer-aided implementation of several of the DFM guidelines given above.

C. Component Design

In addition to reducing assembly cost, a second important DFM goal is to reduce component cost. This is done in a variety of ways. The most direct way is to design components with the manufacturing process in mind. For example, plastic injection molded parts need to be designed to have a uniform wall thickness in order to insure even, distortion-free shrinkage during cooling. Similarly, they require generous draft to allow easy ejection from the mold and reduced production cycle time. These are just a few of many DFM rules and guidelines that the plastic injection molding designer must keep in mind to insure a producible part that is easy to tool and produce with a minimum of scrap.

A multitude of other approaches are also used to reduce component cost. One way is to redesign components to eliminate processing steps. For example, using a through hole instead of a blind hole may eliminate a special machining setup and sequence of operations. Another approach is to standardize components and processes. When done correctly, the standardized component can be used interchangeably in several different products or designs. Some firms, such as automotive companies, develop a systems-level specification for the part that gives the supplier complete freedom to design the component for minimum cost. This allows the supplier to design the component so that it is in complete harmony with the manufacturing process, thereby avoiding unnecessary tooling and fixturing cost as well as special processing and operations.

D. System Cost

Systems cost includes all the day-to-day indirect operations and overhead cost associated with manufacture, distribution, marketing, service, and disposal of the design. Examples include costs associated with purchasing, supplier relations, order processing, inspection, material handling, receiving, shipping, service parts, catalogs, promotional material, worker training, and so forth. Many of these indirect system costs are driven or determined by the design. Hence, perhaps the greatest benefit of DFM is the long-term reduction in systems cost that it produces. Many of the DFM guidelines listed above, such as designing for a minimum number of parts, directly impact system cost. Standardization and approaches that allow the use of similar systems in different designs are also effective approaches for reducing system cost.

Finding the right approach for reducing systems cost can be difficult because of the complex interactions that exist between design and manufacturing at the systems level. For example, consider the sheet metal used in making an automobile. From the standpoint of the design engineer, a wide selection of different sheet metal gages, roll widths, coatings, and material properties are desirable so that the most optimal design can be achieved with respect to structural rigidity, weight and cost. From the manufacturing point of view, however, the ability to process a wide variety of different thickness and types of sheet metal means added cost and complexity in purchasing and supplier relations, material handling and storage, processing equipment and tooling, setups and operation, and so forth. In cases such as this, finding the right balance that insures both design freedom and cost control on the manufacturing floor, can be the key to long-term economic viability and competitiveness.

XI. LIFE CYCLE DESIGN

Life cycle design is a philosophy of design that considers all aspects of the design life cycle as part of the design process. *Life cycle design* may be defined very broadly as the full range of policies, techniques, practices, and attitudes that cause a product or device to be designed for the optimal manufacturing cost, the optimal achievement of manufactured quality, the optimal achievement of operational performance and efficiency, and the optimal achievement of life cycle support (serviceability, reliability, and maintainability). Hence, we see that life cycle design is an extension of the DFM philosophy to consideration of all phases of the life cycle including design, manufacture, distribution, use, and retirement. Life cycle design encompasses a broad range of diverse considerations. We discuss a few of these to illustrate the breath and scope of this important aspect of design.

A. Design for Serviceability

A design that is easy to manufacture and assemble is also, in general, easy to service and maintain. Additional ease of service and maintenance can be achieved by paying close attention to the following considerations [3]: [1] the number and variety of tools, operations, and labor skills required for service and maintenance, [2] ease of accessibility, and [3] ease of reparability. Ideally, only a minimum number of readily available tools should be required and all service and maintenance operations should be easy to perform without special training or skills. The best way to achieve this goal is to clearly define permissible tools and operations early in the design, before detail design decisions are made.

Ease of accessibility implies easily identified parts that are readily visible and easily reached. In particular, critical parts, components, and structural sections, which are likely to have above average failure rates, should be easy to see and inspect. Also, all maintenance items and points should be easily accessible. Most importantly, close attention should be paid to providing sufficient hand and tool manipulation clearance for scheduled maintenance items, adjustments, and so forth, without removal of interfering components. Use of solid modeling capabilities, which permit simulation of assembly and disassembly operations, can be particularly useful for insuring that sufficient clearance is provided.

Repairability is optimized by maximizing commonality and interchangeability of components, providing readily available and easily used trouble shooting aids, minimizing and simplifying adjustments, and designing to minimize repair complexity and turnaround time. In most designs, repairability is best addressed by developing a repairability strategy early in the design. For example, a short list of acceptable common components can be agreed upon prior to detail design and then used exclusively in the design. Such an approach not only simplifies and improves repairability, but also tends to maximize availability of spare parts and enables parts "scavenging" and substitution in emergencies.

B. Design for Recyclability

The ability to recycle easily and efficiently is becoming an increasingly important consideration in many designs. Ease of recyclability can be facilitated by: (1) recycling by repair, (2) recycling by conversion, (3) design for easy separation, and (4) contamination tolerant design [10]. In recycling by repair, items, such as automotive alternators and starter motors, are designed so that they are easy and cost effective to repair and/or rebuild. This makes it economically feasible to repeatedly refurbish and reuse them as opposed to disposing of them. Recycling by conversion involves reusing items in different ways after their intended function has been fulfilled. Henry Ford's use of the wood from shipping containers as automobile floorboards is an excellent example. Embedding a marker or tracer that makes it easy to separate out a wanted material in a sorting process is illustrative of the design for easy separation approach. Finally, an example of contamination tolerant design is to use metalized or painted plastic parts which can accommodate the use of reclaimed material because color degrading can be easily concealed.

C. Human Factors

Human factors relate to the way a design interfaces or interacts with humans throughout its life cycle. In designs that involve human interaction, a consideration of human factors is necessary because problems are likely to develop when humans and machines interact. In general, the design must consider both physical and social human interaction. Physical interaction involves a variety of considerations ranging from feature layout, color, size, shape, and orientation to noise, lighting, vibration, and so on. The social context, on the other hand, involves appearance, safety, and other highly intangible considerations.

Goals of human factor design generally include making the design pleasing and easy to use, insuring the safety, health, and comfort of the user, and insuring worker productivity. To do this, the design must account for the range of differences between people in terms of size, shape, and capabilities. It must also account for the operational and environmental hazards and stresses that may be involved.

D. Safety Considerations

The design must be safe at all stages of its life cycle. Proper consideration of safety in design implies that the design will protect humans from injury, protect property from damage, and prevent harm to the environment. Safety during operation and/or use of a design is generally achieved in three ways [13].

1. Design the product or equipment to be inherently safe such that there is no danger either during normal operation or use of the design, or in case of failure.
2. If inherent safety is not feasible, provide protective devices specifically designed to prevent accidental injury (e.g., a safety screen or guard), interrupt continued operation in the case of a failure (e.g., dead-man switch), and provide features that mitigate against injury in the case of a failure or accident (e.g., air-bag in an automobile).
3. If there is inherent danger in the operation or use of the design, provide warning signs and labels to instruct the user, provide loud sounds, flashing lights, and so forth when a dangerous condition arises or could arise, and provide barriers and other deterrents to keep unsuspecting passersby from being injured.

E. Environmental Considerations

Design decisions determine how the design is manufactured, distributed, used, and ultimately disposed. Most importantly, design decisions constrain downstream processes that often limit the freedom the firm or society has for reducing the undesirable environmental impact of the design. Environmental considerations are therefore important aspect of many designs.

Minimizing undesirable environmental impact involves a coordinated understanding of marketing, design, packaging, and manufacturing issues. It also involves an understanding of governmental regulation and compliance requirements. The following checklist is useful for assessing environmentally related design issues [3].

- What are the environmental issues? Can the design be improved with respect to these issues?
- Can recycled materials be substituted for primary raw materials?
- Which materials in the design are hazardous or regulated? Can they be eliminated?
- Has the cost of safety equipment, training, and exposure monitoring been considered in the material selection process?
- Have all legal and regulatory requirements been considered?
- Will use of the design require excessive energy consumption or generate air, water, or soil pollution?

XII. CONCLUSION

Design is a complex and highly convoluted process. It appears to proceed in a linear fashion starting with a problem definition and ending with a fully specified design that meets customer needs. In reality, the process is highly nonlinear. It involves a variety of iterative cycles during which the design evolves to the ultimate solution. It also involves a multitude of different viewpoints and perspectives including performance, quality, manufacturing, life cycle support, and so forth. Design begins and ends with the customer. A concise and quantified specification is essential to developing suitable concepts and implementing the details of the design. Understanding is gained through "observe–experiment–refine" iterations. The design space is explored by generating numerous product concepts using brainstorming and systematic methods. These concepts are narrowed to the best one or few based on understanding gained through simulation, experiment and testing using analysis and computer-aided techniques as well as physical models and prototypes. The design is further evolved and refined as understanding increases. When understanding is complete, implementation follows naturally through a process of "design–analyze–redesign" cycles that focus on insuring ease of manufacture, long-term performance, quality, and reliability, and a reasonable profit.

REFERENCES

1. Fuchs, H.O. Strategies in design. J. Mech. Design 1980, 102, pp. 1.
2. *Systematic Approach to the Design of Technical Systems and Products*. VDI Society of Product Development, VDI 2221, pp. 7, August 1987.
3. Stoll, H.W. *Product Design Methods and Practices*; Marcel Dekker: New York, 1999.
4. Dixon, J.R.; Poli, C. *Engineering Design and Design for Manufacturing*; Field Strone Publishers, Conway, MA, 1995.
5. Shigley, J.E.; Mischke, C.R. *Mechanical Engineering Design*, 6th Ed.; McGraw-Hill, New York, 2001; 22–30.
6. McLellan, D. Understanding casting factors in aircraft components. Modern Casting October 1994, 21–24.
7. Fuchs, H.O.; Stephens, R. Fatigue Design Methods. *Metal Fatigue in Engineering*; John Wiley & Sons: New York, 1980; 7–12.
8. Junvinal, R.C.; Marshek, K.M. Load analysis. *Fundamentals of Machine Component Design*, 2nd Ed.; John Wiley & Sons: New York, 1991; 51–54.
9. Pahl, G.; Beitz, W. Embodiment Design. *Engineering Design;* Wallace K., Ed.; The Design Council: London, 1988; 196–198.
10. Chow, W. *Cost reduction in Product Design*; Van Nostrand Reinhold Company: New York, 1978.
11. Ettlie, J.; Stoll H. Design for Life-Cycle Manufacturing. *Managing the Design-Manufacturing Process*; McGraw-Hill: New York, 1990; 53–77.
12. Boothroyd, G.; Dewhurst, P.; Knight, W. Product Design for Manual Assembly. *Product Design for Manufacture and Assembly*; Marcel Dekker: New York, 1994; 62–119.
13. Ullman, D. *The Mechanical Design Process*, 2nd Ed.; McGraw-Hill: New York, 1997.

2
Risk-Based Metallurgical Design

Mario Solari
Consejo Nacional de Investigaciones Científicas y Técnicas, Buenos Aires, Argentina

I. INTRODUCTION

A goal of design is to prevent failures from occurring throughout the component life cycle, and situations resulting in severe failure should be avoided. The purpose of this chapter is to present a systematic procedure for minimizing the risks involved in metallurgical design. This procedure is a tool that can be used as a guideline for maximizing availability, reliability and safety, and cost reduction in the review stage of design. Variations of this procedure and the addition of new criteria to assess failure modes should be approached on an individual basis.

Design can be considered as an iterative process, often based on experience, to provide an assessment of the performance of a component during a certain period of time. This idea or concept can be either new or it may be derived based on market demands. Design involves different creative aspects: planning, development, procedures, availability, and fitness concerning the materials and processes used to manufacture the component. The design process culminates in a technical specification resulting in a successful manufacturing process.

In the previous chapter, H.W. Stoll presented basic design principles, pointing out that one of the primary concerns of design is to minimize the occurrence of failures. Under "Failure Prevention", Stoll included several approximations, commonly used to control uncertainties, such as safety factors, design allowable, design codes, fail safe design, safe-life design, damage tolerant design, and FMEA—Failure Mode and Effects Analysis.

The procedures described in this chapter will include some of these failure prevention methods throughout the different stages of component design. A methodology of analysis which combines metallurgical knowledge of failure analysis and fitness in-service criteria, with safety and reliability tools based on risk analysis, will be described here.

II. FAILURES DUE TO DESIGN

A good design incorporates usefulness, low cost, service life, ease of manufacturing, fitness in service, safety, environmental protection, low-cost maintenance, compliance with laws and regulations, originality and esthetic requirements, and market demands [1]. The designed product will be fit to perform the required functions resulting from a suitable selection of materials, shapes, manufacturing processes and procedures, and the implementation of the correct start-up, operation, and maintenance procedures.

Various types of damage may occur throughout the different manufacturing stages of a product during its useful life. These types of damage may lead to failures in the system. The term "failure" refers to the inability of a part of the system or the entire system to perform the functions it was designed for, leading to an unprofitable and technically useless product.

One example is Cr–Mo steel reactors, used in the petrochemical industry and designed to operate at high temperatures, in a hydrogen environment and under high pressures, which may crack primarily in those zones exposed to the welding heat because of the operating failure mechanisms, such as creep and hydrogen attack. However, other reactors which are made of similar materials may easily fracture because of toughness deterioration during service due to "temper embrittlement". These types of failures can be prevented in the design process if there is a suitable technical specification available for the purchase of materials. The consequences, resulting from these failures, can lead to different degrees of product mission loss, with its consequent costs, and damage to users, society and the environment.

Failures may arise during design, manufacturing, start-up, operation, and maintenance stages. Each type of damage can be caused by one or more failure mechanisms or modes. Each failure mode can be operative under a specific material or environmental condition, as well as manufacturing and operating conditions. Surface cracking, for instance, can be caused by fatigue or by stress corrosion, depending on the thermal, mechanical and chemical conditions during use. Deterioration throughout the product useful life should be regarded with special consideration, even when the product has been properly used.

In this chapter, attention will be focused on those potential failures that can be avoided during the design process if proper criteria are applied. Knowledge of potential failure modes, mechanisms that produce them, and their possible consequences is necessary. Materials' experts generally provide this knowledge. However, designers should have simplified criteria that enable them to identify a specific failure mode and corrective actions to avoid it.

The criteria to minimize failures are generally based on empirical data. Knowledge acquired by trial and error has been supported by recent material developments. The use of the scientific method—analysis and synthesis—has led to the development of criteria to reduce failures. A metallurgical knowledge of the nature of failure modes, the criteria to assess them, and the techniques used to avoid them, can be improved by using risk analysis techniques like the ones to be described in this chapter.

III. RISK ANALYSIS AS A DESIGN TOOL

Technical and scientific knowledge which has been assimilated into western culture during the last three centuries is based on Newton's physics, which states that the world can be described and future events estimated on the basis of a certain number of deterministic equations. However, at the present time, knowledge concerning physics, engineering, or human sciences is governed by the notion of uncertainty [2].

Human reaction to uncertainty is that of hesitation or fear. Although deterministic solutions to solve design problems are usually preferred, probabilistic solutions cannot be ignored. The objective is to minimize uncertainties associated with the decisions made during the design process.

A. Risk Assessment

Hazard is a condition that can cause injury or death, damage to or loss of equipment or property, or environmental harm [3]. Risk is the combination of the probability (or frequency of occurrence) and consequence (or severity) of a hazard [4]. Its scope is limited to a specific environment during a certain period of time.

$$\text{Risk(consequences/time)} = \text{Consequences (consequences/event)} \\ \times \text{Likelihood(events/time)}$$

A casual, unexpected, and generally unpleasant event that may occur can be classified in a qualitative manner and, based on its consequences, into four or five categories ranging from catastrophic consequences associated with death, environmental pollution and/or system loss, to insignificant consequences which do not affect the mission of the system.

The likelihood of failure factor can also be determined in a qualitative manner and generally classified into five categories ranging from a high probability for the occurrence of failure ($\chi < 10^{-1}$), this failure being described as "frequent failure" to "unlikely failure" ($\chi < 10^{-6}$). The likelihood of failure can be specified in a quantitative manner in all the cases where specific reliable information is available. The MIL-STD-882C "System Safety Program Requirements" (US Department of Defense, 1993) [5] is a standardized procedure which correlates hazard severity with its likelihood of occurrence. This process provides a ranking system.

Risk analysis methodologies have recently been applied to the oil industry which are based on tools developed by API (American Petroleum Institute), "Risk-Based Inspection" API Publication 581, May 2000 [6] which is based on the earlier work of the ASME (American Society of Mechanical Engineers). These techniques consider a generic likelihood of failure for a given equipment item, which is modified by means of specific factors. Modifying factors account for the environmental damage rate, the effectiveness of the inspection program, and management practices that affect the mechanical integrity of the equipment. The analysis of consequences lies in considering the effect produced by toxic and flammable products, environmental pollution, and business interruption. The consequence and likelihood factors are combined to assess risk for each equipment, section or operating unit.

Other risk analysis methodologies related to safety consider the likelihood of a hazard as the product of two factors: an exposure to risk factor, which considers the likelihood that the first undesired event which initiates the sequence of events leading to an accident may occur, and a second factor which represents the likelihood that, once the risk situation has arisen, the events taking place in the complete sequence of the accident may follow one another in time resulting in the accident and its consequences.

Risk analysis techniques performed during design usually consider risk as the product of three factors: consequences, frequency of occurrence of a specific failure mode, and the possibility of detecting a real or potential failure before the product is used. As in all risk analysis techniques, potential quantified risks are used to establish a criticality rank, permitting focus on those that are most critical.

In all cases, once the method to assess the likelihood of failure and its consequences has been defined, either a risk decision criterion or a risk assessment matrix should be adopted. The allowable risk level adopted will depend not only on technical considerations based on quality but also on social factors related to the protection of human health and the environment. These factors are variable and are strongly driven by environmental awareness.

B. Failure Mode, Effects, and Criticality Analysis

Failure mode analysis is a procedure in which each potential failure mode is analyzed to determine its effects and the criticality of these effects upon the system and rank each potential failure based on its severity. The criticality analysis involves the use of risk analysis techniques based on the assessment of the likelihood of failure and its potential consequences.

Among the most widely used tools in the analysis of design are the FMEA, which is accepted by OSHA (Occupational Safety and Health Administration) [7], and the Failure Mode, Effects and Criticality Analysis (FMECA)—whose scope is wider than that of the FMEA—which focuses on the likelihood of failure which permits the identification of the criticality of the component being analyzed. These methodologies are based on reliability and safety, and are useful to identify single-point failures. A "single-point" failure refers to an individual failure that may cause the entire system to collapse.

By using the FMEA methodology in the analysis of a new design, it is possible to identify single-point failures and redesign the product to avoid them, thus eliminating them completely or to achieve a more robust redesign which is less sensitive to failures. Through the use of FMEA, higher quality designs at lower cost are possible.

An engineer using FMEA should be aware that failure identification does not necessarily mean identifying hazards, since a hazardous situation may be a part of the normal system operation (N. Bahr, personal communication, 1999). Despite the fact that failures may be avoided, a system which often operates at high temperatures and under high pressures, using explosive and flammable fluids, constitutes a highly hazardous system, as in the case of any other systems associated with a high concentration of energy.

The Hazard and Operability Studies (HAZOP) are a primary method used in the petrochemical industry to review a design before its use in identifying hazards and inefficiencies associated with the process. Implementation of these methodologies enables a rational use of resources, minimizing the risk of undesirable consequences.

Risk Assessment Techniques and Risk Acceptance Criteria, as part of Risk Management, are powerful tools within a more extensive: System Safety Program. Risk management is a systematic approach used to define the system, identify and assess risks, ensure that the risk is "a priori" lower than the highest allowable risk level (safe risk level), determine corrective actions and their costs, eliminate or control hazards, verify that the controls are suitable and properly implemented, accept residual risks, and record the actions performed. This procedure results in increased process profitability through cost savings.

Reliability-Centered Maintenance (RCM) is a specific process used to identify policies which must be implemented to manage those failure modes which could cause functional failure of any physical asset in a given operating context [8]. Reliability-Centered Maintenance was initially developed by the commercial aviation industry to improve equipment safety and reliability. However, the RCM process is primarily related to maintenance, some elements of its decision logic may be applied in a risk-based design.

The application of some of these methodologies to review a design for potential failure will be illustrated in the subsequent discussion.

IV. PROCEDURES FOR PREVENTING FAILURES DURING DESIGN

Typical phases during design include: planning and requirements definition, concept design, detail design and test and validation. The iterative nature of design requires a continuous analysis and redesign process as described in the previous chapter.

Regardless of the procedure used for developing the design, at the conclusion of the design process, the required quality of the product should be defined and then described in a technical document which should also include a definitive layout or a final design related to the manufacture of the product. Only the minimum quality needed for the product to perform the function intended should be specified. Overspecifying, including the addition of restrictive features in quality description, may lead to unnecessary delays and increased end-product costs [9].

The design process requires the development of operational definitions (R. López, personal communication, 2000) to assist in communicating the intent of a design concept [10]. An example of an operational definition is a specific test of a material or judgment criteria. Without operational definitions, a specification is meaningless. For example, a casting specification containing the clause: "The casting shall be delivered reasonably clean" requires operational meaning of "reasonably clean" [10].

For this discussion, a design process is divided into PHASE I, which corresponds to the basic definition of the product (including concept and detail design), and PHASE II, which corresponds to the design review to prevent failures and minimize risks.

A. PHASE I: Basic Definition of Design

Figure 1 summarizes the scope of PHASE I which addresses the basic definition of the product. This phase begins with concept definition and ends with a basic description of design. During this phase, all the working loads shall be addressed. Loads, in general, refer to a combination of mechanical, thermal and chemical loads, although other load sources (electrical, magnetic, etc.) may also intervene and vary throughout the product life cycle.

As the first step, the type of material and its geometric configuration should be selected. The properties of the material, such as weldability, machinability, and formability must be suitable for the different manufacturing processes to be used. It is also important to consider the impact of these manufacturing processes and procedures on dimensional, microstructural and compositional variations of the material. The resulting product must be able to perform the required functions in the desired time. In this first phase concerning quality definition and preliminary description, legal regulations, design codes, safety and reliability aspects, maintenance and esthetical appeal requirements must be met while simultaneously meeting the technical requirements at the lowest cost.

As an example, consider a reactor typical of those used in a petrochemical plant which must operate in a certain environment, under certain pressure, at high temperatures during a specified design life. During this first phase of design, it can be established that the plate will be made of $2\frac{1}{4}$ Cr–1 Mo steel (ASTM SA-387 Grade 22) with calculated dimensions consistent for this steel with allowable design stresses and tolerances specified in the ASME code.

PHASE I

BASIC DEFINITION OF DESIGN

Design can be considered as an iterative process which aims to conceive an idea intended to be materialized in a product that will perform a required function during a certain period of time.

During Basic Definition of Design shall be taken into account the following factors

- **Functional requirements and constraints**
    - Mechanical
    - Thermal
    - Chemical
- **Fabricability and producibility**
    - Formability,
    - Cutting and welding
    - Heat Treatments
    - Surface treatments
    - Assembly
- **Conformance to Standards**
- **Energy requirements**
- **Durability**
- **Safety (Health and Environment)**
- **Reliability**
- **Availability**
- **Maintainability**
- **Appearance and resistance to degradation**
- **Economics (Costs)**

↓

A Technical Document or Product Specification must describe the resulting design defined during this phase.

Figure 1 PHASE I, basic definition of design.

In this stage of basic definition, some potential failure modes have been taken into account. Design allowances for corrosion/oxidation resistance were incorporated by the particular corrosion and oxidation-resistant material selected. By calculating the corrosion allowance, considering the material properties exposed to the design temperature, and the design life, ductile failure associated with a plastic collapse of the component under service conditions is avoided. The brittle fracture mode, associated with the lowest temperatures of the metal that may be exposed to, may also be considered in the requirements set forth in the design code. This results in the basic specification of the product.

B. PHASE II: Design Review Oriented to Minimize Risks

Figure 2 shows the actions that are involved in PHASE II design review with a special reference on how to minimize risks. The purpose of this phase is to ensure that the basic design fulfills the requirements and review the design for potential failures. During PHASE II, the following questions are addressed:

- Has product quality been properly defined to fulfill the user's requirements?
- Has product quality been properly described to order the product manufacturing?
- Has product quality been properly described to ensure the desired quality?
- Has susceptibility to failures been minimized?
- Which are the most critical aspects of design?
- Is it possible to detect potential failures before the product is used?
- What are the allowed risks?
- Has the engineering design process been completed?

The following steps should be performed to review the design for failure prevention:

1. Perform a detailed review of the product design to ensure performance; at the time of loading, it will be subjected throughout its life cycle.
2. Determine design criteria, objectives and limitations used in the basic definition of design.
3. Verify the fulfillment of design criteria. These criteria involve combinations between the main properties of a product. The main criteria applied are presented in Sec. 5. During this step, basic types of failure modes have been properly controlled by design.
4. Determine failure modes during the product life cycle should be verified. Section 6 provides a list of potential failure modes and its consequent damage.
5. Determine the types of damages associated with each failure mode in order to implement methods for detection.
6. Determine the likelihood and severity of failure associated with each failure mode. The criteria used to assess susceptibility—described for each failure mode in Sec. 7—may be used as a guideline. Risk level is estimated in a qualitative manner.
7. Describe potential failure effects of the parts and their propagation over the system. Determine whether there are any failures that may affect the entire system, i.e., "single-point failures".
8. Prioritize risks by ranking critical points for the safety and reliability of the product.

9. Determine corrective actions to minimize risk for each failure mode making use of decision logic applied to the evaluation and reduction of risk, Fig. 3, and the criteria to minimize risk for each failure mode are presented in Sec. 7 of this chapter.

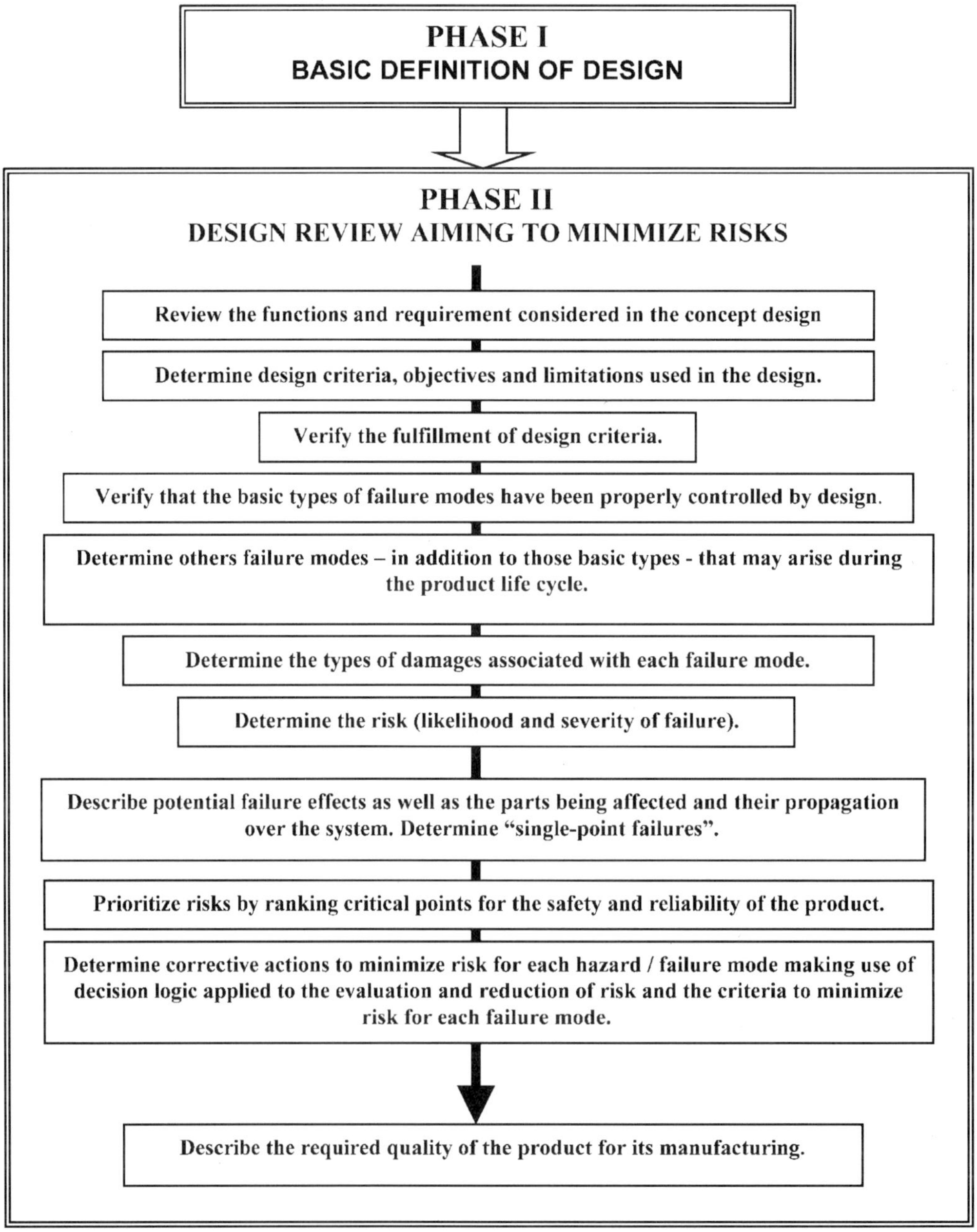

Figure 2 PHASE II, design review process used to minimize the occurrence of failures and its consequences.

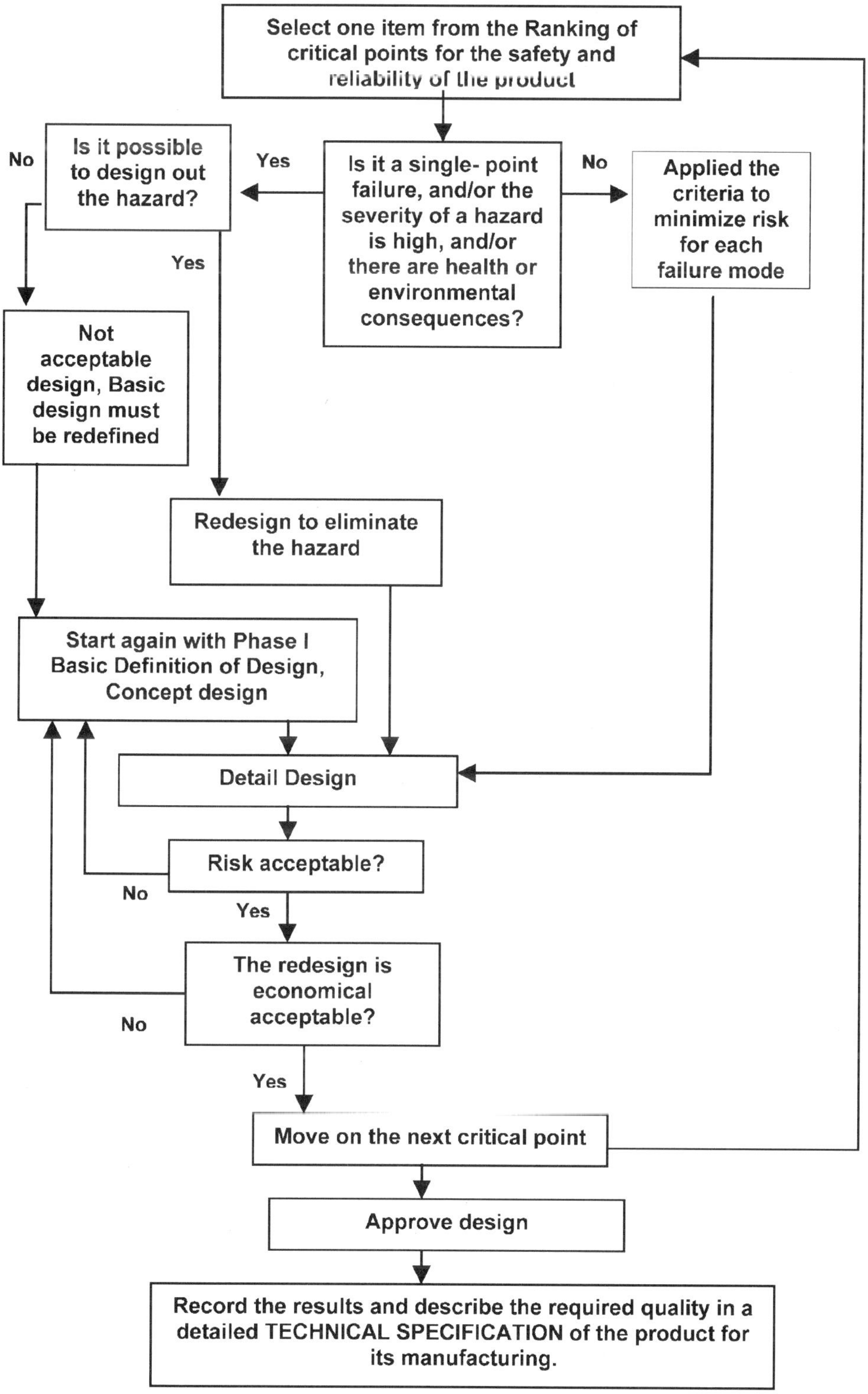

Figure 3 Decision logic applied to the evaluation and reduction of risk.

10. Record the results and describe the required quality criterion in a detailed Technical Specification of the product being manufactured.

Returning to the example of the Cr–Mo reactor provided in Sec. 4, a design review may show that the operating temperature range may produce brittle fracture which may be associated with material degradation due to temper embrittlement. This type of embrittlement may reduce the useful service life and increased risk of a catastrophic collapse of the reactor, especially during a hydraulic test or during shutdown and start-up periods of equipment operation. The decision logic for corrective actions (see Fig. 3) shows that it is necessary to use a Cr–Mo type of steel because of its resistance to both creep and hydrogen attack in addition to cost. Therefore, these design criteria must be applied to minimize temper embrittlement risk.

Purchasing a material with a restricted chemical composition and using welding consumables capable of being exposed to operation temperatures and thermal treatments during the manufacturing process can also minimize risk. Step cooling tests which are intended to assess the material behavior over a long period of time at a high temperature by simulating the expected material degradation during operation. A similar analysis might identify the potential necessary for minimizing the hydrogen attack, or any other more specific failure mode.

The use of this second phase in design permits the verification of the occurrence of the various failure modes and the completion of the definition of quality which contributes to a proper specification for the product to minimize the different types of failure modes.

V. MATERIAL BEHAVIOR

A. Types of Behavior Models

To meet design requirements, materials must have adequate mechanical strength and be able to resist different types of corrosion, wear, etc., so that the various types of failure modes can be avoided. Table 1 is a listing of the main properties and attributes used for material characterization.

Material behavior can be analyzed by developing models related to material attributes, required functions and manufacturing processes. Due to the large number of factors involved, the problem is often simplified by considering blocks of knowledge, corresponding to specific mechanisms and functions. Each block of knowledge represents a simplified model that relates some properties to the required functions. Using the state-of-the-art material engineering criteria which may be based on practical experience, it is possible to optimize performance. The results of this analysis are summarized in the definition of design.

In the next step, two groups of behavior models shown in Fig. 4 will be discussed. The first group is the behavior models used to avoid basic failure modes. These models relate material attributes—generally known as "properties"—which are well defined and individually determined. These properties are density, elastic modulus, tensile strength, elastic limit, elongation, fracture toughness, shape and size, etc. These test and criteria properties have specific operational definitions. Although these basic failure modes are usually present in all the metal designs, only the criteria used to avoid or minimize damage occurrence will be illustrated here.

Table 1 Material Characterization

Material properties	Modulus of elasticity (Young, shear), Poisson's ratio, ultimate tensile strength, yield strength in tension, shear strength, compressive strength, ductility, elongation, impact strength, fracture toughness, modulus of resilience, fatigue endurance limit, hardness, damping, creep strength, creep-rupture strength, thermal conductivity, thermal diffusivity, thermal expansion, specific heat, melting point, density, corrosion rate, electric conductivity
Material parameters	Chemical composition, chemical homogeneity, phases, crystallography, grain size, second-phase particle, non-metallic inclusions, dislocations, vacancies, texture, porosity
General attributes	Shape, size, cost, durability, availability, fabricability, weldability, castability, machinability, formability, hardenability, hot working parameters, heat treatment parameters, producibility, maintainability, surface finish, esthetic appeal, maintainability, biocompatibility, conformance to standards, energy requirements

The second group of behavior models includes attributes that involve the complex interrelation of a number of variables associated with the material, its manufacturing processes, and service conditions.Weldability, wear resistance, localized corrosion resistance, embrittlement, biocompatibility, etc., are among these attributes. Material weldability, for instance, is a broad and relative concept that refers both to the ease with which a material can be joined through some welding processes, and the resulting weld fitness in service.

For the second group of blocks of knowledge, physical metallurgy is used together with the laws of mechanics. Crystalline structure, microstructural and compositional characteristics, grain size and shape, dislocation density, second-phase particles, segregations, non-metallic inclusion shape control are among the concepts applied. In this case, anisotropy and the typical heterogeneities inherent to many materials become evident. These material attributes and criteria have no accurate operational definitions.

B. Behavior Models to Avoid Basic Failure Modes Based upon Well-Defined Material Properties

Behavior models which are used to avoid basic failure modes usually treat materials as being homogeneous, continuous and isotropic, and apply the laws of continuous mechanics. The properties used are those involved in the basic formulas of mechanical design and those typically used to characterize material properties provided in engineering handbooks, specifications, and supplier catalogs. Designers are typically interested in strength, creep strength, ductility, creep ductility, notch toughness, modulus of elasticity, thermal expansion, and fatigue strength. These models state that the combination of some properties may impose performance limitations, in addition to those limitations imposed by individual properties. Using these models, it is possible to develop designs that avoid basic failure modes, such as uniform corrosion, yield, ductile fracture, creep, elastic instability, brittle fracture, and fatigue. The criteria developed in detail by Ashby [11] can be applied in this step in the review process of design.

The primary behavior models intended to avoid some of the basic failure modes will be discussed subsequently.

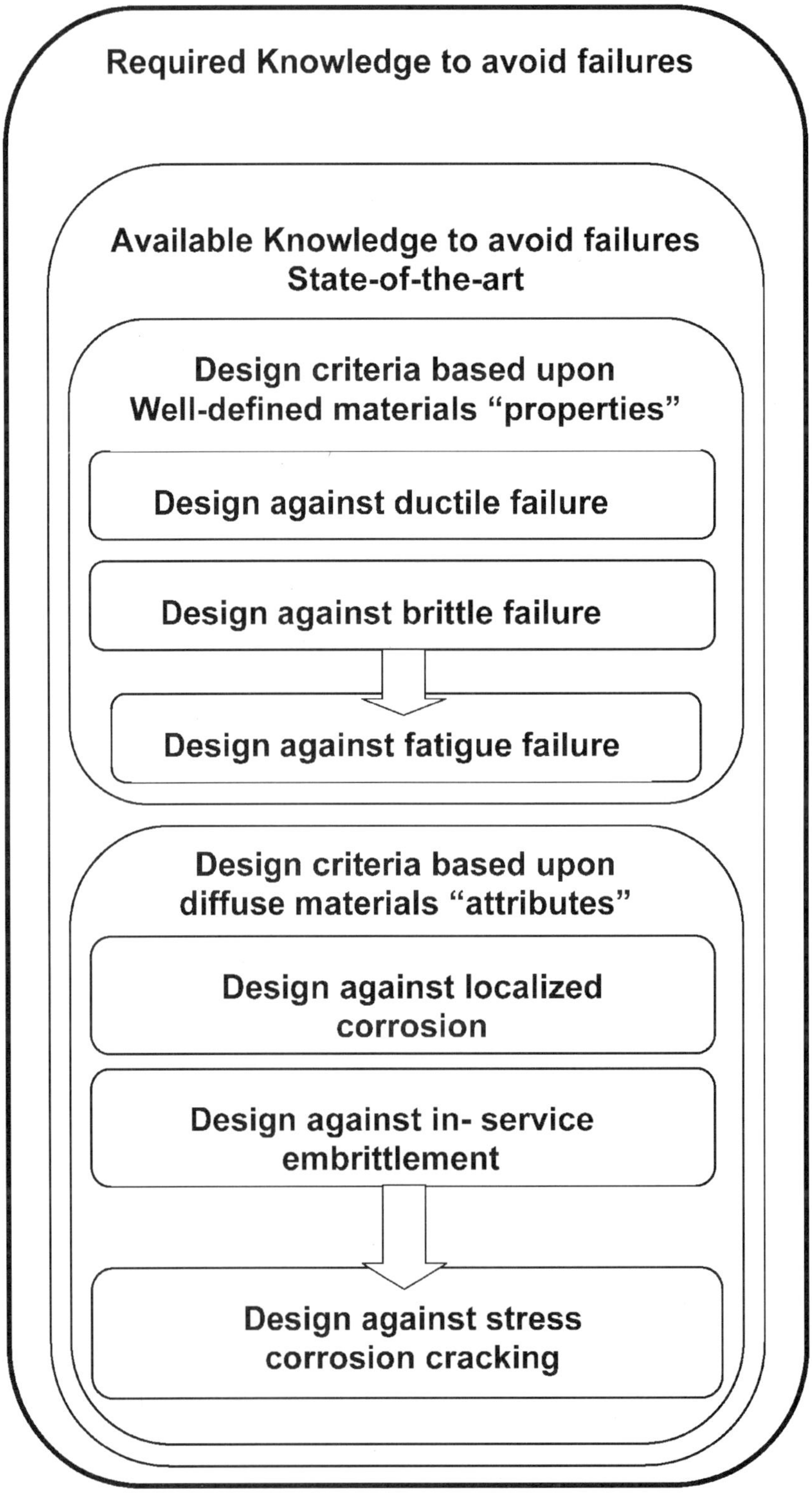

Figure 4 Block of knowledge used in failure design.

1. How to Avoid Uniform Corrosion

The first block of knowledge corresponds to material behavior during exposure to the service environment. In this section, general metal loss due to corrosion/oxidation and/or erosion will be discussed. If the metal loss rate among different points in an area varies by a factor of 4 or less, the damage is considered uniform.

The material selected should have a corrosion rate in the corrosive environment and at the design temperature which provides allowable material loss during the design life of the component. In some cases, design life is determined prior to the design process which may depend on both technical and economic factors. In the petrochemical industry, for example, a design is developed for a 10–15-year life extent, whereas in nuclear projects it is developed for a 40-year life extent.

Corrosion allowance is determined in accordance with the criteria set forth in the design code and the special conditions of the process. Corrosion allowance may be 1.6 or 3 mm; it is usually lower than 6 mm for economic reasons. When selecting the generic material, both corrosion resistance and the cost factor should be optimized. By using this behavior model, the risk of failure modes due to uniform corrosion is minimized. For uniform corrosion upon exposure to strongly corrosive environment, such as strong acids, strong alkali, salt water, and aerated water, the model that is the fraction of wall loss during thinning, due to uniform corrosion, is equivalent to ar/t where time is (a), corrosion rate is (r), and thickness is (t). This model provides a preliminary definition of the generic type of material suitable for design.

2. How to Avoid Ductile Failure Mode

The following block of knowledge refers to both the capacities, the load transfer capacity of the material, and the method to avoid ductile failure. This failure mode is related to "static" stresses and "static" mechanical properties, and is applicable to structures subjected to stresses of approximately constant magnitude, or where the total number of cycles imposed in the course of the working life is comparatively small.

Plastic deformation, which may lead to the component collapse or rupture, should be avoided. Both yield properties which result when the elastic limit of the material is exceeded and potential creep deformation are considered in this model. When a material is exposed to temperatures above $1/3\ T_m$ (melting point) and is mechanically loaded, time-dependent creep occurs. When structures are exposed to the creep temperature range, the material should be measured to avoid unacceptable deformations or rupture throughout its useful design life.

For the selected material, the required shape (casting, forging, plate, pipe) and the possible manufacturing processes should be considered first. Then, in accordance with the design code (e.g., ASME, ANSI) adopted, the specification for the material to be applied is selected and allowable design stresses at the operation temperature are determined [12].

The minimum wall thickness for the size of equipment and its pressure is subsequently calculated with respect to the allowable design stress from the code. The stresses taken into account are the ones known as "primary stresses", which equilibrate the applied loads, and "secondary stresses", which are self-equilibrating. Stresses are caused by any combination of continuous or cyclic loads, of an either mechanical or thermal origin [13]. By assuming that the material is isotropic and applying different criteria (Von Mises's equivalent stress, Tresca's criteria) along with the allowable design stress, the dimension of the product can be calculated. The allowable design stress of the previously selected material exposed to the design temperature is determined by means of the design code. Uniform corrosion allowance—previously determined—should be added to the calculated material thickness.

The criterion applied to avoid plastic deformation states that the calculated stress intensity or effective stress must be lower than the yield and design life creep-rupture stresses of the material:

$$\sigma_e \leq \sigma_y(\text{Von Mises}), \quad \text{where } \sigma_e = (1/\sqrt{2})[(\sigma_1 - \sigma_2)^2 + (\sigma_2 - \sigma_3)^2 + (\sigma_3 - \sigma_1)^2]^{1/2},$$

where $\sigma_1 > \sigma_2 > \sigma_3$ are the principal stresses.

When effective stress is exceeded somewhere within the component it does not necessarily indicate plastic collapse of the entire structure. Primary stresses may locally exceed yield, within certain limits, provided there is enough ductility to allow the material to yield without cracking. Plastic collapse occurs when primary stresses are uniform on the entire structure and exceed effective stress. To prevent an incremental collapse or thermal stress ratchet in each loading cycle, the total elastic stress intensity range, considering residual and applied stresses, should be limited to twice the yield stress.

According to Hooke's law, the elastic deformation is usually calculated as

$$\varepsilon_1 = [\sigma_1 - \nu(\sigma_2 + \sigma_3)]/E,$$

where ε_1 is the principal deformation in direction 1, σ_1, σ_2 and σ_3 are the principal stresses, E represents the elastic constants of the material (Young's modulus), and ν is Poisson's ratio. In addition to the material properties, the resistant section area is also an element of this calculus.

Design codes deal with uncertainties that arise from data, such as properties and calculation methods, by using safety factors to provide a safety margin. Allowable design stresses, used in calculations, are generally 2/3 times lower than yield stress or design life creep-rupture stress. Other restrictive criteria, such as limiting the allowable design stress to 1/4 times the ultimate stress, are applied as well. The safety factor generally ranges from 1.5 to 4.0. When welds are made, allowable design stresses are reduced due to a factor known as "joint efficiencies", depending on the specified non-destructive tests (NDT) level.

The application of this module guarantees an adequate mechanical strength capable of minimizing plastic deformation risk as well as considering thickness loss due to uniform corrosion throughout the design life.

3. How to Control Stiffness

In addition to strength, stiffness must also be calculated. Structure stiffness is based on the maximum elastic deflection under a load. Stiffness depends on both the geometry (properties throughout the section) and elastic modulus of the material. If the product to be designed is an element that must withstand not only tension loads but also bending, torsional, and axial-compressive loads, the combined effect of the applied loads (type, shape, and size) and the material properties associated with stiffness must also be determined. The moment of inertia of the section is used in the calculation. By applying this block of knowledge, failure due to structural distortion is avoided.

4. How to Avoid Elastic Instabilities

Design for those elements subjected to axial compression should be included to avoid not only plastic collapse but also elastic instabilities. Elastic instabilities may cause Euler buckling and local buckling. The occurrence of this failure mode depends on the geometry of the elements and on Young's modulus of the material. When the element subjected to compression is very long or slender, it has a greater tendency to buckling. Slenderness ratio equals l/r, where l is the length of the member and r is the least radius of gyration of the section. The radius of gyration is calculated as

$$r = (I_{\min}/A)^{1/2},$$

where $I_{\min}$ is the moment of inertia and A is the cross-section area. In general, the relationship between Young's modulus (E) and density (ρ), E/ρ should be maximized to increase stiffness; by doing so, yield occurs before buckling, whereas by increasing σ_f/ρ, strength also increases; hence buckling occurs before yield. For example, the critical load needed for a column under compression to become instable is proportional to EI/l^2, where E represents Young's modulus, I the moment of inertia of the section and l the column length.

5. How to Minimize Stress Concentrations

The elastic stresses calculated above are nominal values, which do not take into account local discontinuities, such as holes, notches or section changes. Even on a structure where stress intensity has been limited by yield criteria, there may exist highly localized regions where peak stresses are several times higher than yield. Maximum local stresses on a structure can be determined by considering nominal stresses multiplied by a stress concentration factor, and can be estimated through a detailed stress analysis or by using approximate formulas that account for the most common cases. The design should be verified to confirm whether there are stress concentration points that might activate failure mechanisms due to brittle fracture, corrosion or fatigue.

6. How to Avoid Brittle Fracture

Brittle fracture is associated with very little or no plastic deformation and with a cleavage fracture or intergranular surface, unlike ductile fracture, which is associated with a fibrous surface. A material may fail in an unstable and catastrophic brittle manner under stresses even lower than the allowable design stresses used to avoid ductile failures. This may occur in a given combination of material properties and applied stress levels.

The material property that controls brittle fracture strength is toughness. Other factors that have an effect on brittle fracture are material thickness, local stress level, including nominal stresses, residual stresses and stress concentration factors, temperature, and loading rate. Carbon and low-alloy steels undergo a transition from ductile failure mode to brittle failure mode at low temperatures. Resistance to crack propagation is measured through fractomechanical tests; crack propagation will occur when the stress intensity at crack tip, K, reaches a critical value K_c (MPa m$^{1/2}$). K_c values are experimentally obtained through fatigue precracked test pieces. The critical stress intensity for fracture under plane strain (K_{Ic}) corresponds to mode I loading (Tension) under plane-strain condition. The critical energy-release rate at fracture is

$$G_{Ic} = K_c{}^2(1 - \nu^2)/E(\mathrm{kJ/m^2})$$

where E is Young's modulus and ν is Poisson's ratio. Brittle materials are those that remain elastic until breaking (break occurs before yield).

In the case of brittle metals, the plastic zone associated with the tip of the crack is approximately 0.1–1 mm in diameter; whereas in ductile metals, the diameter of the plastic zone at the tip of the crack can reach 100 mm. For brittle materials, fracture toughness K_{Ic} has relatively low and properly defined values; but when materials are very tough, with significant plastic deformation before break, K_{Ic} is estimated by measuring the critical J-integral (J_c):

$$K_{Ic} = (EJ_c)^{1/2}$$

When design is intended to minimize brittle fracture risk, as in the case of pressure vessels where safety must be maximized, Fracture Mechanics is applied to the design so that yield occurs before break or leak occurs before break, while attempting to keep thickness low for costs and mass reasons. Yield before break is accomplished by maximizing the K_{Ic}/σ_y ratio (fracture toughness/yield stress); whereas leak before break is accomplished by maximizing ${K_{Ic}}^2/\sigma_y$. In both cases, a high yield stress is required to minimize material thickness. The latter ratio can be interpreted as being proportional to the size of the zone close to the tip of a crack (the plastic zone in ductile solids). The global approach of fracture assumes that failure can be described in terms of a single parameter, such as K_{Ic} or J_{Ic}.This approach may yet prove to be questionable in complex situations. This is the reason why local approaches of fracture have also yet to be developed [14].

The material to be selected should have adequate toughness to withstand the lowest design temperature. If the material is welded, both the HAZ and the weld metal shall have adequate toughness in their final condition. The design must be reviewed to minimize the presence of notches and defects that concentrate stresses. For some materials, toughness can deteriorate during service at high temperatures due to metallurgical changes. These aspects will be discussed subsequently.

Both material toughness and uniform corrosion resistance are elements in the selection of the material, through an iterative process, which includes mechanical calculations and costs.

7. How to Avoid Impact Damage Through Design

A material may have good tensile strength and adequate ductility under static loads; however, it may fail when loads are applied at a high velocity. Impact strength, provided the material is stretched below the elastic limit, is associated with the resilience modulus, R:

$$R = {\sigma_y}^2/2E(\mathrm{J/m^3})$$

where E (GPa) is Young's modulus, and represents the slope of the linear elastic part of the stress–strain curve, σ_y (GPa) being the stress to which the stress–strain curve is 0.02% deflected from the elastic line.

Resilience is the measure of the capacity the material has to absorb energy in the elastic field. The resilience modulus represents the maximum energy the material is able to store in an elastic manner, without causing damage, and it can be released when it is discharged/unloaded: in some cases, the element must be designed so that it is sufficiently resilient to absorb the kinetic energy associated with the impact as elastic energy. In other cases, design can be developed in a static manner by increasing loads by a factor of 2 or 4 so that impact loads are accommodated [15].

If the design contains elements that must be capable of storing elastic energy per volume unit, without deformation or breakage due to metal fatigue, as in the case of a spring, the σ^2/E ratio should be maximized regardless of its shape.

When the elastic limit is exceeded, the problem of toughness should be considered.

8. How to Avoid Fatigue Failure

Fatigue failure refers to the deterioration of a component caused by crack initiation and/or crack growth. When mechanical loads alternate, there may be a risk of fatigue failures. In fatigue failures, a crack grows in each loading cycle until the remaining component fails due to ductile or brittle fracture. This phenomenon can occur at stress levels lower than the allowable stresses for static loads. The most important characteristic of a fatigue failure is that some components made from ductile material fracture without revealing any indication of prior plastic deformation, and failure occurs at stress values below the elastic limit of the material. In the past, due to the fracture surface appearance analogous to brittle cleavage fracture, it was incorrectly supposed that a material changed its structure due to the application of cyclic forces and become crystalline and brittle.

To avoid fatigue failure in component design, a stress analysis considering pressure and temperature variations, mechanical vibrations, winds, and fluid pulsation during design life is performed. The results of this analysis, together with stress concentration factors, permit the establishment of the local stress range and the number of cycles. By using the S/N curves set forth in the design code adopted, for example, ASME Code VIII Div 2 [16] or BS 5500 [17], it is possible to calculate fatigue life.

Fatigue life refers to the number of stress cycles of a particular magnitude required to cause the fatigue failure of the component; whereas fatigue limit refers to the magnitude of stress range under constant amplitude loading corresponding to infinite fatigue life or a number of cycles large enough to be considered infinite by design codes.

Cumulative fatigue damage is defined as the $\sum n/N \leq 1.0$ ratio where n is the number of cycles applied for a particular stress range of constant amplitude, and N is the number of allowable material loading cycles or fatigue life for the same stress range of constant amplitude.

When fatigue assessment shows that fatigue life is unacceptable, some changes should be made in design, by either modifying structure stiffness or controlling damping vibrations. In addition, shape control should minimize the stress concentrator effect of welds. Surface flaws should also be eliminated. To this end, stress relief treatments are beneficial.

An alternative design method used to assess the fatigue failure is that used by Fracture Mechanics, which is a branch of mechanics dealing with the behavior and strength of a component containing cracks. The effect of the material properties, stress level and present flaws, characterized by non-destructive tests is combined.

In welded structures, the welds control fatigue life. The International Welding Institute (IWI) has developed "The Fatigue Design of Welded Joints and Components" [18], which contains recommendations for the design and analysis of welded components to avoid fatigue failures. It includes ferritic/perlitic or bainitic structural steels, as well as aluminum alloys. It makes use of approximations based on S/N curves and Fracture Mechanics. It does not include low cycle fatigue (where $\sigma_{\text{nom}} > 1.5\sigma_{\text{y}}$), corrosive conditions or operation temperatures within the creep range.

It is emphasized that fatigue failures strongly depend on design and manufacturing quality, which is accomplished by increasing fatigue strength by minimizing stress concentrators and applying complete joint penetration.

It is generally assumed that fatigue fractures are unrelated to the specific material; however, microstructural fracture mechanics shows that the material itself impacts the fatigue limit [19].

C. Behavior Models to Avoid Specific Failure Modes Based upon Diffuse Material Attributes

Material properties, though sufficient to enable an adequate dimensioning of the product, sometimes are not sufficient to describe the desired quality. Final technical specifications may require the use of other attributes intended to avoid failure modes of very different kinds.

Material suitability for mechanical shaping, welding, or thermal treatments is usually characterized by attributes which are more diffuse and difficult to measure, including formability, weldability, machinability, durability, producibility, etc. The concept of material weldability, for instance, involves both the ease with which materials may be joined by welding and the resulting weld fitness in service. Weldability depends on a complex combination of material properties, processes, welding procedures, and service conditions.

In specific behavior models, grain size is controlled to prevent brittle fracture. The chemical composition of carbon and low-alloy steels is specified through expressions, such as "Carbon Equivalent", to prevent cold cracking during the welding process and to control maximum hardness. In Cr–Mo steels, the impurity content is limited through compositional factors, such as the J or X factor to avoid temper embrittlement. In austenitic stainless steels, the ratio between equivalent chrome and nickel, impurities and delta ferrite contents is limited to avoid hot cracking and minimize sigma-phase embrittlement risk. For stainless steels, pitting corrosion is controlled through a compositional index associated with pitting resistance, and so forth.

Other properties, whose combinations become of interest in certain cases, are those called thermo-physical properties, such as thermal conductivity, λ (W/m K), which relates the heat flow to the thermal jump; and thermal diffusivity, a (m^2/s), which is equal to $a = \lambda/\rho C_p$, where C_p (J/kg K) is the specific heat and ρ (kg/m^3) is density. These properties are a function of temperature. Thermal diffusivity, a, is related to the time-dependent temperature change. These properties define the selection between insulating and thermal-shock protective materials.

Another temperature-dependent property is the thermal expansion coefficient, α (K^{-1}), which measures thermal deformation per degree of temperature increase. By maximizing the λ/α ratio, thermal distortion can be reduced. This is applicable when considering distortion associated with welds.

Owing to the characteristics of these material behavior models which are used to avoid failure modes, the diffuse attributes of the material shall be accommodated by the development of specific criteria. The criteria allow the use of general attribute to be applied to a specific case.

VI. CRITERIA FOR DETERMINING OPERATIVE FAILURE MODES

So far, basic failure modes used for design development have been discussed. In this section, selection of those failure modes that are applicable to each case will be discussed.

In Table 2, material damage has been grouped, based on whether they are associated with the product manufacturing stage or the product service. This list does not represent all the existing failure modes and damage types, but it does provide a guideline for considering the main causes of product failures and relating them, in some cases, to the type of damage they cause.

Damage occurring prior to service can be material flaws from the manufacturing process, and flaws due to processes, such as welding, cutting, shaping, thermal

Table 2 Material Damage [20]

	Pre-service material damage	
	• Material production discontinuities	
	• Fabrication related discontinuities	
	In-service material damage	
Mode of damage		*Type of damage*
General and localized metal loss (corrosion/oxidation/erosion)		Thinning
Hot corrosion	Type I ($T > 760°C$) Transitional hot corrosion (705–760°C) Low temperature hot corrosion type II (650–705°C)	Thinning/surface cracking/pitting
Mechanical fatigue Thermal fatigue Stress corrosion cracking Fatigue cracking Creep fatigue cracking		Surface connected cracking
Hydrogen attack (high temperature) Creep damage (high temperature) Hydrogen damage (low temperature)		Subsurface cracking, microfissuring/microvoid formation
Low temperature embrittlement Strain age embrittlement Sigma-phase embrittlement Hydrogen embrittlement 475°C Embrittlement Graphitization/carburization/nitruration Temper embrittlement Carbide precipitation Overaging		Metallurgical, microstructural and mechanical property changes
Buckling Thermal distortion Overload (plastic deformation)		Dimensional changes

treatments, etc. In many cases, damage prior to service does not cause an immediate failure. Only some gross discontinuities may cause problems during the hydraulic or pneumatic testing.

Sources of damage occurring during use are very diverse and strongly dependent on the working loads and different environments. In the petrochemical industry, for example, the most common sources of damage occurring during service are [20]:

- generalized or localized material loss due to corrosion/erosion,
- surface and subsurface cracking,
- microfissuring/microvoid formation,
- changes in metallurgical and product properties,

- dimensional changes, and
- blistering.

Each type of damage is caused by different failure modes or mechanisms, which involve uniform and localized corrosion, embrittlement, etc., and become operative under a specific combination of materials, environment, and service conditions.

Previously, generalized material loss affecting a significant area of a piece of equipment due to uniform corrosion was addressed and a design process was developed to avoid plastic collapse due to ductile fracture. If the corrosion rate varies considerably within the same region, corrosion is localized, and may be of different types, including pitting corrosion resulting in numerous surface cavities, selective galvanic corrosion on the regions between two electrochemically different materials, selective corrosion on the HAZ, corrosion in crevice due to a localized concentration of the aggressive agent, etc. Usually, the more resistant a selected material is to uniform corrosion, the more likely it is that if corrosion should occur, it would probably be localized.

Crack initiation during service typically starts on the metal surface due to different mechanisms, such as mechanical fatigue, thermal fatigue, fatigue-corrosion, or stress corrosion. Stress corrosion cracking (SCC) on austenitic stainless steels can be caused by polythionic acids and chlorides, by hydrosulfuric acid on carbon and low-alloy steels, and by caustic corrosion on carbon steels. Metallurgical changes associated with creep and/or hydrogen attack can also cause surface and subsurface cracking.

While fatigue cracking growth rate can be calculated, if stresses are known, to develop a design to minimize the risk of this type of failure mode; stress corrosion cracking growth rate, on the other hand, is difficult to estimate. Stress corrosion cracking failures should be prevented during design by appropriate material selection.

When surface cracking and interconnected microfissuring and/or microvoid formations arise within the low temperature range, they may be related to the hydrogen effect on the metal resulting in surface corrosion which allows atomic hydrogen diffusion into the metal. The phenomena related to hydrogen diffusion include blistering, and hydrogen cracking and embrittlement, when the operation temperature is within room temperature range. Operative mechanisms at high temperatures, however, include creep and hydrogen attack which cause microvoids and cracking in the final stages of the damage process. This type of cracking can be related to material surface condition.

Metallurgical properties (ductility, mechanical strength, corrosion resistance, and toughness) can change during service due to microstructural changes in the material caused temperature. Some metallurgical changes can be reversible and thermal treatments may be used so that utility in service can be partially recovered.

Toughness loss is one of the most severe metallurgical changes that may occur during service, since it has a great impact on the structural integrity of a component containing crack-like flaws.

Among the causes leading to the fragile behavior of a material are hydrogen embrittlement on carbon steels operating in H_2S or HF; strain age embrittlement on C and C–Mo steels exposed to a 150°C (300°F)–316°C (600°F) temperature range after cold deformation; graphitization on carbon steels exposed to $T > 427$°C (800°F); various mechanisms that can cause embrittlement on carbon, low-alloy (1/2 Cr–9 Cr) and 12 Cr steels exposed to $T > 700$°C (1290°F); carburization on low-alloy (1/2 Cr a 9 Cr) steels exposed to $T > 593$°C (1100°F); temper embrittlement on $2\frac{1}{4}$Cr–1Mo $T > 399$°C (750°F) steels; embrittlement at 475°C (885°F) on duplex stainless steels exposed to $T > 280$°C (535°F); phase-sigma embrittlement on austenitic stainless steels exposed to $T > 600$°C (1110°F).

Table 3 Determination of a Failure Mode

Material	Screening questions	Failure modes
Austenitic stainless steels	Is the material exposed to chlorides and water in the 35–150°C temperature range?	Chloride stress corrosion cracking
	Is the material exposed to sulfur-bearing compounds?	Stress corrosion cracking by polythionic acid
	Is the operating temperature between 600°C and 900°C?	Sigma-phase embrittlement
	Is the material exposed to temperatures between 600°C and 900°C?	Creep
Ferritic stainless steels	Is the material a high chromium (>13%) ferritic steel and the operating temperature between 350°C and 550°C?	475°C Embrittlement
Cr–Mo steels	Is the material $1\frac{1}{4}$ Cr–1/2 Mo, 21/4 Cr–1/2 Mo, 3Cr–1Mo steel and the operating temperature between 345°C and 565°C?	Temper embrittlement
C and low-alloy steels	Is the operating temperature >200°C, and the operating pressure >0.5 MPa, and the partial pressure of hydrogen high?	Hydrogen attack
High temperature superalloys	Is the operating temperature >650°C and is the material exposed to sulfur and sodium, and/or potassium compounds?	Oxidation–sulfidation phenomenon (hot corrosion)

Table 3 presents some questions that may be used to select the failure modes applicable to each case. This is not a comprehensive listing and should be considered as a guideline for similar questions to be used in other specific applications.

VII. CRITERIA FOR ASSESSING SUSCEPTIBILITY AND MINIMIZING RISKS

The following criteria which are intended to assess susceptibility to the occurrence of a specific failure mode are based on the pattern set below describing each failure mode.

- *Key factors*
  - *Causes*: What are the root causes of this failure mode? Which materials are more susceptible? Which mechanical, thermal and chemical loads initiate this failure? Which manufacturing processes or procedures contribute to this failure?
  - *Type of Damage*: Which type of damage is caused by this failure mode? When can it occur? Where does it occur? What is the propagation rate? What are the possible consequences if damage occurs?
- *Criterion for Assessing Susceptibility*
- *Criterion for Minimizing Risk*: What is the tolerance limit for this damage? Which inspection techniques should be used to detect the damage? Which are the corrective actions to be taken for minimizing risk?

This pattern can be used to describe other failure modes, which have not been included here for space reasons. The following mechanisms will be analyzed:

(A) thinning mechanisms,
(B) cracking mechanisms:
   (a) high temperature hydrogen attack,
   (b) chloride stress corrosion cracking, and
   (c) high temperature hydrogen attack,
(C) metallurgical changes:
   (a) brittle fracture:
      (i) temper embrittlement,
      (ii) sigma-phase embrittlement, and
      (iii) 475°C (885°F) embrittlement,
(D) material processing:
   (a) suitability of general structural steels for fusion welding:
      (i) tendency to brittle fracture,
      (ii) tendency to age,
      (iii) tendency to harden,
      (iv) hydrogen assisted cold cracking, HACC
      (v) tendency to segregate,
      (vi) tendency to lamellar tearing (anisotropy),
      (vii) weld metal solidification cracking,
      (viii) HAZ liquation cracking, and
      (xi) tendency to lamellar tearing.

A. Thinning Mechanisms

1. Causes

There exist several damage modes that result in the type of damage characterized as material thinning. If the rate of thinning is greater than expected, thinning will eventually result in failure due to ductile overload. There are two types of damage: general thinning and localized thinning. General thinning is defined as a relatively uniform metal loss over a significant area of the equipment. This type of damage includes uniform corrosion, such as ammine corrosion, sour water corrosion, hydrofluoric acid corrosion, sulfuric acid (H_2SO_4) corrosion, high temperature (H_2S/H_2) corrosion, high temperature sulfidic/naphthenic acid corrosion, hydrochloric (HCl) corrosion, and high temperature oxidation.

2. Type of Damage

An important step in design is to determine the locations where general or localized thinning is potentially active. Localized thinning includes pitting corrosion resulting in numerous surface cavities, crevice corrosion resulting from the concentration of aggressive chemical species, selective galvanic corrosion in the region between two electrochemically different metals, erosion–corrosion, corrosion-fatigue, wear. Wear is usually defined as the underside removal of material from contacting surfaces through mechanical action. Wear includes abrasive wear, erosion wear, grinding wear, gouging wear, adhesive wear, fretting wear, contact stress fatigue, cavitation fatigue, etc. The damage mechanism must be considered during the review of the material selection, and the material surface treatment specification. Environmental features, such as flow velocity and critical concentration of corroded agents must be considered.

3. Criterion for Assessing Susceptibility

The first step is to identify which damage modes are potentially active and to determine the estimated rate of thinning. The corrosion rate can be predicted by standard corrosion curves. The fraction of wall loss due to thinning can be calculated on the basis of the design life, thinning rate and thickness.

4. Criterion for Minimizing Risk

As discussed previously, uniform corrosion can be avoided through adequate material selection and corrosion allowance. To minimize wear, the type of material, chemical composition, microstructures, and hardness must be considered in the material specification. The susceptibility of the selected material shall be assessed by corrosion or wear test included in the material specification during the detail design stage. The designer can recommend that the equipment be assessed for thinning during service. In-service inspection includes intrusive and non-intrusive inspection with visual examination, ultrasonic thickness measurements or profile radiography; in the case of localized corrosion, the areas to be examined shall be specified by a designer.

B. Cracking Mechanisms

1. High Temperature Hydrogen Attack

a. Causes. Hydrogen attack occurs on carbon and low-alloy steels when they are exposed to temperatures above 200°C (390°F) and subjected to a high partial hydrogen pressure. Hydrogen attack can cause catastrophic fractures on half-pressure units, such as catalytic reformers and becomes a high risk factor for them due to thermal, mechanical, and chemical loads, associated with this type of reactors. This kind of degradation is caused by the presence of hydrogen during service. Hydrogen can dissolve in metal as atomic hydrogen reacting with iron carbides so as to produce methane. This methane cannot diffuse out of the metal. If the methane pressure is sufficient, cracks can arise in the steel. Internal decarburization may occur at high temperatures. The accumulation of methane in discontinuities, usually in grain boundaries, causes high stresses and microvoids, which can lead to cracking or blister formation in the case of steels with a high content of inclusions or laminations. Internal cracking is more frequent in carbon, C–$\frac{1}{2}$ Mo and Cr–Mo steels when the partial hydrogen pressure is high, whereas decarburization is more common in Cr–Mo steels exposed to high temperatures and subjected to low partial hydrogen pressures. In an advanced stage of damage, we can obtain decarburized cracked steel whose strength is lower than half its original strength and its ductility is reduced to zero. Nelson's curves (API RP 941) [21] enable the assessment of susceptibility to this type of damage for specific operating conditions and material composition. Steels containing only Fe_3C carbides are the most susceptible ones, while alloyed steels containing more stable carbides are more resistant. Base metal and postweld heat treatments affect the material resistance to hydrogen attack. A heat treatment at temperatures above

649°C (1200°F) and below the lowest critical temperature is recommended for Cr–Mo steels. The degree of a material cold deformation affects hydrogen service. Hydrogen attack is accelerated with cold deformation.

b. Type of Damage. In the initial stage, the hydrogen attack damage mode causes microvoids in grain boundaries, which collapse forming intergranular microcracks. These interconnected microcracks cause macroscopic cracks that may be connected with the surface. The weld heat-affected zone can be easily attacked, and this attack may result in the catastrophic collapse of the component. Since hydrogen attack is controlled by diffusion, the time for it to occur and the operation temperature are closely related. In some cases, the resulting microcracks are not easily distinguishable from those resulting from the creep mechanism. The presence of tramp elements, such as As, Sb, Sn, and P increases the susceptibility to this phenomenon.

c. Criterion for Assessing Susceptibility. The susceptibility to this phenomenon can be assessed through the parameter

$$Pv = \log(P_{H2}) + 3.09 \times 10^{-4}(T)(\log(t) + 14)$$

where P_{H2} is the partial hydrogen pressure (Kgf/cm^2), T the temperature (K), and t is the time (hours). With the operation temperature and the estimated partial hydrogen pressure, and considering 200,000 hr of service, we can calculate the Pv parameter; in the case of carbon steel, if $Pv = 4.53$, the material exposed to these conditions is considered to be not susceptible to this damage mode [6].

d. Criterion for Minimizing Risk. This type of damage can be detected through "in situ" metallography, and advanced ultrasonic backscatter technique (AUBT). When the damage is advanced and connected with the surface, the magnetic particles or dyed penetrant tests are effective. If operating conditions are characterized by a $Pv > 4.40$, an adequate selection of the material using Nelson's curves is recommended. In addition, Cr–Mo steels should be specified in accordance with the criteria required for avoiding temper embrittlement. Stress relief heat treatments are also recommended after both welding processes and those processes involving cold deformation.

2. Chlorides Stress Corrosion Cracking (SCC)

a. Causes. Stress corrosion cracking may arise when a susceptible material, such as austenitic stainless steel containing 8–10% nickel, is simultaneously combined with certain levels of tensile stresses and chloride ion concentrations in an aqueous environment and within a 35°C (95°F)–200°C (390°F) temperature range. Tensile residual stresses, resulting from manufacturing processes like welds, contribute in causing this type of damage. Also, cold plastic deformation causes hazardous residual stresses.

b. Type of Damage. Chloride stress corrosion cracking causes surface cracks, which are generally transgranular and branched, and initiate on the surface propagating in a direction perpendicular to tensile stresses. The cracking surface is similar to that of brittle fracture, with little or null plastic deformation. Damage can occur during service or shutdown periods in the inner side of a container due to aggressive content conditions, as well as on its external surface under insulations. Hydraulic test conditions should also be considered. Those zones under high mechanical strength are more susceptible. Pitting corrosion can contribute to concentrate chlorides and initiate stress corrosion. It is difficult to estimate a damage rate for stress corrosion cracking, since it can be fast enough, and difficult to notice through visual inspection, to cause significant losses or the brittle collapse of the component.

c. Criterion for Assessing Susceptibility. This failure mode can be operative in the case of an austenitic stainless, 304/316 steel, subjected to tensile stresses arising from the manufacturing process or service loading conditions (tensile stresses >80 MPa are enough, provided that the chloride content is higher than 1 ppm and the temperature is above 35°C (95°F). High susceptibility is associated with aqueous solutions with pH <10, temperatures above 65°C (150°F) and a chloride content > 10 ppm. Chloride contents over 1000 ppm can cause stress corrosion cracking even with pH > 10, provided the temperature is above 90°C (195°F). If austenitic stainless steels covered by insulation operate in a continuous or intermittent manner at temperatures ranging between 35°C (95°F) and 150°C (300°F), they are susceptible to this damage mode.

d. Criterion for Minimizing Risk. The likelihood that this failure mode will become operative must be avoided during design due to the severe consequences of the associated type of damage mode. Tensile stresses must be reduced by controlling manufacturing processes and design. Postweld stress relief heat treatments can be used to minimize susceptibility to stress corrosion cracking, attempting to avoid stainless steel sensitization by using stabilized or low carbon grades. If corrosive environments causing stress corrosion on the selected material cannot be avoided, the selection of another material should be evaluated. The materials to be used can be those with a high Ni content, such as superaustenitic stainless steels or Ni basis superalloys, or alloys with a low Ni content such as duplex stainless steels or ferritic stainless steels of high purity (extra low interstitials). Some titanium alloys can also be selected.

C. Metallurgical Changes

1. Brittle Fracture

a. Causes. Brittle fracture is the sudden failure of a structural component, which usually initiates in a pre-existent flaw. They can be highly hazardous due to the cost resulting from equipment replacement, as well as possible damages to the personnel, environment and facilities. In order to minimize risks associated with brittle fracture, the materials are required to have adequate toughness when they are being used, even when they do not operate during service at low temperatures. Material toughness is the material's ability to absorb energy as plastic deformation without breaking. Brittle fracture depends on the following factors: (a) Stress multi-axiality: high loading levels and their spatial distribution in three directions increase the tendency to brittle fracture. If the loads are too low, brittle fracture does not occur. (b) Residual stresses associated with a weld increase the tendency to brittle fracture. Postweld heat treatments are beneficial. (c) Large thickness is more susceptible. (d) Load application rate: a sudden application of loads may contribute to brittle fracture. (e) Material properties: fine grain structures, such as tempered martensite, with low contents of impurities, are associated with a high degree of toughness. Other microstructural elements, such as precipitates, second-phase particles, dislocations, and solutes in a solid solution contribute to increase the σ_y but deteriorate toughness. (f) Temperature: a wide range of materials, ferritic steels, in particular, present a transition temperature between ductile behavior at high temperatures and brittle behavior below these temperatures. The minimum metal temperature should always be above the material transition temperature. The metal temperature considered is the lowest temperature among the operation, upsets, and the hydraulic test temperatures; in the case of pressurized liquids, the boiling point of the liquid at atmospheric pressure should be considered. The brittle fracture is not very common, since design stresses are low enough to prevent it from occurring. In general, this failure mode is not operative at temperatures above 150°C (300°F). Those flaws that may initiate this fracture mechanism are difficult to detect through conventional non-destructive tests; therefore, it is not possible to develop a strategy for minimizing the risk of brittle fracture in the elimination of crack-like flaws exclusively. In ferritic steels, a ductile-to-brittle transition occurs as the temperature decreases. High-strength steels can undergo a less abrupt ductile-to-brittle transition, although they may register lower values of absorbed energy within the ductile range. Titanium alloys, aluminum, as well as austenitic stainless steels register no transition between ductile fracture and brittle fracture modes depending on the temperature. Fracture toughness decreases with an increase in the strain rate, and increases with the temperature in a manner similar to the variation of absorbed energy with the temperature in the Charpy curve.

b. Type of Damage. Design should be intended to avoid susceptibility to the initiation and propagation of cracks. At low temperatures, the fracture occurs due to completely brittle cleavage mechanisms, absorbing very little energy; thus cracks may initiate and propagate easily. At high temperatures, on the other hand, the fracture occurs due to ductile dimple mechanisms, with a significant absorption of energy. Cracks do not easily initiate and propagate in the upper shelf region of the curve; while in the ductile–brittle transition zone, though they initiate easily, they propagate with difficulty. The damage associated with fracture toughness may occur when the operation temperature is low, or during the equipment shutdown and start-up periods. Stress concentrators prove to be critical regions for the initiation of brittle fractures. A subcritical crack growth rate may result from different failure modes, such as creep, low-cycle fatigue, hydrogen embrittlement, and stress corrosion cracking. When the crack exceeds the critical size, it can propagate in a sudden way causing the catastrophic fracture of the component.

c. Criterion for Assessing Susceptibility. There are several methodologies used to determine a structural component's susceptibility to brittle fracture, such as those presented in the ASME Boiler and Pressure Vessel Code API 581 (6,16) and API RP 579 [20]. These design criteria are intended to prevent brittle fracture due to low temperatures or low toughness, or when taking into consideration the applied loads, the material specification (often qualified for use by performing impact tests), the minimum temperature, residual stresses, postweld heat treatments, and thickness.

d. Criterion for Minimizing Risk. Once a component has proven to be susceptible to brittle fracture, the material with adequate fracture toughness properties should be specified and the performance of mechanical tests, such as the Charpy test and more complex fractomechanical tests should be required. The Charpy V-notch impact test is widely used to characterize the ductile-to-brittle transition in steels. In general, in low strength steels, this test is used as a quality control criterion for structural steels, establishing a minimum level of absorbed energy in order to determine ductile-to-brittle transition temperature, thus steels can be compared. The Charpy test is of a relative value for design, since the notch does not really represent possible present flaws, the test piece size does not match the real thickness, and the loading conditions are not real, either. Other tests (CAT: crack arrest transition temperature, FATT: fracture–appearance, and FTE: fracture-transition elastic) enable the designer to determine not only the ductile-to-brittle transition temperature but also the critical stresses for the propagation of cracks depending on the temperature, thus allowing one to define the temperature above which propagation is difficult. In order to analyze whether a certain flaw (characterized by its size and shape) is critical for the present stress levels on a structure, the Fracture Mechanics approach is used [22]. According to the LEFM (Linear Elastic Fracture Mechanics) approach, the crack propagates when the stress intensification factor, K, which depends on the geometry, mode and magnitude of the applied loads, and on the size, shape and location of defects, reaches a critical K_c value. This approach is valid for those materials in which plastic deformation is restricted to the tip of a crack and the analyzed defects are crack-like defects with zero radius at the tip of the crack. Fracture toughness is determined by means of the ASTM Standard E 399 [23]. To assess those materials whose behavior involves a large plastic deformation around the tip of a crack, the EPFM (Elastic Plastic Fracture Mechanics) has been developed. The concept used is that of the *J*-integral, which leads to a parameter for fracture toughness known as J_{IC}. The method used to determine the J_{IC} parameter is defined in the ASTM 813 [24]. ASME Section VIII Division 2 rules specify toughness in terms of a minimum value of absorbed energy (27 J (20 ft-lb)) at the minimum allowable temperature. Other criteria determine 54 J (40 ft-lb) at 10°C (50°F) or at the minimum operation temperature. This value corresponds to approximately $K_{Ic} = 137$ MPa $m^{1/2}$. In order to meet the specified requirements for toughness in service, the materials should be manufactured through secondary-metallurgy practices so as to ensure an adequate degree of homogeneity, the control of impurities, inclusion shape and size, and sulfur and oxygen contents, and applying fine grain practice and thermal treatments, such as normalizing and tempering or quenching and tempering. In the case of steels, this is accomplished by means of microalloying, thermo-mechanical treatments, and postweld stress relief heat treatments. With respect to impact loads, it is recommended to design the component as an energy absorbing system, with high resilience modulus ($\sigma_y^2/2E$), with sufficient ductility to release stresses in high concentration zones and adequate fatigue resistance. Stress concentrators must be reduced to the minimum, avoiding abrupt section changes. If possible, the material should be used, so that impact load directions coincide with the lamination directions. Inertial forces should be taken into account and flexible supports should be used in order to reduce the acceleration or deceleration of the member.

2. Temper Embrittlement

a. Causes. Temper embrittlement is one of the main causes of toughness degradation in ferritic steels. This degradation may lead to withdrawing the component from service. The problem arises when some types of steels are exposed to temperatures ranging between 345°C (653°F) and 565°C (1050°F). For instance, 2 1/4 Cr–$\frac{1}{2}$ Mo steels with a bainitic structure are the most susceptible to this phenomenon, though temper embrittlement can also occur in C–$\frac{1}{2}$Mo, 1 Cr–$\frac{1}{2}$ Mo, 1$\frac{1}{4}$ Cr–$\frac{1}{2}$ Mo, 3 Cr 1 Mo, 5 Cr–$\frac{1}{2}$ Mo steels. Conversely, 9Cr–1Mo steels are less susceptible. Welded joints (weld metal and HAZ) are the most susceptible zones. In all the cases, the solution to the problem lies in the alloy purity. Exposures within this temperature range may occur during temper or postweld heat treatments or during service, and they should be avoided if at all possible. However, many components operate within the critical temperature range. The segregation of residual elements (antimony (Sb), arsenic (As), phosphorus (P), and tin (Sn)) towards austenitic grain boundaries is the main cause of temper embrittlement. Also, Mn and Si play an important role in this segregation and their content should be limited. Both residual and alloy elements can segregate, but the former can be concentrated up to 300 times their average value in the material. Segregation only occurs in ferrite within a 315°C (600°F)–540°C (1005°F) temperature range, but it never occurs during an austenitization heat treatment. In addition to segregation in grain boundaries, a fine precipitation can occur inside the grains, thus resulting in an increase in its strength (Mo_2C precipitation). The phenomenon associated with changes in grain boundaries causes intergranular brittle fracture. In general, ductility and rupture strength are not affected; nevertheless, both can be reduced under severe conditions. Effect upon toughness: it rises with an up to 100°C (210°F) shift towards the right of the ductile–brittle transition curve resulting from an impact test. The transition temperature used to characterize shifts is the one known as "Fracture Appearance Transition Temperature" (FATT). Sometimes, this effect is accompanied by an up to 25% decrease in the upper shelf energy (the energy absorbed in the ductile fracture field). Since the FATT is related to the K_{Ic}, an increase in the FATT implies a decrease in K_{Ic}. This means that the critical defect size for a given stress level is reduced. Stresses increase temper embrittlement intensity when the P segregation in the grain boundaries increases. Nevertheless, when a material is exposed to a 370°C (700°F)–565°C (1050°F) temperature range, property degradation can become irreversible; in this case, temper embrittlement and creep operate simultaneously. During the operation of a piece of equipment in a hydrogen service environment, hydrogen can diffuse into the metal. During cooling from the operation temperature and associated with a shutdown period, the material becomes oversaturated with hydrogen. The combination between thermal stresses and hydrogen oversaturation may cause hydrogen-induced cracking. If the material toughness has been reduced considerably due to temper embrittlement, the risk of a catastrophic cracking is high.

b. Type of Damage. Material degradation due to temper embrittlement may result not only in brittle fractures with catastrophic consequences, but also in a reduction of the equipment useful life, and in a decrease in the equipment reliability and efficiency, since it may be necessary to operate at lower temperatures so as to avoid temper embrittlement or to depressurize in order to avoid stresses while the equipment is cold. Temper embrittlement is reversible. Heat treatment for a short period of time at temperatures above 565°C (1050°F), followed by quick cooling, can restore the initial properties; however, if material thickness is high and cooling rates required by precipitation kinetics are not achieved, the remaining embrittlement may occur. Moreover, the component may crack during the heat treatment due to the effect of thermal stresses.

c. Criterion for Assessing Susceptibility. In all the cases, susceptibility depends on the alloy purity. The content of impurities is determined by calculating the J or X factors. Embrittlement occurs when some types of steels are exposed to a 345°C (655°F)–565°C (1050°F) temperature range during manufacturing or service. Welded joints (weld metal and HAZ) are the most susceptible zones. Material thickness over 25 mm is more susceptible to brittle fracture due to temper embrittlement. Postweld heat treatments minimize the susceptibility to brittle fracture through this mechanism.

d. Criterion for Minimizing Risk. Metallurgical design criteria to avoid the possible damage. In order to purchase Cr Mo materials to be used during service in environments susceptible to temper embrittlement, it is recommended to: (a) Specify a material with a CHARPY initial value at the lowest operation temperature of 40 J (average) and 30 J min. (b) Restrict the chemical composition through the J and X factors:

forgings and plates: J factor $= (\mathrm{Si} + \mathrm{Mn})(\mathrm{P} + \mathrm{Sn}) \times 10^4$ (% elements) < 120
welded joints: X factor $= (10\ \mathrm{P} + 5\ \mathrm{Sb} + 4\ \mathrm{Sn} + \mathrm{As})/100 < 25$ ppm

After a step-age cooling test, the material should meet the following requirements: $\mathrm{VTr40} + 1.5\ \Delta\mathrm{VTr40}_{sc} \leq 38°\mathrm{C}$, where VTr40 is the impact transition temperature before step cool aging in Charpy V-notch 54 J (40 ft-lb) and $\Delta\mathrm{VTr40}_{sc}$ is the shift on the impact transition temperature after step cool aging, also in Charpy V-notch 54 J (40 ft-lb).

3. Sigma-Phase Embrittlement

a. Causes. When heated within a 600°C (1110°F)–900°C (1650°F) temperature range, austenitic stainless steels (Fe–Cr–Ni), as well as ferritic and martensitic stainless steels (Fe–Cr), can form a brittle intermetallic compound called sigma phase (Fe–46% Cr) [25]. The content increase of ferritizing elements, such as Cr, Mo, Ti, Si, Nb, Al, W, V, increases the tendency to form sigma phase; on the other hand, austenitizing elements, such as C, Ni, Mn, and N tend to delay this phase formation. Carbon combined with Cr, Mo, and W tends to form carbides, thus decreasing its availability to form sigma, whereas nickel decreases the diffusion rate in the austenite of ferrite formers. Sigma phase forms slowly from an entirely austenitic structure; but if delta ferrite is present, it quickly transforms into sigma and austenite phases, since ferrite is a phase richer in Cr than austenite. In addition, delta ferrite formers segregate towards ferrite and many of them accentuate the transformation to sigma. The presence of a high content of delta ferrite in the weld metal constitutes the basic cause of the precipitation of sigma. In an austenitic–ferritic material undergoing a heating process at 700°C (1290°F) for 30 hr, a precipitation of sigma from the existing delta ferrite, and a carbide precipitation of $M_{23}C_6$-type chrome, occur in the ferrite/austenite boundaries. Manufacturing processes, such as welding and some heat treatments, or operating conditions, associated with thermal cycles during which the material is exposed to a 600°C (1110°F)–900°C (1650°F) temperature range, contribute to sigma-phase formation, especially when a high content of delta ferrite is present.

b. Type of Damage. The precipitation of sigma phase causes metal embrittlement. This phase transformation can occur during heat treatments or services at high temperatures. While precipitation kinetics from austenite is slow (since it may take months of service at high temperatures), the precipitation of sigma, when delta ferrite is present, occurs in a few hours, which is compatible with heat treatments. The damage becomes generalized over the entire piece in the case of materials with a high content of chrome, or localized over welds due to their tendency to have delta ferrite. The austenitic structure that contains precipitated sigma has a mechanical strength (tensile strength, hardness, and yield stress), which slightly increases with the content of sigma phase, while the material ductility decreases sharply when the contents of sigma prove to be higher than 5–6%. Toughness is the material property that most easily deteriorates. The precipitation of sigma phase is particularly a problem after long service at high temperatures or even after a post-weld heat treatment at 680°C (1255°F). The presence of sigma phase causes no damage while the material is hot, but it does cause embrittlement when the material gets cold. There is no transition temperature for an austenitic stainless steel containing sigma; ductility is low at room temperature and increases gradually towards high temperatures. Although toughness decreases by the effect of sigma phase, its importance should be estimated considering that the remaining austenite has excellent notch toughness. This phenomenon can also occur at lower temperatures, within a 550°C (1020°F)–600°C (1110°F) temperature range, but it is extremely slow, since the sigma formation is controlled by the diffusion of elements and impurities. As far as fracture mechanisms are concerned, they must be considered as ductile in the case of a weld metal of both ordinary and degraded austenitic stainless steels. However, in ordinary austenitic stainless steels, ductile fracture results from the concentration of microvoids associated with delta ferrite and non-metallic silicon-enriched inclusions, while in materials degraded by service, sigma phase and carbides produce additional sites for the concentration of microvoids, making fracture occur easily. Based on this mechanism, the fracture surface must be characterized by the presence of microvoids, which produce plastically deformed cavities containing precipitated particles inside. Cracks propagate throughout the austenite/ferrite boundaries network, which is present in the austenitic matrix. In addition to the brittle nature of sigma, its plate-like structure makes it break easily.

c. Criterion for Assessing Susceptibility. The following alloys are highly susceptible to sigma embrittlement when subjected to a long service at temperatures above 550°C (1020°F): Fe–Cr alloys with contents of Cr > 16%; Fe–Cr–Ni alloys with a high content of Cr, Si, and relatively low content of Ni; the weld metal susceptibility to sigma embrittlement of an austenitic stainless steel exposed to temperatures above 550°C (1020°F) depends on the content of delta ferrite: delta ferrite <3 FN is susceptible to low embrittlement, since it produces a sigma content <3%. Delta ferrite between 3 and 8 FN is susceptible to medium embrittlement, since it produces a sigma content between 3% and 6%. Delta ferrite >8 FN is susceptible to high embrittlement, since it produces a sigma content >6%.

d. Criterion for Minimizing Risk. Despite the presence of 10% sigma phase, the austenitic stainless steel has adequate toughness during service, but it decreases remarkably below 500°C (930°F) on reaching room temperature. This toughness reduction can be up to 60%. In an austenitic stainless steel undergoing heat treatment of annealed solution, the Charpy impact toughness is above 250 J, and fracture toughness is above 300 MPa $m^{1/2}$, though at room temperature it can be 18 J, and toughness decreases to less than 30 MPa $m^{1/2}$. A small quantity of delta ferrite precipitated sigma phase can be accepted without causing any failures, unless mechanical stresses are too high or stress concentrators are present. The presence of sigma phase can be detected by means of optical and electronic microscopy, EDAX microanalysis techniques, as well as by mechanical bending, Charpy V-notch and fractomechanical (CTOD) tests. If a component is to be subjected to a postweld heat treatment or exposed to temperatures within the critical range during service, it is recommended to specify weld metals without delta ferrite (E-NiCrFe 3 or E 307-type electrodes), or with controlled delta ferrite (E 308H, with <3 FN). In order to enhance oxidation resistance during services at high temperatures, the content of Cr and Si in austenitic steels is generally increased so that a protective oxide film is formed. As a result, the tendency to ductility and toughness losses increases when steels operate within a 650°C (1200°F)–950°C (1740°F) temperature range, due to sigma-phase precipitation embrittlement, as in the case of 310-type alloy (25Cr/20Ni). Nickel enhances this situation, for which 330-type materials (15.5Cr/35 Ni) or 601 alloys (60% Ni) minimize the risk of precipitation of sigma phase. Furthermore, the use of 0.08% carbon contents combined with chrome form $M_{23}C_6$ type carbides that increase creep resistance and delay the precipitation of sigma phase. The specification for a heat treatment with an annealing solution [1050°C (1920°F)/4 hr + water quench] enables the dissolution of both delta ferrite and sigma phase.

4. 475°C (885°F) Embrittlement

a. Causes. Stainless steels as well as their weld metals containing ferrite phase tend to brittle when they are exposed to temperatures within a 350°C (660°F)–550°C (1020°F) range. Toughness reduction results from the precipitation of brittle intermetallic phases within the above temperature range. Embrittlement is induced by the segregation processes that occur in ferrite and produce an iron-enriched ferromagnetic (alpha, α) phase and a chrome-enriched (80% Cr) paramagnetic phase known as alpha prime (α'). This transformation is known as the spinodal decomposition of the ferrite. The alpha prime toughness is much lower than that of ferrite. Ferritic and austenitic-ferritic (Duplex) stainless steels, castings containing ferrite, as well as austenitic stainless steels containing ferrite, are susceptible to this failure mode. The degree of embrittlement depends not only on the content of chrome and ferrite but also on the temperature. With a 12% Cr content, 105 hr within the critical range are required for embrittlement to occur. At 475°C (885°F), precipitation occurs in short periods of time. The addition of Cr, Si, Al, and Mo increases and accelerates embrittlement, whereas the addition of carbon decreases the tendency to embrittlement, though in the presence of Ti and/or Nb, this carbon effect is not considerable. The microstructure of duplex stainless steels contains austenite and 45–50% ferrite. The ferrite phase is susceptible of transforming into alpha prime, which is brittle even within the ductile austenitic phase. This phenomenon can occur even at about 280°C (535°F), depending on the type of alloy. Susceptibility to embrittlement is increased by a Cr percentage >27%. Ferritic stainless steels of a 405, 430, 442, and 446-type are susceptible. The family of 17% Cr is, in general, more susceptible than that of 13% Cr. Soft martensitic stainless steels with 13% Cr are susceptible to this phenomenon, though due to its low content of Cr, embrittlement occurs within long exposure periods (of about 20,000 hr) at high temperatures. Entirely austenitic stainless steels are immune to this phenomenon. Slow cooling within the critical temperature range must be avoided during manufacturing processes involving highly susceptible materials. A heat treatment performed at a temperature above 600°C (1110°F) with air cooling, if applicable, enables us to dissolve precipitates and recover toughness [2].

b. Type of Damage. When the Cr content proves to be high (26–30% Cr), this damage mode causes a considerable increase in hardness, tensile strength and yield strength, along with a decrease in ductility and impact strength in particular. In the case of duplex stainless steels, only delta ferrite is transformed, increasing its hardness and taking part in the embrittlement process, which occurs in very short periods of time (20-min exposure periods at a critical temperature). As a result of this change in the mechanical properties of the material, the component may break when it operates at temperatures lower than the transition temperature. The transition temperature of susceptible steels is around room temperature, depending on the alloys and their thickness. Low thickness (2 mm) 409-type steels register a transition temperature lower than 0°C (30°F). High thickness (10 mm) steels with a high content of Cr (26-1S-type) register a transition temperature up to 100°C (210°F). Although embrittlement occurs while the material is exposed to high temperatures, its brittle behavior becomes evident when the material is exposed to room temperature. In general, the resulting fracture initiates as flaws associated with stress concentrators, and propagates in a sudden way. This results in the catastrophic breaking of the component.

c. Criterion for Assessing Susceptibility. It should be assessed whether the above-mentioned susceptible steels will stay within the critical temperature range [350°C (660°F)–550°C (1020°F)] during manufacturing and/or service. If so, the component is considered to be susceptible to this failure mode. Harmful precipitation is avoided through quick cooling from heat treatment temperatures.

d. Criterion for Minimizing Risk. A structural component that becomes susceptible to this damage mode during its life cycle cannot be designed when the involved thickness is higher than 2 mm. Precipitation increases the material tensile strength and hardness. By measuring the material hardness, the degree of embrittlement can be controlled. If the component is susceptible to embrittlement during manufacturing, a heat treatment involving the dissolution of precipitates within an adequate temperature range [760°C (1400°F)–815°C (1500°F)] for each type of steel should be performed, and air cooling for thin sections as well as water cooling for higher thickness should follow. Furthermore, since these types of steels, particularly those with a high content of Cr, can be susceptible to sigma-phase embrittlement, which occurs exactly within that temperature range; a heat treatment ensuring the dissolution of both phases [1010°C (1850°F)/air cooling] should be performed. If the component is susceptible to embrittlement during service, a non-susceptible type of steel should be selected. In order to avoid this damage mode in duplex stainless steels, the operation temperature is limited up to 280°C (535°F).

D. Material Processing Suitability

In order to control the material processing and heat treatment, the designer will take into account the service requirement properties, the relationship between properties and microstructural and compositional parameters, and some design criteria to optimize these parameters and produce the best combination of properties within a realistic economic framework. The relationship between properties and microstructural and compositional parameters includes the strengthening mechanisms in steel. The main effects to be considered are substitutional and interstitial solid solution, grain size, second-phase particle, dislocations, and transformation. Several excellent reviews present different relationships obtained by using the multiple-regression technique [1], for example, the well-known Hall–Petch relationship by means of which yield stress may be related linearly to the reciprocal of square root of the grain size. Similar relationships describe toughness in terms of the impact transition temperature. In addition, relationships between strengthening and the square root of the solute element concentrations, dislocation density or second-phase parameters are employed. Steels are designed on the basis of the optimum combination of strength and toughness. It is important to notice that the grain size is unique by the fact that, as it becomes smaller, both strength and toughness increase.

5. Suitability of General Structural Steels for Fusion Welding

The behavior of structural steels during welding, as well as their suitability for service within the 30°C (85°F)–400°C (750°F) temperature range, depends on the following factors.

a. Tendency to Brittle Fracture. The risk of brittle fracture increases depending on the multi-axial tensile stress (high thickness), low temperature and high stress rate. The criteria to avoid brittle fracture presented above can be applied. Furthermore, Table 4 presents a guideline for the determination of steel type to avoid fracture, the procedure considers the factors affecting brittle fracture. It allows to determine the steel type including deoxidation treatment, Charpy-V impact test requirements, and heat treatment. The choice of material strength must be executed previously when the thickness is determined to avoid ductile fracture.

b. Tendency to Age. The tendencies to age after cold working increase hardness and tensile strength with a simultaneous reduction in ductility. Therefore, the risk of embrittlement due to welding in cold worked areas increases. The killed steels show enhanced resistance to aging in normalized conditions. In case the degree of deformation is high, the material should be thermally treated (normalized or stress relief) before the deformed zone is welded. The allowable risk of embrittlement due to welding in cold worked areas admit to weld, without special requirements, under the following conditions: the relationship between the inner bending radius (r) and the plate thickness (t) shall be $r/t \geq 1.0$ and $t \leq 4.0$ mm, or $r/t \geq 1.5$ and $t \leq 8.0$ mm, or $r/t \geq 2.0$ and $t \leq 12.0$ mm, or $r/t \geq 3.0$ and $t \leq$

Table 4 Suitability of Structural Steels for Fusion Welding: Avoiding Brittle Fracture

Guideline for the determination of steel type (deoxidation treatment, Charpy-V impact test requirements, and heat treatment condition) to avoid brittle fracture (valid for hot rolled C steels with service temperature between −30°C and 450°C)

1	Select appropriate value for structural complexity (restraint intensity)	Low	1
		Medium	2
		High	3
2	Select appropriate value for stress level	Low	1
		High	2
3	Select appropriate value for type of loads	Compression	1
		Tensile	3
4	Select appropriate value for rate of stressing	Low	0
		High	2
5	Select appropriate value for the minimum allowable metal temperature (operation, shutdown, etc.)	$T > -10°C$	1
		$-30°C < T < -10°C$	4
6	The structural factor will be the sum of points 1–5		
7	Plate thickness	Thickness in mm	
8	Multiply point 6 by point 7		
9	Select appropriate value for damage consequences	Low	0
		Medium	25
		Catastrophic	125
10	The *susceptibility factor* will be the sum of points 8 and 9		

The type of steel is determined from the susceptibility factor (point 10) using the following:

Type of steel	*Susceptibility factor*
Semi-kilned steel	<90
Semi-kilned steel and Charpy-V (28 J @ +20°C)	91–179
Kilned steel and Charpy-V (28 J @ 0°C)	180–299
Kilned steel + Normalized HT and Charpy-V (28 J @ −20°C)	>300

The minimum average Charpy-V impact energy of three specimens value of 28 J and minimum individual energy of three specimens of 21 J corresponds to steel with UTS < 450 MPa; an average value of 41 J and minimum individual value of 30 J are recommended for steel with UTS > 450 MPa.

24.0 mm, or $r/t \geq 10.0$ and all thickness. These conditions correspond to an elongation, $\varepsilon = 1/(2r/t + 1)$, in the cold worked areas of 33%, 25%, 14%, and 5%, respectively.

c. Tendency to Harden. The maximum hardness in the heat-affected zone depends on the chemical composition (carbon equivalent) and the cooling rate during welding ($t_{8/5}$). Hardenability of steel is not necessarily an indicator of HAZ hardness. It is important to control the maximum hardness in the HAZ in order to avoid two main problems: (a) cold cracking (Hydrogen Assisted Cold Cracking, HACC) during welding fabrication, in which the limit of maximum hardness usually is 350 Hv, and (b) in-service cracking in hydrogen environments, where the maximum hardness shall be less than 220 Hv. Maximum HAZ hardness can be accurately calculated as a function of chemical composition and the cooling time from 800°C to 500°C ($t_{8/5}$) [26].

d. Hydrogen Assisted Cold Cracking. CAUSES. Cold cracking can occur if three factors are present together: (a) local hydrogen concentration, (b) susceptibility weld metal or HAZ, local metal hardness, and (c) high level of residual stress remaining after welding; they cause cracking at temperature less than 100°C (210°F). The HACC causes are related to many factors including H_o (initial weld metal hydrogen content), H_r (residual hydrogen content at 100°C), CE (steel carbon equivalent), σ_y (yield stress of steel or weld metal), Q (heat input), T_p (preheat temperature), d (material thickness), R_F (joint restraint intensity), K_t (notch concentration factor), η (welding process thermal efficiency), and others. It is a diffusion-controlled phenomenon that does require days or weeks at ambient temperature to develop cracks.

TYPE OF DAMAGE. HAZ hydrogen induced cracking can be parallel to the fusion boundary adjacent to a fillet weld, or in the form of toe cracks. Weld metals are by no means immune when the steels have high yield strength. So, the HACC induce surface connected cracking or subsurface cracking, which may provide initiation points for further cracking by brittle fracture or fatigue.

CRITERION FOR ASSESSING SUSCEPTIBILITY. Recently, Yurioka [27] reviewed existing methods for prevention of HACC. The conclusion was that many predictive methods are available for prevention and control of hydrogen-assisted cold cracking in general welding construction. However, its limitations and applicable ranges must be understood when employing a certain method in welding construction. It is important to understand the metallurgical meaning of carbon equivalent to be referred to in welding practice. The significance of the effect of welding residual stress on HACC should be recognized and this effect must be taken into account in any predictive methods. Coupling the predictive methods for prevention of HAZ and weld metal HACC is urgently needed. It is considered necessary to develop criteria with operational definition in the blocks of knowledge that relate not well-defined material attributes, as is the case of the HACC. These statements were confirmed by the Yurioka [27] conclusion.

CRITERION FOR MINIMIZING RISK. Methods of avoidance cracking consist of adopting procedures that remove or attenuate one of the three factors which are necessary for it to occur. Thus, the problem can be avoided in some cases by the choice of material that does not harden in the HAZ or weld metal with the particular welding process employed. The material specification can include CE and C content restrictions. Benefit may be gained by controlling cooling time from 800°C to 500°C ($t_{8/5}$), and also from 800°C to 100°C ($t_{8/1}$). The thermal cycle depends upon the heat input, preheat temperature, plate thickness, thermo-physical properties of the material, thermal efficiency of the welding process and joint configuration. Properly controlled hydrogen procedure must be employed. If the above expedients are insufficient by themselves, preheating and welding operation must be followed by a postweld heat treatment at 250°C for 4–6 hr in order to avoid hydrogen concentration.

e. Tendency to Segregate. The segregation of elements (phosphorus, sulfur, carbon, etc.) impairs weldability.

HAZ HARDENABILITY. The chemical heterogeneity can contribute to increase locally the hardenability; so, in spite of a normal chemical composition of the heat, hardness can be higher than the maximum allowable hardness in a certain part of the HAZ. Chemical heterogeneity can contribute to increase hardenability locally so that hardness can be higher than the maximum allowable hardness in a certain part of the HAZ, despite a normal chemical composition of the heat. From the point of view of segregation behavior, semi-killed and killed steels are better than rimmed steels. If segregation zones are in fact involved—as in butt-welding—care should be taken to limit penetration and hence minimize weld metal dilution. In addition, a suitable filler metal, low hydrogen basic electrodes, shall be used.

WELD METAL SOLIDIFICATION CRACKING (HOT CRACKING). Segregations of solute and impurities can occur during weld pool solidification; so, if the ductility of the weld metal is lowered by the presence of residual liquid films to a level where it cannot accommodate the strain, hot cracking or solidification cracking will occur. Contamination with sulfur or phosphorus will be avoided. The tolerance of the weld metal to impurities also depends on the carbon level and alloy balance. Chemical composition restriction to avoid hot cracking can be included into the material specification.

HAZ LIQUATION CRACKING. In ferritic steels in the region adjacent to the fusion boundary, where high peak temperatures are reached during welding, local melting may occur at the grain boundaries due to the formation of a sulfide eutectic. The phenomenon is also known as "burning". Thus, heat-affected zone liquation cracking may appear when the plastic strains accompanying welding can open up microcracks at liquated boundaries. Furthermore, the residual liquid solidifies to give a brittle sulfide film, and the presence of these brittle films may reduce the fracture toughness of the HAZ to unacceptable levels. The material specification must ensure low carbon content and high manganese-to-sulfur ratio in order to control hot tearing and burning. For example, Mn/S of about 20 : 1 will ensure freedom from liquation cracking in carbon–manganese steels, with a carbon content of above 0.2%.

f. Tendency to Lamellar Tearing (Anisotropy). CAUSES. Generally, complex joint designs and 30–150 mm thick rolled products, certain types of welded details, including "T"-joint, multi-pass fillet or butt welds, corner welds and "T" fillet welds are likely to induce high restraint stresses perpendicular to the plate surface. Lamellar tearing in these circumstances has often occurred after welding. This tearing results from opening of lamellar segregations in a rolled plate at local temperatures of 200°C. This opening is due to the restrain stresses caused by passes and/or hydrogen embrittlement. The decrease in the elongated inclusions, manganese-sulfide, in particular, in a rolled plate is very effective in preventing lamellar tearing. Large inclusions, such as laminations are obviously harmful. However, arrays of very small inclusions can also be responsible. Silicates, sulfides, alumina and duplex inclusions may all be a part of the problem as they have been reported in many different types of mild and low-alloy steels, whether semi-killed or fully killed or vacuum degassed steels. It is also desirable to increase the value of reduction of area (Z-contraction) of the Z-tensile specimen. In the thickness direction, the toughness can be lower than parallel to the rolled plate plane as a result of the anisotropy. With decreasing deformation during rolling, this anisotropy becomes less.

TYPE OF DAMAGE. The cracks are parallel to the plate surface and almost invariably associated with planar inclusions. Lamellar tearing frequently initiates in or near the HAZ, and propagates with a characteristic steeped morphology.

CRITERION FOR ASSESSING SUSCEPTIBILITY. All pressure vessel steels must be considered potentially liable to the problem. However, the possibility of its occurrence can generally be neglected except where high reaction stresses exist in the short transverse direction or where design stresses are to be transmitted through the thickness of a rolled plate. Injurious factors affecting the susceptibility of laminar tearing are high plate thickness, complex joint configuration (T-joints), high weld fillet dimension, high degree of restraint, and material susceptibility; preheat temperature > 100°C is advantageous.

CRITERION FOR MINIMIZING RISK. Avoidance of lamellar tearing includes: (a) improved design avoiding joints in which load is transmitted through plates in the short transverse direction; where this is not possible, the design must allow deformation which relieves the material of its stresses in the thickness direction, (b) the use of forged products, as a replacement for those fabricated from plate, (c) the material specification including ultrasonic examination, vacuum degassed plate (O $<$ 20 ppm), low sulfur (S $<$ 0.010%), shape non-metallic inclusion control (rare earth metals (REM), Zr or Ca), and a short transverse tensile test ensuring that the reduction in area value exceeds 20%, (d) the welding procedure specification considering low strength weld metal, basic electrodes, and the "buttering" or "grooving and buttering" techniques.

VIII. CONCLUSION

In this chapter, a systematic procedure for minimizing the risks involved in design was presented. This procedure constitutes a tool that can be used as a guideline for maximizing availability, reliability and safety, and reducing costs in the review stage of design. The analysis combines metallurgical knowledge, failure analysis, fitness in-service criteria, with safety and reliability tools based on risk analysis. The use of failure mode analysis to rank each potential failure based on its risks, identify single-point failures and redesign the product to avoid them, or, at least, achieving a redesign less sensitive to failures was described. The knowledge required to avoid failures and hazards is partially covered by the state-of-the-art material engineering criteria. This available knowledge can be analyzed, divided into blocks of knowledge that relate material attributes, required functions and manufacturing processes. The first group of the material behavior models involves well-defined "properties", and they consider materials as being homogeneous, continuous and isotropous, according to the continuous mechanics laws. Usually, these properties, test and criteria have accurate operational definitions. Furthermore, a second group of material behavior models includes attributes that usually have no accurate operational definitions. Therefore, it is necessary to develop specific criteria, which are applicable to avoid failures on a case-by-case basis.

The design review procedure presented in this chapter is used to develop a suitable product specification, based upon operational definitions, where the likelihood and consequences of failure during the product life cycle result less than the allowable risk level.

REFERENCES

1. Pickering, F.B. *Physical Metallurgy and the Design of Steels*; Applied Science Publishers: London, 1978.
2. Prigogine, I. *El Fin de las Certidumbres*; Editorial Andres Bello: Ed. Chile, 1996.
3. Roland, H.E.; Moriarty, B. *System Safety Engineering and Management*; John Wiley & Sons: New York, 1990.
4. Bahr, N. *System Safety Engineering and Risk Assessment*; Taylor & Francis: Philadelphia, 1997; 7–10.
5. US Department of Defense. *Military Standard, System Safety Program Requirement*, MIL-STD-882C; US Department of Defense: Washington, DC, 1993.
6. *Risk-Based Inspection*, API Publication 581, 1st Ed.; American Petroleum Institute: Washington, DC, 2000.
7. US Department of Labor. *Occupational Safety and Health Administration. Process Safety Management of Highly Hazardous Chemicals*, 29 CFR Part 1919.119; Bureau of National Affairs: Washington, DC, 1992.
8. SAE Surface Vehicle Aerospace Standard. *Evaluation Criteria for Reliability-Centered Maintenance (RCM) Processes*, SAE JA 1011; SAE International, The Engineering Society For Advancing Mobility Land Sea Air and Space: Warrendale, 1999.
9. Lee, L., Jr.; Dobler, D.W. *Purchasing and Materials Management*; McGraw-Hill Book Company: New York, 1977.

10. Deming, W.E. *Out of the Crisis*, 23rd Ed.; Massachusetts Institute of Technology: Cambridge, MA, 1995.
11. Ashby, M.F. *Materials Selection in Mechanical Design*, 2nd Ed.; Butterworth Heinemann: Great Britain, 1999.
12. Breen, A.J. Low alloy steels in oil refinery service 1. In *Material Selection. Metal Construction*; The Welding Institute: Cambridge, England, November 1984, 671–676.
13. Glendinning, I.S.; Vincent, J.R. Low alloy steels in oil refinery service 2. In *Design Against Failure. Material Selection. Metal Construction*; The Welding Institute: Cambridge, England, January 1985, 23–29.
14. Pineau, A. Global and local approaches of fracture. In *Topics in Fracture and Fatigue*; Argon, A. S., Ed.; Springer-Verlag: New York, 1992, 97–234.
15. Blodgett, O.W. *Design of Welded Structures*; The James F. Lincoln Arc Welding Foundation: Cleveland, OH, 1966.
16. American Society of Mechanical Engineers. *Boilers and Pressure Vessel Code*, Section VIII, Div 2 – Alternatives Rules; ASME: New York, 1999.
17. *BS* 5500 : 1982 *Specification for Unfired Fusion Welded Pressure Vessels*; British Standards Institution: London, 1982.
18. *The International Welding Institute (IIW). The Fatigue Design of Welded Joints and Components*; Abington Publishing: Cambridge, England, 1996.
19. Miller, K.J. Metal fatigue—a new perspective. In *Topics in Fracture and Fatigue*; Argon, A. S., Ed.; Springer-Verlag: New York, 1992, 309–330.
20. *API Recommended Practice* 579. *Fitness-For-Service*, 1st Ed.; American Petroleum Institute: Washington, DC, 2000.
21. Garverick, L. *Corrosion in the Petrochemical Industry*; ASM International: Materials Park, OH, 1994.
22. Viswanathan, R. *Damage Mechanisms and Life Assessment of High-Temperature Components*; ASM International: Materials Park, OH, 1989.
23. *Standard Method of Test for Plane Strain Fracture Toughness of Metallic Materials*, ASTM 399; American Society for Testing and Materials: Philadelphia, 1985.
24. *Standard Test Methods for* J_{Ic}, *A Measure of Fracture Toughness*, ASTM 813; American Society for Testing and Materials, Philadelphia, 1985.
25. Folkhard, E. *Welding Metallurgy of Stainless Steels*; Springler-Verlag: New York, 1994.
26. Duren, C. *Equations for the Prediction of Cold Cracking in Field-Welding Large Diameter Pipes*, IIW Doc. IX-1356–85; The International Institute of Welding, Cambridge, England.
27. Yurioka, N. *Predictive Methods for Prevention and Control of Hydrogen Assisted Cold Cracking*, III Doc.IX-1938–99, Proceedings of the First International Conference on Weld Metal Hydrogen Cracking in Pipeline Girth Welds, Welding Technology Institute of Australia, Wollongong, March, 1999.

3

Designing with Carbon-, Low-, and Medium-Alloy Steels

Guoxun Liu
University of Science and Technology, Beijing, China

I. INTRODUCTION

Carbon-, low- and medium-alloy steels cover a wide range of standardized commercial steel grades. Some of them are produced and supplied as strips, sheets, plates, beams of various sections with different profiles and dimensions, rails, wires, linepipes, tubes, etc. Most of these steels are used as structural steels and engineering steels; however, a minor portion of these may be used to manufacture various tools. The specifications of these three different kinds of steel can be found in Ref. [1], which includes a complete, comprehensive and a comparative compilation of the different kinds of steel produced worldwide.

When the steels which have been standardized and manufactured over a long period time cannot meet: a new property requirement for certain applications, necessary requirement, cost reduction, weight reduction, energy conservation, substitution or if a reduction of a certain alloying element is necessary, a new grade must be developed. To meet this challenge and opportunity, a material designer would try to identify a material with adequate performance at minimum cost to satisfy the new requirements. The general philosophies may be tentatively suggested as follows:

(A) Based on the new property requirements put forward by engineering analysis, one may find a base material from the standardized steels from which least changes may be required as a starting point. Usually, a new grade steel is a derivative, or a variation of the base material. But on certain occasions, radical change or risk-based design may be encouraged.
(B) Typically experimentations to rectify the steel chemistry and the processing parameters design is necessary. The experimentations may be conducted through three stages: laboratory, pilot trial and full-scale operation.
(C) Experiments should be supplemented by computer-aided technique and the database of related alloy systems should be utilized as fully as possible. The material engineer should use more quantitative approaches.
(D) What one attempts to do in an experiment should follow the underlying metallurgical principles both in process metallurgy and in physical metallurgy, and the capability of the facilities available.

(E) The working direction should be guided by the relationship between properties and microstructures that have been established.
(F) Utilization of the scientific achievements in inter-related fields, such as in chemistry, physics, applied mechanics and applied mathematics.
(G) Characterization of the microstructure and the properties involved in the investigation and establishing new relationships between them. Sieving out the best ones.
(H) Evaluation of economics and energy conservation.
(I) Standardization if production tonnage is sufficient.

Every material designer has his creative ideas and his personalized working style; there is no unified working procedure to approach the aim of the design. But the underlying principles are shared in common. Now we will discuss the underlying principles in this chapter.

II. UNDERLYING PRINCIPLES IN THE EMPIRICAL APPROACH TO DESIGNING CARBON-, LOW- AND MEDIUM-ALLOY STEELS

Several general reviews about this topic have been published by several authors [2].

A. Mechanical Properties

The mechanical properties of steels are determined by their constitutional and structural parameters. The important mechanical properties considered in this section are:

- Yield strength σ_s (if there is no yield phenomenon during deformation test, the flow stress, or proof stress σ_p is used instead).
- Ultimate strength or tensile strength σ_u.
- Ductility, characterized by two parameters, ε_t, the strain to fracture, and ε_u, maximum uniform strain prior to plastic instability.
- Toughness, usually characterized by the Charpy Shelf Energy (CSE) value and by the impact transition temperature (ITT) ; in fracture mechanics, it is measured by crack opening displacement (COD).
- Fatigue, a failure mode under cyclic stress.

1. Strength, Ductility and Toughness

a. Microstructure Composed of Polygonal Ferrite (Annealed and Normalized Low-Carbon Steels). For mild strip steels, IF steels, micro-alloyed high strength low alloy (HSLA) steels, polygonal ferrite is the basic and the predominant microstructural constituents. The mechanical properties of the steel are basically determined by the properties of the ferrite intrinsically.

The yield strength of ferrite is determined by a number of factors, namely, its lattice friction stress (σ_o), solid solution strengthening (σ_s), precipitation strengthening (σ_p), dislocation and substructure strengthening (σ_d), texture strengthening (σ_t) and last, but the most influential, grain refining strengthening (σ_g).

(1) STRENGTH. (a) *Intrinsic Lattice Friction Stress* (σ_o). This term is the origin of the strength of steels. It comes from the bonding force between atoms in the lattice. Friction

stress of ferrite is determined by the shear modulus (μ) of iron, it measures the resistance to the movement of dislocations in the periodic field of the crystal lattice due to the interatomic interaction potential barriers. Assuming a sinusoidal force–displacement relationship, σ_o may be expressed approximately as

$$\sigma_o \cong 2\mu \exp(-2\pi w/b) \cong 10^{-3} \qquad (1)$$

in which $w=a/(1-\nu)$, and is the width of dislocation, b is Burgers' vector. σ_o is basically determined by the shear modulus of the base metal of the alloy. For ferrite, $\mu_{Fe}=8.3\times10^{11}$ dyne/cm^2, then σ_o can be calculated as about 83 MN/m^2. For all kinds of steels, it is nearly of the same value. It changes slightly either with the microstructure resulting from different processing parameters, or from alloying or heat treatment.

(b) *Solid Solution Strengthening* (σ_s). Unlike Al-, Ti- and Cu- alloys, solid solution strengthening is not the main method to increase the strength of carbon-, low alloy- and medium-alloy steels. Generally speaking, the mechanism which plays the role in solid solution strengthening is different in different alloys and at different temperatures, namely, it may be due to the misfit of atomic size, chemical locking, different in elastic modulus, change in electronic structure of the dislocation core, local ordering or clustering, long range order in intermetallic compounds, etc. For carbon- and low- and medium-alloy steels, solid solution strengthening is caused by the interaction of solute atoms and dislocations.

For both substitutional and interstitial solid solutions, solutes of different atomic size dissolved in body centered cubic iron produce localized strain fields which interact with those of the dislocations. The size misfit can be represented by

$$\delta_s = (1/a)(da/dc) \qquad (2)$$

where a is the lattice parameter of ferrite, in which the solute atoms are distributed randomly, the misfit produces interaction forces on dislocation that are to be summed statistically. Taking the flexibility of the dislocation into account, Mott and Nabarro first carried out the calculation and the result predicts the strengthening effects of magnitude

$$\mu(\delta_s)^{4/3} c^{1/2-2/3} \qquad (3)$$

$c^{1/2}$ applies to very dilute solutions, $c^{2/3}$ applies to more concentrated solutions. μ is the shear modulus of the alloy.

For steel, the effect of solid solution strengthening is less effective than that of precipitation strengthening or that of transformation strengthening. The lattice distortion caused by solid solution usually impairs the toughness. It is also limited by the solid solubility limit of the solute in the matrix metals.

The solid solution strengthening coefficients of the common alloying elements used in steel are listed in Table 1 [3–5]. The elements most commonly used in carbon-, low- and medium-alloy steels are phosphorus, manganese and silicon, though boron may also be used in IF steels. Carbon and nitrogen are the most potent solid solution strengthening elements but substantial quantities of these elements in solution are not normally used because they lead to deleterious room temperature strain aging. Very small quantities in solution are used to give bake hardening.

Solid solution strengthened steels usually have a minimum yield stress in the range 220–300 N/mm^2, though higher strength may be obtained depending on the processing facilities available. Phosphorus is the element most commonly used for fairly low increase in strength, since it is relatively cheap and gives a higher increase in strength per unit

Table 1 Solid Solution Strengthening Coefficient

Solute	Solution strengthening coefficient (A), MNm^{-2} $(wt\%)^{-1}$
C and N	5544
P	680
Sn	120
Si	83
Cu	38
Mn	37
Mo	11
Ni	0
Cr	−30

addition than manganese or silicon. Phosphorus addition is usually restricted to well below 0.1% to avoid problems with welding and because phosphorus may lead to secondary cold work embrittlement (SCWE) of IF steel. SCWE is a phenomenon which leads to brittle fracture after the steel has previously received strain. A very small addition of boron, such as 0.0005%, is sufficient to reduce the tendency to SCWE [6]. The strengthening by phosphorus is often supplemented first by manganese and then by silicon.

An advantage of solid solution strengthening steels is that they retain the general characteristics of the mild steel from which they are derived, good formability and the strain ratio of thickness to width. Furthermore, there is less loss in strength from the hot-rolled to the cold-rolled and annealed condition.

For structural steels containing up to 0.25% and 1.5% Mn, the microstructure is ferrite + pearlite. When pearlite is less than 1/5 by volume fraction in the microstructure, pearlite has no effect on σ_y of the steel. The solid solution principle is also valid for the ferrite in these steels. The solid solution effects of the common alloying elements are illustrated in Fig. 1 [7].

Additions of up to about 0.1% P are incorporated in the higher strength rephosphorized grades that are used in the automotive body panel. But phosphorus is not used as a strengthening agent per se in structural steels. Phosphorus is added to the weathering grades because of its beneficial effect on atmospheric corrosion resistance. Silicon and manganese are cost effective as solid solution strengtheners in low-carbon structural steel, but silicon is added to steel primarily as a deoxidizing agent.

The following equation indicates approximately how the tensile strength of a batch-annealed solid solution strengthened AK strip steels varies with the steel composition (wt%) [8]:

$$\sigma_u = 270 + 441[O] + 64[Mn] + 98[Si] + 930[P] \tag{4}$$

Another equation is as follows [9]:

$$\sigma_u = 292 + 563[C] + 678[P] + 18[Mn] + 90[Si] - 1534[S] \tag{5}$$

An equivalent equation for a continuously annealed steel is as follows [10]:

$$\sigma_u = 477 + 918[P] + 48[Mn] + 127[Si] - 0.019\ (T°C) \tag{6}$$

where T°C is the annealing temperature.

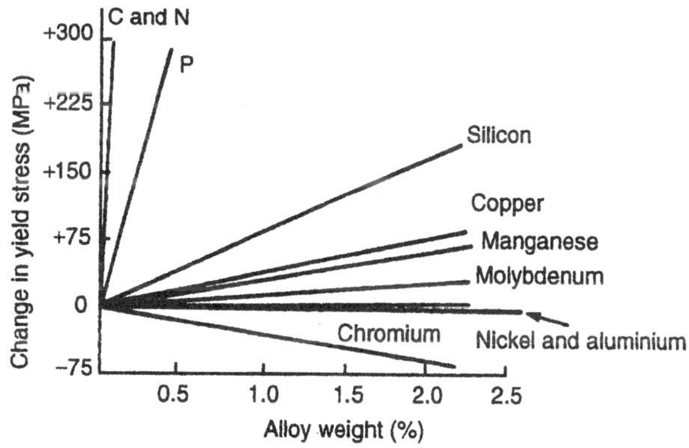

Figure 1 Solid solution strengthening effects in ferrite–pearlite steels. (From Ref. 7.)

Solid solution strengthened steels are usually based on titanium, niobium or titanium plus niobium IF ultra-low-carbon steels but with sufficient phosphorus, manganese, silicon or boron to give the strength required. Figure 2 is a plot of properties vs. [Si] + [Mn] + 10[P] for niobium bearing ULC IF steels. These curves suggest that the increase in strength from manganese is equivalent to that from silicon and that the effect of phosphorus is 10 times as great. The ratio of the true strain in width direction to the true strain in the thickness (*r* value) decreases most for manganese and least for phosphorus. Figure 3 gives the effect of solid solution additions to titanium steel, in this case, the effect of silicon is greater than that of manganese as it is in an AK steel [10]. Figure 4 shows that the decrease in elongation is greatest for phosphorus addition and least for silicon. Boron imparts solid solution strengthening to titanium treated IF steel, even though boron has high affinity for nitrogen, in this case, nitrogen is precipitated as titanium nitride.

(c) *Precipitation Strengthening* (σ_p). The effect of precipitation strengthening is stronger than that of solid solution strengthening for steels. Both strip and structural grades of steel may be precipitation strengthened by the addition of elements such as titanium, niobium, and vanadium, and these elements may have an additional effect on strength by leading to grain refinement. These elements are strong carbide and nitride formers which may be partially or completely dissolved at the slab reheating stage prior to hot rolling and which may then be reprecipitated into a fine form on subsequent cooling and transform to ferrite. The degree of strengthening is dependent on both fraction and size of the precipitates; finer precipitates produce a greater effect. Coarse precipitates, which are not dissolved at the slab reheating stage, are ineffective as strengthening agents.

For quench aged carbides and precipitated carbonitrides in Nb, V and Ti added steels, the strengthening effect can be expressed by the modified Ashby–Orowan relationships [12,13]:

$$\sigma_\mathrm{p}(\mathrm{MN/m^2}) = 5.9\frac{\sqrt{f}}{\bar{x}}\ln\left(\frac{\bar{x}}{2.5\times10^{-4}}\right) \tag{7}$$

where f = the volume fraction of precipitate, $\bar{x}$ = the mean planar intercept diameter of the precipitates. Figure 5 shows the dependence of precipitation strengthening on precipitate size ($\bar{x}$) and volume fraction (f) according to the Ashby–Orowan model, compared with

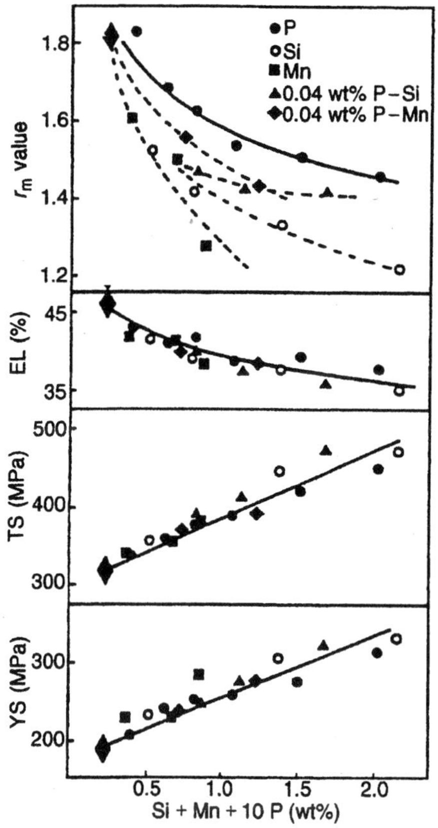

Figure 2 Mechanical properties of ULC continuously annealed Nb bearing steel vs. [Si]+[Mn]+10[P]. (From Ref. 11.)

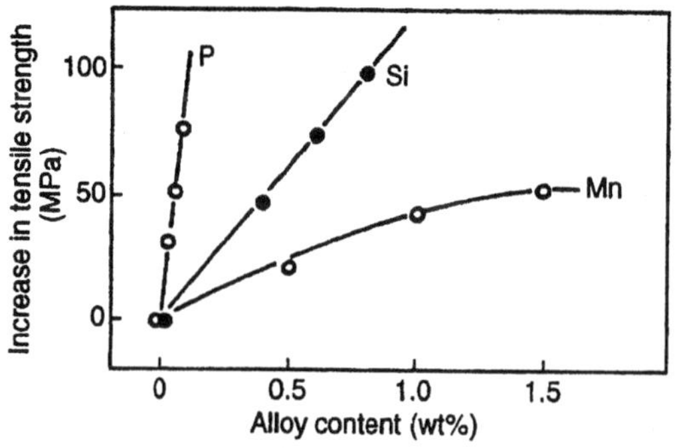

Figure 3 Increase in tensile strength of ULC Ti bearing steel vs. content of solid solution elements. (From Ref. 10.)

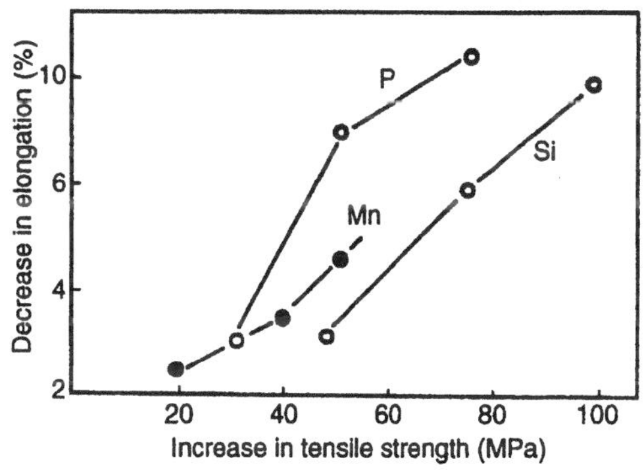

Figure 4 Deterioration of elongation by solid solution elements. (From Ref. 10.)

experimental observations for a given micro-alloying addition [14]. A simplified approach leads to an easily manipulated relationship:

$$\sigma_{\text{precipitate}} = B(\%\ \text{solute}) \tag{8}$$

The values of B in Table 2 have been published in Ref. [3].

Variation of the value of B below the maximum depends on:

(i) Whether some precipitation occurs in the austenite prior to ferrite transformation, especially during controlled rolling; if it occurs, it will decrease the value of B.
(ii) The temperature at which the precipitates are formed.
(iii) The extent to which the precipitates may overage during subsequent cooling.
(iv) Between 750°C and 650°C, with decreasing transformation temperature σ_p increases linearly with a decrease in transformation temperature; it amounts to $20\,\text{MN/m}^2/10°\text{C}$.
(v) σ_p increases to a maximum value with increasing cooling rate; if precipitation is suppressed by faster cooling, σ_p will decrease.
(vi) Tempering may be used to reactivate the precipitate process if such suppression occurs.

The solubility product for the precipitate is important, since it determines the amount of precipitate that can be taken into the solution at any temperature and consequently the volume fraction that may be subsequently reprecipitated in a fine form. The temperature dependence of the solubility product is generally represented by an equation of the form:

$$\log[x][y] = -A/T + B \tag{9}$$

where $[x]$ and $[y]$ are the weight percentage of elements in solution, T is the temperature in degree Kelvin, and B is a constant. Table 3 gives the solubility products for a number of compounds in austenite.

Figure 6 shows the solubility of NbC and VN in austenite at various temperatures [17].

A number of precipitates are solid soluble in each other and the precise composition of such precipitates depends on the composition of the austenite matrix with which they are in equilibrium, as well as on the temperature. For example, niobium carbonitride

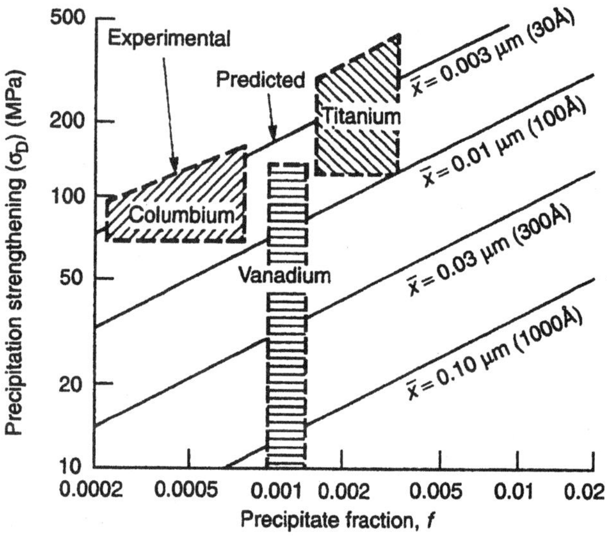

Figure 5 Ashby–Orowan model compared with experimental observations. (From Ref. 14.)

has a wide range of solid solubility; the ratio of carbon to nitrogen in the precipitate in equilibrium at a given temperature could be calculated for different amounts of carbon, nitrogen, and niobium in the steel [18].

It is assumed that a precipitate with the formula NbC_xN_{1-x} is in equilibrium with a matrix based on the reaction:

$$Nb_{sol} + xC_{sol} + (1-x)N_{sol} = NbC_xN_{1-x\ ppt} \tag{10}$$

The solubility product K for this reaction is written as

$$[Nb][C]^x[N]^{1-x}/[NbC_xN_{1-x}] = K \tag{11}$$

where $[NbC_xN_{1-x}]$ is the activity of the carbonitride which is taken as unity. It is also assumed that the activity of NbC in the precipitate is x and the effective activity of NbN in the precipitate was $1-x$ and the separate solubility product equations for the equilibrium between the matrix and NbC and NbN, respectively, are:

$$[Nb][C]/x = K_1 \tag{12}$$

$$[Nb][N]/1-x = K_2 \tag{13}$$

Table 2 The Coefficient (B) of Precipitation Strengthening

Solute and precipitate	B_{max} $(MN/m^2/wt\%)$	B_{ave} $(MN/m^2/wt\%)$	Solute concentration (wt%)
V as V_4C_3	1000	500	0–0.15
V as VN	3000	1500	0–0.06
Nb as Nb (CN)	3000	1500	0–0.05
Ti as TiC	3000	1500	0.03–0.18

Table 3 Temperature Dependence of Solubility Product (wt%) for Carbides, Nitrides, Carbonitride [15], Sulfides and Carbosulfides [16]

Solubility product (K)	$\text{Log}_{10} K$	Solubility product (K)	$\text{Log}_{10} K$
[B][N]	$-13970/T+5.24$	[V][N]	$-7700/T+2.86$
[Nb][N]	$-10150/T+3.79$	$[V][C]^{0.75}$	$-6500/T+4.45$
$[Nb][C]^{0.87}$	$-7020/T+2.81$	[Ti][S]	$-16550/T+6.92$
$[Nb][C]^{0.7}[N]^{0.2}$	$-9450/T+4.12$	$[Ti][C]^{1/2}[S]^{0.5}$	$-15350/T+6.32$
[Ti][N]	$-15790/T+5.40$	[Mn][S]	$-9020/T+2.93$
[Ti][C]	$-7000/T+2.75$		

where K_1 and K_2 are the solubility product for the pure carbide and nitride, respectively, at the desired temperature. For a given steel composition and temperature, these equations could be solved to give a value of x. The model was later developed further to involve more than one precipitate [19] and a number of computer programs able to predict the

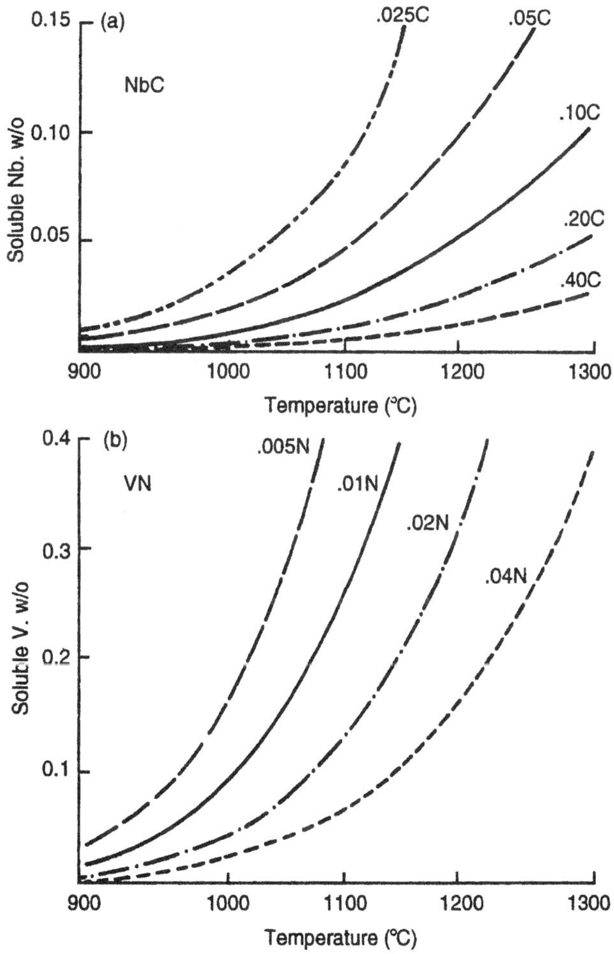

Figure 6 Solubility of NbC and VN in austenite. (From Ref. 17.)

equilibrium conditions for a series of multi-component systems have now been developed. These programs can be used to predict the amounts of titanium, niobium, carbon, nitrogen, and sulfur, etc., retained in solution in equilibrium at any temperature as well as the amounts and composition of the precipitates.

Micro-alloyed steels are essentially low-carbon–manganese steels alloyed with the addition of the strong carbide or nitride-forming elements niobium, titanium or vanadium, separately or together, and are often known as high strength low alloy (HSLA) steel. In the hot-rolled condition, they usually have a yield stress in the range of 300 up to 500 or 600 N/mm^2. The upper limit of the potential yield stress range is usually lower for a cold-rolled and annealed product, depending on the processing given.

The alloying elements have widely differing effects [20] due to the different solubility of their carbides and nitrides in both austenite and ferrite, due to their different precipitation kinetics. They increase the strength both by precipitation when sufficient carbon (nitrogen for vanadium steels) is present in the steel and grain refinement; but the grain refinement itself may arise from several mechanisms. In ferrite–pearlite steels, the systems of commercial significance are those involving niobium, vanadium, and titanium. These elements have only a limited solubility in steels due to their strong affinity for carbon and nitrogen. They are added to steels in small amounts, e.g., up to about 0.06% Nb or 0.15% V. At the slab or bloom reheating temperature of 1250°C, a substantial amount of niobium will be taken into solution. Nb(C, N) will precipitate at the austenite–ferrite interface during transformation, called interphase precipitation, which leads to a substantial strengthening. But when reheating to a typical normalizing temperature of 920°C, very little Nb(C, N) will dissolve and virtually no precipitation strengthening can take place. The undissolved particles will act as pinning agents, restricting austenite grain growth and leading to the formation of a fine ferrite grain size. Therefore, the reheating temperature controls the potential for precipitation strengthening and the strength increases progressively as the temperature raises from 920°C to 1250°C. Vanadium dissolves more easily than niobium and a complete solution of V_4C_3 would be expected to occur in commercial grades of structural steel at a typical normalizing temperature, e.g. 920°C. For the dissolution of VN, slightly higher temperatures are required; at 920°C, VN can act as a grain refining agent. Aluminum is a powerful nitride former, in the presence of 0.04% Al, significant vanadium will go into the solution at 920°C and be available for the precipitation of V_4C_3 on transformation to ferrite. Vanadium steels provide significant precipitation strengthening effect, i.e., up to 150 N/mm^2 per 0.10% V.

Another instance of the application of precipitation strengthening is the micro-alloying forging steel. Historically, after the methods of alloy reduction and substitution had largely been exhausted, attention was turned to potential savings in manufacturing costs and particularly in the area of heat treatment. Traditionally, components such as crank shaft and connecting rods are cooled to room temperature after the forging operation, only to be reheated to a temperature of about 850°C prior to oil quenching. Tempering at 550–650°C then produces tensile strengths in the range 800–1100 N/mm^2. In the mid-1970s, a micro-alloy-medium carbon steel (49MnVS3) was manufactured in Germany; after air cooling from the forging operation, heat treatment was eliminated. Since then, a major effort has been devoted to the development of micro-alloy forging steels in Europe, Japan, and China.

As indicated earlier, niobium, titanium, and vanadium are used as micro-alloying elements in low-carbon steels; high soaking temperature must be employed in order to achieve a substantial solution of Nb(CN) , TiC, and TiN. Vanadium is the most suitable micro-alloying element for medium-carbon steels due to its high solubility in austenite

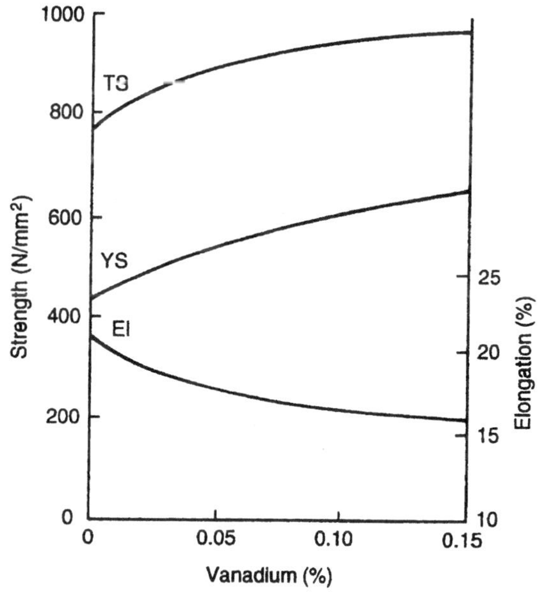

Figure 7 Effect of V on the tensile properties of air cooled 0.45% C, 0.9% Mn steel. (From Ref. 5.)

regardless of carbon content. Vanadium carbonitride precipitates in both the proeutectoid ferrite and ferrite lamellae of the pearlite on cooling from the solution treatment temperature [21]. Vanadium content in the range 0.05–0.2% is employed depending upon the levels of strength required. Figure 7 illustrates the tensile properties of these grades that increase progressively with the vanadium content. Nitrogen content influences the level of precipitation strengthening. The tensile strength of these steels can be expressed as a function of $(V + 5 \times N)$ % [22]. In order to intensify the strengthening effect up to 0.02%, Ni is incorporated in the steels.

One of the disadvantages of these micro-alloy steels is that they display significantly lower levels of toughness than the traditional quenched and tempered martensitic grades. The low impact strength is related to the coarse pearlitic structure but this effect is exacerbated by precipitation strengthening. Whereas this problem has been overcome in structural steel plates with the use of low temperature finishing (controlled rolling), there is little scope for the adoption of this practice in dropping the forging operation due to the metal flow/die filling problems that occur at low forging temperatures.

The impact strength of these grades can be improved by lowering the carbon content and increasing the manganese, vanadium, and nitrogen content to compensate the loss in strength. Its practice in Sweden and Germany has shown that an improvement in toughness can also be obtained by increasing the silicon content. But grain refining by the addition of titanium in pursuit of higher impact strength has been given attention. The stoichiometric level required for reaction with nitrogen in TiN and the growth of particles are also minimized by rapid solidification from the liquid state; then the finely dispersed particles of TiN present at the normal soaking temperature for forging, namely 1150°C, will refine the austenite grain size [21]. In practice, the titanium additions are restricted to levels of about 0.01% and the need for rapid solidification is generally satisfied by continuous casting.

The formations of TiN for grain refinement can reduce the level of available nitrogen for precipitation strengthening by V(CN). However, this problem can be overcome by adjusting nitrogen such that free nitrogen exceeds 0.006%.

(d) *Dislocation and Substructure Strengthening* (σ_d). Dislocations are introduced into ferrite by cold working, by controlled rolling at low temperatures, or by decreasing the transformation temperature. The complete processing will result in a substantial increase in strength. The strengthening effect due to dislocation is related to the dislocation density ρ,

$$\sigma_d = \alpha\mu b\sqrt{\rho} \tag{14}$$

σ_d is the stress increase due to cutting forest dislocations, μ shear modulus, b Burgers' vector of dislocation, and α is a constant.

The measurement of dislocation density is a sophisticated and time consuming work using the transmission electron microscope. As subgrain boundaries formed after recovery, the σ_d may be expressed in a form as

$$\sigma_d = c + kd^{-n} \tag{15}$$

where c, k and n are constants and n has a value in the range of 0.5–1.0.

An increase of 60 MN/m^2 as the transformation temperature decreases from 800°C to 650°C is produced by dislocation strengthening. However, an increase in the strength in the range of 9–15 MN/m^2 per unit increase in $d^{-1/2}$ (mm$^{-1/2}$) was produced by controlled rolling of ferrite–pearlite steels.

A steel with mixed microstructure consisting of recovered subgrains and recrystallized grains, with its σ_y, may be described by the relationship in the following form:

$$\sigma_y = \sigma_o + K[d^{-1/2}(f_r) + d_c^{-1/2}(1 - f_r)K_c/K] \tag{16}$$

d=recrystallized grain size, d_c=recovered subgrain size, f_r=fraction of recrystallized grain. K and K_c are experimentally determined values for fully recrystallized and fully cellular subgrain structures, respectively.

Due to the fact that all the boundaries act as slip barriers, for the purpose of specifying an effective grain size, Eq. (16) can be changed to the form:

$$\Delta_s = K\,[(f_r)/d_e + (1 - f_r)/d_e]^{1/2} \tag{17}$$

where d_e is the effective grain size.

Figure 8 illustrates the effect of "effective" grain size on yield strength of controlled rolled steels [14,23]. In the range of very fine cell sizes of $d_c^{-1/2}$=50 mm$^{-1/2}$, the strengthening by the dislocations in the subgrain boundaries is of the order of 9 MN/mm^2 per unit increase in $d_c^{-1/2}$ (mm$^{-1/2}$). The dislocation substructure strengthening due to controlled rolling has been shown to amount to some 80 MN/mm^2 for a finishing temperature of 600°C, as shown in Fig. 9 [14,24].

(e) *Grain size Refining Strengthening* (σ_g). The most influential factor in the strengthening of ferrite is grain size refining. Grain refinement produces a simultaneous improvement in toughness, while other strengthening effects usually lead to a decrease in toughness; this effect is unique to grain refinement.

According to the Hall–Petch equation [25,26],

$$\sigma_y = \sigma_o + kd^{-n} \tag{18}$$

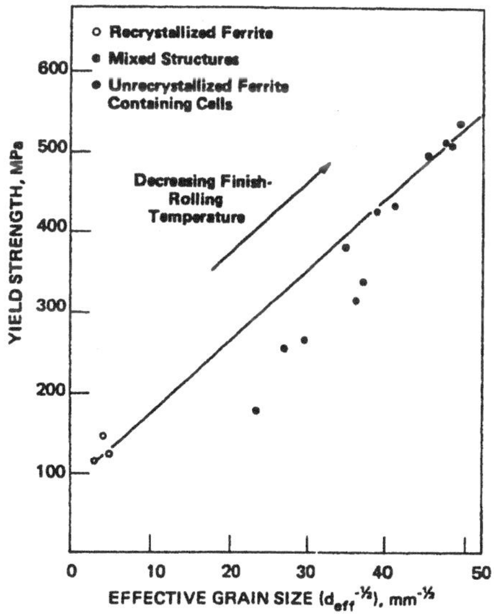

Figure 8 The effect of "effective" grain size on yield strength. (From Refs. 14,23.)

the exponent n is approximately 0.5, σ_o is the friction stress, k is a material constant; for ferrite, it is about 15, d is the ferrite grain size.

Refinement of grain size can be achieved in a number of ways: controlled rolling, controlled cooling, micro-alloying and others. Traditionally, fine grained steels contain about 0.03% Al which is soluble at normal slab or bloom reheating temperatures of around 1250°C and which remains in solution during rolling and after cooling to ambient temperature. However, on subsequent reheating through the ferrite range to normalizing or solution treatment temperature, Al combines with nitrogen in the steel to form a fine dispersion of AlN. The particles pin austenite grain boundary at normal heat treatment temperatures just above Ac3, typically 850–920°C, and result in the formation of a fine austenite grain size. In turn, a fine austenite grain size results in the formation of a fine ferrite grain size on cooling to room temperature.

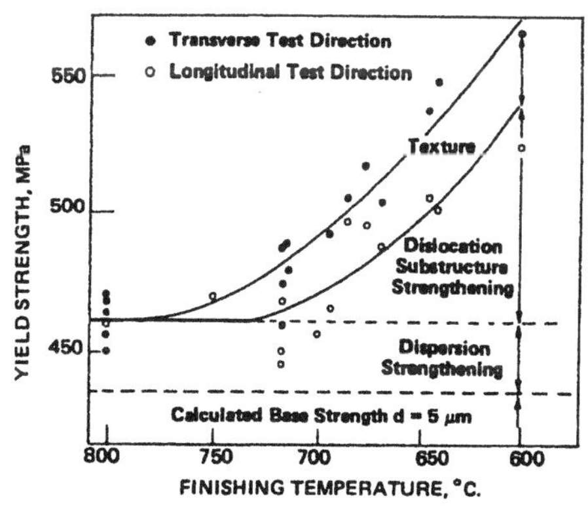

Figure 9 Effect of controlled rolling finishing temperature on contribution of various mechanisms to the yield strength of low temperature rolled steels. (From Refs. 14,24.)

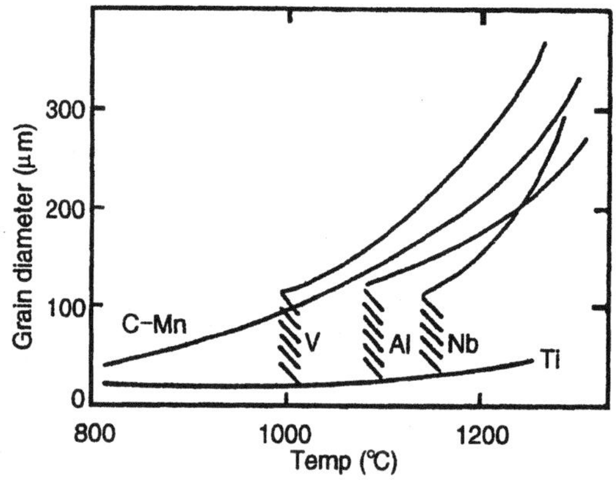

Figure 10 Austenite grain growth characteristics in steels containing various micro-alloy additions. (From Ref. 28.)

Titanium or niobium micro-alloyed in steels plays an important role in controlling the kinetics of recrystallization during hot rolling. Elements in solid solution may retard recrystallization through the processes of solute drag, but strain-induced precipitation of Nb(CN) is also very effective in retarding or stopping recrystallization. The final effect depends on the initial austenite grain size, the temperature, the amount of deformation given after the last previous recrystallization and the level of alloy addition [27].

Figure 10 shows the austenite grain growth characteristics in steels containing various micro-alloy additions [28]. Figure 11 shows the change of recrystallization stop temperature with content of micro-alloy solutes in a 0.07% C 1.40% Mn, 0.25% Si steel [29]. Figure 12 illustrates the variation of ferrite grain size with austenite grain boundary area per unit volume for deformed and recrystallized austenite [30]. For the deformed austenite, deformation band and boundaries also may act as nucleation sites. Figure 13 shows the effects of recrystallized austenite grain size and total reduction below recrystallization temperature on ferrite grain size [31]. As an example, Fig. 14 shows how the critical strain for

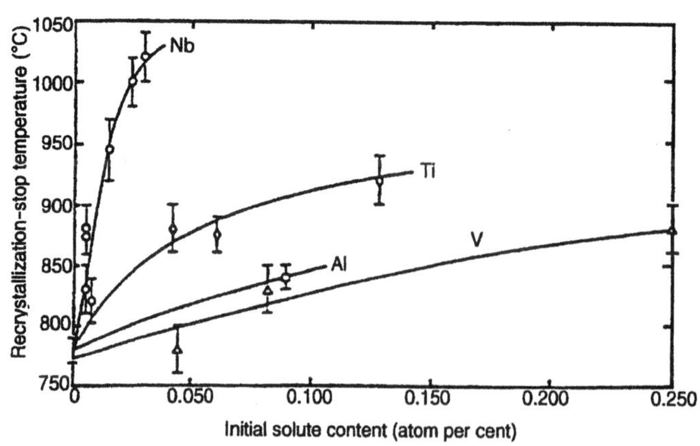

Figure 11 The recrystallization stop temperatures change with the content of solutes in a 0.07% C, 1.4% Mn, 0.25% Si steel. (From Ref. 29.)

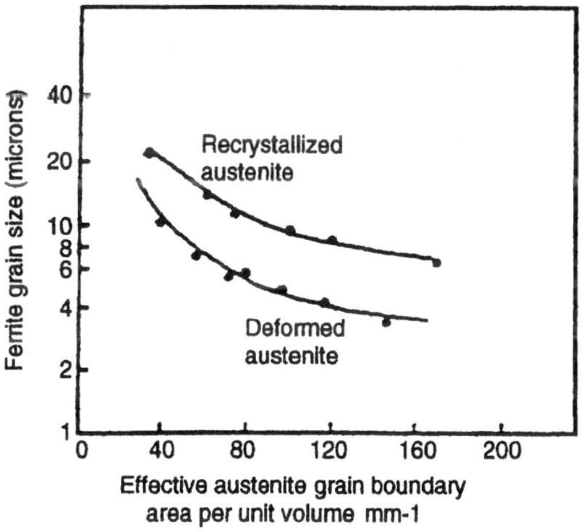

Figure 12 Variation of ferrite grain size with austenite grain boundary area. (From Ref. 30.)

austenite recrystallization varies with prior austenite grain size and rolling temperature for a steel containing 0.03% Nb [32].

The processing parameters after cold work also have a large effect on the final grain size of ferrite.

Depressing the transformation temperature of austenite to ferrite by the addition of elements such as Mn or increase of cooling rate from the austenite range will also lead to refinement of ferrite grains.

With an unrecrystallized austenite, the austenite grain boundary area per unit volume is increased during grain elongation; the nucleation frequency on grain boundary per unit volume of austenite is higher and nucleation may also occur on deformation bands; the overall effect is that a finer ferrite grain size is produced from a deformed unrecrystallized austenite than from a recrystallized austenite. Figure 15 shows that for a given grain boundary area per unit volume, S_v, the ferrite grain size decreases with increasing cooling rate [33].

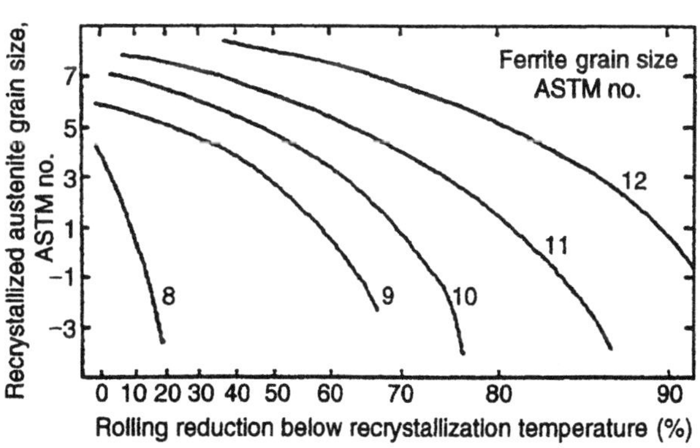

Figure 13 Effect of recrystallized austenite grain size and total reduction below recrystallization temperature on ferrite grain size. (From Ref. 31.)

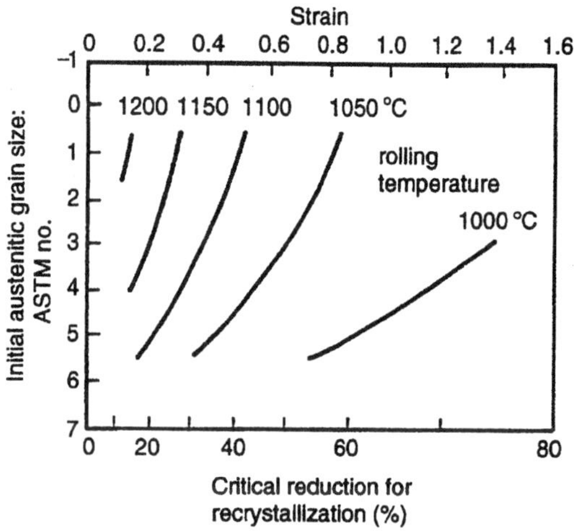

Figure 14 Critical strain varies with prior austenite grain size and rolling temperature. (From Ref. 32.)

(f) *Texture Strengthening* (σ_t). Preferred orientation in ferrite grains can be inherited from austenite or can be the result from deformation of ferrite in the controlled rolling. The steel will become anisotropic in mechanical behavior to a certain extent depending upon the intensity and crystallographic feature of the texture, i.e., the yield strength in the longitudinal direction is different from those in the transverse direction. In general, pronounced texture may result in those steels which have been rolled at a low temperature. The texture usually observed has been shown to cause higher σ_y in the direction transverse to the rolling direction than that in the longitudinal direction. For rolling finishing temperature of 600°C, about 30 MN/mm^2 accounts for the strengthening in the transverse direction, see Fig. 9 [14,24].

In summary, assuming that the linear superposition principle is valid for these strengthening mechanisms, i.e., all these mechanisms are independent of each other, the

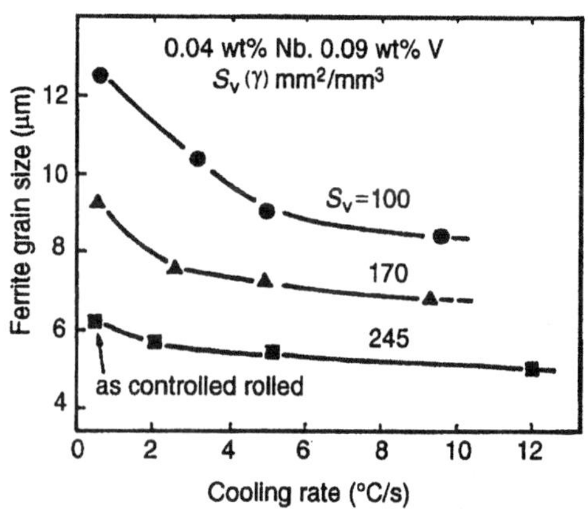

Figure 15 The change of ferrite grain size with increasing cooling rate in controlled rolling. (From Ref. 33.)

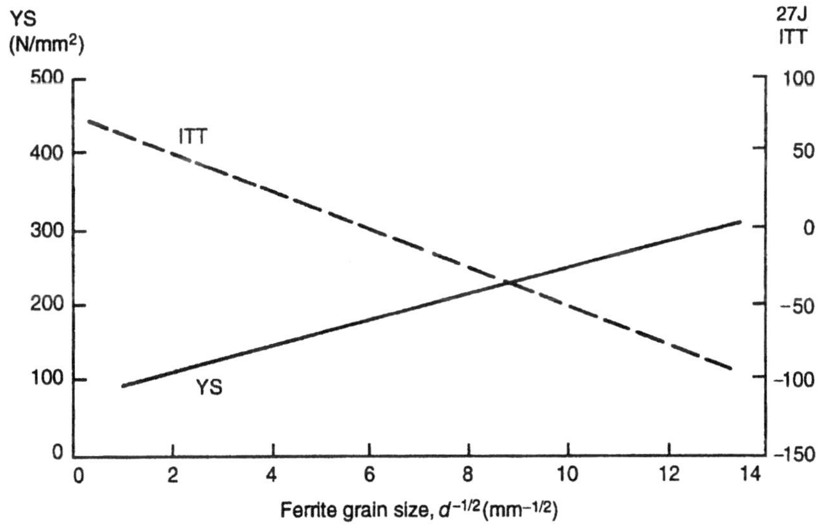

Figure 16 Effect of ferrite grain size on impact properties. (From Ref. 5.)

yield strength of ferrite can be analytically expressed as follows:

$$\sigma_y = \sigma_o + \sigma_s + \sigma_p + \sigma_d + \sigma_t + k_y d^{-1/2} = \sigma_o + A(\%) + B(\%) + kd^{-n} + \sigma_t + k_y d^{-1/2} \quad (19)$$

The yield strength of silicon killed carbon–manganese steels composed of polygonal ferrite containing up to 0.25% C can be expressed as

$$\sigma_y(\mathrm{MN/mm^2} \pm 31\,\mathrm{MN/mm^2}) = 88 + 37(\%\ \mathrm{Mn}) + 83(\%\ \mathrm{Si}) + 2918(\%\ \mathrm{N_{free}}) + 15.1d^{-1/2} \quad (20)$$

(2) TOUGHNESS (ITT AND CSE). Firstly ITT.

(a) *Grain Refinement*. Strengthening effect usually leads to a decrease in toughness, but the refinement of ferrite grain size also produces a simultaneous improvement in toughness. The Petch equation linking toughness to grain size is given below:

$$\beta T = \log \beta - \log C - \log d^{-1/2} \quad (21)$$

where β and C are constants, T is ductile-to-brittle transition temperature and d is ferrite grain size. As illustrated in Fig. 16, the impact transition temperature decreases as ferrite grain size decreases.

(b) *Solid Solution Strengthening*. Other than grain size refining, all the other strengthening mechanisms, such as solid solution strengthening, precipitation strengthening, dislocation strengthening, etc., impair toughness, i.e. raise ITT.

For low-carbon and structural steels, an impact transition temperature has been described by the following equation:

$$\mathrm{ITT}(^\circ\mathrm{C}) \pm 30(^\circ\mathrm{C}) = -19 + 44(\%\ \mathrm{Si}) + 700(\%\ \mathrm{N_{free}})^{1/2} + 2.2(\%\ \mathrm{pearlite}) - 11.5d^{-1/2} \quad (22)$$

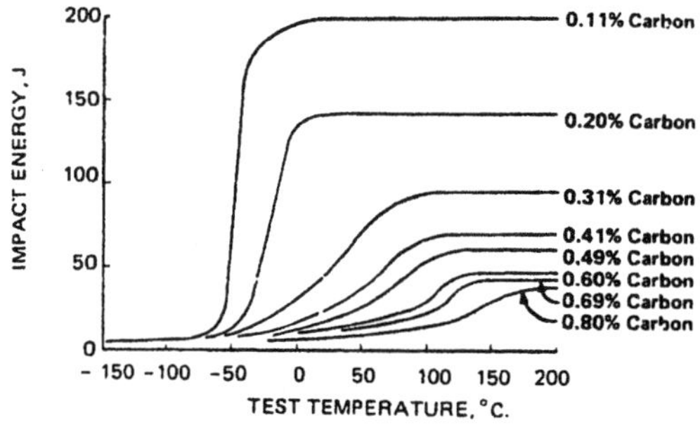

Figure 17 Effect of pearlite on the ITT of normalized carbon steel. (From Refs. 34,35.)

This equation suggests that:

(i) Grain refinement produces both high strength and low ITT.
(ii) Grain refinement is a critical means to produce low ITT.
(iii) Direct effect of manganese on ITT is not explicitly included in the equation; its effect is implied in the pearlite and grain size term.
(iv) Pearlite is detrimental to ITT, see Fig. 17.
(v) Filamental grain boundary carbides also have a detrimental effect in increasing the ITT, particularly when the carbide thickness exceeds about 2 μm [32,33], see Fig. 18.
(vi) Solid solution strengthening would be expected to raise the ITT with the exception of Mn and Ni which reduce the ITT. Manganese makes the thickness of grain boundary carbide thinner and Ni possibly increases the ease of dislocation cross-slip. A small amount of Al depresses the ITT by moving N from solution, but larger amount of Al raises the ITT. With the exception of Ni, Mn and Al, the magnitude of the effect of solution strengthening elements on the ITT should be proportional to the influence on the strength, but the exact value should be determined by experiments. For silicon, the calculated value is +25/+40, the experimental result is +17/+44 per wt% of the element. For Mn and Ni, the decrements in ITT are 20/40°C and 13°C, respectively, per wt% of the element [34].

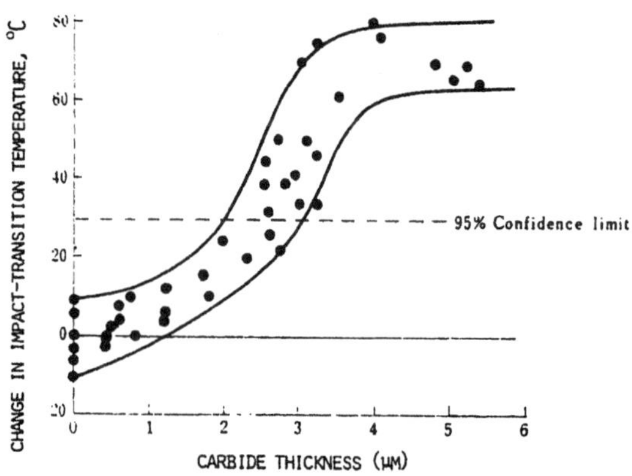

Figure 18 Effect of carbide thickness on ITT (From Refs. 14,36.)

(c) *Precipitation Strengthening*. Precipitation strengthening invariably raises the ITT values of +0.2/0.5°C per MN/mm^2 increment of σ_y as reported [3,12,35]. A reasonable value is 0.3°C per MN/mm^2. This value seems to apply for a wide range of precipitates in micro-alloy steels, for Cu precipitation and quench aging.

(d) *Dislocation Strengthening*. Dislocation strengthening raises the ITT lying in the range of $0.2°C/MN/mm^2$, and increases in σ_y [3,12,37].

(e) *Crystallographic Texture*. Attempts have been made to relate the impact transition temperature to a simplified crystallographic texture parameter [23,38]. The texture parameter is

$$p = [(I_{1\,1\,1})_{Rp}(I_{1\,1\,0})_{Tp}] - 1 \tag{23}$$

where $(I_{111})_{Rp}$ is relative intensity of the {1 1 1} planes in the rolling plane, and $(I_{110})_{Tp}$ is relative intensity of {1 1 0} plane in the transverse plane. The ITT was correlated with this texture parameter by the following equation:

$$\text{ITT}(°\text{C}) = 75 - 13d^{-1/2} + 0.63(\text{P}) \tag{24}$$

the confidence interval seems to be of the order of ±20°C.

Based on the experimental analysis of commercial micro-alloyed steels, an additional term, carbide thickness t (mm), has been incorporated into ITT equation 3

$$\text{ITT}\,(\pm 29°\text{C}) = 46 + 0.45\,\sigma_p + 131t^{1/2} - 12.7d^{-1/2} \tag{25}$$

where σ_p is the precipitation strengthening in MN/m^2.

(f) *Non-metallic Inclusions*. Up to ~0.05% S, the elongated sulfides raise both the longitudinal and transverse ITT. It seems fairly well established, however, that the transverse ITT is 20/40°C higher than the longitudinal ITT [36], probably due to structural fibering by elongated inclusions.

Due to the occurrence of lamellar tearing in which plate materials separate along planar arrays of non-metallic inclusions under the forces generated in highly restrained welds, production of steels which are isotropic with regard to toughness and ductility has been given major attention. The need for higher levels of impact strength in structural steels and the requirement for better cold formability in strip products have also focused attention on the development of cleaner steels and inclusion shape control. For the reduction of sulfide and oxide inclusions, and for modifying the shape and size of these inclusions, the practice of adding calcium to steels is being used world wide now. One of the pioneers in this field was Thyssen Niederrhein in Germany [39]. After tapping, calcium in the form of calcium silicide or calcium carbide is injected deep into the ladle, using argon as a carrier. The calcium vaporizes and forms bubbles floating through the molten bath. The bubbles combine with sulfur and oxygen in the steel, and the reaction products go into the slag. Prior to this treatment, the steel was deoxidized with aluminum to make the oxygen level in the range 20–100 ppm. After calcium treatment, the oxygen content reduced to 10–20 ppm. The desulfurizing effect of calcium is determined largely by the amount of calcium added and also influenced greatly by the type of ladle refractories employed. With dolomite ladles, the addition of 1 kg of calcium as a desulfurizing agent per ton reduces the sulfur content from an initial level of 0.02% to a final level of 0.003%, but sulfur less than 0.001% can be achieved by this technique. Figure 19 illustrates that magnesium is also very effective as a desulfurizing agent, but calcium is generally preferred because it is cheaper and more controllable [39]. Both manganese sulfide and

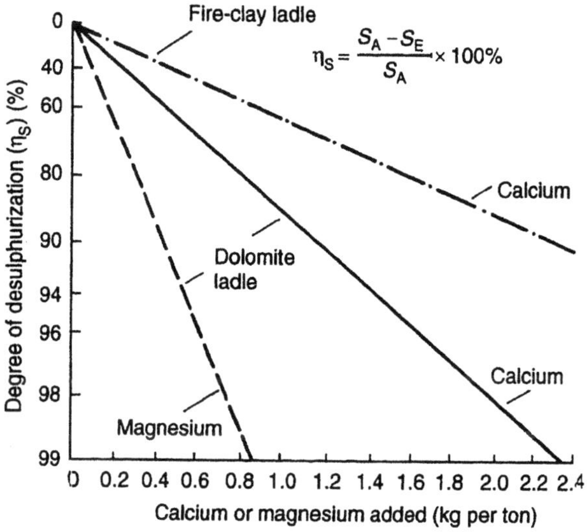

Figure 19 Effect of Ca and Mg on desulfurization. (From Ref. 39.)

strings of oxide will be elongated into planar arrays which cause the anisotropy of toughness and ductility. This effect will damage the steel. If the inclusions are present as small, isolated and non-deformed particles, the problem will be eliminated. Therefore, major attention has been devoted to inclusion shape control, in addition to the reduction in the volume fraction of inclusions. The elements used as inclusion modifiers are calcium, zirconium, tellurium, and the rare earth metals.

In aluminum deoxidized steels, the inclusion population will generally include elongated type II manganese sulfides, alumina, and some silicates. After calcium treatment, the inclusions are restricted to calcium aluminate of type $CaO{\cdot}Al_2O_3$. The sulfur in the steel is also associated with these inclusions, either as calcium sulfide or as sulfur in solution.

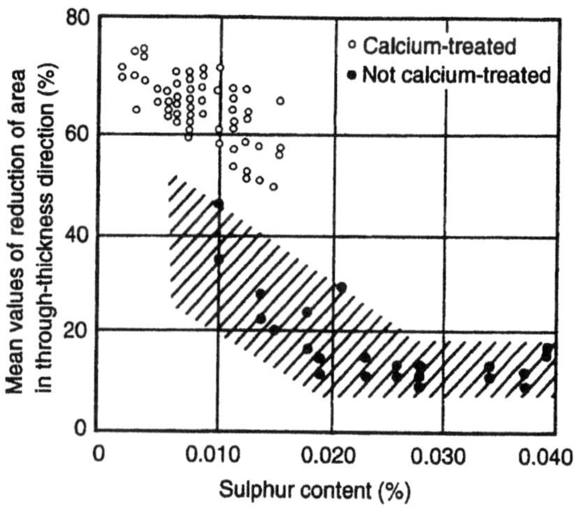

Figure 20 S content vs. mean values of reduction of area in the through-thickness direction of steel grade FG36. (From Ref. 39.)

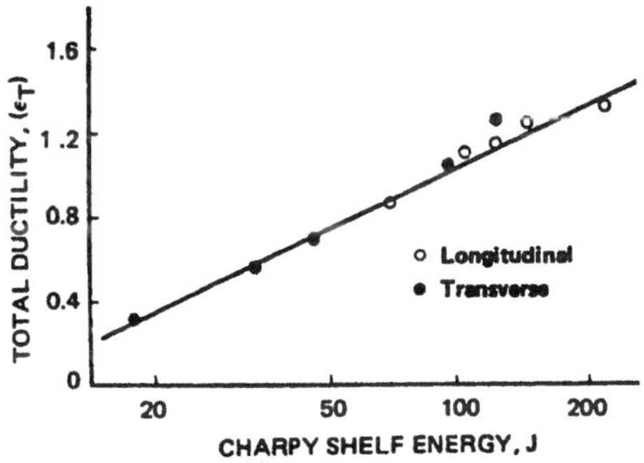

Figure 21 Relationship between CSE and total ductility at fracture. (From Ref. 37.)

The calcium aluminate particles are globular in nature and tend to retain their shape on hot rolling. The beneficial effect of the reduction in sulfur content and calcium treatment on the ductility (reduction in area) in the through-thickness direction is shown in Fig. 20.

Secondly CSE values.

It has been shown that there is a linear relationship between ε_t and CSE values, as shown in Fig. 21. The influence of various factors on the CSE may be inferred from their effect on ε_t. ε_t which may be expressed as

$$\varepsilon_t = C - f_1(V_f)f_2(S) \tag{26}$$

where C is a constant, f_1 and f_2 are different functions, V_f=volume fraction of second phase particles, S=a second phase particle shape factor.

The factors which influence the CSE may be summarized as:

(i) Pearlite decreases the CSE value. Some more systematic work to quantify this effect is required.
(ii) Spheriodizing the pearlite will increase the CSE value.
(iii) The effect of ferrite grain size on the CSE value is less quantitatively known. It may be as small as the effect of ferrite grain size on ε_T. The published data have confirmed that the coefficient of $d^{-1/2}$ is only -2.8 J [40].
(iv) The effects of solid solution strengthening on the CSE value are also less quantified.
(v) Precipitation strengthening decreases the transverse CSE value by 0.3 J per MN/m^2 increase in strength; the effect is probably not linear [37].
(vi) Dislocation strengthening is known to decrease the CSE value more or less progressively by a factor of -35 J per MN/m^2 increase in σ_d [43].
(vii) CSE value decreases linearly with increasing texture parameter P, but some discrepancy between different authors exists.
(viii) Increase of the volume fraction of non-metallic inclusions markedly decreases the CSE value; this relationship is true for both sulfide and oxide inclusions, but the effect is not linear, refer to Fig. 22.

The effect of inclusion elongation on the toughness is illustrated by some of the fracture toughness data available, Fig. 23 [34,41]. Due to the elongation of the sulfide and

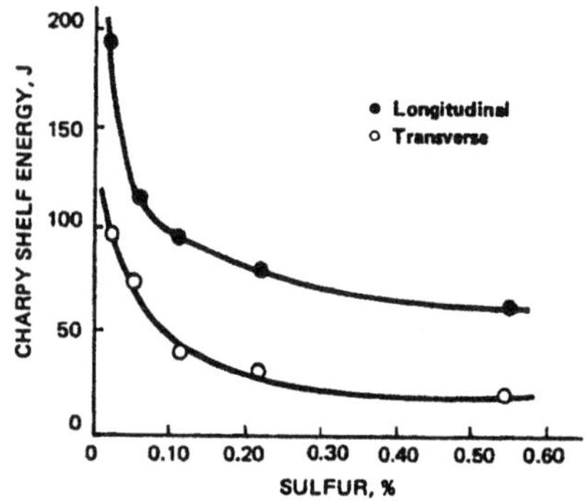

Figure 22 Anisotropic effect of S content on the CSE (From Ref. 36.)

oxide inclusions, the difference between longitudinal and transverse CSE value exists. As discussed in the ITT section, the anisotropy of toughness is one of the reasons which leads to the technology of inclusion control in which the MnS inclusions are converted to non-plastic sulfide by Zr, Ca, or rare earth additions [41,42].

By Zr sulfide modification, the equation for transverse CSE value [43] is

$$\begin{aligned} \text{CSE (J)} = {} & 112 - 2.8d^{-1/2} - 0.18\,\sigma_p - 832(\%\ \text{S}) - 43(\%\ \text{Al}) \\ & - 0.76(\%\ \text{pearlite}) + 107(\%\ \text{Zr}) \end{aligned} \tag{27}$$

It can be seen that the effect of grain size is very small, the negative effect of sulfur reflects the effect of sulfide volume fraction, and the benefits of Zr inclusion shape control increase the CSE value.

b. Microstructure Composed of Ferrite Plus Pearlite. In the polygonal ferrite section, we have extended to ferrite plus pearlite in some respects. It was proposed that the Hall–Petch equation could be extended to take into account the strengthening effects of alloying elements. Thus:

$$\sigma_y = \sigma_i + k(\%\ \text{alloy}) + k_y d^{-1/2} \tag{28}$$

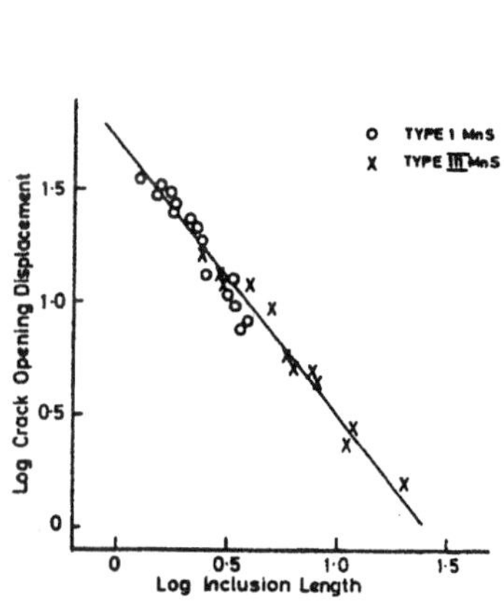

Figure 23 Effect of inclusion elongation on fracture toughness parameter (COD). (From Ref. 41.)

The following quantitative relationships for σ_y, σ_t, and ITT were developed:

$$\sigma_y(\text{N/mm}^2) = 53.9 + 32.3(\%\ \text{Mn}) + 83.2(\%\ \text{Si}) + 354(\%\ \text{N}_f) + 17.4d^{-1/2} \tag{29}$$

$$\sigma_t(\text{N/mm}^2) = 294 + 27.7(\%\ \text{Mn}) + 83.2(\%\ \text{Si}) + 3.83(\%\ \text{pearlite}) + 7.7d^{-1/2} \tag{30}$$

$$\text{ITT}(^\circ\text{C}) = -19 + 44(\%\ \text{Si}) + 700\text{N}(\%\ \text{N}_f) + 2.2(\%\ \text{pearlite}) - 11.5d^{-1/2} \tag{31}$$

where d is mean ferrite grain size in millimeter, and N_f is free (soluble) nitrogen. Each of these equations illustrates very clearly the beneficial effects of fine ferrite grain size in increasing the yield and tensile strength and depressing the ITT. It is interesting to note that the pearlite content has no significant effects on the yield strength of these low-carbon predominantly ferritic steels. On the other hand, pearlite increases the tensile strength and has a detrimental effect on toughness [61]. The solid solution strengthening effects of manganese, silicon, and free nitrogen are also highlighted in the above equations and it will be noted that free nitrogen is particularly detrimental to the impact properties. Figure 24 illustrates the effect of manganese content on the components contributing to σ_y of steel.

c. Microstructure Composed of Pearlite. Steels with predominantly pearlitic structure are used for rails, wheels, axles, tires, reinforcing rods, drill rods, and other applications. Pearlite itself begins to play an increasingly important role in controlling the toughness, yield strength and tensile strength as the proportion of pearlite in microstructure increases up to 100%. A relationship has been established between the microstructural and constitutional parameters and the yield strength, tensile strength and ITT in such steels with pearlitic structures.

(1) *Tensile Properties of Pearlite.* The following basic equation is usually used to analyze the yield stress of materials with mixed structural constituents:

$$\sigma_y = f_\alpha^n \sigma_\alpha + (1 - f_\alpha^n)\sigma_\pi \tag{32}$$

where f_α^n is volume fraction of ferrite, σ_y, σ_α and σ_π are the yield strength of ferrite–pearlite aggregates, of the ferrite and of the pearlite, respectively. The value of σ_α obeys the normal Hall–Petch equation. σ_π was first shown to be linearly related to logarithm of the mean free path through the pearlitic ferrite, but more recent work on fully pearlitic structure [60] showed that σ_π obeys a Hall–Petch type relationship of the following form:

$$\sigma_\pi = \sigma_i + k(S_o)^{-1/2} \tag{33}$$

where σ_i is an apparent friction stress, S_o is the pearlite interlamellar spacing, and k is a constant. The index n in Eq. (32) makes the strength non-linear with pearlite content and the value of n=1/3 gives least residual error.

The pearlite colony size, p, and plate thickness of Fe_3C, t, were found to have no influence on the yield or the tensile strength, but they do influence ITT, as will be shown later. The equations derived to describe the yield strength σ_y and tensile strength σ_t are

$$\sigma_y(\pm 45\,\text{MN/mm}^2) = f_\alpha^{1/3}[35 + 58(\text{Mn}\ \%) + 17.4d^{-1/2}] + (1 - f_\alpha^{1/3})[178 + 3.8S_o^{-1/2}] + 63(\text{Si}\ \%) + 42(\sqrt{N_f} \tag{34}$$

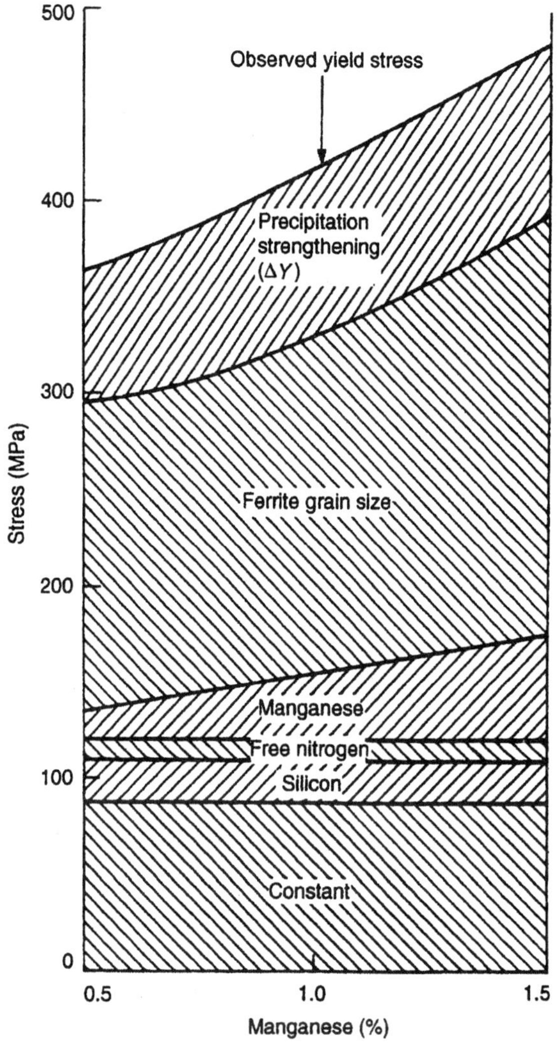

Figure 24 The effect of manganese on σ_y. (From Ref. 14.)

$$\sigma_t(\pm 45\ \mathrm{MN/mm^2}) = f_\alpha^{1/3}[246 + 1140\sqrt{N_f} + 18.2d^{-1/2}] + (1 - f_\alpha^{1/3})[720 + 3.5S_o^{-1/2}] + 97(\mathrm{Si}\ \%) \tag{35}$$

These equations show that the ferrite grain size gives a smaller contribution to the strength as the pearlite content increases, refer to Fig. 25. Pearlite contributes more to the strength as the composition approaches the eutectoid; in this case, S_o plays an important role in its contribution to strength. As discussed previously, the flow stress at certain strain σ_f, the work-hardening rate, the uniform ductility ε_u, the total ductility ε_t, decrease with increasing pearlite content [62]; however, these relationships are not really linear [8,36,43].

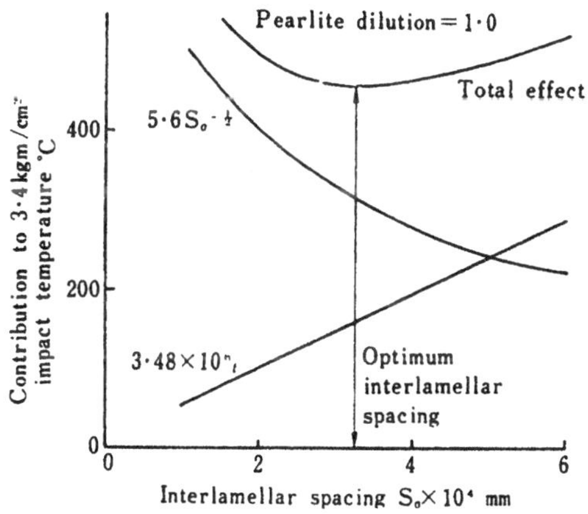

Figure 25 Effect of interlaminar spacing on the ITT (From Ref. 36.)

For a fully pearlite structure, Eqn. (34) and (35) reduce to

$$\sigma_y(\mathrm{MN/mm^2}) = 178 + 3.8S_o^{-1/2} + 63(\mathrm{Si}\ \%) + 425(\sqrt{N_f}) \tag{36}$$

$$\sigma_t(\mathrm{MN/mm^2}) = 720 + 3.5S_o^{-1/2} + 97(\mathrm{Si}\ \%) \tag{37}$$

The above relationships were confirmed for a eutectoid steel with interlamellar spacing varying from 2500 to 5000 Å.

The flow stress and the work-hardening rate of a full pearlite structure decreased as S_o increases:

$$\sigma_y(\mathrm{MN/mm^2})_{\varepsilon=0.5} = 1310 - 0.06S_o \tag{38}$$

$$d\sigma/d\varepsilon(\mathrm{MN/mm^2})_{\varepsilon=0.5} = 1560 - 0.09S_o \tag{39}$$

where S_o is expressed in Å.

ε_t slightly decreases as S_o increases, this effect is probably due to the thicker Fe_3C plates cracking more readily and initiating cavities at smaller strain values.

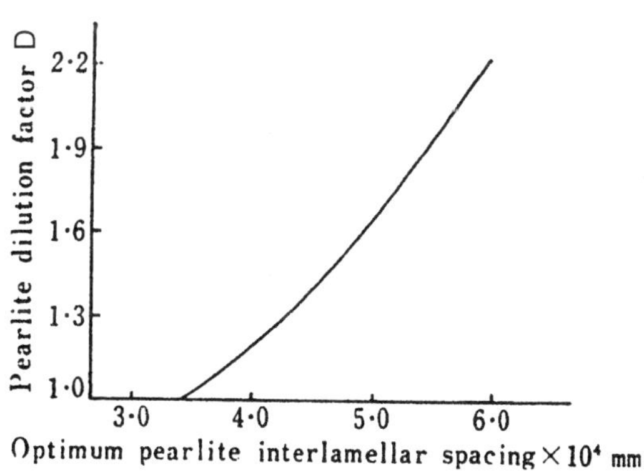

Figure 26 Dilution factor vs. optimum pearlite interlaminar spacing. (From Ref. 36.)

(2) *Toughness*. As discussed previously, increasing pearlite content decreases the CSE value of normalized steels. Using a very empirical approach, to a first approximation, the effect of pearlite content on CSE can be described by

$$\begin{aligned} \text{CSE (J)} = {} & 112 - 2.8d^{-1/2} - 0.18\,\sigma_p - 832(\%\ \text{S}) - 43(\%\ \text{Al}) \\ & - 0.76(\%\ \text{pearlite}) + 107(\%\ \text{Zr}) \end{aligned} \tag{40}$$

but the validity of this linear equation over wide variations in pearlite content is very doubtful.

Pearlite elevates ITT as described by the following equation:

$$\text{ITT } (\pm 30^\circ\text{C}) = -19 + 44(\%\ \text{Si}) + 700(\%\ \text{N}_f)^{1/2} + 2.2(\%\ \text{pearlite}) - 11.5d^{-1/2} \tag{41}$$

However, the effects are not as simple as the equation implies when wide variations in the pearlite content and microstructure are examined. Using a simple law of mixture [36,61], the 27 J transition temperature is given by

$$\begin{aligned} \text{ITT } (\pm 30^\circ\text{C}) = {} & f_\alpha[-46 - 11.5d^{-1/2}] + (1 - f_\alpha)[-335 + 5.6S_o^{-1/2} - 13.3p^{-1/2} \\ & + 3.48 \times 10^6 t] + 48.7(\%\ \text{Si}) + 767(\sqrt{N_f}) \end{aligned} \tag{42}$$

where d=ferrite grain m.l.i. in mm, p=pearlite colony m.l.i. in mm, S_o = interlamellar spacing of pearlite in mm, t=pearlite carbide lamellar thickness in mm, and f_α=volume fraction of ferrite.

It can be seen that the size of pearlite colony, interlammellar spacing of pearlite and pearlitic carbide lamellar thickness have a profound effect on ITT. A small pearlite colony size is beneficial to ITT. Metallographic evidence has shown that a pearlite colony boundary impedes a cleavage crack propagation. Solid solution strengthening is harmful to ITT. Interlamellar spacing S_o and carbide plate thickness t act on ITT in opposite directions; refinement of S_o is detrimental but a decrease in t is beneficial. A compromised condition may be reached in order to have an optimum total effect on ITT, as shown in Fig. 25.

(3) *Ductility*. The effect of interlamellar spacing of pearlite S_o on ε_u is not observed. Perhaps this is due to $S_o^{-1/2}$ having the same coefficient in the equations of σ_s and of $d\sigma/d\varepsilon$. Consequently, they can compensate each other. But slight increase in S_o can decrease ε_t; this may be due to t relating with S_o by the equation

$$t = 0.14S_o \tag{43}$$

in eutectoid steel, i.e., the pearlitic carbide thickness increases if S_o increases; as we have described previously, thicker carbides crack at a small strain value and voids form more easily. This is the reason why a wire with an extremely fine pearlite structure can undertake very heavy loads.

Increasing pearlite percentage can decrease ε_t and ε_u exponentially [2].

(4) *Dilution of Pearlite*. Pearlite dilution factor, D, may be defined as

$$D = 0.8f_p/\%\ \text{C} \tag{44}$$

where f_p is the pearlite volume fraction, % C is the carbon content of the steel. D will be unity for fully pearlitic eutectoid steel, and greater than unity for hypoeutectoid steel in fully pearlitic structure. The optimum value of S_{opt} for a minimum ITT increases as the

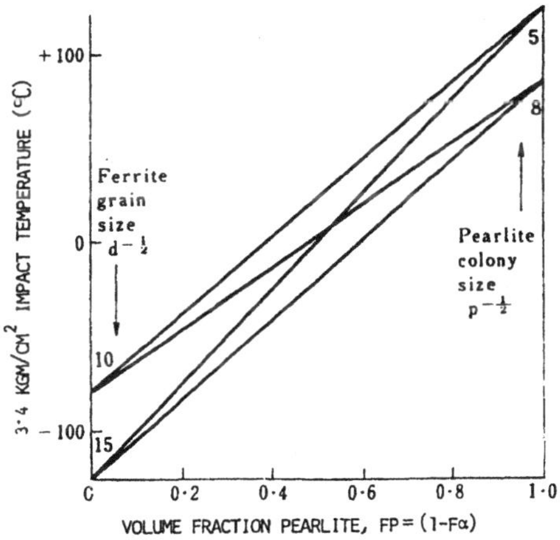

Figure 27 Effect of ferrite grain size and pearlite colony size on ITT (From Ref. 36.)

pearlite dilution factor increases [67]. Figure 26 illustrates this relationship and is given by

$$S_{\text{opt}}(\text{mm}) = 6.7 \times 10^{-6}(D - 0.12)^{2/3} \tag{45}$$

Thus a low-carbon dilute pearlite is conducive to a low ITT for a fully pearlitic structure. Increasing the pearlite proportion at a given carbon content will lower ITT. But as dilution pearlite is achieved by the sacrifice of fine grained ferrite, it would be detrimental to ITT.

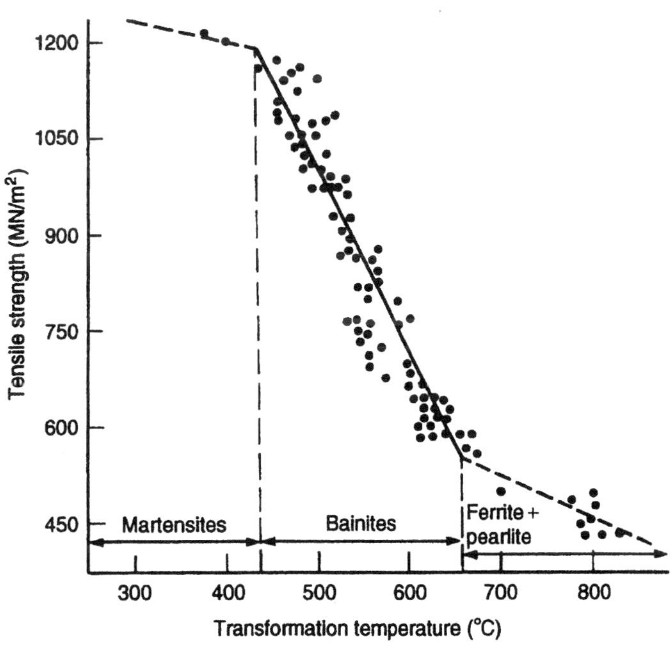

Figure 28 Relationship between 50% transformation temperature and tensile strength. (From Ref. 64.)

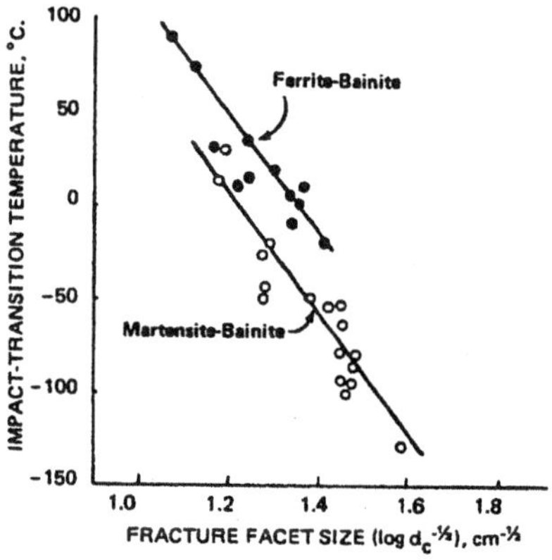

Figure 29 Effect of fracture facet size on ITT of martensite–bainite structure. (From Refs. 36,69.)

Some general inter-relationships between the ferrite grain size and pearlite colony size in terms of ITT are shown in Fig. 27. Replacing coarse ferrite grains by fine pearlite colonies in a structure comprising pearlite and ferrite can be beneficial. Pearlite morphology is most important as carbon content increases in the steels.

In general, decreasing the transformation temperature that can be brought about either by increased austenite grain size or alloy content, or rate of cooling, will be detrimental to impact toughness. The decreased value of S_o generated by too much depression of the transformation temperature will prove most detrimental.

A predominantly pearlitic steel for optimum combination of strength and toughness will be achieved by using a minimum carbon content that is consistent with the strength requirements and by grain refinement to produce fine ferrite grain and pearlite colony size, and by a minimum solid solution strengthening.

d. Microstructure Composed of Acicular Ferrite and Bainite. Both alloying elements and faster cooling rates depress the temperature of transformation of austenite to ferrit; ultimately, the effect will be sufficient to cause a transformation to bainite or martensite. The consequence of this progression is illustrated in Fig. 28, which relates to steels containing 0.05–0.20% carbon. The strength is increased progressively with the introduction of lower temperature transformation product with the sacrifice of toughness and ductility.

The structure of bainites/upper bainite, lower bainite, acicular bainite (very low-carbon bainite), and other modifications of bainite are relatively complex and comprises the following features:

(i) Acicular ferrites have lath-like grains with low angle misorientations between neighboring lath-like grains. The boundaries between packets composed of acicular ferrite are high angle misorientation boundaries; within a prior austenite grain, there may be several such pockets. The ferrite boundaries coincident with the prior austenite grain boundaries are high angle boundaries. In one word, the structure of the grain boundaries in bainite structure is very complex. This complexity makes the measurement of the grain size which is a very important parameter to characterize the bainite structure quantitatively difficult.

(ii) Dislocation density and dislocation arrays vary with transformation temperature and carbon content; the density increases with decreasing transformation temperature and increasing carbon content.

(iii) In the microstructure of conventional upper bainite, there are quite large carbides at lath boundaries within a packet, at packet boundaries, at prior austenite boundaries, the quantity of such carbides increases with the increasing carbon content.
(iv) In conventional lower bainite, carbides disperse within the individual ferrite laths in the packets. These carbides increase in number with increasing carbon content and decrease in size with decreasing transformation temperature.
(v) Perhaps due to the presence of enriched austenite, regions of alloy enriched ferrite [64,65,66,67] exist within the lath ferrite.
(vi) Due to both carbon and alloy enrichment in the austenite during its transformation, regions of high carbon martensite, higher carbon bainite or retained austenite exist between ferrite laths. Retained austenite is particularly prevalent in bainite that is formed during continuous cooling.
(vii) Solid solution strengthening both by interstitial or by substitutional solutes is associated with dislocations.
(viii) The B_s and B_f temperature is related with the composition approximately by the following equation:

$$B_s(^\circ\mathrm{C}) = 830 - 270[\mathrm{C}] - 90[\mathrm{Mn}] - 37[\mathrm{Ni}] - 70[\mathrm{Cr}] - 83[\mathrm{Mo}] \tag{46}$$

$$B_f(^\circ\mathrm{C}) = B_s - 120^\circ\mathrm{C} \tag{47}$$

In view of the complexity of the microstructure of bainite, quantitative relationships between microstructure and properties have been less well developed.

(1) STRENGTH. The factors contributing to the strength of bainite are:

(a) *Grain Refining Strengthening* (*Bainitic Ferrite Grain Size or Lath Width*). The Hall–Petch relationship $\sigma_y = \sigma_o + k_y d^{-1/2}$ is still valid; if bainitic ferrite lath width is used for d in the Hall–Petch equation, strength contribution is overestimated; but by using the prior austenite grain size, a more reasonable estimate is given. It is probable that a more realistic estimate may result from using bainite ferrite "packet" dimensions.

(b) *Dislocation Strengthening* (σ_d)

$$\sigma_d(\mathrm{MN/m^2}) = 1.2 \times 10^{-3}\sqrt{\rho} \tag{48}$$

where ρ is dislocation density in lines/cm^2.

(c) *Carbide Dispersion Strengthening* (σ_p). Over a wide range of carbon content, for bainite, σ_p can be described by

$$\sigma_d(\mathrm{MN/m^2}) = 600 + \frac{1.25\mu b}{\lambda} \tag{49}$$

where μ=shear modulus, b=Burgers' vector of slip dislocations, λ=interparticle spacing (μm) [66].

There are two parameters describing the dispersion of precipitates, one is interparticle spacing λ, and the other one is the number of the particles per unit section of the structure, n. λ and n are inter-related by

$$\lambda^{-1/2} = 0.94n^{1/4} \tag{50}$$

A multiple regression analysis gives the following equation for the 0.2% proof stress, which combines the grain size and particle density together:

$$\sigma_{ps}(\mathrm{MN/m^2}) = -1.91 + 17.2d^{-1/2} + 14.9n^{1/4} \tag{51}$$

(d) *Transformation Temperature*. There is a linear relationship between the tensile strength and the transformation temperature, see Fig. 28. It has been possible to relate the strength to the composition of steels with bainitic structure. This is consequent to the well-known linear dependence of the B_s temperature on composition. The relationship is

$$\sigma_t(\pm 60\,\mathrm{MN/m^2}) = 246 + 1900(\%\,\mathrm{C}) + 230(\%\,\mathrm{Mn} + \%\,\mathrm{Cr}) + 185(\%\,\mathrm{Mo}) + 90(\%\,\mathrm{N}) + 125(\%\,\mathrm{Ni}) + 65(\%\,\mathrm{Cu}) + 385(\%\,\mathrm{V} + \%\,\mathrm{Ti}) \tag{52}$$

(e) A more recent analysis of the strength of low-carbon bainite and/or acicular ferrite [14,67] has reverted to the typical yield stress equation for ferrite–pearlite structure, but modified to include precipitation and dislocation strengthening:

$$\sigma_y(\mathrm{MN/m^2}) = 88 + 37(\%\,\mathrm{Mn}) + 83(\%\,\mathrm{Si}) + 2900(\%\,\mathrm{N_f}) + 15.1(d_l^{-1/2}) + \sigma_d + \sigma_p \tag{53}$$

where d_l=bainite ferrite lath size in mm, σ_d=strengthening due to dislocations and given by $\sigma_d = \alpha\mu b\rho^{1/2}$, σ_p=strengthening from dispersed carbide particles within the ferrite laths but not at the lath boundaries [66].

It is clear that considerably more systematic work needs to be done before reliable quantitative expressions of a less empirical nature are available for the prediction of yield (or proof) stress of bainite, particularly over a wide range of carbon content.

(2) TOUGHNESS. ITT and CSE values of bainite are less well understood in terms of quantitative relationships with microstructure, than that of the strength. In low-carbon steels, upper bainite has a higher ITT [36,68]. There is a linear relationship between the ITT and logarithmic of $d^{-1/2}$, d representing unit crack path of ferrite size, which is basically equivalent to the bainitic ferrite packet size (Fig. 29).

By refining the prior austenite grain size, and thereby the bainitic ferrite packet size, controlled rolling is beneficial to the ITT (Fig. 30). A recent analysis of the factors influencing the ITT of bainite or acicular ferrite is based on the equation for ferrite–pearlite structures, but modified as follows:

$$\mathrm{ITT}(^\circ\mathrm{C}) = -19 + 44(\%\,\mathrm{Si}) + 700\left(\sqrt{\%\,\mathrm{N_f}}\right) + 0.26(\sigma_p + \sigma_d) - 11.5d^{-1/2} \tag{54}$$

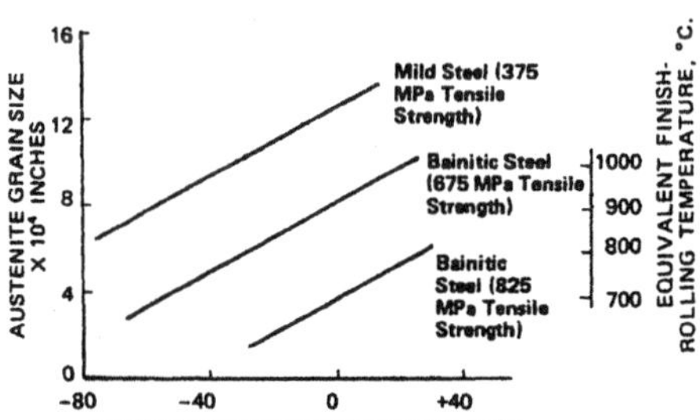

Figure 30 Effect of finishing rolling temperature on ITT of low-carbon bainite. (From Refs. 34,68,70.)

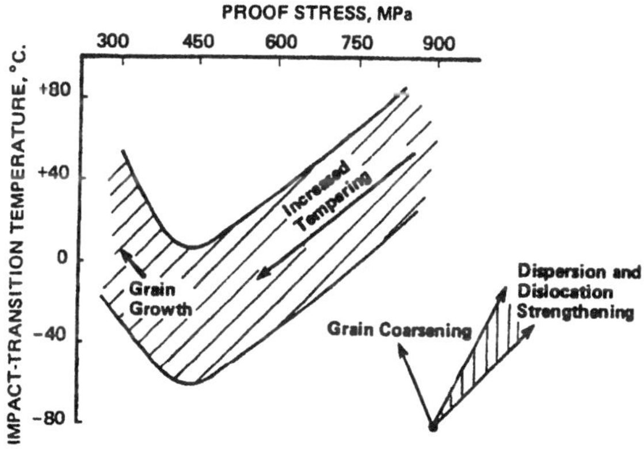

Figure 31 Effect of tempering on the relationship between proof stress and ITT of lower bainite. (From Refs. 36,68.)

where σ_p=precipitation strengthening as described by the Ashby–Orowan relationship, σ_d=strengthening caused by random dislocation and dislocations in low angle bainitic ferrite lath boundaries, d is the mean spacing between high angle boundaries, i.e., packet or prior austenite grain boundaries. The value of σ_d can be decomposed into two terms: σ_{df}, a strengthening term contributed by forest dislocations and σ_b, a term for low angle boundary strengthening:

$$\sigma_d = \sigma_{df} + \sigma_b \tag{55}$$

σ_{df} can be described by Eq. (14), $\sigma_{df}=\alpha\mu b\sqrt{\rho}$, whereas σ_b is more difficult to assess. It may be estimated by the difference in yield strength between structures with grain of size d, and those with bainite ferrite lath size of d_l, i.e.

$$\sigma_b = 15.1(d_l^{-1/2} - d^{-1/2}) \tag{56}$$

The reason is that high angle boundaries increase strength and lower the ITT by acting as barriers to crack propagation, the vector being –0.7°C per MN/m^2 increase in yield stress. On the other hand, low angle boundaries increase the strength and by so doing raise the ITT, because they do not act as barriers to crack propagation.

The relationship of σ_y and ITT agrees well with the best observed relationship over a wide range of strength, but for some low-carbon bainite structures, anomalously high values of ITT impinge. This embrittlement is due to the coarse carbides at the ferrite lath boundaries in upper bainite, just as the filament carbides at ferrite grain boundaries which raise the ITT.

Increasing carbon content markedly increases the ITT of bainitic structures, although carbon effect has entered into the term σ_p; however, some extra terms would be needed in order to predict the ITT over a wide range of carbon content in bainite. Retained austenite in bainite would decrease the ITT of bainite structure, but the effects have not been described quantitatively.

The application of Eq. (54) to acicular ferrite structure in high manganese steels has shown that the prior austenite grain size instead of bainite ferrite packet size gives more reliable predictions.

If a preferred orientation texture parameter has been included, a more improved prediction of the ITT of acicular ferrite after controlled rolling may be achieved.

Very little work has been done on the CSE value of bainitic steels. But there is evidence that:

(i) Increasing the carbon content decreases the CSE value [36,68].
(ii) Decreasing both σ_p and σ_d increases the CSE value.
(iii) Tempering of bainitic structure has been shown to decrease the ITT [36,64,68] due to the decrease in σ_p and σ_d (Fig. 31). But when tempering at a very high temperature, the ITT will increase due to grain growth of the ferrite despite a further decrease in strength. Equation (54) would apply to ITT of bainite after tempering.

One must be aware of the fact that embrittlement may be introduced in tempering due to:

(i) Temper brittleness.
(ii) The precipitation of fine alloy carbides during tempering steels which show secondary hardening [70]
(iii) The formation of large grain carbides during tempering at high temperature.
(iv) In the high manganese acicular ferrite steels, martensite forms during cooling when tempering temperature above the A_e.

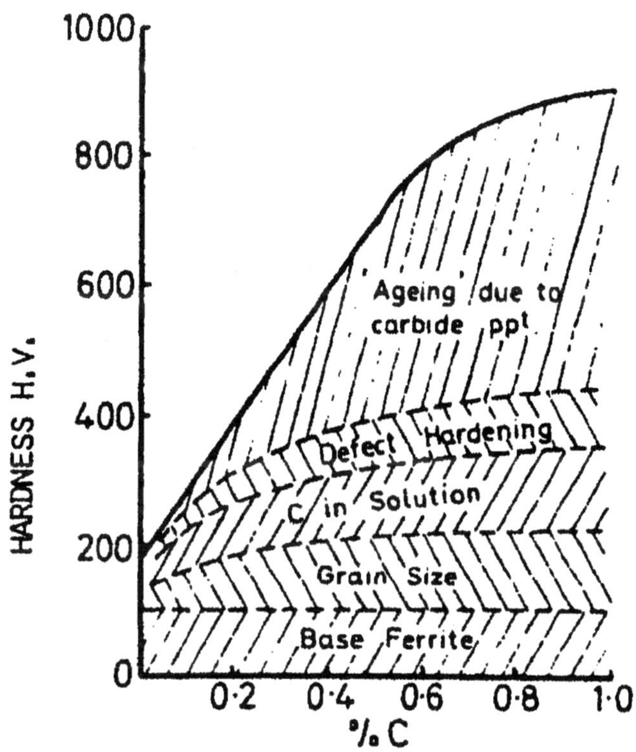

Figure 32 Possible contributions of various mechanisms to hardness of martensite. (From Ref. 42.)

(v) The transformation characteristics of any austenite formed during tempering are clearly of importance. Structures, comprising islands of matensite and bainite of varying carbon content (or even austenite) in a polygonal or acicular ferrite matrix, are reported to have good combinations of strength, ductility and formability, however, toughness may vary depending on the detailed structure produced.

For an optimum combination of high yield strength, good weldability, good formability, low ITT and high CSE, the following structural and compositional factors are recommended:

(i) Carbon content as low as possible.
(ii) A low B_s or transformation temperature.
(iii) A high M_s temperature.
(iv) A very fine austenite grain size or bainitic ferrite packet size.
(v) A low non-metallic inclusion content.
(vi) A high value of the ratio of depression of the B_s per wt% to the depression of M_s per wt% is recommended for good strength and weldability.

e. Microstructure Composed of Martensite. Martensite is the hardest structure that can be generated in steels; the strength increases with carbon content. There are many attempts to develop an equation to relate the M_s temperature to the composition of steels. Table 4 lists the formulae for M_s calculations.

The symbols in the brackets again refer to weight percentage and M_f, the martensite finish temperature is usually about 200°C below the M_s temperature

$$M_f = M_s - 200°C \tag{57}$$

For athermal transformation kinetics, the relationships between the volume fraction transformed to martensite and the degree of undercooling below M_s have been described by [129,130]

$$f = 1 - 6.96 \times 10^{-15}[455 - \Delta T]^{5.32} \tag{58}$$

Table 4 Formulae for M_s [128]

Equations	Authors	Date
M_ε (°F) = 930 − 570C − 60Mn − 50Cr − 30Ni − 20Si − 20Mo − 20W	Payson and Savage	1944
M_s (°F) = 125×(1 − 0.62C) (1 − 0.092Mn) (1 − 0.033Si) (1 − 0.045Ni) (1 − 0.070Cr) (1 − 0.029Mo) (1 − 0.018W) (1 + 0.020Co)	Caragealla	1944
M_s (°F) = 930 − 600C − 60Mn − 50Cr − 30Ni − 20Si − 20Mo − 20W	Rowland and Kyle	1946
M_s (°F) = 1000 − 650C − 70Mn − 70Cr − 35Ni − 50Mo	Grange and Stewart	1946
M_s (°F) = 930 − 540C − 60Mn − 40Cr − 30Ni − 20Si − 20Mo	Nehrenberg	1946
M_s (°C) = 561 − 474C − 33Mn − 17Cr − 17Ni − 21Mo	Steven and Haynes	1956
M_s (°C) = 539 − 423C − 30.4Mn − 12.1Cr − 17.7Ni − 7.5Mo	Andrews	1965
M_s (°C) = 512 − 453C − 71.5(C)(Mn) + 15Cr − 16.9Ni − 9.5Mo + 217(C)^2 − 67.6(C)(Cr)	Andrews	1965

and

$$f = 1 - \exp -(1.10 \times 10^{-2}\Delta T) \tag{59}$$

The strength and toughness of martensite and its tempered products in relation to microstructure are less quantitatively available.

(1) STRENGTH. The following mechanisms are operative in martensite strengthening:

(i) Interstitials in solid solution, most commonly carbon and nitrogen.
(ii) The fine martensite plate or lath size.
(iii) The high dislocation density in low-carbon lath martensite.
(iv) The obstruction of dislocations movement by twin boundaries in high carbon lenticular martensite.
(v) The precipitation of carbides, especially in tempered and auto-tempered structure, and the clustering of carbon atoms.
(vi) The interaction between interstitially dissolved carbon (and nitrogen) atoms and dislocations.

(a) *Solid Solution Strengthening by Carbon*. It is often reported that the hardness value of martensite is proportional to $C^{1/2}$ and a function of $C^{1/3}$ is also reported.

The hardness values of alloy martensite are usually higher than those of plain carbon martensite; this is due to the fact that M_s is depressed and less auto-tempering occurs.

(b) *Precipitation Strengthening*. Carbon as a carbide precipitation in martensite may be twice as effective as that in the state of solid solution. At temperatures above –60°C, precipitation effects occur in all the martensites. It is difficult to differentiate precipitation from clustering and atmosphere effect, but it has been estimated that carbon as precipitates gives an increment of hardness of 500 HV in 0.8% martensite compared with a maximum hardness increment of 350 HV from solid solution strengthening.

(c) *Grain Size*. It is difficult to define martensite grain size. It was estimated an increase in yield stress of 450–600 MN/m^2 is effected by fine lath size. Some workers have shown that the strength of martensite depends on the prior austenite grain size. However, the prior austenite grain size only confines the martensite crystals formed immediately below M_s; when the temperature goes down below M_s austenite grain was partitioned into compartments and the martensite crystal decreases as the partition process goes on; thus, the grain size represents the average.

(d) *Dislocation Strengthening*. The effect of dislocation density in low-carbon martensite has estimated that the contribution to the strength is of the order of 150–300 MN/m^2.

(e) *Substructure Strengthening*. A systematic study of the internal structure of martensite in Fe–Ni alloys showed that twinning in martensite more or less provides strengthening.

The effect of the different strengthening mechanisms as a function of carbon content of martensite is shown in Fig. 32. Attempts have been made to relate quantitatively the yield strength of martensite to the carbon content. Typical equations are

$$\sigma_y(MN/m^2) = 290 + 1800(\%\,C)^{1/3} \text{ valid up to } 0.4\%\,C \tag{60}$$

and

$$\sigma_y(\mathrm{MN/m^2}) = 815 + 3850(\%\,\mathrm{C}) \text{ valid up to } 0.25\%\,\mathrm{C} \tag{61}$$

Up to 0.2%, these equations give very similar strength values. But the results diverge considerably at higher carbon content.

The strength value of tempered martensite in terms of dislocation density and carbide interparticle spacing has been given by several investigators [71–73].

The general form of the formula may be described by

$$\sigma_y = \text{constant} + f(\rho) + f(\lambda) + f(d) \tag{62}$$

ρ is dislocation density, λ is carbide interparticle spacing, d is mean linear intercept of the grain size. $f(d)$ should be included in the equation when the interparticle spacing is less than the mean linear intercept of the grain size.

An equation for yield stress of low-carbon lath martensite takes the following form:

$$\sigma_y(\mathrm{MN/m^2}) = 88 + 37(\%\,\mathrm{Mn}) + 83(\%\,\mathrm{Si}) + \alpha\mu b\rho^{1/2} + 28d^{-1/2} + \sigma_{\mathrm{lb}} \tag{63}$$

where μ = shear modulus, b = Burgers' vector of slip dislocation, ρ = dislocation density in lines/mm^2, d = mean linear intercept of packet size, σ_{lb}=strength contribution from lath boundaries.

There is no carbide term in (63); this would imply that carbides were largely confined to the boundaries and thus would not influence the strength.

In the case of tempered martensite, particularly in alloy steels in which the carbide type may change with increasing tempering temperature, multi-modal carbide distributions may occur. Two extreme cases may exist: one is that the carbides are found exclusively at the ferrite grain boundaries, and the other case is that the carbides are located within the ferrite grains. In the former case, the carbides do not contribute to strengthening effect, the grain size is the major factor

$$\sigma_y(\mathrm{MN/m^2}) = 108 + 18.2d^{-1/2} \tag{64}$$

In the latter case, carbides contribute to

$$\sigma_y(\mathrm{MN/m^2}) = 71 + (0.015/\lambda)\ln(D/2b) + 23.9d^{-1/2} \tag{65}$$

where D = carbide particle diameter, Å, b = Burgers' vector of slip dislocations, λ = carbide particle spacing (μm), d=the mean linear intercepting grain size (mm).

A similar expression has also been used to describe the flow stress.

(2) TOUGHNESS. The quantitative analysis of the effect of microstructure of tempered martensite on ITT and CSE values is very sparse.

It is complex in the case of tempered martensite structure and compounded by various embrittling phenomena, such as:

(i) Low temperature embrittlement during tempering at 250–300°C.
(ii) Temper embrittlement.
(iii) Embrittlement due to secondary hardening.
(iv) Upper nose embrittlement during tempering at temperature just below Ac_1. Equation (54) may also be used to express the ITT of low-carbon lath martensite.

> In this case, d is the austenite grain size or the packet size. The fracture facet size is determined by the austenite grain size in turn. There is evidence to indicate that the ITT of martensite and its temper product increases with increasing carbon content. It might be inferred that for tempered martensite the ITT increases more or less linearly with strength by a factor of 0.2/0.3°C per MN/m^2.

There are very few quantitative data available on the effect of composition and microstructure on the CSE value of martensite and its tempered products. Some evidence indicates that the prior austenite grain size has no effect on CSE.

There is a direct relationship between the CSE value and the total ductility at fracture, it may be concluded that the CSE value decreases as the carbon content increases for tempered martensite; however, this may be in an exponential manner rather than a linear one.

The factors influencing the design of martensitic steels may be summarized as follows:

(i) The lowest carbon content to achieve the requisite strength.
(ii) The requisite hardenabilty for the section sizes involved.
(iii) Adequate ductility which will depend on the strength level required.
(iv) Maximum toughness commensurate with the required strength.
(v) Some measures for adequate fatigue resistance.
(vi) Some measures to improve weldability.
(vii) Resistance to quench cracking.

For good ductility, toughness and fatigue resistance, the greatest possible freedom from non-metallic inclusions is required. In general, the higher the tempering temperature which can be used, the better is the toughness and the ductility, provided no embrittlement is encountered. Consequently, there is a need to increase the tempering temperature at which requisite strength can be achieved. For example, in order to develop a steel which will have a high strength at tempering temperature 350°C, and with a minimum tendency to quenching cracking, together with adequate ductility, toughness and weldability, the ratio of retardation of tempering to the depression of M_s per wt% alloy addition should be maximum. Typical values for this ratio are listed in Table 5.

Based on the above figures, carbon is detrimental and should be as low as possible consistent with the strength requirements. Cr should be limited except to the economic

Table 5 Ratio of Retardation of Tempering to ΔM_s

Elements	Ratio
C	−12
Cr	0
Mn	0.24
Ni	0.48
Mo	0.80
W	0.90
V	1.0
Si	1.8
Co	8.0

level for hardenability. Silicon and cobalt are very effective and Si is by far the most economical. Hardenability will be discussed in another section.

2. Fatigue

Fatigue is a failure mode which occurs in components and structures under the action of cyclic stress or vibration condition. This kind of failure could occur at a lower stress level than that could cause failure by the unidirectional application of the stress. Fatigue damage usually occurs at critical positions in a component or in a structure, such as holes, corners, places where sectional dimension abruptly changes, junctions or joints, interfaces between inclusions and matrix, i.e., at the either geometrical or microstructural discontinuities.

Total fatigue failure consists of two stages, i.e., crack nucleation and crack growth. The total fatigue life, i.e., the number of strain cycles to failure, may be estimated by the summation of the crack nucleation life N_n and the crack growth life N_g, i.e. N_n+N_g. The motor industry is more concerned with crack nucleation. The first stage fatigue life may be estimated using data obtained from smooth specimens with a gauge length of only a few millimeters. Buckling should be avoided in the tests involving both tension and compression. The cyclic stress during each test varies gradually and reaches a constant value. As the first crack is initiated, the stress falls rapidly; at this point, the test is terminated. A strain life curve could be obtained by plotting the magnitude of the strain with the number of circles to crack initiation. Figure 33 shows an example for two steels. At higher values of strain, the number of cycles to give crack nucleation becomes relatively small, depending on the strength and other characteristics of the steel. At very low applied strain, crack nucleation can never happen. For practical purpose, N_g may be considered as infinite.

In order to estimate the fatigue life of the critical part of the structure, it is necessary to know the stress history to which the component has been subjected. Using the cyclic stress–strain curve, the local stress history may be calculated. The so-called cyclic stress–strain curve involves the use of several specimens; however, alternative methods have been proposed to obtain a cyclic curve from a single specimen.

The cyclic stresses developed under the application of a constant cyclic strain often vary during the test. In the stable region, stresses and strains may be plotted against each

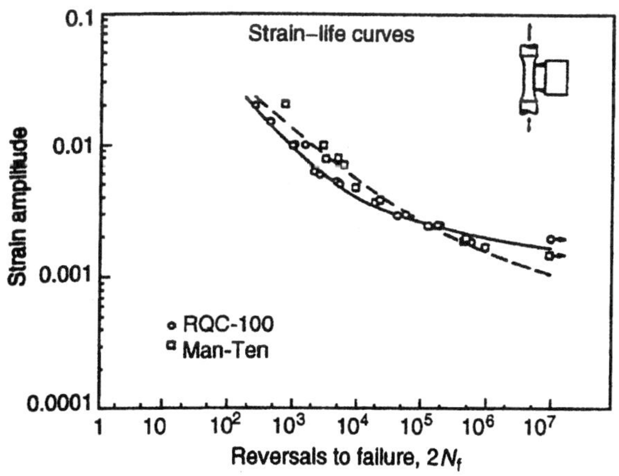

Figure 33 Strain–life curves for two structural steels. (From Ref. 133.)

other to form the so-called stress–strain curve. The monotonic curve obtained from a conventional tensile test would correspond with the first part of the first cycle in a fatigue test. In a fatigue test, cyclic softening, cyclic hardening and cyclic stability, or a mixture of them have been observed for various types of steel when plastic strain is involved in the test, but the stress is stable when the deformation is substantially elastic. The stability tends to be associated with the strengthening mechanisms used in the steel, strengthening by solid solution, by grain refinement and precipitation tend to be stable in the presence of plastic deformation, whereas strengthening by temper rolling or baking hardening tends to be unstable.

The method will enable calculating the contribution to the fatigue damage that would arise from each of the strains in the expected strain history. It is assumed that the total fatigue damage is cumulative and is a linear superposition of each in the sequence of the strains [131,132]. If the proportion of the various strains m in the strain history is p_1, $p_2, p_3, \ldots, p_n$, etc., and the number of circles to crack nucleation for these strains taken from the strain life curve is N_1, N_2, $N_3, \ldots$, etc., then the cumulative damage is given by

$$\text{Damage} = p_1 N_i/N_1 + p_2 N_i/N_2 + p_3 N_i/N_3 + \cdots \tag{66}$$

and failure is expected to occur when $\sum p_m N_i/N_m = 1$. N_i is the fatigue life to crack nucleation to be calculated. The difficulty with this method is that calculation of p_m is not straight-forward. Several methods have been proposed, of which the rainflow method is probably the most accurate one [133].

An alternative way of conducting a fatigue test is to develop a stress–life curve by using a series of constant cyclic stresses. A stress may be identified below which the number of cycles to crack is never reached. This stress is regarded as the fatigue strength of the steel. The relationship between the fatigue strength and the monotonic tensile strength depends on the yield to tensile strength ratio of the steel [134].

A fatigue limit which relates to crack growth may be measured using a standard test method [135]. Figure 34 illustrates a type of specimen used for the test. Before the test, a fatigue crack is performed at the notch tip. A stress intensity factor at the crack tip is defined and this depends on the difference between the minimum and the maximum loads applied. The crack length is monitored by a traveling microscope and based on a plot of crack length vs. number of cycles, the crack growth rate per cycle can be obtained. A fatigue threshold is defined as the maximum stress intensity below which an existing crack does not propagate. Contact fatigue failure is a special failure mode occurring in bearings. Bearings are required to have long endurance life; laboratory evaluation test has to be accelerated such that a meaningful result can be obtained in a matter of days. Various bearing fatigue tests have been developed which are capable of assessing the performance of bearing steels in the form of balls, washers, cones, and cylinders. Figure 35 shows a Unisteel washer test machine which can provide a maximum calculated Hertizian stress of $3725\,\text{N/mm}^2$ at a relatively low load.

For steel cylinders, the influence of hydrogen on the rate of fatigue crack growth is an important topic which has been under investigation both theoretically and practically. Figure 36 illustrates that hydrogen accelerates the fatigue crack growth tested on Cr–Mo and C–Mn steels. Figure 37 shows the acceleration factor at a stress intensity value of $20\,\text{MNm}^{-3/2}$ as a function of hardness. For transformable gas containers, the hardness is restricted to the range of 230–290 HB. The yield strength is not allowed to exceed $680\,\text{N/mm}^2$ and the tensile strength must be within the range $800–930\,\text{N/mm}^2$.

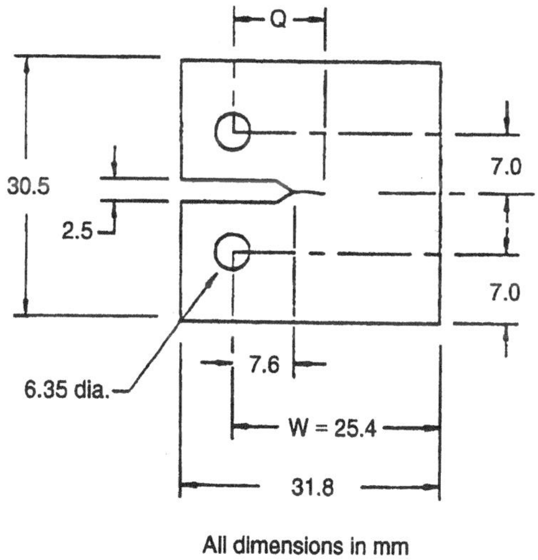

Figure 34 Compact tension specimen for crack growth test. (From Ref. 5.)

B. Formability

Cold formability is one of the most important requirements for low-carbon strip steel grades. For many applications, the main requirement is to be able to form the part without splitting, necking or wrinkling. For structural applications, the strength of the steel is more

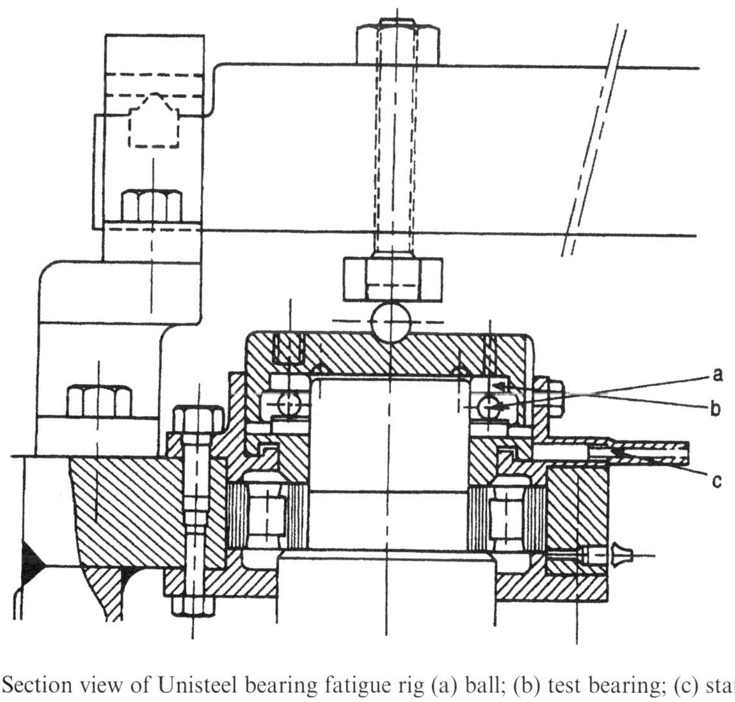

Figure 35 Section view of Unisteel bearing fatigue rig (a) ball; (b) test bearing; (c) standard thrust race. (From Ref. 5.)

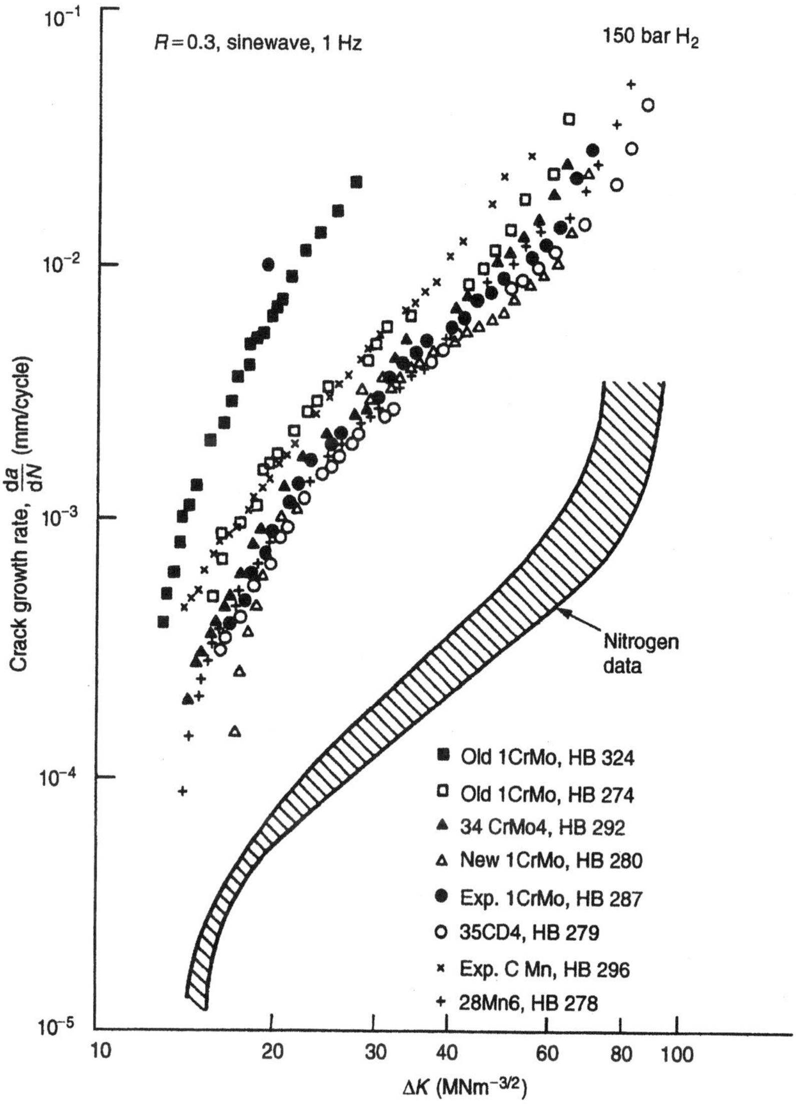

Figure 36 The influence of hydrogen on the rate of fatigue crack growth in the gas container steel. (From Ref. 136.)

important and must be above a given minimum value. There is a general tendency for the cold formability of any type of steel to reduce as the strength increases. Many developments of higher strength steels have been specifically aimed at providing higher strength while minimizing the loss in formability that would otherwise have taken place.

The cold formability of sheet steel may be resolved into two separate but related components, namely drawability and stretchability. The drawability of a steel is the ability to be drawn in, to make a component without local necking or splitting, whereas its stretchability is the ability to be stretched to form a component without local necking or splitting. Stretching involves both positive major and minor strains in the plane of

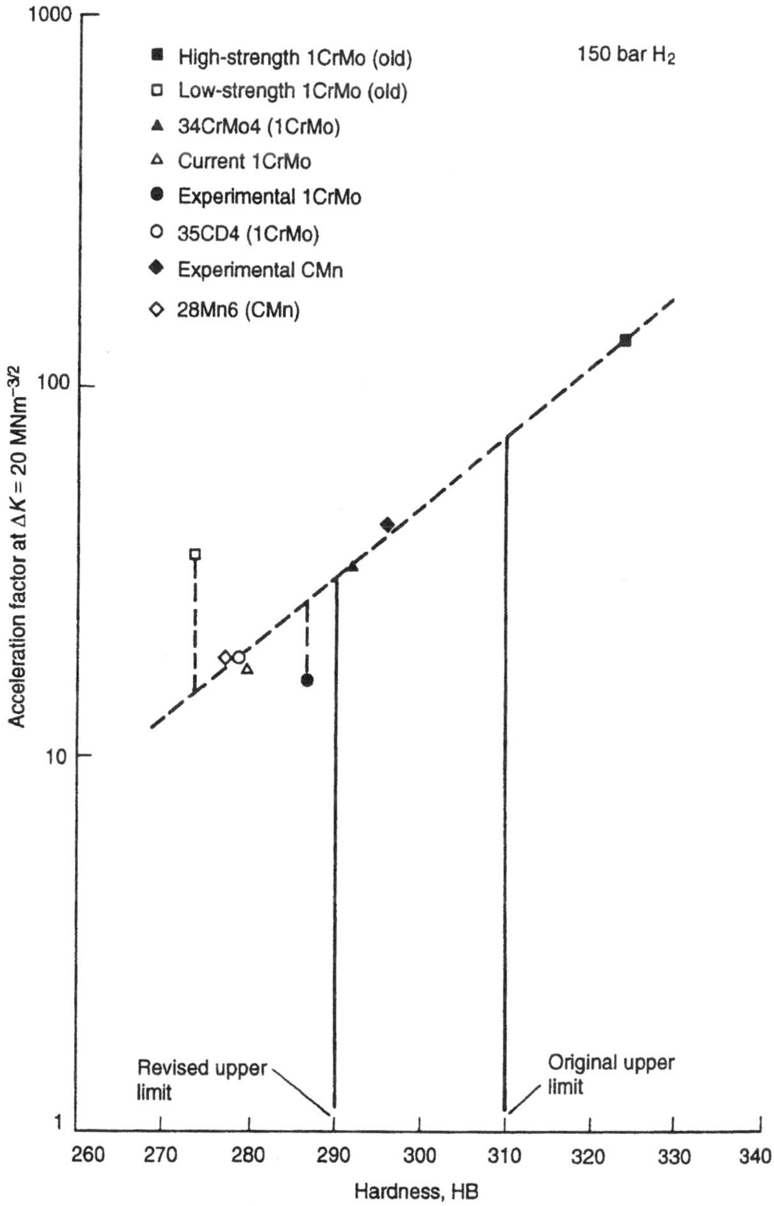

Figure 37 The influence of hardness on the enhancement of fatigue growth by hydrogen. (From Ref. 136.)

the sheet, whereas drawing involves major and minor strains, one of which is positive and the other is negative.

A simple measure of the deep drawability of a steel may be obtained by forming flat bottomed cylindrical cups from circular blanks; the maximum ratio between the circular blank diameter and the punch diameter that may be drawn in a single stage to form a cylindrical cup without necking or splitting is a measure of the drawability of the steel.

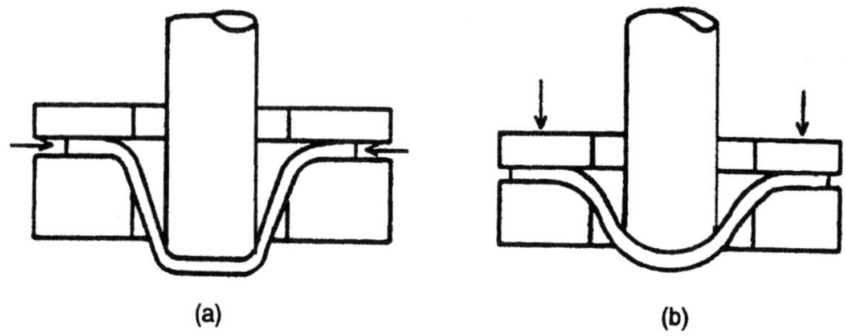

Figure 38 Schematic drawing to show (a) deep drawing; (b) stretch forming. (From Ref. 5.)

The ratio is called the limiting ratio and may vary up to about 2.5 for very good deep drawing steels.

A simple measure of the stretchability of steel is by using a hemispherical punch to form a circular dome. The maximum ratio of dome height to dome diameter at the moment of necking or splitting is a measure of the stretchability. In Fig. 38 are two schematic drawings to show these two tests. Hydraulic bulge test may be used as a similar test, i.e., oil under pressure being employed to form a dome. In these tests, the shape of the dome is circular; the strain at the top of the dome is said to be balanced biaxial strain. The result would not be influenced by friction.

The formability of sheet steels may also be assessed using parameters that may be measured directly from a conventional universal tensile test. The first parameter is strain ratio, and is usually called the r value. It gives a measure of the drawability and also a measure of the resistance to thinning, resulted from the orientation of the slip system that is active during drawing. The second parameter is the work-hardening coefficient, designated as n, which is closely related to the stretchability. The uniform and local elongation measured in the tensile test is also related to the stretchability.

1. *The Strain Ratio r*. In a tensile test, the strain ratio r is defined as the ratio of the true strain in the width direction, ε_w, to the true strain in the thickness direction, ε_t, at a particular moment during the test, usually when the length engineering strain is either 15% or 20%, corresponding to true length strain of with 0.1398 or 0.1823, thus

$$r = \varepsilon_w/\varepsilon_t \tag{67}$$

Constant volume is assumed during plastic deformation, which implies that the sum of the true width, thickness and length strains would be equal to 0; this means that only width w_f at true length strain of either 0.1398 or 0.1823 needs to be measured. Then

$$r = \varepsilon_w/\varepsilon_t = \frac{\ln w_f/w_0}{(\ln w_0/w_f - 0.1823)} \tag{68}$$

Usually, the strain ratio has different values for different directions in the plane of the sheet. A mean value, r_m or $\bar{r}$, is defined by measuring the value in the rolling, transverse and diagonal directions and calculating a mean, giving the diagonal value double weight as follows:

$$r_m = (r_0 + 2r_{45} + r_{90})/4 \tag{69}$$

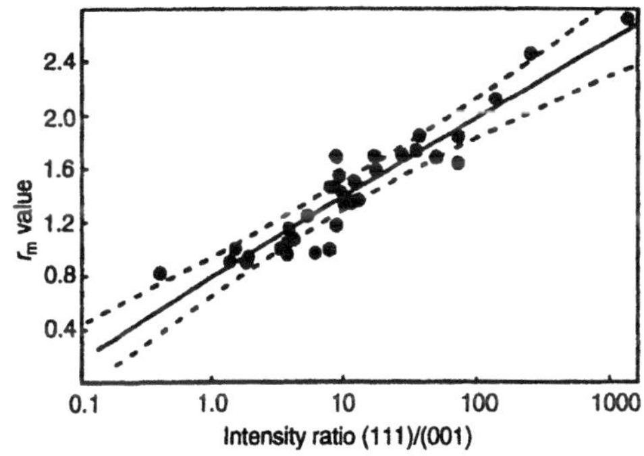

Figure 39 Ratio of the intensity of the (1 1 1) components to that of (0 0 1) vs. r_m of low-carbon steel sheets. (From Ref. 44.)

The variation in r value around the rolling plane, Δr, is defined as

$$\Delta r = (r_0 - 2r_{45} + r_{90})/4 \tag{70}$$

Crystallographic texture is the main structural factor which influences the r value. However, the relationship between r value and crystallographic texture is a complicated one.

The γ fiber texture, i.e., a high proportion of grains with (1 1 1) planes parallel to the surface, leads to high r values. A high proportion of grains with (1 0 0) planes parallel to the surface, which is one component of the α fiber texture, tends to lead to a low r value. Various empirical relationships between the strength of texture components and r value have been established. Figure 39 shows a relationship between the r_m value and the ratio of the intensity of (1 1 1) to (1 0 0) texture components.

The r_m value affects the distribution of strain in a pressing by influencing the thinning tendency and gives a measure of the deep drawability of the steel. Figure 40 gives a plot of r_m value vs. limiting drawing ratio for a number of steels drawn to form cylindrical cups.

The r_m value also influences the strain distribution during stretching. The ratio of Δr to the r_m value determines the height of ears that would be obtained in cups prepared using

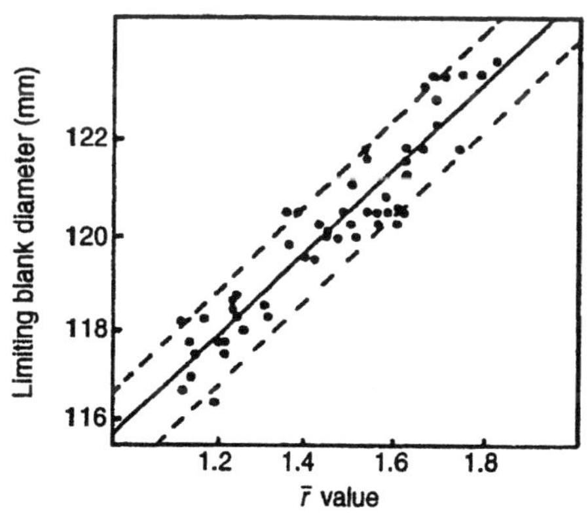

Figure 40 Effect of r_m on limiting blank diameter for carbon steels drawn using polythene sheet lubrication. (From Ref. 45.)

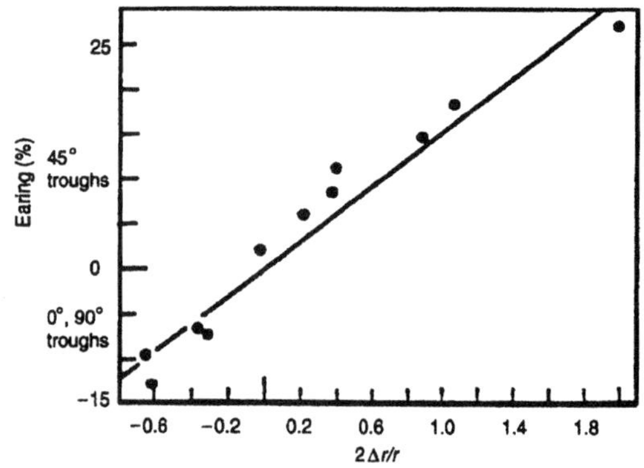

Figure 41 The relationship between earing and Δr. (From Ref. 46.)

a cylindrical punch and a circular blank. Figure 41 illustrates the relationship between $\Delta r/r_{\rm m}$ and earing. The ear height is also influenced by the punch and blank diameters (p and B). The following equation gives the relationship [95]

$$\text{ear height} = M(B^2 - p^2)(B - p)^{1.5}/p^{2.5} \tag{71}$$

where M is a dimensionless earing parameter related mainly to r and Δr values of the steel, and also influenced a little by cup drawing conditions.

Ferrite grain size is closely related to the drawing properties of steel; $\bar{r}$ values greater than unity are essential for good deep drawing properties. Figure 42 illustrates the effect of grain size on $\bar{r}$ value.

In addition to the normal anisotropy due to crystallographic texture, it should also be mentioned that there is another anisotropy in mechanical properties due to periodic banding, i.e., the pearlite and ferrite were arranged in alternate bands due to the dendritic segregation of P, Mn, etc. This has been investigated by several authors. The severity of the periodic banding may be reduced by the bridging effect of pearlite which can be achieved by proper C/Mn ratio adjustment under the same processing parameters [48].

2. *The Strain Hardening Coefficient n.* Any true stress–strain curve measured throughout a tensile test may be fitted to an equation of the following type:

$$\sigma = k\varepsilon^n \tag{72}$$

where k and n are constants.

Taking logarithm of both sides in Eq. (72):

$$\ln \sigma = \ln k + n \ln \varepsilon \tag{73}$$

The coefficient n is the slope of the true stress–strain curve plotted on a logarithmic scale. It determines how quickly the true stress builds up with increasing true strain, and is referred to as a work-hardening coefficient. In practice, the true stress–strain curve plotted on a logarithmic scale is not always a perfect straight line, i.e., in this case n value varies with the strain and takes more than one value, which corresponds to a different range of strains.

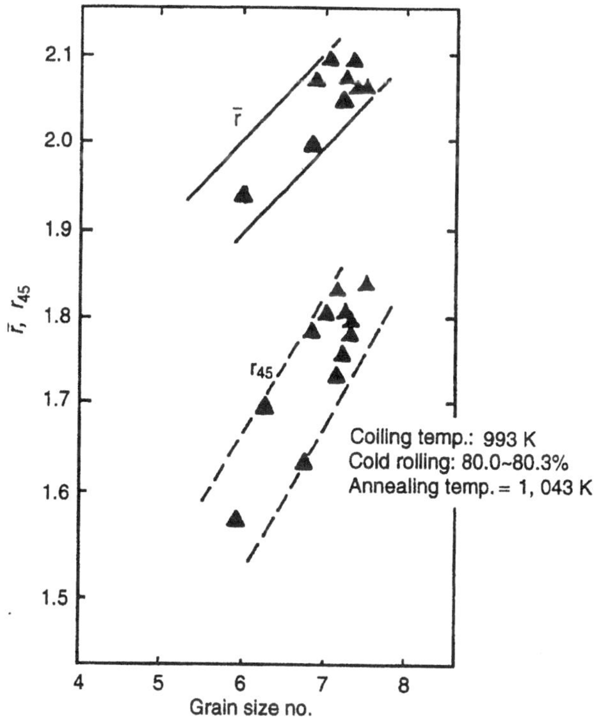

Figure 42 Effect of grain size on r_m values of IF steel sheets. (From Ref. 47.)

The uniform and the total elongation measured in tensile test closely relate to the n value of a steel. Figure 43 illustrates this relationship for several types of steels. If the n value is truly constant throughout a tensile test, then it would be numerically equal to true strain at the maximum load which is the uniform elongation.

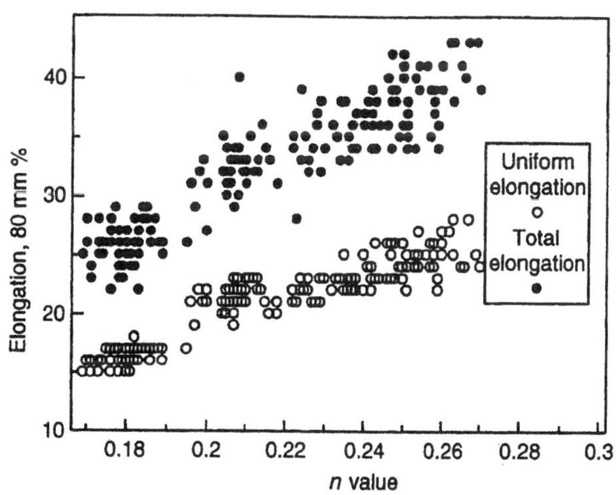

Figure 43 n values vs. ε_t and ε_u. (From Ref. 5.)

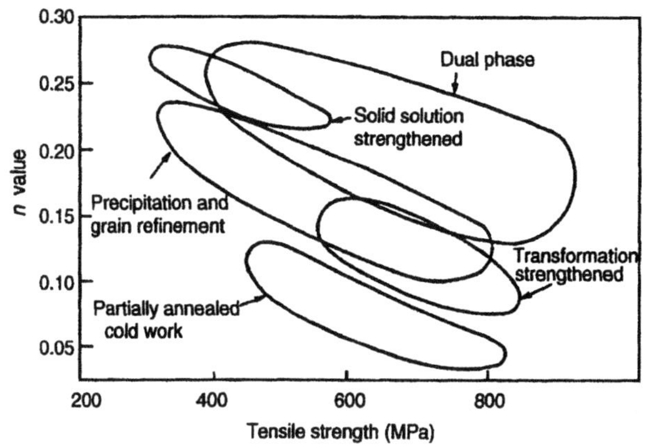

Figure 44 Relationship between n value and tensile strength. (From Ref. 49.)

Strength and strengthening mechanisms used to develop the strength are the most important factors that influence the n value of the steels. Figure 44 illustrates the relationship between n value and tensile strength and the corresponding strengthening mechanisms. It can be seen that for each type of steel, the n value decreases with increasing strength. It is clear that the highest possible n value is obtained when the strength of any type of steel is as low as possible and this is the reason why low strength mild steels are always more formable than higher strength steels. During the development of higher-strength steels, one should choose or develop strengthening mechanisms that give the highest possible n value and hence elongation for the strength needed.

The n value is the second main parameter that determines the distribution of strain across any pressing. In general, a high n value for steel leads to a more uniform strain distribution than low n value steel. A higher n value enables the forming strains to be distributed across a pressing with less likelihood of local necking and failure than for a low n value. In practice, it is found that success with any particular pressing may be dependent on a high value of one or other of the parameters, i.e., r or n, whereas success with another pressing may depend on the high value of both parameters [50].

The n values of freshly solidified steels in simulation with the solid shell structure in continuously casting had been measured for several low-carbon steels. The results show that n value changes with deformation temperature strain rate and with the plastic deformation mechanism [51].

In summary, beside r and n derived from the tensile test which defines the formability of the material, the following properties derived from tensile test have been paid attention to: σ_f, flow stress at a given strain; the actual strain hardening rate $d\sigma/d\varepsilon$, the maximum uniform strain prior to plastic instability ε_u, and the strain at fracture

$$\begin{aligned}\sigma_f(\pm 35\,\mathrm{MN/m^2})_{\mathrm{at}\,\varepsilon=0.2} &= 246 + 45(\%\,\mathrm{Mn}) + 138(\mathrm{Si}\,\%) + 920(\mathrm{P}\,\%) + 70(\mathrm{Sn}\,\%)\\ &\quad + 3750(\%\,\mathrm{N_f}) + 4.2(\%\,\mathrm{pearlite}) + 15.0d^{-1/2}\end{aligned} \tag{74}$$

$$\begin{aligned}\sigma_f(\pm 55\,\mathrm{N/m^2})_{\mathrm{at}\,\varepsilon=0.8} &= 385 + 57(\%\,\mathrm{Mn}) + 185(\mathrm{Si}\,\%) + 920(\mathrm{P}\,\%) + 215(\mathrm{Sn}\,\%)\\ &\quad + 4150(\%\,\mathrm{N_f}) - 508(\mathrm{S}\,\%) + 418(\%\,\mathrm{pearlite}) + 27.7d^{-1/2}\end{aligned} \tag{75}$$

$$d\sigma/d\varepsilon(\pm 52\,\mathrm{MN/m^2})_{\mathrm{at}\,\varepsilon=0.8} = 385 + 57(\%\mathrm{Mn}) + 110(\mathrm{Si}\%) + 460(\mathrm{P}\%) + 215(\mathrm{Sn}\%) + 4150(\%\mathrm{N_f}) - 508(\mathrm{S}\%) + 4.8(\%\mathrm{pearlite}) + 23.1d^{-1/2} \quad (76)$$

$$d\sigma/d\varepsilon(\pm 52\,\mathrm{MN/m^2})_{\mathrm{at}\,\varepsilon=0.8} = 169 + 37(\%\,\mathrm{Mn}) + 99(\mathrm{Si}\,\%) + 123(P\,\%) - 950(\mathrm{S}\,\%) \quad (77)$$

$$\varepsilon_u(\pm 0.026) = 0.27 - 0.016(\%\,\mathrm{pearlite}) - 0.025(\%\,\mathrm{Mn}) - 0.044(\mathrm{Si}\,\%) - 0.039(\mathrm{Sn}\,\%) - 1.2(\%\,\mathrm{N_f}) \quad (78)$$

$$\varepsilon_t(\pm 0.21) = 1.3 - 0.020(\%\,\mathrm{pearlite}) + 0.3(\%\,\mathrm{Mn}) + 0.2(\mathrm{Si}\,\%) - 3.4(\mathrm{S}\,\%) - 4.4(\mathrm{P}\,\%) - 0.29(\mathrm{Sn}\,\%) + 0.015d^{-1/2} \quad (79)$$

Based on the above equations, it can be summarized that:

(i) High pearlite content increases σ_f and $d\sigma/d\varepsilon$ but reduces ε_u and ε_t.
(ii) Decreasing the pearlite interlamellar spacing also increases $d\sigma/d\varepsilon$.
(iii) Pearlite decreases ε_u by virtue of its effect on reducing the yield elongation, thus raising the flow stress curve.
(iv) An increase in $d\sigma/d\varepsilon$ relative to σ_f increases ε_u which is beneficial for stretch forming requirements.
(v) ε_u is the most influencing factor on the limitation of formability by plastic instability; high ε_u is essential for good stretch forming.
(vi) ε_t is the limiting criteria with regard to fracture; a high ε_t value together with a high ε_u is required for good bendability. Pearlite is detrimental, but spherodizing can improve the ductility in comparison with the lamellar pearlite structure.

Solid solution strengthening effect is similar to that on σ_y and varies with its atomic size. In general, solid solution strengthening increases σ_y but has less effect on the work-hardening rate [52]. Substitutional solutes decrease n [47], Table 6 illustrates the changes per wt%.

Interstitial solutes do not significantly affect n.

Grain refinement will increase $d\sigma/d\varepsilon$; the opposite effect has been reported also but not to a great extent [52,53].

Table 6 Change in n per wt% of Solutes

Alloying element	Change in n/wt%
Cu	−0.06
Si	−0.06
Mo	−0.05
Mn	−0.04
Ni	−0.04
Co	−0.04
Cr	0.02

The work-hardening exponent, n, is decreased by decreasing the grain size [54] based on the equation

$$n = 5/(10 + d^{-1/2}) \tag{80}$$

Grain refinement has the same effect on σ_f and $d\sigma/d\varepsilon$ at a given strain; it has little effect on ε_u and thus cannot be used to offset the detrimental effect of pearlite.

Grain refinement is beneficial in terms of ε_t.

There are little quantitative data about the effect of precipitation and dislocation strengthening on the formability of ferrite and ferrite–pearlite structure. The following relationships are reported:

(i) Precipitation strengthening increases $d\sigma/d\varepsilon$ at low strains [52].
(ii) Precipitation strengthening decreases n [47,54].
(iii) Strengthening by dislocation reduces $d\sigma/d\varepsilon$. However, the relationship between n and $d\sigma/d\varepsilon$ is not simple; an increase in n decreases $d\sigma/d\varepsilon$ at low strains but the reverse effect occurs at high strain [55].

The effect of second phase particles:

(i) ε_t is decreased due to the existence of second phase particles such as MnS, oxide inclusions and carbides. Eq. (79) shows the effect of sulfur.
(ii) The effect is not linear, an exponential decrease in ε_t with increasing volume fraction of second particles (Fig. 45).
(iii) The shape of the second phase particle is influential; as the length/width ratio of the particle increases, it gives a higher ductility in the longitudinal direction compared with plate-like particles tested parallel to their minor axis. Figure 46 illustrates this anisotropy of ductility.
(iv) Unlike inclusions, carbides do not crack or decohese at low strains, although carbides affect ε_t similar to inclusions.
(v) Carbide cracking is stress dependent. [56,57] and the strain to nucleate voids decreases with increasing carbide fraction.

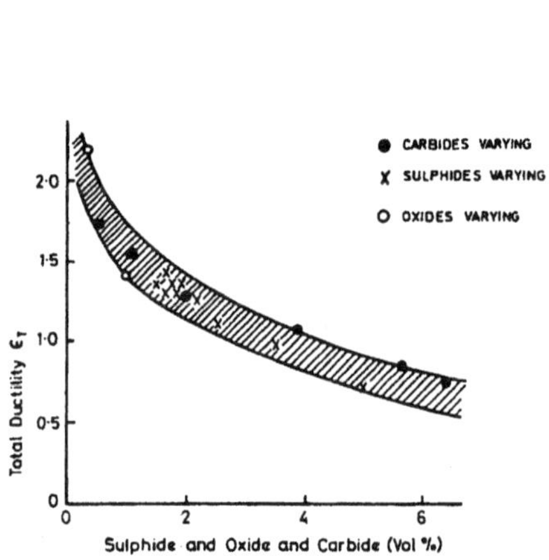

Figure 45 Effect of volume fraction of second phase on ε_t. (From Refs. 12,37.)

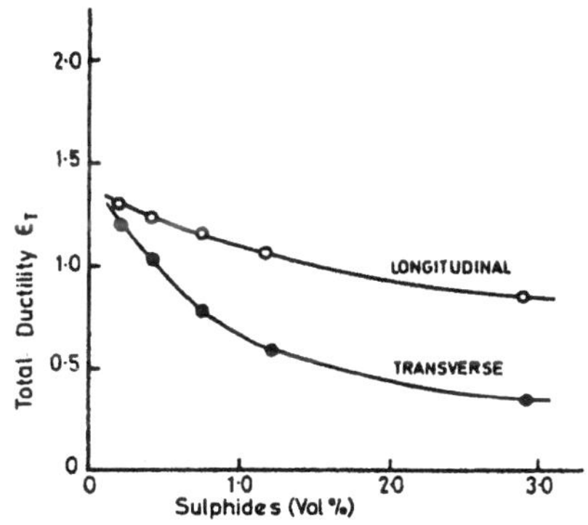

Figure 46 Anisotropic effect of sulfide particle on ε_t. (From Refs. 12,37.)

(vi) Carbide cracking and void nucleation can occur at smaller strains with increasing flow stress and work-hardening rate.

(vii) A high $d\sigma/d\varepsilon$ can be beneficial in inhibiting void growth by developing high σ_f values in strain concentrative regions around voids and confining the spread of the plastic zone.

(viii) The anisotropy of ductility is effectively minimized by inclusion shape control, in the same way that the anisotropy of CSE value is minimized.

C. Weldability

For high-strength low alloy steels, high yield strength with good toughness and good weldability is required.

Weldability of steel may be defined as the ease with which it can be welded by a normal welding process to produce a joint free from defects which would adversely affect its service performance.

The main influence of the parent steel on the soundness of the welded joint is associated with the behavior of the heat effective zone (HAZ), a narrow zone on either side of the fusion zone, which is heated sufficiently during welding to cause ferrite plus pearlite transform to austenite in the solid state. The behavior of the HAZ on cooling depends on the composition of the steel, the cooling rate, the hydrogen content of the weld metal, and the stress imposed on the weld either by external loads or as a result of thermal expansion and contraction. Higher carbon and alloy content in the steel and faster cooling rate, higher hydrogen content and stresses in the HAZ will increase the susceptibility to cracking in HAZ.

In order to achieve weldability, the M_s temperature must not be too low, and the quantitative effect of alloy elements on the M_s temperature is known. By regression methods, several equations for calculation of the M_s as a function of chemical composition have been listed in Table 4. The last one in the table was based on a total of 184 carefully selected sets of M_s measurements and chemical analysis of steels from several countries. The application range is up to 0.6% C, 4.9% Mn, 5% Cr, 5% Ni and 5.4% Mo. Si, S, and P did not give significant correlations [58].

The combined effects of C and alloying elements on the likely hardness in HAZ result from a particular thermal cycle and hence the index to assess the weldability of the steel is usually in terms of the "C equivalent" of steels, the most widely used definition of C.E. given by the International Institution Of Welding (IIW) :

$$\text{C.E.} = \text{C} + 1/6\text{Mn} + (\text{Cr} + \text{Mo} + \text{V})/5 + (\text{Ni} + \text{Cu})/15 \tag{81}$$

If the maximum hardness of HAZ could be kept below 350 DPH (diamond pyramid hardness), the HAZ cracking is unlikely to occur. In practice, steels of low C.E. less than 0.3 are readily welded by most processes.

With regard to pipe fabrication and to the girth welds particularly in low temperature environment that are used in the field for pipeline construction, the weldability of pipeline steels is important. The use of C.E. formula appears to be adequate in ensuring crack-free weld. There is a general feeling that the IIW formula is not adequate to define the behavior of modern steels with low C content and the following relationship by Ito and Bessyo is sometimes preferred.

$$\text{C.E.} = \text{C} + \text{Si}/30 + (\text{Mn} + \text{Cu} + \text{Cr})/20 + \text{Ni}/60 + \text{Mo}/15 + \text{V}/10 + 5\text{B} \tag{82}$$

The thermomechanical processing has permitted the development of high-strength steels with low-carbon content and this has contributed to improve the weldability in line-pipe steels.

D. Machinability

Machining is an important stage in the production of most engineering components. Machining of automotive components can account for up to 60% of the total cost. Steels, with improved and consistent levels of machinability whilst still maintaining other properties that ensure good service performance, have been called now and then. Turning, milling, grinding, drilling, broaching, and other operations might be carried out on an automatic lath in the production of a single component. Each of these operations differs in terms of the metal cutting action and involves different conditions of temperature, strain rate, stress and strain conditions, chip formation, and heat generation. Therefore, the machining performance of a steel is a multi-variable function and cannot be defined by means of a single parameter. However, some of the features that are often regarded to be measures of machinability are: tool life, product rate, power consumption, service finish of components, chip form, ease of swarf removal, dimensional tolerance. Many different methods for evaluating machinability have been developed. However, reproducibility is of absolute importance, regardless of the type of test employed, the cutting speed, the rate of feeding and the depth of cut must be carefully controlled.

Taylor tool life test is a widely adopted laboratory test for machinability, in which the life of the tool is determined at various speeds. Figure 47 shows a plot of log T (tool life) against cutting speed (V) that follows the equation

$$VT^n = C \tag{83}$$

where n and C are constants. $V20$, the cutting speed that will provide a tool life of 20 min, is one particular parameter of tool life which is featured prominently in machining evaluations. However, steel users often call for longer term tests involving multi-machining operations. Thus up to 2 tons of bright drawn bar stocks might be consumed over a period of about 7 hr in test. The machinability rating of a given steel sample can then be expressed in terms of components per hour or production of time per component. In order to

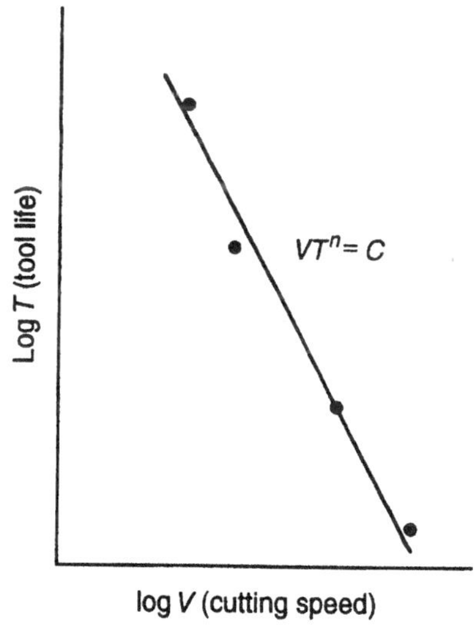

Figure 47 Taylor tool life curve. (From Ref. 5.)

improve the machining performance, various elements are added to steels. The main free cutting additives are as follows:

(1) *Sulfur*. Sulfur is the cheapest and most widely used free cutting additive in steels. Most specifications for engineering steels restrict the sulfur content to 0.05% max, but levels up to about 0.35% are incorporated in free cutting steels. At the same time, sufficient manganese is added to such steel to ensure that the sulfur is present as MnS rather than FeS, which may cause hot shortness during hot working. The MnS inclusions deform plastically during chip formation into planes of low strength which facilitate deformation in the primary shear zones. The MnS inclusions also exude into the tool–chip interface, acting as a lubricant and also forming a protective deposit on the tool. The net effect is a reduction in cutting forces and temperatures and a substantial reduction in the tool wear rate. Figure 48 shows a marked decrease in flank wear rate with increases in the volume fraction of MnS inclusions. In the region of MnS greater than 1.5% volume, the chip form and surface finish continue to improve with further addition of sulfur.

It has been postulated that large globular inclusions are far more effective than thin elongated inclusions. However, sulfide morphology is influenced markedly by the state of deoxidation of the steel, a heavily killed steel promoting the formation of MnS inclusions that are easily deformed into elongated inclusions during hot working. At the same time, killed steels tend to contain hard, abrasive oxide inclusions which have an adverse effect on machinability. Therefore, it is difficult to differentiate between the effects that might be due to sulfide morphology and those that are clearly due to the presence of hard, abrasive oxide inclusions.

High sulfur content will impair the transverse ductility of steels, particularly when MnS is present as elongated inclusions. Inclusion modifying agents can be added to high sulfur steels to promote a more favorable sulfide morphology.

(2) *Lead*. Lead is the next most common additive and additions of 0.15–0.35% Pb are incorporated in free cutting steels. Such levels are soluble in molten steel but are precipitated as discrete particles of lead during solidification. Segregation due to lead's high density should be avoided [63]. Lead reduces the frictional effects at the tool–chip interface

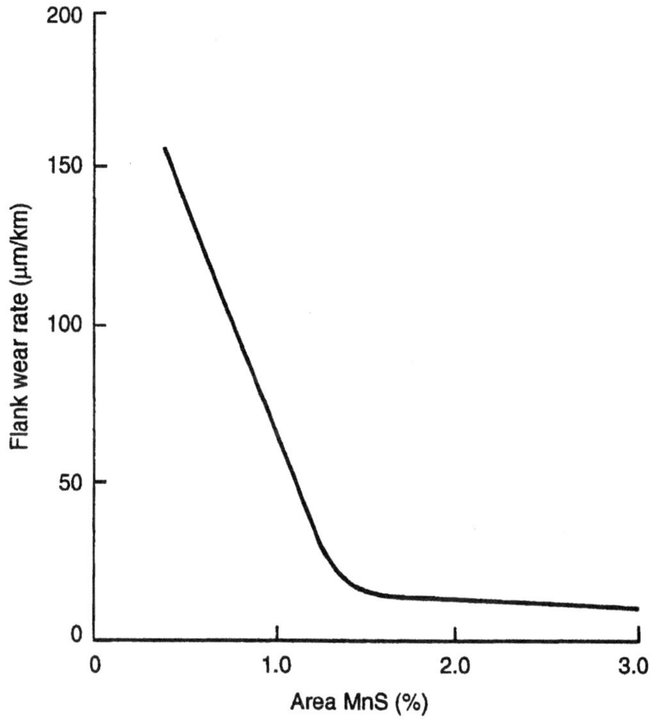

Figure 48 Effect of sulfide content on machinability. (From Ref. 74.)

where it becomes molten. An embrittlement in the primary shear zone is effected, thereby shortening the chips and improving the surface finish. The lead particles are often present as tails to the MnS inclusions. Care should be taken to contain lead fumes avoiding its toxic effects.

(3) *Tellurium*. Tellurium is an efficient but relatively expensive addition in free cutting steels and is generally restricted to a maximum level of 0.1%. It leads to hot shortness if large amounts are added. Tellurium is generally present as manganese telluride which is a low melting point compound and acts in a manner similar to lead. Tellurium has a high surface activity. More globular MnS inclusions and better transverse properties can be produced by 0.01% additions of tellurium to engineering steels. Fume extraction system is also required.

(4) *Selenium*. To low alloy steels, 0.05–0.1% selenium have been added, but in free machining stainless steels, many specifications call for a minimum of 0.15% Se. Selenium is present as a mixed sulfide–selenide, and globular sulfide inclusions are promoted by a small addition of this element.

(5) *Bismuth*. Bismuth is closely related chemically to lead. Levels of up to 0.1% Bi are typical in free cutting grades. Bismuth is present as tails to the MnS inclusions. Its action in promoting improved machining characteristics appears to be similar to that of lead.

(6) *Calcium*. Oxide inclusions are hard and abrasive and act as a negative factor in the machinability of steels. Alumina inclusions are formed in engineering steels as a product of deoxidation process by the addition of aluminum or as a grain refining element. However, the adverse effects of alumina can be reduced with the addition of calcium,

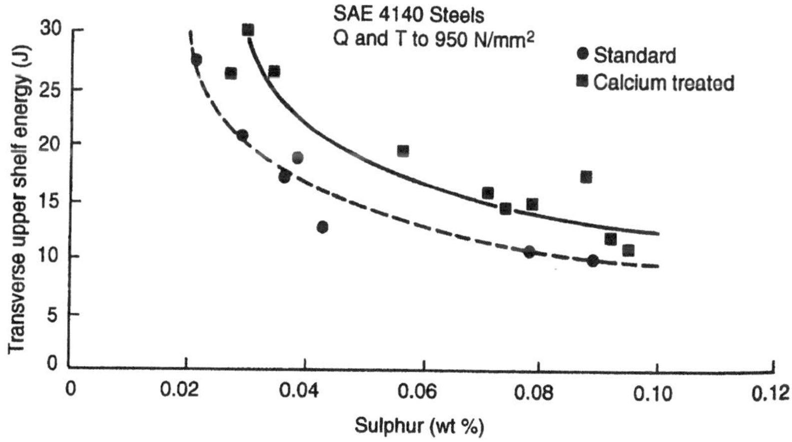

Figure 49 Effect of calcium treatment on transverse upper shelf energy of SAE4140 steel. (From Ref. 75.)

which results in the formation of calcium aluminate. These inclusions soften and form protective layers on the surface of carbide tools during high speed machining. Calcium is also effective in reducing the projected length of MnS inclusion, thereby improving the transverse properties of resulfurized steels. Figure 49 illustrates these effects. Where extremely high rates of machining are required, a low-carbon free cutting steel might be treated with sulfur, lead and bismuth, e.g., 0.25% S, 0.25% Pb, 0.08% Bi.

E. Hardenability and Resistance to Tempering

Hardenability is used to describe the heat treatment response of steels. It is often defined as the capacity of a steel to harden in depth under a given set of heat treatment conditions. It should be emphasized that hardenability concerns the depth of hardening, or the hardness profile in a machine component, rather than a specific level of hardness attained. Much of the information that is available today on the hardenability concepts and the metallurgical factors affecting hardenability was generated in the 1930s with names such as M.A. Grossmann, E.C. Bain, R.A. Grange, W.E. Jominy and J.L. Lamont featuring prominently in the literature. The period also coincided with the introduction of isothermal transformation diagrams which paved the way to the detailed understanding of the decomposition of austenite and a quantitative indication of hardenability.

(1) MEASUREMENT OF HARDENABILITY. The hardenability of a steel can be measured from continuous-cooling transformation diagrams. But a number of experimental methods have been developed, the best known and most widely used is the Jominy end-quench test. The Jominy specimen is a cylinder 102 mm long × 25.4 mm diameter with a flange at one end. The material is normalized, prior to machining, in order to eliminate variations in microstructure in as rolled condition. The specimen is heated to the appropriate austenitizing temperature and then transferred quickly to a fixture which suspends the specimen above a tube through which a column of water is directed against the bottom face. The water flow is tightly specified and controlled in order to produce a consistent quenching effect. After the quenching operation, flats are ground at diametrically opposed positions on the specimen to a depth of 0.38 mm to remove decarburized material and provide a suitable surface for hardness testing.

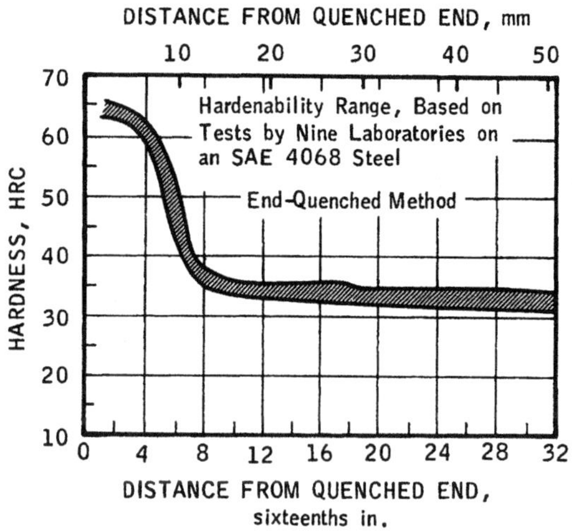

Figure 50 End-quench hardenability band of a single heat of steel, tested in nine laboratories, showing reproducibility. (From Ref. 76.)

Vickers (HV) or Rockwell (HRC) hardness at intervals of about 1.5 mm for alloy steels or 0.75 mm for carbon steels is measured in succession. Figure 50 shows a Jominy hardenability curve tested in nine laboratories, and Fig. 51 shows a standard form of Jominy end-quench test specimen, fixture, and water quench requirement. In a steel of reasonable hardenability, martensite will be formed at the quenched end and other transformation products will be formed at slower rates of cooling. But for shallow hardening steels, the SAC test is deemed to be more appropriate. It involves a cylinder specimen of the measure of 140 mm long×25.4 mm diameter. After normalizing and austenitizing at a suitable temperature above Ac_3, the specimen is quenched overall in water. A cylinder 25 mm long is cut from the test specimen and the end faces are ground very carefully to remove any possible tempering effects that might have been introduced during cutting. HRC hardness measurements are then made at four positions on the original cylinder surface and the average value provides the surface (S) value. Rockwell testing is then carried out along the cross-section of the specimen from surface to center. The total area under the curve provides the (A) value in units of Rockwell-inch and the hardness at center gives the (C) value.

(2) FACTORS AFFECTING HARDENABILITY. (a) *Grain Size*. The larger the grain boundary surface area, the greater are the nucleation sites for pearlite formation; thus the hardenability of a given composition will increase with increasing austenitizing temperature and austenite grain size. Figure 52 illustrates a series of end-quench hardenability curves, when employing the successively higher austenitizing temperature shown, on a SAE/AISI 4150 steel (0.55% C, 0.84% Mn, 0.30% Si, 0.13% Ni, 0.92% Cr, 0.21% Mo).

However, grain coarsening is rarely adopted as a method to achieve high hardenability because toughness and ductility are impaired. Instead, most commercial engineering steels are subjected to an aluminum-killed treatment to make a fine grained steel that will provide a good combination of strength and toughness.

(b) *Alloying Elements*. The common elements alloyed in engineering steels worked in quenched and tempered conditions are Cr, Mn, Mo, Si, Ni, V, B, and others.

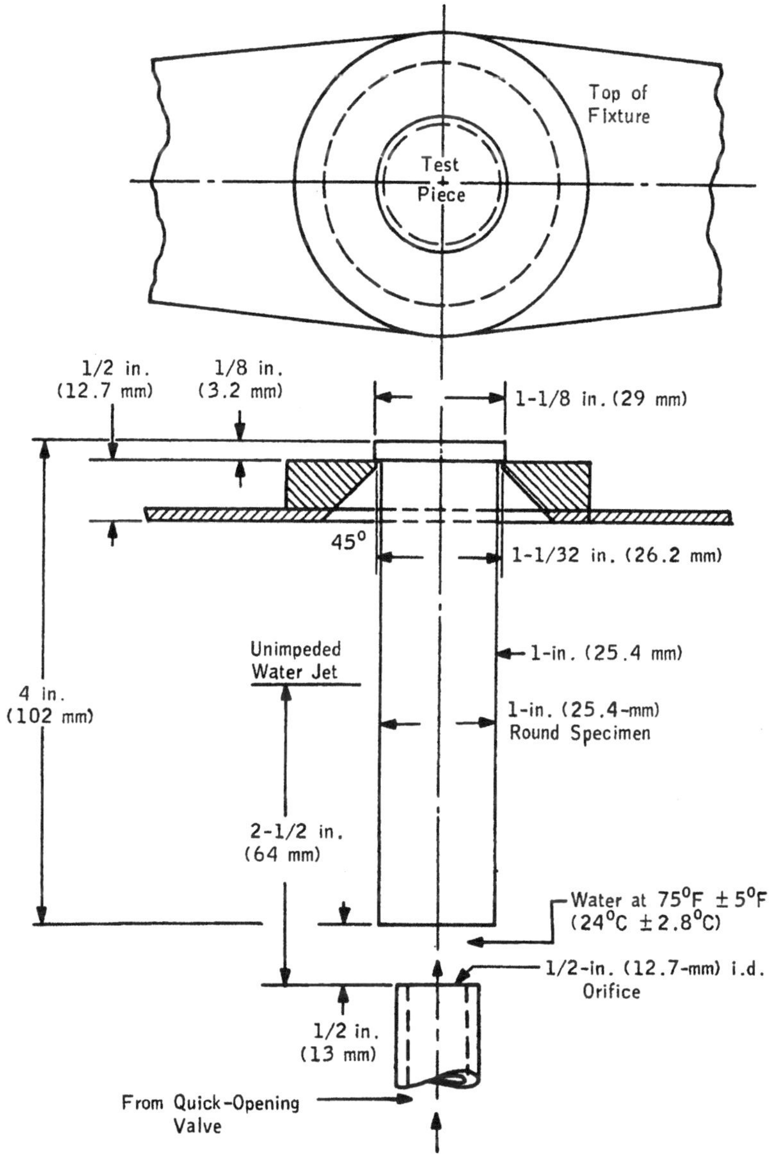

Figure 51 Standard form of Jominy end-quench test specimen, support and water-quench requirements. (From Ref. 76.)

First, we will discuss carbon, even though it was not included in the list of alloying elements. Carbon defines the maximum attainable hardness in the steel of a given chemical composition. The relationship between carbon content and the hardness of martensite is shown in Fig. 53; it shows that the hardness declines after reaching a maximum plateau at the carbon content about 0.6% C, depending upon the alloy content of the steel. This is due to the presence of retained austenite caused by the depression of the M_s temperature below the ambient temperature. This is important when attempting to compare the hardenability of steels of different carbon content. Over and above, its influence on hardness,

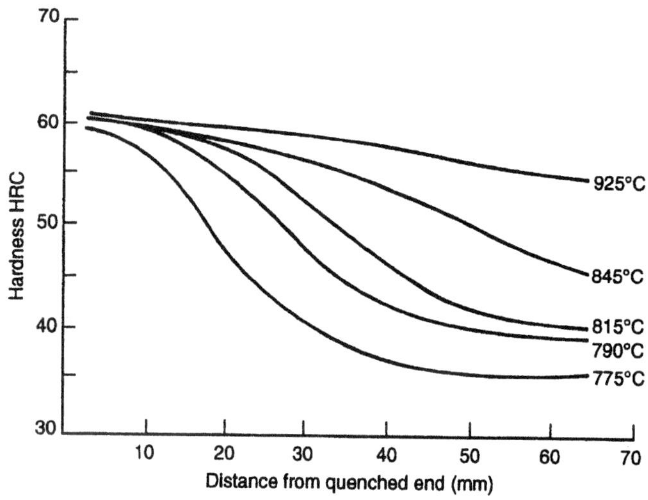

Figure 52 Effect of austenitizing temperature on the Jominy hardenability of 0.55% C, 0.84% Mn, 0.92% Cr, 0.2% Mo steel. (From Ref. 5.)

carbon contributes to hardenability. Figures 54 and 55 illustrate the progressive effect of carbon in several base materials: iron–carbon alloy, 0.8% Mn steel, and SAE8600 (0.5% Ni, 5% Cr, 0.2% Mo), respectively. The effect on hardenability is much more pronounced in the Ni–Cr–Mo steel. However, carbon is rarely used as a hardenability agent because of its adverse effect on toughness and its tendency to promote distortion cracking. High carbon steels also are hard and difficult to cut or shear in the annealed condition.

With the exception of cobalt, small additions of all the alloying elements will retard the transformation of austenite to pearlite and thereby increase hardenability. The ele-

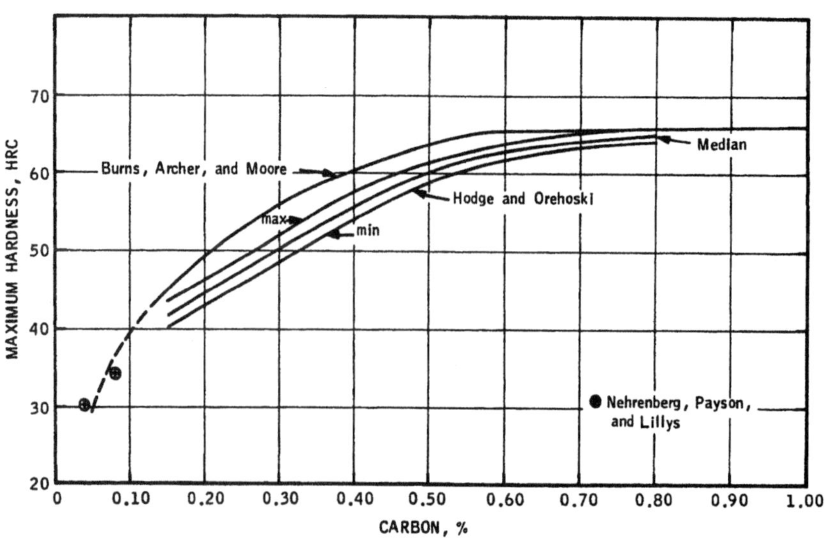

Figure 53 Hardness of martensite as a function of carbon content. (From Ref. 76.)

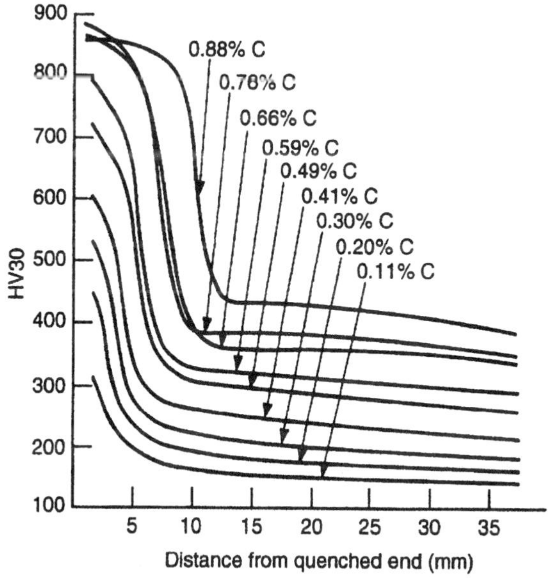

Figure 54 Effect of carbon on hardenability of 0.8% Mn steel. (From Ref. 77.)

ments most commonly used for the promotion of hardenability are manganese, chromium, and molybdenum, but nickel and vanadium are frequently incorporated for additional purposes. The effects of Cu, W, P, as well as the five alloying elements and the alloy interactions in carbon–manganese steels, medium-carbon low alloy steels, carburizing steels, and high hardenability steels have been summarized in Ref. [76].

The effects of the major alloying elements on the hardenability of low-carbon steels of the type used for case carburizing have been evaluated [78]. The effect on hardenability is expressed by means of a multiplying factor, defined as multiplying factor=hardenability of (base steel+alloying elements) /hardenability of base steel.

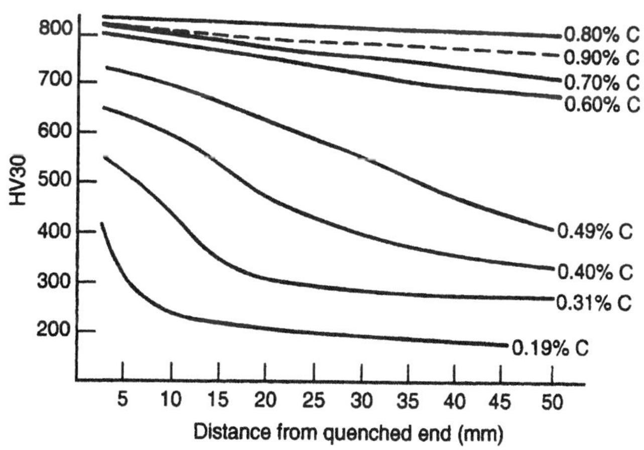

Figure 55 Effect of carbon on hardenability of SAE8600 steel. (From Ref. 77.)

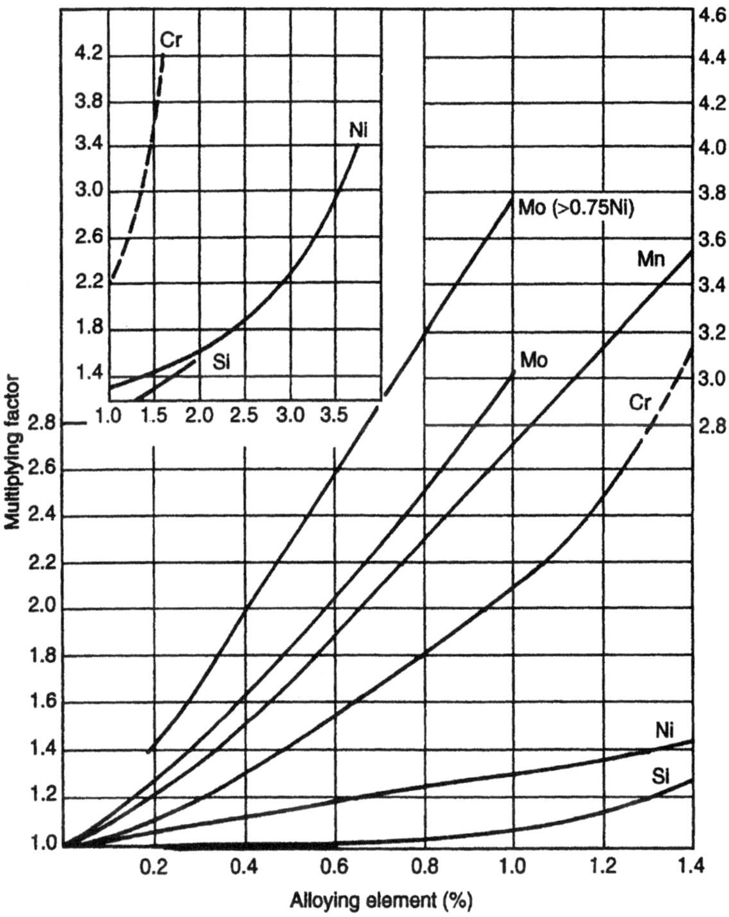

Figure 56 Average multiplying factors for several elements in alloy steels containing 0.15–0.25%C. (From Ref. 78.)

Figure 56 shows the data derived from Jominy tests on a variety of commercial and experimental casts of carburizing grades in which only one element was varied in the initial part of the work. The multiplying factors were tested subsequently in multi-element steels and modified empirically to provide more widely applicable averaged factors. But for molybdenum, due to its interactive effect, separate factors have to be used in low and high nickel steels. Although, the effects can vary significantly with carbon content and base composition, the sequence of the elements in promoting hardenability in decreasing order is: (C, P, N), V, Mo, Cr, Mn, Si, Cu, Ni. Vanadium has a powerful effect on hardenability, but it has a low solubility in steel due to the formation of vanadium carbide. The addition of vanadium is usually for the purpose of retarding the tempering process rather than for promoting hardenability.

Boron is a unique alloying element in low alloy engineering steels for the following reasons: (1) The addition of 0.002–0.003% B to a suitably protected base composition produces a hardenability effect comparable with that obtained from 0.5% Mo, 0.7% Cr, or 1.0% Ni. (2) The effect of boron on hardenability is relatively constant provided a minimum level of soluble boron is present in the steel. (3) The potency of boron is related to the carbon content of the steel, being very effective at low-carbon content but decreasing to 0 at the eutectoid carbon level.

Boron has a high affinity for oxygen and nitrogen; in order to produce a metallurgical active, soluble boron, it is added to steel in conjunction with even stronger oxide- and nitride-forming elements. Electric arc steel making involves the addition of about 0.03% Al and 0.03% Ti, either separately or in the form of proprietary compounds containing the required level of boron, aluminum, and titanium. Without this addition, boron would react with oxygen and nitrogen in the steel and form insoluble boron compounds which have no positive effects on hardenability.

The location of boron in steels has been investigated by the tracer method; there are several modes depending on the applied heat treatment : (1) During austenitizing, boron is uniformly solved through the austenite grains. (2) Boron segregates to grain boundaries in air cooled samples. (3) After segregation to grain boundaries, it is then precipitated at the grain boundaries based on a typical "C" curve pattern in a steel with base composition 0.1% C, 3.0% Mn [79]. (4) The incubation period for segregation and precipitation decreases when boron content increases. (5) Under normal heat treatment conditions, i.e., oil quenching from temperature of 820–920°C, boron segregates to austenite grain boundaries and suppresses the formation of high temperature transformation product.

Based on a detailed investigation of boron treated engineering steels with a base composition of 0.2% C, 0.5% Ni, 0.5% Cr, 0.2% Mo (SAE 8620) containing a wide range of carbon content, the following conclusions have been drawn: (1) Boron effect on hardenability reaches a maximum at a soluble content of about 0.0007%. (2) Further additions produce a reduction in hardenability but a reasonable steady state condition is achieved at boron content in excess of 0.0015%. (3) In commercial practice, it is usual to aim for soluble boron content of 0.002–0.003%, accepting a slight loss in hardenability in favor of a consistent hardenability effect.

The interaction between boron and carbon was investigated in SAE 8600 base steel. The result shows that boron produces a marked increase in hardenability at a carbon content of 0.2%, but the effect is steadily reduced to 0 as the carbon content is increased to 0.53–0.54%. Above this carbon level, boron has a detrimental effect on hardenability. A similar trend of boron effect varying with carbon content was obtained in experiments carried out in base steels containing 0.8% Mn, but in this case, the effect becomes 0 at a carbon content of about 0.85%.

It is evident that the boron effect varies not only with carbon content but also with the alloy content of the base steel. It has been proposed that the critical (zero effect) carbon content may be around the eutectoid level concerned. A consequence of the above effect is that the case hardenability of a carburized boron treated steel is lower than that of a boron free steel of comparable core hardenability.

After quenching, components are tempered at an elevated temperature to obtain a better combination of properties. As applied to low-carbon carburizing steel for gears, the tempering treatment might be carried out at a temperature as low as 180°C for the purpose of relief internal stresses without causing any significant softening in either the case or core of components. Medium-carbon steels containing about 0.4% C might be tempered at temperatures up to 650°C to produce adequate toughness and ductility with a sacrifice of strength.

The mechanism of tempering involves a partial dissolution and decomposition of martensite through the diffusion of carbon atoms out of solution to form fine carbides.

Table 7 Retardation Effect of Alloying Elements

Element	Retardation/1% addition
C	–40
Cr	0
Co	8
Ni	8
W	10
Si	20
V	30

Alloying elements affect the tempering process by retarding or suppressing the formation of Fe_3C, either by stabilizing the epsilon carbide ($Fe_{2.4}C$) which is formed initially in the breakdown of martensite or by forming carbides growing more slowly than Fe_3C. Alloying elements provide the possibility of higher temperature tempering in order to obtain a higher toughness for a given strength level. Alternatively, improved tempering resistance might be employed to allow a component to operate at a higher temperature without softening. Table 7 lists the retardation effect per 1 wt% addition of several alloying elements [80]. Carbon has a negative value; this indicates an acceleration effect which is presumably due to the increased supersaturation, i.e. driving force effect.

The design of composition in engineering steels is dictated as much by costs as property requirement. Vanadium and molybdenum are effective in promoting both hardenability and tempering resistance, but they are expensive. Hardenability is achieved very cheaply through the addition of manganese, chromium, and boron but with little contribution to tempering resistance. Such a condition may be acceptable in the case of carburized component where tempering is carried out at low temperature. Molybdenum and vanadium are added when high tempering temperature is required for the properties' requirement or operating conditions, but at relatively moderate levels. Molybdenum is also added to steels to suppress temper brittlement. It involves the co-segregation of alloying elements, such as manganese and silicon, and trace impurity elements, such as antimony, arsenic, tin and phosphorus, to the prior austenite grain boundaries in steels with bainitic or martensitic microstructures. Temper embrittlement occurs during exposure in the temperature range 325–575°C, either from operating within this range or by slow cooling through the range from a higher tempering temperature. Steels for steam power turbines (turbine casings, HP and IP and LP rotors) are alloyed with Cr, Mo, V and with low levels of silicon (0.1% maximum) and manganese (typical 0.2%) together with restricted levels of arsenic, antimony, tin, and phosphorus. Nickel is expensive and is not effective in enhancing either hardenability or temper resistance. It is perceived to be a solid solution strengthener in ferrite and a toughness promoter in engineering steels.

III. COMPUTATIONAL APPROACH TO THE DESIGN OF CARBON-, LOW-, AND MEDIUM-ALLOY STEELS

Computational approach to metallurgical alloy design to realize optimum material properties required through microstructural and constitutional selection and control has made significant strides in the last decade. Computational techniques, such as first principle calculations and the Monte Carlo simulation, have been increasingly utilized in the calculation of phase diagrams. Several databases for calculating phase diagrams of commercial alloy systems using CALPHAD approach have been developed. A wealth of information concerning various aspects of phase equilibria have been extracted, such as stable and metastable phase equilibria, the temperature of T_0 as a function of composition, volume fractions of the phase constituents under specific conditions, value of activity and enthalpy of solutions, Gibb's energy of formation, driving force for phase transformations, etc. Several instances of the computational approach which is related to designing with C-, low alloy and medium-alloy steels are briefly reviewed.

(A) SEGREGATION AND PERITECTIC REACTION. The degree of segregation of a solute is quantitatively analyzed by the distribution (or partition) coefficient of the solute. It is defined by $k_x^{s/l}=x_x^s/x_x^l$, where x_x^s and x_x^l are the concentrations of solute in the solid and liquid phases, respectively. $1 - k$ is defined as segregation coefficient which can be used to quantify the rejection of the solute at the interface during the solidification process [81]. As a result of solute additions, changes in the relative positions of liquidus and solidus lines are directly related to the segregation coefficients. Thermodynamic calculations have been utilized in calculating these coefficients for several solutes and such calculations have been extended for estimating the partition coefficients of γ-loop forming elements, even though there exists no direct phase equilibrium between austenite and liquid phase in the steel. Table 8 lists some of the equilibrium partition coefficients of alloying elements between austenite and liquid or between ferrite and liquid in steel, and the effect of adding 1 wt% alloying elements on the variation in the position of the liquidus and the solidus lines of the steel [82]. Sulfur, phosphorus, and carbon have low values of k in steel and, therefore, have a strong tendency to segregate during solidification.

The characteristics of the peritectic reaction and thereby the solidification behavior are also affected by the addition of the alloying elements. Figure 57 shows the results from

Table 8 Effect of Alloying Elements on Solid/Liquid Equilibria in Iron Base Alloys

			Change in liquidus and solidus/% element			
	Partition coefficient		Liquidus		Solidus	
Element	$K_x^{\delta/L}$	$K_x^{\gamma/L}$	$[\Delta T]^{L/\delta}$	$[\Delta T]^{L/\gamma}$	$[\Delta T]^{S/\delta}$	$[\Delta T]^{S/\gamma}$
S	0.05	0.02	32.2	27.9	596	1300
P	0.14	0.11	30.4	26.3	223	236
C	0.15	0.31	75.0	51.5	493	169
Nb	0.28	0.20	8.6	8.0	31.0	41.0
N	0.38	0.47	47.3	31.0	125	72.7
Ti	0.46	0.32	12.3	13.0	26.8	41.0
Si	0.65	0.61	13.8	12.8	21.3	21.0
Mn	0.69	0.70	6.3	5.0	9.2	7.1
Cu	0.7	0.72	5.2	4.1	7.1	5.6
V	0.78	0.62	4.7	6.8	6.0	10.9
Mo	0.79	0.61	2.4	3.7	3.0	6.1
Ni	0.81	0.87	3.6	2.1	4.5	2.4
W	0.86	0.70	0.8	1.5	1.0	2.2
Co	0.87	0.91	2.4	1.4	2.7	1.6
Cr	0.90	0.88	2.1	2.2	2.3	2.5
Al	0.99	0.80	0.2	6.6	0.2	8.5

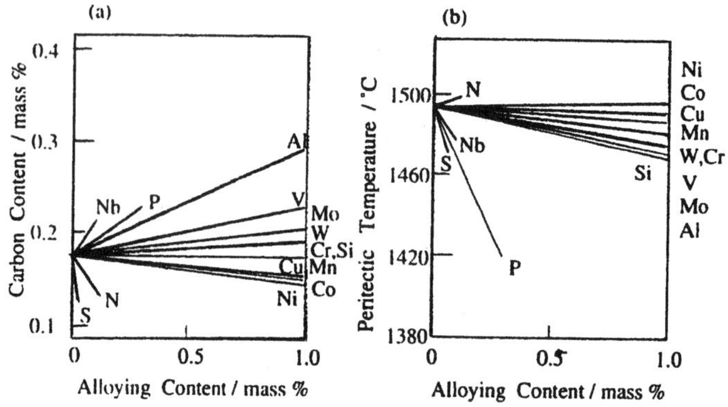

Figure 57 Effect of alloying elements on (a) the peritectic composition, and (b) temperature of Fe–C system. (From Ref. 82.)

thermodynamic calculations predicting the effect of alloying elements on the temperature and composition of the peritectic reaction in Fe–C system.

(B) CONTROL OF INCLUSION MORPHOLOGY. The morphology of MnS in steels can be broadly classified into three types: (1) randomly dispersed globular sulfides, (2) rod-like fine sulfides, and (3) angular sulfides. Globular or droplet MnS would form only when the solidification path involves the monotectic reaction (L = Fe(s) + L). However, based on the stable equilibrium phase diagram of Fe–MnS pseudo-binary system (Fig. 58(a)) the primary Fe phase would be expected to solidify first, then followed by eutectic reaction $L = \mathrm{Fe(s)} + \mathrm{MnS(s)}$. This would preclude the formation of any globular type MnS. But in practice the globular MnS has always been observed in Fe–Mn–S ternary alloys cooled from liquid state in the normal way. This dilemma can be resolved now by the calculated meta-stable phase diagram represented by the dotted line in Fig. 58(b). The calculated difference between the eutectic temperature (*e*) and the monotectic temperature (*m*) is so small (only 3°C) that a monotectic reaction seems possible. The deciding factor that tips the balance in favor of globular MnS is the magnitude of interfacial energy between Fe(L_1) / MnS(L_2) boundary which is estimated to be lower than that of the Fe(L_1) /MnS(s) bound-

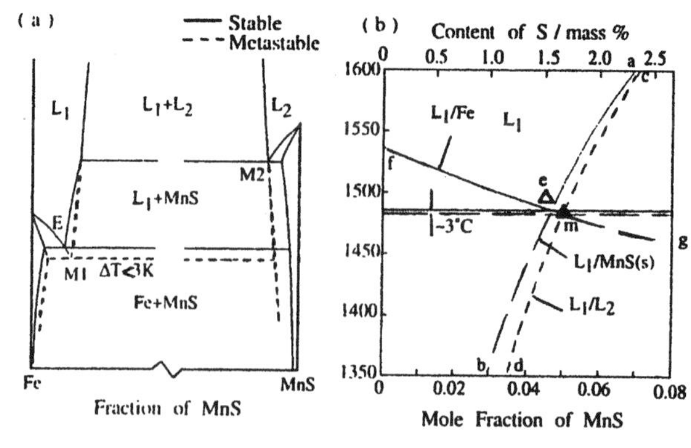

Figure 58 Pseudo-binary Fe–MnS phase diagrams in stable and metastable systems: (a) schematic, and (b) calculated one in Fe-rich portion. (From Ref. 82.)

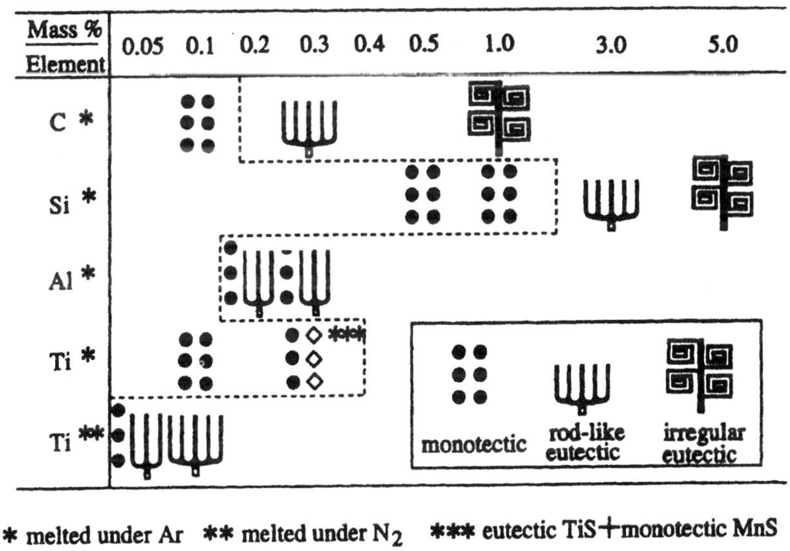

Figure 59 Morphology of MnS in hypoeutectic (monotectic) alloys containing C, Si, and Ti. (From Refs. 83,84.)

ary; the energy barrier for the nucleation of $MnS(L_2)$ is energetically more favored and the metastable monotectic reaction predominates over the stable eutectic reaction.

Alloying element addition in steel strongly affects the morphology of MnS inclusions [84,84]. The changes in morphology of MnS inclusions brought about by the additions of C, Si, Al, and Ti in hypo and hyper eutectic Fe–MnS system are summarized in Figs. 59 and 60. TiN crystallites act as nucleation catalysts for MnS(s) in Fe–MnS alloys containing Si (melted under N_2) and Al_2O_3 crystallites act as nuclei in Fe–MnS alloys containing Al (melted under Ar), promoting primary crystallization of the MnS(s) phase and the eutectic reaction associated with the stable system. When the same alloys containing Ti

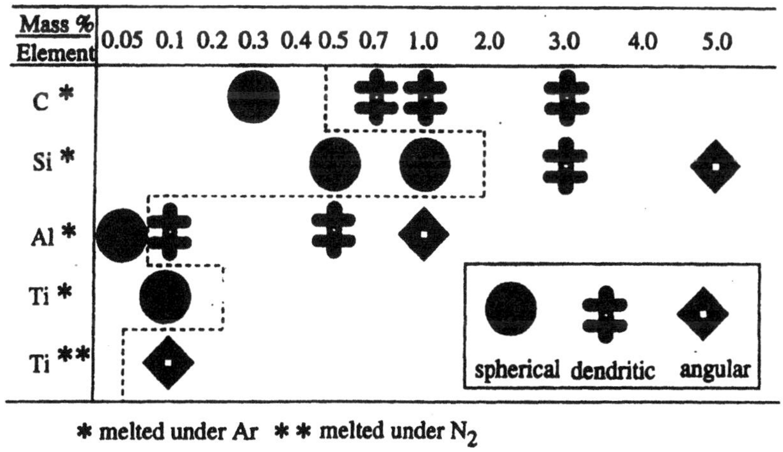

Figure 60 Morphology of primary MnS in hypereutectic (monotectic) alloy. (From Refs. 83,84.)

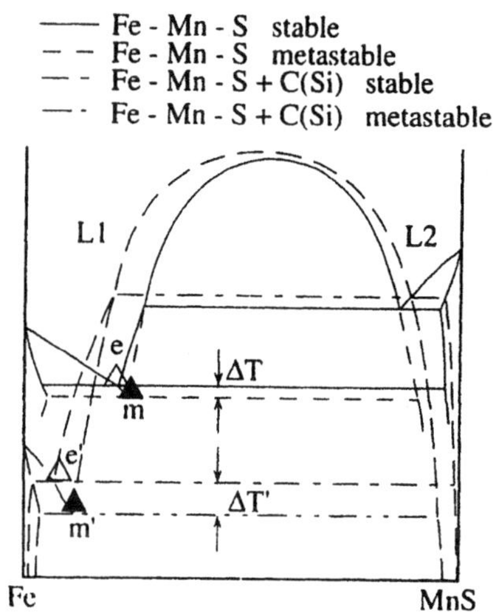

Figure 61 Change in phase equilibria in the pseudo-binary system by alloying with carbon and silicon. (From Refs. 137,138.)

are melted under Ar atmosphere, only oxides TiO_2 and SiO_2 are formed which tend to react with MnO forming compounds with lower melting temperatures, although its own melting temperature is about 100°C higher than that of Fe. As a result, liquid oxide droplets are formed first and act as nucleants for MnS (L_2), thereby facing the metastable monotectic reaction leading to monotectic morphology of MnS. The addition of C and Si lowers the melting point of iron and raises the activity of sulfur, resulting in an expansion of the two phase region and an increase in the slope of the liquidus line for iron, leads to an increase in the temperature difference between the metastable monotectic and the stable eutectic points, as shown in Fig. 61. The magnitude of this difference is estimated to be about 10°C in the Fe–1% Mn–0.3% S–1% Si alloy, which could explain the preference of stable eutectic over the metastable monotectic reaction in these alloys.

(C) SURFACE FISSURE IN CU-CONTAINING STEELS. During hot rolling, surface fissure defects occur in steels containing residual copper which comes from the ferrous scrap. Copper has a weaker oxidation tendency as compared to that of iron in steel. Cu is not oxidized during the heating stage prior to hot working and remains as a dissolved solute in iron-based solution. When the concentration of Cu in the surface layer exceeds the solubility limit of Cu in solid Fe, liquid Cu forms at the interface between steel and the oxide. This liquid penetrates the grain boundaries of Fe matrix during hot rolling and generates surface crack. Figure 62 is a calculated 1200°C isothermal section of Fe–Cu–Sn phase diagram [85] and superimposes some experimental points obtained for the composition of the liquid phase formed at the interface between the matrix and oxide in specimens containing Cu and Sn heated at 1200°C in air. It should be noted that the liquid phase in the specimen is located on the boundary of the two phase field ($\gamma + L$), i.e., the liquid is in equilibrium with austenite. It is evident that the surface fissure is closely related to the austenite/liquid phase equilibrium in the Fe–Cu system. The solubility of Cu in Fe is effected by alloying additions. Figure 63 shows the calculated and the experimental results. It can be seen that Ni and Co increase the solubility of Cu while Al decreases up to ~2.5% and then increases it in ferrite. All the other elements such as Si, V, Mn, Cr, and Sn decrease the solubility of Cu. Sn drastically decreases the solubility of Cu. The calculated results

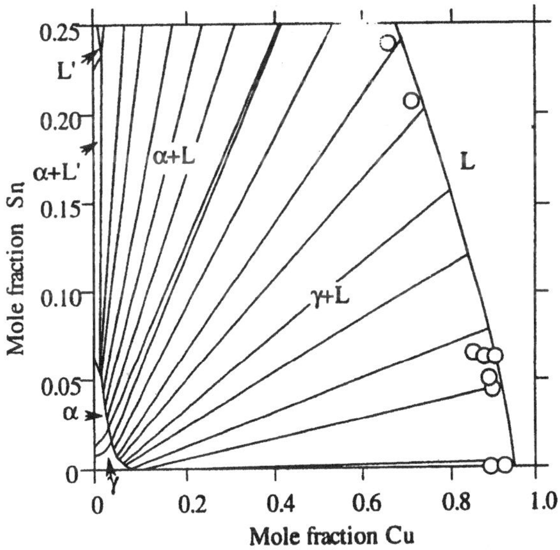

Figure 62 Composition of liquid phase formed at the interface of matrix and oxide in the Fe–Cu–Sn system projecting the calculated phase at 1200°C. (From Ref. 85.)

can explain the origin of the surface fissure due to liquid embrittlement in steel containing residual Cu, which is suppressed by Ni and promoted by Sn.

(D) EFFECT OF ALLOYING ELEMENT ON THE RELATIVE STABILITY OF FERRITE AND AUSTENITE. Various types of high-strength low alloy steels with good ductility have been developed by thermo-mechanical treatment, i.e., controlled rolling combined with controlled cooling. Dual phase steel and low alloy TRIP-type steels (a group of C–Mn–Si), are processed by intercritical treatment, consisting of heating the steel to a temperature range in the $(\alpha + \gamma)$ region followed by austempering. During the austempering treatment, Si suppresses the formation of cementite in bainite and promotes carbon enrichment of austenite, which in turn depresses the M_s temperature of the austenite. In the case of both

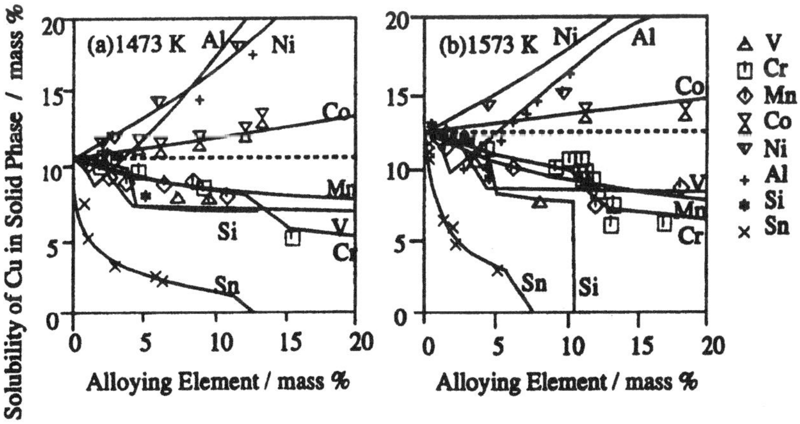

Figure 63 Effect of alloying elements on the solubility of Cu in solid phases. (From Ref. 86.)

Table 9 Data Base of Calculated Phase Diagrams of Low Alloy Steels

Steels	Systems	Phases
Low alloy steels	Fe(C, N, Si, Mn, Cr, Mo, Ni, Co, Al, Nb, V, Ti, W)	L, α, γ, carbide, nitride
Micro-alloying steels	(C, N, S, Mn, Si, Al, Cr, Ti, Nb, V)	L, α, γ, carbide, nitride, sulfide

dual phase and low alloy TRIP-type steels, the intercritical treatment in the (α+γ) two phases field is a decisively important step in controlling the final microstructure. The extent of the dual phase field, volume fraction of the phases and their equilibrium compositions and Ae_3 temperature are critical processing parameters for a successful intercritical treatment. All these parameters can now be accurately predicted using information from thermodynamic databases shown in Table 9.

The partition coefficient $k_x^{\alpha/\gamma}=x_x^{\alpha}/x_x^{\gamma}$ is one of the most important factors that determine the α/γ phase equilibrium. It directly relates to the partial molar Gibb's energy change at infinite dilute solution of the element x accompanying the α/γ transformation [87]. Figure 64 shows the calculated $k_x^{\alpha/\gamma}$ as a function of temperature [88]. Most of the elements exhibit strong temperature dependence.

The calculated Ae_3 is illustrated in Fig. 65. There is a significant agreement between the calculated and the observed value [89].

(E) CALCULATION OF M_S TEMPERATURE. The effect of the alloying element addition on the martensitic transformation start temperature can be calculated by the thermodynamic approach [90]. This has been done by dividing the free energy change accompanying the martensitic transformation into chemical and non-chemical parts. The chemical free energy is estimated from the change in partial molar Gibb's energy associated with the $\alpha \rightarrow \gamma$ transformation. The non-chemical part arises mainly as a result of the change in friction stress required to move dislocations which is related to the strengthening mechanism in austenite. The calculated chemical and mechanical contributions of alloying elements to

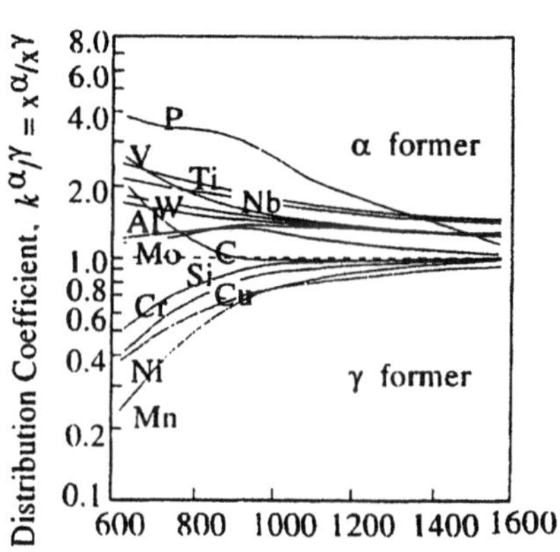

Figure 64 Temperature dependence of the distribution of alloying elements between α and γ phases. (From Ref. 88.)

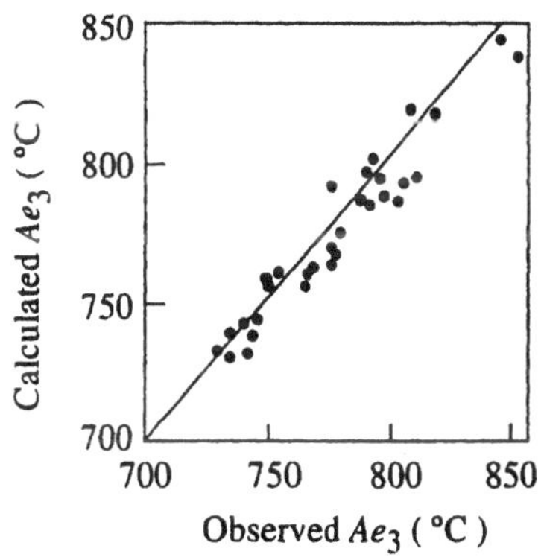

Figure 65 Comparison of calculated Ae3 temperatures and observed ones. (From Ref. 89.)

M_s are shown in Fig. 66. Based on such a calculation, M_s temperature relates to the chemistry of low alloy steels:

$$M_s(^\circ C) = 545 - 330C + 2Al + 7Co - 14Cr - 13Cu - 23Mn \\ - 5Mo - 4Nb - 13Ni - 7Si + 3Ti + 4V + 0W \tag{84}$$

(F) CARBIDES AND NITRIDES IN MICRO-ALLOYED STEELS. The solubility of carbides, nitrides and carbonitrides in both liquid and solid state of steel is of fundamental importance for exercising microstructural control in micro-alloyed steels.

Figure 67 shows the solubility products of various carbides, nitrides and sulfides in austenite and ferrite. It can be seen that the overall solubilities of nitrides are smaller than those of carbides, while both the compounds are more soluble in austenite than in ferrite.

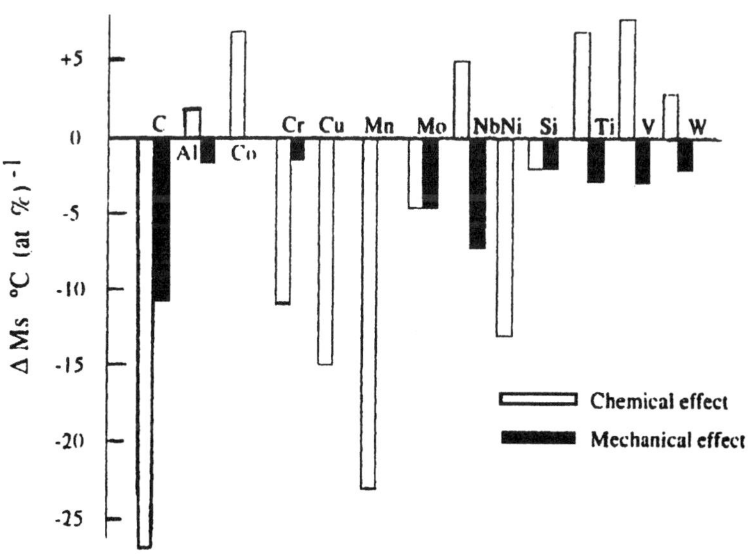

Figure 66 Effect of alloying elements on M_s temperature. (From Ref. 90.)

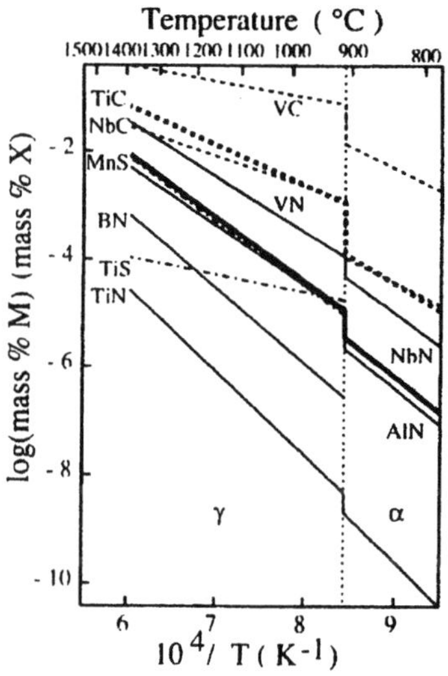

Figure 67 Solubility products of carbides, nitrides, and sulfides in γ and α. (From Ref. 91.)

Figure 68 illustrates the effect of micro-alloying elements on the γ field in Fe–C alloys [91], a very small additional amount significantly changed the extent of the phase field. Figure 69 shows the calculated phase equilibria between austenite and (Nb, Ti) (C, N) at 1200°C in a steel where the total Nb + Ti content in austenite is kept at 0.02% [92]. A small increase in the content of Ti (from 0.00001 to 0.0001) would result in a phase separation of carbonitride and the three phase field (γ + TiN + NbC) appears.

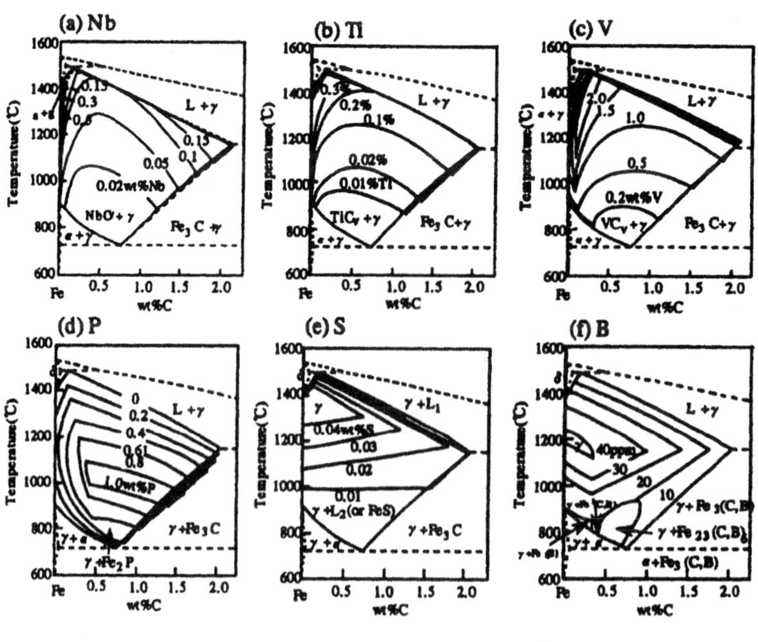

Figure 68 Effect of micro-alloying elements on the austenite region of Fe–C alloy. (From Ref. 91.)

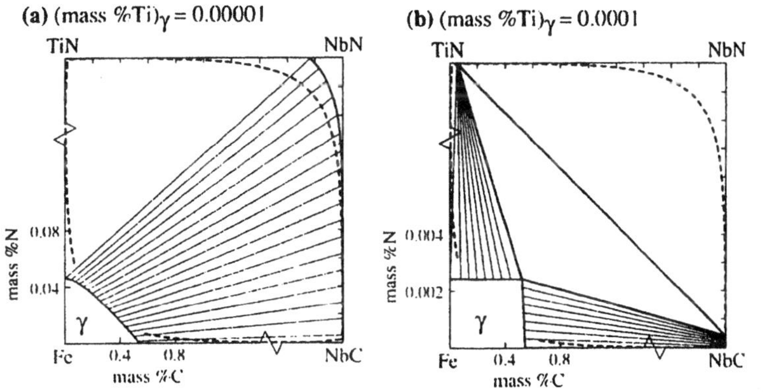

Figure 69 Phase equilibria between γ and (Nb, Ti)(C, N) at 1200°C keeping (Nb+Ti)=0.02% by mass percentage. (From Ref. 92.)

Figure 70 shows the calculated miscibility gap contours in carbonitrides (NaCl type) consisting of Ti, Nb, and V. The miscibility gap in NbC–VC pseudo-binary system and the miscibility island in double-pseudo-binary system of (Nb, Ti) (C, N), (Nb, V) (C, N) and (V, Ti) (C, N) are formed due to the large difference in lattice parameters between NbC and VC and the difference in Gibb's energy of formation of the compounds, respectively [93]. These calculated phase diagrams may be utilized in predicating phase separation in steels.

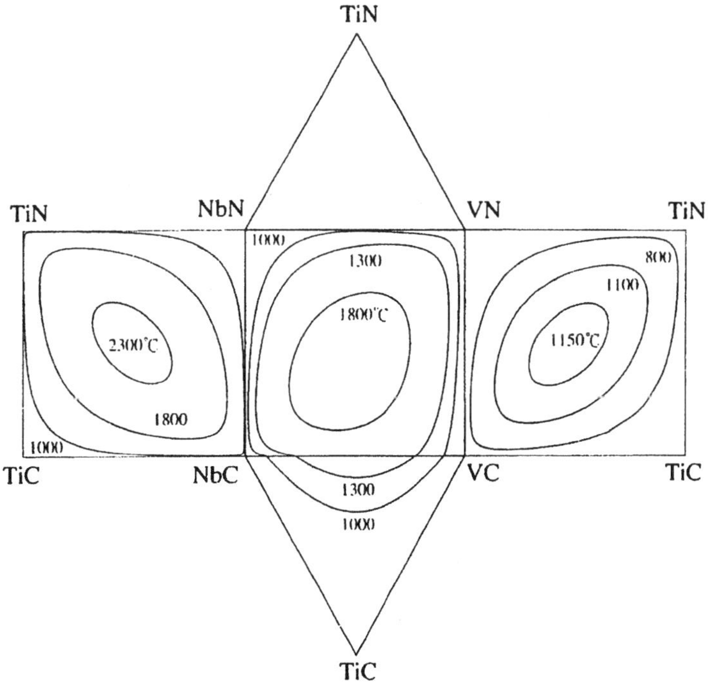

Figure 70 Miscibility gaps of complex carbonitrides. (From Ref. 93.)

IV. CASE ANALYSIS AND APPLICATION OF THE METALLURGICAL PRINCIPLES

(A) ALUMINUM-KILLED (AK) STEEL. AK steel is a traditional mild steel. Mild steel implies that the yield strength is below 280 N/mm^2. The AK steels usually have a carbon content ranging from 0.02 (> 0.015) to 0.12 (maximum to 0.2). It may be manufactured by both batch-annealed and continuously annealed processes. For batch-annealed AK steel, the annealing temperature is close to 700°C. The main influencing elements are Al, N, C, and Mn. Figrue 71 shows the r_m value in relation to Al and N content. The optimum nitrogen and the optimum soluble aluminum content are 0.005–0.01% and 0.025–0.04%, respectively. Normally, with carbon content close to 0.04%, the carbon reprecipitates completely onto the undissolved carbides during cooling. The nitrogen is combined as aluminum nitride and the steel becomes perfectly non-aging. At lower carbon content, the number of undissolved carbides reduces and eventually becomes 0 at about 0.02% C depending on the annealing temperature. Less than this carbon content, the reprecipitation process becomes progressively more difficult. It results in a rise in strength and the steel ceases to be completely non-aging. Such a steel exhibits bake hardening.

The distribution of AlN plays a decisive role in the processing of an AK steel having a high r_m value. The main requirement is that the AlN is kept in solution in the hot band structure before cold reduction. Then, it can lead to an enhancement of the (1 1 1) γ fiber texture components and a decrease in the deleterious (1 0 0) texture components during annealing. A pancake grain structure develops at the same time, which is caused by the preferential precipitation of AlN on the prior cold worked grain boundaries that provide

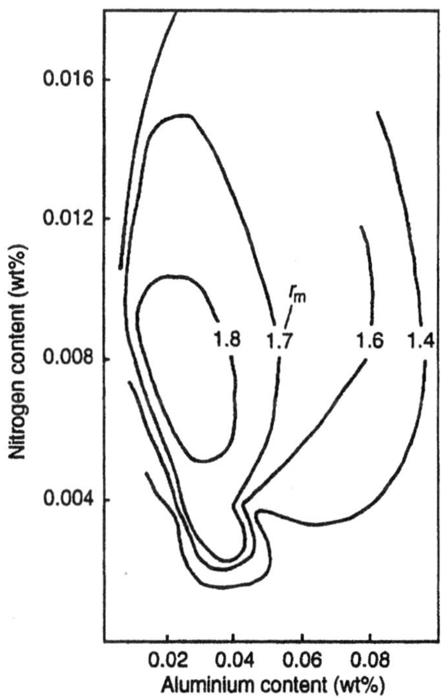

Figure 71 Effect of N and Al on r_m values of batch-annealed AK steel. (From Ref. 94.)

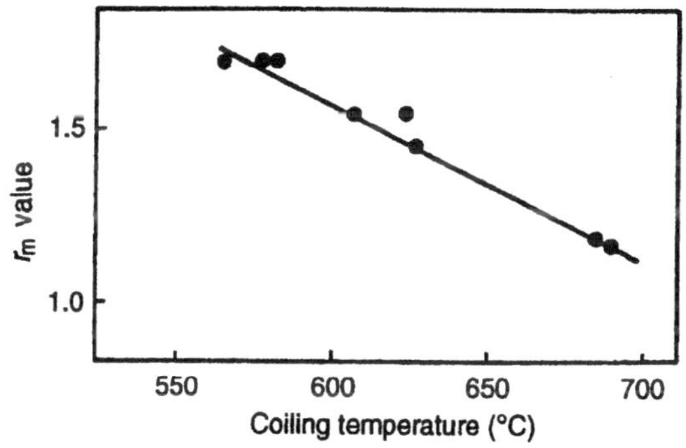

Figure 72 Variation of r_m value of batch-annealed AK steel with coiling temperature, Al=0.036%. (From Ref. 98.)

an anisotropic barrier to grain boundary movement. It is generally agreed that the precipitation of AlN or the clustering of Al and N on subboundary sites during slow heating to the annealing temperature contributes to this beneficial effect. This effect is associated with an inhibition of recovery and subgrain growth and an increase in recrystallization temperature compared with the same steel treated to precipitate the AlN prior to cold reduction. The peak r_m value is obtained when there is just sufficient inhibition of recrystallization and when the drag on grain boundary movement caused by the AlN decays slowly as a result of Ostwald ripening of the particles [96]. The slow coarsening of AlN particles leads to the preferential development of (1 1 1) component and a coarser grain size due to restricted nucleation of other components. The AlN must be dissolved in solution at the slab reheating stage so that the solution holds enough Al and N in the hot band structure prior to coiling and cold rolling.

The solubility of AlN in austenite is given by [97]:

$$\log[\mathrm{Al}][\mathrm{N}] = -6770/T + 1.033 \quad (85)$$

where T is in Kelvin.

So, a relatively high slab reheat temperature is required to take the Al and N into solution, the most commonly used temperature being 1250°C. To avoid substantial precipitation of AlN in hot band, low coiling temperature below 600°C (Fig. 72) and the finishing temperature in the single phase region (Fig. 73) must be used. Mn has a detrimental effect on r_m value depending markedly on the carbon content (Fig. 74). Cold reduction and heating rate during annealing also have an effect on r_m value. The heating rate which

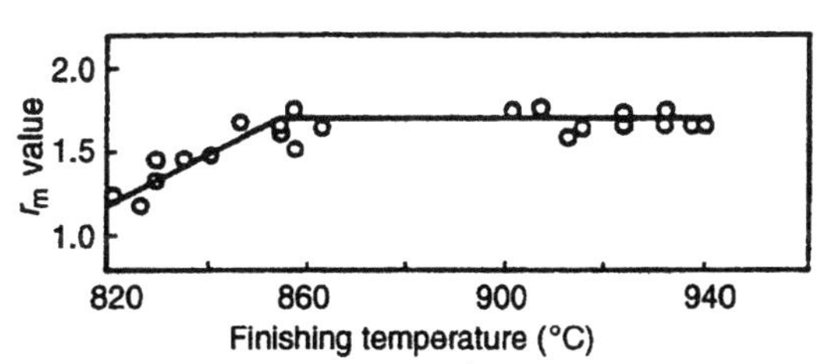

Figure 73 r_m value vs. finishing temperature, batch-annealed AK steel coiled at 525–565°C. (From Ref. 99.)

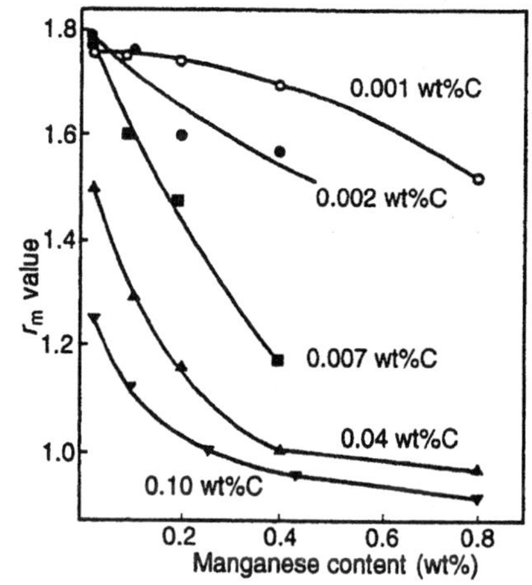

Figure 74 Effect of Mn content on the r_m values of cold-rolled and annealed steels with various C%, heating rate 100°C/hr, held at 700°C/hr. (From Ref. 100.)

gives a maximum r_m depends on the Al content (Fig. 75). These optimum heating rates also generate a maxima in the grain size and a minima in the yield strength. The equation below gives the optimum heating rate in °C/hr:

$$\log(\text{optimum heating rate °C/hr}) = 18.3 + 2.7\log[\text{Al}][\text{N}][\text{Mn}]/\text{CR} \tag{86}$$

where [Al], [N], and [Mn] represent the weight percentages of these elements in solid solution and CR is the cold reduction.

For continuously annealed AK(CAAK) steel, in order to match the grain size, the crystallographic texture and no strain markings on subsequent pressing, the steel chemistry, prior processing and annealing temperature have to be adjusted to compensate for the lack of time on continuous line.

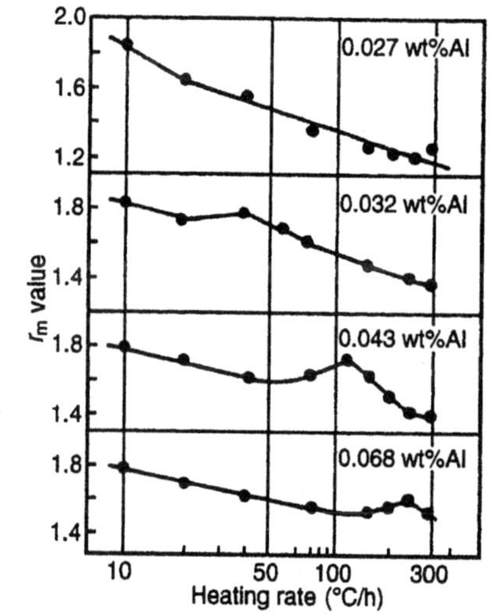

Figure 75 r_m vs. heating rate, batch-annealed AK steels with different Al%. (From Ref. 101.)

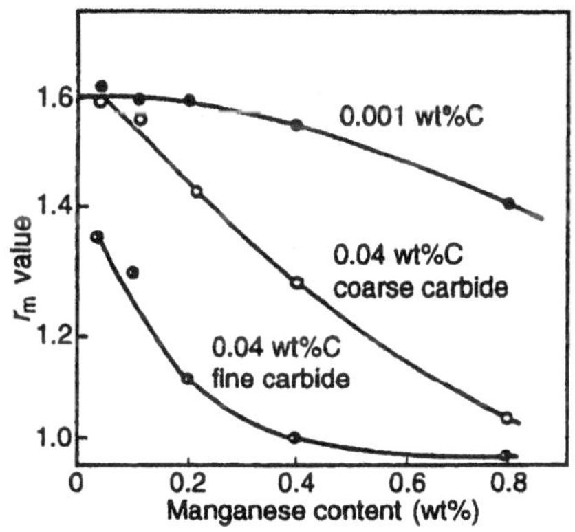

Figure 76 r_m vs. Mn% of steels with different C% heated at 50°C/sec. (From Ref. 103.)

Suitable large grain size and suitable crystallographic texture were obtained by using a number of scavenging reactions involving mainly Al, N, C, Mn, and S to purify the ferrite matrix and a suitable high annealing temperature. Taking out the interstitials from the solution was carried out mainly by the incorporation of holding the steel at temperatures in the range 350–450°C for up to a few minutes or allowing the steel to cool slowly from such temperatures.

Controlling Mn in relation to the S content had a marked effect on grain size and hence on properties. A parameter K was defined by

$$K = [\text{Mn}] - 55/32[\text{S}] - 55/16[\text{O}] \quad (87)$$

where [Mn], [S], and [O] represent the weight percentage of these three elements in steel. For rimming steel, grain size, elongation and r_m reach a maximum when the K value was in the range 0–0.15%, and a corresponding minimum in yield stress [102]. If the term involving oxygen is removed, a similar relationship would be applied to AK steel.

Figure 76 illustrates the effect of Mn depending on the carbon content and also shows that the detrimental effect of Mn was greatly reduced when C in the steel is present as coarse carbides arising from a high coiling temperature such as 750°C. Figure 77 illustrates the effect of carbon content on mechanical properties. The maximum in yield stress below 0.01% C coincided with a maximum in aging index and was associated with additional carbon being retained in solution. The grain size of CAAK steel increases with the coarsening of mainly carbides and AlN in the hot band. This may be achieved using high coiling temperature which depends on the carbon content. The slower dissolution rate of coarse carbides during heating enables recrystallization to occur in a matrix that is relatively free from interstitial carbon; hence, higher r_m values are achieved.

The effect of nitrogen content on the properties of the rimming steel without Al addition is illustrated in Fig. 78, which shows an increase in yield stress with increasing nitrogen content. Figure 79 shows the influence of nitrogen content on yield stress depending on the coiling temperature. High coiling temperatures lead to the combination of Al and N. The combined nitrogen is then no longer able to have any further major deleterious influences on the annealing process. Low total nitrogen content is important for highest formability.

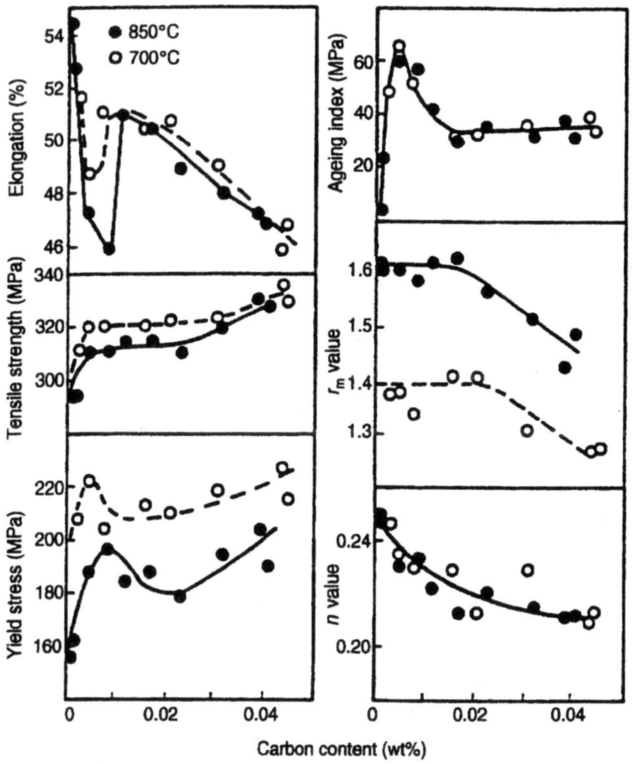

Figure 77 Mechanical properties of AK steel vs. C%, continuously annealed at 700°C and 850°C. (From Ref. 104.)

An adequate amount of boron is added to combine with nitrogen to improve the formability of CAAK steel. The combining reaction takes place during hot rolling. A high coiling temperature is not necessary as in the case of nitrogen combining with aluminum. Higher yield stress loss and more difficult pickling can result from a high coiling temperature; therefore, it should be avoided. The formability of a boron bearing steel which was coiled at low temperature is inferior from that of a non-boron bearing steel coiled at high temperature due to the presence of fine carbides which dissolve quickly on annealing. Just opposite to batch annealing, a low slab reheat temperature is required for continuous annealing. A low slab reheat temperature leaves undissolved AlN particles present in the structure prior to hot rolling and these particles act as nucleation sites for precipitation after hot coiling. The yield stress decreases with decreasing slab reheat temperature, as illustrated in Fig. 80.

Finishing temperature, coiling temperature, cold reduction, annealing temperature, coiling rate, each part of the annealing cycle needs to be carefully designed if the most favorable properties are required. Figures 81, 82 and 83 show the effect of coiling temperature, cold reduction, annealing temperature on the mechanical properties of the corresponding steels. The most formable properties are usually obtained when the finishing temperature is in the mono-phase austenite region. However, for many applications, the best practices were not required and cheaper processes were adopted. Ferritic rolling is one of these methods. The roughing sequence is carried out at lower temperature in the

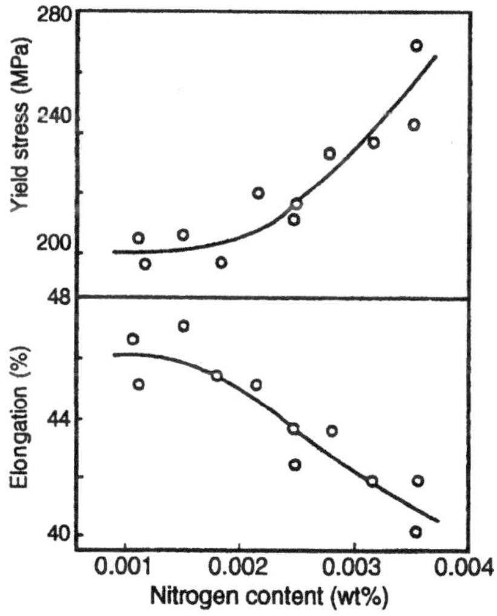

Figure 78 Effect of nitrogen content on the yield stress and elongation of subcritical annealed rimming steel. (From Ref. 102.)

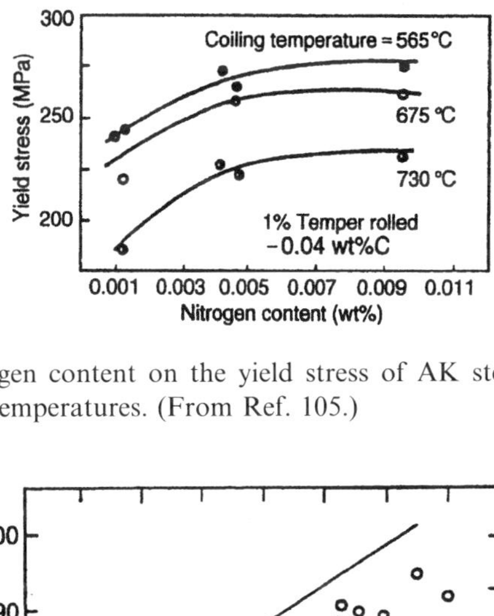

Figure 79 Effect of nitrogen content on the yield stress of AK steel continuously annealed at 760°C for different coiling temperatures. (From Ref. 105.)

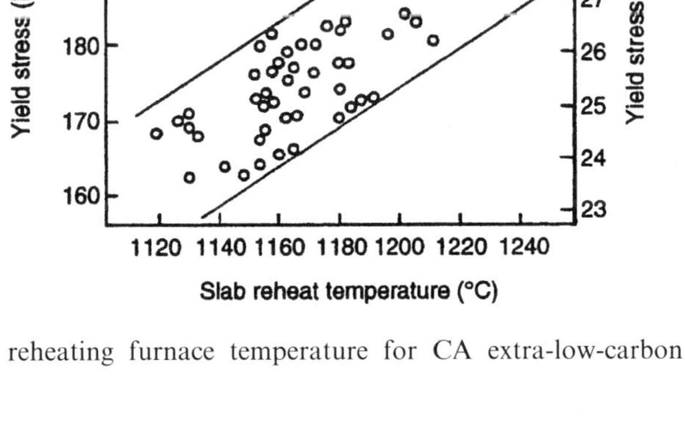

Figure 80 YS vs. reheating furnace temperature for CA extra-low-carbon AK steel. (From Ref. 106.)

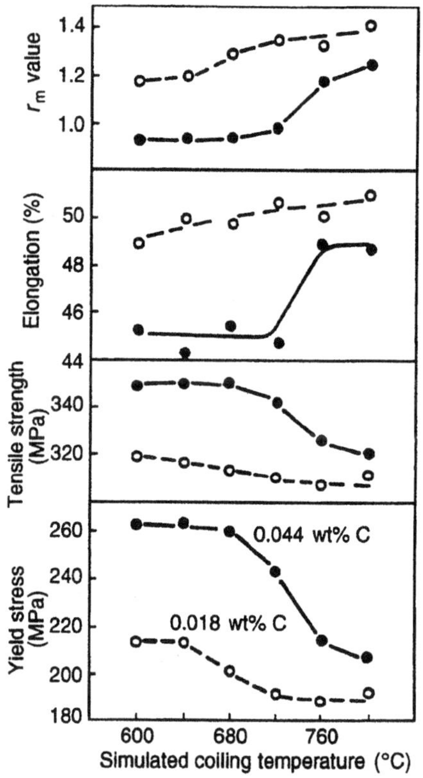

Figure 81 Effects of coiling temperature on the mechanical properties of AK steel annealed at 700°C for 1.5 min. (From Ref. 104.)

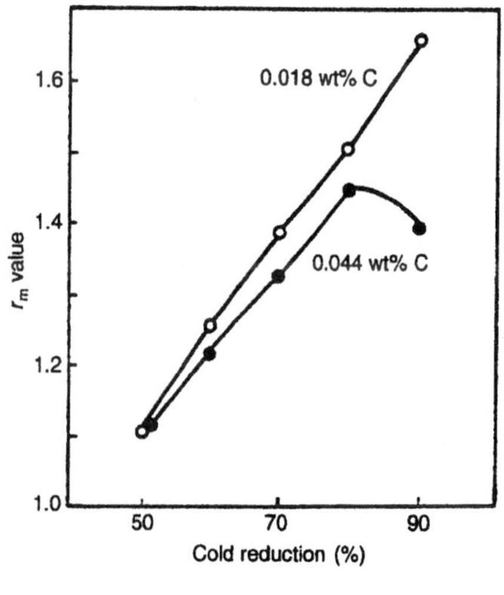

Figure 82 r_m values vs. cold reduction for AK steel annealed at 700°C for 1.5 min. (From Ref. 104.)

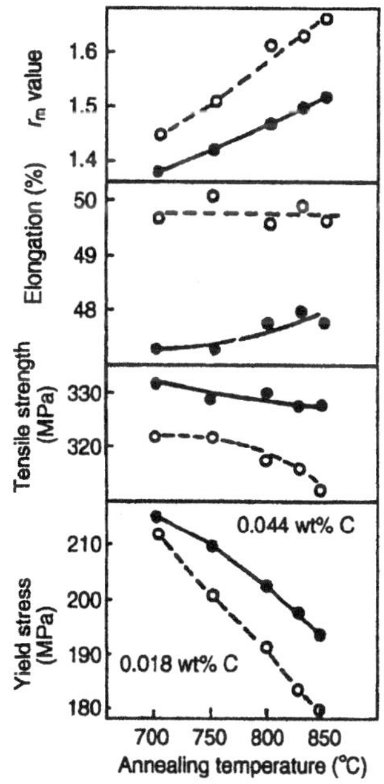

Figure 83 Mechanical properties vs. annealing temperature for AK steel annealed for 1.5 min after 71% cold reduction. (From Ref. 104.)

austenite region and the finish rolling is carried out in the single phase region of ferrite. This process takes the advantage of the fact that the resistance to hot deformation is similar in the ferrite region below 800°C to the resistance in the austenite region above 900°C.

(B) TRANSFORMATION STRENGTHENED STEELS. Steels containing a proportion of transformation product such as martensite or bainite and retained austenite in their microstructure thereby to develop high strength are transformation strengthened steels. When these relatively hard transformation products up to 20% or more are distributed throughout a matrix of ferrite, favorable combinations of strength and ductility may be developed. The ultra-high-strength steels contain a high proportion of bainite, martensite or retained austentite. The dual phase steels exhibit continuous yielding followed by a high work-hardening rate. They possess a low yield stress to tensile strength ratio and good cold formability for their strength. TRIP steels containing ferrite, bainite and/or martensite, also retained austenite, and provided even higher formability than dual phase steels. The benefits of TRIP effect were first recognized in steels containing high levels of chromium, nickel, and molybdenum [107], but it is now well known that low alloy steels containing manganese and silicon or aluminum are also suitable.

Both dual phase and TRIP steels, with tensile strength up to above 600 and 800 MN/m^2, respectively, may be manufactured in the hot-rolled condition by controlling the cooling conditions in relation to the chemistry; however, they may also be obtained by cold milling and continuous annealing to give higher strength. Ultra-high-strength steels with tensile strength of up to 1000 or 1200 N/mm^2 or more may only be manufactured at present by cold rolling followed by continuous annealing.

1. *Dual Phase Steel.* The steels are firstly annealed in the intercritical region of the phase diagram to produce a structure of ferrite and relatively high carbon concentration

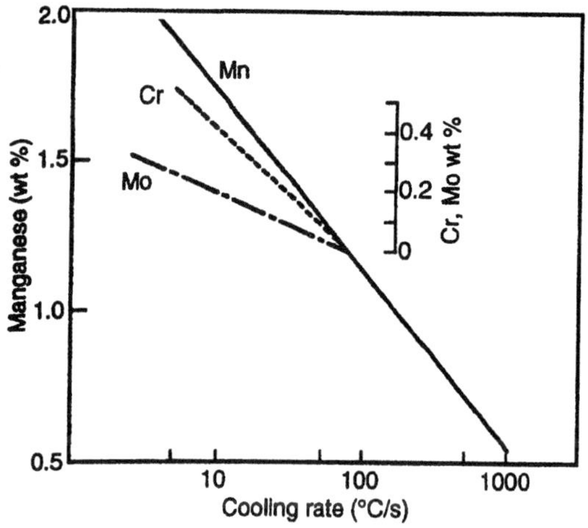

Figure 84 Relation between critical cooling rate and amount of alloying element present. (From Ref. 59.)

austenite and ferrite enriched with carbon on cooling through the intercritical region. The subsequent cooling rate must continue to be sufficiently fast at least at the M_s temperature to avoid the formation of pearlite or bainite. Then sufficient martensite must form to be able to eliminate the yield point elongation. This critical cooling rate depends on the manganese content of the steel and is also influenced by the Cr and Mo content (Fig. 84). Taking into account the effect of Mo and Cr, the equation to calculate the equivalent Mn content is given as [59]:

$$[\mathrm{Mn}]_{\mathrm{equ}} = [\mathrm{Mn}] + 1.3[\mathrm{Cr}] + 2.67[\mathrm{Mo}] \tag{88}$$

The S and Si content and holding time at temperature may also affect the critical cooling rate. The tensile strength of a classical, ferrite–martensite dual phase steel increases with volume fraction of martensite. But the yield stress first decreases and then increases as shown in Fig. 85. This is associated with the gradual removal of the yield point elongation at low volume martensite fractions. For a given volume fraction of martensite, a decrease in the mean island diameter had little influence on tensile strength, and has a marked effect in increasing the uniform elongation. However, an increase in tensile strength has also been reported [109]. The effect of the martensite volume fraction on the yield stress depends on the relative strength of the ferrite and the martensite which depends primarily on the martensite carbon content [110] and on the solid solution strengthening of the ferrite. The martensite carbon content depends on the annealing temperature and the initial cooling rate, but this carbon content is also influenced by silicon and manganese [111]. The optimum combination of strength and formability is obtained by a very fine distribution of martensite island and a very fine grain size matrix [109]. Increasing carbon, manganese and silicon leads to an increase in strength [112]. Phosphorus by solid solution strengthening and vanadium by increasing the martensite volume fraction and also by decreasing the ferrite grain size increase the strength. Silicon provides solid solution strengthening, increases the carbon content of martensite and suppresses pearlite formation.

Variations in processing parameters has great effects on the microstructure and properties of dual phase steel. Hot rolling conditions may affect the hot band microstruc-

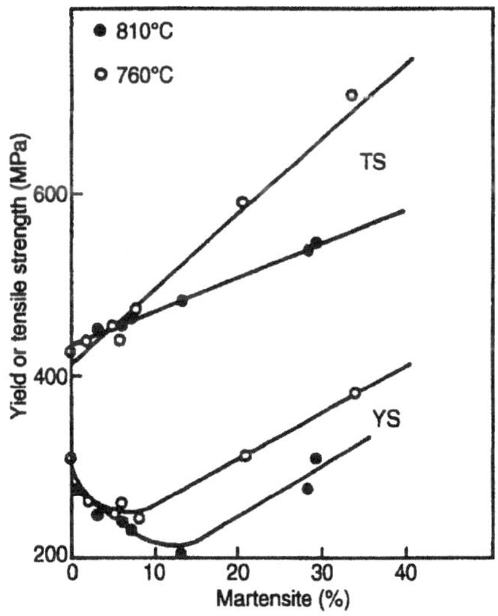

Figure 85 Yield stress and tensile stress vs. martensite% by volume for a dual phase steel containing 06% C and 1.29% Mn annealed at 760–810°C. (From Ref. 108.)

ture, thereby influencing the strength of cold-rolled and annealed steel [113]. Figure 86 illustrates the tensile strength–elongation balance for dual phase steel developed from different band structures and different cold reductions. No specific microstructure was reported to give a superior strength/ductility balance. A higher coiling temperature may lead to a coarser grain size and lower fraction of martensite [114]. In some cases, minimizing cold mill loads by low hot band strength rather than optimizing product properties determines the selection of hot rolling conditions. Annealing temperature has a major influence in determining the volume fraction of martensite. Figure 87 shows how the yield stress and tensile strength vary with annealing temperature for steels with different carbon, manganese, and silicon content. Sufficient time is needed at annealing temperature for the formation of sufficient austenite. Figure 88 shows that the volume fraction of second phase continues to increase after holding up to 10 min at 755°C. The yield point elongation

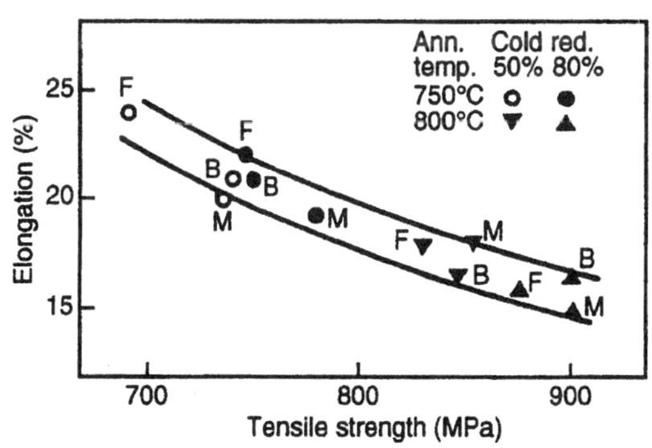

Figure 86 Tensile stress–elongation balance, dual phase steels, the strength developed from different hot band structure and different cold reductions. (From Ref. 113.)

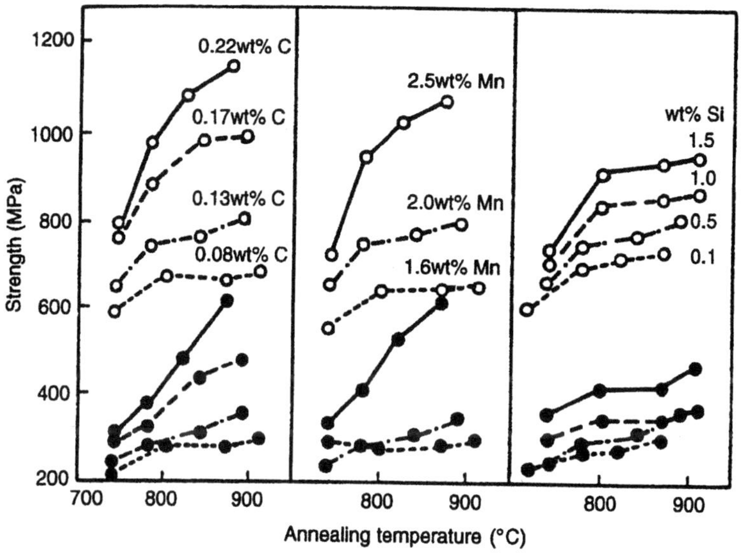

Figure 87 Effect of annealing temperature and chemical composition on the yield (closed cycle) and tensile strength (open cycle) of gas jet cooled steel. (From Ref. 115.)

is eliminated after 1 min at this temperature as Fig. 89 shows. Figure 90 shows the effect of cooling rate. Dual phase steel tends to possess high n values for the strength at low degree of strain in a tensile test, but the n values reduce at higher strain. Dual phase steels usually have a poor hole expansion value. The hole expansion value may be improved by titanium addition to provide precipitation strengthening to the ferrite matrix which reduces the hardness difference between the two phases present.

2. *Multi-phase Steel* (*TRIP Steel*). These steels are based on carbon–manganese compositions, usually also with silicon or possible aluminum additions to inhibit carbide precipitation. After intercritical annealing, the TRIP steels are cooled to austempering temperature, i.e., reasonably close to 400°C, to transform to the necessary products. The details of the micro-structural changes and their effects on final properties depend critically on the precise chemistry and processing used. The steels are processed on a continuous annealing line that would give a sufficient cooling rate to develop the necessary

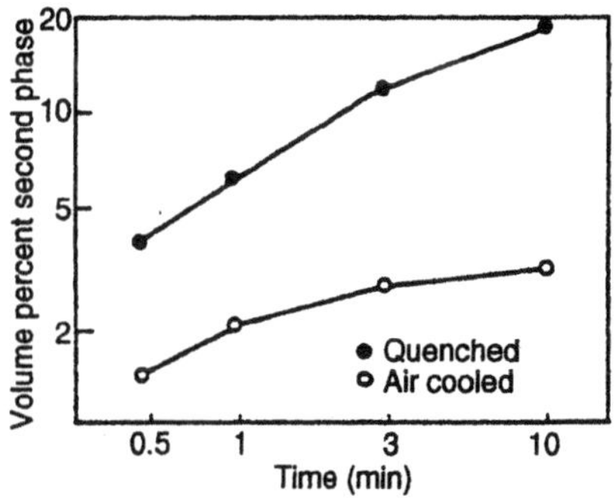

Figure 88 Second phase% by volume vs. time of annealing at 775°C. (From Ref. 116.)

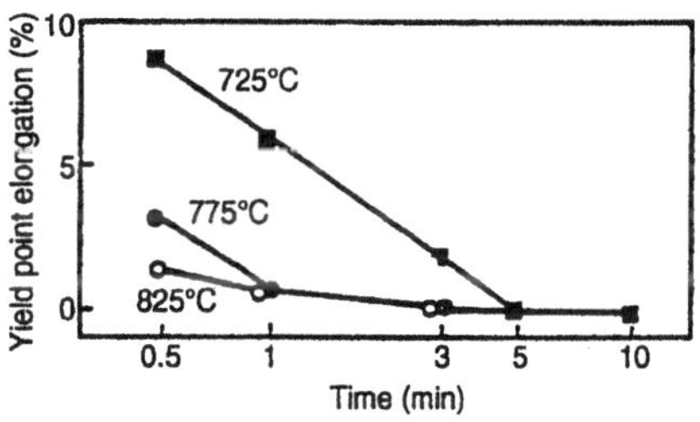

Figure 89 Variation of yield point elongation with annealing time for a 0.06% C, 1.23% Mn steel for different annealing temperatures. (From Ref. 116.)

structures. At the highest annealing temperature, ferrite and austenite are formed; as the temperature is increased, the proportion of ferrite decreases. Partitioning of carbon, manganese, and silicon gives a higher carbon and manganese content in austenite, and a little lower silicon content in austenite than in ferrite. During cooling, the amount of austenite may decrease and its carbon content may increase, particularly in the case of slow cooling, as illustrated in Fig. 91. In some cases, the partition process only occurs within the periphery of the austenite grains; then only a carbon rich layer may be developed. At the austempering temperature, some of the austenite transforms gradually to bainite and leaves a mixture of bainite and some retained austenite located between the laths of bainite or

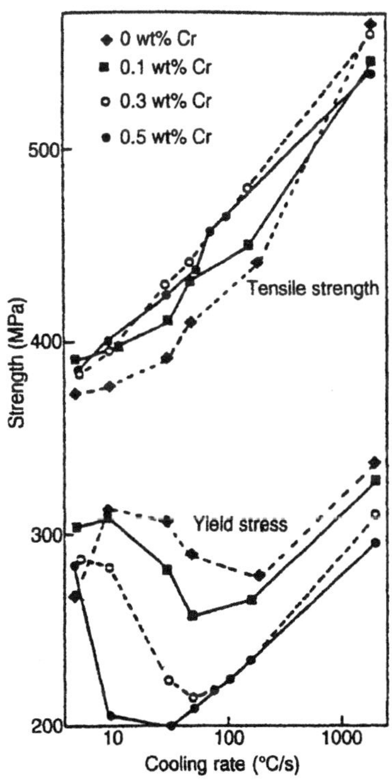

Figure 90 Yield stress and tensile stress vs. cooling rate for steels containing 1.25% Mn and different Cr content. (From Ref. 59.)

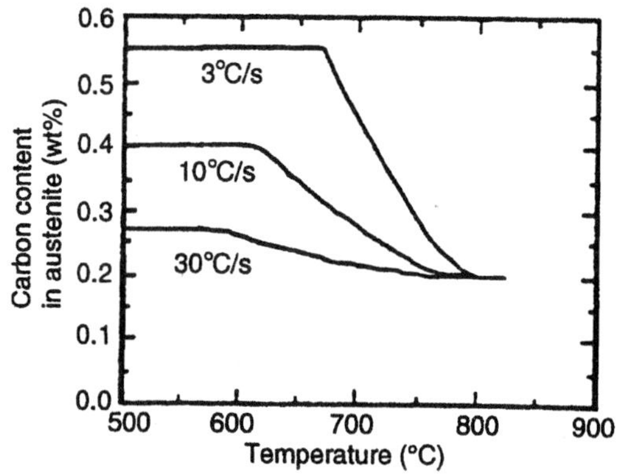

Figure 91 Evaluation of carbon enrichment in untransformed austenite during cooling from 825°C. (From Ref. 117.)

along ferrite boundaries. When the holding time is short and the austempering temperature is relatively low, the carbon in the austenite remains low. Much of this austenite transforms to martensite. As the carbon builds up in the austenite, it is stable at room temperature. On further holding, carbide may form, but the tendency of carbide formation is inhibited by the presence of silicon; thus a silicon addition leads to the retention of a higher volume fraction of austenite with a higher carbon content. The volume fraction of austenite retained at room temperature in a steel containing 1.2% Mn and 1.2% Si varies with holding time at different temperature, and the corresponding variations of carbon content of the retained austenite are given in Figs. 92 and 93, respectively. Figure 94 shows an example of the effect of holding time at 400°C on the nature of the stress–strain curve. It can be seen that for 1 min holding, the curve shows continuous yielding and changes to an increasing yield point elongation with longer holding times. An increasing hold time leads to a progressive decrease in tensile strength and an increase in yield stress;

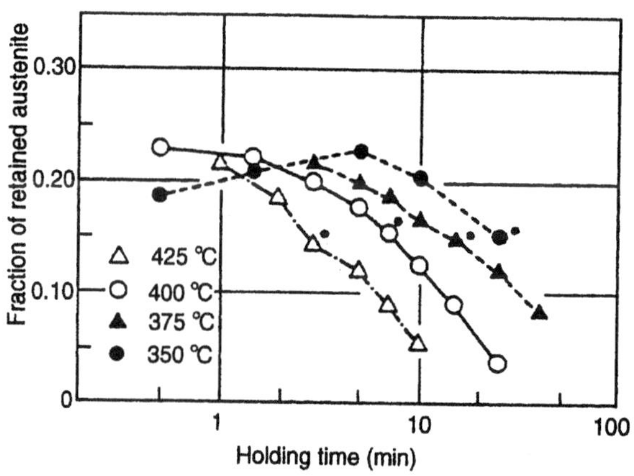

Figure 92 Change in retained austenite volume fraction with isothermal holding time for a steel containing 1.2% Si for different holding temperature. (From Ref. 118.)

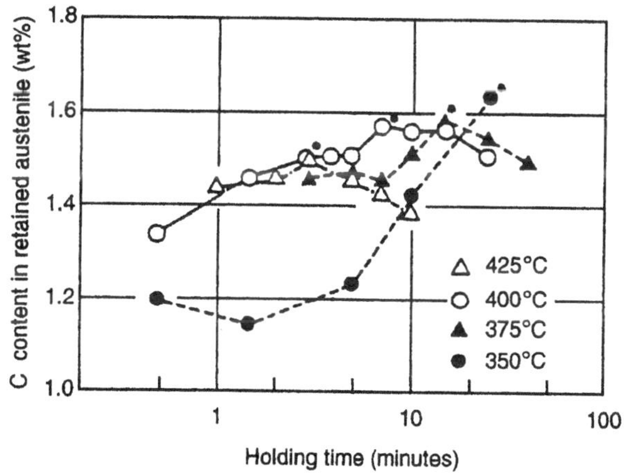

Figure 93 Carbon content in retained austenite for different isothermal holding time for a steel containing 1.2% Si for different holding temperature. (From Ref. 118.)

the elongation values pass through a peak for an intermediate holding time of 6 min as shown in Fig. 95.

Retained austenite has a tendency to transform to martensite during forming, and this process will lead to a high work-hardening rate and consequently a larger range of uniform elongation. The mechanical stability of retained austenite against deformation changes with carbon content of the austenite. Austenite with a very high carbon content may be so stable that most of them remain untransformed even after a high level of strain. Austenite containing lower levels of carbon content could transform completely to martensite under the action of relatively small amounts of strain. The mechanical stability

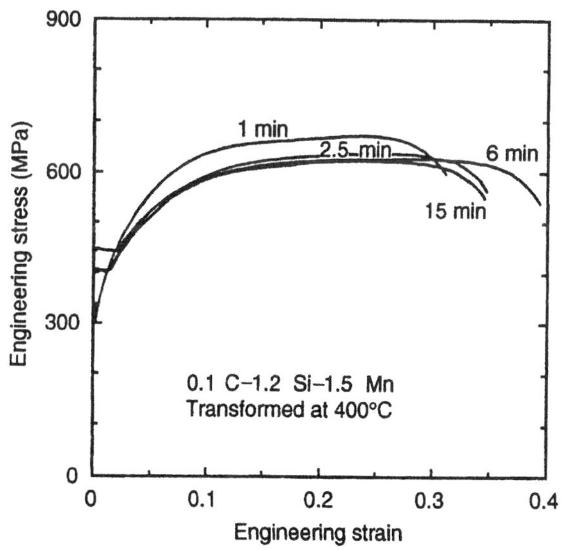

Figure 94 Engineering stress–strain curves of specimen of a 0.1% C, 1.2% Si, 1.5% Mn steel transformed at 400°C for different time. (From Ref. 119.)

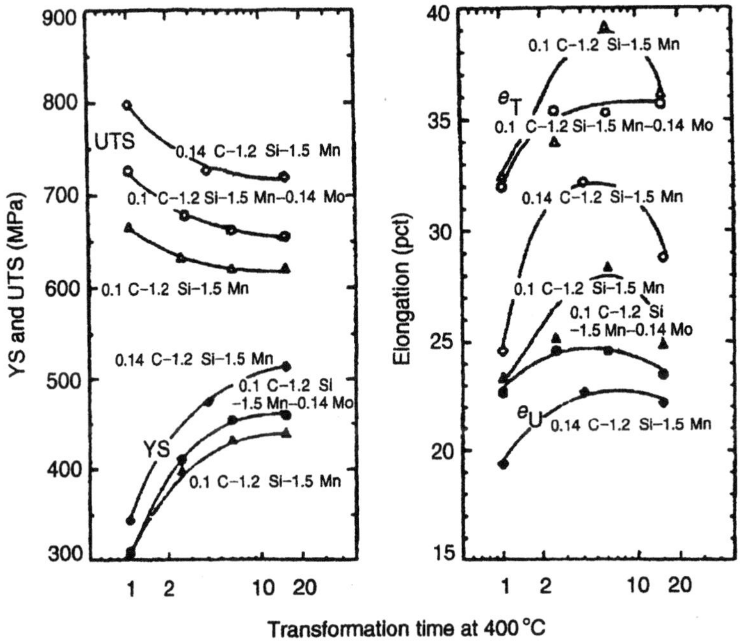

Figure 95 Changes in ambient temperature mechanical properties as a function of isothermal transformation time of three steels. (From Ref. 119.)

of retained austenite is critical for best ductility of TRIP steels [117]. The austenite should have an intermediate stability to transform gradually over the complete strain range which is needed for the forming process. Figure 96 shows an example of the effect of holding time

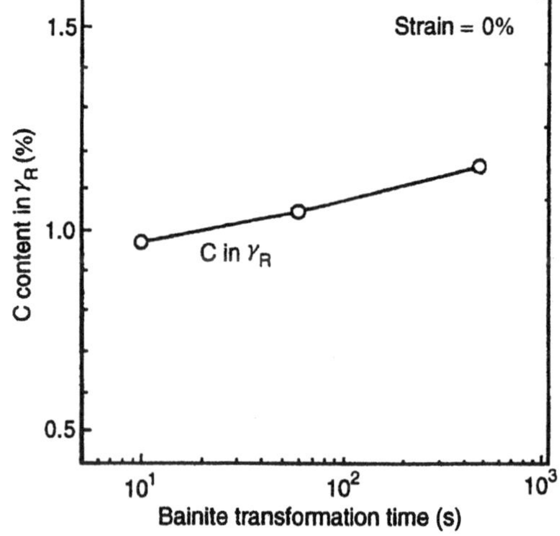

Figure 96 Effect of bainite transformation time at 400°C on carbon content in retained austenite. (From Ref. 120.)

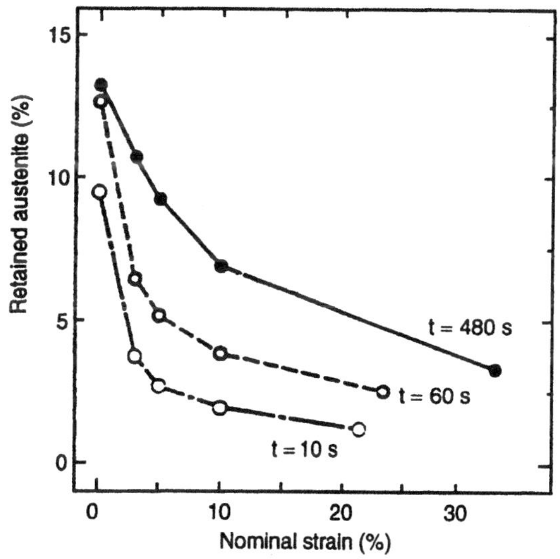

Figure 97 Effect of bainite transformation time on change in volume fraction of retained austenite with deformation. (From Ref. 120.)

on the bainitic region on the build up of carbon in the retained austenite in a steel containing 0.14% C, 1.94% Si, and 1.66% Mn. The effect of hold time on the stability of retained austenite during deformation is illustrated in Fig. 97. The retained austenite having highest carbon content has the highest yield stress as well as the highest elongation; it reflects the gradual transformation of austenite to martensite over the complete strain range.

There are steels with tensile strength about 1000 N/mm^2; sometimes steels with a tensile strength above 800 or 900 N/mm^2 may be regarded as ultra-high-strength steels. There are two basic types, one containing a microstructure of a high proportion of martensite, and one containing a high proportion of bainite. Each may contain a smaller proportion of other phases, together with significant proportions of ferrite or retained austenite. One approach to produce a steel with a tensile strength above 980 N/mm^2 is to produce a ferrite–martensite dual phase steel comprising a high hardness martensite in a ductile ferrite containing a minimum of dissolved carbon. The method involves intercritical, continuous annealing at the lowest possible temperature, cooling and then quenching from just above the B_s temperature to give a transformation product enriched as far as possible with carbon and the highest proportion of retained austenite. The highest elongation was obtained with the highest silicon content used (1.9%) which led to a homogeneous distribution of fine martensite islands [121].

Steels containing niobium and a high carbon equivalent expressed as $C_{eq} = C + Si/24 + Mn/6$ can obtain a tensile strength up to above 1600 N/mm^2 by quenching from single phase austenite region to get a substantially complete martensite structure. Figure 98 shows the effect of tempering temperature on the balancing of bendability and tensile strength of various steels. Figure 99 illustrates the relationship between the time for delayed fracture and C_{eq} for a given tensile strength close to 1350 N/mm^2. It was considered that controlling the carbide distribution within the grain is important to achieve good bendability, whereas avoiding the carbides formed at the grain boundaries at higher temperatures was necessary to give good delayed fracture characteristics. Both

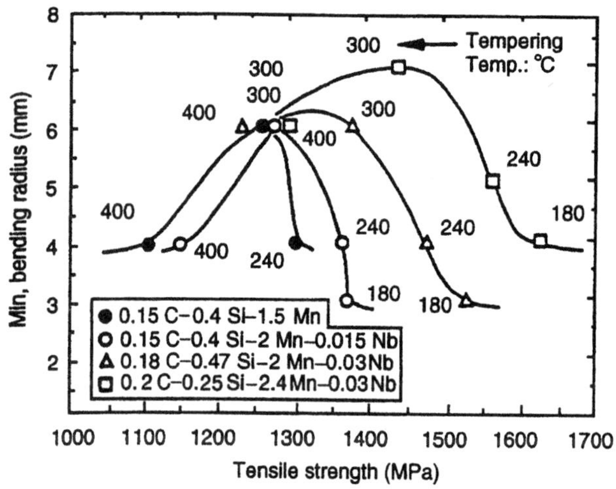

Figure 98 Effect of tempering temperature on balance of bendability and tensile strength of various steels. (From Ref. 122.)

factors are controlled by tempering at temperatures below 240°C. Ultra-high-strength TRIP steels containing a high proportion of bainite may be obtained by rapidly cooling a steel with a suitable composition to temperature close to 400°C. Optimization of the strength/elongation balance depends on retaining a sufficient proportion of retained austenite. Figure 100 illustrates an example how the strength, elongation and structure vary with annealing temperature for a steel containing 0.4% C, 1.55% Si, and 0.8% Mn. The highest values of elongation were obtained for low annealing temperature, for which the carbon in the austenite would have been high enough leading to the highest proportion

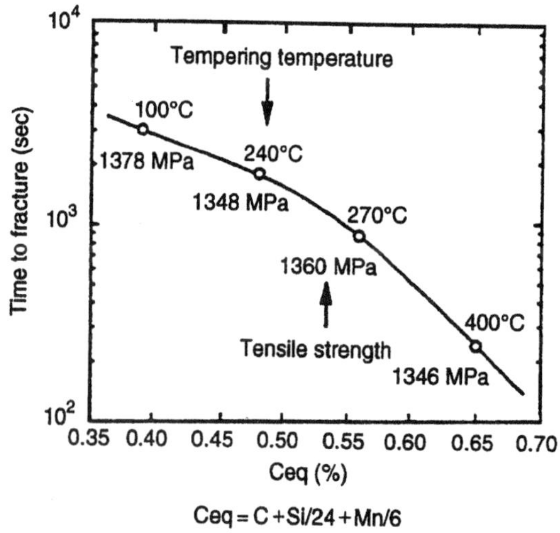

$$Ceq = C + Si/24 + Mn/6$$

Figure 99 Effect of carbon equivalent and tempering temperature on time to fracture of four-point bent samples. (From Ref. 122.)

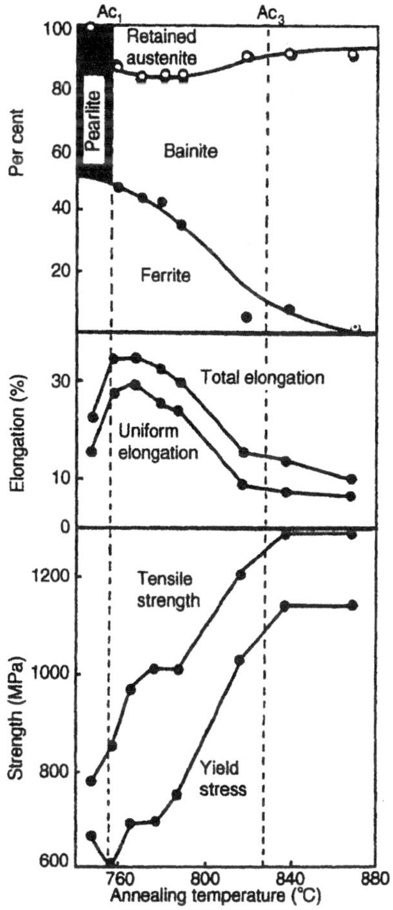

Figure 100 Strength, elongation and structure vs. annealing temperature, samples quenched to 400°C and held for 5 min. (From Ref. 123.)

of retained austenite. Quenching from the single phase austenite region was able to give a strength above 1200 N/mm^2 but with a lower elongation. Tensile strength up to above 1400 N/mm^2 could be obtained using a similar type of annealing cycle but with a richer chemistry.

(C) WEATHERING STEELS. Here, we take weathering steels as an example to illustrate how atmospheric corrosion resistance can be achieved in structural steels.

Weathering steels are structural grades containing small amounts of Cu, P, Si, and Cr to improve the resistance to atmospheric corrosion. Under favorable climate conditions, e.g., relatively long dry summer period, these steels can develop a relatively stable layer of hydrated iron oxide which retards further attack, thus rust at a rate lower than plain carbon steels. Weathering steels can provide cost savings by eliminating the initial painting operation and subsequent maintenance work.

These steels were first introduced in 1937 by the United States Steel Corporation under the brand name Cor-Ten. The mid-specifications of the chemical composition in the Cor-Ten series are given in Table 10. Cor-Ten A is the original high phosphorus grade,

Table 10 Mid-specification of Chemical Composition in Cor-Ten Series

Percent	Cor-Ten A	Cor-Ten B	Cor-Ten C
C	0.08	0.14	0.16
Si	0.50	0.20	0.20
Mn	0.25	1.10	1.20
P	0.11	0.04 max	0.04 max
Cr	0.75	0.50	0.50
Ni	0.35	—	—
Cu	0.40	0.35	0.35
V	—	0.06	0.07

whereas Cor-Ten B and Cor-Ten C have normal levels of phosphorus and are microalloyed with vanadium to provide high strength.

The mechanism of the superior performance of the weathering steels is still relatively obscure. Studies by the USSC [124] showed that Cor-Ten steels rusted faster than carbon steel in the initial stage and it was only after a period of eight days that carbon steel showed a greater gain in weight. After this period, both materials had developed a continuous covering of rust mounts, but those on carbon steel grew to a layer size and eventually spalled from the surface. With Cor-Ten A, splitting was less frequent and no spalling was observed. X-ray diffraction work showed that the rust on both types of steel consisted essentially of $\gamma Fe_2O_3{\cdot}H_2O$ in the initial stage, but after 30 days, $\alpha Fe_2O_3{\cdot}H_2O$ was detected. In addition to iron oxides, iron sulfates ($FeSO_4{\cdot}3H_2O$, $FeSO_4{\cdot}7H_2O$ and $Fe_2(SO_4)_3$) have been detected in the rust layers formed on steel in polluted atmospheres. It is suggested tentatively that the beneficial alloying elements render these sulfates less soluble and thereby retard the penetration of air and moisture through the oxide layer to the steel interface.

The effect of individual alloying element on the corrosion resistance of Mayari R steel was examined [125]. The base composition used is: 0.08% C, 0.28% Si, 0.70% Mn, 0.10% P, 0.03% S, 0.60% Cr, 0.40% Ni, 0.60% Cu. The results are summarized in Fig. 101(a), which shows the effect of variations in a single alloying element in the above base. The order of effectiveness is: P (most beneficial), Cr, Si, Ni, Mn (no effect), S (detrimental). Another result, shown in Fig. 101(b,c), involved a base composition: $<$ 0.10% C, 0.22% Si, 0.25–0.40% Mn, 0.10% P, $<$ 0.02% S, 0.63% Si, ~0% Ni, 0.42% Cu give the order of effectiveness as follows: P (most beneficial), Si, Cu (up to 0.3%), Cr, Ni, Cu ($>$0.3%). But copper exerts a more powerful effect at the industrial site and would be promoted to a higher ranking than that shown above.

It is generally acknowledged that phosphorus produces a marked and progressive benefit up to at least 0.1%. However, P is very detrimental to both toughness and weldability and therefore, it is not included in some grades of weathering steel. Copper is regarded as an essential constituent in weathering steels but little benefit is gained by increasing the copper content above 0.3%. Silicon and chromium are mildly beneficial and the greatest benefit is obtained at levels up to about 0.25 and 0.6, respectively. Manganese is neutral, whereas sulfur is detrimental. A European standard for weathering grades is BS EN 10155.

(D) STEELS FOR SOUR GAS SERVICE. A fluid containing greater than 0.0035 atm partial pressure of H_2S is designated as sour by the National Association of Corrosion Engineers (NACE). Both H_2S and CO_2 become corrosive in the presence of moisture and the

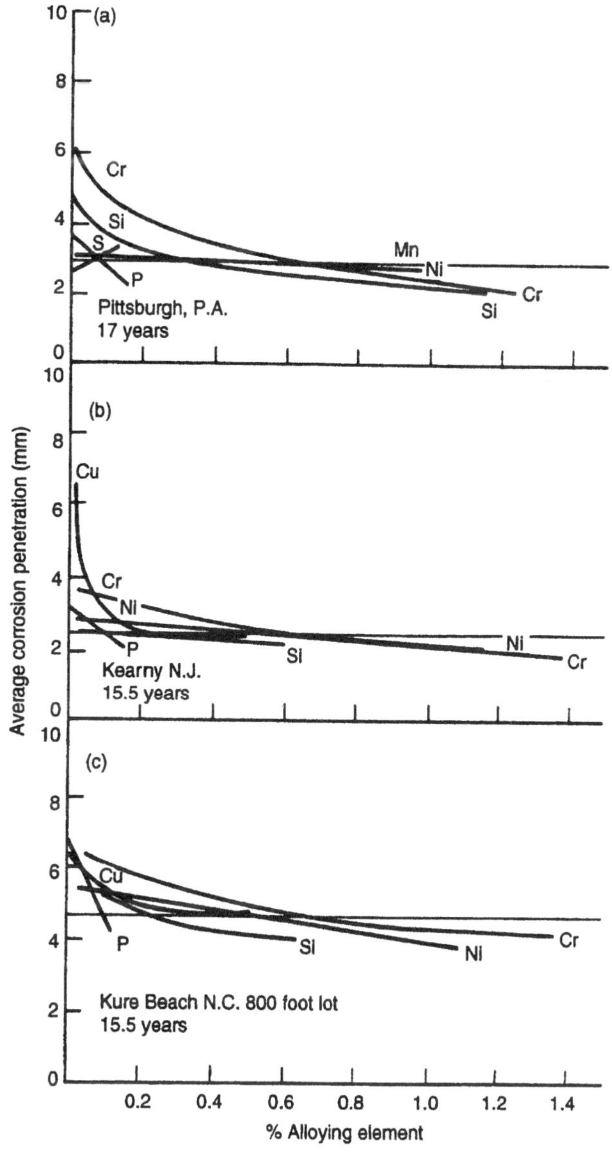

Figure 101 Corrosion behavior of weathering steels: (a) Bethlehem data on Mayari R; (b) USS Cor-Ten A in industrial atmosphere; and (c) USS Cor-Ten A in marine atmosphere. (From Refs. 124,125.)

corrosivity of natural gas is determined solely by the levels of these compounds. There is a need for steels with inherent resistance to sour environments.

Hydrogen-induced cracking (HIC) and sulfide stress corrosion cracking (SSCC) are two types of fractures which can be introduced by H_2S. Sulfide stress corrosion cracking failure is generally confined to steels with yield strengths greater than about 550 N/mm^2 and therefore do not feature prominently in linepipe steels. However, in the high hardness heat affected zones of welds, SSCC should not be overlooked since it can result in catastrophic brittle failure. Hydrogen-induced cracking results in a form of blistering or delamination and can take place in the absence of stress. Atomic hydrogen is generated at cathodic sites under wet sour conditions and diffuses into the steel, forming molecular hydrogen at the interface between non-metallic inclusions and the matrix. When the internal pressure due to the build-up of molecular hydrogen exceeds a critical level, HIC is initiated. Elongated inclusions such as TypeII MnS are particularly favorable sites for crack initiation but planar arrays of globular oxides are also effective. Cracking can proceed along segregated bands containing lower transformation products such as bainite and martensite. Cracking tends to be parallel to the surface but can be straight or stepwise.

The susceptibility to HIC is assessed by the immersion of unstressed coupons in synthetic solution of sea water saturated in H_2S with a pH of 5.1–5.3 (BP test) or in the more aggressive solution of 0.5% CH_3COOH + % NaCl + H_2S saturated at a pH of 3.5–3.8 (NACE test). In either case, the test duration is 96 hr and the test parameters include crack length, crack width, or blister formation.

Linepipe steels with resistance to HIC embody the following features: (1) Keep levels of C, S, and P as low as possible to minimize segregation. (2) A minimum level of Mn must be maintained, particularly in linepipe for arctic service, Mn is beneficial in improving toughness by refining the ferrite grain size. (3) Additions of calcium or rare earth metals to globularize the sulfide inclusions. (4) Low levels of oxide inclusions. (5) Free from periodic banding so as to avoid bainitic or martensitic bands. (6) Addition of small amounts of Cu or Cr to form protective film that prevents the diffusion of hydrogen into the steel.

(E) BEARING STEELS. Bearing steel is a member belonging to the family of engineering steels. Engineering steels cover a wide range of chemical compositions that are generally heat treated to produce tensile strengths greater than 750 N/mm^2. Diversified machine components are made from these steels. The concept of hardenability was primarily introduced in engineering steels to describe the generation of a particular level of strength in a ruling section.

Bearings are vital components constituting a machinery. SAE52100, a high carbon low chromium steel (1.0% C, 1.5% Cr), is an internationally adopted steel for through-hardened bearings. It is generally austenitized under about 860°C, followed by oil quenching and tempering in the temperature range 180–250°C. This treatment results in a microstructure of lightly tempered martensite, undissolved carbides and up to about 5% retained austenite. Carburizing grades such as SAE4720 (2% Ni, 1% Mo) are adopted for large bearings. Fatigue performance is a decisive consideration in the properties of bearing steels. Oxides, such as alumina and silicates, have an adverse effect on fatigue performance of bearings. Figure 102(a) shows the relationship between the average life and the total oxide inclusion content. It is believed that the oxides act as stress raisers and are brittle and are also surrounded by tensile stresses due to the difference in the thermal expansion coefficients between the oxides and the matrix [127], so they are liable to form incipient cracks which then propagate under stress reversals until a fatigue pit (spall) is formed on the surface of the components. Sulfides appear to be beneficial (Fig. 102(b)) which are ascribed to the fact that they tend to encapsulate the more angular oxide inclusions, thereby reducing the tessellated tensile stress. However, there is another view that sulfide is neutral, i.e., non-detrimental rather than positively beneficial. TiN inclusions have a relatively small effect on bearing fatigue performance, the tessellated stress around TiN is small. It is accepted that the undissolved carbides increase the wear resistance and are instrumental in inhibiting grain growth during austenitizing. However, it is considered that large undissolved carbides, particularly when they are present in the segregated bands, act as stress raisers and are detrimental to fatigue life. The retained austenite will

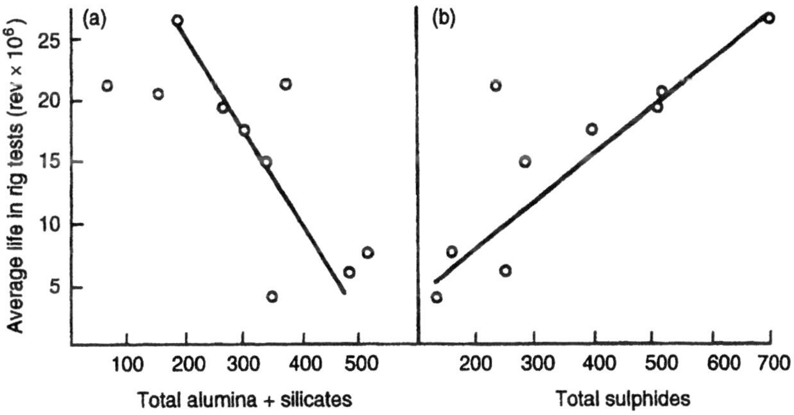

Figure 102 Relation between average life and inclusion content; counts based on tital inclusions observed 750 magnification in 516 fields representing a total area of about 9 mm. (From Ref. 126.)

transform to martensite under the high Hertzian stresses induced in service and will introduce dimensional changes in bearing component. For this reason, retained austenite is generally limited to a maximum of about 5%. The compressive residual stress created by the strain-induced transformation of retained austenite should be beneficial to the fatigue performance. The fatigue performance of bearing steels has been improving progressively over the years through the improvement of steel cleanness and the reduction of non-metallic inclusions by means of the adoption of facilities such as vacuum degassing (1950s), argon shrouding of the molten stream (1960s) and vacuum induction melting (VIM) and vacuum arc remelting (VAR) (1970s), and particularly with the introduction of secondary steel making facilities in bulk steel making processes for bearing steels.

REFERENCES

1. Stahl Schlüssel, Verein Eisenhüttenleute, Düsseldorf, 1989.
2. Pickering, F.B. Structure-Property Relationship in Steels. Mater. Sci. Technol. 1992, *7*, 45–90.
3. Morrison, W.B. Controlled processing of H.S.L.A. steel. In British Steel Corporation Conference, York, 1976.
4. Pickering, F.B.; Gladman, T. *Metallurgical Developments in Carbon Steels*; I.S.I. Special Report, No. 81; Williams Lea & Company: London, 1963; 10–20.
5. Llewellyn, D.T.; Hudd, R.C. *Steel: Metallurgy and Application*, 3rd Ed.; Butterworth-Heinmann: Woburn, MA, 1998; 137–145 and 280–281.
6. Takahashi, N.; Shibata, M.; Furuno, Y.; Hayakawa, H.; Kubuta, K.; Yamamoto, K. Boron-Bearing Steels for Continuous Annealing to Produce Deep-Drawing and High-Strength Steel Sheets. *Metallurgy of Continuous Annealed Sheet Steel*, Bramfitt, B.L.; Manganon, P.L., Eds.; AIME: Dallas, 1982; 133–153.
7. Pickering, F.B. *Physical Metallurgy and the Design of Steels*; Applied Science Publishers: London, 1978; 10–36.
8. Hayami, S.; Furukawa, T. A Family of High-Strength, Cold-Rolled Steels. *Microalloying '75*, Korchynsky, M., Ed.; Union Carbide Corporation: New York, 1977; 311–321.
9. Gawne, D.T.; Lewis, G.M. Strengthening Mechanisms in High-Strength Microalloyed Steels. Mater. Sci. Technol. 1985, *1*, 183.

10. Katoh, H.; Takechi, H.; Takahashi, N.; Abe, M. Proceedings of Symposium on Metallurgy of Continuously Annealed Cold Rolled Sheet Steel; Pradhan, R., Ed.; AIME: Detroit, 1984; 37–58.
11. Ohashi, N.; Irie, T.; Satoh, S.; Hashimoto, D.; Takahashi, I. SAE Paper 810027, 1981.
12. Kocks, U.F. A Statistical Theory of Flow Stress and Work-Hardening. Phil. Mag. 1996, *13*, 541–566.
13. Ashby, M.F. Oxide dispersion strengthening. In AIME Conference, New York, 1966; 143.
14. Gladman, T.; Dulieu, D.; McIvor, I.D. Structure-Property Relationships in High-Strength Microalloyed Steels. *Microalloying '75*, Korchynsky, M., Ed.; Union Carbide Corporation: New York, 1977; 32–58.
15. Turkdogan, E.T. Causes and Effects of Nitride and Carbonitride Precipitation During Continuous Casting. Iron Steel Making, 1989, *16*, 61–75.
16. Copreux, J.H.; Gage, H.; Henry, Lanteri, S. ECSC Report, 7210, EC/303, 1994.
17. Irvine, K.J.; Pickering, F.B.; Gladman, T. Grain-Refined C-Mn Steels. JISI 1967, *205*, 161–182.
18. Hudd, R.C.; Jones, A. A Method for Calculating the Solubility and Composition of Carbonitride Precipitates in Steel with Particular Reference to Niobium Carbonitride. JISI 1971, *209*, 121–125.
19. Liu, W.J.; Jonas, J.J. Calculation of the $Ti(C_y N_{1-y})$-$Ti_4 C_2 S_2$–MnS-Austenite Equilibrium in Ti-Bearing Steels. Met. Trans. A 1989, *20*, 1361–1374.
20. Paules, J.R. Microalloyed Long Products-High Temperature and Processing. *Microalloyed Vanadium Steels*, Korchynsky, M.; Gorczuca, S.; Blicharski, M., Eds.; Association of Polish Metallurgical Engineers: Krakow, 1990; 19.
21. Gladman, T. Physical Metallurgy of Microalloyed Medium-Carbon Engineering Steels. Iron Steel Making 1989, *16*(4), 241–245.
22. Langenborg, R. Proceedings on Fundamentals of Microalloying Forging Steels; Krauss, G., Banerji, S.K., Eds.; TMS: 1987; 39–54.
23. Bramfitt, B.L.; Marder, A.R. *Processing and Properties of Low Carbon Steels*; AIME: New York, 1973; 191.
24. Little, J.M. Proceedings of Third International Conference on Strength of Metals and Alloys; The Institute of Metals/Iron and Steel Institute: Publ. No. 36, Vol. 1, 1973; 80.
25. Hall, E.O. The Deformation and Aging of Mild Steel. Proc. Phys. Soc. B 1951, *64*, 742–747.
26. Petch, N.J. The Cleavage Strength of Polycrystals. Proceedings of Swampscott Conference, MIT Press: 1953, 174, 25–28.
27. Kozaku, I. Thermomechanical Controlled Processing. Mater. Sci. Technol.,1992, *7*, 183–216.
28. Cuddy, L.J.; Raley, J.C. Austenite Grain Coarsening in Microalloyed Steels. Met. Trans. A 1989, *14*, 1983.
29. Cuddy, L.J. The Effect of Microalloy Concentration on the Recrystallisation of Austenite During Hot Deformation. *Plastic Deformation of Metals*; Academic Press: New York, 1975; 129–140.
30. Kasper, R.; Lotter, U.; Biegus, C. The Influence of Thermomechanical Treatment on the Transformation Behavior of Steels. Steel Res. 1994, *65*(6), 242–247.
31. Sckine, H. Fundamental Research on Manufacturing of High-Tough, High-Tension Steel by Controlled Rolling. Seitetsu Kenkyu, 1976, *43*, 209.
32. Kosasu, I.; Ouchi, C.; Sampei, T.; Okita, T. Hot Rolling as a High Temperature Thermomechanical. *Microalloying'75*; Union Carbide Corporation: New York, 1977; 120–135.
33. Ouchi, C. High strength low alloy steel. In Proceedings of Wollongong International Conference, University of Wollongon: Australia, 1985; 17.
34. Pickering, F.B. High Strength Low-Alloy Steels—A Decade of Progress. Proceedings of International Conference on HSLA Steels; Union Carbide Corporation: 1977; 9–31.
35. Burns, K.W.; Pickering, F.B. Deformation and Fracture of Ferrite-Pearlite Structures. JISI 1964, *202*, 899–906.
36. Pickering, F.B. The Effect of Composition and Microstructure on Ductility and Toughness, Climax Molybdenum Company: Kyoto, 1971; 9–31.

37. Heslop, J.; Petch, N.J. The Ductile-Brittle Transition in Fracture of α-iron II. Phil. Mag. 1958, *36*, 1128–1136.
38. Melloy, G.E. Proceedings of Third International Conference on Strength of Metals and Alloys; The Institute of Metals/Iron and Steel Institute: Publ. no. 36, 1973, 60.
39. Pircher, H.; Klapdar, W. Controlling Inclusions in Steel by Injecting Calcium into Ladle. *Microalloying '75*; Union Carbide Corporation: New York, 1977; 232–239.
40. Bucher, J.H. Strength and Toughness of Hot-Rolled Ferrite-Pearlite Steels. 1969, *6*, 247.
41. Hilty, D.C.; Farrell, J.W. *Inclusions and Their Effects on Steel Properties*; British Steel Corporation: Leeds, 1974; 4.
42. Pickering, F.B. The optimization of microstructures in steel and their relationship to mechanical properties In *Hardenability Concepts with Applications to Steel*, Doane, D.V.; Kirkaldy, J.S., Eds.; Metallurgical Society of AIME: Warrendale, PA, 1978; 179–228.
43. Mintz, B. *Low Carbon Structural Steels for Eighties*; The Institute of Metallurgists: Chameleon Press, London, Spring Residential Course, No. 10, I47–53.
44. Held, J.F. Proceedings of Eighth Conference on Mechanical Working and Steel Processing IV; Edgecombe, D.A., Ed.; AIME: Warrendale, 1965; 3.
45. Atkinson, M.; Maclean, I.M. The Measurement of Normal Anisotropy in Sheet Steel. Sheet Met. Ind. 1965, *42*, 290 298.
46. Wilson, D.V.; Butler, R.D. J. Inst. Met. 1961, *90*, 473.
47. Kino, N.; Matsumura, Y.; Tsuchiya, H.; Furukawa, Y.; Akagi, H.; Sanagi, S. CAMP-JISI 1990, *3*, 785.
48. Liu, G.X.; Guan, Z.M.; Notis, M.R.; Marder, A.R. Periodic Banding Structure in Fe-Mn-C Systems. J. Univ. Sci. Technol. Beijing 1993, *15*(1), 67–72.
49. Dasarathy, C.; Goodwin, T.J. Recent Developments in Automotive Steels. Met. Mater. 1990, *6*(1), 21–28.
50. Pearce, R. *Sheet Metal Forming*; Adam Hilger: 1991.
51. Liu, G.X.; Dahl, W. The Influence of Temperature and Strain Rate on the Stress-Strain Behavior of "In-Situ Solidified" Steel during Tensile Test. Steel Res. 1989, *5*, 221–229.
52. Dillamore, I.L. Controlled processing of H.S.L.A. steel. In British Steel Corporation Conference, York, 1976 (Paper 3).
53. Armstrong, R. Discontinuous Twinning during Essentially Elastic Compression of Steel at 4.2°K. Phil. Mag. 1972, *25*(2), 519–522.
54. Morrison, W.B. Trans ASM 1966, *59*, 824.
55. Gladman, T. Work Hardening of Low-Carbon Steels. JISI 1970, *208*, 172–183.
56. Barnby, J.T. The Initiation of Ductile Failure by Fractured Carbides in Austenitic Steel. Acta Met. 1967, *15*, 903–909.
57. Smith, E. Int. J. Fract. Mech. 1970, *4*, 131.
58. Andrews, K.W. Empirical Formulae for the Calculation of Some Transformation Temperatures. JISI 1965, *203*, 721–727.
59. Irie, T.; Satoh, S.; Hashiguchi, K.; Takahashi, I.; Hashimoto, O. Metallurgical Factors Affecting the Formability of Cold-Rolled High-Strength Steel Sheets. Trans. Iron Steel Inst Jpn. 1981, *21*, 793–801.
60. Embury, J.M.; Fisher, R.M. The Structure and Properties of Drawn Pearlite. Acta Met. 1966, *14*, 147–159.
61. Gladman, T. Some Aspects of the Structure-Property Relationships in High-Carbon Ferrite-Pearlite Steels. JISI 1972, *210*, 916–930.
62. Pickering, F.B. Some Aspects of Relationship between Microstructure and Mechanical Properties. Iron & Steel 1965, *38*(3), 110–117.
63. Gladman, T. Effect of Steel Composition on Lead Embrittlement. JISI 1972, *210*, 938–940.
64. Habraken, L.J.; Economopoulos, M. Bainitic Microstructures in Low-Carbon Alloy Steels and Their Mechanical Properties. Climax Molybdenum Company: Ann Arbor, 1967, 69–108.
65. Brownrigg, A. J Met. Sci. 1975, *9*, 313.

66. Bush, M.E.; Kelly, P.M. Strengthening Mechanisms in Bainitic Steels. Acta Met. 1971, *19*, 1363–1371.
67. Whiteman, J.A. Low Carbon Structural Steels for the Eighties,1997, *3*(6), I 1–7.
68. Irvine, J.K.; Pickering, F.B. The Impact Properties of Low-Carbon Bainitic Steels. JISI 1963, *201*, 518–531.
69. Ohtani, H. The Microstructure and Toughness of High-Tensile Strength Steels. Trans I.S.I. Jpn. 1972, *12*, 118–127.
70. Pickering, F.B. High-Strength Steels and Their Uses Today. Pergamon Press: 1965; 199–258.
71. Cox, A.R. Metallography and Strengthening Mechanisms of 0.3%C and 0.3%–1.5% Cu Steels. JISI 1967, *205*, 51–57.
72. Malik, L.; Lund, J.A. A Study of Strengthening Mechanisms in Tempered Martensite From a Medium-Carbon Steel. Met Trans. 1972, *3*, 1403–1406.
73. Leslie, W.C.; Solers, R.J. Trans ASM 1967, *60*, 459.
74. Wannel, P.H.; Blank, J.R.; Naylor, D.J. Proceedings of International Symposium on the Influence of Metallurgy on the Machinability of Steel, September, ISIJ/ASM: Tokyo, 1977.
75. Pickett, M.L.; Cristinacce, M.; Naylor, D.J. Proceedings of High Productivity Machining, Materials and Processing, May, ASM: New Orleans, 1985.
76. Siebert, C.A.; Doane, D.V.; Breen, D.H. *The Hardenability of Steels—Concepts, Metallurgical Influences, and Industrial Applications*; ASM:Metal Park, OH, 1977; 22.
77. Llewellyn, D.T.; Cook, W.T. Metallurgy of Boron-treated Low-Alloy Steels. Met Technol. 1974, *12*, 517–529.
78. de Retana, A.F.; Doane, D.V. Predicting Hardenability of Carburizing Steels. Met Prog. 1971, *100*, 65–69.
79. Ueno, M.; Inoue, J. Distribution of Boron at Austenite Grain Boundaries and Bainitic Transformation in Low-Carbon Steels. Trans ISI Jpn. 1973, *3*, 210–217.
80. Smallman, R.E. Tempering and Heat Treatment. 4th Ed. *Modern Physical Metallurgy*; Butterworth: London.
81. Chipman, J. Ingot Structure and Segregation. Physical Chemistry of Liquid Steel. *Basic Open Hearth Steel Making*; AIME: New York, 1951, *444*, 632.
82. Ishida, K. *Fundamentals and Application of Phase Diagrams*; Seminar Text; The Japan Institute of Metals: 1987.
83. Oikawa, K.; Ohtani, H.; Ishida, K.; Nishizawa, T. The Control of the Morphology of MnS Inclusions in Steel during Solidification. ISIJ Int. 1995, *35*, 402–408.
84. Oikawa, K.; Ishida, K.; Nishizawa, T. Effect of Titanium Addition on the Formation and Distribution of MnS Inclusions in Steel during Solidification. ISIJ Int. 1997, *37*, 332–338.
85. Akamatzu, S.; Senuma, T.; Akisue, O. Effect of Cu and other Elements on Steel Properties; ISI Japan: 1997; 126.
86. Ohtani, H.; Suda, H.; Ishida, K. Solid/Liquid Equilibria in Fe-Cu Based Ternary Systems. ISIJ Int.: Japan 1997, *37*, 207–216.
87. Hillert, M.; Wada, T.; Wada, H. The α/γ Equilibrium in Fe-Mn, Fe-Mo, Fe-Ni, Fe-Sb, Fe-Sn, and Fe-W Systems. JISI 1967, *205*, 539–542.
88. Ishida, K.; Wakakuwa, T.; Nishizawa, T. Computation of Dual Phase Region in Low-Alloy Steels. *HSLA Steels, Metallurgy and Applications*, Gray, J.M.; Zhang, S.H.; Ke, J., Eds.; CSM/ASM International, 1987; 851–859.
89. Hashiguchi, K.; Kirkaldy, J.S.; Fukuzumi, T.; Pavaskar, V. Prediction of Equilibrium, Para-Equilibrium and No Partition Local Equilibrium Phase Diagrams for Multi-Components Fe-C Base Alloys. Calphad 1984, *8*, 173–186.
90. Ishida, K. Calculation of the Effect of Alloying Elements on the Ms Temperature in Steels. J Alloys Compd. 1995, *220*, 126–131.
91. Ohtani, H.; Ishida, K.; Nishizawa, T. Trans. Mater. Res. Jpn. 1992, *9*, 199.
92. Inone, K.; Ohtani, H.; Ishida, K. Calculation of Phase Equilibria between Austenite and (Nb,Ti,V)(C,N) in Steel. Calphad XXVI Florida 1997, 21.
93. Rudy, E. Boundary Phase Stability and Critical Phenomena in High Order Solid Solution System. J Less Common Met. 1973, *33*, 43–70.

94. Gawne, D.T. *Steel Research*; British Steel Cooperation: London, 1976.
95. Hudd, R.C.; Lyons, L.K. Effect of Cup Geometry on Ear Height and Thickening in Cylindrical Flat-Bottomed Steel Cups. Met. Technol. 1975, *2*, 428–432.
96. Meyzand, Y.; Michant, B.; Parniere, P. *Texture and Properties of Materials*. Fourth International Texture Conference, Cambridge; The Metals Society: London, 1975; 255–265.
97. Leslie, W.C.; Ricket, R.L.; Dotson, C.L.; Walton, C.S. Solution and Precipitation of Aluminium Nitride in Relation to the Structure of Low-Carbon Steels. Trans. ASM 1954, *46*, 1470–1499.
98. Whitely, R.L.; Wise, D.E. Relationship among Texture, Hot-Mill Practice and the Deep Drawability of Sheet Steel. In *Flat Rolled Products III*, Earhart, E.W., Ed.; Interscience: New York, 1962; 47–63.
99. Parayil, T.R.; Gupta, I. *Mechanical Working and Steel Processing XXVII*. Proceedings of 31st Conference, Kuhn, L.G., Hultgren, F.A. Eds.; AIME: Warrendale, 1989; 314.
100. Osawa, K.; Matsudo, K.; Kurihara, K.; Suzuki, T. Effects of Carbon and Manganese on the Deep Drawability of Cold-Rolled Steel Sheets. Tetsu-to-Hagane 1984, *70*, 522.
101. Shimisu, M.; Matsukura, K.; Takahashi, N.; Shinagawa, H. The Effect of Heating Rates on the Recrystallisation Behavior of the Cold-Rolled Sheet Steel. Tetsu-to Hagane 1964, *50*, 2094–2097.
102. Toda, K.; Gondoh, H.; Takeuchi, H.; Abe, M.; Uehara, N.; Komita, K. Metallurgical Investigations on Continuous Annealing of Low-Carbon Capped Steel Sheets. Trans. Iron Steel Inst. Jpn. 1975, *15*, 305–313.
103. Osawa, K.; Matsudo, K.; Kurihara, K.; Susuki, T. Effects of Carbon and Manganese on the Deep Drawability of Cold-Rolled Steel Sheets. Tetsu-to-Hagane 1984, *70*, 552.
104. Ono, S.; Nozoe, O.; Shimomura, T.; Matsudo, K. Effect of Carbon on the Mechanical Properties of Continuous Annealed Drawing Quality Steels. Proceedings of Symposium on Metallurgy of Continuous Annealed Sheet Steel; Bramfitt, B.L., Mangonon, P.L., Eds.; AIME: Dallas, 1982; 99–115.
105. Pradhan, R. Effect of Nitride Formers (B,Zr,Ti) on the Mechanical Properties of Continuously Annealed Low-Carbon Steel Sheet. Proceedings of Symposium on Metallurgy of Continuously Annealed Cold Rolled Sheet Steel, Pradhan, Ed.; AIME: Detroit, 1984; 185–202.
106. Prum, N.; Meers, U.; Marthy, H.; Lerog, V. Proceedings of Symposium on Hot and Cold Rolled Sheet Steels; Pradhan, R., Ludkovsky, G., Eds.; The Metallurgical Society: Cincinnati, 1988; 8.
107. Zackay, V.F.; Parker, E.R.; Fahr, D.; Busk, R. The Enhancement of Ductility in High-Strength Steels. Trans. Am. Soc. Met. 1967, *60*, 252.
108. Lawson, P.D.; Matlock, D.K.; Krauss, G. The Effect of Microstructure on the Deformation Behavior and Mechanical Properties of a Dual Phase Steel. Proceedings of Conference on Fundamentals of Dual Phase Steel; Kot, R.A., Bramfitt, R.L., Eds.; AIME: Warrendale, 1981; 347–381.
109. Gladman, T. *Advances in the Physical Metallurgy and Applications of Steel*, Eyre, B.L., Ed.; The Metals Society: London, 1981; 55.
110. Speich, G.R.; Miller, R.L. Tempering of Ferrite-Martensite Steels. Proceedings of Conference on Fundamentals of Dual Phase Steel; Kot, R.A.; Bramfitt, B.L., Eds.; AIME: Warrendale, 1981; 279–304.
111. Messien, P.; Herman, J.C.; Greday, T. The Influence of Cooling Rates after Continuous Annealing during the Production of D.D.Q. and D.P. Steel Grades. Proceedings of Symposium on Metallurgy of Continuous Annealed Sheet Steel; Bramfitt, B.L., Mangonon P.L., Eds.; AIME: Dallas, 1982; 271–286.
112. Speich, G.R.; Demarest, V.A.; Miller, R.L. Formation of Austenite during Intercritical Annealing of Dual Phase Steels. Metal. Trans. A 1981, *12*, 1419–1428.
113. Shirasawa, H.; Thomson, J.G. Effect of Hot-Band Microstructure on Strength and Ductility of Cold-Rolled Dual-Phase Steels. Trans. Iron Steel Inst. Jpn. 1987, *27*, 360–365.

114. Sakuma, O.; Matsumura, O.; Akisue, O. Influence of C Content and Annealing Temperature on the Microstructure and Mechanical Properties of 400°C Transformed Steel Containing Retained Austenite. ISIJ Int. 1991, *31*, 1348–1353.
115. Okamoto, A.; Nagao, N.; Takahashi, M.; Saiki, K. Strength and Bendability of Ultra-High Strength Cold-Rolled Steel Sheet. Proceedings of Symposium on Metallurgy of Continuous Annealed Sheet Steel; Bramfitt, B.L., Mangonon, P.L., Eds.; AIME: Dallas, 1982; 287–300.
116. Sudo, M.; Kokubo, I. Microstructure-Mechanical Property Relationships in Multi-Phase Steel Sheet. Scand. J. Met. 1984, *13*, 329–342.
117. Minote, T.; Torizuka, S.; Ogawa, A.; Nikura, M. Modeling of Transformation Behavior and Compositional Partitioning in TRIP Steel. ISIJ Int. 1996, *36*, 201–207.
118. Matsumura, O.; Sakuma, Y.; Takechi, H. Retained Austenite in 0.4C-Si-1.2Mn Steel Sheet Intercritically Heated and Austempered. ISIJ Int. 1992, *32*, 1014–1020.
119. Sakuma, Y.; Matlock, D.K.; Krauss, G. *Developments in the Annealing of Sheet Steels*, Pradhan, R., Ed.; TMS: 1992; 321.
120. Itami, A.; Takahashi, M.; Ushioda, K. Plastic Stability of Retained Austenite in the Cold-Rolled 0.14C%-1.9%Si-1.7%Mn Sheet Steel. ISIJ Int. 1995, *35*, 1121–1127.
121. Rigsbee, J.M.; Vander Arend, P.J. Laboratory Studies of Microstructure and Structure-Property Relationship in D. P. HSLA Steels. *Formable HSLA and Dual-Phase Steel*, Davenport, A.T., Ed.; AIME: Warrendale, 1979; 56–86.
122. Hosoya, Y.; Tsuyama, S.; Nagataki, S.; Kanetoh, S.; Izuishi, T.; Takada, Y. Microstructure Control of Ultra High Strength Steel Sheets. NKK Technical Review 1995, *72*, 20.
123. Matsumura, O.; Sakuma, Y.; Takechi, H. Enhancement of Elongation by Retained Austenite in Intercritical Annealed 0.4C-1.5Si-0.8Mn Steel. Trans. Iron Steel Inst. Jpn. 1987, *27*, 570–579.
124. United States Steel Technical Report—A Study of the Initial Atmospheric Corrosion of Carbon Steel, USS Cor-Ten A Steel and USS Cor-Ten B Steel, May, 1966, Project No. 47004–002 (1).
125. Horton, J.B. The rusting of low alloy steels in the atmosphere. In Paper to Pittsburgh Regional Technical Meeting of AISI, November, 1965.
126. Johnson, R.F.; Sewell, J. The Bearing Properties of 1%C-Cr Steel as Influenced by Steel Making Practice. JISI 1960, *12*, 414–444.
127. Brooksbank, D.; Andrews, K.W. Stress Fields Around Inclusions and Their Relation to Mechanical Properties. JISI 1972, *210*, 246–255.
128. Krauss, G. Martensitic Transformation, Structure, and Properties in Hardenable Steels. *Hardenability Concept with Application to Steels*, Doane, D.V.; Kirkaldy, J.S., Eds.; 1977; 229–248.
129. Harris, W.J.; Cohen, M. Stabilization of the Austenite-Martensite Transformation. Trans. AIME 1949, *180*, 447–470.
130. Koistinen, D.P.; Marburger, R.E. A General Equation Prescribing the Extent of Austenite-Martensite Transformation in Pure Iron-Carbon Alloys and Plain Carbon Steels. Acta Met. 1959, *7*, 59–60.
131. Palmgren, A.Z. Die Lebens Dauer von Kugellagern. Zeitschrift Verein Deutscher Ingenieur 1924, *68*, 339–341.
132. Miner, M.A. Cumulative Damage in Fatigue. J. Appl. Mech. 1945, *12*, A159–164.
133. Proceedings of the SAE Fatigue Conference, Dearborn, MI, April 1982; 109.
134. Sperle, J.O.; Trogen, H. Influence of Yield Ratio on the Fatigue Strength of Steel Sheet. Scand. J. Met. 1989, *18*, 147–154.
135. *Annual Book of ASTM Standards*, Vol. 03.01, E674–93
136. Harris, D.; Priest, A.; Davenport, J.; McIntyre, P.; Almond, E.A.; Rowbuck, B. Proceedings of Integrity of Gas Cylinders; Materials Technology, NPL, 69, 1985
137. Oikawa, K.; Ohtani, H.; Ishida, K.; Nishijawa, T. The Control of the Morphology of MnS Inclusions in Steel during Solidification. ISIJ Int. 1995, *35*, 402–408.
138. Oikawa, K.; Ishida, K.; Nishijawa, T. Effect of Titanium Addition on the Formation and Distribution of MnS Inclusions in Steel during Solidification. ISIJ Int. 1997, *37*, 332–338.

4

Designing with Tool Steel

Lin-Jiang Yang
Nanyang Technological University, Singapore, Republic of Singapore

David N. Collins
University College Dublin, Dublin, Ireland

I. INTRODUCTION

International Standard EN ISO 4957:1999 defines tool steels as: "Special steels suitable for working or processing of materials, for handling and measuring workpieces and, for this purpose, exhibiting high hardness and wear resistance and/or toughness" [1]. The most commonly used materials for tool, die, and mold applications are wrought tool steels, which are either carbon, alloy, or high-speed steels capable of being hardened and tempered to a high hardness. Other tool materials used for metal working applications include powder metallurgy (P/M) steels, martensitic and precipitation-hardened stainless steels, beryllium copper alloys, and cemented tungsten carbides [2,3].

The earliest tool steels were simple plain-carbon steels. Steel swords from Damascus, made in AD 540 and Japan in AD 900, were produced by layer welding of high- and low-carbon steels by multiple forging steps, to obtain a high hardness from the high-carbon steel and a high toughness from the low-carbon steel [4,5]. However, modern tool steels depend strongly on alloying and heat treatment for their high hardness and toughness. Robert Mushet introduced the alloying technique in 1868, by adding tungsten to a high-carbon steel to produce Mushet's Special tool steel with a nominal composition of 2% C, 2.5% Mn, 7% W, 0.5% Cr, and 1.10% Si [2,6]. Many complex, highly alloyed tool steels were developed in the early 20th century. These tool steels contain, among other elements, relatively large amounts of molybdenum, tungsten, vanadium, chromium, and manganese, and they make it possible to meet increasingly severe service demands and to provide greater dimensional control and freedom from cracking during heat treatment. Many alloy tool steels are also widely used for machinery components and structural applications in which particularly stringent requirements must be met, such as high-temperature springs, ultrahigh-strength fasteners, special-purpose valves, and bearings of various types for elevated-temperature service.

Tool steels are a very special class of steels that are uniquely manufactured for tool applications and are free from internal porosity, sizable undesirable non-metallic inclusions, serious chemical segregation, and surface defects. Hence, for typical wrought tool steels, raw materials (including scrap) are carefully selected not only for alloy content but also for qualities that ensure cleanliness and homogeneity in the finished product. Benjamin Huntsman was reported to have melted pieces of blister steel in a crucible to produce a much more homogenous and high-quality steel for clock springs in 1740 [2]. Modern tool steels are also generally melted in relatively small-tonnage electric arc furnaces to economically achieve composition tolerances, cleanliness, and precise control of melting conditions. Semi-finished and finished bars are given rigorous in-process and final inspections such as macroetching, ultrasonic testing, hardness testing, grain size determination, microscopic examination, and magnetic particle inspection. The finished tool steel bars must be free from decarburization, surface and internal discontinuities.

Modern steel making process involves at least two stages. The first stage consists of melting with little or no refining by using an electric arc furnace. In the second stage, hot metal from the electric arc furnace is transferred to a ladle or converter vessel for refining, using the argon–oxygen decarburizatoin (AOD) or vacuum oxygen decarburization (VOD) process [7]. Alloy additions of chromium, tungsten, molybdenum, vanadium, silicon, and manganese are usually made in the AOD in the form of ferroalloys. Molybdenum, cobalt, and vanadium can also be added as oxides, which can be reduced in the AOD. By the injection of controlled ratios of O_2 and argon or N_2 by means of a submerged tuyere, excellent control of chemistry, including carbon, sulfur, phosphorus, O_2, and H_2, is possible. The principal benefits associated with secondary refining are reduced furnace time, improved yields, quality, consistency, and reproducibility in steel production. Although the AOD process has been used extensively for the production of stainless steels, however it is not clear to what extend it is used for the production of tool steels.

Special secondary remelting processes have also been used for tool steel manufacturing, to satisfy particularly stringent demands regarding tool steel quality and performance. These include electroslag remelting and vacuum arc remelting for producing premium tool and mold steels with uniformity in composition and structure and high quality. Electroslag remelting or refining (ESR) is a progressive melting process in which the end of a consumable electrode is submerged in a reactive, molten slag contained in a water-cooled copper mold [8,9]. By passing a high current, usually alternating current, through the molten slag, the end of the electrode melts, droplets of molten metal fall through the slag are refined, and collect in a pool in the mold to form a sound ingot, with no pipe or porosity. Uniform, relatively rapid solidification leads to an improvement in the uniformity of macrostructure and a reduction in the tendency for center segregation. The ESR process can also reduce the sulfur content substantially, improve hot workability and processing yields, reduce the degree of carbide banding, carbide size, and grain size, increase transverse tensile ductility and fatigue properties. However, ESR is an expensive process due to its slow remelting rate and is most often applied to certain specialized types of tool steel. In the vacuum arc remelting (VAR) which is a companion process to ESR, an arc in a high-vacuum environment is used to replace the resistance-heated molten slag as the heat source. Vacuum arc remelting is capable of removing O_2, N_2, and volatile elements such as manganese, lead and bismuth, but not sulfur. Its use is hence more limited in tool steel refining.

Powder metallurgy (P/M) is used extensively for production of high-carbon, high-chromium, and high-speed tool steels [10,11]. These steels are difficult to process by the conventional ingot metallurgy route. The relatively slow cooling of the conventional static cast ingot allows the formation of coarse eutectic carbide structures which are difficult to break down during hot working, resulting in non-uniform and excessive grain growth, non-uniform hardening, poor transverse properties, and low toughness. The P/M process can produce a fine, uniform distribution of carbides as a result of rapid solidification during atomization. Hence, P/M tool steels have also improved machinability in the annealed condition, faster response to hardening heat treatment, improved toughness and grindability in the fully hardened condition. In the P/M process, molten metal is poured through a small-diameter nozzle into jets of high-pressure gas, usually nitrogen that breaks the stream into small droplets. Hot isostatic pressing (HIP), with a pressure of about 1000 bar and a temperature of about 1150°C, sometimes with an intermediate cold isostatic pressing (CIP) with a pressure of about 4000 bar, is the most common technique used to consolidate the powders produced by the gas atomizing process to form P/M tool steels. In the Osprey process, molten alloy is poured from an induction furnace through a nozzle that directs the stream into high-pressure gas atomization jets, fragmenting it into a large number of small droplets on a rotating collector [12]. However, P/M tool steels constitute only a small market. The bulk of the tool steel market is still supplied by product made by the highly cost-competitive conventional ingot metallurgy approach.

Most tools in service are subjected to extremely high loads that are supplied rapidly. The tools must withstand these loads without breaking and without undergoing excessive wear or deformation, and in many applications, under conditions which produce high temperatures in the tool. No single tool material combines maximum wear resistance, toughness, and resistance to softening at elevated temperatures. Consequently, the selection of the proper tool material for a given application often requires a tradeoff to achieve the optimum combination of properties. Furthermore, the performance of a tool in service depends on: (i) the proper design of the tool; (ii) the selection of the proper tool steel; (iii) the accuracy with which the tool is made, (iv) the application of proper heat treatment; and (v) its proper setup, use, and maintenance. It is therefore important for tool designers to have a good knowledge of: (i) the type(s) of tool steel normally specified in a similar application; (ii) the tool steels available from the market; (iii) the facility available to fabricate it; and (iv) the heat-treatment procedures and the availability of heat-treatment facility.

This chapter is divided into four main sections. Section I will introduce the definition of tool steels, a brief development history, and the various topics to be covered in this chapter. Section II will introduce the latest standard for tool steels, the International Standard EN ISO 4957:1999. The types of tool steel, their compositions, and hardness obtainable according to this standard will be mentioned. The more comprehensive AISI (American Iron and Steel Institute, known as Iron and Steel Society recently) classification system of tool steels will also be described [13]. It will cover: (i) the composition limits of each type of tool steel under this classification; (ii) their major properties such as wear resistance, toughness, and hot hardness; their tempered hardness and annealed hardness; (iii) their processing and service characteristics, such as resistance to decarburization, depth of hardening, amount of distortion, safety in hardening, machinability, and grindability; and (iv) their applications.

Section III will discuss the procedure and the main property requirements in the selection of tool steels. Some typical selection examples based on AISI designations will also be given to cover the selection of carbon tool steels, alloy cold-work tool steels, hot-work tool steels, and high-speed tool steels for: (i) molds and dies; (ii) tools and dies for cold-working applications; (iii) tools and dies for hot-working applications; (iv) cutting tools; and (v) structural component parts. Cold-work tool steels and hot-work tool steels are generally used for applications in which the surface temperature is below 200°C and over 200°C, respectively. High-speed tool steels are used mainly for machining and for forming processes and which, because of their chemical composition, have the highest high-temperature hardness and temper resistance up to about 600°C [1].

Section III will also mention briefly the applications of some other tool materials, like stainless steels, beryllium copper alloys, and cemented carbides; and the applications of hubbing and electrical-discharge machining for tool, mold, and die manufacturing. Some precautions to prevent premature tool failures due to poor design, poor finishing in fabrication, and from cracking and distortion in heat treatment will also be provided.

Section IV will cover the heat treatment of tool steels. To facilitate machining, most of the tool steels are supplied in the soft annealed condition. To obtain their optimum properties, to meet the various demands such as high hardness, high toughness, high wear resistance, and high hot hardness, the selected tool steel must be properly heat treated, using a correct procedure, a suitable temperature and time selection, and suitable equipment. As it is critical to have a good metallurgical knowledge of the heat treatment of tool steels, considerable space is devoted in this section to cover the heat-treatment principles and the effect of alloying elements. The selection of heat-treatment equipment and heat-treatment processes for the heat treatment of tool steels, with some examples, will also be given.

II. CLASSIFICATION AND PROPERTIES OF TOOL STEELS

A. Introduction

Tool steels are divided into groups according to their properties and uses, rather than by chemical composition alone. The approach is useful to assist the user in understanding the many types and their eventual selection to meet certain property requirements. In this section, the latest international standard on tool steel, the ISO 4957:1999 classification system, will be introduced. A list of comparable steels, in EN10027-2:1992 (European Standards) and JIS (Japanese Industrial Standards), is also included for reference. Next, the most comprehensive AISI (American Iron and Steel Institute) tool steel classification system will be described [13–15]. In addition to their compositions, major properties, hardening, and service characteristics, the applications of each type of tool steel will also be mentioned. Products according to these international standards may be either wrought or produced by powder metallurgy. They can be in the form of hot- and cold-finished bars, special shapes, forgings, hollow bar, hot extrusions, wire, drill rod, plate, sheet, strip, tool bits, and castings.

1. General Properties of Alloying Element

To compare the properties, hardening, and service characteristics of different types or grades of tool steels, a good knowledge of the metallurgical effects of common alloying elements and heat-treatment principles is required, as will be discussed in Sec. IV. However, to facilitate understanding of both Secs. II and III, the general properties of the following nine alloying elements should be noted here [16].

Carbon. Carbon is the most important alloying element that affects the properties of tool steels. It is carbon that enables a tool steel to harden by martensite formation. In general, increased carbon contents provide higher hardness after heat treatment and improved wear resistance in service, accompanied by some sacrifice in toughness.

Silicon. Silicon improves the toughness of low-alloy tool steels of the shock-resisting group. When added to hot-work tool steels, silicon raises the critical points and reduces scaling tendencies. Silicon also increases hardenability and resistance to tempering. This element is added to the graphitic free-machining steels to promote the formation of free carbon.

Manganese. The addition of manganese increases the hardenability of tool steels. Even small amounts have significant effects on depth of hardening in carbon tool steels.

Chromium. Chromium, a moderately strong carbide former, contributes to wear resistance in the cold-work die steels. This is especially true for the high-carbon, high-chromium types. Chromium also promotes resistance to tempering and hot hardness in hot-work and high-speed tool steels. Addition of chromium also improves the hardenability of tool steels.

Molybdenum and tungsten. Both molybdenum and tungsten are crucial alloying elements for hot-work and high-speed steels. This is because they provide hot hardness, increased resistance to tempering, and wear-resistant carbides that are harder than chromium carbides. Molybdenum has about double the potency of tungsten in its effect on hot hardness. Relatively small amounts of molybdenum are frequently added to low-alloy tool steels for improved hardenability.

Nickel. Nickel goes into solid solution and does not form carbides. Improved toughness and lower critical points normally result from the addition of nickel. In general, high nickel content (above 2%) is not desirable in tool steels because of the element's strong tendency to stabilize austenite. It should also be noted in Table 4 that the Ni content for a number of steels is specified as 0.20% or 0.30% max. In those cases, Ni is not a deliberate alloying addition. The values merely indicate the restriction on the Ni content.

Vanadium. Vanadium, a very strong carbide former, is added to hot-work and high-speed tool steels for increased wear resistance. This alloying element also improves hot hardness and tempering resistance, particularly in high-speed steels, and promotes grain refinement.

Cobalt. Cobalt is added to improve hot hardness and resistance to tempering in both high-speed and hot-work tool steels. This alloying element remains entirely in solid solution in the steels and does not form carbides.

2. Basic Heat-Treatment Principles

Next, as will be mentioned in Sec. IV, the final heat treatment of tool steels consists of a three-stage process: heating for austenite formation, cooling to transform the austenite to martensite, and tempering to eliminate retained austenite and form carbides within the martensite. The hardened microstructure of a typical tool steel consists of matrix of tempered martensite containing various dispersions of iron and alloy carbides. The following points should be noted here:

- High carbon and high alloy content promote hardenability or the ability to form martensite on cooling.
- Any alloy carbides that remain undissolved in the austenite during hot work and austenizing will still remain undissolved in the martensite produced on quenching. These carbides are retained as components of the microstructure in addition to those formed in the martensite during tempering.
- The higher the carbon content of the martensite and the higher the density of carbides, the higher the hardness and wear resistance but the lower the toughness of a tool steel microstructure.

B. ISO 4957:1999 Classification of Tool Steels

The International Standard EN ISO 4957:1999 has divided tool steels into four categories: (i) non-alloy cold-work tool steels; (ii) alloy cold-work tool steels; (iii) alloy hot-work tool steels; and (iv) high-speed tool steels, as shown in Fig. 1. Table 1 shows the chemical composition of tool steels according to ISO 4957:1999, with a list of comparable steels, in

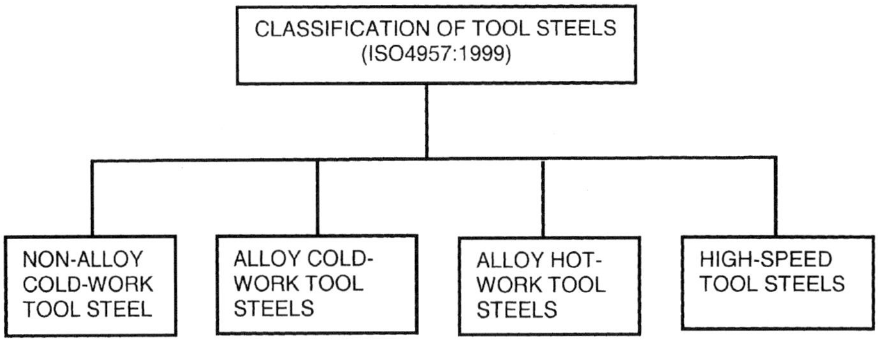

Figure 1 Classification of tool steels according to ISO 4957:1999.

EN10027-2:1992 (European Standards) and JIS (Japanese Industrial Standards). Table 2 shows the hardness values of tool steels, according to ISO 4957:1999, in their annealed and hardened conditions [1]. The hardening method used to achieve the hardness values, i.e. hardening temperature, quenching medium and tempering temperature, are also indicated in the table. Hardness is an important property of a tool material. Without adequate hardness, the steel would not be able to withstand the loads imposed upon it. The annealed hardness and hardened hardness may also provide some indication of its machinability and grindability, respectively.

1. Non-alloy Cold-Work Tool Steels

There are six grades of non-alloy cold-work tool steel, with composition ranges from 0.42% to 1.25% C, 0.10% to 0.40% Si, and 0.10% to 0.80% Mn. These are all plain-carbon steels, and only iron carbides are produced by heat treatment. Steel C45U is used in the non-heat-treated condition. Steels C70U to C120U are shallow hardening steels. For diameters of 30 mm, the hardness penetration depth will be approximately 3 mm. Through-hardening may only be achieved in diameters up to 10 mm. As the hardenability of these steels is low, water quenching is required to form martensite. The hardness value obtainable from the various types of non-alloy cold-work steel, with water quenching, ranges from 54 to 62 HRC as indicated in Table 2.

2. Alloy Cold-Work Tool Steels

There are seventeen grades of tool steels classified under this group. Their carbon, silicon and manganese content varies from 0.18% to 2.30%, 0.10% to 1.00%, and 0.10% to 2.5%, respectively. The alloying elements added are Cr (from 0% up to 17.5%), Mo (from 0% up to 1.4%), Ni (from 0% up to 4.3%), V (from 0% up to 1.0%), and W (from 0% up to 2.2%). Steel 105V has a composition similar to C105U, except with an addition of 0.10–0.20% V. Its hardened hardness of 61 HRC, with water as the quenching medium, is also similar to that obtainable with C105U. Steel 21MnCr5 contains 0.18–0.24% C. However, with 1.10–1.40% Mn and 1.00–1.30% Cr, the steel should achieve a surface hardness of 60 HRC when it is carburized, quenched and tempered. Steels 35CrMo7, 40CrMnNiMo8-6-4 and X38CrMo16 are normally supplied in the quenched and tempered condition with a hardness of approximately 300 BHN. Steel X40Cr14 may also be supplied in the prehardened condition with a similar hardness value.

Table 1 Chemical Composition of Tool Steels According to ISO 4957:1999, with a List of Comparable Steels [1]

Steel name	Chemical composition (%)									Comparable steels	
	C	Si	Mn	Cr	Mo	Ni	V	W	Co	EN10027-2:1992	JIS
Non-alloy cold-work tool steels[a]											
C45U	0.42–0.50	0.15–0.40	0.60–0.80	—	—	—	—	—	—	1.1730	—
C70U	0.65–0.75	0.10–0.30	0.10–0.40	—	—	—	—	—	—	1.1520	SK7
C80U	0.75–0.85	0.10–0.30	0.10–0.40	—	—	—	—	—	—	1.1525	SK6
C90U	0.85–0.95	0.10–0.30	0.10–0.40	—	—	—	—	—	—	1.1535	SK5, SK4
C105U	1.00–1.10	0.10–0.30	0.10–0.40	—	—	—	—	—	—	1.1545	SK3
C120U	1.15–1.25	0.10–0.30	0.10–0.40	—	—	—	—	—	—	1.1555	SK2
Alloy cold-work tool steels[a]											
105V	1.00–1.10	0.10–0.30	0.10–0.40	—	—	—	0.10–0.20	—	—	1.2834	SKS43
50WCrV8	0.45–0.55	0.70–1.00	0.15–0.45	0.90–1.20	—	—	0.10–0.20	1.70–2.20	—	1.2549	—
60WCrV8	0.55–0.65	0.70–1.00	0.15–0.45	0.90–1.20	—	—	0.10–0.20	1.70–2.20	—	1.2550	—
102Cr6	0.95–1.10	0.15–0.35	0.25–0.45	1.35–1.65	—	—	—	—	—	1.2067	—
21MnCr5	0.18–0.24	0.15–0.35	1.10–1.40	1.00–1.30	—	—	—	—	—	1.2162	—
70MnMoCr8	0.65–0.75	0.10–0.50	1.80–2.50	0.90–1.20	0.90–1.40	—	—	—	—	1.2824	—
90MnCrV8	0.85–0.95	0.10–0.40	1.80–2.20	0.20–0.50	—	—	0.05–0.20	—	—	1.2842	—
95MnWCr5	0.90–1.00	0.10–0.40	1.05–1.35	0.40–0.65	—	—	0.05–0.20	0.40–0.70	—	1.2825	—
X100CrMoV5	0.95–1.05	0.10–0.40	0.40–0.80	4.80–5.50	0.90–1.20	—	0.15–0.35	—	—	1.2363	SKD12
X153CrMoV12	1.45–1.60	0.10–0.60	0.20–0.60	11.00–13.00	0.70–1.00	—	0.70–1.00	—	—	1.2379	—
X210Cr12	1.90–2.20	0.10–0.60	0.20–0.60	11.00–13.00	—	—	—	—	—	1.2080	—
X210CrW12	2.00–2.30	0.10–0.40	0.30–0.60	11.00–13.00	—	—	—	0.60–0.80	—	1.2436	—
35CrMo7	0.30–0.40	0.30–0.70	0.60–1.00	1.50–2.00	0.35–0.55	—	—	—	—	1.2302	—
40CrMnNiMo8-6-4[b]	0.35–0.45	0.20–0.40	1.30–1.60	1.80–2.10	0.15–0.25	0.90–1.20[b]	—	—	—	1.2738	—
45NiCrMo16	0.40–0.50	0.10–0.40	0.20–0.50	1.20–1.50	0.15–0.35	3.80–4.30	—	—	—	1.2767	—
X40Cr14	0.36–0.42	≤1.00	≤1.00	12.50–14.50	—	—	—	—	—	1.2083	—
X38CrMo16	0.33–0.45	≤1.00	≤1.00	15.50–17.50	0.80–1.30	≤1.00	—	—	—	1.2316	—
Hot-work tool steels[c]											
55NiCrMoV7[d]	0.50–0.60	0.10–0.40	0.60–0.90	0.80–1.20	0.35–0.55	1.50–1.80	0.05–0.15	—	—	1.2714	SKT4
32CrMoV12-28	0.28–0.35	0.10–0.40	0.15–0.45	2.70–3.20	2.50–3.00	—	0.40–0.70	—	—	1.2365	SKD7
X37CrMoV5-1	0.33–0.41	0.80–1.20	0.25–0.50	4.80–5.50	1.10–1.50	—	0.30–0.50	—	—	1.2343	SKD6
X38CrMoV5-3	0.35–0.40	0.30–0.50	0.30–0.50	4.80–5.20	2.70–3.20	—	0.40–0.60	—	—	1.2367	—

(Continued)

Table 1 (Continued)

Steel name	Chemical composition (%)									Comparable steels	
	C	Si	Mn	Cr	Mo	Ni	V	W	Co	EN10027-2:1992	JIS
X40CrMoV5-1	0.35–0.42	0.80–1.20	0.25–0.50	4.80–5.50	1.20–1.50	—	0.85–1.15	—	—	1.2344	SKD61
50CrMoV13-15	0.45–0.55	0.20–0.80	0.50–0.90	3.00–3.50	1.30–1.70	—	0.15–0.35	—	—	1.2355	—
X30WCrV9-3	0.25–0.35	0.10–0.40	0.15–0.45	2.50–3.20	—	—	0.30–0.50	8.50–9.50	—	1.2581	SKD5
X35CrWMoV5	0.32–0.40	0.80–1.20	0.20–0.50	4.75–5.50	1.25–1.60	—	0.20–0.50	1.10–1.60	—	1.2605	SKD62
38CrCoWV18-17-17	0.35–0.45	0.15–0.50	0.20–0.50	4.00–4.70	0.30–0.50	—	1.70–2.10	3.80–4.50	4.00–4.50	1.2661	SKD8
High-speed tool steels[a]											
HS0-4-1	0.77–0.85	≤0.65	≤0.40	3.90–4.40	4.00–4.50	—	0.90–1.10	—	—	1.3325	—
HS1-4-2	0.85–0.95	≤0.65	≤0.40	3.60–4.30	4.10–4.80	—	1.70–2.20	0.80–1.40	—	1.3326	—
HS18-0-1	0.73–0.83	≤0.45	≤0.40	3.80–4.50	—	—	1.00–1.20	17.20–18.70	—	1.3355	SKH2
HS2-9-2	0.95–1.05	≤0.70	≤0.40	3.50–4.50	8.20–9.20	—	1.70–2.20	1.50–2.10	—	1.3348	SKH58
HS1-8-1	0.77–0.87	≤0.70	≤0.40	3.50–4.50	8.00–9.00	—	1.00–1.40	1.40–2.00	—	1.3327	—
HS3-3-2	0.95–1.03	≤0.45	≤0.40	3.80–4.50	2.50–2.90	—	2.20–2.50	2.70–3.00	—	1.3333	—
HS6-5-2	0.80–0.88	≤0.45	≤0.40	3.80–4.50	4.70–5.20	—	1.70–2.10	5.90–6.70	—	1.3339	SKH51
HS6-5-2C[e]	0.86–0.94	≤0.45	≤0.40	3.80–4.50	4.70–5.20	—	1.70–2.10	5.90–6.70	—	1.3343	—
HS6-5-3	1.15–1.25	≤0.45	≤0.40	3.80–4.50	4.70–5.20	—	2.70–3.20	5.90–6.70	—	1.3344	SKH53
HS6-5-3C	1.25–1.32	≤0.70	≤0.40	3.80–4.50	4.70–5.20	—	2.70–3.20	5.90–6.70	—	1.3345	—
HS6-6-2	1.00–1.10	≤0.45	≤0.40	3.80–4.50	5.50–6.50	—	2.30–2.60	5.90–6.70	—	1.3350	SKH52
HS6-5-4	1.25–1.40	≤0.45	≤0.40	3.80–4.50	4.20–5.00	—	3.70–4.20	5.20–6.00	—	1.3351	SKH54
HS6-5-2-5[e]	0.87–0.95	≤0.45	≤0.40	3.80–4.50	4.70–5.20	—	1.70–2.10	5.90–6.70	4.50–5.00	1.3243	SKH55
HS6-5-3-8	1.23–1.33	≤0.70	≤0.40	3.80–4.50	4.70–5.30	—	2.70–3.20	5.90–6.70	8.00–8.80	1.3244	—
HS10-4-3-10	1.20–1.35	≤0.45	≤0.40	3.80–4.50	3.20–3.90	—	3.00–3.50	9.00–10.00	9.50–10.50	1.3207	SKH57
HS2-9-1-8	1.05–1.15	≤0.70	≤0.40	3.80–4.50	9.00–10.00	—	0.90–1.30	1.20–1.90	7.50–8.50	1.3247	SKH59

[a]For all steels; phosphorus ≤0.030% and sulfur ≤0.030%.
[b]By agreement, sulfur may be increased to between 0.050% and 0.100% and Ni may be omitted.
[c]For all steels (unless otherwise specified), phosphorus ≤0.030% and sulfur ≤0.020%.
[d]The sulfur content for this grade is ≤0.030%.
[e]A sulfur range of 0.060–0.150% may be agreed at the time of enquiry and order for this grade. In this case, a max. of 0.8% Mn applies.

Table 2 Hardness of Tool Steels According to ISO 4957:1999 [1]

Steel name	Hardness (Annealed)[a], HB Max	Hardening method: Hardening temperature °C (±10 °C)	Quenching medium[b]	Tempering temperature °C (±10 °C)	Hardness HRC min.
Non-alloy cold-work tool steels					
C45U	207[c]	810	W	180	54
C70U[d]	183	800	W	180	57
C80U[d]	192	790	W	180	58
C90U[d]	207	780	W	180	60
C105U[d]	212	780	W	180	61
C120U[d]	217	770	W	180	62
Alloy cold-work tool steels					
105V	212	790	W	180	61
50WCrV8	229	920	O	180	56
60WCrV8	229	910	O	180	58
102Cr6	223	840	O	180	60
21MnCr5	217	[e]	[e]	[e]	[e]
70MnMoCr8	248	835	A	180	58
90MnCrV8	229	790	O	180	60
95MnWCr5	229	800	O	180	60
X100CrMoV5	241	970	A	180	60
X153CrMoV12	255	1020	A	180	61
X210Cr12	248	970	O	180	62
X210CrW12	255	970	O	180	62
35CrMo7	[f]				[f]
40CrMnNiMo8-6-4	[f]				[f]
45NiCrMo16	285	850	O	180	52
X40Cr14[g]	241	1010	O	180	52
X38CrMo16	[f]				[f]
Hot-work tool steels					
55NiCrMoV7	248[h]	850	O	500	42[i]
32CrMoV12-28	229	1,040	O	550	46
X37CrMoV5-1	229	1,020	O	550	48
X38CrMoV5-3	229	1,040	O	550	50
X40CrMoV5-1	229	1,020	O	550	50
50CrMoV13-15	248	1,010	O	510	56
X30WCrV9-3	241	1,150	O	600	48
X35CrWMoV5	229	1,020	O	550	48
38CrCoWV18-17-17	260	1,120	O	600	48
High-speed tool steels					
HS0-4-1	262[j]	1,120	O[k]	560	60
HS1-4-2	262[j]	1,180	O[k]	560	63
HS18-0-1	269[j]	1,260	O[k]	560	63
HS2-9-2	269[j]	1,200	O[k]	560	64
HS1-8-1	262[j]	1,190	O[k]	560	63
HS3-3-2	255[j]	1,190	O[k]	560	62
HS6-5-2	262[j]	1,220	O[k]	560	64

(Continued)

Table 2 (Continued)

Steel name	Hardness (Annealed)[a], HB Max	Hardening method: Hardening temperature °C (±10 °C)	Quenching medium[b]	Tempering temperature °C (±10 °C)	Hardness HRC min.
HS6-5-2C	269[j]	1,210	O[k]	560	64
HS6-5-3	269[j]	1,200	O[k]	560	64
HS6-5-3C	269[j]	1,180	O[k]	560	64
HS6-6-2	262[j]	1,200	O[k]	560	64
HS6-5-4	269[j]	1,210	O[k]	560	64
HS6-5-2-5	269[j]	1,210	O[k]	560	64
HS6-5-3-8	302[j]	1,180	O[k]	560	65
HS10-4-3-10	302[j]	1,230	O[k]	560	66
HS2-9-1-8	277[j]	1,190	O[k]	550	66

[a]Hardness in the cold drawn condition may be 20 HB higher than in the annealed condition.
[b]Quenching medium: A=Air, O=Oil, W=Water.
[c]This grade is used in the non-heat-treated condition.
[d]Steels C70U to C120U are due to their chemical composition shallow hardening steels. For diameters of 30 mm, the hardness penetration depth will be approximately 3 mm. Through-hardening may only be achieved in diameters up to 10 mm.
[e]This material when carburized, quenched, and tempered should achieve a surface hardness of 60 HRC.
[f]This steel is normally supplied in the quenched and tempered condition with a hardness of approximately 300 HB.
[g]This steel may also be supplied in the pre-heat-treated condition with a hardness of approximately 300 HB.
[h]For greater dimensions, this steel is normally supplied in the quenched and tempered condition with a hardness of approximately 380 HB.
[i]This value applied for smaller dimensions only.
[j]Hardness in the annealed plus cold drawn condition may be 50 HB and hardness in the annealed plus cold rolled condition may be 70 HB higher than in the annealed condition.
[k]For the reference, hardening test either oil or salt bath. In cases of dispute, however, only oil. Usual quenching media in practice are air, gas, or salt bath.

All the steels within this group, except steel 105V, contain Cr and other alloying elements to improve their hardenability and mechanical properties. For a number of steels, a good depth of hardening by oil quenching can be obtained. For steels X100CrMoV5 and X153CrMoV12, due to their high carbon and high Cr contents, hardenability is high enough to permit martensite formation on air quenching, to obtain a hardness of 60 and 61 HRC, respectively. Other similar steels may also be air quenched e.g. steel 70MnMoCr8 with a high Mn content. The advantage of air quenching is to minimize distortion and promote good dimensional stability during heat treatment. Table 2 also shows the hardness values obtainable with various grades of alloy cold-work tool steel, ranging from 58 to 61 HRC with air quenching, and from 52 to 62 HRC with oil quenching.

3. Alloy Hot-Work Tool Steels

Nine grades of tool steels are classified under this group. Their carbon, silicon and manganese content varies from 0.25% to 0.60%, 0.10% to 1.20%, and 0.15% to 0.90%, respectively. The medium carbon content promotes toughness by limiting the carbon concentration of the martensite and by limiting the size of alloy carbide particles. The alloying elements added are Cr (0.80–5.5%), Mo (0–3.2%), Ni (1.50–1.8% for 55NiCrMoV7 only), V (0.05–2.1%), W (0–9.5%), and Co (4.00–4.5% for 38CrCoWV18-17-17 only). Chromium, molybdenum, vanadium, and tungsten are strong alloy carbide-forming elements, capable of producing a large volume fraction of alloy carbides, which results in stable microstructures when the steels are exposed to high-temperature forming operations. The hardened hardness values obtainable with the various grades of hot-work tool steel, with oil quenching, varies from 42 to 56 HRC, as can be seen from Table 2. However, the high-alloy content of the steels also provides excellent hardenability and permits hardening by air quenching.

4. High-Speed Tool Steels

Sixteen grades of high-speed tool steels are classified under this group. Their carbon, silicon, and manganese contents varies from 0.73% to 1.40%, $\leq$0.45% to $\leq$0.70%, and $\leq$0.40%, respectively. The alloying elements added are Cr (3.5–4.5%), Mo (0–10.00%), V (0.9–4.2%), W (0–18.70%), and Co (0–10.50%). Nickel is not added to these steels. Type HS18-0-1 does not contain Mo, but with the highest content of W (17.20–18.70%). Steels HS6-5-2-5, HS6-5-3-8, HS10-4-3-10, and HS2-9-1-8 contain Co to increase red hardness, and their hardened hardness values, with oil quenching, vary from 64 to 66 HRC. Type HS2-9-1-8 also has the highest Mo content of 9.00–10.00%.

High-speed steels are used for high-speed cutting tool applications. By virtue of their relatively high carbon and very high alloy contents, these steels have high hardenability and are processed to develop microstructures with large volume fractions of high-temperature stable carbides, which create excellent wear resistance and red hardness. Usual quenching media used for high-speed steels are air, gas, or salt bath. Table 2 shows the hardness values obtainable with the various grades of high-speed tool steel, from 60 to 66 HRC, with oil quenching.

C. AISI Classification of Tool Steels

AISI (American Iron and Steel Institute) has divided tool steels into 10 types, each assigned a prefix letter as shown in Table 3 [13].

The groupings are based on prominent characteristics such as (i) alloying element, (ii) application, and (iii) heat-treatment method. In case of (i), T and M are used for tungsten and molybdenum high-speed steels, respectively. In case of (ii), H is used for hot-work tool steels, L for low-alloy special-purpose tool steels, P for mold steels, and S for shock-resisting tool steels. In case of (iii), A is used for air-hardening tool steels, O for oil-hardening tool steels, and W for water-hardening tool steels.

To make comparisons easier, and in line with the classification method used in ISO 4957:1999, AISI tool steels are re-arranged into the following four major groups in this section, namely (i) carbon tool steels; (ii) alloy cold-work tool steels; (iii) hot-work tool steels, and (iv) high-speed tool steels. Figure 2 shows the classification of AISI tool steels into the four major groups [13,15].

Table 4 shows the chemical composition of principal types of AISI tool steels, which are also identified by designations in the Unified Numbering System (UNS) for Metals and Alloys, by the Society of Automotive Engineers (SAE), and the American Society for Testing and Materials (ASTM) [14,15].

Table 5 shows the major properties and hardness of AISI tool steels. The major properties considered are wear resistance, toughness, and hot hardness. The rating for

Table 3 Prefix Letters for AISI Tool Steel Groups [13]

Prefix letter	AISI tool steel group
A	Air-hardening medium-alloy, cold-work steels
D	High-carbon, high chromium cold-work steels
H	Hot-work tool steel
L	Low-alloy special-purpose steels
M	Molybdenum high-speed steels
O	Oil-hardening cold-work steels
P	Low-carbon mold steels
S	Shock-resisting steels
T	Tungsten high-speed steels
W	Water-hardening steels

major characteristics ranges from 1 (low rating) to 9 (high rating). The numbers are primarily used for comparative ranking purposes only, as will be illustrated in examples presented in Sec. III. They are not intended to indicate absolute quantitative differences. For example, several tool steels given the same rating were found to have different abrasion wear resistance [2]. The wear resistance increases with increasing carbon content, while toughness decreases with increasing carbon content and depth of hardening. A higher hardness and the presence of hard undissolved carbide particles will also improve the wear resistance of tool steel. The hardening temperature, quenching medium, tempering temperature, and approximate tempered hardness obtainable for each grade of steel are also listed. The quenching media used can be air (A), brine (B), oil (O), salt bath (S), or water (W). The annealed hardness of each type of tool steel is also listed in Table 5.

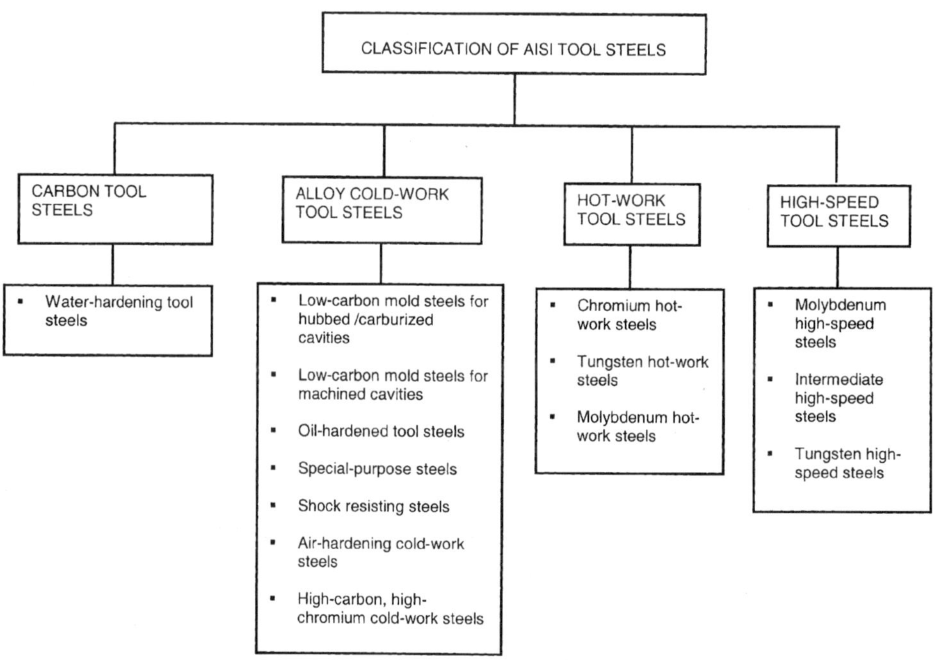

Figure 2 Classification of AISI tool steels.

Table 4 Chemical Composition of Principal Types of AISI Tool Steels [14,15]

Designation		Composition[a], %								
AISI	UNS	C	Si	Mn	Cr	Mo	Ni	V	W	Co
Group I—Non-alloy cold-work tool steel										
I (i) Water-hardening tool steels										
W1	T72301	0.70–1.50[b]	0.10–0.40	0.10–0.40	0.15 max	0.10 max	0.20 max	0.10 max	0.15 max	—
W2	T72302	0.85–1.50[b]	0.10–0.40	0.10–0.40	0.15 max	0.10 max	0.20 max	0.15–0.35	0.15 max	—
W5	T72305	1.05–1.15	0.10–0.40	0.10–0.40	0.40–0.60	0.10 max	0.20 max	0.10 max	0.15 max	—
Group II—Alloy cold-work tool steels										
II (i) Low-carbon mold steels for hubbed and/or carburized cavities										
P6	T51606	0.05–0.15	0.10–0.40	0.35–0.70	1.25–1.75	—	3.25–3.75	—	—	—
II (ii) Low-carbon mold steels for machined cavities										
P20	T51620	0.28–0.40	0.20–0.80	0.60–1.00	1.40–2.00	0.30–0.55	—	—	—	—
P21	T51621	0.18–0.22	0.20–0.40	0.20–0.40	0.50 max	—	3.90–4.25	0.15–0.25	—	1.05–1.25 Al
II (iii) Oil hardening tool steels										
O1	T31501	0.85–1.00	0.50 max	1.00–1.40	0.40–0.60	—	0.30 max	0.30 max	0.40–0.60	—
O2	T31502	0.85–0.95	0.50 max	1.40–1.80	0.50 max	0.30 max	0.30 max	0.30 max	—	—
O6	T31506	1.25–1.55[c]	0.55–1.50	0.30–1.10	0.30 max	0.20–0.30	0.30 max	—	—	—
O7	T31507	1.10–1.30	0.60 max	1.00 max	0.35–0.85	0.30 max	0.30 max	0.40 max	1.00–2.00	—
II (iv) Special-purpose tool steels										
L2	T61202	0.45–1.00[d]	0.50 max	0.10–0.90	0.70–1.20	0.25 max	—	0.10–0.30	—	—
L6	T61206	0.65–0.75	0.50 max	0.25–0.80	0.60–1.20	0.50 max	1.25–2.00	0.20–0.30[e]	—	—
II (v) Shock-resisting steels										
S1	T41901	0.40–0.55	0.15–1.20	0.10–0.40	1.00–1.80	0.50 max	0.30 max	0.15–0.30	1.50–3.00	—
S2	T41902	0.40–0.55	0.90–1.20	0.30–0.50	—	0.30–0.60	0.30 max	0.50 max	—	—
S4	T41904	0.50–0.65	1.75–2.25	0.60–0.95	0.10–0.50	—	—	0.15–0.35	—	—
S5	T41905	0.50–0.65	1.75–2.25	0.60–1.00	0.50 max	0.20–1.35	—	0.35 max	—	—
S6	T41906	0.40–0.55	2.00–2.50	1.20–1.50	1.20–1.50	0.30–0.50	—	0.20–0.40	—	—
S7	T41907	0.40–0.55	0.20–1.00	0.20–0.90	3.00–3.50	1.30–1.80	—	0.20–0.30[e]	—	—
II (vi) Air-hardening cold-work steels										
A2	T30102	0.95–1.05	0.50 max	1.00 max	4.75–5.50	0.90–1.40	0.30 max	0.15–0.50	—	—
A4	T30104	0.95–1.05	0.50 max	1.80–2.20	0.90–2.20	0.90–1.40	0.30 max	—	—	—

(Continued)

Table 4 (Continued)

Designation		Composition[a], %								
AISI	UNS	C	Si	Mn	Cr	Mo	Ni	V	W	Co
A6	T30106	0.65–0.75	0.50 max	1.80–2.50	0.90–1.20	0.90–1.40	0.30 max	—	—	—
A7	T30107	2.00–2.85	0.50 max	0.80 max	5.00–5.75	0.90–1.40	0.30 max	3.90–5.15	0.50–1.50	—
A8	T30108	0.50–0.60	0.75–1.10	0.50 max	4.75–5.50	1.15–1.65	0.30 max	—	1.00–1.50	—
A9	T30109	0.45–0.55	0.95–1.15	0.50 max	4.75–5.50	1.30–1.80	1.25–1.75	0.80–1.40	—	—
A10	T30110	1.25–1.50[c]	1.00–1.50	1.60–2.10	—	1.25–1.75	1.55–2.05	—	—	—
II (vii) High-carbon, high-chromium, cold-work steels										
D2	T30402	1.40–1.60	0.60 max	0.60 max	11.00–13.00	0.70–1.20	0.30 max	1.10 max	—	—
D3	T30403	2.00–2.35	0.60 max	0.60 max	11.00–13.50	—	0.30 max	1.00 max	1.00 max	—
D4	T30404	2.05–2.40	0.60 max	0.60 max	11.00–13.00	0.70–1.20	0.30 max	1.00 max	—	—
D5	T30405	1.40–1.60	0.60 max	0.60 max	11.00–13.00	0.70–1.20	0.30 max	1.00 max	—	2.50–3.50
D7	T30407	2.15–2.50	0.60 max	0.60 max	11.50–13.50	0.70–1.20	0.30 max	3.80–4.40	—	—
Group III—Hot-work tool steels										
III (i) Chromium hot-work steels										
H10	T20810	0.35–0.45	0.80–1.20	0.25–0.70	3.00–3.75	2.00–3.00	0.30 max	0.25–0.75	—	—
H11	T20811	0.33–0.43	0.80–1.20	0.20–0.50	4.75–5.50	1.10–1.60	0.30 max	0.30–0.60	—	—
H12	T20812	0.30–0.40	0.80–1.20	0.20–0.50	4.75–5.50	1.25–1.75	0.30 max	0.50 max	1.00–1.70	—
H13	T20813	0.32–0.45	0.80–1.20	0.20–0.50	4.75–5.50	1.10–1.75	0.30 max	0.80–1.20	—	—
H14	T20814	0.35–0.45	0.80–1.20	0.20–0.50	4.75–5.50	—	0.30 max	—	4.00–5.25	—
H19	T20819	0.32–0.45	0.20–0.50	0.20–0.50	4.00–4.75	0.30–0.55	0.30 max	1.75–2.20	3.75–4.50	4.00–4.50
III (ii) Tungsten hot-work steels										
H21	T20821	0.26–0.36	0.15–0.50	0.15–0.40	3.00–3.75	—	0.30 max	0.30–0.60	8.50–10.00	—
H22	T20822	0.30–0.40	0.15–0.40	0.15–0.40	1.75–3.75	—	0.30 max	0.25–0.50	10.00–11.75	—
H23	T20823	0.25–0.35	0.15–0.60	0.15–0.40	11.00–12.75	—	0.30 max	0.75–1.25	11.00–12.75	—
H24	T20824	0.42–0.53	0.15–0.40	0.15–0.40	2.50–3.50	—	0.30 max	0.40–0.60	14.00–16.00	—
H26	T20826	0.45–0.55[d]	0.15–0.40	0.15–0.40	3.75–4.50	—	0.30 max	0.75–1.25	17.25–19.00	—
III (iii) Molybdenum hot-work steels										
H42	T20842	0.55–0.70[d]	—	0.15–0.40	3.75–4.50	4.50–5.50	0.30 max	1.75–2.20	5.50–6.75	—

Group IV—High-speed tool steels										
IV (i) Molybdenum high-speed steels										
M1	T11301	0.78–0.88	0.20–0.50	0.15–0.40	3.50–4.00	8.20–9.20	0.30 max	1.00–1.25	1.40–2.10	—
M2	T11302	0.78–0.88; 0.95–1.05	0.20–0.45	0.15–0.40	3.75–4.50	4.50–5.50	0.30 max	1.75–2.20	5.50–6.75	—
M3-Class 1	T11313	1.00–1.10	0.20–0.45	0.15–0.40	3.75–4.50	4.75–6.50	0.30 max	2.25–2.75	5.00–6.75	—
M3-Class 2	T11323	1.15–1.25	0.20–0.45	0.15–0.40	3.75–4.50	4.75–6.50	0.30 max	2.75–3.75	5.00–6.75	—
M4	T11304	1.25–1.40	0.20–0.45	0.15–0.40	3.75–4.75	4.25–5.50	0.30 max	3.75–4.50	5.25–6.50	—
M6	T11306	0.75–0.85	0.20–0.45	0.15–0.40	3.75–4.50	4.50–5.50	0.30 max	1.30–1.70	3.75–4.75	11.00–13.00
M7	T11307	0.97–1.05	0.20–0.55	0.15–0.40	3.50–4.00	8.20–9.20	0.30 max	1.75–2.25	1.40–2.10	—
M10	T11310	0.84–0.94; 0.95–1.05	0.20–0.45	0.10–0.40	3.75–4.50	7.75–8.50	0.30 max	1.80–2.20	—	—
M33	T11333	0.85–0.92	0.15–0.50	0.15–0.40	3.50–4.00	9.00–10.00	0.30 max	1.00–1.35	1.30–2.10	7.75–8.75
M34	T11334	0.85–0.92	0.20–0.45	0.15–0.40	3.50–4.00	7.75–9.20	0.30 max	1.90–2.30	1.40–2.10	7.75–8.75
M36	T11336	0.80–0.90	0.20–0.45	0.15–0.40	3.75–4.50	4.50–5.50	0.30 max	1.75–2.25	5.50–6.50	7.75–8.75
M41	T11341	1.05–1.15	0.15–0.50	0.20–0.60	3.75–4.50	3.25–4.25	0.30 max	1.75–2.25	6.25–7.00	4.75–5.75
M42	T11342	1.05–1.15	0.15–0.65	0.15–0.40	3.50–4.25	9.00–10.00	0.30 max	0.95–1.35	1.15–1.85	7.75–8.75
M46	T11346	1.22–1.30	0.40–0.65	0.20–0.40	3.70–4.20	8.00–8.50	0.30 max	3.00–3.30	1.90–2.20	7.30–8.80
IV (ii) Intermediate high-speed steels										
M50	T11350	0.78–0.88	0.20–0.60	0.15–0.45	3.75–4.50	3.90–4.75	0.30 max	0.80–1.25	—	—
M52	T11352	0.85–0.95	0.20–0.60	0.15–0.45	3.50–4.30	4.00–4.90	0.30 max	1.65–2.25	0.75–1.50	—
IV (iii) Tungsten high-speed steels										
T1	T12001	0.65–0.80	0.20–0.40	0.10–0.40	3.75–4.50	—	0.30 max	0.90–1.30	17.25–18.75	—
T4	T12004	0.70–0.80	0.20–0.40	0.10–0.40	3.75–4.50	0.40–1.00	0.30 max	0.80–1.20	17.50–19.00	4.25–5.75
T5	T12005	0.75–0.85	0.20–0.40	0.20–0.40	3.75–5.00	0.50–1.25	0.30 max	1.80–2.40	17.50–19.00	7.00–9.50
T6	T12006	0.75–0.85	0.20–0.40	0.20–0.40	4.00–4.75	0.40–1.00	0.30 max	1.50–2.10	18.50–21.00	11.00–13.00
T8	T12008	0.75–0.85	0.20–0.40	0.20–0.40	3.75–4.50	0.40–1.00	0.30 max	1.80–2.40	13.25–14.75	4.25–5.75
T15	T12015	1.50–1.60	0.15–0.40	0.15–0.40	3.75–5.00	1.00 max	0.30 max	4.50–5.25	11.75–13.00	4.25–5.25

[a]All steels except group W contain 0.25 max Cu, 0.03 max P, 0.03 max S; group W steels contain 0.20 max Cu, 0.025 max P, 0.025 max S. Where specified, sulfur may be increased to 0.06–0.15 to improve machinability of groups A, D, H, M, and T steels.

[b]Specified carbon ranges are designated by suffix numbers.

[c]Contains free graphite in the microstructure.

[d]Available in several carbon ranges.

[e]Optional.

Table 5 Major Properties and Hardness of AISI Tool Steels [14,15]

	Major properties[a]			Hardening method and hardness				
AISI designation	Wear resistance[b]	Toughness [c]	Hot hardness	Hardening temperature, °C	Quenching medium[d]	Tempering temperature, °C	Approximate tempered hardness, HRC	Annealed hardness, BHN
Group I—Non-alloy cold-work tool steel								
I (i) Water-hardening tool steels								
W1	2–4	3–7	1	760–843	B or W	177–343	64–50	202
W2	2–4	3–7	1	760–843	B or W	177–343	64–50	202
W5	3–4	3–7	1	760–829	B or W	177–343	64–50	202
Group II—Alloy cold-work tool steels								
II (i) Low-carbon mold steels for hubbed and/or carburized cavities								
P6	1[e]	9	3[e]	788–816	A or O	177–232	61–58[e]	212
II (ii) Low-carbon mold steels for machined cavities								
P20	1[e]	8	2[e]	816–871	O	482–593	37–28	—
P21	1	8	4	704–732 (Solution HT)	A or O	510–552 (Aging)	40–30	—
II (iii) Oil-hardening tool steels								
O1	4	3	3	788–816	O	177–260	62–57	212
O2	4	3	3	760–802	O	177–266	62–57	217
O6	3	3	2	788–816	O	177–316	63–58	229
O7	5	3	3	788–829	W	177–288	64–58	241
				816–885	O			
II (iv) Special-purpose steels								
L2	1	7	2	788–843	W	177–538	63–45	197
				843–927	O			
L6	3	6	2	788–843	O	177–538	62–45	235
II (v) Shock-resisting steels								
S1	4	8	5	899–954	O	204–649	58–40	229
S2	2	8	2	843–899	B or W	177–427	60–50	217

S4	2	8	2	843–899 871–927	W or B O or W	177–427	60–50	229
S5	2	8	3	871–927	O	177–427	60–50	229
S6	2	8	3	913–954	O	204–316	56–54	229
S7	3	8	5	927–954	A or O	204–621	57–45	229
II (vi) Air-hardening cold-work steels								
A2	6	4	5	927–982	A	177–538	62–57	248
A4	5	4	4	816–871	A	177–427	62–54	241
A6	4	5	4	829–871	A	149–427	60–54	248
A7	9	1	6	954–982	A	149–538	67–57	269
A8	4	8	6	982–1010	A	177–593	60–50	241
A9	4	8	6	982–1024	A	510–621	56–35	248
A10	3	3	3	788–816	A	177–427	62–55	269
II (vii) High-carbon, high-chromium, cold-work steels								
D2	8	2	6	982–1024	A	204–538	61–54	255
D3	8	1	6	927–982	O	204–538	61–54	255
D4	8	1	6	968–1010	A	204–538	61–54	255
D5	8	2	7	982–1024	A	204–538	61–54	255
D7	9	1	6	1010–1066	A	149–538	65–58	262
Group III—Hot-work tool steels								
III (i) Chromium hot-work steels								
H10	3	9	6	1010–1038	A	538–649	56–39	229
H11	3	9	6	996–1024	A	538–649	54–38	235
H12	3	9	6	996–1024	A	538–649	55–38	235
H13	3	9	6	996–1038	A	538–649	53–38	235
H14	4	6	7	1010–1066	A	539–649	47–40	235
H19	5	6	7	1093–1204	A	538–704	57–40	241
III (ii) Tungsten hot-work steels								
H21	4	6	8	1093–1204	A or O	593–677	54–36	235
H22	5	5	8	1093–1204	A or O	593–677	52–39	235

(Continued)

Table 5 (Continued)

AISI Designation	Major Properties[a]			Hardening Method and Hardness				Annealed Hardness, BHN
	Wear Resistance[b]	Toughness[c]	Hot Hard-ness	Hardening Temperature, °C	Quenching Medium[d]	Tempering temperature, °C	Approximate Tempered Hardness, HRC	
H23	5	5	8	1204–1260	O	649–732	47–34	255
H24	5	5	8	1093–1232	O	566–649	55–45	241
H26	6	4	8	1177–1260	S, O, or A	566–677	58–43	241
III (iii) Molybdenum hot-work steels								
H42	6	4	7	1121–1218	O, A, or S	566–649	60–50	235
Group IV—High-speed tool steels								
IV (i) Molybdenum high-speed steels								
M1	7	3	8	1177–1218	O, A, or S	538–593	65–60	248
M2	7	3	8	1191–1232	O, A, or S	538–593	65–60	248/255
M3, Class 1	8	3	8	1204–1232	O, A, or S	538–593	66–61	255
M3, Class 2	8	3	8	1204–1232	O, A, or S	538–593	66–61	255
M4	9	3	8	1204–1232	O, A, or S	538–593	66–61	255
M6	8	3	9	1177–1204	O, A, or S	538–593	66–61	277
M7	8	3	8	1177–1218	O, A, or S	538–593	66–61	255
M10	7	3	8	1177–1218	O, A, or S	538–593	65–60	248/255
M33	8	1	9	1204–1232	O, A, or S	538–593	65–60	269
M34	8	1	9	1204–1232	O, A, or S	538–593	65–60	269

M36	7	1	9	1218–1246	O, A, or S	538–593	65–60	269
M41	8	1	9	1191–1216	O, A, or S	538–593	70–65	269
M42	8	1	9	1163–1191	O, A, or S	510–593	70–65	269
M46	8	1	9	1191–1218	O, A, or S	524–566	69–67	269
IV (ii) Intermediate high-speed steels								
M50	6	3	6	1039–1121	O, A, or S	524–593	63–58	248
M52	6	3	6	1121–1177	O, A, or S	524–593	64–58	248
IV (iii) Tungsten high-speed steels								
T1	7	3	8	1260–1302	O, A, or S	538–593	65–60	255
T4	7	2	8	1260–1302	O, A, or S	588–593	66–62	269
T5	7	1	9	1273–1302	O, A, or S	538–593	65–60	285
T6	8	1	9	1273–1302	O, A, or S	588–593	65–60	302
T8	8	2	8	1260–1302	O, A, or S	538–593	65–60	255
T15	9	1	9	1204–1260	O, A, or S	538–649	68–63	277

[a]Rating range from 1 (low) to 9 (high).
[b]Wear resistance increases with increasing carbon content.
[c]Toughness decreases with increasing carbon content and depth of hardening.
[d]A, air cool; B, brine quenched; O, oil quenched; S, salt-bath quench; W, water quench.
[e]After carburizing.

Table 6 Processing and Service Characteristics of AISI Tool Steels [13]

		Hardening and tempering			Fabrication and service				
AISI designation	Resistance to decarburization	Depth of hardening	Amount of distortion[a]	Safety in hardening	Machinability	Grindability	Toughness	Resistance to softening	Resistance to wear
Group I—Non-alloy cold-work tool steel									
I (i) Water-hardening steels									
W1	Highest	Shallow	High	Low to medium	Highest	Highest	High[b]	Low	Low to medium
W2	Highest	Shallow	High	Low to medium	Highest	Highest	High[b]	Low	Low to medium
W5	Highest	Shallow	High	Low to medium	Highest	Very high to highest	High[b]	Low	Low to medium
Group II—Alloy cold-work tool steels									
II (i) Low-carbon mold steels for hubbed and/or carburized cavities									
P6	High	—	A, very low; O, low	High	Medium	Medium	High	Low	Medium
II (ii) Low-carbon mold steels for machined cavities									
P20	High	Low to medium	Low	High	Medium to high	Medium	High	Low	Low to medium
P21	High	Deep	Lowest	Highest	Medium	Medium	Medium	Medium	Medium
II (iii) Oil-hardening tool steels									
O1	High	Medium	Low	Medium to high	High	High	Medium	Low	Medium
O2	High	Medium	Low	Medium to high	High	High	Medium	Low	Medium
O6	High	Medium	Low	Medium to high	Highest	High	Medium	Low	Medium
O7	High	Medium	O, low; W, high	W, low; O, high	High	Medium to high	Medium	Low	Medium
II (iv) Special-purpose steels									
L2	High	O, medium; W, high	O, medium; W, low	O, medium; W, low	High	High	Very high	Low	Low to medium
L6	High	Medium	Low	Medium	Medium	High	Very high	Low	Medium

II (v) Shock-resisting steels									
S1	Low	Medium	Medium	High	Medium	Medium to high	Very high	Medium	Low to medium
S2	Low	Medium	High	Low	Medium to high	Medium to high	Highest	Low	Low to medium
S4	Low	Medium	W, high; O, medium	W, low; O, high	Medium to high	Medium to high	Highest	Low	Low to medium
S5	Low	Medium	Medium	High	Medium to high	Medium to high	Highest	Low	Low to medium
S6	Low	Medium	Medium	High	Medium	Medium to high	Very high	Low	Low to medium
S7	Medium	Deep	A, lowest; O, low	O, high; A, highest	Medium to high	Medium to high	Very high	High	Low to medium
II (vi) Air-hardening cold-work steels									
A2	Medium	Deep	Lowest	Highest	Medium	Medium	Medium	Medium to high	High
A4	Medium to high	Deep	Lowest	Highest	Low to medium	Medium	Medium	Medium	Medium to high
A6	Medium to high	Deep	Lowest	Highest	Low to medium	Medium	Medium	Medium	Medium to high
A7	Medium	Deep	Lowest	Highest	Lowest	Lowest	Lowest	High	Highest
A8	Medium	Deep	Lowest	Highest	Medium	Medium	High	High	Medium to high
A9	Medium	Deep	Lowest	Highest	Medium	Medium	High	High	Medium to high
A10	Medium to high	Deep	Lowest	Highest	Medium to high	Low to medium	Medium	Medium	High
II (vii) High-carbon, high-chromium, cold-work steels									
D2	Medium	Deep	Lowest	Highest	Low	Low	Low	Medium to high	High to very high
D3	Low to medium	Deep	Low	Medium to high	Very low	Very low	Very low	Medium to high	Very high
D4	Low to medium	Deep	Lowest	Very high	Very low	Very low	Very low	Medium to high	Very high
D5	Medium	Deep	Lowest	Highest	Low	Low	Low	High	High to very high
D7	Low to medium	Deep	Lowest	Highest	Lowest	Lowest	Lowest	High	Highest

(Continued)

Table 6 (Continued)

AISI designation	Resistance to decarburization	Hardening and tempering: Depth of hardening	Amount of distortion[a]	Safety in hardening	Fabrication and service: Machinability	Grindability	Toughness	Resistance to softening	Resistance to wear
Group III—Hot-work tool steels									
III (i) Chromium hot-work steels									
H10	Medium to high	Deep	Very low	Highest	Medium to high	Medium to high	High	High	Medium
H11	Medium to high	Deep	Very low	Highest	Medium to high	Medium to high	Very high	High	Medium
H12	Medium to high	Deep	Very low	Highest	Medium to high	Medium to high	Very high	High	Medium
H13	Medium to high	Deep	Very low	Highest	Medium to high	Medium to high	Very high	High	Medium
H14	Medium to high	Deep	Very low	Highest	Medium	Medium to high	High	High	Medium
H19	Medium	Deep	A, low; O, medium	High	Medium	Medium	High	High to very high	Medium to high
III (ii) Tungsten hot-work steels									
H21	Medium	Deep	A, low; O, medium	High	Medium	Medium	High	High to very high	Medium to high
H22	Medium	Deep	A, low; O, medium	High	Medium	Medium	High	High to very high	Medium to high
H23	Medium	Deep	Medium	High	Medium	Medium	Medium	Very high	Medium to high
H24	Medium	Deep	A, low; O, medium	High	Medium	Low to medium	Medium	Very high	High
H26	Medium	Deep	A, or S, low; O,medium	High	Medium	Low to medium	Medium	Very high	High
III (iii) Molybdenum hot-work steels									
H42	Medium	Deep	A or S, low; O, medium	Medium	Medium	Low to medium	Medium	Very high	High

Group IV—High-speed tool steels									
IV (i) Molybdenum high-speed steels									
M1	Low	Deep	A or S, low; O, medium	Medium	Medium	Low to medium	Low	Very high	High to very high
M2	Medium	Deep	A or S, low; O, medium	Medium	Medium	Low	Low	Very high	Very high
M3(Class 1 and Class 2)	Medium	Deep	A or S, low; O, medium	Medium	Medium	Very low	Low	Very high	Highest
M4	Medium	Deep	A or S, low; O, medium	Medium	Low to Medium	Lowest	Low	Very high	Highest
M6	Low	Deep	A or S, low; O, medium	Medium	Medium	Low	Low	Highest	Very high
M7	Low	Deep	A or S, low; O, medium	Medium	Medium	Low	Low	Very high	Very high
M10	Low	Deep	A or S, low; O, medium	Medium	Medium	Low	Low	Very high	Very high
M33	Low	Deep	A or S, low; O, medium	Medium	Medium	Low	Low	Highest	Very high
M34	Low	Deep	A or S, low; O, medium	Low to medium	Medium	Low	Low	Highest	Very high
M36	Low	Deep	A or S, low; O, medium	Low to medium	Medium	Low	Low	Highest	Very high
M41	Low	Deep	A or S, low; O, medium	Low to medium	Medium	Very low	Low	Highest	Highest
M42	Low	Deep	A or S, low; O, medium	Low to medium	Medium	Low	Low	Highest	Very high to highest
M46	Low	Deep	A or S, low; O, medium	Medium	Medium	Very low	Low	Highest	Very high
IV (ii) Intermediate high-speed steels									
M50	Low	Deep	A or S, low; O, medium	Medium	Medium	Medium	Low	High	High
M52	Low	Deep	A or S, low; O, medium	Medium	Medium	Low	Low	High	High
IV (iii) Tungsten high-speed steels									
T1	High	Deep	A or S, low; O, medium	High	Medium	Low to medium	Low	Very high	Very high
T4	Medium	Deep	A or S, low; O, medium	Medium	Medium	Low	Low	Highest	Very high
T5	Low	Deep	A or S, low; O, medium	Medium	Medium	Low	Low	Highest	Very high
T6	Low	Deep	A or S, low; O, medium	Medium	Low to medium	Low	Low	Highest	Very high
T8	Medium	Deep	A or S, low; O, medium	Medium	Medium	Low	Low	Highest	Very high
T15	Low	Deep	A or S, low; O, medium	Low to medium	Lowest	Lowest	Lowest	Highest	Highest

[a]A, air cool; B, brine quench; O, oil quench; S, salt bah quench; W, water quench.
[b]Toughness decreases with increasing carbon content and depth of hardening.

Table 6 shows the processing and service characteristics of AISI tool steels. The hardening and tempering factors considered are resistance to decarburization, depth of hardening, amount of distortion and safety in hardening. The fabrication and service factors considered are machinability, grindability, toughness, resistance to softening, and resistance to wear [13]. The qualitative ratings used for resistance to decarburization, safety in hardening, machinability, grindability, toughness, resistance to softening, and resistance to wear are: highest, very high, high, medium, low, very low, and lowest. The depth of hardening is rated as shallow, medium or deep; and the amount of distortion in heat treatment is rated as: lowest, very low, low, medium, or high. The amount of distortion and the safety in hardening depend on the quenching medium used, whether it is air (A), brine (B), oil (O), salt bath (S), or water (W).

1. Carbon Tool Steel

This group contains only one type of carbon tool steel, that is, the water-hardening tool steels. The following steels are excluded as they are no longer in common use: W3, W4, W6, and W7 [13].

a. Water-hardening Tool Steels (W1–W5). This group contains three grades (W1–W5). Their carbon, silicon, and manganese content varies from 0.70% to 1.50%, 0.10% to 0.40%, and 0.10% to 0.40%, respectively. The alloying elements added are Cr (0.15–0.60%), Mo (0.10% max.), V (0.10–0.35%), and W (0.15% max.). They are essentially carbon steels, and are among the least expensive of tool steels. Due to the low-alloy contents, they are very shallow hardening steels that must be water quenched to attain the necessary hardness, ranging from 50 to 64 HRC, while the toughness is high, the resistance to softening is low, and the wear resistance is medium, as indicated in Table 6. Typical applications include: axes, blanking dies, cold heading dies, cold striking dies, countersinks, drills, files, forming dies, jewelers' dies, reamers, taps, and woodworking tools. Wrought plain-carbon steels are also used in the fabrication of structural components for fixtures and auxiliary die components, to be described in Sec. III.C.

2. Alloy Cold-Work Tool Steel

This major group contains: (i) one grade of low-carbon mold steels for hubbed and/or carburized cavities (P6); (ii) two grades of low-carbon mold steels for machined cavities (P20 and P21); (iii) four grades of oil-hardening tool steels (O1–O7); (iv) two grades of (low alloy) special-purpose tool steels (L2 and L6); (v) six grades of shock-resisting steels (S1–S7); (vi) seven grades of air-hardening cold-work steels (A2–A10); and (vii) five grades of high-carbon, high-chromium, cold-work steels (D2–D7). The following steels are excluded as they are no longer in common use: P1, P2, P3, P4, P5, L1, L3, L4, L5, L7, F1, F2, F3, S3, D1, D6, A3, and A5 [13].

a. Low-Carbon Mold Steels for Hubbed and/or Carburized Cavities (P6). The carbon, silicon, and manganese content varies from 0.05% to 0.15%, 0.10% to 0.40%, and 0.35% to 0.70%, respectively. The alloying elements added are Cr (1.25–1.75%) and Ni (3.25–3.75%). The low-carbon content allows the mold or die steel cavities to be shaped by hubbing (hobbing) while the steel is in a soft annealed condition. The molds are then carburized and hardened to obtain high surface hardness, and range from 58 to 61 HRC. The toughness of these steels is high, resistance to softening is low and wear resistance is medium. They are used for injection and compression molds for plastics.

b. Low-Carbon Mold Steels for Machined Cavities (P20 and P21). Their carbon, silicon, and manganese content varies from 0.18% to 0.40%, 0.20% to 0.80%, and

0.20% to 1.00%, respectively. The alloying elements added are Cr (0.5% to 2.00%), Mo (0.30–0.55% for P20 only), Ni (3.90–4.25% for P21 only), V (0.15–0.25% for P21 only), and Al (1.05–1.25% for P21 only). P20 and P21 have slightly higher carbon contents than P6 mold steel. They are usually supplied in the prehardened condition so that the cavity can be machined and the mold placed directly in service. The hardness of these steels, after tempering in temperature range normally recommended for them, ranges from 28 to 40 HRC, while the toughness ranges from medium to high, resistance to softening from low to medium, and wear resistance from low to medium. They may be used for plastic molds, zinc die-casting dies, and holder blocks.

c. Oil-Hardening Tool Steels (O1–O7). Their carbon, silicon, and manganese content varies from 0.85% to 1.55%, 0.50% to 1.50% and 0.30% to 1.80%, respectively. The alloying elements added are Cr (0.35–0.85%), Mo (0–0.30%), V (0–0.40%), and W (0–2.00%). Due to the high carbon and moderate alloy contents, high hardness and good depth of hardening can be obtained with oil hardening to provide the high wear resistance required under cold-working conditions. Grade O7 can promote graphite formation to enhance machinability and die life. The hardened hardness of these steels ranges from 57 to 64 HRC, while the toughness is medium, the resistance to softening is low, and the wear resistance is medium. Their safety in hardening is ranked medium to high with oil hardening. Typical applications include: blanking dies, cams, coining dies, drawing dies, forming dies, gauges, machine ways, plastic molds, shear blades, and trim dies.

d. Special-Purpose Tool Steels (Low Alloy) (L2 and L6). Their carbon, silicon, and manganese content varies from 0.45% to 1.00%, 0.50% max and 0.10% to 0.90%, respectively. The alloying elements added are Cr (0.60–1.20%), Mo (0.25 max. to 0.50% max.), Ni (0–2.00%), and V (0.10–0.30%). Moderate hardenability is obtainable with these steels which can also be hardened by oil quenching. The medium-carbon grades will also give a higher toughness as compared with that from the high-carbon oil-hardening tool steels. Hence, these steels can be used in a variety of applications with characteristics overlapping those of water- and oil-hardening grades. The hardness of these steels ranges from 45 to 63 HRC, while the toughness is very high, the resistance to softening is low, and the wear resistance is low to medium. Typical applications for the medium-carbon types include: arbors, brake dies, cams, chucks, collets, drift pins, jigs, machine tool parts, pinions, punches, rivet sets, rolls, spindles, and wrenches. Typical applications for the high-carbon types include: arbors, dies, drills, gauges, knife edges, knurls, rolls, and taps.

e. Shock-Resisting Steels (S1–S7). Their carbon, silicon, and manganese content varies from 0.40% to 0.65%, 0.15% to 2.50% and 0.10% to 1.50%, respectively. The alloying elements added are Cr (0–3.50%), Mo (0–1.80%), V (0.15–0.50%), and W (1.50–3.00% for S1 only). With a medium-carbon content, these steels will have low-carbon martensite and fine carbide dispersions after hardening to give high toughness and fracture resistance in combination with high strength and wear resistance under conditions of impact loading. With a higher alloy content than that of the water-hardening tool steels, the hardenability is also higher. Depth of hardening of these steels varies from medium to deep. The hardness of these steels ranges from 40 to 60 HRC, while the toughness ranges from very high to highest, the resistance to softening from low to high, and the wear resistance from low to medium. Types S1 and S7 are also used for moderately elevated-temperature service. Typical applications include: chipping chisels, circular pipe cutters, drift pins, grippers, hand chisels, knock out pins, mandrels, punches, rivet sets, screw driver blades, shear blades, stamps, and track tools.

f. Air-Hardening Cold-Work Steels (A2–A10). Their carbon, silicon, and manganese content varies from 0.45% to 2.85%, 0.50% to 1.50% and 0.50 max. to 2.50%, respectively.

The alloying elements added are Cr (0–5.75%), Mo (0.90–1.80%), Ni (0.30 max.–2.05%), V (0–5.15%), and W (0–1.50%). Due to their high alloy contents, these steels can be hardened with air quenching. However, they have various combinations of hardness and toughness due to ranges of carbon and alloy content used for the various grades. The low-carbon grades A8 and A9 offer greater shock resistance than the other steels in this group, but are lower in their resistance to wear. Grade A7 exhibits maximum abrasion resistance, but should be restricted to applications where toughness is not a prime consideration. Grades A4, A6, and A10 are characterized by relatively low hardening temperatures. Grade A10 also contains graphite in its microstructure to improve machinability and die life. Depth of hardening of these steels is deep. The hardened hardness and the annealed hardness of these steels range from 35 to 67 HRC and 241 to 269 BHN, respectively. Typical applications include: blanking dies, coining dies, cold forming tools, forming dies, knurls, mandrels, master hubs, molds, rolls, shear knives, slitter knives, and spindles.

g. High-Carbon, High-Chromium, Cold-Work Steels (*D*2–*D*7). Their carbon, silicon, and manganese content varies from 1.40% to 2.50%, 0.60% max. and 0.60% max., respectively. The alloying elements added are Cr (11.00–13.50%), Mo (0–1.20%), V (1.00–4.40%), W (1.00% max. for D3 only), and Co (2.50% to 3.50% for D5 only). The grades containing molybdenum are air hardening. With high-carbon martensite and large volume fractions of alloy carbides formed after hardening, these steels have extremely high wear and abrasion resistance. However, they are very difficult to machine and grind during manufacture, with a lowest to low machinability and grindability. Depth of hardening of these steels is deep. The hardened hardness and annealed hardness of these steels range from 54 to 65 HRC and 255 to 262 BHN, respectively, while the toughness is from lowest to low, the resistance to softening is from medium to high, and the wear resistance from high to the highest. Typical applications include: blanking dies, burnishing tools, cold forming dies, drawing dies, gauges, knurls, lamination dies, punches, rolls, shear knives, slitter knives, and thread rolling dies.

3. Hot-Work Tool Steel

This major group contains: (i) six grades of chromium hot-work steels (H10–H19); (ii) five grades of tungsten hot-work steels (H21–H26) and (iii) one grade of molybdenum hot-work steel (H42). The following steels are excluded as they are no longer in common use: H15, H16, H20, H25, H41, and H43 [13].

Chromium Hot-Work Steels (*H*10–*H*19). Their carbon, silicon, and manganese content varies from 0.30% to 0.45%, 0.20% to 1.20% and 0.20% to 0.70%, respectively. The alloying elements added are Cr (3.00–5.50%), Mo (0–3.00%), V (0–2.20%), W (0–5.25%), and Co (4.00–4.50% for H19 only). With medium-carbon contents and relatively high concentrations of chromium and other strong carbide-forming elements, these steels have excellent resistance to high-temperature impact loading, to softening during high-temperature exposure, and to thermal fatigue. The molybdenum-containing grades H10, H11, H12, and H13 are the most widely used of all the hot-work steels, and are characterized by high hardenability to allow air hardening, and excellent toughness. They offer the best availability. The chromium–tungsten grade H14 and the chromium–tungsten–cobalt type H19 offer greater resistance to softening, but are less ductile in service, and are not as readily available from stock. Depth of hardening of these steels is deep. The hardness of these steels ranges from 38 to 57 HRC. Typical applications include: bolsters, die-casting dies, dummy blocks, extrusion dies, forging dies, gripper dies, hot header dies, mandrels, punches, and piercing tools.

b. Tungsten Hot-Work Steels (*H*21–*H*26). Their carbon, silicon, and manganese content varies from 0.25% to 0.55%, 0.15% to 0.60% and 0.15% to 0.40%, respectively. The alloying elements added are Cr (1.75% to 12.75%), V (0.30% to 1.25%), and W (8.50% to 19.00%). However, Mo and Co are not used with these steels. The addition of substantial amounts of tungsten which, in addition to the chromium, produce large volume fractions of stable alloy carbides in the microstructure to increase significantly the resistance to softening during high-temperature exposure of tungsten hot-work tool steels, as compared with that of the chromium hot-work steels. However, the toughness of tungsten hot-work steels is lower. Depth of hardening of these steels is deep. The hardness of these steels ranges from 34 to 58 HRC, while the toughness ranges from medium to high, the resistance to softening from high to very high, and the wear resistance from medium to high. Hence these steels are intended for those hot-work applications where resistance to the softening effect of elevated temperature is of greatest importance and a lesser degree of toughness can be tolerated. Typical applications include: hot blanking dies, dummy blocks, extrusion dies, gripper dies, mandrels, piercer points, punches, hot shear blades, and hot trim dies.

Molybdenum Hot-Work Steel (*H*42). H42 contains 0.55–0.70% C, 0.15–0.40% Mn, 3.75–4.50% Cr, 4.50–5.50% Mo, 0.30% (max) Ni, 1.75–2.20% V and 5.50–6.75% W. This grade is modified from low-carbon molybdenum high-speed steels. It has excellent resistance to the softening effect of elevated temperature, comparable to that of the tungsten hot-work steels. However, it is not readily available from stock.

4. High-Speed Tool Steels

This major group contains: (i) 14 grades of molybdenum high-speed steels (M1–M46); (ii) two grades of intermediate high-speed steels (M50 and M52); and (iii) six grades of tungsten high-speed steels (T1–T15). All high-speed steels are rated low or lowest in toughness in Table 6, relative to other types of tool steels. However, small differences exist in this property among the high-speed steel grades; and these differences may be important when selecting a particular high-speed steel for a cutting application, as indicated in Table 5. The following steels are excluded as they are no longer in common use: M8, M15, M30, M35, M43, M44, M45, M47, T2, T3, T7, and T9 [13].

a. Molybdenum High-Speed Steels (*M*1–*M*46). Their carbon, silicon, and manganese content varies from 0.75% to 1.30%, 0.15% to 0.65% and 0.10% to 0.40%, respectively. The alloying elements added are Cr (3.50–4.75%), Mo (3.25–10.00%), V (0.95–4.50%), W (0–7.00%), and Co (0–13.00%). Molybdenum high-speed steels, with high-carbon and high alloy contents, have high hardenability and high wear resistance. Grades M2 and M10 are available in both the regular carbon and high-carbon versions. Those grades with higher carbon and vanadium contents generally offer improved abrasion resistance, but machinability and grindability may be adversely affected, ranking low to medium and low to lowest, respectively. The cobalt-bearing M30 series and M40 series are premium grades, with the latter capable of being heat treated to a high hardness of 70 HRC. With the presence of stable alloy carbides and the addition of cobalt to increase red hardness, the resistance to high-temperature softening of these steels is also high. As compared with the tungsten high-speed steels, molybdenum high-speed steels have slightly higher toughness and better cost advantage. However, the M steels more readily decarburize and require more care in hardening. For most of the grades, the rating for susceptibility to decarburization is low. Depth of hardening of these steels is deep. The hardened hardness and annealed hardness of these steels range from 60 to 70 HRC and 248 to 269 BHN, respectively. The toughness of these steels is low, the resistance to softening and the wear resistance are from high to highest. Typical applications include: broaches, chasers, drills, end mills, hubs, lathe tools, milling cutters, planer tools, punches, reamers, routers, saws, taps, and woodworking tools. In addition to the above cutting tool applications, some of these high-speed steels are used for cold-work applications such as cold header die inserts, thread rolling dies, punches, and blanking dies. For such applications in order to increase toughness, the high-speed steels frequently are hardened from a lower temperature than that used for cutting tools.

b. Intermediate High-Speed Steels (M50 and M52). Their carbon, silicon, and manganese content varies from 0.78% to 0.95%, 0.20% to 0.60% and 0.15% to 0.45%, respectively. The alloying elements added are Cr (3.50–4.50%), Mo (3.90–4.90%), V (0.80–2.25%), and W (0.75–1.50% for M52 only). These steels contain no cobalt. With a lower Mo and W content, as compared with those with molybdenum high-speed steels, the hardness of these steels is lower, ranges from 58 to 64 HRC. These grades normally are limited to less severe service conditions. Typical applications include: bearings, chasers, drills, hydraulic pump components, punches, router bits, taps, and woodworking tools.

c. Tungsten High-Speed Steels (T1–T15). Their carbon, silicon, and manganese content varies from 0.65% to 1.60%, 0.15% to 0.40% and 0.10% to 0.40%, respectively. The alloying elements added are Cr (3.75–5.00%), Mo (0–1.25%), V (0.80–5.25%), W (11.75–21.00%) and Co (0–13.00%). Grade T1 does not contain molybdenum and cobalt. Cobalt-bearing grades range from T4 through T15 and contain various amounts of cobalt. Tungsten is a very strong carbide-forming element. The addition of substantial amounts of tungsten, which in addition to the chromium, vanadium, and molybdenum, again produces large volume fractions of high-temperature stable alloy carbides in the microstructure to increase significantly the wear resistance and red hardness of tungsten high-speed steels. Cobalt has also the ability to increase red hardness. Except for the slightly lower toughness, the performance characteristics of tungsten high-speed steels are similar to those of molybdenum high-speed steels. They are more resistant to decarburization in heat treatment than the molybdenum types, and usually are hardened from higher temperatures. The hardness of these steels ranges from 60 to 68 HRC.

III. SELECTION AND APPLICATIONS OF TOOL STEELS

A. Introduction

As mentioned earlier, essential requirements for any tool, die, or mold, with respect to optimum performance and economy, include: (i) proper design; (ii) proper material; (iii) accurate manufacture; (iv) correct heat treatment; and (v) proper setup, use, and maintenance. As indicated in Sec. II, there are many types and grades of tool steels that can be used, and many tools, dies, and molds also contain several materials. Furthermore, in addition to both ISO 4957:1999 and AISI specifications, tool steels are also manufactured to meet other international or industrial standards. Hence, the selection of a proper tool, die, or mold material for a specific application needs to be considered carefully. The following factors should be evaluated [16]:

- The operations to be performed, including their severity, forces applied, temperature encountered, and lubricants used.
- The workpiece material, including its hardness, thickness, and size.

- The production rate and quantity, accuracy, and finish requirements.
- The machine to be used, including its type and condition.
- The design of the tool, die, or mold.
- The accuracy and rigidity of the setup.
- The cost per part produced, based upon the material, manufacturing, heat treatment, and maintenance costs, as well as the life of the tool, die, or mold.
- The current availability of the tool, die, or mold material.
- The major characteristics of the tool material and its fabrication or heat-treatment properties.

Table 7 shows the general grouping of common tool steel applications and their requirements. The tool steel applications can be divided into five groups: (i) molds/dies; (ii) tools/dies for cold-working; (iii) tools/dies for hot-working; (iv) cutting tools for machining; and (v) structural component parts. As indicated in Table 5, wear resistance, toughness, and hot hardness are the major characteristics (factors) for a tool material. As mentioned previously, wear resistance increases as the carbon and alloy contents of a tool steel increase. A higher hardness and the presence of hard undissolved carbide particles will also improve its wear resistance. For hot-work steels, wear resistance includes resistance to erosion at elevated temperatures. In tool steels, the concept of toughness is best expressed as the ability to resist chipping and breaking. High-carbon steels generally have lower toughness than low-carbon steels, as indicated by the high toughness ratings given to types S and H steels. However, carbon tool steels and other shallow hardening steels, which harden with a case and core, may be rated high in toughness. The hardness of the tool itself also has a considerable effect on the toughness of the steel.

Hot hardness or red hardness, is the resistance to the softening effect of elevated temperature, that is, the resistance of the tool to tempering. It is an important selective factor especially for tools used for hot working. The resistance to softening increases as the molybdenum, tungsten, vanadium, cobalt, and chromium content increases. Hence, carbon tool steels are lowest in this respect and high-speed steels are highest. However, an additional major property required for tool steels in hot-work applications, e.g., die-casting dies or forming punches, is resistance to heat checking or thermal fatigue cracking. Heat checking is characterized by a network of fine cracks that appear on the working surfaces of tooling as a result of stresses associated with alternate rapid heating and cooling during service.

Polishability, grindability, safety, and low distortion in heat treatment are some of the factors to be considered in fabrication/heat-treatment requirements. Safety in heat treatment is related to the resistance of a tool to cracking during heat treatment, especially when complicated or intricate sections are involved. The amount of distortion will depend on the tool steel selected and the quenching medium used. The steels rated lowest or very low can usually be machined very close to finished size prior to heat treatment so that little grinding will be required after the hardening operation. Water-hardening steels generally exhibit the most distortion and air-hardening steels the least. However, carbon tool steels and other shallow hardening types may distort very little in heat treatment when the hardened case is small in comparison with the unhardened core for a particular tool design. As most of the tools need to be machined and ground, it is important to bear in mind their machinability and grindability. In addition to the hardness in the annealed condition, the quantity of hard excess carbides present and the microstructure of the steel itself can also affect significantly the machinability of tool steels. Grindability and wear resistance have an inverse relationship, as higher heat-treated hardness and the presence of larger and

Table 7 General Grouping of Tool Steel Applications and Their Requirements

Applications	Major characteristics required	Fabrication/heat-treatment characteristics required
(A) Molds/dies		
Cold	Wear resistance	Polishability, low distortion
Hot	Heat resistance, wear resistance and resistance to heat checking	Low distortion
(B) *Tools/dies for cold working*		
(C) *Tools/dies for hot working*		
(i) Bending		
Cold	Wear resistance	Low distortion
(ii) Blanking/shearing/trimming		
Cold- light duty	Wear resistance	Low distortion
Cold-heavy duty	Wear resistance and toughness	Safety in hardening
Hot-light duty	Wear and heat resistance	Safety in hardening
Hot-heavy duty	Heat resistance and toughness; resistance to heat checking	Safety in hardening
(iii) Coining/embossing		
Cold	Wear resistance and toughness	Low distortion
(iv) Drawing		
Cold	Wear resistance	Polishability, low distortion
(v) Extrusion/heading		
Cold	Wear resistance	Low distortion
Hot	Heat resistance and toughness; resistance to heat checking	Low distortion
(vi) Forging/swaging		
Cold	Wear resistance and toughness	Low distortion
Hot	Heat resistance and toughness; resistance to heat checking	Low distortion
(vii) Punching		
Cold	Wear resistance	Low distortion
Hot	Heat resistance and toughness; resistance to heat checking	Low distortion
(viii) Rolling/roll forming		
Cold	Wear resistance	Polishability, low distortion
Hot	Heat resistance and toughness; resistance to heat checking	Polishability, low distortion
(ix) Spinning		
Cold	Wear resistance	Low distortion
(D) *Cutting tools for machining*		
Hot	Wear resistance, heat resistance and toughness	Grindability
(E) *Structural parts*		
Cold	Wear resistance and toughness	Safety in hardening
Hot	Heat resistance and wear resistance	Low distortion

more numerous primary alloy carbides decrease grindability while increasing wear resistance.

As mentioned previously, cold-work tool steels and hot-work tool steels are generally used for applications in which the temperature is below 200°C and over 200°C, respectively. Hence, it is useful to take note of this operating temperature to facilitate tool steel selection. Wear resistance and toughness are generally considered to be the major characteristics required for tools to be used for cold-working; while heat resistance, wear resistance, toughness, and resistance to heat checking are to be considered for tools to be used for hot-working.

Weldability should be taken into consideration in the following cases: (i) assembly of components into a single tool or die; (ii) overlay of tool steel weld metal on specific areas of a carbon steel or low-alloy steel base metal; (iii) rebuilding of worn surfaces and edges; (iv) alteration of a tool or die to meet a different application requirement; and (v) repair of cracked, broken, or otherwise damaged tools, dies, or other components.

The selection of a tool material begins with the identification of the group or groups of tool steels that best satisfy the requirements of a given tool or die application. Selection within a group then proceeds with a more detailed evaluation of the major performance properties and fabrication properties of the grades of steel within that group. In evaluating the properties of the various grades of tool steels, it also becomes apparent that some compromising is required. For example, a particular tool steel may have a greater wear resistance and hot hardness, but lower in toughness as compared with another. Moreover, it may be more difficult to machine, heat treat, and grind. Hence, the challenge of selecting tool steels is to get an optimum blend of both major and fabrication/heat-treatment properties, and to match these properties to the requirements of the tool. In many instances, the choice is not limited to a single material that can be used for an application. It is desirable to select the one material that will provide the most economical overall performance. However, it should be noted that the relative evaluation of properties for various grades is only qualitative in nature, and the proper choice for a specific application cannot always be made with assurance. One approach is to select a material that has proved successful in the past for a similar operation. Consultations with the material supplier; tool, mold or die producer; and heat treater are also highly recommended.

B. Application Examples of AISI Tool Steels

In this section, some tool steel application examples for plastic molds, die-casting dies, tools and dies for cold-working, tools and dies for hot-working, cutting tools for machining and structural component parts, will be given for reference. These examples will be very useful to a tool designer in two aspects. For designing a new tool, he needs to know what grades of tool steel to select; or what grades of tool steel are usually specified for a similar tool. For an existing tool already used in a production plant, the tool designer should know how it is performing. However, he may want to change it to another hardness or even another grade of tool steel to improve its wear resistance, or toughness, etc. in order to improve its overall performance in his plant. Hence, wherever possible, more than one material from more than one source will be provided. The major properties for each tool material, based on values given in Table 5, will also be listed to facilitate comparisons. As different factors were apparently taken into account for each case, it will be difficult to find two recommendations exactly identical, as one expert may emphasize more on the wear resistance and life of the tool while the other may emphasize more on its toughness or safety in heat treatment. Even if the same grade of material is recommended by both experts, the hardness values recommended may also be different as one of them may prefer a lower hardness value to obtain a slightly higher toughness to suit his operations. Similarly, one expert may prefer air-hardening steels to have a better safety or lower distortion in hardening than water- or oil-hardening steels. It is therefore important to understand the rationale of each recommendation. Furthermore, it should also be noted that the selection of tool steels should not be limited to AISI grades only, as many tooling components are manufactured from non-standard proprietary grades that may be superior for specific applications. It is therefore worthwhile to discuss application requirements with tool steel specialist suppliers whenever possible.

1. Tool Steel Selection Examples for Plastic Molds

As indicated in Table 8, P6 is used for hubbed and carburized cavities, P20, P21, and H13 for machined cavities; and A2, D2, S1, S7 for inserted cavities. P21 is a prehardened tool steel which has good polishability. H13 has excellent resistance to tempering, less tendency to crack and very little or no distortion with air hardening. A2 and D2 will provide a better wear resistance. O1, A2, D2, and H13 are also used for making the master hubs [16–19].

It should be noted that a significant number of molds are nitrided. Hence, one should bear this requirement in mind when selecting a mold material, as some grades are much more suitable for nitriding than others. For example, H13 is an excellent material for this purpose, as it responds very well to nitriding, producing an excellent nitrided surface with hardness of about 1100 HV which is considerably harder than that obtainable with P20, for instance. H13 also retains good core hardness at the nitriding temperature, whereas many other grades soften during nitriding. Steels with high Ni contents are much less suitable for nitriding and one should note the significant difference between P20 and P20 + Ni grades in making an integrated decision to meet both the material choice and processing need. It should also be noted that stainless steels are widely used for plastic molds, as will be discussed in Sec. III.C.2.

2. Tool Steel Selection Examples for Die-Casting Dies

Table 9 shows some tool steel selection examples for die-casting dies. As die-casting dies are exposed to high casting temperatures, hot-work tool steels with good heat resistance and resistance to heat checking are generally used. H11, H12, and H13 are used for dies for casting zinc, aluminum, and magnesium; while H19, H21, and H23 for dies for casting brass. For short runs, P20 and S7 are also used for dies for casting zinc, which has a relatively lower melting temperature [2,16]. An example of the heat treatment of H13 is given in Sec. IV.P.

3. Tool Steel Selection Examples for Tools and Dies for Cold-Working Applications

Table 10 shows some tool steel selection examples for cold bending, blanking, coining, embossing, drawing, extrusion, heading, forging, swaging, punching, rolling, roll forming, and spinning. Cold-work tool steels, like A2, D2, D4, O1, O2, O6, W1, and W2, are used mainly for their wear resistance; while S1, S5, S7, L3, and L6 for their wear resistance and toughness. T1, T15, M2, M4, M42 are high-speed steels which have even better wear resistance, but may have lower toughness. Low distortion and safety in hardening are also factors to be considered in the tool steel selections [2,16,20].

It should also be noted that P/M M3:2 and indeed other P/M grades are also commonly used for cold-working applications. The toughness of P/M3:2 is better than D2,

Table 8 Examples of AISI Tool Steel Selection for Plastic Molds

Steel	Wear resistance	Toughness	Hot hardness	Applications	Remarks
Hubbed and carburized cavities					
P6	1	9	3	Hubbed and carburized, high core strength, 56–60 HRC	Ref. [17]
Machined cavities					
P20	1	8	2	Carburized, 56–60 HRC; or prehardened to 300 HB	Ref. [17]
P21	1	8	4	Prehardened to 28–30 HRC	
H13	3	9	6	Excellent resistance to tempering, high core strength, no distortion; 48–54 HRC	
H13	3	9	6	Excellent resistance to tempering, good toughness, no distortion; 54–56 HRC	Ref. [18]
P20	1	8	2	Carburized, 54–61 HRC; or prehardened to 300 HB	
P21	1	8	4	Prehardened to 300–330 HB	
Inserted cavities					
A2	6	4	5	Best for small molds, long wearing; 59–61 HRC	Ref. [17]
D2	8	2	6	For small molds where maximum resistance to abrasion is required; also for molds used up to 750°F; 58–60 HRC	
S1	4	8	5	For small and medium sized molds; tougher than D2; 53–56 HRC	
S7	3	8	5	For medium and large molds, tough and stable; 53–56 HRC	
Molds					
A2	6	4	5	Inserts, 50–60 HRC	Ref. [16]
H13	3	9	6		
S7	3	8	5		
P20	1	8	2	Insert holders, 28–32 HRC	Ref. [16]
H13	3	9	6	Ejector pins	Ref. [16]
Master hubs					
O1	4	3	3	O1: good wear resistance, compressive strength and dimensional stability in heat treating; A2 and D2: very good wear resistance, compressive strength and dimensional stability in heat treating	Ref. [19]
A2	6	4	5		
D2	8	2	6		
H13	3	9	6	Heat-treated to 56 HRC, highly polished and plated with copper	Ref. [18]

and its abrasion resistance is much better. Examples of the heat treatment of P/M M3:2 and D2 are given in Sec. IV.P.

In addition to the various surface hardening and coating techniques, titanium nitride coating is also widely used on cold-work tool steels to improve their wear resistance. The physical vapor deposition (PVD) process is employed at a temperature range of 200–500°C to minimize the effect of tempering.

Table 9 Examples of AISI Tool Steel Selection for Die-Casting Dies

Steel	Wear resistance	Tough-ness	Hot hardness	Applications	Remarks
Die-Casting					
(i) Aluminum and magnesium					
H13	3	9	6	Dies, 42–52 HRC; inserts and cores, 46–52 HRC; plungers, 46–50 HRC; slides, 46–52 HRC; shot sleeves, 44–48 HRC; nozzles, 32–42 HRC	Ref. [16]
H11	3	9	6	General purpose for aluminum die casting; 47 HRC preferred hardness	
H11	3	9	6		Ref. [2]
H13	3	9	6		
H12	3	9	6		
P20	1	8	2	Die holders, 28–32 HRC	Ref. [16]
H13	3	9	6	Ejector pins, 42–46 HRC	Ref. [16]
(ii) Brass					
H19	5	6	7	Dies, inserts and cores, 40–44 HRC	Ref. [16]
H21	4	6	8		
H13	3	9	6		
H21	4	6	8	Short runs on brass die casting; 38–42 HRC	Ref. [2]
H19	5	6	7	Medium runs on die casting; 38–43 HRC	Ref. [2]
H23	5	5	8	Medium runs on die casting; 32–39HRC; Highest tempering resistance	Ref. [2]
H13	3	9	6	Plungers, 40–44 HRC	Ref. [16]
H21	4	6	8		
H13	3	9	6	Ejector pins, 42–46 HRC	Ref. [16]
M2	7	3	8		
P20	1	8	2	Die holders	Ref. [16]
(iii) Zinc alloys					
P20	1	8	2	Dies and inserts, 28–48 HRC	Ref. [16]
H13	3	9	6		
S7	3	8	5		
P20	1	8	2	Pre-hardened to 300 HB; used for zinc casting dies; short to medium runs.	Ref. [2]
H11	3	9	6	Long runs for zinc die casting. 47 HRC preferred hardness	Ref. [2]
H13	3	9	6		
H12	3	9	6		
H13	3	9	6	Plungers, 42–48 HRC; ejector pins, cores and slides, 39–44 HRC	Ref. [16]
H11	3	9	6		

4. Tool Steel Selection Examples for Tools and Dies for Hot-Working Applications

Table 11 shows some tool steel selection examples for hot shearing, extrusion, heading, forging, swaging, punching, and rolling. Hot-work tool steels H11, H12, H13 are used for hot extrusion and forging of aluminum, magnesium, and steel, while H19, H21, H24, H26 are used for steel, brass, and high-temperature alloys. Cold-work tool steels S1, S2, S5, S7 are used for light-duty hot shearing tools while A8 and D2 are used for dies for hot extrusion of zinc, with a better wear resistance [2,16].

5. Tool Steel Selection Examples for Cutting Tools

Table 12 shows some tool steel selection examples for turning tools, twist drills, end mills, taps, and dies, reamers and broaches, milling cutters, and saws. High-speed tool steels are generally used for making these cutting tools that require wear resistance, heat resistance, toughness, and grindability. M1, M2, and T1 are popular choices for normal machining applications; while M3:2, M4, T4, T5, and T8 for high-speed cutting; M42 and T15 for heavy-duty applications [3,21]. Again, it should be noted that P/M grades are also commonly used for making cutting tools, being considerably tougher than normal wrought grades, and having much more isotropic properties, which can be a distinct advantage for some applications.

Again PVD titanium nitride coating is widely used to improve wear resistance of high-speed steel cutting tools, to minimize heat buildup and to prevent welding of the workpiece material.

6. Tool Steel Selection Examples for Structural Component Parts

Table 13 shows some tool steel selection examples for structural component parts. Depending on the requirements of the component part, it can be made of cold-work tool steel, hot-work tool steel, or high-speed steel. Cold-work tool steels O1, A2, and D2 are selected to provide for better wear resistance; S5 and S7 for their better resistance to chipping; O2 and O6 for their extreme dimensional stability; A2, O2, O6, A7, and L6 for an increased degree of safety in heat treatment. Hot-work tool steels H11, H12, and H13 have high tensile strength, high fatigue strength, and fracture toughness. High-speed steels M50 and M2 give a better heat resistance, wear resistance, and strength [2,3].

C. Applications of Other Types of Tool Materials

1. Carbon Tool Steels

Wrought carbon tool steels are often used in the fabrication of fixtures and auxiliary die components in which strength and weldability, rather than wear resistance, are the primary requirements. Hot-rolled low-carbon steels are relatively inexpensive and they are used extensively for components for which machining/welding is required. These materials can be purchased in standard-size bars and plates from stock. Large plates may be cut to required sizes and contours, to reduce machining costs during die construction. On the other hand, cold-rolled low-carbon steels have smoother surface finishes, closer dimensional tolerances, and higher strengths than hot-rolled low-carbon steels. They are generally used for die components for which hardening is not required, but wear

Table 10 Examples of AISI Tool Steel Selection for Cold Working

Steel	Wear resistance	Tough-ness	Hot hardness	Applications	Remarks
Bending dies					
A2	6	4	5	Bending dies, 58–64 HRC	Ref. [16]
O2	4	3	3		
O1	4	3	3	Bending dies, 58–62 HRC	Ref. [20]
A2	6	4	5		
D2	8	2	6		
Blanking dies					
A2	6	4	5	Blanking dies, 58–64 HRC	Ref. [16]
D2	8	2	6		
S7	3	8	5		
O1	4	3	3		
W2	2–4	3–7	1	Blanking dies and punches (short runs), 57–65 HRC for W2; 58–62 HRC for O1 and A2	Ref. [20]
O1	4	3	3		
A2	6	4	5		
A2	6	4	5	Blanking dies and punches (long runs), 58–62 HRC	Ref. [20]
D2	8	2	6		
M4	9	3	8		
Coining dies					
A2	6	4	5	Coining dies, 58–62 HRC	Ref. [16]
D2	8	2	6		
S1				Coining dies, 52–55 HRC for S1; 58–62 HRC for W1, A2, D2, and D4	Ref. [20]
W1	2–4	3–7	1		
A2	6	4	5		
D2	8	2	6		
D4	8	1	6		
Deep drawing					
A2	6	4	5	Deep drawing dies for steel, 62–65 HRC	Ref. [16]
M4	9	3	8		
D2	8	2	6		
A2	6	4	5	Deep drawing dies for aluminum, 62–64 HRC	Ref. [16]
D2	8	2	6		
M4	9	3	8	Deep drawing dies for brass, 62–65 HRC	Ref. [16]
D2	8	2	6		
D2	8	2	6	Drawing punches, 62–65 HRC	Ref. [16]
M2	7	3	8		
A2	6	4	5		
W1 or W2	2–4	3–7	1	Drawing dies, 58–64 HRC for W1 or W2; 58–62 HRC for O1, O6, A2, D2, and D4	Ref. [20]
O1	4	3	3		
O6	3	3	2		
A2	6	4	5		
D2	8	2	6		
D4	8	1	6		

(Continued)

Table 10 (Continued)

Steel	Wear resistance	Tough-ness	Hot hardness	Applications	Remarks
Embossing					
O1	4	3	3	Embossing dies, 59–61 HRC	Ref. [20]
O2	4	3	3		
A2	6	4	5		
D2	8	2	6		
S1	4	8	5	Embossing punches, 59–61 HRC	Ref. [20]
S5	2	8	3		
Extrusion (cold-work)					
T15	9	1	9	Die insert: 65–67 HRC for T15; 64–66 HRC for M4; 62–64 HRC for D4; 60–62 HRC for D2	Ref. [2]
M4	9	3	8		
D4	8	1	6		
D2	8	2	6		
T15	9	1	9	Punches: 64–66 HRC for T15 and M4; 62–64 HRC for D4; 60–62 HRC for D2; or 60–63 HRC under conditions requiring slightly higher toughness for T15, M4, and D4	Ref. [2]
M4	9	3	8		
D4	8	1	6		
D2	8	2	6		
A2	6	4	5	Punch shank or mandrel, 56–58 HRC	Ref. [2]
O1	4	3	3		
S1	4	8	5		
A2	6	4	5	Die base; 54–56 HRC	Ref. [2]
S1	4	8	5		
L6	3	6	2		
H11	3	9	6	Die holder; 48–50 HRC	Ref. [2]
H11	3	9	6	Knockout pin; 53–55 HRC for H11; 54–56 HRC for S5, L6 and S1	Ref. [2]
S5	2	8	3		
L6	3	6	2		
S1	4	8	5		
M4	9	2	8	Die anvil; 60–62 HRC for M4; 58–60 HRC for D2 and A2	Ref. [2]
D2	8	2	6		
A2	6	4	5		
M2	7	3	8	Extrusion of steel, die, and punches, 60–66 HRC	Ref. [16]
M4	9	3	8		
O2	4	3	3		
A2	6	4	5	Extrusion of aluminum, die, and punches, 56–62 HRC	Ref. [16]
D2	8	2	6		

(Continued)

Table 10 (Continued)

Steel	Wear resistance	Tough-ness	Hot hardness	Applications	Remarks
Heading (cold-work)					
W1	2–4	3–7	1	Heading dies, 58–62 HRC	Ref. [16]
M4	9	3	8	Heading die inserts, 60–64 HRC	Ref. [16]
D2	8	2	6		
M2	7	3	8	Shearing dies, 58–63 HRC	Ref. [16]
D2	8	2	6		
M4	9	3	8		
M2	7	3	8	Upsetting dies, 58–64 HRC	Ref. [16]
A2	6	4	5		
D2	8	2	6		
M2	7	3	8	Trimming dies, 60–64 HRC	Ref. [16]
T1	7	3	8		
M4	9	3	8	Chamfering cutters, 62–65 HRC	Ref. [16]
M42	8	1	9		
W1	2–4	3–7	1	Heading punches, 58–60 HRC	Ref. [16]
M2	7	3	8		
M4	9	3	8		
M2	7	3	8	Indenting punches, 60–64 HRC	Ref. [16]
M4	9	3	8		
D2	8	2	6	Blanking punches, 58–62 HRC	Ref. [16]
A2	6	4	5		
S5	2	8	3		
M2	7	3	8	Piercing punches, 58–62 HRC	Ref. [16]
D2	8	2	5		
M2	7	3	8	Upsetting punches, 58, 64 HRC	Ref. [16]
A2	6	4	5		
S5	2	8	3		
H11	3	9	6	Die-insert holders, 45–50 HRC	Ref. [16]
H13	3	9	6		
M2	7	3	8	Cutoff blades, 60–64 HRC	Ref. [16]
W1	2–4	3–7	1		
O1	4	3	3		
A2	6	4	5	Quills and knockout pins, 59–63 HRC	Ref. [16]
W1	2–4	3–7	1		
M2	7	3	8		
Punching					
S5	2	8	3	Light-duty punches, 58–62 HRC	Ref. [16]
D2	8	2	6		
S5	2	8	3	Heavy-duty punches, 58–62 HRC	Ref. [16]
M2	7	3	8		
S7	3	8	5		
Rolling-mill rolls (cold-work)					
01	4	3	3	Cold rolls 6.4–100 mm (0.25–4 in.) round. Through-hardening in oil; 60–62 HRC	Ref. [2].
A2	6	4	5	Cold rolls 6.4–250 mm (0.25–10 in.) round. Through-hardening in air; 58–62 HRC; work rolls and backup rolls	Ref. [2]

(Continued)

Table 10 (Continued)

Steel	Wear resistance	Tough-ness	Hot hardness	Applications	Remarks
D2	8	2	6	Cold rolls 6.4–300 mm (0.25–12 in.) round. Through-hardening in air; 58–62 HRC; Sendzimir mill work rolls	Ref. [2]
M4	9	3	8	Cold rolls 6.4–150 mm (0.25–6 in.) round; high-speed steels; 62–65 HRC; Sendzimir mill work rolls	Ref. [2]
Shearing (cold-work)					
O1	4	3	3	Cold-work shears for thin or medium-thickness stock	Ref. [2]
A2	6	4	5		
D2	8	2	6		
S2	2	8	2	Cold shearing heavy stock; scrap shears	Ref. [2]
S5	2	8	3		
S1	4	8	5		
M3	8	2	8	Long-life cold shears for thin or medium-thickness stock; highest wear resistance	Ref. [2]
M4	9	3	8		
D2	8	2	6	Light-duty shear blades, 56–68 HRC	Ref. [16]
A2	6	4	5		
O1	4	3	3		
S5	2	8	3	Heavy-duty shear blades, 55–60 HRC	Ref. [16]
S7	3	8	5		
Spinning					
A2	6	4	5	Mandrels, 50–58 HRC	Ref. [16]
H11	3	9	6		
P20	1	8	2		
M2	7	3	8	Rollers, 58–62 HRC	Ref. [16]
M4	9	3	8		
Swaging dies					
S5	2	8	3	Cold swaging dies, 56–60 HRC	Ref. [16]
S7	3	8	5		
A2	6	4	5		
D2	8	2	6		
Trimming					
W2	2–4	3–7	1	Dies and punches, 57–60 HRC	Ref. [20]
A2	6	4	5		
D2	8	2	6		
D4	8	1	6		
M4	9	3	8		

surfaces are often case hardened. The major advantage of cold-rolled steels is that they can often be used without machining, thus reducing costs. Cold-rolled steels, however, have internal stresses that are relieved by machining or welding. This can cause sufficient bowing, warping, or twisting to require additional costly operations.

Table 11 Examples of AISI Tool Steel Selection for Hot Working

Steel	Wear resistance	Tough-ness	Hot hardness	Applications	Remarks
Hot extrusion					
(i) Aluminum and magnesium					
H11	3	9	6	Dies, 48–50 HRC; dummy and backer blocks; 44–48 HRC; liners, 42–46 HRC; rams, 40–42 HRC; mandrels, 48–52 HRC	Ref. [16]
H12	3	9	6		
H13	3	9	6		
H11	3	9	6	Dies, dummy blocks, rams, liners for magnesium and aluminum; 42–45 HRC	Ref. [2]
H13	3	9	6		
H12	3	9	6		
(ii) Brass/copper alloys					
H21	4	6	8	Extrusion of brass-dies for tubes and round shapes, 39–42 HRC; dies for other shapes, 31–37 HRC	Ref. [16]
H19	5	6	7		
H11	3	9	6	Rams and liners for copper alloy extrusion. Die holders for extrusion. Dummy blocks for copper, brass; 42–45 HRC. Mandrels for copper alloys; 50–52 HRC	Ref. [2]
H13	3	9	6		
H12	3	9	6		
H21	4	6	8	Dies and dummy blocks for copper alloy extrusion. Dies for brass and copper; 42–45 HRC. Dummies for cupro-nickel; 41–44 HRC	Ref. [2]
H26	6	4	8	Dies for copper alloy extrusion. Dies for brass and copper; 40–44 HRC	Ref. [2]
H11	3	9	6	Extrusion of brass-dummy and backer blocks, 40–45 HRC	Ref. [16]
H21	4	6	8		
H12	3	9	6	Extrusion of brass-liners, rams, 40–44 HRC; mandrels, 40–46 HRC	Ref. [16]
H11	3	9	6		
(iii) Steel					
H21	4	6	8	Dies, 43–47 HRC	Ref. [16]
H11	3	9	6		
H21	4	6	8	Dummy blocks, 40–44 HRC	Ref. [16]
H19	5	6	7		
H11	3	9	6	Mandrels, 42–46 HRC	Ref. [16]
H13	3	9	6		
H21	4	6	8	Billet shears, 46–52 HRC	Ref. [16]
H12	3	9	6		
H11	3	9	6		
A2	6	4	5	Trimmer dies, 56–58 HRC	Ref. [16]
H13	3	9	6		

(Continued)

Table 11 (Continued)

Steel	Wear resistance	Tough-ness	Hot hardness	Applications	Remarks
(iv) Refractory/high-temperature alloy					
T15	9	1	9	Dies for refractory metal extrusion and high-temperature alloy extrusion	Ref. [2]
(v) Zinc					
A2	6	4	5	Dies for zinc extrusion	Ref. [2]
D2	8	2	6		
Hot forging					
(i) Aluminium					
H11	3	9	6	Punches and dies, 44–48 HRC; die inserts, 46–50 HRC	Ref. [16]
H12	3	9	6		
H13	3	9	6		
(ii) Brass					
H21	4	6	8	Punches, dies and inserts, 48–52 HRC	Ref. [16]
H11	3	9	6		
H13	3	9	6		
(iii) Steel					
H13	3	9	6	Punches, dies and inserts, 44–48 HRC	Ref. [16]
H12	3	9	6		
H19	5	6	7		
D2	8	2	6	Trim dies, 58–60 HRC	Ref. [16]
A2	6	4	5		
(iv) General					
S1	4	8	5	Short runs; die temperature in service not over 480°C; dies can be water cooled	Ref. [2]
H12	3	9	6	Long runs; die temperature in service not over 480°C; dies can be water cooled; H12 is widely used; A8 provides better resistance to wear	Ref. [2]
H11	3	9	6		
H13	3	9	6		
A8	4	8	6		
H21	4	6	8	Long or short runs; die temperatures above 480°C. H21 is widely used; H24 and H26 have best wear resistance. H26 is used for press forging of nickel and cobalt-base high-temperature alloys	Ref. [2]
H20	4	6	8		
H24	5	5	8		
H26	6	4	8		
Hot punching					
H13	3	9	6	Punches, 48–58 HRC	Ref. [16]

(Continued)

Table 11 (Continued)

Steel	Wear resistance	Tough-ness	Hot hardness	Applications	Remarks
Hot rolling-mill rolls					
H11	3	9	6	Hot rolls; short runs; backup rolls; low temperatures; 40–48 HRC	Ref. [2]
H21	4	6	8	Hot rolls; medium runs; medium temperatures; 40–50 HRC	Ref. [2]
H26	6	4	8	Hot rolls; long runs; special materials; high temperatures; 45–58 HRC	Ref. [2]
Hot shearing					
S5	2	8	3	Light-duty shear blades, 55–60 HRC	Ref. [16]
H13	3	9	6		
S7	3	8	5		
S2	2	8	2	Hot shears for light-duty and moderate life	Ref. [2]
S5	2	8	3		
S1	4	8	5		
H21	4	6	8	Heavy-duty shear blades, 45–50 HRC	Ref. [16]
H13	3	9	6		
H12	3	9	6		
H11	3	9	6	Hot shears for heavy sections	Ref. [2]
H13	3	9	6		
A8	4	8	6		
Hot swaging					
H13	3	9	6	Hot swaging dies, 48–52 HRC	Ref. [16]
H12	3	9	6		
H11	3	9	6		

2. Stainless Steels and Beryllium Copper Alloys

Corrosion resistance is a major consideration in selecting mold steels. Typical cases are: (i) when potentially corrosive plastics (e.g. polyvinyl chloride) are being processed; (ii) where cooling water is of poor quality; (iii) where a mold will be used intermittently, with or without proper drainage and drying between uses, and stored "on the shelf" for long periods; and (iv) where plants are required to work under high-humidity conditions. Hence martensitic and precipitation-hardening (e.g. 15-5 PH and 17-4 PH) stainless steels are often used for plastic molds.

Beryllium copper alloys have a high thermal conductivity, and hence they are often selected for mold components to optimize productivity [3,17–19].

3. Tungsten Carbides

Tungsten carbides, which have a higher softening temperature than that of high-speed steels (HSS), are used widely in machining when the cutting temperature exceeds about 1000–1100°F (538–593°C). However, HSS tools possess better impact toughness and are more capable of taking the shock loading of interrupted cuts than carbide tools. When HSSs are in the annealed state, they can be hot worked or machined to produce the final cutting tool shape readily and economically. Obviously most drills, reamers, taps, thread chasers, end mills, and form tools are made from HSS.

Table 12 Examples of AISI Tool Steel Selection for Cutting (Machining) Tools

Steel	Wear resistance	Tough-ness	Hot hardness	Applications	Remarks
General cutting (machining)					
M1	7	3	8	Normal duty; M2 is the most popular choice for most purposes	Ref. [21]
M2	7	3	8		
T1	7	3	8		
T4	7	2	8	High-speed cutting	Ref. [21]
T6	8	1	9		
M42	8	1	9	Higher hardness	Ref. [21]
Turning tools					
M42	8	1	9	For the most difficult cutting operations	Ref. [21]
T15	9	1	9	Medium- and heavy-duty applications	Ref. [21]
M1	7	3	8	Regular all-purpose tool bits	Ref. [3]
M2	7	3	8		
T1	7	3	8		
M3:2	8	3	8	Higher wear resistance than regular tool bits due to higher carbon and vanadium contents	Ref. [3]
M4	9	3	8		
T4	7	2	8	Higher wear resistance due to higher carbon and vanadium contents; higher hot hardness due to the cobalt content	Ref. [3]
T5	7	1	9		
T8	8	2	8		
T15	9	1	9	Exceptional wear resistance; most popular tool steel for lathe tools	Ref. [3]
Twist drills					
M2	7	3	8	Universal choice for drills for general application	Ref. [21]
M7	8	3	8	Popular choice	Ref. [21]
M42	8	1	9	Heavy-duty applications	Ref. [21]
M41	8	1	9		
M1	7	3	8	General-purpose drills	Ref. [3]
M2	7	3	8		
M7	8	3	8		
M10	7	3	8		
M50	6	3	6	For lower-cost hardware-quality drills	Ref. [3]
M52	6	3	6		
M33	8	1	9	Drills with high hot hardness for more difficult-to- machine alloys, such as nickel-base or titanium products	Ref. [3]
M42	8	1	9		
T15	9	1	9		
End mills					
M1	7	3	8	End mills for general-purpose applications	Ref. [3]
M2	7	3	8		
M7	8	3	8		

(Continued)

Table 12 (Continued)

Steel	Wear resistance	Tough-ness	Hot hardness	Applications	Remarks
M10	7	3	8		
M33	8	1	9	For milling work-pieces made from hardened materials (> 300 HB); improved hot hardness with their cobalt contents	Ref. [3]
M42	8	1	9		
T15	9	1	9		
Taps and dies					
M7	8	3	8	Light-duty applications	Ref. [21]
M2	7	3	8		
M3:2	8	3	8	Medium duty applications	Ref. [21]
Reamers and broaches					
M2	7	3	8	Light-duty applications	Ref. [21]
M3:2	8	3	8	Medium duty applications	Ref. [21]
M41	8	1	9	Heavy duty applications	Ref. [21]
M1	7	3	8	General-purpose grades	Ref. [3]
M2	7	3	8		
M7	8	3	8		
M10	7	3	8		
M3	8	3	8	For applications required greater wear resistance	Ref. [3]
M4	9	3	8		
T15	9	1	9		
Milling cutters					
M2	7	3	8	Light-duty applications	Ref. [21]
M3:2	8	3	8	Medium-duty applications	Ref. [21]
M42	8	1	9		
T15	9	1	9	Heavy-duty applications	Ref. [21]
M2	7	3	8	For machining free-machining types of materials or materials under 30 HRC	Ref. [3]
M3	8	3	8	More wear resistance; for machining materials up to about 35 HRC	Ref. [3]
M4	9	3	8		
M42	8	1	9	For machining materials with a hardness above 35–50 HRC	Ref. [3]
T15	9	1	9		
Saws					
M2	7	3	8	The usual first choice	Ref. [21]
M2	7	3	8	General-purpose saw material; lower hardness is used to increase the toughness	Ref. [3]
M42	8	1	9	Often used to machine stainless steels, aluminium, and brass, with better production life and higher speeds	Ref. [3]
T15	9	1	9	Very special applications only	Ref. [3]

Table 13 Examples of AISI Tool Steel Selection for Structural Component Parts

Steel	Wear resistance	Tough-ness	Hot hardness	Applications	Remarks
Normal temperatures					
O1	4	3	3	For parts requiring abrasion resistance, e.g. guides, plain sliding bearings, bushings, cams and cam followers.	Ref. [3]
A2	6	4	5		
D2	8	2	6		
S5	2	8	3	Shock-resisting steels noted for their resistance to chipping, for applications involving impact loading, e.g. clutch teeth and ratchets.	Ref. [3]
S7	3	8	5		
H11	3	9	6	For components to withstand repeated flexing, e.g. split collets, leaf and reed springs, torsion bars, and critical aircraft parts such as landing-gear components and helicopter rotors. These materials have high tensile strength, high fracture toughness and fatigue strength.	Ref. [3]
H12	3	9	6		
H13	3	9	6		
O2	4	3	3	For gages, squares, straight edges, templates, and micrometer anvils requiring extreme dimensional stability.	Ref. [3]
O6	3	3	2		
A2	6	4	5	To provide an increased degree of safety in heat treatment for component parts, e.g. manifold plates with a complicated geometry. A2 can be air-hardened; O2, O6, A7 and L6 have a relatively low austenitizing temperature of about 790 °C.	Ref. [3]
O2	4	3	3		
O6	3	3	2		
A7	9	1	6		
L6	3	6	2		
L2	1	7	2	Machine elements, cams, shafts, spindles, gears, sprockets	Ref. [2]
S1	4	8	5		
P21	1	8	4		
H11	3	9	6	Ultrahigh-strength steel used for aircraft and missile structures, bolts at 1515–2070 MPa (220–300 ksi) ultimate tensile strength (UTS)	Ref. [2]
H10	3	9	6		
A8	4	8	6	For strengths of 1790–2205 MPa (260–320 ksi) UTS. Less ductility than H11.	Ref. [2]
A9	4	8	6		

(Continued)

Table 13 (Continued)

Steel	Wear resistance	Tough-ness	Hot hardness	Applications	Remarks
High-temperatures					
M50	6	3	6	For components to operate continuously at elevated temperatures, e.g. roller bearings in gas turbine engines, valves and valve seats in plastic injection molding equipment. Both M50 and M2 have good hot hardness.	Ref. [3]
M2	7	3	8		
M50	6	3	6	High-temperature bearings; strengths up to 2585 MPa (375 ksi), low ductility.	Ref. [2]
M10	7	3	8		
M2	7	3	8		

Cemented carbides are also employed in metal-forming applications because of their combination of high compressive strength, good abrasion resistance, high elastic modulus, good impact, and shock resistance, and ability to take and retain excellent surface finish. Typical applications include drawing dies, hot and cold rolling of strips and bars, cold heading dies, forward and back extrusion punches, swaging hammers and mandrels, and can body punches and dies [3].

D. Special Fabrication Processes

1. Hubbing Process

The hubbing process consists of pressing a master hub made of hardened tool steel into an annealed steel blank made of hubbing steel (e.g. P6) by a slow-acting hydraulic press. The hubbed cavities are mostly heat treated by carburizing. Duplicates of identical cavities can be made easily with this process. However, it is not usually possible to hub cavities with undercuts and hence this process is applicable only to relatively simple shapes.

2. Electrical-Discharge Machining Process

Electrical-discharge machining (EDM) is a popular method for producing plastic molds, die-casting dies, forging dies, etc. It is a spark erosion method involving electrical discharges between an anode (graphite or copper) and a cathode (tool steel or other tooling material) in a dielectric medium. The discharges are controlled in such a way that erosion of the tool or work piece takes place. In this process, the surface of the steel is subjected to very high temperatures, causing the tool steel to melt or vaporize. Consequently, a melted and re-solidified layer (white zone) is produced on the surface. Both a re-hardened layer, with as-quenched martensite, and a tempered layer are formed on a spark-machined surface. This may result in cracking and shortened tool life, unless suitable precautions are taken [3,22].

E. Precautions in Tool Design and Fabrication

To avoid premature failures, all tools should be designed to avoid sharp corners, notch effects and large differences in section thickness. Sharp corners concentrate and magnify applied stresses, stresses that arise in tool and die manufacturing such as during liquid quenching, and stresses that occur during service. They promote cracking during liquid quenching, and also promote buildup of residual stresses that may not be fully relieved by tempering and can therefore reduce service life. A change in section size is often the site of premature failure. Hence one should choose air-hardening steels for complex tools, and suitable hardness value to avoid any temperature range that can reduce toughness after tempering. There is also a need to bear in mind the risk of distortion and the presence of de-carburization in hot-rolled wrought products or in heat treatment, in specifying machining allowances. A stress relieving operation should also be carried out after rough machining. Sec. IV.H and IV.I will discuss more about surface integrity, dimensional change, and distortion in heat treatment.

Great care should also be taken in machining, as rough machining marks and identification stamp marks are common source of failures in heat treatment and in service. If the EDM process is chosen, then the following steps should be taken: (i) Finish the EDM operation by fine sparking, that is, using low current and high frequency. (ii) The affected surface layer should be removed by grinding or polishing, (iii) If the spark eroded surface texture is to be used in the finished mold, it should be re-tempered at a temperature 15–25°C (25–45°F) below that used previously [3,22].

IV. HEAT TREATMENT OF TOOL STEELS

A. Introduction

The properties of tool steels, even more than most other classes of steel, result from a combination of material selection and treatment, particularly heat treatment. The subject of heat treatment is of such vital importance in the area of tool steels that any discussion of these alloys without relation to their heat treatment would be incomplete. The selection of a tool steel and its heat treatment have to be seen as an integral decision. The availability of suitable specialized heat-treatment facilities may in some cases be a deciding factor in the selection of a particular grade.

B. General Heat-Treatment Considerations

1. Crystal Structure of Iron

The main reasons why steels are so amenable to heat treatment lie in the crystal structure of iron, and in the behavior of carbon in relation to that crystal structure. Metal atoms are normally arranged in a regular, repeating 3D pattern or *space lattice*. Iron is perhaps the most notable example of a metal that can have more than one crystal structure, that is, *polymorphic*. Up to about 910°C, pure iron is body centered cubic, but when heated above this temperature, it changes to face centered cubic. If it is heated to about 1400°C, it reverts to its body centered cubic structure again, and this structure remains until the iron melts. In steels, the body centered cubic (bcc) structure is called *ferrite*, and the face centered cubic (fcc) structure is called *austenite*.

2. Solubility of Carbon in Iron

Carbon atoms are very small compared with iron atoms, and when present, they fit in between the iron atoms (*interstitial*). The spaces in the fcc austenite structure are a more suitable size and shape for the carbon atoms than those in the bcc ferrite. The solubility of carbon in ferrite is extremely low (less than 0.01% at room temperature), but austenite can dissolve much more (about 2% at about 1130°C). When austenite containing significant amounts of carbon is cooled slowly, and is ready to transform to ferrite, it must first get rid of the surplus carbon that would not dissolve in the ferrite. This carbon is rejected (*precipitated*) normally as iron carbide (Fe_3C) or *cementite*, which contains about 6.7% carbon.

This polymorphic transformation, accompanied by the associated change in carbon solubility, is the key to our ability to manipulate the structure and properties of steels. We can greatly influence these changes by controlling heating and cooling of the steel; in other words by heat treatment.

3. Transformation of Austenite (Slow Heating/Cooling)

In a plain-carbon steel under conditions of slow heating or cooling, we can predict what will happen from the *Iron/Carbon phase diagram* (Fig. 3). This shows what phase or mixture of phases should be stable at a given composition/temperature. The addition of carbon to iron changes the temperature at which ferrite changes to austenite, and at which other changes take place. From the diagram, it will be seen that with about 0.8% carbon, austenite can be stable down to about 723°C, instead of about 910°C in pure iron.

If a plain-carbon steel with about 0.8% carbon is cooled slowly from say 850°C, at which temperature it is completely austenitic, nothing happens until a temperature of about 723°C is reached. On cooling through this temperature, the austenite becomes unstable, and needs to be transformed to ferrite; however, ferrite cannot hold 0.8% carbon. Before the transformation can take place, the carbon atoms must move (*diffuse*) out of the way. They collect together in such quantities as to form cementite, and the areas they vacate now have such low-carbon content that they can transform to ferrite.

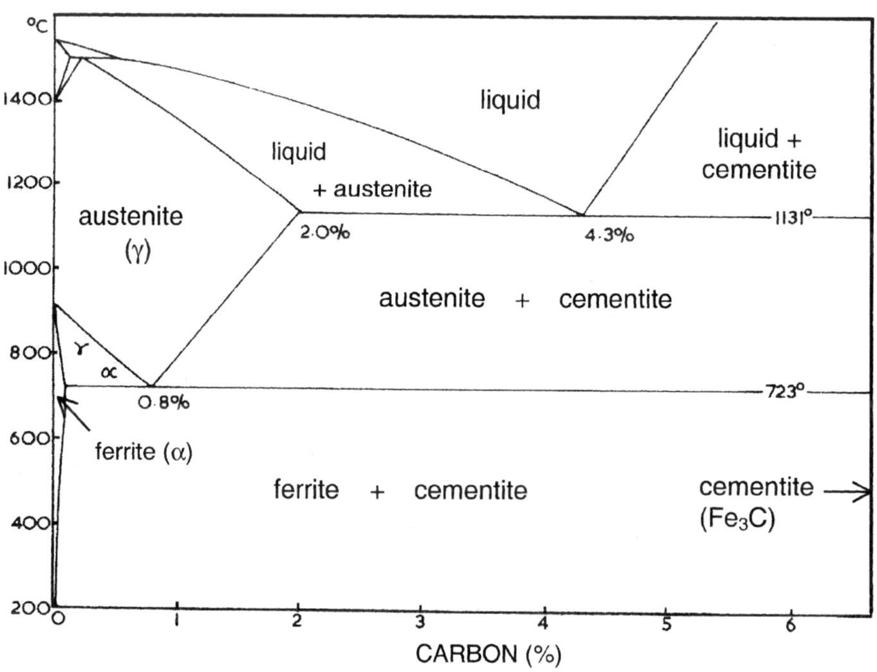

Figure 3 Iron/iron carbide phase diagram. (From Engineering Metallurgy by R.A. Higgins. Reprinted by permission of Butterworth-Heinemann.)

The result is that the steel will have alternate layers of high carbon (cementite) and low carbon (ferrite). This *lamellar* structure of ferrite/cementite is given the name *pearlite*. This whole reaction, of austenite breaking down to a mixture of two different phases, is an example of a *eutectoid* reaction. A plain-carbon steel with about 0.8% carbon is called a eutectoid steel.

Plain-carbon steels with less than 0.8% C (*hypoeutectoid* steels) will begin to transform at a higher temperature (called the upper critical temperature), depending on carbon content. At this temperature, ferrite begins to form, rejecting carbon as it does so. The carbon that is rejected from the ferrite goes back into the remaining austenite, so that as cooling proceeds, more and more ferrite forms, and the remaining austenite becomes progressively richer in carbon, until at the eutectoid temperature, it will have 0.8% C. At this point, the steel will be composed of some ferrite (with no more than about 0.03% C), and austenite (with 0.8% C). On further cooling, the austenite transforms, as before, by the eutectoid reaction, so that the final microstructure will be a mixture of ferrite and pearlite, the proportions of which depend on the overall carbon content of the steel.

A plain-carbon steel with more than 0.8% C (a *hypereutectoid* steel) also begins to transform at the upper critical temperature; this time it is cementite that first forms. Cementite contains about 6.7% C, and so in order to form, must absorb some carbon from its surroundings. The remaining austenite therefore becomes progressively depleted in carbon, until at the eutectoid temperature, it contains 0.8% C, as in previous cases. After undergoing the eutectoid reaction, the final structure in this case is a mixture of pearlite and cementite.

One important point to note here is that there is a volume change associated with the transformation of austenite. There is a net volume contraction on heating, as austenite is formed, and an expansion on cooling, as austenite transforms. This volume change is superimposed on the normal thermal expansion on heating (and contraction on cooling). This volume change is the root cause of some of the distortion and cracking problems encountered during heat treatment.

4. Effect of Cooling Rate on Austenite Transformation

a. Pearlite. At the eutectoid temperature (723°C in a plain-carbon steel), ferrite and austenite are equally stable, and transformation would take an infinitely long time to occur. As the temperature falls below the eutectoid temperature, austenite becomes more unstable and hence there is an increased thermodynamic driving force for the transformation to occur, resulting in a faster rate of transformation. At the same time, as temperature falls, the diffusion rate of carbon also falls. These two effects, faster transformation and slower carbon diffusion, lead to a finer structure. This is demonstrated in the upper part of the *Time–Temperature-Transformation* (TTT) diagram (Fig. 4), also known as the Isothermal Transformation diagram. At temperatures progressively lower than the eutectoid temperature, the pearlite formed will be finer.

When a steel is *normalized* (air cooled, instead of slow cooled or annealed), the cooling rate is such that the temperature falls well below the eutectoid temperature before the transformation can begin, so that the reaction takes place at a lower temperature than the theoretical eutectoid temperature. This results in a larger amount of pearlite in the final structure, and the pearlite will be finer. All this results in the steel being stronger and harder than an annealed steel of the same composition.

b. Martensite. If the austenite is cooled very rapidly, the diffusion of carbon is suppressed, and the carbon atoms remain trapped in the iron crystal structure. Austenite is now unable to transform to bcc ferrite, but instead transforms by a process of

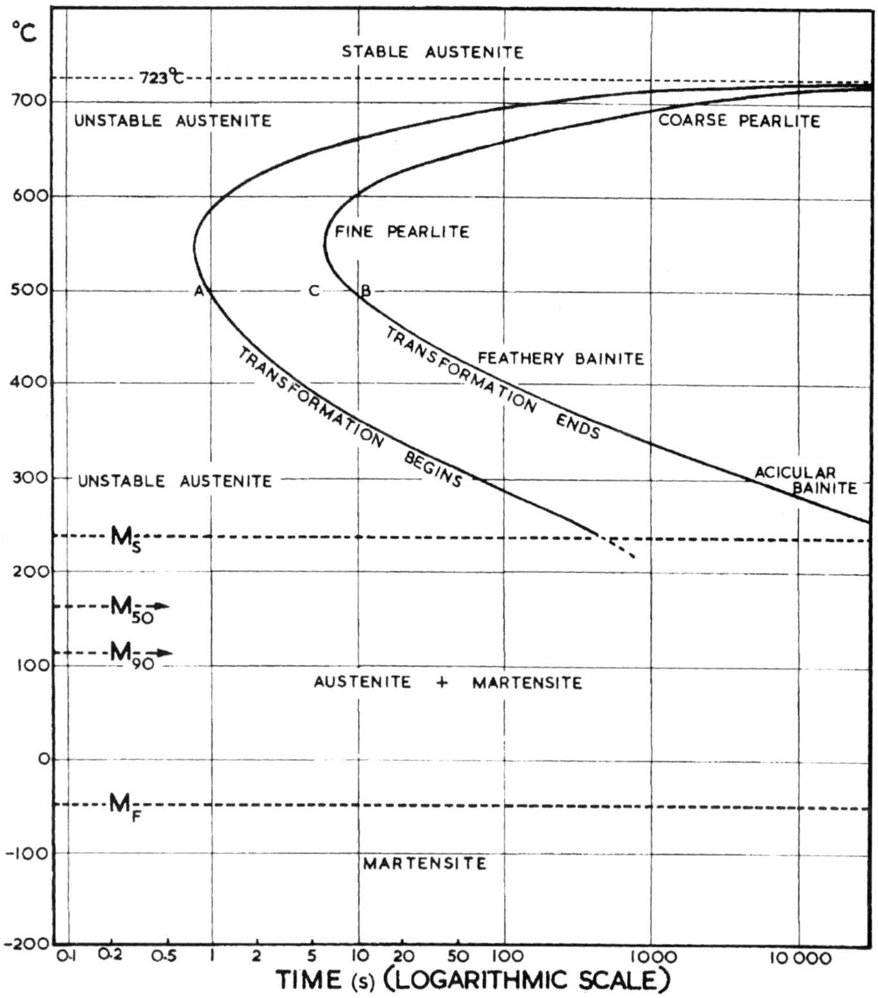

Figure 4 Time–temperature-transformation diagram for eutectoid plain-carbon steel. (From Engineering Metallurgy by R.A. Higgins. Reprinted by permission of Butterworth-Heinemann.)

crystallographic shear to body centered tetragonal (bct) *martensite*. Because carbon does not diffuse out to form carbides, the transformation to martensite is said to be a *diffusionless* transformation.

This transformation takes place, not at a fixed temperature like pearlite, but over a range of temperature from a starting temperature (denoted M_s) to a finishing temperature (denoted M_f). It is said to be an *athermal* reaction. The temperature must be continuously lowered to produce more martensite. If the cooling is stopped somewhere above the M_f the transformation will be incomplete, and some *retained austenite* will be present with the martensite. This transformation is extremely rapid, depending on temperature rather than time. As such, the martensite transformation should not theoretically be shown on the TTT diagram, but for convenience it is usually shown as horizontal lines (M_s and M_f), often together with other lines representing proportions of austenite transformed to martensite at different intermediate temperatures.

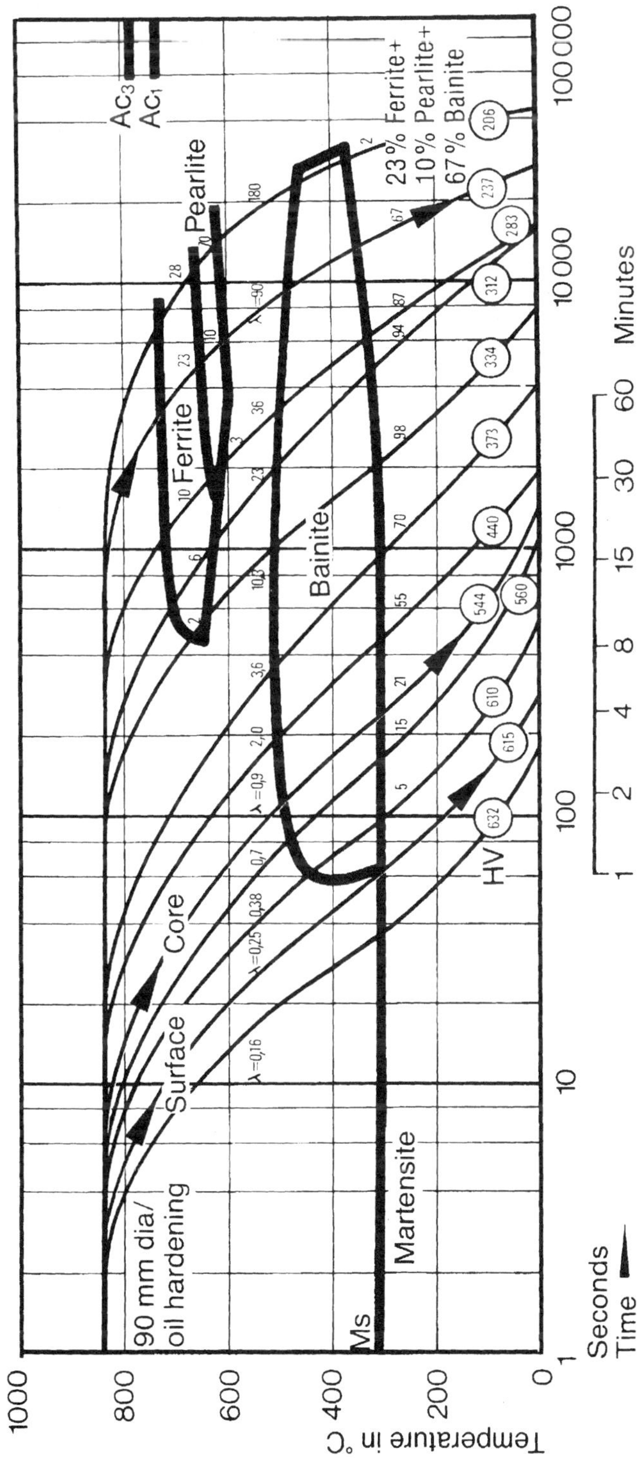

Figure 5 Continuous cooling transformation diagram. (From Ref. 23.)

The presence of a great deal of lattice strain, and a high density of crystal defects such as dislocations in the martensite structure, caused by the crystallographic shear mechanism of its formation, results in its great hardness, the hardness increasing with the degree of lattice strain and dislocation density, which increases with carbon content.

There is a significant volume expansion as austenite transforms to martensite: the higher the carbon content of the steel, the greater the volume change. The high strains resulting from this, coupled with the inherent hardness and brittleness of martensite, are the reason why high-carbon steels are so prone to cracking and distortion when rapidly quenched. The volume change in itself is not the sole cause of the problem; if the whole workpiece changed volume throughout, all at the same time, then there would be no stress set up. The problem arises because there will be a temperature distribution in the piece as it cools rapidly, and transformation takes place in different regions of the piece at different times, causing localized differential volume changes.

One method of minimizing this problem is to minimize temperature differentials during the martensite transformation. This can be done by *martempering* (also known as *marquenching* or *step hardening*). In this process, the workpiece is first rapidly quenched (to avoid transformation to pearlite or bainite) to a temperature near to but preferably just above the M_s, holding at this temperature to equalize the temperature throughout the piece, and then cooling more slowly to minimize temperature gradients during the transformation. The first quench (to the M_s) is normally done in molten salt or hot oil, depending on the required temperature, and further cooling is normally in air (since the martensite transformation is not time dependent).

c. Bainite. At intermediate temperatures, below those for pearlite transformation but above martensite transformation temperatures, austenite will transform by a mechanism that can for simplicity be visualized as a combination of crystallographic shear and short-range diffusion of carbon, into a structure known as *bainite*, the nature of which depends on its formation temperature, more closely resembling fine pearlite when formed at higher temperatures (*upper bainite*), and more closely resembling martensite when formed at temperatures closer to the M_s (*lower bainite*).

The mechanical properties of bainite are between those of pearlite and martensite, and are often not dissimilar from tempered martensite. Bainite is sometimes deliberately produced by cooling to an intermediate temperature (e.g. in a molten salt bath) and allowing sufficient time for the transformation to be complete. This is called *austempering*.

5. Critical Cooling Rate and Mass Effect

How fast must cooling be in order to form martensite? The question needs re-phrasing, because the transformation to martensite is temperature dependent, not time dependent. The question should really be how fast must cooling be to avoid other transformations, such as pearlite or bainite.

To accurately assess this, another diagram is needed: a *continuous cooling transformation* (CCT) diagram (Fig. 5). Cooling must be sufficiently fast to avoid the pearlite or bainite "noses" of the CCT diagram. The cooling rate that just avoids the pearlite nose is called the *critical cooling rate*; cooling at this rate or faster will avoid pearlite formation, and result in complete hardening, but slower than this will result in some pearlite and consequent reduction in overall hardness.

When quenched, even in identical conditions, it is obvious that a small workpiece will cool faster than a large one, and also the outside of a large piece will cool faster than the center. This is called the *mass effect*. There will be a size where the cooling rate in the center will fall below the critical cooling rate, even with a "perfect" (or *ideal*) quench (which cools the surface instantaneously). This size is called the *ideal critical diameter*. With a "real" quenching medium, such as oil or water, the maximum diameter that can be fully hardened will be smaller, depending on the severity of the quench; the less severe the quench, the smaller the diameter that can be fully hardened. The maximum size of workpiece that can be hardened to achieve specified properties throughout is called the *ruling section*.

To harden larger pieces in "real" quenchants, and especially if we wish to reduce distortion and cracking due to very severe quenching, we need steels with lower critical cooling rates, or better *hardenability*. This is achieved by adding appropriate amounts of alloying elements.

6. Tempering

Martensite is very hard, but also very brittle, because of its highly distorted and stressed crystal structure. In order to make workpieces at all useful, their toughness must normally be improved. This is done by tempering, which involves heating the martensite up to a suitable temperature to modify its structure, and hence its properties.

The results of tempering depend primarily on temperature (and to some extent time). With plain-carbon steels, at low temperatures (up to about 200°C), some of the stresses are relieved, and there is some carbon diffusion, leading to the precipitation of very fine carbides (not cementite, but epsilon carbide). This process improves toughness with little reduction in hardness.

Tempering at progressively higher temperatures results in other changes taking place, with progressive reduction in hardness, until at very high tempering temperatures (about 650°C) the structure reverts to the "equilibrium" structure of ferrite and cementite, which is very tough, but considerably less hard than martensite. An example of the effect of tempering temperature on properties is illustrated in Fig. 6.

C. Alloying Elements and Heat Treatment

The effects of alloying elements on steel are many and varied. Only those elements and effects specific to the heat treatment of tool steels will be considered here.

Alloying elements can be classified in several ways. Two of these are particularly relevant in this context.

1. Solution Strengtheners and Carbide Formers

Elements may be classified depending on their relationship either with the iron or with the carbon in the steel. The solution strengtheners do not normally form stable carbides in steel, but dissolve in the ferrite (or austenite) to form solid solutions, either interstitial, if their atoms are small (e.g. N, B, P), or substitutional, if their atoms are large (e.g. Ni, Co, Mn). The general effect of these, especially the substitutional elements, is to increase the strength of the steel whilst maintaining good toughness and ductility. The carbide formers predominantly react with the carbon in the steel to form alloy carbides, which are more stable than cementite, and may be either simple or complex in structure and composition. These carbides (e.g. those of Cr, Mo, W, V, Ti, Nb) normally increase the wear resistance and hardness of the steel. It should be noted that some elements, notably Cr, act in both of these ways; some Cr will dissolve in ferrite, and it will also form carbides.

Many tool steels, because of their requirement for high hardness and wear resistance, contain large amounts of these carbide-forming elements. Some of these complex alloy

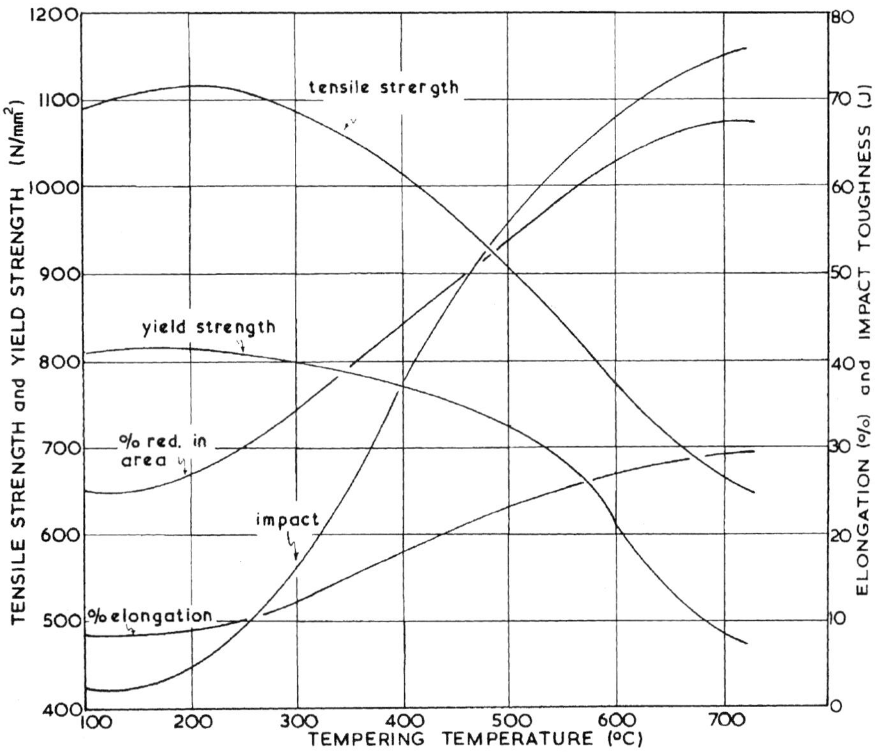

Figure 6 Effect of tempering temperature on properties. (From Engineering Metallurgy by R.A. Higgins. Reprinted by permission of Butterworth-Heinemann.)

carbides are very stable, and are quite resistant to dissolution in austenite, thus necessitating higher austenitizing temperatures than would be needed to dissolve simpler, less stable carbides like cementite. The use of very high austenitizing temperatures brings with it the danger of grain coarsening. This effect of complex carbides can be used to advantage, however, in some of these particles, "insoluble" carbides can lodge in grain boundaries and thus help to prevent grain growth during austenitizing. An example of this is vanadium in high-speed steels.

2. Ferrite Stabilizers and Austenite Stabilizers

Some elements, especially Cr, Mo, W, preferentially stabilize ferrite at the expense of austenite. This has the effect of shrinking the austenite phase field in the iron/carbon phase diagram, raising critical temperatures, and reducing maximum solubility of carbon in austenite. These elements also reduce the eutectoid carbon content. Higher austenitizing temperatures are therefore necessary in order to promote sufficient carbon dissolution to achieve sufficiently high hardness in the martensite formed after quenching. This situation is particularly characteristic of many tool steels, which often contain large amounts of these ferrite-stabilizing elements. An example of a pseudo-binary phase diagram for a high-speed steel is shown in Fig. 7.

Some elements, especially N, Ni, Co, Mn, increase the stability of austenite in preference to ferrite. This has the effect of enlarging the austenite phase field on the

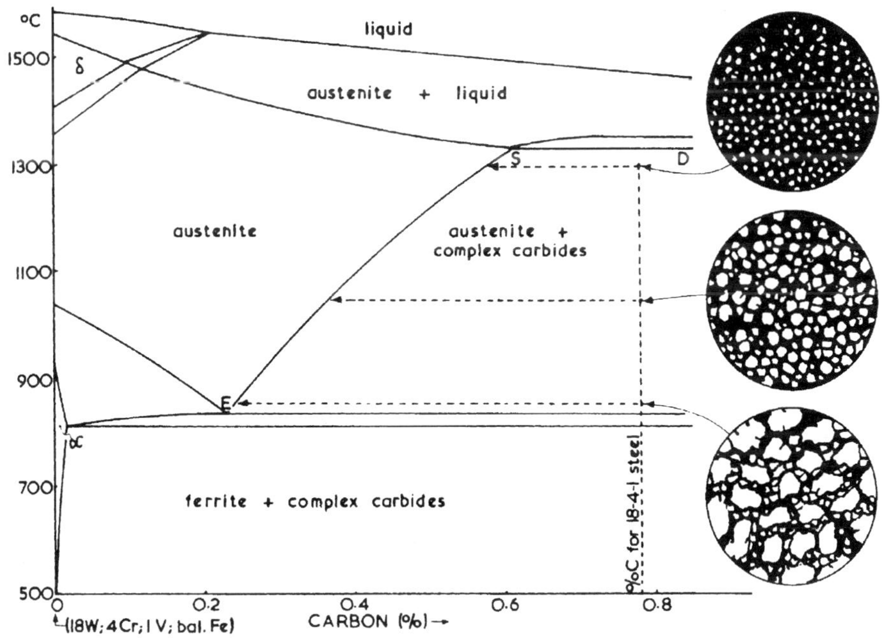

Figure 7 Pseudo-binary phase diagram for T1 high-speed steel. (From Engineering Metallurgy by R.A. Higgins. Reprinted by permission of Butterworth-Heinemann.)

iron/carbon phase diagram and lowering critical temperatures. These lower heat treatment and transformation temperatures will help to promote a finer grain size and microstructure in the heat-treated parts.

The effect of widening the austenite phase field can be used to advantage as, for example, in the case of martensitic stainless steels, where there is a restriction on the allowable Cr content (and hence corrosion resistance) for any given carbon content; exceeding this makes the steel ferritic. This problem can be overcome by adding small amounts of Ni, as is the case with type 431 stainless, thus allowing higher Cr and higher corrosion resistance at a given carbon content. Another example is the use of Co in some high-speed steels. This extends the austenite phase field to greater carbon contents, counteracting the opposite effects of Mo, W, and Cr. This allows a greater degree of carbon dissolution in the austenite during hardening, and a consequently higher martensite hardness after quenching and tempering.

3. Effect on Transformation Rates

As carbon content increases up to about eutectoid, there is a slowing down of the austenite transformation kinetics. A low-carbon steel cannot be hardened because the critical cooling rate is so fast that it cannot be practically achieved, whereas a eutectoid steel can be relatively easily hardened. As carbon content is further increased beyond the eutectoid, austenite transformation again speeds up. Almost all other alloying elements, with the notable exception of Co, slow down austenite transformation rates, shifting the TTT and CCT diagrams to longer times, thus reducing critical cooling rates and increasing hardenability, meaning that milder quenches can be used. Many tool steels fall into this category, being of sufficiently high alloy content as to render them air-hardening, or at least easily hardened with sufficiently high-pressure gas quenching.

In heat treatment, it must be remembered that more than one transformation needs to be considered. The magnitude of the effect on pearlite and bainite transformation kinetics can be substantially different. Mo, for example, has a much larger effect on slowing down the pearlite reaction than it does on the bainite reaction, thus effectively pushing the pearlite "nose" over to a much greater extent than the bainite nose. It thus becomes possible to have a direct cooling rate in a Mo-containing steel that will result in a bainite structure, whereas this is impossible in a plain-carbon steel, in which a cooling rate fast enough to avoid pearlite always results in a martensitic structure.

4. Effect on Transformation Temperatures

The effect of alloying elements on critical temperatures has already been mentioned. This effect (ferrite stabilizers moving temperatures up and austenite stabilizers moving them down) explains their effect on pearlite transformation temperatures.

In the case of the bainite transformation range, most alloys (Si being an exception) move the bainite transformation temperature range downwards. Thus, any element (such as Cr, Mo, V) that moves the pearlite temperature range up and the bainite temperature range down, effectively separates the pearlite and bainite transformation curves. This is in

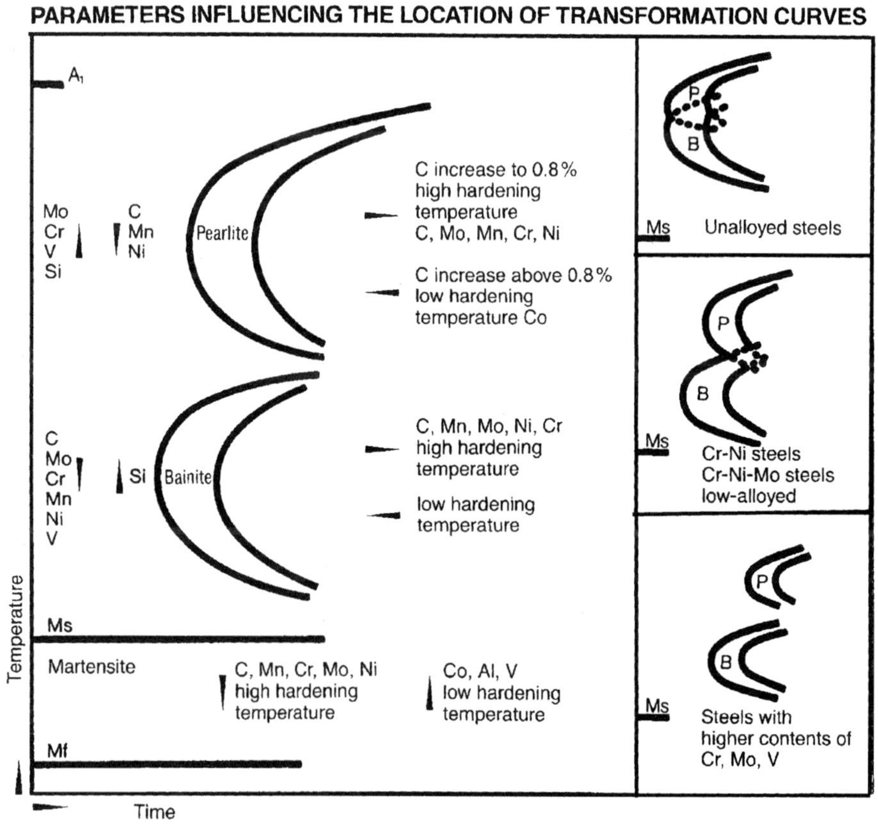

Figure 8 Effect of alloying elements on transformation temperatures and rates. (From Ref. 23.)

addition to the effect of these elements on the transformation rates of the two reactions. The general effects of alloys on the CCT diagram are illustrated in Fig. 8.

Most alloying elements (Co, Al, V being exceptions) move the martensite transformation temperature range (M_s to M_f) down. Lowering the M_s means that transformation stresses are likely to be higher, since the steel will at this lower temperature have a higher yield strength, and will thus be less compliant to transformation strains. This will increase the danger of quench cracking. These elements do, however, increase hardenability, and render the steel more amenable to martempering, allowing a lower temperature initial quench.

Lowering the M_f to below normal ambient temperature will result in incomplete transformation of austenite to martensite; the lower the M_f, the greater the proportion of retained austenite in the as-quenched microstructure.

5. Effect on Tempering

The general effect of most alloying elements is to slow down the tempering reactions, and move them to higher temperatures. This is particularly the case with the strong carbide-forming elements. The overall result of retarding the tempering reactions is that alloy steels retain higher levels of hardness at higher tempering (and therefore service) temperatures. This is known as *hot hardness*. Tempering curves for a range of tool steels are illustrated in Fig. 9.

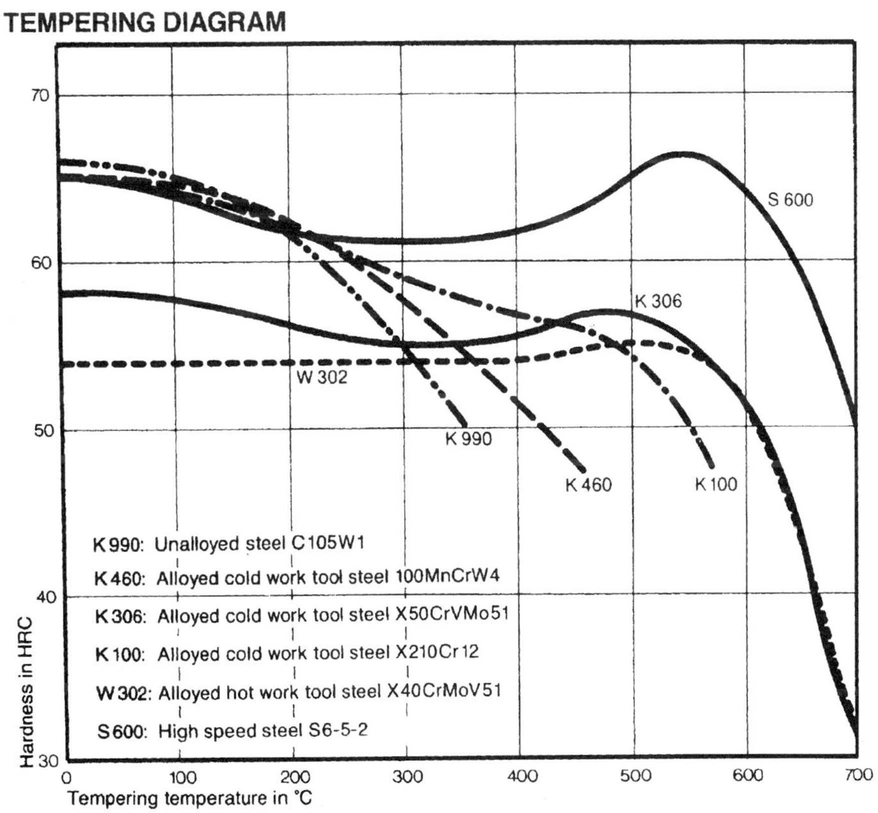

Figure 9 Tempering curves for various alloy steels. (From Ref. 23.)

In plain-carbon steels and lean low-alloy steels, any retained austenite in the as-quenched microstructure will transform, normally to bainite, during tempering at temperatures in the range 250–350°C. Increasing amounts of alloying elements, as well as increasing the amount of retained austenite present, also causes the austenite to be more stable. In some high-alloy steels, tempering temperatures in excess of 500°C are often employed. In this temperature range, the austenite becomes "*conditioned*" by precipitation, and transforms to martensite on cooling after the temper. This freshly formed martensite, if in significant quantities, will then need to be tempered; hence the necessity to *multiple temper* some high-alloy steels.

6. Alloy Carbides and Wear Resistance

The wear resistance of a steel depends not only on its hardness, but primarily on its microstructure, particularly the type and distribution of carbides present. Whilst a hardened plain carbon or low-alloy steel may be almost entirely martensitic in microstructure, many tool steels will have a significant proportion of alloy carbides in their heat-treated microstructure. Most alloy carbides are considerably harder than cementite, and the presence of some of them (e.g. VC) can dramatically improve wear resistance.

The type, size, and distribution of alloy carbides is strongly influenced by heat treatment, primarily during austenitizing, which governs the extent of dissolution of primary carbides, and tempering, which governs the precipitation of carbides from the martensite (and austenite) present after hardening. Both the composition of carbides, and their size and distribution, will be influenced by tempering temperature, and to a lesser extent time. In particular, some complex alloy carbides will only form during tempering at temperatures of the order of 550°C or higher. This will be discussed in a later section.

7. Secondary Hardening

The phenomenon known as *secondary hardening*, observed particularly in high-speed steels, results in an increase in hardness after tempering in the temperature range around 550°C. It is the result of a combination of two different mechanisms: firstly, the transformation of retained austenite present in the as-quenched microstructure, and secondly an effect similar to precipitation hardening as a result of the precipitation of an array of fine, hard complex alloy carbides during the tempering. These carbides are often referred to as "secondary hardening carbides", and are often of the form MC, M_2C, or sometimes M_6C, depending on the alloying elements present and tempering temperature ("M" denoting metal atoms).

D. Precipitation Hardening

For most steels, the most effective hardening mechanism is martensitic hardening, as described in earlier sections. Martensite containing significant amounts of carbon is hard because the presence of lattice strain and large numbers of crystal defects act as barriers to dislocation movement, and hence inhibit plastic deformation. An alternative mechanism, used in some low-carbon steels and many non-ferrous alloys, is precipitation hardening, in which strain fields are set up round a very large number of sub-microscopic non-equilibrium "embryo precipitates".

An alloy composition is chosen that contains alloying elements which at high temperature would be soluble in the matrix, but with falling temperature, under equilibrium conditions, would become insoluble and precipitate out. Large, widely dispersed precipitates that are totally distinct from the matrix (said to be *incoherent*) do not have a very significant influence on hardness. If the normal precipitation reaction is suppressed, for example by quenching, then the alloy atoms will remain in supersaturated solid solution. This is a non-equilibrium condition, and given sufficient time and thermal energy these solute atoms would cluster together and form precipitates. The temperature and time are controlled, depending on the particular alloy, in order to allow the atoms to cluster together into zones, but not to have sufficient energy to form distinct (incoherent) precipitates. These zones or clusters of atoms, still crystallographically part of the matrix (said to be *coherent*), result in the setting up of strain fields around them, caused by the mismatch in atomic size between the alloying elements and the matrix. These strain field inhibit dislocation movement, and thus increase hardness.

Some mold steels, mostly stainless but some not, fall into this category. They are normally low in carbon content, and are usually supplied in the "solution treated" (i.e. quenched) condition. Their matrix may often be either austenitic or martensitic (low carbon, hence soft), or a mixture of the two. In this condition, they are relatively soft, and can be machined, etc. After fabrication, they are subjected to a precipitation heat treatment, often in the temperature range 450–550°C. Normal hardness achieved varies with alloy and treatment, but is usually in the range 40–50 HRc. The advantage of this treatment is that the dimensional change during the precipitation treatment is negligible, there being no quenching involved, and no phase change.

E. Surface Hardening

1. Thermal Methods

In some cases, only the surface, or even only a specific area of the surface, is required to be hardened. In such cases, a high intensity heat source (oxy-gas flame, induction, laser, electron beam) can be used to heat up the surface so rapidly that the heat does not have time to conduct far into the core, and then rapidly quench. Such processes have the advantages again of minimal distortion, as only the surface layer is affected, and retain a tough inner core. With careful control of process parameters, case depth can be controlled. Hardenability is not normally an important issue in these processes, so they are often used on carbon and lower-alloy grades.

2. Carburizing

As previously explained, the hardness of martensite depends on carbon content. If a low-carbon steel is heated to its austenitizing temperature, and subjected to an environment containing high levels of carbon, then the surface layers will become enriched with carbon. On quenching, there will be a gradation in hardness, with the high-carbon surface being harder than the low-carbon core. Careful selection of process parameters, particularly to control surface carbon content and depth of carbon diffusion, can control surface hardness and case depth. This process is widely used in engineering steels, but only rarely for tooling.

3. Nitriding

In certain alloy steels, surface properties can be improved by nitriding. Nitriding is normally performed by exposing the steel for a prolonged time, at a temperature usually in the range 500–540°C, to an environment rich in atomic nitrogen, often derived from ammonia gas, which "cracks" on the metal surface and releases nitrogen in its atomic form. Ordinary molecular nitrogen (N_2) is effectively inert to steel.

The nitrogen that diffuses into the steel combines with certain alloying elements, notably Al, Mo, Cr, and others, to form very hard, stable alloy nitrides, and also forms a diffusion layer, increasing the strength of the matrix, and introducing compressive residual stresses in the surface layer. The resultant surface has high hardness, which can be up to the equivalent of 70 HRC, depending on alloy content, and has a reduced coefficient of friction, thus further improving wear properties. The presence of the compressive residual stresses in the surface improves fatigue resistance. The properties are fairly stable almost up to the original nitriding temperature. The case depth depends on steel composition and on nitriding temperature and time, but is usually in the range 0.1–0.3 mm.

The fact that there is no phase change and no quenching involved, together with the relatively low process temperature, means that dimensional change is very small (there is a very slight, predictable growth in dimensions in the surface). Nitriding is normally performed on finished products. It is strongly recommended that prior to nitriding, components have a microstructure that is stable at the process temperature, so tempering temperatures should be in excess of the nitriding temperature. In the case of some processes, plasma nitriding for example, the workpiece is not deliberately heated to above 500°C, but nevertheless, the surface does become heated during the process, so the same recommendation regarding tempering still applies.

4. Coatings

A number of processes are available for overlaying or implanting hard coatings of various compositions onto or into the surface of finished workpieces. Description of these processes is outside the scope of this section, but many processes are based on physical vapour deposition (PVD), chemical vapor deposition (CVD), and ion implantation. These are in addition to the more traditional electro (or electroless) plating and thermal spraying. Just one example is the now widely used PVD of Titanium Nitride (TiN), especially on cold-work and high-speed steels.

It should be remembered that during some of these processes the workpiece becomes heated, and therefore, as in the case of nitriding, the prior heat treatment, especially tempering temperature, must be selected with this in mind.

Another important consideration is that most of these coatings are very thin, and therefore need a relatively hard supporting surface underneath, if they are to perform well, without becoming cracked and spalled off because of plastic deformation in the underlying piece. This is why some of these coatings have proved most successful when overlaid onto inherently hard cold-work and high-speed steels.

F. Special Considerations for Tool Steel Heat Treatment

Although the heat treatment of tool steels is not fundamentally different in principle from that of other types of steel, certain special characteristics of tool steels mean that control of the heat-treatment process is more critical than with plain-carbon and low-alloy steels. Some of these considerations can be summarized as follows:

- These steels are normally used for very demanding applications, often using the steel at the limits of its potential, making optimum heat treatment of vital importance.

- Although some tool steels may be used in mass production, they will perhaps be more commonly used for expensive "one-off" applications than other steels, making it even more critical than ever to get it "right first time".
- Many of the tool steel grades are high alloy, containing considerable proportions of strong carbide-forming elements, resulting in significant amounts of complex carbides in the microstructure. This has an influence on austenitizing temperatures, which may need to be significantly higher than for ordinary grades of steel in order to dissolve these carbides. In some cases, however, not all the carbides are required to go into solution during austenitizing.
- The significantly higher austenitizing temperatures used for many high-alloy tool steels mean that precautions to protect the surface during the treatment are of increased importance.
- Many dies are heat treated in the finished condition, so that maintaining surface integrity is of vital importance.
- The high levels of both carbon and alloying elements in many of these steels mean that the M_s and M_f temperatures are often quite low, significantly increasing the risk of quench cracking. Lowering the M_f increases the likelihood of having significant amounts of retained austenite in the as-quenched microstructure.
- The demands on dimensional tolerances are normally more exacting than with most ordinary steels. This, together with the greater difficulty of machining most of these grades in the hardened condition, makes distortion control even more critical than with most steels.
- The high-carbon contents of many tool steel grades mean that there is an increased volume change during the martensite transformation, further exacerbating the dimensional tolerance problem, and also further increasing the risk of quench cracking.
- Some high alloy grades have a lower thermal conductivity than low-alloy grades, thus increasing the importance of control of heating rates, and appropriate preheating, many grades requiring at least two preheats prior to austenitizing.
- Some of the higher alloy grades have a higher coefficient of thermal expansion, and most have higher yield strengths, especially at elevated temperatures, resulting in higher thermal stresses during heating, again emphasizing the importance of heating rates and preheating.

From the foregoing it will be apparent that optimum heat treatment of tool steels is a somewhat more specialized procedure than that for ordinary grades of carbon and low-alloy steels, and hence special procedures, equipment, and controls are employed.

G. Choice of Equipment and Processes

The optimum choice of equipment and processes for heat treatment of tool steels, like other similar decisions, depends on a number of factors specific to the needs of the organization. Some of these considerations are:

- Is it only a single grade of steel to be treated, for one particular purpose (e.g. mass production of HSS drill bits), or has a variety of grades to be catered for?
- What range of steel grades needs to be catered for, and in particular what hardenabilities (i.e. have water- or oil-hardening grades to be accommodated)?
- Amount of each broad type in the workload—whether it is essential to cover all possible grades, or just the majority and contract-out the minority.

- Availability of suitable sub-contract facilities for some or all production.
- Whether installed capacity could be utilized for sub-contract work for others.
- Maximum piece size and weight.

These considerations are in addition to the normal economic and resource-related issues (capital, personnel, etc.).

Except in the unusual case of a single grade or product, which would allow for an optimum fully dedicated equipment/process, in most cases, tool steel heat treatment is much more "jobbing" in nature, and thus flexibility is perhaps the keyword in choosing equipment and processes. If there is no sufficient volume of work to justify the not inconsiderable expense of installing the specialized equipment and process control measures needed for these steels, it is much better to sub-contract the heat treatment to a reputable specialized supplier of such services. It is never sensible to "cut corners" and try to get by with marginally suitable equipment or processes.

For heat treating tool steels to an adequately high standard, "ordinary" heat-treatment equipment, either batch or continuous, designed for treating carbon and low-alloy steels (such as sealed-quench furnaces) are unlikely to be suitable. Higher austenitizing temperatures, accurate control of preheating and heating rates, and a greater flexibility in quenching are needed, as well as consideration of the often much smaller job-lots involved in tool steels. It is unlikely that a single type of furnace will be suitable for all grades, and therefore a combination of furnace types is usually found to be preferable. These may be summarized as follows.

1. Vacuum Furnaces

These are the general "work-horses" for tool steel heat treatment, providing a full range of temperatures, optimum surface integrity, and excellent process control of most parameters, including quench severity. Modern furnaces with high-pressure efficiently designed gas quenching can achieve cooling rates approaching, but not equal to oil quenching, and also have the ability to achieve marquenching and also more gentle cooling rates when desired.

The main limitations on vacuum furnaces are that they are not suitable for the normal water or oil hardening grades, and unless equipped with forced convection heating facilities, are very inefficient for tempering and other low-temperature work. Even with forced-convection heating, the economics of using a vacuum furnace for tempering, unless overnight or when unmanned, when nothing else could otherwise be processed, is not normally attractive.

There are many design variations in vacuum furnaces, and the choice of a specific model for a given requirement is best discussed between prospective purchaser and supplier, starting with the overall processing requirements.

2. Air-Recirculating Furnaces

For most processes up to about 650°C, including tempering, stress relieving, precipitation hardening, preheating, etc., unless very rigid requirements on surface finish must be met, forced recirculation air furnaces provide a simple and economic solution. There are numerous variations in design, and it is sensible to make the furnace compatible in size and shape with vacuum or other furnaces installed. There is a discoloration when treating at temperatures above about 500°C, but this is rarely a problem, unless a mirror-bright finish is essential, in which case a vacuum furnace may be employed (or fluidized-bed furnace), if the very slight, but much reduced, discoloration can be tolerated.

3. Salt Baths

A combination of a number of salt-bath furnaces provides a very flexible and rapid heat treating facility, especially when "jobbing" or treating small batches to different requirements. With suitable salts, surface hardening also becomes a possibility. The problems associated with toxic salts and effluents have in recent years been reduced (but not completely eliminated) by some of the low-toxicity salts now available. There do still remain the problems of ensuring total freedom from moisture on the workpieces, and drag-out of salt, especially in "blind holes", and possible contamination of quenching media, together with the washing requirements to ensure clean workpieces. To minimize the drag-out and contamination problems, wherever possible salt-bath quenching should be considered, provided the work has sufficient hardenability.

4. Fluidized Beds

Fluidized-bed furnaces are an attractive alternative to salt baths, and have become more popular in recent years. They have most of the advantages of salt baths, including rapid heat transfer and flexibility, and provided a number of units are installed, allow almost total flexibility, especially if some are provided with a range of active gases, such as hydrocarbons and ammonia, in addition to inert gas such as nitrogen. They also have the advantage that the bed material is inert and non-toxic, and does not need washing off.

There is inevitably a small amount of drag-out of bed material, even when the work is vibrated on removal, and for this reason, it is advisable, wherever hardenability will permit, to quench in a fluidized bed. Tempering can also be done in another fluidized bed, as can preheating. Bed material (usually alumina grit) washed into liquid quenching media will tend to sink to the bottom, but some filtration is advisable to prevent recirculation through pumps, etc.

Provided suitable gas equipment is installed, all the normal thermochemical surface hardening processes may be accomplished in any unit. The only real limitation is that at very high austenitizing temperatures, such as those used for some high-speed steels, retort life is likely to be limited, and a vacuum furnace may be a better alternative. The work, although "semi-bright", will be discolored during transfer from preheating to austenitizing, or from austenitizing bed to quenching. The finish achieved in a vacuum furnace is superior to either salt bath or fluidized bed.

5. Overall Choice

The overall combination of facilities depend, as mentioned previously, on the mix of steel types to be processed, and some typical possibilities are as follows:

- If only small quantities of a wide range of grades is to be processed, it may be wise to sub-contract, provided a reputable source of such service is available locally.
- If oil- and water-hardening grades do not have to be accommodated, then a vacuum furnace plus air-recirculating tempering furnaces provides the best solution.

- If high-speed steels do not need to be accommodated, but all other grades, including water and oil hardening grades, plus the facility to surface harden, then fluidized-bed furnaces may be the best choice.
- If all grades need to be accommodated, and complete flexibility with optimum quality is needed, then a combination of vacuum and fluidized-bed, or possibly vacuum and salt bath, may be the best combination.

H. Surface Integrity

When exposed to air at an elevated temperature, steel will react with its environment. There are two possibilities:

1. The iron atoms will react with oxygen to form iron oxide. This is called *scaling*.
2. Carbon in the steel will react with oxygen to form carbon dioxide, leading to *decarburization*.

Other reactions are of course possible in different atmospheres. It is obvious that both surface oxidation and decarburization are undesirable, and steps need to be taken to reduce or eliminate them.

It should also be remembered that not only atmospheric constituents, but also surface contaminants, can cause undesirable reactions. Heating a piece with its surface covered in oxide, even in a vacuum, will cause decarburization, as the surface oxide reacts with carbon in the underlying steel. This emphasizes the need for effective cleaning of workpieces before treatment, removing all oil, grease, oxides, and any other contaminants.

Some methods of promoting surface integrity are as follows.

Vacuum. Provided the steel surface is clean, no undesirable reaction is possible, so treatment in a vacuum is ideal. There is a remote possibility of loss of volatile alloying elements, but with steels in a normal vacuum furnace, this is not usually a serious problem.

Atmosphere Control. Atmospheres containing oxidizing gases (including carbon dioxide and water vapor) will decarburize, whilst atmospheres containing carbon monoxide and hydrocarbons will carburize. Careful control of atmosphere to achieve the optimum balance of carburizing and decarburizing gases, depending on the grade of steel being treated, can ensure neither carburizing nor decarburizing conditions prevail. The atmosphere composition to maintain this correct level of *carbon potential* will vary with temperature, and therefore needs careful control.

Inert Atmospheres. Nitrogen, although not truly chemically inert like argon, is inactive with steel, and so can be considered "inert" in this context. Nitrogen-based atmospheres can be used for many heat-treatment processes, for gas quenching in vacuum furnaces and also in atmosphere furnaces and fluidized beds. Another commonly used "inert" (so far as the steel is concerned, though obviously not inert) is cracked ammonia, comprising a nitrogen/hydrogen mixture. Unlike nitrogen, which is safe at all temperatures, cracked ammonia can only safely be used at austenitizing temperatures, otherwise there is a risk of explosion.

Salt Baths. Immersion in a molten salt excludes the surrounding atmosphere. The composition of the salt bath can be maintained to be either neutral or carburizing, so in addition to preventing undesirable reactions, salt baths can also be used for *carbon restoration*, adding carbon back into a previously decarburized surface.

Fluidized Beds. Essentially, any gas can be used as the fluidizing medium. At low temperatures, air is satisfactory, even at temperatures in excess of 500°C, such as for tempering high-speed and hot-work steels. The agitation and (very) mildly abrasive conditions in the bed promote a brighter surface than is achieved at a similar temperature in an air-recirculating furnace. At higher temperatures, nitrogen-based atmospheres are commonly used, but the full range of active gases to promote any thermochemical reaction may be used.

Wrapping. In the absence of atmosphere control, workpieces can be wrapped in stainless steel foil or envelopes, thus excluding the outside air. It can be advantageous to have an inner layer of paper, which on heating will partially burn and use up any oxygen inside the package. It should be remembered that this wrapping constitutes an insulating layer (especially if it includes paper) which will reduce heat transfer, especially during quenching. This method should only be used as a last resort—it is much better to use more suitable furnace equipment, or sub-contract if no such facilities are available in-house.

Packing. Pieces can be packed in sealed boxes containing spent carburizing granules, cast iron chips or the like. This is really suitable only for annealing, as it is almost impossible to effectively quench in such conditions. Like wrapping, it is not recommended except as a last resort.

I. Dimensional Change and Distortion

Dimensional change may be defined as a change in volume as a result of the treatment, whereas distortion may be defined as a change in shape (which may include a change in volume).

1. Unavoidable Dimensional Changes

Thermal expansion and contraction occur on heating and cooling. Superimposed on this, there is also a volume change as a result of phase transformations (ferrite/austenite, and austenite/martensite). Volume changes also occur during tempering of martensite. These changes are totally unavoidable, and must be allowed for in the component design.

On heating and slow cooling, the thermal expansion and phase change on heating are reversed on cooling, so theoretically there should be no permanent volume change. On hardening, however, the expansion as austenite transforms to martensite is greater than the contraction as ferrite transformed to austenite on heating, so in consequence, there will be a net overall expansion as a result of the treatment (Fig. 10).

The expansion on martensite formation increases with increasing carbon content, so the dimensional change would normally be expected to be greater with high-carbon steels. This is also true of increasing austenitizing temperatures, resulting in higher carbon dissolution. However, to counteract this, the presence of retained austenite results in a volume reduction (the transformation from ferrite not having been reversed). The overall effect of increasing carbon content and austenitizing temperature are therefore complex, depending on the amount of retained austenite in the microstructure, which might or might not compensate for the expansion due to martensite formation. Careful control of austenitizing temperature can therefore help to at least control the overall volume change (see Fig. 11).

On tempering, there is a volume reduction as carbides precipitate out form martensite, and the crystal structure changes. Again, the higher the carbon content, the greater the volume change. Once again, however, retained austenite can play a part, since there is an expansion as it transforms. These changes occur progressively, and at different temperatures, so careful choice of tempering temperatures can also influence overall dimensional change. The relationship between tempering temperature and dimensional change naturally varies with steel grade (Fig. 12); details for any particular grade should be available with the technical literature provided by steel suppliers.

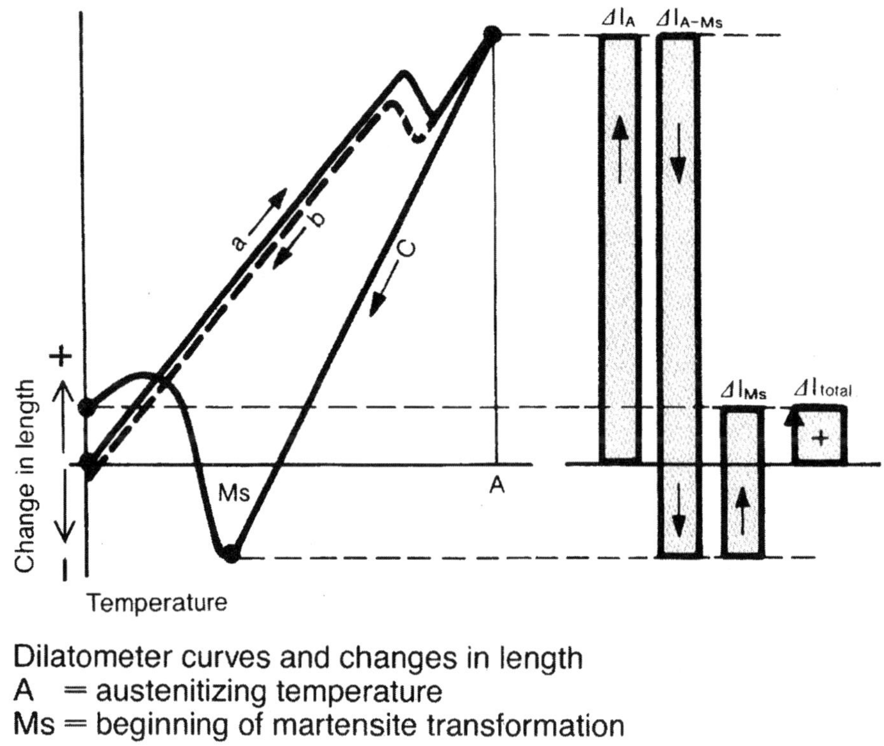

Figure 10 Volume changes on heating and cooling: a represents heating, b represents slow cooling, and c represents quenching. (From Ref. 23.)

It should be remembered that dimensional change may not be isotropic in some grades, depending on the microstructure developed during the steel processing, especially rolling and forging operations. It is important in such cases to know the orientation of the workpiece in relation to the rolling direction, and take this into account when allowing for dimensional changes as a result of heat treatment (Fig. 13). This anisotropy will also vary with steel quality and supplier, and can be minimized by using premium grades such as Electro-Slag-Refined (ESR) grades, or Powder Metallurgy (P/M) grades.

From the above, it will be clear that to achieve excellent dimensional accuracy, even with a "perfect" heat-treatment operation, a detailed knowledge of the steel and its source, and close liaison between the tool designer and heat treater are needed.

2. Controllable Distortion

Some distortion problems may be caused by poor heat-treating operations, and are therefore "avoidable". In other cases (due to difficult tool design for instance), in any real heat-treatment procedure, it may be difficult or impossible to avoid some distortion. Steps may be taken to at least minimize the problems. Distortion may result from a number of causes.

Residual Stresses. These may be present in the incoming steel stock or induced during heavy machining or forming operations. On heating, these will relieve themselves by plastic deformation, resulting in a shape change (distortion). The only real solution to this is to undertake a stress relief anneal after the bulk of the machining is complete, but before final machining and hardening.

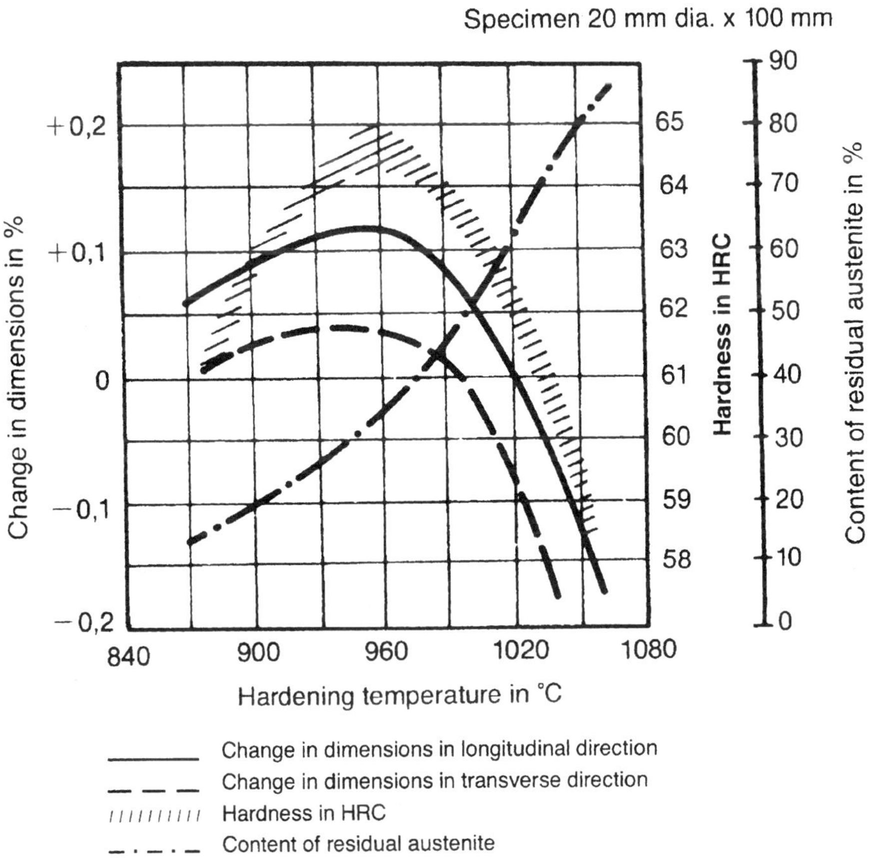

Figure 11 Effect of austenitizing temperature on dimensional change in D2 tool steel. (From Ref. 23.)

Segregation or Decarburization in the Steel. This can result in differential volume changes in different parts of the piece during heat treatment, thus causing it to distort. There is little either the heat treater or tool designer can do about this, except choose a better quality material. It is often worth paying the extra for premium grade material for critical work.

Uneven Heating and Cooling. If all parts of the workpiece could be heated or cooled uniformly, thermal expansion and contraction would occur uniformly through the piece, and no distortion would result. Rapid or uneven heating or cooling results in temperature differentials (for example between thin and thick sections, and inevitably between the case and core of the piece). This results in thermal stresses, as different regions of the piece try to conform to the differential volume change. If these stresses exceed the yield strength of the material (which is obviously much lower at high temperatures), then plastic deformation will occur in the weaker region, resulting in a permanent distortion. This is the reason for the importance of control of heating rates and preheating, and also for choosing the mildest possible quench that will avoid undesired transformations.

Sagging and Creep. At elevated temperatures, the yield strength of the steel will be severely reduced, so much so that slender sections may be unable to provide enough support for the weight of the piece. Over a period of time, plastic deformation (*creep*) may

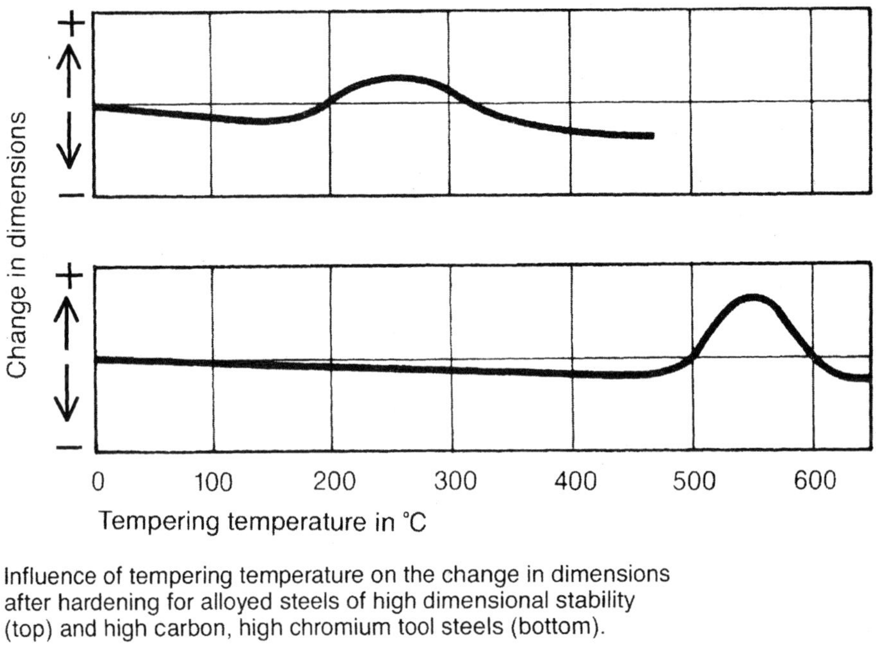

Figure 12 Effect of steel composition and tempering temperature on dimensional change. (From Ref. 23.)

occur. To avoid this, careful attention is needed when loading the work, and providing adequate support for delicate or slender pieces (for example, supporting long thin pieces vertically, hanging them rather than standing them up).

Quenching. During quenching, there is likely to be considerable temperature variations in the piece (thin/thick sections, case/core), and thus transformation to martensite, with its associated volume change, will take place at different times in different regions of the workpiece. The volume change is unavoidable; all that can be done is to minimize the magnitude of the temperature variation, and thus reduce both the difference in yield strength between hottest and coolest parts, and also the degree of differentiation of martensite formation in different regions. This explains the advisability of choosing the mildest quench that will just harden the piece, and also the advantage of marquenching.

Delays in Tempering. It is important to temper hardened tool steels without delay. Failing to do this may result in uncontrolled volume changes, either due to the spontaneous transformation of austenite, if the temperature falls well below the specified endpoint of the quench, or alternatively delays can result in the stabilization of retained austenite, which then fails to transform as expected during the subsequent tempering.

J. Austenitizing

1. Temperatures

The first stage of both soft annealing and hardening processes is austenitizing, that is converting the microstructure to austenite, after which the desired final microstructure can be developed by controlling the transformation of austenite on subsequent cooling.

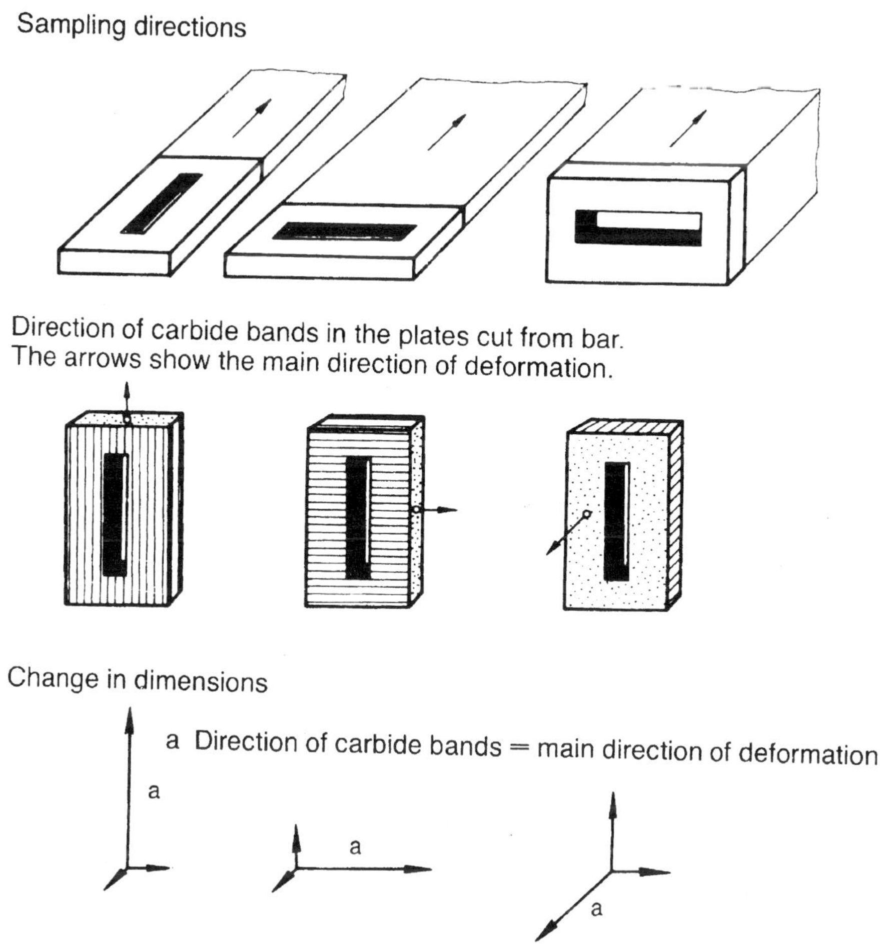

Figure 13 Anisotropy in dimensional change. (From Ref. 23.)

In lower alloy grades, the object is to form a homogenous austenite structure, but without excessive grain growth. Dissolution of carbon and alloying element atoms in the austenite is both temperature and time dependent. The temperature and time must therefore be controlled so as to effect the required dissolution and homogenization of carbon and alloying elements, but not to allow undue grain coarsening. This can be demonstrated by the use of *TTA diagrams* (Time-Temperature-Austenitizing—Fig. 14). Unfortunately, these are not so readily available as TTT or CCT diagrams. So in general, the best policy is to use temperatures and times recommended by manufacturers in their technical literature, or in standard reference works [24].

Unlike lower alloy grades, with high alloy tool steels, especially high-speed steels and the high-carbon high-chrome grades (the "D" series), it is neither possible nor desirable to effect complete dissolution of all carbon and alloying elements. Sufficient carbon and alloy dissolution must be achieved in order to develop the required hardness in the martensite formed after quenching (only the carbon and alloying elements dissolved in the austenite prior to quenching take part in the hardening process). The remaining carbon and alloys left undissolved during austenitizing remain in the microstructure as primary carbide

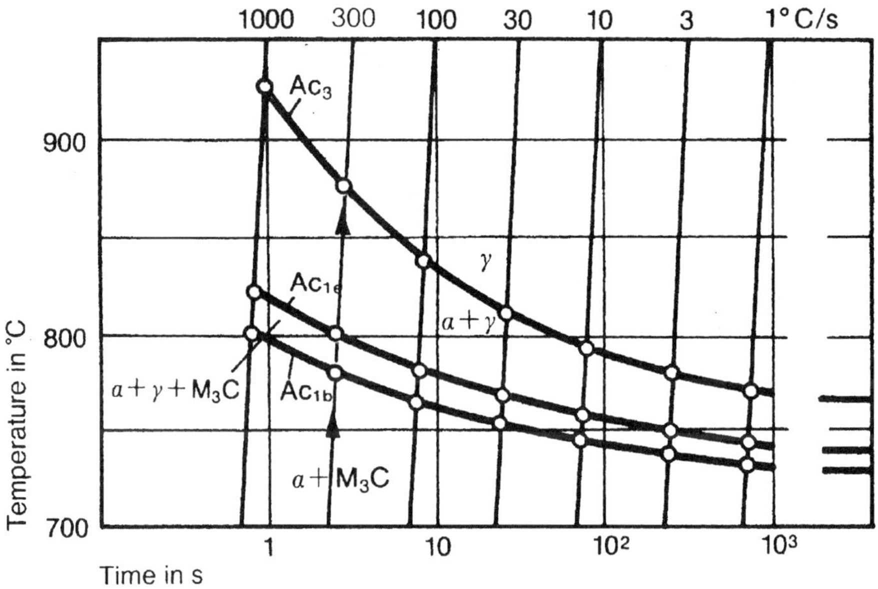

Figure 14 TTA diagram for conditions of continuous heating. (From Ref. 23.)

particles, and are an essential part of the steel's final microstructure. There is often a wide range of possible austenitizing temperatures for any given grade, and it is important to utilize this variable to optimize final properties, rather than controlling final hardness only by varying the tempering temperature. This will be discussed later, in relation to tempering.

2. Temperature Control

Accurate control of austenitizing temperature is critical to the success and quality of the result of the heat-treatment process, and therefore the performance of the tool. In the case of treatments in salt baths or fluidized beds, where temperature uniformity and heat transfer are excellent, the temperature of the bed or bath, once it has stabilized after introducing the work, can be taken as an accurate reflection of the workpiece temperature. The process control can therefore be based on bed or bath temperature.

In the case of an atmosphere or vacuum furnace, however, it is essential to have an accurate measure of the actual workpiece temperature. The best way of doing this is to insert a thermocouple into a suitably sized hole in the workpiece (many tool steel workpieces will have such holes in them). If there is no suitable location in the workpiece that can represent the section of the piece, then it is necessary to use a "dummy block". The heat treater will require a range of different sized dummy blocks to represent different sizes of workpiece. There can be considerable variation in temperature during heating up, especially in a vacuum or atmosphere furnace, between thin and thick workpieces, and in different locations in the furnace. It is therefore highly desirable to have at least two thermocouples, one to register the anticipated hottest part, and one to register the anticipated coolest part, and control the furnace so that the hottest part does not become overheated, whilst at the same time ensuring that the coolest part reaches the required temperature.

3. Preheating

As mentioned in an earlier section, some high alloy tool steels require quite high austenitizing temperatures, many in excess of 1000°C, many high-speed steels in excess of 1100°C, and some in excess of 1200°C. Introducing a cold tool steel workpiece into a furnace at such a high temperature, and holding until the whole piece is heated through, is totally impractical. Firstly, the thermal shock involved would almost certainly crack or severely distort the piece, and secondly, the long holding time needed to through heat any large section would almost certainly result in severe grain growth in thinner sections. This demonstrates the absolute necessity for controlled rates of heating and preheating.

It is recommended that all tool steels, even the lower alloy water hardening and oil hardening grades, should have at least one preheat. This first preheat should be at a temperature just below the eutectoid. Its purpose is to take the whole load up to a uniform temperature prior to any phase change. For lower alloy grades, this is commonly at about 650°C, but for some high alloy grades with higher eutectoid temperatures, it can be 700°C or 750°C. If thermal shock is considered to be a problem, especially when introducing a high-speed steel into a salt bath or fluidized bed at 750°C, then a preliminary preheat in the temperature range 500–550°C may be used. This preliminary preheat is normally unnecessary in a vacuum or atmosphere furnace, as the heating rate is much slower in such a furnace.

Having achieved temperature uniformity below the eutectoid, the next stage is to take the temperature up into the austenitic range, and achieve uniformity. For the lower alloy grades, this will normally be the hardening temperature (often in the range 790–830°C), but for some high alloy grades, and especially for high-speed steels, this will be an intermediate preheat, often at between 850°C and 900°C. The object of this is to achieve a homogenous austenitic structure throughout the section, at a temperature well below that at which significant grain growth is likely, so that a longer soak time can be used at this intermediate temperature, to minimize the time at the much higher true austenitizing (hardening) temperature.

K. Annealing and Stress Relieving

Most tool steels are supplied in the soft-annealed condition, so annealing is normally only necessary when re-working pieces. Full annealing should always be undertaken before re-hardening. Specific annealing procedures for each particular grade will normally be supplied with the technical literature provided by the steel supplier. In general, there are two commonly used procedures: either slow cool from the austenitizing temperature (cooling rate depends on grade) to a temperature below the transformation range, or alternatively an "isothermal anneal", in which the work is cooled at a controlled rate to a temperature below the austenitizing temperature, and held at that temperature for transformation to proceed.

Stress relieving is an important part of the production process, and should not be neglected. On completion of the bulk of the machining operations, but before finishing, it is likely that there will be residual stresses in the workpiece, either from the incoming stock, if machining has removed material asymmetrically, or induced by the machining operations themselves. On heating, these stresses will relieve themselves by plastic deformation, thus causing dimensional movement in the piece. This distortion is unavoidable. The stress relieving process, heating up the steel to a temperature normally about 650°C, and holding for at least 2 hr, will allow this distortion to occur before final machining and hardening, thus minimizing the distortion on hardening, and also minimizing the machining required to correct the distortion in the hardened condition.

Because most tool steels have high hardenability, and many are air hardening, normalizing is not usually practiced on them.

L. Quenching

The general principles of quenching are not different from ordinary grades of steel. The mildest quench that effectively hardens the given section size of the grade of steel should be used, and step hardening may often be used to advantage, even in a vacuum furnace. In the case of oil-hardening grades, oil temperature should in all cases be not less than 80°C, and where possible martempering should be used. This will minimize distortion on quenching, which is an inherent danger with these grades. Where hardenability permits, it is advantageous to use a fluidized-bed quench for steels treated in a fluidized-bed furnace, eliminating the problem of carry-over of grit into liquid quenching baths. For similar reasons, it is usually advantageous to quench salt-bath treated work in a molten salt bath.

Continuous cooling transformation diagrams and recommendations for quenching are normally provided in technical literature, or standard references. The exact choice, for example of quench pressure in a vacuum furnace, has to be based on trials and actual experience with a given furnace, as cooling rate is influenced not only by pressure but by flow rate and efficiency of heat exchangers, which will be specific to a particular furnace.

Quenching is deemed to be complete when a specified end temperature has been achieved. Bearing in mind that the M_f temperature may be substantially below room temperature, it is important to observe the end temperature recommended for the particular grade of steel. On completion of quenching, workpieces should be transferred to the next process without delay. There are two main reasons for this, firstly the steel will be vulnerable to cracking, especially if the temperature falls uncontrolled, and secondly any retained austenite may become stabilized over a period of time, making it very difficult to subsequently transform.

M. Cold Treatment

Cold treatment normally refers to the process of reducing the temperature of the as-quenched pieces to a temperature of about −80°C, often using a dry-ice based process or mechanical refrigeration. This temperature is near or below the M_f temperature for most steels, and thus retained austenite will be eliminated or substantially reduced, increasing hardness and improving dimensional stability. For some high-alloy grades, such as high-speed steels, the M_f may be well below −80°C, and treatment down towards −140°C may be necessary to eliminate all retained austenite. This latter process is normally achieved using vaporized liquid nitrogen. Neither of these processes is time dependent; so, provided the required temperature has been achieved throughout the section, long soak times are not required.

N. Deep Cryogenic Treatment

Deep cryogenic treatment involves subjecting the work to liquid nitrogen temperatures (−196°C), and soaking for a prolonged period, often 24 hr or more. This causes changes to occur in the as-quenched martensite, and results in the precipitation of a fine array of carbides after tempering, increasing the wear resistance of the steel, but having little effect on hardness. It should be remembered that it is impossible to do deep cryogenic treatment without retained austenite being transformed as the temperature falls through the cold treatment temperature range, so the net effect will be a combination of cold treatment and deep cryogenic treatment.

O. Tempering

Tempering is an extremely important part of the heat-treatment cycle. It could even be argued that it is the most important element, since it is the tempering that develops the final microstructure (and therefore properties) in the workpiece. Incorrect tempering can negate the whole of the heat-treatment cycle, necessitating beginning again, starting with a soft anneal before re-hardening. Even if the desired microstructure can finally be achieved, there is likely to be considerable loss of dimensional accuracy.

The purposes of tempering are mainly threefold:

- To improve toughness, thus reducing the tendency to crack or chip in service.
- To develop the desired carbide type and distribution in the finished piece, thus maximizing wear resistance.
- To stabilize the microstructure, thus avoiding uncontrolled transformations during any subsequent processing, or in service. Examples of this may be the removal or complete stabilization of any retained austenite, and high-temperature tempering of pieces destined for nitriding or PVD surface treatments, or for hot-work or high-speed cutting applications.

It may be at first surprising to note that in none of these purposes is hardness mentioned. The reduction in hardness that normally accompanies tempering (except in the case of secondary hardening grades) is a consequence of the tempering process, not a purpose of it. It may therefore be questioned why "tempering curves" normally relate hardness to tempering temperature, and it is the final hardness that is specified, not toughness or microstructure. The reason for this is simply a matter of convenience; it is easy to measure hardness, but not so easy to specify or measure microstructure or toughness. It is therefore an error to assume that hardness is the important criterion for tempering, or that reducing hardness necessarily results in an improvement in toughness.

P. Illustrative Examples

The approach to optimizing final structure and properties must involve optimizing the whole heat-treatment cycle; austenitizing, quenching, cryogenic treatment (if used), and tempering. This can perhaps best be explained by using typical examples.

Example 1—H13

H13 is a very commonly used steel, for a wide range of applications, including injection molding dies, die-casting dies, hot forging and extrusion dies, and it also has an excellent response to nitriding, so making it useful for highly stressed machine parts, especially where high toughness coupled with excellent fatigue resistance and surface wear resistance is needed. The normal hardening temperature for H13 is about 1030°C, and except for special purposes, this is not varied significantly.

A wide range of heat treatments on H13 can yield very similar finished hardness (see the tempering curve, Fig. 15). If almost any normal tempering temperature will result in a very similar final hardness, does it matter what tempering temperature is used? This demonstrates the limitations of basing tempering on the normal "tempering curve". The properties resulting from different heat-treatment cycles can be quite different, and therefore the heat-treatment cycle needs to be tailored to the particular end-use of the component. To optimize properties, the purposes of tempering must be recalled (toughness, carbides, and stabilizing microstructure). In the case of H13, the relatively low-carbon content means that there is not a large proportion of carbides in the final microstructure, so the tempering decision is based on toughness and microstructural stability.

To optimize toughness, a curve relating toughness to tempering temperature is required (Fig. 16). From this, it will be seen that there is a toughness peak at tempering temperatures around 250°C. Higher tempering temperatures result in lower toughness, and not until temperatures in excess of about 600°C are used (resulting in much lower hardness values) does toughness exceed that developed at 250°C. For any application not involving exposure to high temperatures, such as plastic injection molds, tempering in the range 200–250°C gives the optimum combination of properties.

In situations where the piece will subsequently be exposed to high temperatures, either during processing (e.g. nitriding) or in service (e.g. hot forming or die-casting dies), it is important to stabilize the microstructure by tempering at a temperature preferably

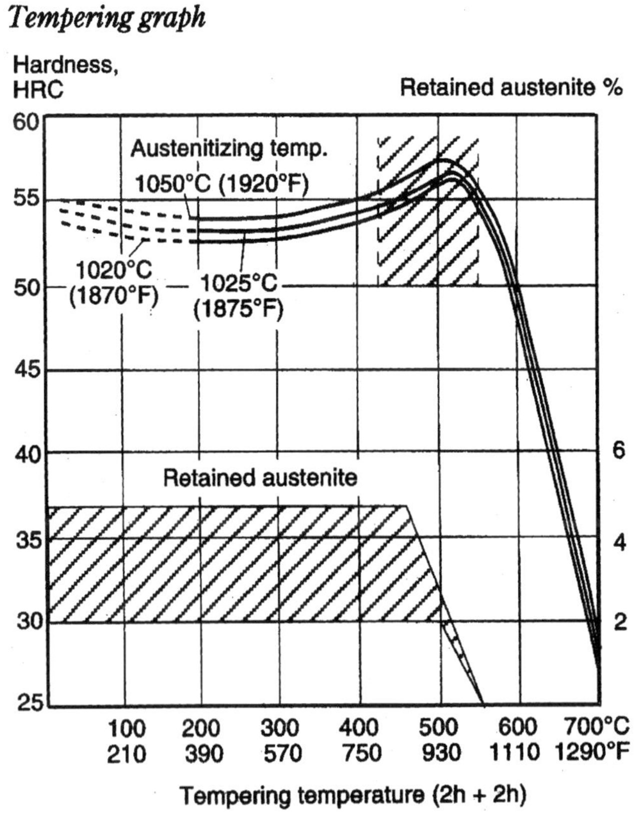

Figure 15 Tempering curve (hardness) for Premium Grade H13 tool steel. (From Ref. 25.)

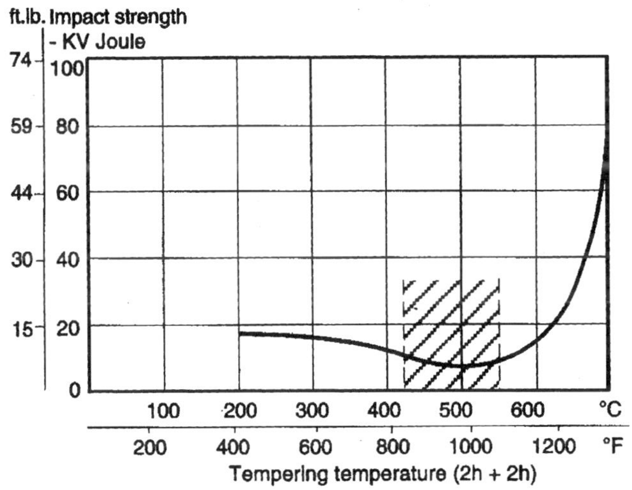

Tempering within the range 425–550° C (800–1020° F) is normally not recommended due to the reduction in toughness properties.

Figure 16 Tempering curve (toughness) for Premium Grade H13 tool steel. (From Ref. 25.)

slightly higher than that to which it will be exposed, and thus tempering temperatures in the range 540–620°C are often used.

The results of either of these tempering treatments, in terms of hardness, are very similar. It is thus important when specifying the heat treatment to ensure that the heat treater is aware of subsequent processing or the end-use of the piece, so that it can be properly treated.

Example 2—High-Speed Steel M3:2 (P/M)

Powder metallurgy versions of this grade are widely used, and are very versatile, for cutting tools, and also for hot-forming tools, fine blanking, and many other hot- and cold-forming applications. The grade is sold by several major suppliers under a number of trade names.

The final microstructure of this steel consists of tempered martensite and an array of well-dispersed hard, complex "secondary hardening" carbides, that give the steel its combination of hardness, toughness, and excellent wear resistance. To develop this microstructure, particularly the carbide type and distribution, tempering in the secondary hardening range is essential. The manufacturers normally recommend 560°C. Use of considerably lower temperatures would fail to develop the desired carbides, and radically higher temperatures would result in carbide coarsening.

With an effectively "fixed" tempering temperature range, control of properties, especially toughness (for cold-forming operations, to reduce susceptibility to chipping), is achieved by varying the austenitizing temperature, which can be anywhere between 1000°C and 1200°C, to give final hardness in the range 58–66 HRC (Fig. 17). Note here, as mentioned in a previous section, that although hardness is the quoted control variable,

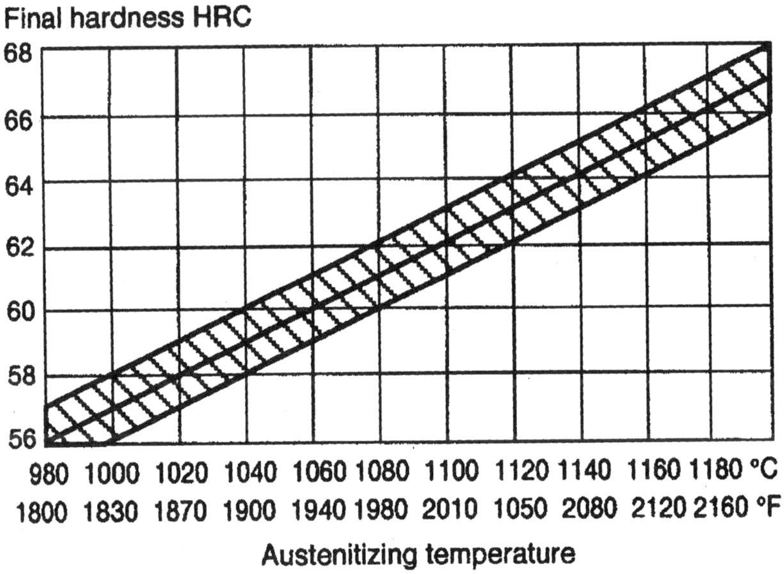

Figure 17 Final hardness vs. austenitizing temperature for P/M M3:2 high-speed steel. (From Ref. 26.)

what is actually being aimed at is to control toughness. If the toughness achieved at the lowest of these hardness values could be achieved whilst maintaining the highest hardness value, no one would complain! Hardness is here being used as a control measure to imply a desired toughness value (Fig. 18).

Example 3—Cold-Work Steel D2

This is perhaps one of the most commonly used of the high-carbon, high-chromium cold-work steels. It has excellent wear resistance, good dimensional stability, and has good corrosion resistance. Its main limitation is that it is somewhat brittle. It is available from most suppliers, and often in variant grades. It is widely used for cold-forming dies of many types, and also for precision gauges, etc.

There are a variety of heat-treatment procedures available, to achieve different combinations of final properties, and even differing schools of thought as to the optimum heat-treatment procedure for a given desired result. The specified finished hardness is normally in the range 58–60 HRC, but can be anywhere in the range 56–62 HRC. A tempering curve for this grade is shown in Fig. 19. As well as achieving the desired final hardness, the main criteria for the heat treatment are:

- To make the best of the limited toughness of this grade.
- To minimize dimensional change during the treatment, and maintain dimensional stability in service.
- To stabilize the final microstructure for subsequent processing (e.g. PVD or any other high-temperature processing; some would also include EDM in this. This steel is not frequently used for hot-work applications.)

Depending mainly on austenitizing temperature, there may be significant proportions of retained austenite in the as-quenched microstructure, and a major consideration

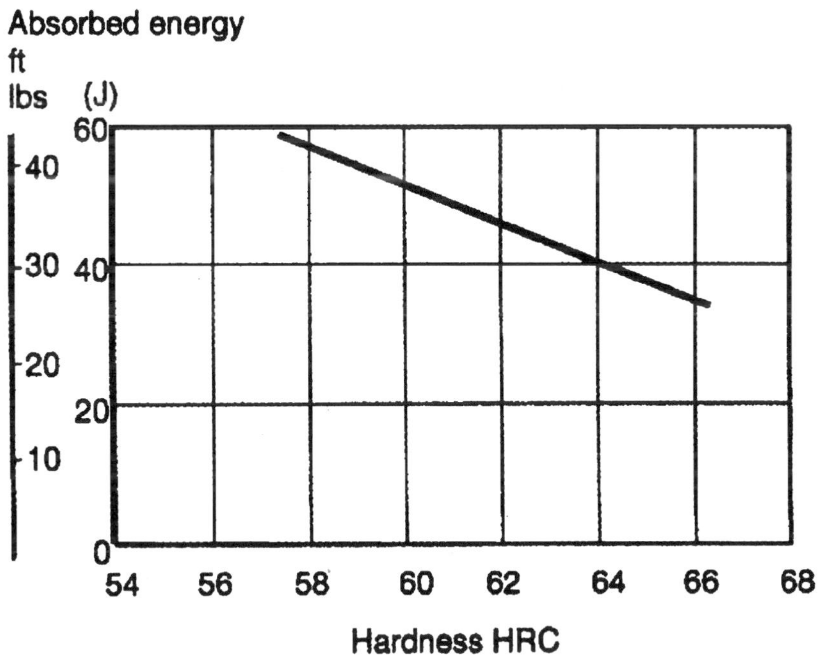

Figure 18 Hardness vs. toughness for P/M M3:2 high-speed steel. (From Ref. 26.)

is what to do about this, either to eliminate it or to stabilize it. The decision on this has implications both for toughness, dimensional change, and microstructural stability.

Microstructural Stability. Any austenite still retained in the microstructure is likely to transform when the steel is subjected to high temperatures. If high temperatures are likely to be experienced, it is strongly advisable to eliminate the retained austenite completely from the microstructure during heat treatment. In this case, a high-temperature (double) temper is needed, usually of the order of 550°C. In order to achieve sufficient hardness after this temper, an austenitizing temperature of the order of 1050–1060°C is often employed. This combination of hardening and tempering temperatures is also acceptable from the point of view of dimensional change.

Toughness. The presence of small, controlled amounts of retained austenite in the microstructure imparts a significant improvement in fracture toughness. It is, however, essential that the proportion of retained austenite be controlled, and that it be relatively stable (or at least metastable) in the conditions prevailing during any subsequent processing or service. Hardening from about 1010°C to 1020°C, and double tempering at 180 250°C will achieve this. This range of austenitizing temperatures leaves some retained austenite in the structure (but not excessive amounts), and the double temper effectively stabilizes the austenite, so that in normal circumstances, it will not subsequently transform in service. The net result is a finished hardness in the required range, a toughness at least 25% better than that achieved after high temperature tempering, and minimal dimensional change, the austenite shrinkage effectively counteracting the martensite expansion.

In cases where the normal toughness obtainable from D2, even when it contains stabilized retained austenite, is not sufficient, it is sometimes mistakenly assumed that by lowering the hardness requirement still further, a substantial increase in toughness can be

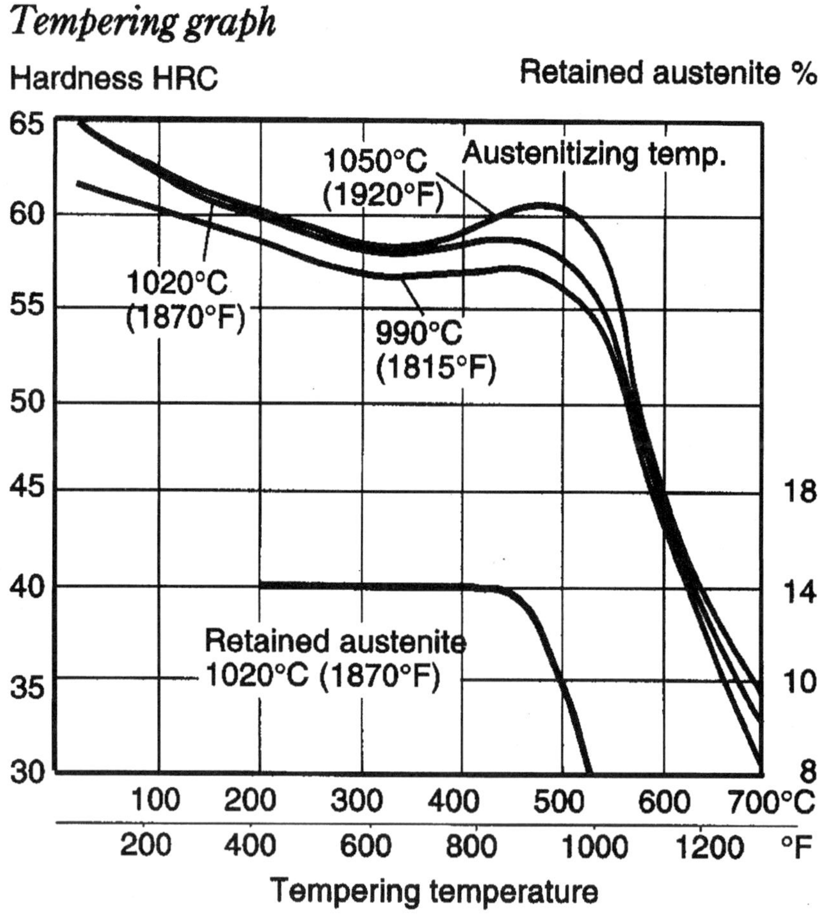

Figure 19 Tempering curve for D2 tool steel. (From Ref. 27.)

achieved. This is not a sensible option, as the toughness improvements by further lowering the austenitizing temperature are not large (there will be less retained austenite in the as-quenched microstructure), and also control of hardness at very high tempering temperatures is difficult, and again the gain in toughness is not large. If significant increases in toughness are needed, then it is much better to use a different grade of steel, with inherently better toughness, than to try to force D2 into a "property envelope" for which it is not designed.

Wear Resistance and Hardness. Increases in wear resistance can be achieved in this grade by deep cryogenic treatment, which results in an increase in the number of fine carbides in the as-tempered microstructure. During this treatment, all retained austenite will also be converted to martensite, so a lower hardening temperature can be used to achieve the same finished hardness (e.g. 960–980°C instead of 1010–1020°C). This gives toughness values intermediate between workpieces containing stabilized retained austenite and those tempered at high temperature, but closer to the former. Increases in hardness can be achieved by increasing austenitizing temperature and deep cryogenic treatment, followed by low temperature tempering. Hardness values as high as 65 HRC have been achieved in this way.

V. CONCLUSION

From these three examples, it can be seen that there are a variety of different strategies for obtaining a given final hardness value, by controlling both hardening temperatures and tempering temperatures:

- As was the case with H13, with a (more or less) fixed hardening temperature, tempering temperature may be the controlling variable, either to control hardness, or to control microstructure and other properties at a given hardness value.
- As with the M3:2, with a fixed tempering temperature, properties can be controlled by varying the hardening temperature.
- In some cases, as with D2, hardening and tempering temperatures have to be chosen in conjunction with each other, and only certain combinations of the two will achieve optimum properties.

It should also be clear that hardness alone is not a sufficient guide to the required heat treatment, since in many cases, the same hardness can be achieved by different process variations, but with a different mix of other properties.

Finally, it should also be clear that in order to optimize the properties of a tool steel by heat treatment, it is essential to have a knowledge of the steel, its grade, quality (and preferably supplier), and a detailed knowledge of any subsequent processing of the pieces, as well as its final end-use. Only by close collaboration between the tool-maker, user, and heat treater can properties be optimized *for a given specific application.*

ACKNOWLEDGMENTS

The authors would like to thank British Standards Institution, for permission to use their data from BS EN ISO 4957:2000; ASM International, for permission to use their data from GEM 2001—Guide to Engineering Materials; Böhler GMBH for their kind permission to use a number of diagrams from their publication "Böhler Shop Service", and to Uddeholm AB, for permission to use diagrams from their technical literature.

REFERENCES

1. British standard BS EN ISO 4957:2000, Tool Steels; British Standards Institution, European Standard EN ISO 4957:1999E (English version). Tool Steels; European Committee for Standardization: Brussels.
2. Roberts, G.; Krauss, G.; Kennedy, R. Tool Steels, 5th Ed. ASM International: Materials Park, OH, 1998.
3. Davis, J.R. Tool Materials—ASM Specialty Handbook; ASM International: Materials Park, OH, 1995.
4. Figiel, L.S. On Damascus Steel; Atlantis Arts Press: Atlantis, FL, 1991.
5. Yumoto, J.M. The Samurai Sword; Charles E. Tuttle Company: Tokyo, Japan, 1958.
6. Townsend, A.S. Alloy tool steels and the development of high-speed steel. Trans. Am. Soc. Steel Treat. 1933, 21 (Jan–Dec).
7. Lange, K.W. Thermodynamic and kinetic aspects of secondary steel-making processes Int. Mater. Rev. 1988, 3 (2).
8. Philips, T.V. ESR: a means of improving transverse mechanical properties in tool and die steels Met. Technol. 1975, (Dec), 2(12).

9. Hoyle, G. Electroslag Processes, Principles and Practice, Applied Science: New York, 1983.
10. Tengzelius, J.; Grinder, O. P/M in Sweden. Int. J. Powder Metall. 1996, 32 (3).
11. White, D.G. P/M in North America. Int. J. Powder Metall. 1996, 32 (3).
12. Leatham, A.G.; Elias, L.G. The Osprey Process. Spray Forming: Science Technology, and Application; Metal Powder Industries Federation: Princeton, NJ, 1992.
13. Steel Products Manual-Tool Steels; American Iron and Steel Institute: Washington, DC, September 1981.
14. GEM 2001—Guide to Engineering Materials: Advanced Materials & Processes. ASM International: Materials Park, OH, 2000.
15. A600-92a standard specification for tool steel high speed; A681-94 standard specification for tool steels alloy; A686-92, standard specification for tool steel carbon. ASTM Annual Book of ASTM Standards, Vols. 01–05, The American Society for Testing and Materials, Philadelphia, PA, 1998.
16. Wick, C. Tool and Manufacturing Engineers Handbook; Forming, Society of Manufacturing Engineers, Dearborn, MI, 1984.
17. Schwartz, S.S.; Goodman, S.H. Plastics Materials and Processes; Van Nostrand Reinhold Company: New York, 1982.
18. Dym, J.B. Injection Molds and Molding—A Practical Manual, 2nd Ed. Van Nostrand Reinhold Company. New York, 1987.
19. Dubois, J.H.; Pribble, W.I. Plastic Mold Engineering Handbook, 4th Ed. Van Nonstrand Reinhold Company: New York, 1987.
20. Suchy, I. Handbook of Die Design; McGraw-Hill: New York, 1998.
21. Hoyle, G. High Speed Steels; Butterworths: London, 1988.
22. ASSAB, EDM of Tool Steel, Publication # 201; ASSAB Pacific Pte Ltd: Singapore, 1995.
23. Böhler Shop Service, A Practical Guide to Successful Processing of Special Steels; Böhler GMBH: Kapfenberg, Austria.
24. Chandler, H. Heat Treater's Guide, Practices and Procedures for Irons and Steels; ASM International: Materials Park, OH, 1995.
25. Orvar Supreme. Uddeholm Tooling AB; SE-68385; Hagfors: Sweden.
26. Vanadis 23. Uddeholm Tooling AB; SE-68385; Hagfors: Sweden.
27. Sverker 21. Uddeholm Tooling AB; SE-68385; Hagfors: Sweden.

5

Designing with High-Strength Low-Alloy Steels

Lin Li and Luoping Xu
Shanghai University, Shanghai, China

I. GENERAL CONCEPT FOR THE DESIGNING OF HIGH-STRENGTH LOW-ALLOY (HSLA) STEEL

The structural steel industry experienced interesting and technologically important changes during the last four decades. Prior to the mid-1960s, there were essentially two types of steels available for use in the transportation, energy and construction industries [1]:

(a) low-strength ($250 \leq YS \leq 400$ MPa) hot rolled or normalized plain carbon steels, and
(b) high-strength ($YS \geq 560$ MPa) quenched and tempered (QT) low-alloy steels.

Since both the ferrite–pearlite microstructure and the tempered martensite (or bainite) microstructure in steels derive their strengths from the carbon content, these steels have not always exhibited the desired levels of weldability, formability, the ratio of YS to UTS, and resistance to brittle fracture. These shortcomings have rendered these traditional steels less cost effective than they might be otherwise and became serious when challenges of high level requirements to materials came from the rapidly developing industries such as automobile industry.

During the last 30 years, significant progress has taken place in the development of high-strength and tough microalloyed steels. Important steps have been added to the technology of thermo-mechanically treated microalloyed steels:

(a) rolling process improvement (controlled rolling),
(b) production of clean steels (very low content of C, P, S, and N),
(c) development of new alloy designs (microstructure comprising either traditional ferrite–pearlite or lower temperature transformation products such as bainite or martensite), and
(d) application of accelerated cooling (AC).

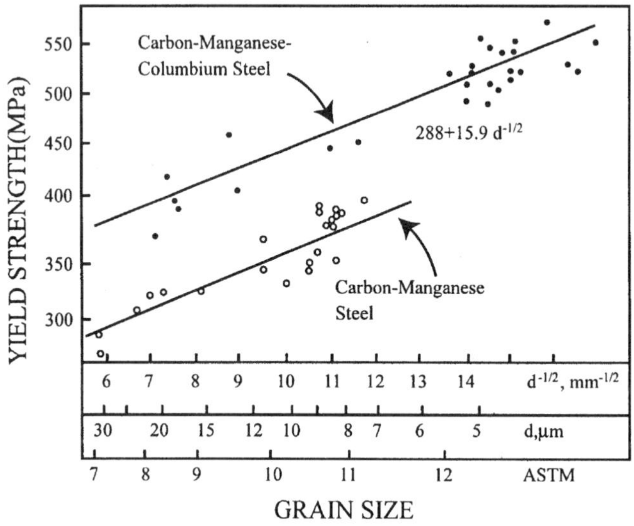

Figure 1 Experimental Hall–Petch relationships determined by quantitative metallography for C–Mn and C–Mn–Nb steels. (From Ref. 2.)

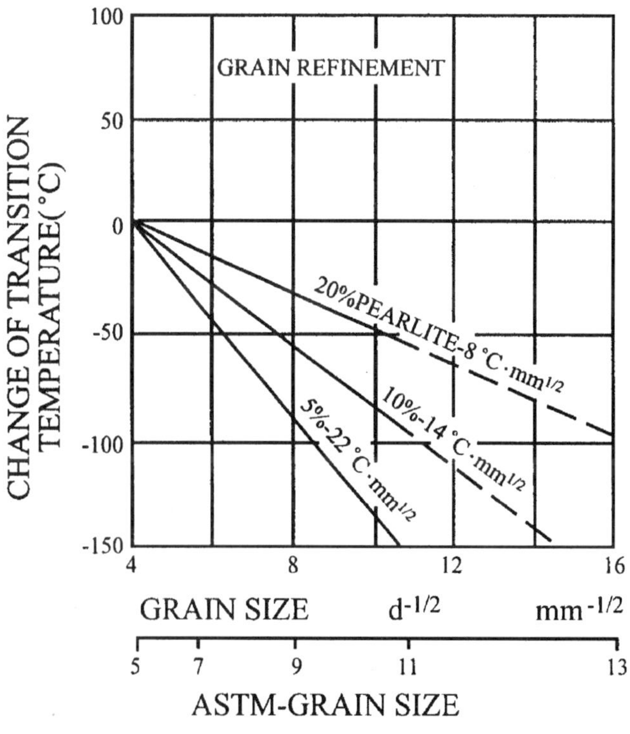

Figure 2 Change in transition temperature through variation of the structure of the steel. (From Ref. 3.)

Starting with a standard Nb or Nb–V steel, or C–Mn–Si steel, combinations of other microalloys have been developed in a number of alloy design approaches. Among alloying and microalloying elements, Nb, V, Ti, B, Mo, Cu, and Cr have found their place in this product area. The importance, or usage, of one or combinations of these and other elements in each project depends on property requirements and also on preferences from steel makers. It is generally understood that these elements exert their influence mainly through their behavior during and subsequent to the hot deformation processing of austenite. The microalloying elements (MAEs) are well known to impede the motion of dislocation, grain boundaries, and recrystallization boundaries when the MAE is present either as solute or as precipitate. Therefore, the MAE can suppress grain coarsening, static recrystallization, and the motion of dislocations. Moreover, the dissolved MAE in austenite can increase hardenability, and thus lower the transformation temperature. The suppression of both grain coarsening and recrystallization contributes to the control of the austenite microstructure (i.e., conventional controlled rolling or recrystallization controlled rolling). The combination of this effect and hardenability controls the final microstructure. Superimposed on this is the potential for the precipitation during or after transformation, which can lead to the additional strengthening of the final microstructure [2]. The increase in strength was a result of grain refinement and precipitation hardening of the ferrite as shown in Fig. 1 [2]. These original MA steels were not only stronger but also tougher than the C–Mn–Si steels. The improved toughness was caused by an increased resistance to cleavage fracture which resulted from both the finer ferrite grain size and the reduction in the amount of the existed pearlite as shown in Fig. 2 [3]. The strengthening resulted from pearlite which was no longer required in this case. Moreover, the lower carbon content in MA steels also

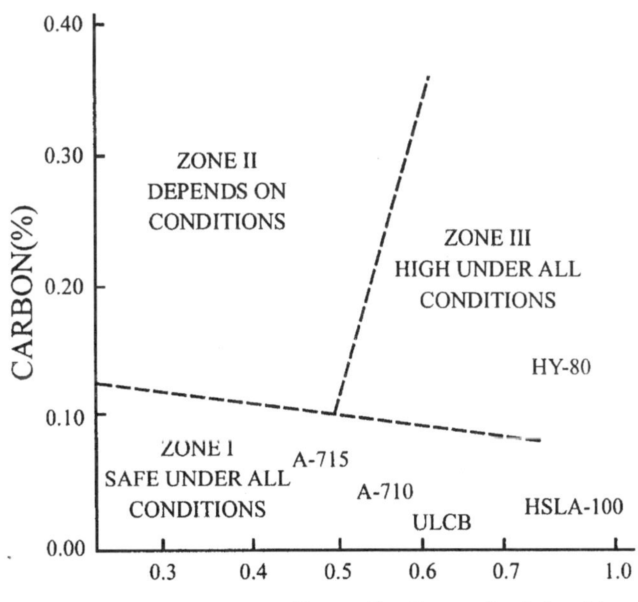

$$CE = C + \frac{Mn+Si}{6} + \frac{Ni+Cu}{15} + \frac{Cr+Mo+V}{5}$$

Figure 3 Influence of steel composition on the susceptibility of the heat affected zone to cold cracking. Carbon concentration and carbon equivalent (CE) values are represented by their weight percent. (From Ref. 4.)

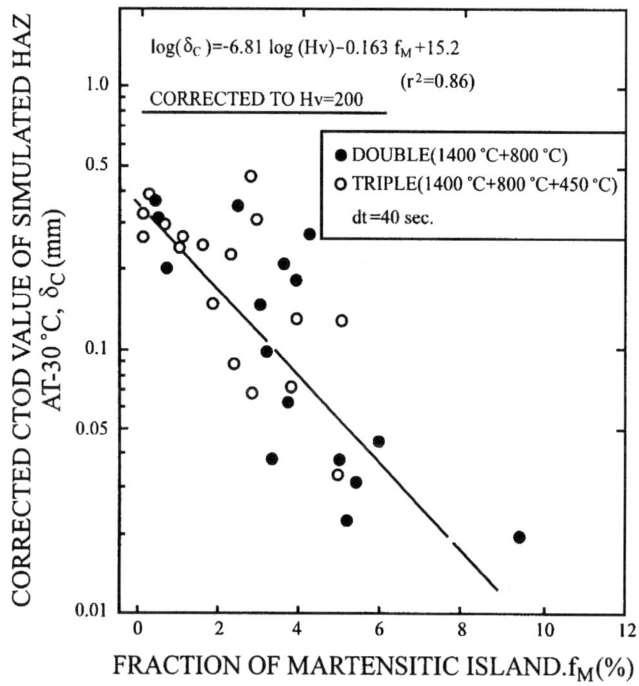

Figure 4 Relation between CTOD value and fraction of martensitic islands of simulated HAZ. (From Ref. 5.)

led to improved weldability and weldment properties. Graville [4] has shown in Fig. 3 that the susceptibility to cold cracking of the weld heat affected zone (HAZ) is directly related to the composition of the base plate. Haze et al. [5] has shown the relationship between the HAZ toughness and the amount of untempered martensite with good accuracy in Fig. 4. In order to get impressive improvements in both weldability and weldment toughness, it was shown that it was not necessary to rely on high carbon content even if high-strength in the base plate is required.

A. Linepipe Steels

Although the mechanical properties associated with controlled rolled ferrite–pearlite microstructures were attractive and found wide use in structural steel applications, there existed some limitations since the maximum strength which can be developed in ferrite–pearlite structures was not always high enough to meet all of the requirements from industry. For example, in order to increase productivity, the oil and natural gas industries continue to request pipelines of larger diameter and higher operating pressure. Obviously, larger diameters and higher pressures correspond to higher strength of steels, and this request led pipe producers to investigate new approaches other than the conventional controlled rolled pearlite-reduced micro alloyed (MA) steels. In addition, the pipe used in cold areas such as in the arctic region still requires excellent low temperature toughness. The evolution of high performance linepipe steel over the past three decades is shown in Fig. 5 [6]. The hot rolled and normalized steel with ferrite–pearlite microstructure shown represents Western European technology of the mid-1960s. The controlled rolled MA plus

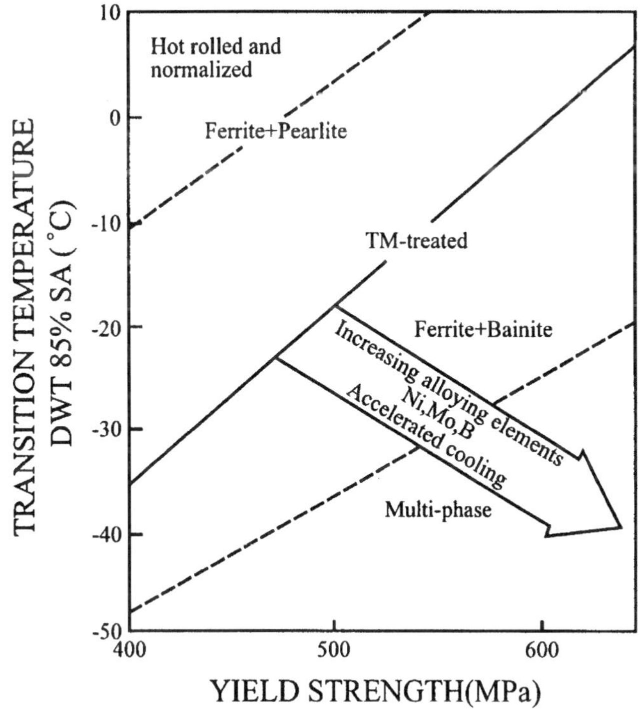

Figure 5 Influence of microstructure and production process on mechanical properties. (From Ref. 6.)

thermomechanically (TMP) ferrite–pearlite line represents the technology in the mid-1970s, while the multi-phase property line represents that in the mid-1980s.

B. Strip Steels

1. Dual-Phase Steel

During the past several years, dual-phase ferrite–martensite steels have received increasing attention, because of its high-strength and ductility combined with excellent cold formability. The strengthening and toughening principle of the dual-phase structure involves the incorporation of inherently strong martensite and a soft ferrite matrix. Besides, the high work hardening rates and the high strain rate sensitivity contribute to suppress the formation of the localized neck during cold deformation. Thus, these alloys can be considered similar to fiber composites, where coherent phases are obtained through solid-state phase transformation. However, it should be emphasized that the properties of dual-phase steels are versatile and largely dependent on the following factors [7]:

(a) morphology (size, shape, and distribution of ferrite and martensite phases),
(b) microstructure and properties of the individual phase, and
(c) volume fraction of martensite (or bainite).

Morphology plays the most important role in controlling the mechanical properties (see in Fig. 6) in the three factors stated above. In Fig. 6, A indicates globular shape of martensite within ferrite matrix in Fe–0.1C–4Cr duplex steel. B represents in Fe–0.1C–0.5Cr

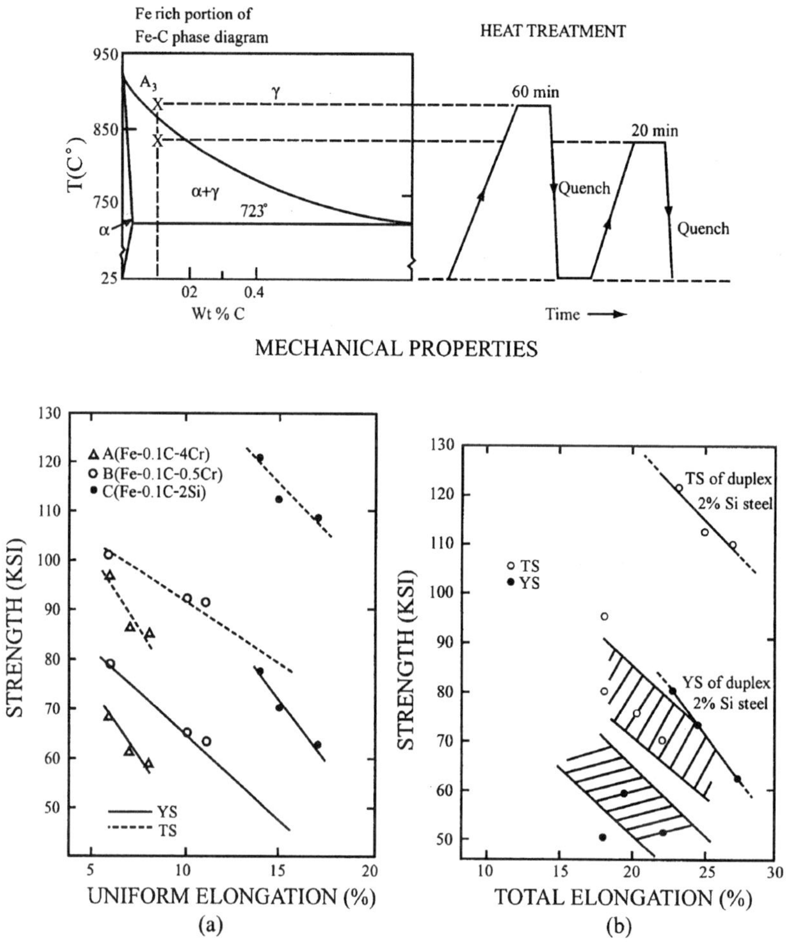

Figure 6 Scheme showing the effects of composition on microstructure of dual-phase steels. Although tensile strength is not markedly affected, morphology critically affects fracture and fatigue. (a) Depending on the annealing temperature in the range, various volume fraction of martensite and a wide range of strength and elongation are obtained. (b) Tensile properties of the duplex 2% Si steel are compared with those of commercial HSLA steels shown in the area with hatched lines. (From Ref. 7.)

steel needle-like martensite within prior austenite grain, which is connected with continuous austenite phase and C exhibits the fine aligned and acicular martensite in Fe–0.1C–2Si steel surrounded by prior austenite.

It has been shown that a coarse dual-phase structure exhibits very poor tensile elongation ductility while a fine structure acts in an opposite way. This is because the maximum stress concentration easily takes place in the coarse grain of ferrite matrix, which leads to the initiation and propagation of cleavage crack. However, when the microstructure is refined, the degree of stress concentration decreases. Fracture is primarily initiated by void nucleation and growth of the crack starts around the ferrite/martensite interfaces.

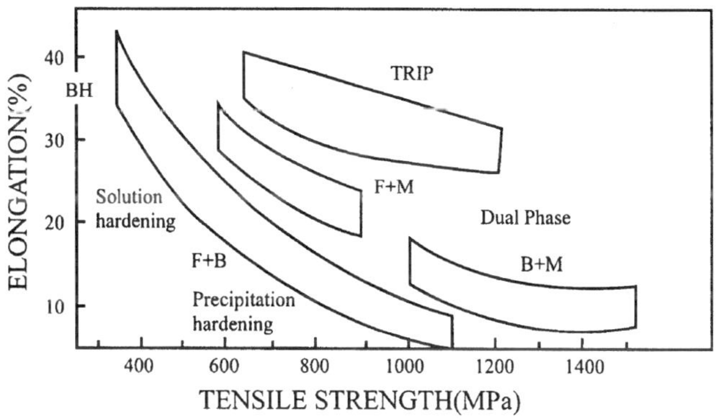

Figure 7 Relation between tensile strength and elongation in terms of strengthening mechanism in the case of HSLA produced by Nippon Steel Corp. (From Ref. 8.)

2. Transformation-Induced Plasticity Steel

Figure 7 shows the relationship between tensile strength and total elongation in high-strength low-alloy (HSLA) sheets strengthened by different strengthening mechanisms [8]. It is obvious that transformation-induced plasticity (TRIP) steel has the best balance between strength and ductility. Transformation-induced plasticity sheet steel has been successfully developed for automobile parts using cheaper and effective elements such as C, Si, and Mn. In the steel, C atoms are ejected from ferrite during the $\gamma \rightarrow \alpha$ transformation under the critical temperature and transferred to austenite and stabilize as retained austenite in the course of baintic transformation. Silicon accelerates the $\gamma \rightarrow \alpha$ transformation, increases the activity of C, and it inhibits the formation of pearlite and then stabilizes austenite. Manganese is also an austenite stabilizing element and can suppress the intercritical temperature to let the transformation to take place in lower temperatures.

Factors to enhance the formation of retained austenite are as follows:

(a) Adequate supply of C atoms to austenite through control of ferrite volume during $\gamma \rightarrow \alpha$ transformation.
(b) Austenite grain refining before $\gamma \rightarrow \alpha$ transformation.
(c) Suppression of pearlite formation.

Comparison of mechanical properties between cold rolled TRIP steel and Al alloy is shown in Table 1 [8], where TRIP steel exhibits elongation more than Al alloy sheet. Unlike the ordinary cases, the $\bar{r}$ value of TRIP steel in Table 1 is not high but less than 1.0 though its deep drawability is fairly good. This is because retained austenite does

Table 1 The Comparison of Mechanical Properties Between TRIP Steel and Al Sheet

Steel	YS (MPa)	TS (MPa)	Elongation (%)	n value	$\bar{r}$ value	Young's modulus (MPa)
Al(5182-0)	150	300	25	0.3	0.7	70,000
TRIP	800	1,000	30	0.3	0.9	210,000

Source: Ref. 8.

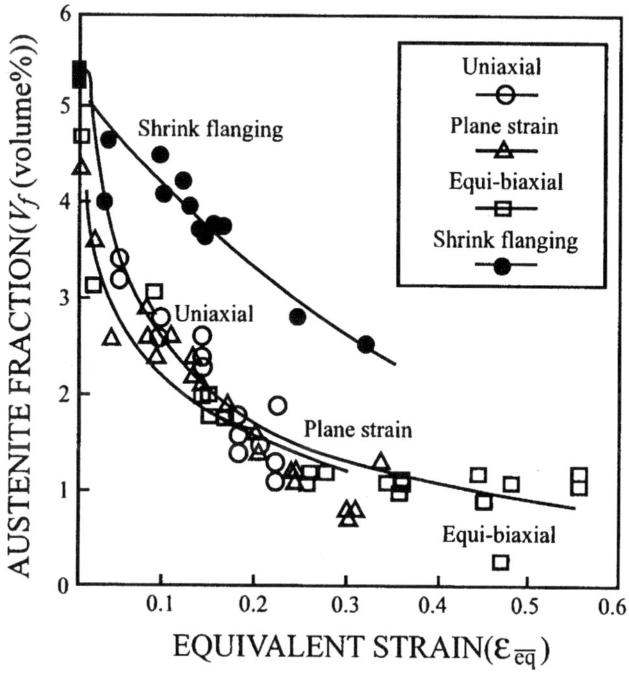

Figure 8 Changes in volume fraction of austenite vs. equivalent plastic strain under the four deformation conditions. (From Ref. 8.)

not tend to transform into martensite by compressive stress, which occurs at flanging area of blank, but transforms much easily by tensile stress applied at the shoulder of cup during drawing as shown in Fig. 8. Softer flange incorporating harder cup shoulder improves the deep drawability in the light of martensite transformation. Actually, TRIP steel could be considered as the most prospective HSLA steel against the challenge of Al alloy in automobile industry.

The precipitation of carbides or carbonitrides in microalloyed steels, which contain one or more of the transition metals Nb, Ti, and V, is of primary importance in the controlling of the properties of HSLA steel, especially for TRIP steels. The promoting of Nb(CN) precipitation by the change in TMP (thermo-mechanical process) conditions results in a decrease of retained austenite volume fraction [9], which is the primary factor governing the TRIP effect.

3. Interstitial-Free Based Precipitation Hardening Steel

Based on the development of steel-making technology in vacuum degassing and continuous annealing, new series of deep drawing automobile steel sheet with basic character of extra-low carbon was designed and produced in 1980s. The traditional rimmed and Al-killed low carbon sheet steels were then by extra-low carbon sheet steel replaced. The typical product of these steels is interstitial-free (IF) sheet steel, a steel with the addition of elements as Ti and Nb, to facilitate formation of carbides and nitrides in the steel even with extra-low form carbon content ($C \leq 80\,ppm$). The C and N atoms are therefore fully fixed as a carbonitride (Ti(C, N), Nb(C, N)), and same less interstitial atoms can exist in

Table 2 Properties of Cold-Rolled Steel Sheets

Type	σ_s (MPa)	σ_b (MPa)	δ_{10} (%)	$\bar{r}$	$\bar{n}$
Rimmed steel	180–230	290–390	30–34	1.0–1.2	0.19–0.21
Al-killed steel	160–210	270–390	32–44	1.4–1.8	0.20–0.23
IF steel	120–180	250–330	45–55	1.8–2.8	0.23–0.28

Source: Ref. 10.

the steel. The properties of three generations of cold-rolled sheet steels for automobile applications are shown in Table 2 [10].

Interstitial-free steels possess the following characteristics:

(a) low impurity (Si, Mn, S, P, etc.) content,
(b) low yield point ($\leq$180 MPa) and low yield strength to tensile strength ratio ($\leq$ 0.55),
(c) high ductility ($\delta_{10} \geq 45\%$),
(d) high $\bar{r}$ and $\bar{n}$ value, and
(e) no aging required.

According to the alloy element addition, IF steel can be classified into three kinds, i.e., Ti-IF steel, Nb-IF steel and (Ti + Nb)-IF steel.

Recently, the same kind of HSLA steel has been developed by using conventional IF steel with slightly larger C content and the modification of Ti/Nb which kill C in solution sufficiently. An example is shown in Table 3 [8] and Fig. 9 [8]. The steel has tensile strength greater than 450 n mm^{-2}, excellent $\bar{r}$ value and deep drawing behavior. When it is produced by hot dip Zn coating process, the content of Si is reduced and Nb is preferable to Ti.

4. Bake Hardening Steel

The bake hardening steel is also one of the most important steels in automobile industry since the dent resistance of the steel will be appropriately raised after baking. The bake hardening (BH) value of the steel is dominated by two factors as follows:

(a) adequate number of C and N interstitial atoms of solid solution, and
(b) adequate number of free dislocations in crystal lattices.

Before baking, the free dislocations are easy to move under stress, which leads to a low yield strength. But during baking, the interstitial C, N atoms diffuse rather quickly into the free dislocations and fix them, thereby raising the yield strength of the steel and obtaining baking hardening.

During pre-strain the number of the dislocations increases. When the pre-strain level is low, the BH value of the steel increases with the increase of the number of the

Table 3 Chemical Composition (wt%) and Mechanical Properties of Developed Steel

Steel	C	Si	Mn	P	Ti	Nb	TS (MPa)	YS (MPa)	El (%)	$\bar{r}$
Developed	0.007	0.50	0.81	0.076	0.061	0.016	467	289	36.0	1.76

Source: Ref. 8.

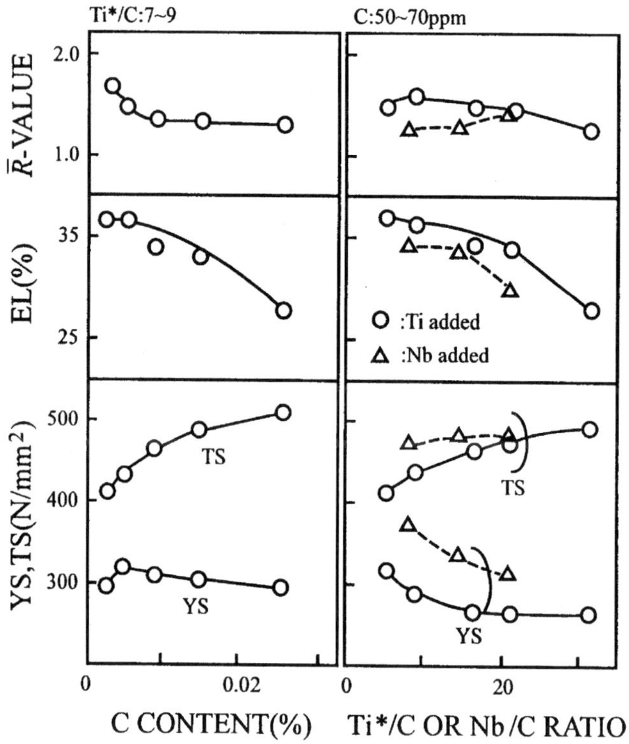

Figure 9 Effect of C, Ti, and Nb on mechanical properties, where $Ti^* = Ti - 48/14N$; when Nb added, Ti equivalent to N% is also existed. (From Ref. 8.)

dislocation. However, large number of dislocations and the shift of new and old dislocations can lead to the interweaving and fixing of dislocations and thus diminish the number of free dislocations and reduce the BH value. This is revealed in Fig. 10(a).

As the bake temperature increases, the ability of diffusion of C and N interstitial atoms increases accordingly, more dislocations are fixed by the diffused atoms, and then higher BH value is obtained. At 160°C for BH 340 steel, the diffusion of C and N atoms is strong enough, further increase in temperature leads to less increase in BH value as shown in Fig. 10(b).

The length of the baking time has also an effect on the number of the diffusing C and N atoms. The relation is obvious: the longer the baking time, the more the diffusing atoms. As more dislocations are fixed by diffused interstitial atoms, the BH value becomes higher and higher together with the increasing of the baking time, as shown in Fig. 10(c).

The composition and mechanical properties of a typical bake hardening steel (BH340) are listed as an example in Table 4 [11].

C. Bar Steels

Bar steels for structural uses are only rarely used in the as-hot rolled condition, but are used widely as the feed stock for hot or cold forging applications, which is discussed separately in the following.

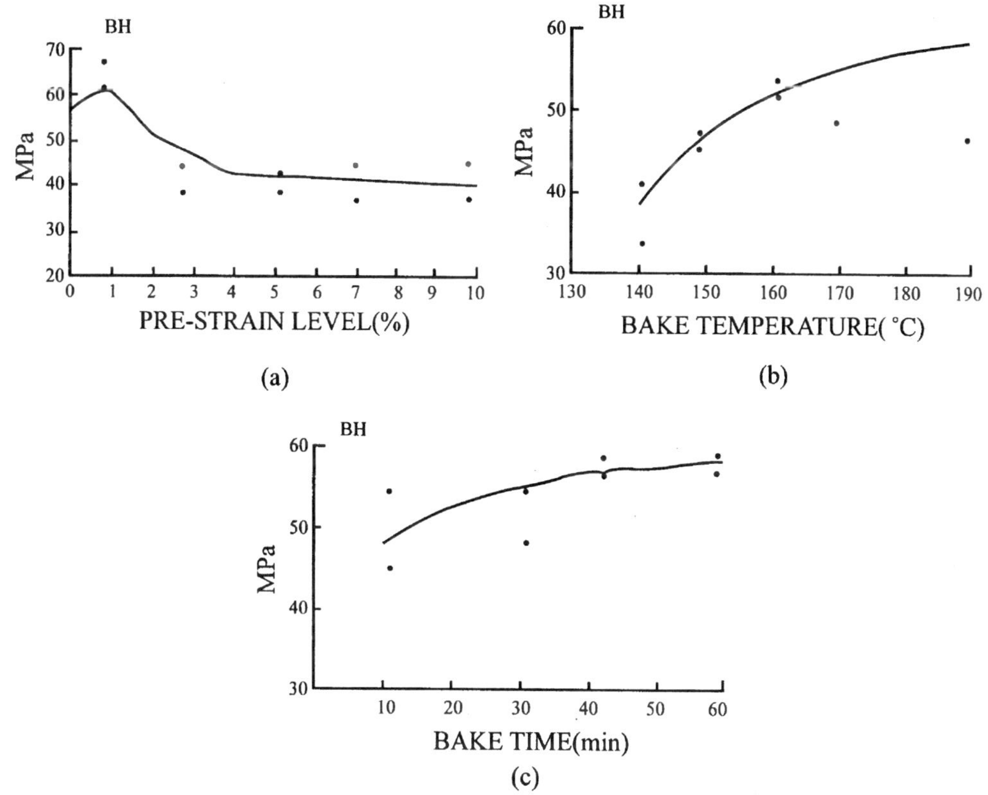

Figure 10 Relation between BH value and treatment parameters: (a) relation between BH value and pre-strain level; (b) relation between BH value and bake temperature; and (c) relation between BH value and bake time. (From Ref. 11.)

1. Bar Steels for Hot Forging Applications

Traditionally, steels for hot forging applications have been medium carbon (0.4 wt% C) steels. When high-strength is required in the final component, the as-forged part is reheated, quenched, and tempered (QT). Until 1970, these low-alloy steels in the QT condition were the only route to high-strength forgings. However, this additional treatment declined because of the high cost, the problems of quenching cracking, distortion, and susceptibility to hydrogen embrittlement that accompany the QT process and the very wide range in properties that is often obtained.

Table 4 Chemical Composition (in wt%) and Properties of BH340 Steel

C	Si	Mn	P	S	Al	YS (MPa)	TS (MPa)	El[a] (%)	IE (mm)	n	r	BH (MPa)
0.018	0.03	0.20	0.063	0.004	0.056	250	370	33	10.5	1.49	0.20	51.8

[a] L_0=80 mm.
Source: Ref. 11.

The major advance was made in mid-1970s when microalloying was commercially applied to forgings in Germany. This first MA forging steel was designated as 49MnVS3 and was essentially 1144 microalloyed with V. Later work showed that similar results could also be obtained with 1144 + Nb [12]. This steel was to be used in the as-forged condition without final QT heat treatment. In this condition, the steels would exhibit a structure consisting of perhaps 75% pearlite and 25% ferrite. The role of the MAE was principally to induce precipitation hardening of the proeutectoid and pearlitic ferrite during cooling from the forging operation. The yield strengths and tensile strengths of these steels were 500 and 900 MPa, respectively, without heat treatment. While these strength levels are impressive, they do not approach those of most QT steels, e.g., yield and tensile strengths in excess of 800 and 950 Mpa, respectively. These early MA forging steels also suffered from very low notched-impact toughness levels. It is common for these steels to exhibit the toughness value at room temperature of less than 20 J in a Charpy V-notch test [13]. Efforts have been made to improve the properties of these medium carbon, microalloyed, pearlite–ferrite steels for decades, however, they continue to be troubled with the unacceptably low levels of strength and toughness. This leads to components which must either be overdesigned or used in safetynon-critical applications.

A radically new approach has been developed in the production of high-strength forgings since 1985. In this new approach, the steel forging is directly quenched from the trim dies after forging has been completed. This new process is called direct quenching (DQ). In order to ensure high-strength and toughness in the as-quenched forging without the need of subsequent tempering or other heat treatments, three different families of MA steels: Mn–Mo–Nb [14], Mn–Mo–V–Ti–N [15], and Mn–Cr–Ti–B [16] have been developed for DQ forging application. These steels have microstructures ranging from fully low carbon martensite to mixtures of martensite and bainite in the DQ condition, which result in minimum yield strengths from 700 to 930 MPa and minimum tensile strengths from 950 to 1170 MPa, respectively. In addition, the minimum values of Charpy V-notch impact toughness values at room temperature for the steels range from 40 to 80 J. The comparative benefit of using one of the DQ steels relative to a standard QT steel was shown by Garcia et al. [17]. As shown in Table 5, the arms forged of BHS (Mn–Mo–Nb) steel, whatever treated by oil quench (OQ) or polymer quench (PQ), were at least as strong and tough as those forged from 1541H/(QT). Furthermore, the full-component fatigue resistance of the DQ arm was much higher. Clearly, these new MA multi-phase steels offer opportunity to supply components, which not only do not have to be heat treated but also exhibit better properties to assemblers and end-users.

Table 5 Mechanical Properties of Lower Control Arms

Sample	YS (MPa)	UTS (MPa)	RA (%)	CVN (J) 25°C	CVN (J) −40°C	Average kilocycles to failure
1541H QT	800	930	60	75		80
BHS (OQ)	841	1,047	66	74	43	280
BHS (PQ)	779	1,013	65	74	32	205.6

Source: Ref. 17.

2. Bar and Rod Steels for Cold Forming

As known, steels are subjected to cold deformation either to increase the strength of the hot rolled bar or to forge the hot rolled bar into a complex final stage. Cold drawing can be considered as an alternative to QT heat treatment as a method of increasing the strength of hot rolled bars. It has been known that the addition of MA elements to hot rolled bars always results in much higher strengths[18]. For example, the yield strength of 1144 after a 20% cold draw could be increased from 750 to 1000 MPa by the presence of an MA addition of V + N to the 1144. The combination of MA and cold drawing offers flexibility in processing and in final properties, however, little deterioration is found in machinability of the cold drawn bars caused by the MA elements [19].

In general, cold forgings have traditionally been either heat-treated coarse grain plain carbon steel or low-alloy steel, depending upon the section size and strength level required. There are at least three requirements of a steel to be successful in cold forging application [1]:

(a) Sufficient ductility to survive the cold forming operation without fracture, and to completely fill the die cavity.
(b) Sufficient strength in the final form for adequate load bearing capacity and fatigue resistance.
(c) Sufficient toughness in the final form to resist fracture under service stresses in the presence of unavoidable or undetectable flaws.

Since traditional medium carbon steels were not specifically designed as cold forgings, the cold forging process scheme that has been developed over time for these steels was rather complicated and expensive and is outlined as follows:

Hot rolled bar/rod → spheroidizing anneal → cold draw to size → cold form to fine size and shape → hardening heat treatment → stress relieving and/or hydrogen baking anneal.

The use of traditional steels in these applications leads to serious economic problems because of the large number of steps required in deformation conditioning, the large number of deformations used in forming the part, and the post-deformation treatments. The final cost reflecting both material and process would be thus considerably increased.

It is also obvious that the use of traditional steels causes metallurgy problems: the high-strength, which results from the final QT heat treatment, is derived from the tempered martensite microstructure. Though this microstructure is capable of exhibiting adequate strength, it does have shortcomings in the areas of uniformity, quench cracking, distortion, toughness, and delayed failure under static loading conditions resulting from hydrogen pick-up during plating. These problems lead to low variability in component evaluation and service performance.

Therefore, an ideal new steel for cold forming applications is one which does not rely on carbon content but rather on the microstructure for strength. A good example of this type of microstructure is the multi-phase microstructure, which derives its strength from the low carbon ferrite, bainite, and martensite and their interactions in steel. Many benefits, which are largely independent of the carbon content, are: no need for spheroidizing anneal, possible elimination of deformation control steps, high-strength where actually required, no need for post-forming heat treatments, and less prone to hydrogen cracking due to the multi-phase microstructure.

An example of a multi-phase steel of this type, which is designated as BHS-1, is similar to the Mn–Mo–Nb steel described above for hot forging and DQ applications. The microstructure of this steel is dependent upon the composition (hardenability), rolling practice (austenite conditioning), and cooling rate. For the case of the controlled rolled multi-phase

Table 6 Mechanical Properties of Controlled Rolled BHS-1 (19 mm ϕ Bar) Subjected to Cold Deformation Trials

				CVN (J)	
Condition	YS (MPa)	UTS (MPa)	RA (%)	RT	−50°C
AC	462	751	65	152	112
AC + 5% Red.	772	854	62	148	145
AC + 10% Red.	834	882	58	112	63
AC + 15% Red.	882	944	56	93	47
AC + 20% Red.	971	1,013	53	84	

Source: Ref. 20.

steel, the microstructure consists of ferrite, upper bainite, and lower bainite. The mechanical properties of controlled rolled BHS-1 are shown in Table 6 [20]. Also shown are the properties after various cold reductions. For example, a cold reduction of 10% results in an 80% increase in YS with very little loss in ductility and toughness. The high rate of work hardening shown in Table 6 is typical for multi-phase microstructures. Table 6 indicates that the BHS-1 in the continuous cold rolled (CCR) condition is an excellent material for cold forming operations since it responds well to cold deformation without the need for prior spheroidizing annealing. Table 6 also show that the strength levels associated with SAE Grade 8 and 8.2 (YS $\geq$ 130 ksi or 896 MPa and TS $\geq$ 150 ksi or 1034 MPa) bolts can be achieved simply by forming of BHS-1 in the as-CCR condition.

The excellent austenite conditioning obtained after controlled rolling contributes to the superior toughness of the as-rolled bar/rod, which is required to avoid cracking during cold forming and to resist fracture in service in the final component. For a material to be considered in a cold forming operation, it must be capable of surviving the cold forming operation without fracture. In this case, the tensile ductility parameters are often used to access the cold formability of a given material. A more valid test is considered, such as the upset test using flange specimens, where the material is compressed until failure is found on the surface of the flange. The strain to failure is then measured and recorded and can be used to compare materials.

II. COMPOSITIONS AND MECHANICAL PROPERTIES OF HIGH-STRENGTH LOW-ALLOY STEELS

A. Compositions and Mechanical Properties

High-strength low-alloy steel grades are generally available in all standard wrought forms such as sheet, plate, strip, bars, structural shapes, and special sections. Heat-treated grades are normally available as plate and bars and sometimes, as sheet and structural shapes. However, if an application involving hot forming is required, a non-heat-treated condition can also be ordered. Chemical compositions of typical HSLA steels are shown in Table 7 [21]. Specifications of ASTM and SAE which are applicable to HSLA steels are listed in Table 8 [21]. Tables 9–11 [21] list alloying elements, mill forms, special characteristics, intended uses, grades, and mechanical properties as described in the ASTM specifications. Tables 12–14 [21] list composition limits, descriptions, mechanical properties, and other characteristics of steels as designated in SAE J410c. Because HSLA steels are sold on the basis of minimum mechanical properties and because the chemical compositions of a lot of steels are considered proprietary, specific alloy content is left to the discretion of the steel producer. A widely used standard of HSLA steels is SAE J410c, which provides a series of minimum standards for tensile strength. Chemical composition requirements are also minimal; broad limits have been set only for the elements C, Mn, P, S, and Si.

B. Miscellaneous Aspects

Properties which do not appear in the tables listed above are very simply described in the following

1. The Control of Inclusion Shape in Steels

One of the more significant developments in microalloyed HSLA steels is to control the shape of inclusion. Small additions of rare earth elements, Zr, Ti, or V together with TMCP (thermo-mechanical control process) can change the shape of the sulfide inclusions from elongated strings to small, dispersed particles. This change substantially improves formability and increases transverse impact energy, which is a measure of resistance to ductile fracture. This technology, i.e., inclusion shape control, was introduced with the development of hot rolled sheet and light plate with the yield strength of 550 MPa (or 80 ksi) in the as-rolled condition. This technology has also been extended to include lower yield strength grades ranging from 310 to 550 MPa (or 45–80 ksi) and specific product specifications are given in ASTM A715.

2. Heat-Treated High-Strength Low-Alloy Steels

The high-strength heat-treated grades include: (a) steels with more than 0.75% Cu attain yield strength of about 485 MPa through precipitation hardening, and (b) steels of other compositions that attain yield strength of 620 MPa or more in quenched and tempered condition.

3. Notch Toughness

The notch toughness of HSLA steels, whether evaluated by Charpy impact or drop weight tests, is superior to that of structural carbon steels. The Charpy V-notch impact values for a structural carbon steel (ASTM A7), a HSLA steel and a heat-treated constructional alloy steel are shown in Fig. 11 [21]. The transition temperatures in the as-rolled or normalized conditions are principally controlled by chemical composition and ferrite grain size. Notch toughness is mainly affected by microstructure for both precipitation hardening steel and quenched and tempered steel.

The effects of individual elements, as well as ferrite grain size on the 20-J Charpy V-notch transition temperature of hot rolled or normalized pearlitic steels are listed in Table 15 [21].

Microstructure governs the notch toughness of quenched and tempered steels. In general, there exists a relationship which shows the lower the phase transformation temperature of austenite, the lower its transition temperature after a suitable tempering treatment. Low carbon tempered martensite and lower bainite lead to optimum values. Theeffects of individual elements on notch toughness are mainly through their effects on microstructure.

4. Directionality of Properties

In microalloyed HSLA steels, the changes in mechanical properties through the addition of Nb and V, together with TMCP, result in improved yield strength, weldability, and

Table 7 Compositions of HSLA Steels

Steel	Composition, %[a]										
	C	Mn	P	S	Si	Cr	Ni	Mo	Cu	V	Others
Heat-treated high-strength low-alloy steels											
1	0.28	1.50	0.025	—	0.25	—	—	0.15	—	—	—
2	0.28	0.75	0.025	—	0.75	0.50	—	0.16	—	—	0.12 Zr
3	0.12	0.50	0.025	—	0.25	—	1.85	—	1.00	—	—
4	0.18	1.00	0.025	—	0.25	—	1.35	0.25	0.65	—	—
5	0.12	0.75	0.025	—	0.10	—	1.25	0.25	1.15	—	—
6	0.18	0.75	0.025	—	0.10	—	1.50	0.25	1.25	—	—
7	0.15	0.80	0.025	—	0.25	0.60	0.85	0.50	0.35	0.05	0.004 B
8	0.12–0.21	0.45–0.70	0.035	—	0.20–0.35	—	—	0.50–0.65	—	—	0.001–0.005 B
9	0.12–0.21	0.45–0.70	0.035	—	0.20–0.35	—	1.20–1.50	0.45–0.60	—	—	0.001–0.005 B
10	0.12–0.21	0.45–0.70	0.035	—	0.20–0.35	0.85–1.20	1.20–1.50	0.45–0.60	—	—	0.001–0.005 B
11	0.10–0.20	0.60–1.00	0.035	—	0.15–0.35	0.40–0.65	0.70–1.00	0.40–0.60	0.15–0.50	0.03–0.08	0.0005–0.006 B
12	0.12–0.21	0.70–1.00	0.035	—	0.20–0.35	0.40–0.65	—	0.15–0.25	—	0.03–0.08	0.01–0.03 Ti; 0.0005–0.005 B
13	0.12–0.21	0.95–1.30	0.035	—	0.20–0.35	0.40–0.65	0.30–0.70	0.20–0.30	—	0.03–0.08	0.0005 B
14	0.21	0.60–1.10	0.04	—	0.40–0.90	0.40–0.90	—	0.28	—	—	0.05 Zr; 0.0025 max B
15	0.18	0.10–0.40	0.025	—	0.15–0.35	1.00–1.80	2.00–3.25	0.20–0.60	—	—	—

16	0.20	0.10–0.40	0.025	—	0.15–0.35	1.00–1.80	2.20–3.50	0.20–0.60	—	—	—
17	0.12	0.60–0.90	0.015	—	0.15–0.35	0.40–0.70	4.75–5.25	0.30–0.65	—	0.05–0.10	—
18	0.16	0.60–0.90	0.015	—	0.15–0.35	0.40–0.70	4.75–5.25	0.30–0.65	—	0.05–0.10	—
Microalloyed high-strength low-alloy steels											
19	0.09	0.35	0.010	0.012	0.01	—	—	—	—	< 0.01	0.05 Al; 0.005 N; 0.015 Nb
20	0.09	1.25	0.010	0.012	0.30	—	—	—	—	0.09	0.05 Al; 0.005 N; 0.09 Nb
21	0.14	1.25	0.03	0.03	0.30	—	—	—	—	0.02	0.01 Nb
22	0.16	1.40	0.03	0.03	0.30	—	—	—	—	0.02	0.01 Nb
23	0.18	1.50	0.03	0.03	0.30	—	—	—	—	0.02	0.01 Nb
24	0.18	1.60	0.03	0.03	0.60	—	—	—	—	0.05	—
25	0.06	0.45	0.01	0.02	0.10	—	—	—	—	—	0.02 Nb
26	0.06	0.75	0.01	0.02	0.10	—	—	—	—	—	0.04 Nb
27	0.06	0.95	0.01	0.02	0.10	—	—	—	—	—	0.10 Nb
28	0.10	0.40	—	—	—	—	—	—	—	—	0.05 Ti; 0.01 Al
29	0.10	0.40	—	—	—	—	—	—	—	—	0.05 Ti; 0.01 Al
30	0.10	0.40	—	—	—	—	—	—	—	—	0.05 Ti; 0.01 Al

[a]Single values are typical values unless otherwise noted.
Source: Ref. 21.

Table 8 List of Specifications Applicable to HSLA Steels

Alloy group	Applicable specifications	
	ASTM	SAE
Niobium or vanadium	A572	J410c
	A607	—
	A656	—
	A715	—
Low manganese–vanadium–titanium	A606	J410c
	A633	—
	A715	—
Manganese or manganese–copper	A440	[a]
	A537	—
	A633	—
	A662	—
	A678	—
Manganese–vanadium–copper multiple-alloy plus copper	A441	J410c
	A242	J410C
	A588	—
	A606	—
Multiple-alloy plus copper and phosphorus	A242	J410c
	A606	—
Precipitation hardening	[a]	[a]
Constructional[b]	A514	J410c
	A517	—

[a]None apply.
[b]Extra high-strength steels.
Source: Ref. 21.

toughness. The reduced ferrite grain size brings about the increase in yield strength. Because of this increase, the reduction in toughness due to precipitation strengthening can be tolerated. The remaining properties, however, are dependent on the test direction of samples. In the transverse direction, toughness is reduced considerably because of the characteristic shape of non-metallic inclusions which become elongated in the rolling direction during rolling. For Al-killed steels, this is mainly caused by elongated sulfide inclusions. To prevent sulfides from becoming excessively elongated in the hot rolling process, elements such as Zr, Ti or rare earth elements can be added to form sulfides with high melting points. These sulfides are less plastic and cannot be deformed easily at hot rolling temperature and the existence of these compounds leads to the enhanced transverse toughness remarkably.

5. Fatigue Characteristics of High-Strength Low-Alloy Steels

Many structural applications for HSLA steels are under the condition of cyclic loading. The fatigue behavior of these steels therefore becomes important. Some of the fatigue characteristics of HSLA steels are compared to those of hot rolled low-carbon steels in the following [21]:

(a) High-strength low-alloy steels possess fatigue properties equivalent or superior to those of hot rolled low-carbon steel.
(b) High-strength low-alloy steels possess greater notch fatigue resistance than hot rolled low-carbon steels.
(c) Large plastic prestrains tend to impair the fatigue life of both HSLA and hot rolled low-carbon steels.
(d) Most of the work hardening effects are not retained under cyclic loading for both HSLA and hot rolled low-carbon steel.

III. WELDING DESIGN OF HIGH-STRENGTH LOW-ALLOY STEELS

Drastic reduction in material thickness and weight can be achieved by using HSLA steels. Some properties are in common among HSLA steels for welding and welded components. A low-carbon content and the maximized grain refinement are fundamentally important the HSLA steels for welding. Actually, a chemical composition with less than the 0.09% C threshold value is preferred to avoid the peritectic reaction during solidification, which is responsible for microsegregation and thus a deterioration of the heat affected zone (HAZ) toughness. The toughness in the HAZ is largely improved with a microstructure of low carbon acicular bainite or ferrite.

A. Weldability of Cast Steel for Offshore Structures

The important requirements for cast offshore steels are weldability, toughness, and fatigure resistance. The carbon content in these steels is usually decreased below 0.15–0.18%, or even below 0.12% to fit the weldability requirements [22]. Lowering carbon content decreases the hardness of HAZ and reduces the formation of grain boundary carbides, which are known as local brittle zones reducing the fracture resistance of HAZ.

The sulfur level has been reduced below 0.005% or less to result in greatly improved toughness, weldability, and fatigue properties. Other critical impurities to be reduced are N, O, H, and P. Because an extremely low impurity level is required for high-strength offshore steels, which has yield strength above 450 MPa, secondary melting processing is necessary to increase the purity level.

Microalloying elements such as Ti, Nb, and V have been added to provide grain refinement and precipitation hardening in these steels. On the other hand, microalloying exhibits a deleterious effect on the toughness of the HAZ, especially in thick sections. Nickel is the most common alloying element in offshore steels. It increases strength through solid-solution strengthening and improves low temperature toughness, which is important to offshore steels.

There are a number of methods to improve the mechanical properties of cast offshore steels. Grain refinement and precipitation strengthening, as stated above, are a traditional way and have been applied to wrought materials together with the sophisticated rolling treatments. Another method is to modify the microstructure after austenitizing from ferrite to low carbon lath martensite and bainite by the addition of hardening agents Cr and Mo and efficient water quenching. The addition of Cr and Mo, however, increases the carbon equivalent and this effect should be reduced by minimizing the carbon content

Table 9 Summary of Characteristics and Intended Uses of HSLA Steels Described in ASTM Specifications[a]

ASTM Specification	Title	Alloying elements[b]	Available mill forms	Special characteristics	Intended uses
A242	High-strength low-alloy structural steel	Cr, Cu, N, Ni, Si, Ti, V, Zr	Plates, bars, and shapes up to 102 mm (4 in.) thick	Atmospheric corrosion resistance four times that of carbon steel	Structural members, in welded, bolted, or riveted constructions
A440	High-strength structural steel	Cu, Si	Plates, bars and shapes up to 102 mm (4 in.) thick	Atmospheric corrosion resistance twice that of carbon steel	Structural members, primarily in bolted, or riveted constructions
A441	High-strength low-alloy structural manganese–vanadium steel	V, Cu, Si	Plates, bars and shapes up to 203 mm (8 in.) thick	Atmospheric corrosion resistance twice that of carbon steel	Welded, bolted or riveted structures, but primarily welded bridges and buildings
A572	High-strength low-alloy niobium–vanadium steels of structural quality	Nb, V, N	Plates, bars, shapes and sheet piling up to 152 mm (6 in.) thick	Yield strengths of 290–450 MPa (42–65 ksi), in six grades	Welded, bolted or riveted structures, but mainly bolted or riveted bridges and buildings
A588	High-strength low-alloy structural steel with 50 ksi minimum yield point to 4 in. thick	Nb, V, Cr, Ni, Mo, Cu, Si, Ti, Zr	Plates, bars and shapes up to 203 mm (8 in.) thick	Atmospheric corrosion resistance four times that of carbon steel; nine grades of similar strength	Welded, bolted or riveted structures, but primarily welded bridges and buildings where weight savings or added durability is important
A606	Steel sheet and strip, hot rolled, and cold-rolled, high-strength low-alloy with improved corrosion resistance	Not specified	Hot rolled and cold rolled sheet and strip	Atmospheric corrosion resistance twice that of carbon steel (type 2) or four times that of carbon steel (type 4)	Structural and miscellaneous purposes where weight savings or added durability is important
A607	Steel sheet and strip, hot rolled and cold rolled, high-strength low-alloy niobium and/or vanadium	Nb, V, Cu, N	Hot rolled and cold rolled sheet and strip	Atmospheric corrosion resistance twice that of carbon steel, but only when copper content is specified; yield strengths of 310–485 MPa (45–70 ksi) in six grades	Structural and miscellaneous purposes where greater strength or weight savings is important

A618	Hot-formed welded and seamless high-strength low-alloy structural tubing	Nb, V, Si, Cu	Square, rectangular, round, and special shape structural tubing, welded, or seamless	Three grades of similar yield strength; may be purchased with atmospheric corrosion resistance twice that of carbon steel	General structural purposes, including welded, bolted, or riveted bridges, and buildings
A633	Normalized high-strength low-alloy structural tubing	Nb, V, Cr, Ni, Mo, Cu, N, Si	Plates, bars, and shapes up to 152 mm (6 in.) thick	Enhanced notch toughness; yield strengths of 290–415 MPa (42–60 Ksi) in five grades	Welded, bolted, or riveted structures for service at temperatures down to −45°C (−50°F)
A656	High-strength low-alloy hot rolled structural vanadium–aluminum–nitrogen and titanium–aluminum steels	V, Al, N, Ti, Si	Plates, normally up to 15.9 mm (5/8 in.) thick	Yield strength of 552 MPa (80 ksi)	Truck frames, brackets, crane booms, rail cars, and other applications where weight savings is important
A690	High-strength low-alloy steel H-piles and sheet piling for use in marine environments	Ni, Cu, Si	Structural quality H-piles and sheet piling	Corrosion resistance two to three times greater than carbon steel in splash zone of marine structures	Dock walls, sea walls, bulkheads, excavations, and similar structures exposed to seawater
A715	Steel sheet and strip, hot rolled, high-strength low-alloy with improved formability	Nb, V, Cr, Mo, N, Si, Ti, Zr, B	Hot rolled sheet and strip	Improved formability[c] compared to A606 and A607; yield strengths of 345–550 MPa (50–80 ksi) in four grades	Structural and miscellaneous applications where high strength, weight savings, improved formability, and good weldability are important

[a]For grades and mechanical properties, see Table 10
[b]In addition to C, Mn, P, and S, a given grade may contain one or more of the listed elements, but not necessarily all of them; for specified composition limits, see Table 11.
[c]Obtained by producing killed steel, made to fine grain practice, and with microalloying elements such as Nb, V, Ti, and Zr in the composition.
Source: Ref. 21.

Table 10 Mechanical Properties of HSLA Steel Grades Described in ASTM Specifications[a]

ASTM Specification	Type, Grade or Condition	UNS Designation	Min Tensile Strength[b]		Min yield strength[b]		Min Elongation, %[b]		Impact Energy[c]		Bend Radius[b]	
			MPa	ksi	MPa	ksi	In 200 mm or 8 in.	In 50 mm or 2 in.	J	ft lb	Longi-tudinal	Trans-verse
A242	Type 1	K11510	435–480	63–70	290–345	42–50	18	21	—	—	[d]	—
	Type 2	K12010	435–480	63–70	290–345	42–50	18	21	—	—	[d]	—
A440	—	K12810	435–485	63–70	290–345	42–50	18	21	—	—	[d]	—
A441	—	K12211	415–485	63–70	275–345	40–50	18	21	—	—	[d]	—
A572	Grade 42	—	415	60	290	42	20	24	—	—	[d]	—
	Grade 45	—	415	60	310	45	19	22	—	—	[d]	—
	Grade 50	—	450	65	345	50	18	21	—	—	[d]	—
	Grade 55	—	485	70	380	55	17	20	—	—	[d]	—
	Grade 60	—	520	75	415	60	16	18	—	—	[d]	—
	Grade 65	—	550	80	450	65	15	17	—	—	[d]	—
									—	—		—
A588	Grades A–J	[e]	435–485	63–70	290–345	42–50	18	21	—	—	[d]	—
A606	Hot rolled	—	480	70	345	50	—	22	—	—	t	$2t$–$3t$
	Hot rolled and annealed or normalized	—	450	65	310	45	—	22	—	—	t	$2t$–$3t$
	Cold rolled	—	450	65	310	45	—	22	—	—	t	$2t$–$3t$

A607	Grade 45	—	410	60	310	45	—	22–25	—	—	t	$1.5t$
	Grade 50	—	450	65	345	50	—	20–22	—	—	t	$1.5t$
	Grade 55	—	480	70	380	55	—	18–20	—	—	$1.5t$	$2t$
	Grade 60	—	520	75	415	60	—	16–18	—	—	$2t$	$3t$
	Grade 65	—	550	80	450	65	—	15–16	—	—	$2.5t$	$3.5t$
	Grade 70	—	590	85	485	70	—	14	—	—	$3t$	$4t$
A618	Grade I	K02601	483	70	345	50	19	22	—	—	t–$2t$	—
	Grade II	K12609	483	70	345	50	18	22	—	—	t–$2t$	—
	Grade III	K12700	448	65	345	50	18	20	—	—	t–$2t$	—
A633	Grades A and B	[e]	430–570	63–83	290	42	18	23	[f]	[f]	[d]	—
	Grades C and D	[e]	450–620	65–90	315–345	46–50	18	23	[f]	[f]	[d]	—
	Grade E	K12202	520–690	75–100	380–415	55–60	18	23	[f]	[f]	[d]	—
A656	Grades 1 and 2	[e]	655–793	95–115	552	80	12	—	—	—	[d]	—
A690	—	K12249	485	70	345	50	18	—	—	—	$2t$	—
A715	Grade 50	—	415	60	345	50	—	22–24	—	—	0	t
	Grade 60	—	485	70	415	60	—	20–22	—	—	0	t
	Grade 70	—	550	80	485	70	—	18–20	—	—	t	$1.5t$
	Grade 80	—	620	90	550	80	—	16–18	—	—	t	$1.5t$

[a]For characteristics and intended uses, see Table 9; for specified composition limits, see Table 11.
[b]May vary with product size and mill form; for specific limits, see the article in this handbook on the appropriate mill form.
[c]Charpy V-notch.
[d]Optional supplementary requirement given in ASTM A6, section S14.
[e]See Table 11.
[f]Optional supplementary requirement given in ASTM A6, section S1.
Source: Ref. 21.

Table 11 Composition Limits for HSLA Steel Grades Described in ASTM Specification[a]

ASTM Specification	Type or Grade	UNS Designation	Heat Composition Limits, %[b]									
			C	Mn	P	S	Si	Cr	Ni	Cu	V	Other
A242	Type 1	K11510	0.15	1.00	0.45	0.05	—	—	—	0.20 min	—	—
	Type 2	K12010	0.20	1.35	0.04	0.05	—	—	—	0.20 min[c]	—	—
A440	—	K12810	0.28	1.10–1.60	0.04	0.05	0.30	—	—	0.20 min	—	—
A441	—	K12211	0.22	0.85–1.25	0.04	0.05	0.30	—	—	0.20 min	0.02 min	—
A572	Grade 42	—	0.21	1.35[d]	0.04	0.05	0.30[d]	—	—	0.20 min[e]	—	[f]
	Grade 45	—	0.22	1.35[d]	0.04	0.05	0.30[d]	—	—	0.20 min[e]	—	[f]
	Grade 50	—	0.23	1.35[d]	0.04	0.05	0.30[d]	—	—	0.20 min[e]	—	[f]
	Grade 55	—	0.25	1.35[d]	0.04	0.05	0.30	—	—	0.20 min[e]	—	[f]
	Grade 60	—	0.26	1.35[d]	0.04	0.05	0.30	—	—	0.20 min[e]	—	[f]
	Grade 65	—	0.23[d]	1.65[d]	0.04	0.05	0.30	—	—	0.20 min[e]	—	[f]
A588	Grade A	K11430	0.10–0.19	0.90–1.25	0.04	0.05	0.15–0.30	0.40–0.65	—	0.25–0.40	0.02–0.10	—
	Grade B	K12043	0.20	0.75–1.25	0.04	0.05	0.15–0.30	0.40–0.70	0.25–0.50	0.20–0.40	0.01–0.10	—
	Grade C	K11538	0.15	0.80–1.35	0.04	0.05	0.15–0.30	0.30–0.50	0.25–0.50	0.20–0.50	0.01–0.10	—
	Grade D	K11552	0.10–0.20	0.75–1.25	0.04	0.05	0.50–0.90	0.50–0.90	—	0.30	—	0.04 Nb; 0.05–0.15 Zr
	Grade E	K11567	0.15	1.20	0.04	0.05	0.15–0.30	—	0.75–1.25	—	0.05	0.10–0.25 Mo
	Grade F	K11541	0.10–0.20	0.50–1.00	0.04	0.05	0.30	0.30	0.40–1.10	0.30–1.00	0.01–0.10	0.10–0.20 Mo
	Grade G	K12040	0.20	1.20	0.04	0.05	0.25–0.70	0.50–1.00	0.80	0.30–0.50	—	0.10 Mo; 0.07 Ti
	Grade H	K12032	0.20	1.25	0.035	0.040	0.25–0.75	0.10–0.25	0.30–0.60	0.20–0.35	0.02–0.10	0.15 Mo; 0.005–0.030 Ti
	Grade J	K12044	0.20	0.60–1.00	0.04	0.05	0.30–0.50	—	0.50–0.70	0.30 min	—	0.03–0.05 Ti
A606	—	—	0.22	1.25	—	0.05	—	—	—	—	—	—
A607	Grade 45	—	0.22	1.35	0.04	0.05	—	—	—	0.20 min[e]	—	[f]
	Grade 50	—	0.23	1.35	0.04	0.05	—	—	—	0.20 min[e]	—	[f]
	Grade 55	—	0.25	1.35	0.04	0.05	—	—	—	0.20 min[e]	—	[f]
	Grade 60	—	0.26	1.50	0.04	0.05	—	—	—	0.20 min[e]	—	[f]
	Grade 65	—	0.26	1.50	0.04	0.05	—	—	—	0.20 min[e]	—	[f]
	Grade 70	—	0.26	1.65	0.04	0.05	—	—	—	0.20 min[e]	—	[f]
A618	Grade I	K02601	0.22	1.25	—	0.05	—	—	—	[g]	—	—
	Grade II	K12609	0.22	0.85–1.25	0.04	0.05	0.30	—	—	0.20 min	0.02 min	—
	Grade III	K12700	0.23	1.35	0.04	0.05	0.30	—	—	—	0.02 min	0.005 min Nb[h]

A633	Grade A	K01802	0.18	1.00–1.35	0.04	0.05	0.15–0.30	—	—	—	—	0.05 Nb
	Grade B	K01803	0.18	1.00–1.35	0.04	0.05	0.15–0.50	—	—	—	0.10	—
	Grade C	K12000	0.20	1.15–1.50	0.04	0.05	0.15–0.50	—	—	—	—	0.01–0.05 Nb
	Grade D	K02003	0.20	0.70–1.60[d]	0.04	0.05	0.15–0.50	0.25	0.25	0.35	—	0.08 Mo
	Grade E	K12202	0.22	1.15–1.50	0.04	0.05	0.15–0.50	—	—	—	0.04–0.11	0.01–0.05 Nb[e]; 0.01–0.03 N
A656	Grade 1	K11804	0.18	1.60	0.040	0.050	0.60	—	—	—	0.05–0.15	0.020 min Al; 0.005–0.030 N
	Grade 2	K11503	0.15	0.90	0.040	0.050	0.10	—	—	—	—	0.05–0.50 Ti; 0.01 min Al
A690	—	K12249	0.22	0.60–0.90	0.08–0.15	0.05	0.10	—	0.40–0.75	0.50 min	—	—
A715	Type 1	—	0.15	1.65	0.025	0.035	0.10	—	—	—	—	0.05 min Ti
	Type 2	—	0.15	1.65	0.025	0.035	0.60[j]	—	—	—	—	0.005 min N[j]
	Type 3	—	0.15	1.65	0.025	0.035	0.60[j]	—	—	—	0.08[j]	0.005 min Nb; 0.020 N[j]
	Type 4	—	0.15	1.65	0.025	0.035	0.90	0.80[j]	—	—	—	0.005–0.06 Nb[k]; 0.10 Ti[j]; 0.05 min Zr; 0.0025 B[j]
	Type 5[m]	—	0.15	1.65	0.025	0.035	0.30	—	—	—	—	0.20 min Mo; 0.03 min Nb
	Type 6	—	0.15	1.65	0.025	0.035	0.90	—	—	—	—	0.005–0.10 Nb
	Type 7	—	0.15	1.65	0.025	0.035	—	—	—	—	0.005 min[n]	0.020 N

[a]For characteristics and intended uses, see Table 9; for mechanical properties, see Table 10.
[b]Where a single value is shown, it is maximum unless otherwise stated.
[c]Applicable only if Si and/or Cr content is below 0.5%.
[d]Values may vary, or minimum value may exist, depending on product size and mill form.
[e]Optional or when specified.
[f]May be purchased as type 1 (0.005–0.05 Nb), type 2 (0.01–0.15 V), type 3 (0.05 max Nb plus 0.02–0.15 V) or type 4 (0.015 max N plus V $\geq$ 4 N).
[g]As agreed between producer and purchaser.
[h]May be substituted for all or part of the V.
[j]Not added to Grades 50 and 60.
[k]Might not be added to Grade 50.
[m]Available as Grade 80 only.
[n]0.005 min Nb may be added in place of or in addition to the V.
Source: Ref. 21.

Table 12 Compositions, Mill Forms and Characteristics of HSLA Steels Described in SAE J410c [a]

Grade[c]	Heat composition limits, %[b] C, max	Mn, max	P, max	Other elements[d]	Available mill forms	Special characteristics
942X	0.21	1.35	0.04	Nb, V	Plates, bars, and shapes up to 102 mm (4 in.) thick	Similar to 945X and 945C except for better weldability and formability
945A	0.15	1.00	0.04	—	Sheet, strip, plates, bars, and shapes up to 76 mm (3 in.) thick	Excellent weldability, formability, and notch toughness
945C	0.23	1.40	0.04	—	Sheet, strip, plates, bars, and shapes up to 76 mm (3 in.) thick	Similar to 950C except that lower carbon and manganese improve weldability, formability and notch toughness
945X	0.22	1.35	0.04	Nb, V	Sheet, strip, plates, bars, and shapes up to 38 mm (3 in.) thick	Similar to 945C except for better weldability and formability
950A	0.15	1.30	0.04	—	Sheet, strip, plates, bars, and shapes up to 76 mm (3 in.) thick	Good weldability, notch toughness, and formability
950B	0.22	1.30	0.04	—	Sheet, strip, plates, bars, and shapes to 76 mm (3 in.) thick	Fairly good notch toughness, and formability
950C	0.25	1.60	0.04	—	Sheet, strip, plates, bars, and shapes to 76 mm (3 in.) thick	Fair formability and toughness
950D	0.15	1.00	0.15	—	Sheet, strip, plates, bars, and shapes to 76 mm (3 in.) thick	Good weldability and formability; phosphorus should be considered in conjunction with other elements
950X	0.23	1.35	0.04	Nb, V	Sheet, strip, plates, bars, and shapes to 38 mm (1.5 in.) thick	Similar to 950C except for better weldability and formability

955X	0.25	1.35	0.04	Nb, V, N	Sheet, strip, plates, bars, and shapes to 38 mm (1.5 in.) thick	
960X	0.26	1.45	0.04	Nb, V, N	Sheet, strip, plates, bars, and shapes to 38 mm (1.5 in.) thick	Similar to 945X and 950X except that progressively higher strengths are obtained by increasing the carbon and manganese contents, or by adding nitrogen up to 0.015%; formability and weldability generally decrease with increased strength; toughness varies with composition and mill practice
965X	0.26	1.45	0.04	Nb, V, N	Sheet, strip, plates, bars, and shapes to 19 mm (0.75 in.) thick	
970X	0.26	1.65	0.04	Nb, V, N	Sheet, strip, plates, bars, and shapes to 19 mm (0.75 in.) thick	
980X	0.26	1.65	0.04	Nb, V, N	Sheet, strip and plates to 10 mm (0.38 in.) thick	

[a]For mechanical properties, see Table 13.

[b]0.05 max P and 0.90 max Si, all grades.

[c]Fully killed steel made to fine grain practice may be specified by adding a second suffix "K"; for instance, 945XK. Steels made to "K" practice are normally specified only for applications requiring better toughness at low temperatures than steels made to normal semikilled practice.

[d]Elements normally added singly or in combination to produce the specified mechanical properties and other characteristics. Other alloying elements such as copper, chromium, and nickel may be added to enhance atmospheric corrosion resistance.

Source: Ref. 21.

Table 13 Mechanical Properties of HSLA Steel Grades Described in SAE J410c[a]

Grade	Min Tensile Strength[b]		Min Yield Strength[b,c]		Min Elongation, %[b]		Bend diameter[b,d]
	MPa	ksi	MPa	ksi	In 200 mm or 8 in.	In 50 mm or 2 in.	
942X	415	60	290	42	20	24	t–$3t$
945A	415–450	60–65	275–310	40–45	18–19	22–24	t–$3t$
945C	415–450	60–65	275–310	40–45	18–19	22–24	t–$3t$
945X	415	60	310	45	19	22–25	t–$2.5t$
950A	430–483	63–70	290–345	42–50	18–19	22–24	t–$3t$
950B	430–483	63–70	290–345	42–50	18–19	22–24	t–$3t$
950C	430–483	63–70	290–345	42–50	18–19	22–24	t–$3t$
950D	430–483	63–70	290–345	42–50	18–19	22–24	t–$3t$
950X	450	65	345	50	18	22	t–$3t$
955X	483	70	380	55	17	20	t–$3.5t$
960X	520	75	415	60	16	18	$1.5t$–$3t$
965X	550	80	450	65	15	15	$2t$–$3t$
970X	590	85	485	70	14	14	$3t$
980X	655	95	550	80	10	12	$3t$

[a]For compositions, available mill forms and special characteristics of these steels, see Table 12 .
[b]May vary with product size and mill form; for specific limits, refer to SAE J410c.
[c]0.2% offset.
[d]180° bend test at room temperature. Used for mill acceptance purposes only; not to be used as a basis for specifying fabricating procedures.
Source: Ref. 21.

to maintain excellent weldability. Further increasing of strength could be obtained through solid-solution strengthening.

New secondary steel-making facilities (argon oxygen decarburization converter (AODC) and vacuum oxygen decarburization converter (VODC) are being installed in

Table 14 Listing of SAE J410c Grades in Approximate Order of Decreasing Toughness, Formability, and Weldability

Toughness	Formability	Weldability
945A	945A	945A
950A	950A	950A
950B	945C, 945X	950D
950D	950B, 950X, 942X	945X
945X, 950X	950D	950B, 950X
945C, 950C, 942X	950C	945C
955X	955X	955X, 950C, 942X
960X	960X	960X
965X	965X	965X
970X	970X	970X
980X	980X	980X

Source: Ref. 21.

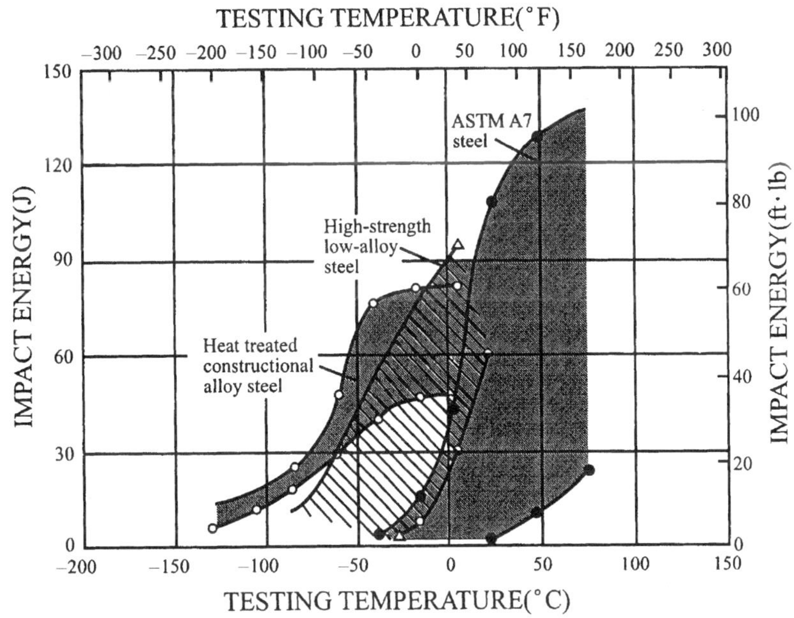

Figure 11 Notch toughness of HSLA steel compared to structural carbon steel and hardened alloy steel. (From Ref. 21.)

Table 15 Effects of Individual Elements on Charpy V-notch Transition Temperature of Pearlitic Steel

Element	Range Covered, %	Transition temperature change[a]	
		°C	°F
C	0.12–0.32	+1.7	+3.0
Mn	0.20–1.50	−0.6	−1.0
P	0.01–0.06	+5.5	+10.0
S	0.02–0.05	None	None
Si	0.02–0.20	−1.7	−3.0
Al	0.00–0.06	Beneficial	Beneficial
N	0.003–0.022	+18	+33
Cu	0.03–2.00	+0.22	+0.40
Ni	0.04–2.50	−0.25	−0.45
Cr	0.005–1.00	+0.28	+0.50
Mo	0.00–0.29	+1.0	+1.8
Ti	0.00–0.08	+2.8	+5.0
V	0.00–0.05	Nil	Nil
Zr	0.05–0.15	+3.3	+6.0
Ferrite grain size		14[b]	−25[b]

[a]Change in 20-J Charpy V-notch transition temperature for each 0.01% increase in element concentration.
[b]For each increase of one ASTM number (decrease in grain size).
Source: Ref. 21.

Table 16 The Nominal Chemical Analysis of Lokomo Steel Works' Cast Offshore Steels

Element	OS-340	OS-540	OS-690
C	0.10	0.08	0.09
Si	0.30	0.25	0.25
Mn	1.00	0.40	0.40
P	0.007	0.008	0.008
S	0.002	0.002	0.002
Cr	<0.20	1.25	0.80
Mo	<0.08	0.40	0.55
Ni	0.90	2.50	4.50
Ti	<0.01	<0.01	<0.01
Nb	<0.01	<0.01	<0.01
V	<0.01	<.01	<0.01
Al	0.035	0.035	0.035
N	0.008	0.008	<0.008
O	<30 ppm	<30 ppm	<30 ppm
H	< 2 ppm	< 2 ppm	< 2 ppm

Source: Ref. 22.

Table 17 Mechanical and Fracture Toughness Properties of Lokomo Steel Works' Cast Offshore Steels

	OS-340	OS-540	OS-690
Re	>340MPa	>540MPa	>690MPa
Rm	>450MPa	>650MPa	>760MPa
As	>16%	>16%	>16%
Z	>50%	>50%	>50%
CVN(−40°C)	200J	175J	175J
NDT*	−55°C	−90°C	−125°C
CTOD[a]	0.602 mm	0.718 mm	0.560 mm

*NDT nil ductility temperature
[a]200 × 200 mm cross-section test speciment at −40°C.
Source: Ref. 22.

Table 18 The Weldability Properties of Lokomo Steel Works' OS grades

Grade	Preheating	Heat input	PWHT
OS-340	Not necessary	10–40 kJ/cm	Not necessary
OS-540	Not necessary	10–35 kJ/cm	Not necessary
OS-690	In difficult circumstances	10–35 kJ/cm	Not necessary

Source: Ref. 22.

many cast steel foundries to allow production of greatly improved cast steels, even for subzero temperatures.

As an example, the chemical composition, mechanical properties, and weldability of several OS grade of cast offshore steels developed in Lokomo Steel Works [22] are shown in Tables 16, 17, and 18, respectively. The strength properties, which are shown in Table 17, have been achieved only by solid-solution strengthening and fine grain size.

B. Design and Development of Welding Wire for High-Strength Low-Alloy Steel

Welding electrodes are traditionally developed using "trial-and-error" techniques, which employ iterative procedures to identify appropriate alloy systems. However, new metallurgical models and approaching methods [23–29] based on chemical composition, processing, structure, and property relationship were developed to design the composition of bare solid electrodes. The carbon equivalent number (CEN) equation was analyzed and carbon content there was found to have the largest effect on CEN [23] and therefore a substantial reduction of carbon in weld metal is necessary to reduce the CEN as well as the preheat level. Moreover, it is also desirable to design weld metal containing reduced levels of elements with the highest coefficients in CEN equation (i.e., B, Cr, Mo, V, and Nb) but increased levels of those with the lowest coefficients in CEN equation (i.e., Si, Ni, Cu, and Mn) to further reduce the CEN value and sensitivity of the weld metal to preheat controls. Since excess additions of Si and Cu may promote fusion-zone solidification cracking, it is preferable that a substantial reduction in carbon content is compensated with appropriate increases in Ni or Mn to achieve adequate strength and toughness without increasing the necessity of preheat [23]. It can also be expected that significant reductions in carbon content may bring about the effect which allows a substantial increase in the upper limit of CEN, and facilitates welding without preheat and interpass temperature controls.

As stated above, the low-carbon bainitic steels [30,31], which exhibit high tensile strength (100–200 ksi range), good toughness at low temperature and show resistance to hydrogen assisted cracking (HAC) can be designated as possible candidates to meet the strength and toughness requirements of high-strength weld metals. These steels contained about 0.08 wt% to 0.17 wt% carbon [31] and other alloy additions that retard the transformation of austenite to blocky proeutecoid ferrite, and instead facilitate the transformation to bainite. The alloy additions also suppress the martensite transformation starting temperature (M_s). The B_{50} and M_s temperatures of these steels are related to their chemical composition as shown below [30]:

$$B_{50}\ (^\circ\mathrm{C}) = 770 - (270 \times \mathrm{C}) - (90 \times \mathrm{Mn}) - (37 \times \mathrm{Ni}) - (70 \times \mathrm{Cr}) - (83 \times \mathrm{Mo}) \quad (1)$$

$$M_s\ (^\circ\mathrm{C}) = 561 - (474 \times \mathrm{C}) - (33 \times \mathrm{Mn}) - (17 \times \mathrm{Ni}) - (17 \times \mathrm{Cr}) - (21 \times \mathrm{Mo}) \quad (2)$$

Thus welding electrodes designed to provide weld metal characterized by a B_{50} temperature in the range of 400–500°C and a M_s temperature below the B_{50} temperature will likely exhibit high-strength and acceptable CVN toughness, thereby meeting the all-weld metal mechanical property requirements for MIL-100S and MIL-120S bare electrodes.

Attempts have been made to provide calculated quantities to characterize the weldability of steels for many years. Based on the early work of Dearden and O'Neill [32], the International Institute of Welding (IIW) gave the IIW formula for "carbon equivalent"

(CE_{IIW}):

$$CE_{IIW} = C + Mn/6 + (Cr + Mo + V)/5 + (Ni + Cu)/15 \tag{3}$$

or the following expression for CE [46]:

$$CE = C + Mn/6 + Si/24 + Ni/40 + Cr/5 + Mo/4 + V/14 \tag{4}$$

where all the elements are expressed in weight percent. Several other formulas have also been developed. The P_{cm} equation was developed and has been applied to low carbon low-alloy steel [33], for which CE_{IIW} is not entirely suitable

$$P_{cm} = C + Si/30 + (Mn + Cu + Cr)/20 + Ni/60 + Mo/15 + V/10 + 5B \tag{5}$$

However, a general criticism of using such weldability expression is that they consider only the chemical composition of the material and not the cooling rate. P_{cm} and CE_{IIW} can only be effectively used for correlation of alloy content to hardenability if a constant heat input is being used. More recent efforts in HAZ studies have developed weldability expressions including cooling rate. To design the composition ranges of weld metals that would eliminate or reduce the necessity for preheat and interpass temperature controls and achieve enough resistance to HAC, the CEN formula developed by Yurioka et al. is used [34]:

$$CEN=C+A(C)\times\{Si/24+Mn/6+Cu/15+Ni/20+(Cr+Mo+V+Nb)/5+5B\} \tag{6}$$

where $A(C) = 0.75 + 0.25\tanh\{20 \times (C - 0.12)\}$ is preferable.

The flux cored are welds, which meet the mechanical requirements for MIL-100S and MIL-120S, have CEN in the range of 0.29–0.37, calculated B_{50} temperature in the range of 445–515°C and calculated M_s temperature in the range of 415–451°C [23].

The metal cored arc welds that meet the MIL-100S and MIL-120S mechanical property requirements have CEN in the range of 0.31–0.35, calculated B_{50} temperature and M_s temperature ranging from 453°C to 474°C and 428°C to 439°C, respectively [23].

The formulation of the base composition of a shielded metal arc (SMA) welding electrode was developed for joining HSLA-100 steel plates [24]. To achieve successful SMA welds of HSLA-100 steel with properties comparable to those of the original plate, a consumable with excellent properties is required. The electrode must produce weld metal with proper microstructure to have both sufficient strength and toughness. Acicular ferrite is known as the weld metal constituent which best promotes toughness in HSLA steels with around 500 MPa yield strength due to the random orientation of the ferrite laths and their ability to deflect the crack propagation direction. Factors in the nucleation and growth of acicular ferrite are considered as the following [35]:

(a) chemical composition of the weld deposit,
(b) chemical composition of inclusions (mainly oxides),
(c) size distribution of inclusions, and
(d) crystallographic and thermal disregistry between inclusions and the matrix.

However, in order to be aware of weld metal transformation from austenite to ferrite, it is essential to know the weld metal oxygen content. Oxygen has a very low solubility in iron; however, it occurs as inclusions with a variety of sizes and compositions in the weld metal. The majority of these inclusions form as a result of different deoxidation steps in the liquid weld pool. It is these oxide inclusions that effect the formation of weld metal ferrite. While assessing the factors which control weld metal oxygen, alloying elements in the weld metal play an important role both on the reaction with oxygen to form inclusions and on the formation of the final microstructure.

Due to the variance of welding environment with changing welding conditions, the chemical reactions in the weld pool can be difficult to characterize. Chemical equilibrium is not attained in a weld pool; however, a trend toward equilibrium is generally observed that can be estimated with basic thermodynamic principles. Oxidation in the pool can be generalized as [24]:

$$x\underline{\mathrm{M}} + y\underline{\mathrm{O}} \Leftrightarrow \mathrm{M}_x\mathrm{O}_y \tag{7}$$

According to the law of mass action, the equilibrium constant k is:

$$k = \frac{[a_{\mathrm{M}_x\mathrm{O}_y}]}{[a_{\underline{\mathrm{M}}}]^x[a_{\underline{\mathrm{O}}}]^y} \tag{8}$$

where $[a_{\underline{\mathrm{M}}}]$ and $[a_{\underline{\mathrm{O}}}]$ represent the activities of weld metal alloying element M and oxygen, respectively, $[a_{\mathrm{M}_x\mathrm{O}_y}]$ is the activity of the metal-oxide inclusion in the weld metal and can be taken as unity. Thus, yields:

$$k = \frac{1}{[a_{\underline{\mathrm{M}}}]^x[a_{\underline{\mathrm{O}}}]^y} \tag{9}$$

Competition between chemical reactions within the weld pool also affects the final composition of the weld metal. As local compositions vary in solidification, the local activity of alloying elements will change, which alters the amount of alloying additions that will be oxidized. As a result, these competitive reactions alter the predicted weld metal chemical composition.

One parameter that was introduced to quantify the role of a flux in determining the final weld metal chemical reaction is the "delta quantity" [24], which is the amount of a specified element either gained or lost during the process and is written as

$$[\text{Delta Quantity}] = [\text{Analytic Content}] - [\text{Estimated Content}] \tag{10}$$

The analytic content is determined by the chemical analysis, while the estimated content is determined by adding the contribution of elements from the weld wire to those from the base plate by dilution calculation. A negative delta quantity implies that the alloying element in consideration is lost to slag and a positive value implies that flux contributes the element to the pool. Zero delta quantity means no transfer of alloying element from any of the sources. The most desirable flux systems are those that have delta quantities, at least for the major alloying elements, that remain relatively constant with variations in flux composition and welding condition.

Figure 12 is a schematic in Ref. [24] showing the general changes in electrode coating as the formulation progressed. The initial electrodes had a high rutile and silica content and their flux composition can be located in the lower portion of the diagram. As the silica and rutile were replaced with more basic components, it moved the coating composition toward the top portion of the ternary diagram. Although CaO and CaF_2 are not expected to behave the same, they are both considered as the primary constituents of a basic flux. Following the arrow indicated in Fig. 12, an initial rutile-based electrode coating was

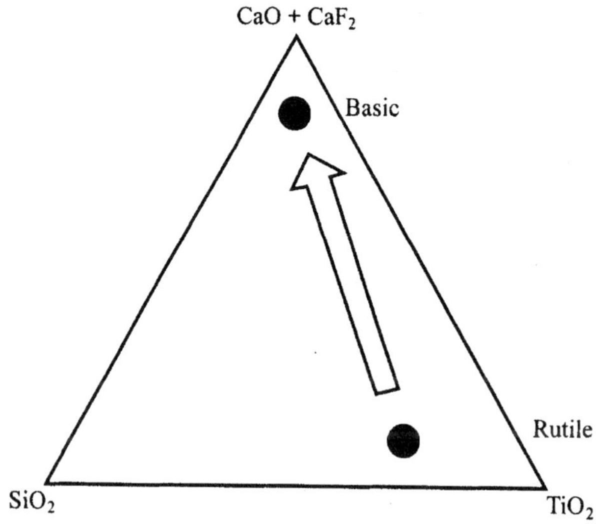

Figure 12 Diagram of the formulation concept: moving from a flux that is rutile in nature (high SiO_2 plus TiO_2content) to a flux that is more basic in nature (high CaO plus CaF_2 content). (From Ref. 24.)

modified to improve its metallurgical characteristics in weld pool oxygen and alloying elements control. The changes were made gradually such that the good electrode performance of a rutile-based electrode would not be radically altered.

Based on the knowledge stated above, for example, an experimental matrix was devised in Ref. [24] to systematically vary two components simultaneously in the flux. The experimental electrode designation has the format Axxx, with x representing numbers from 0 to 9. The first digit is the series number, the second is the sequence number within each series, and the third is used to designate later modifications in each series.

Fleming et al. [24] changed two components in the coating simultaneously (one decreasing a specific amount while the other is increased by the same amount). All electrodes within each series were evaluated and the one that produced the best results for the flux system was used as the starting formulation for the next series. Careful selection of each substitution component was made with the primary function of each component in mind. Figure 13 show schematically the electrode "genealogy" for the test matrix. The major constituents of each electrode were placed at the beginning of each line, and the electrode selected as the starting point for the next series appeared highlighted (circled by a bold circle). For example, electrode A230 performed most satisfactorily in A2xx series, therefore it would be chosen to be further modified in the next series A3xx.

The first electrode (A1) was formulated from an experimental rutile-based electrode for its good welding performance such as slag formability, slag detachability, arc stability, etc. The initial electrode series was designated as A1xx. These electrodes had high rutile content and were comparable to a commercial E6013 steel welding electrode. Continuous addition of FeTi in the coating was the character of this series, which supplied the possibility to study the influence of titanium on the microstructure in an electrode with high oxygen potential.

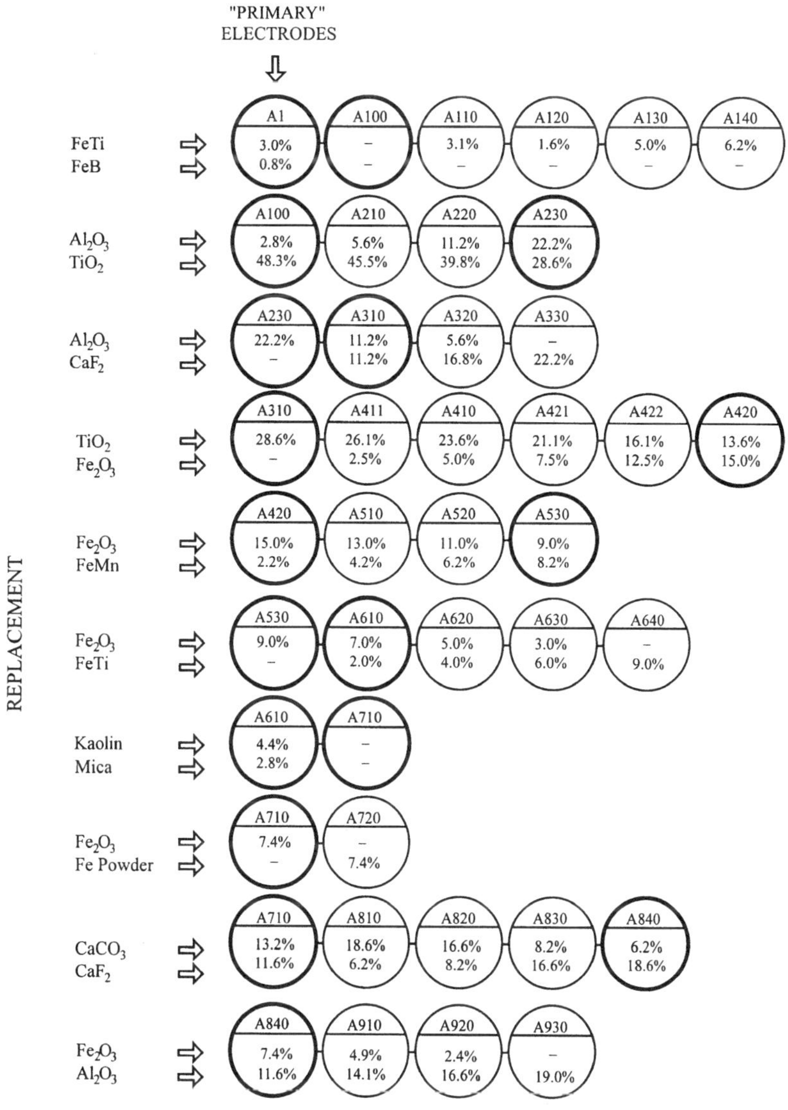

Figure 13 Representation of electrode formulation sequence. Primary changes made in each series are designated above the corresponding line. Electrodes used as the starting composition for each new series are placed at the beginning of each new line. (From Ref. 24.)

In the A2xx series, TiO_2 was substituted by Al_2O_3, with Al_2O_3 increasing from 2.8 to 22.2 wt%. Alumina is amphoteric and should reduce the oxygen in the weld metal while decreasing the viscosity of the slag. This series was also the first step to characterize the effect of oxygen on weld metal microstructure.

For the A3xx series, Al_2O_3 was substituted with CaF_2 so as to further decrease the oxygen in the weld metal and the viscosity of the slag. According to Donchenko et al. [36], CaF_2 can reduce the free hydrogen in the weld metal by the reaction

$$CaF_2 + H_2O \Leftrightarrow CaO + 2HF \tag{11}$$

However, high content of CaF_2 causes arc instability and leads to extremely high mean voltage to sustain the arc.

In A4xx series, there existed the substitution of rutile with Fe_2O_3, which resulted in a further increase of the basicity index and improved arc stability. However, there would be a subsequent increase in the weld metal oxygen.

In order to remove hematite, the A5xx series consisted of additions of Fe–Mn at the expense of Fe_2O_3. The manganese content in the weld metal was increased to improve the hardenability of the weld metal, to promote lower transformation temperature products and to decrease oxygen content and oxide inclusions.

A6xx series consisted of the addition of Fe–Ti and the reduction of ferric-oxide. Titanium was added to promote the formation of acicular ferrite together with the formation of titanium oxides in the weld metal.

For the A7xx series, mica and kaolin, both containing water of crystallization, were removed and also replaced was ferric oxide with iron powder to further decrease oxygen.

In the A8xx series, the CaF_2 to $CaCO_3$ ratio was altered to control the weld metal oxygen, as well as slag viscosity. Calcium fluoride is known to reduce the melting temperature of the slag and thus lower the viscosity. Calcium carbonate decomposes during welding into CO_2 gas and CaO. The former in the products will serve to shield the weld pool, while the latter, incorporated in the slag, will affect the melting temperature and viscosity of the slag, and the amount of oxygen in the weld metal.

In the A9xx series, ferric oxide was replaced with alumina to improve slag viscosity for out-of-position welding. However, alumina is amphoteric and alters the amount of oxygen in the weld metal. If other ingredients are more basic than alumina, the weld pool will pick up oxygen from the dissociation of alumina. But if the others are more acid than alumina, alumina will limit the amount of oxygen pickup in the weld pool. The reduction of ferric oxide would also lead to the decrease of oxygen content in the weld metal. Miscellaneous functions of the flux ingredients are listed in Table 19 [24].

Graphs of the weld metal delta quantity as a function of weld metal oxygen content for manganese, silicon, and titanium are shown in Fig. 14A, B, and C, respectively [24],

Table 19 Components and Primary Functions of the Flux Ingredients Used in the Experiments

Component	Formulae	Primary function
Alumina	Al_2O_3	Arc stabilizer
Calcium carbonate	$CaCo_3$	Shielding gas
Calcium fluoride	CaF_2	Slag former
Feldspar	$K_2O \cdot Al_2O_3 \cdot 6SiO_2$	Shielding gas
Ferroboron	Fe–B	Alloying
Ferromanganese	Fe-Mn	Deoxidizer
Ferrotitanium	Fe–Ti	Alloying
Iron oxide	Fe_2O_3	Slag former
Kaolin	$Al_2O_3 \cdot 2SiO_2 \cdot H_2O$	Slipping agent
Mica	$K_2O \cdot 3Al_2O_3 \cdot 6SiO_2 \cdot H_2O$	Slipping agent
Potassium silicate	K_2SiO_2	Binder
Potassium titanate	$2K_2O \cdot 2TiO_2$	Arc stabilizer
Silica	SiO_2	Slag former
Sodium silicate	Na_2SiO_2	Binder
Titanium dioxide	TiO_2	Slag former

Source: Ref. 24.

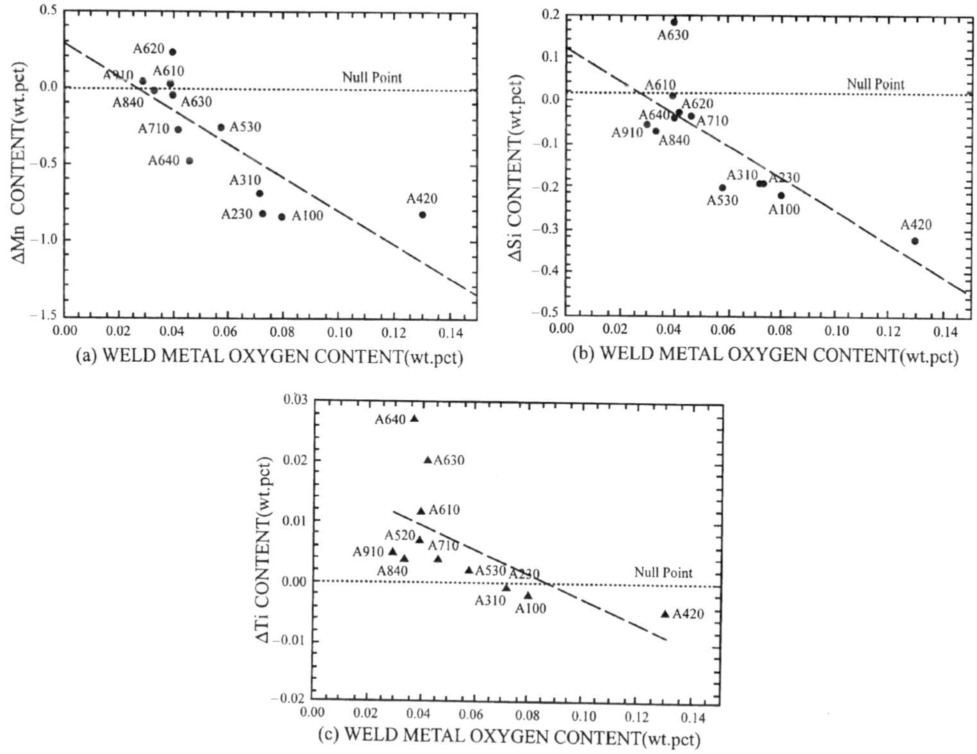

Figure 14 Weld metal delta quantity of: A—manganese; B—silicon; C—titanium as a function of weld metal oxygen content for selected electrodes. (From Ref. 24.)

Table 20 Results of Point Count for the Welds Deposited with the A1xx, A6xx, and A9xx Series Electrodes (in vol.%)

Electrode designation	PF	AF	FS (A)	FS (NA)	M
Al	12	18	33	37	—
A100	28	—	54	18	—
All0	18	12	49	21	—
A120	15	6	36	43	—
A130	16	14	47	23	—
A140	15	24	40	21	—
A610	1	59	9	—	31
A620	3	57	—	—	40
A630	3	57	—	—	40
A640	—	—	—	31	69
A910	—	54	12	6	28
A920	—	44	13	3	40
A930	—	26	7	—	67

PF, primary ferrite; AF, acicular ferrite; FS(A), ferrite with aligned second phase; FS(NA), ferrite with second phase non-aligned; M, martensite.
Source: Ref. 24.

where the rutile electrode (e.g. A100) led to the greatest loss of alloying elements to the slag, shown as the negative delta quantity.

As the value of weld metal oxygen content decreased, the delta quantity increased, implying that the flux coating removed less of the alloying elements from the metal.

The resulted microstructure of the welds deposited with the A1xx, A6xx, and A9xx series electrodes was point counted and listed in Table 20 [24] showing effective control of the microstructures.

C. Standard for Offshore and Arctic Regions of Canadian Standards Association

The Canadian Standards Association (CSA) has completed a preliminary version of the first standard in Canada to stress at the design, fabrication, installation, and operation of steels, which include HSLA steels for oil and gas production in offshore and arctic regions [37] . The toughness requirements for base metal and welded joints need considerable flexibility to allow the wide variety of designs because of the large difference in environment. A combination of the crack arrest and initiation toughness approach was applied. The approach holds that the welded joint, including the HAZ, is the mostly likely location of geometric discontinuities, defects and fatigue cracks, and also the location of low tough-

Table 21 Classification of Structural Elements for Fracture Control and Associated Strategy for Deriving Toughness Requirements

	Safety class of structural element[a]		
		Safety class 1	
Susceptibility to fracture initiation	Safety class 2	Highly redundant structure	Limited redundancy of structure
Low For example, element in compression or low tension	Box 1 No toughness requirements	Box 2 Nominal toughness	Box 3 Some initiation toughness Moderate crack arrest toughness
Moderate For example, element in tension; no high stress concentrations; no hot spots; no plastic straining	Box 4 Nominal toughness	Box 5 Moderate control of initiation Moderate crack arrest toughness	Box 6 Moderate control of initiation Good crack arrest toughness
High For example, element in tension; high stress concentration; hot spots; plastic straining	Box 7 Some initiation toughness Some crack arrest toughness	Box 8 Good control of initiation Moderate crack arrest toughness	Box 9 Good control of initiation Good crack arrest toughness

[a]The box number is a 3×3 matrix grid which denotes the level of risk for fracture. The greater the box number, the greater the risk.
Source: Ref. 37.

ness in comparison with the base metal. The general framework for base metal, HAZ, and weld metal toughness requirements is listed in Table 21 [37] showing a 3×3 matrix of fracture initiation risk and failure consequence. Various levels of susceptibility to fracture initiation were divided into the following conditions of local stress:

(a) *Low*

Mainly for compression or low tension stress which was less than $0.3\sigma_y$; no fatigue load and significant stress concentration factor (<1.2).

(b) *Moderate*

Moderate stress concentration factor (<1.67) and tension stress ($<0.6\sigma_y$); no dynamic loading and fatigue life limitation.

(c) *High*

High stress concentration factor (>2.5) and tension stress; dynamic load possible and fatigue life approaching to that of the designed structure.

In Table 21, the susceptibility to fracture initiation generating from a flaw in the weld joint depends on the flaw size, the magnitude of the local stress, and its application rate. It is obvious that large flaw size, high local stresses, and dynamic loading all increase the risk of fracture initiation.

Some examples for the fixed structures (their corresponding box number in Table 21) are listed in Table 22 [37].

In order to establish the fracture toughness requirements, a flaw size was assumed which could escape detection and exist in the weld joint to develop during service. Thus, the flaw size is 10% of the member thickness up to a maximum depth of 5 mm and keeps the same as in UK Department of Energy Guidance Notes toughness requirements [38]. In determining the levels of fracture initiation toughness, several assumptions were made concerning applied stress levels, stress concentrating factors, and defect growth from fatigue or dynamic loading. Moreover, the fracture mechanics approach was used to determine levels of crack tip opening displacement (CTOD) for weld metals and HAZ.

The specific toughness requirements in the standard are listed in Table 23 [37], where each box has a corresponding box number in Table 21. The CVN energy requirement at the specified test temperature is equal to 1/10 of the yield strength of the base steel.

IV. CORROSION PROTECTION DESIGNING OF HIGH-STRENGTH LOW-ALLOY STEELS

A. Atmosphere Corrosion Resistance of High-Strength Low-Alloy Steels

Because of the worldwide oil crisis in 1970s, weight reduction of automobile became one of the main projects in automobile and iron–steel industries, which led to a surge of the substitution of mild steel with HSLA steel. However, because automobiles are exposed to various atmospheric conditions, the corrosion of automobiles is of serious concern especially for the suspension members, exhaust pipes, floor panels, etc. The application of HSLA steel leads to the reduction of thickness of sheets in components in automobile and also the reduction of the safety margin against corrosion. It is thus of importance to address the corrosion resistance to the atmosphere in HSLA steels. Complicated reactions occur during atmospheric corrosion [39], Fe is the first one resolving into Fe^{++} in the formula of $FeOH^{+}$, and then is oxidized in air to form

Table 22 Some Examples of Structural Elements and Corresponding Box Number in Table 21

Box No.	Element Description	Susceptibility to Fracture Initiation	Safety Class of Structural Element
3	Deck leg on three-leg well protector platform	Low Compressive stresses No fatigue or dynamic stresses	1 Limited redundancy No risk to life-platform unmanned High risk of pollution-uncontrolled flow from well after collapse of deck possible
5	Skin plating and stiffeners of caisson type structure with oil storage	Moderate Bending stresses Some stress concentrations Low fatigue	1 Highly redundant Low risk to life; one element failure will not lead to overall collapse Low risk of pollution; local failure will limit leakage to one compartment
6	Living quarters cantilever support	Moderate Tensile stresses No fatigue or dynamic stresses	1 Limited redundancy High risk to life; failure will cause loss of living quarters module No risk of pollution; no liquids stored in living quarters
7	Skin plating and stiffeners of caisson type structures	High Plastic catenary design	2 No risk to life or pollution
7	Cantilever flare boom tieback	High Tension member Dynamic wind loads High fatigue Stress concentrations	2 No risk to life; structure unmanned No risk of pollution; no liquids in flare

8	Node in a redundant (e.g. eight leg) jacket structure	High	1 Highly redundant
		Tensile stresses	Low risk to life; multiple load paths
		High fatigue loading	Low risk of pollution; no liquids stored in jacket, process fluids only
		Dynamic stresses	
		High stresses concentrations (hot spots)	
9	Transition piece in three legged GBS type structure	High	1 Limited redundancy
		High complex stresses	High risk to life; failure of non-redundant element will cause loss of deck
		High fatigue stresses	High risk of pollution; immediate failure of overall structure may prevent shutdown of wells and process.
		Stress concentrations	
		Dynamic loading	
9	Monocone ice structure neck section	High	1 Limited Redundancy
		High complex stresses	High risk to life; failure will cause loss of complete deck
		Dynamic ice loads	High risk of pollution; wells pass through neck section; loss of wells could result in uncontrolled well flow
		Risk of accidental damage	
		Fatigue	

Source: Ref. 37.

Table 23 Toughness Requirements and Test Temperatures (see notes 1,2,3, and 4)

Thickness, t (mm)	Base metal Test	Base metal Test Temperature	HAZ and weld metal Test	HAZ and weld metal Test Temperature HAZ	Weld metal AW	Weld metal SR	Base metal Test	Base metal Test Temperature	HAZ and weld metal Test	HAZ and weld metal Test Temperature HAZ	Weld metal AW	Weld metal SR	Base metal Test	Base metal Test Temperature	HAZ and weld metal Test	HAZ and weld metal Test Temperature HAZ	Weld metal AW	Weld metal SR
	Box 1						Box 2						Box 3					
$t < 26$	Steel Grade in		Welding consumables				Base metal		Welding consumables				NDT, CVN	T	CVN	T	$T-5$	T
$26 < t < 38$	accordance		in accordance with				toughness to		in accordance with				NDT, CVN	$T-10$	CVN	T	$T-10$	T
$38 < t < 52$	with Clause		Clause S 7.1.2 of the				CSA G40.21		Clause S7.1.2 of the				NDT, CVN	$T-20$	CVN	T	$T-20$	T
$t > 52$	5.1.2		Supplement to W59				Category 1 and		Supplement to W59				NDT, CVN	$T-35$	CVN	T	$T-20$	T
							Clause 6.2.1.4											
	Box 4						Box 5						Box 6					
$t < 26$	Base metal		Welding consumables				NDT, CVN	T	CVN	T	$T-10$	$T-5$	NDT, CVN	$T-10$	CVN	T	$T-10$	$T-5$
$26 < t < 38$	toughness to		in accordance with				NDT, CVN	$T-10$	CVN	T	$T-20$	$T-5$	NDT, CVN	$T-20$	CVN	T	$T-20$	$T-5$
$38 < t < 52$	CSA G40.21		Clause S7.1.2 of the				NDT, CVN	$T-20$	CVN	$T-10$	$T-30$	$T-10$	NDT, CVN	$T-30$	CVN	$T-10$	$T-30$	$T-10$
$t > 52$	Category 1 and		Supplement to W59				NDT, CVN	$T-35$	CVN	$T-10$	$T-30$	$T-10$	NDT, CVN	(Note 4)	CVN	$T-10$	$T-30$	$T-10$
	Clause 6.2.1.4																	
											HAZ and weld metal						HAZ and weld metal	
	Box 7						Box 8				AW	SR	Box 9				AW	SR
$t < 26$	NDT, CVN	T	CVN	T	$T-5$	T	NDT, CVN	T	CVN		T	T	NDT, CVN	$T-10$	CVN		$T10$	$T-10$
									CTOD		0.10	0.07			CTOD		0.10	0.07
$26 < t < 38$	NDT, CVN	T	CVN	T	$T-10$	T	NDT, CVN	$T-10$	CVN		T10	$T-10$	NDT, CVN	$T-20$	CVN		$T20$	$T-20$
									CTOD		0.15	0.11			CTOD		0.15	0.11
$38 < t < 52$	NDT, CVN	T	CVN	T	$T-10$	T	NDT, CVN	$T-20$	CVN		T20	$T-20$	NDT, CVN	$T-30$	CVN		$T30$	$T-30$
									CTOD		0.20	0.14			CTOD		0.20	0.14
$t > 52$	NDT, CVN	$T-20$	CVN	T	$T-20$	T	NDT, CVN	$T-35$	CVN		T35	$T-35$	NDT, CVN	(Note 4)	CVN		(Note 4)	
									CTOD		0.20	0.14			CTOD		0.20	0.14

Notes:

[1]Box numbers in this table correspond to box number in Table 21.

[2]T is the toughness design temperature of the structural element (in degree Celsius).

[3]CTOD in mm at T.

[4]Requirements to be established and documented on a case-by-case basis. Legned: AW = as-welded; SR = stress-relieved; HAZ = heat-affect-zone; CVN = Charpy V-notch; CTOD = crack tip opening displacement; NDT = nil ductility transition.

Source: Ref. 37.

γ-FeOOH. If the pH value decreases in the film of water or moisture on the steel surface, the γ-FeOOH will precipitate as non-crystalline hydroxy iron oxide or α-FeOOH to fill the gap and pores of the rusty layer to form a compact and protective layer against atmosphere corrosion. In addition to its large effect of solid-solution strengthening, phosphorus is also an anode depolarization agent. It accelerates the dissolution rate of Fe and the oxidation rate of Fe^{++} in the corrosion process, which is favorable to the building up of a homogeneous and compact rust layer on the surface of steel, increasingly the resistance of atmospheric corrosion of the steel. The addition of phosphorus to the steel for both protection from atmospheric corrosion and strengthening of the steel was discussed by Gao and Wang [40]. The well-known Hall–Petch equation is used to estimate the strength of the steel

$$\sigma_y = \sigma_0 + \sigma_s + \sigma_p + \sigma_d + \sigma_t + kyd^{-0.5} \tag{12}$$

where σ_y indicates the yield strength, σ_0 the friction force of lattice, σ_s the solution strengthening, σ_p the precipitation strengthening, σ_d the dislocation strengthening, σ_t the texture strengthening, and $kyd^{-0.5}$ represents grain refinement strengthening.

In Eq. (12), σ_d was omitted because this term was low in cold rolled and annealed sheet steels [40]. σ_t was considered as zero due to the fact that the sheet steels were used for automobile components, which were manufactured by ordinary rolling and stamping, the formation of a strong cold rolled annealing texture was not necessary [40]. Therefore, the equation listed above was simplified as

$$\sigma_y = \sigma_0 + \sigma_s + \sigma_p + kyd^{-0.5} \tag{13}$$

σ_0 in the above equation was closely related to the linkage force between atoms and was difficult to calculate theoretically. Experiments exhibited these values varying from 27, 51, 70 to 105 MPa [41]. A mean value, i.e., $\sigma_0 = 63.25$ MPa of them, was taken in Ref. [40]. σ_p was taken as 60 MPa after Ref. [41] for the low carbon cold rolled and annealed sheet steel. σ_s, the solution strengthening factor, was taken as the most common one used in low carbon low-alloy steels [40]:

$$\sigma_s = 32\text{Mn} + 84\text{Si} + 680\text{P} \quad (\text{MPa}) \tag{14}$$

ky was experimentally determined and was reported as 19.5, 22.0, 23.0, and 25.8 $\text{MPa}^{1.5}$ for annealed steels [41]. The mean value $ky = 22.58\ \text{MPa}^{1.5}$ was adapted in calculation [40]. By substituting all the data above and the grain size $d = 0.031$ mm, say, the size corresponding to ASTM No. 7, into the Hall–Petch equation, one obtained

$$\sigma_y = 251.55 + 32\text{Mn} + 84\text{Si} + 680\text{P} \tag{15}$$

According to Fig. 15 [39] and the equation listed above, suitably high content of phosphorus and other elements were added into the steel to attain both good resistance of atmospheric corrosion and enough strengthening of the steel (yield strength set to 345 MPa). The steel was designed with the composition in wt% as follows [40]:

$$\text{C} \leq 0.01\%, \quad \text{Si} = 0.2\%, \quad \text{Mn} = 0.8\%, \quad \text{P} = 0.80\%, \quad \text{and Ti} = 0.10\%$$

where the addition of titanium is for obtaining the precipitation strengthening effect as suggested in the following [42]:

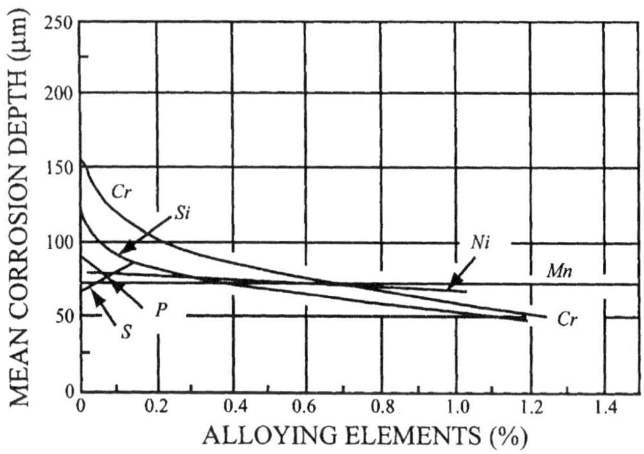

Figure 15 The effect of alloy element content on atmospheric corrosion of steel in industrial air of exposing test for 17 y. (From Ref. 39.)

$$Ti^* = Ti_{total} - 48N/14 - 48S/32 \tag{16}$$

when $Ti^* > 0$, TiC was formed and ferrite strengthened [42].

Experimental results [40] proved the design successful.

Table 24 Corrosion Zones in Sea Water

Zone	Environment	Corrosion behavior
Atmosphere	Weather conditions, minute particles of sea water	Surface washed by rain corrode less than sheltered surfaces
Splash	Wet, well aerated surface, no biofouling	Most aggressive zone, difficult to protect
Tidal	Biofouling, pollutants, ample oxygen available	May act cathodically related to zone below, isolated panels attacked, coatings useful
Shallow	Saturated with oxygen, biofouling pollutants	More active than atmosphere, mineral scale and biofouling reduce the available content of oxygen; control by coating or cathodic protection
Continental shelf depth	Less oxygen and biofouling, lower temperature	
Deep ocean	Oxygen varies, low temperature, low PH	Corrosion often less, no protective mineral scale, cathodic protection
Mud	Sulfate reducing bacteria may be present	Cathodic protection

Source: Ref. 43.

B. Sea Weather Corrosion and Protection

In general, HSLA steels have similar corrosion resistance with ordinary mild steels. Design corrosion allowance will normally be the same for HSLA steel. However, because of the weight reduction, the percentage corrosion allowance in HSLA steel is larger than that in the mild steels. In order to meet the requirements of the operating conditions, protection measures must be defined. When considering the protection of sea weather corrosion, it is useful to refer to Table 24 [43], which divided the environment into several different zones.

Corrosion can be divided into general corrosion and local corrosion. General corrosion includes a comparatively uniform attack over the entire exposed surface. In stagnant ocean water, the corrosion rate is between 0.075 and 0.125 mm/y in the immersed area, however, in the splash zone, it may reach 0.6 mm/y.

Local corrosion is normally far more severe than general corrosion. The most common damage encountered is pitting, which is caused by restricted liquid exchange in the metal surface and may reach a corrosion rate of 1 mm/y.

Corrosion of welded joints, belonging to local corrosion, is also a severe problem in arctic circumstance. Coatings and passivating oxide layers are not effective to protect the metal and will be easily eroded by ice. In order to protect production platforms or artificial islands, which cannot be moved for maintenance, a combination of protective methods of cathodic protection, coatings, and the increasing of section size should be applied.

Another kind of local corrosion is the corrosion in the tidal splash zones. Though there exists a variety of solutions, the most common for protection is to increase the corrosion allowance by increasing section size with 10 or 12 mm. Cladding the critical region with sheaths of highly resistant material such as copper nickel alloy is effective.

Corrosion in external surfaces in the immersed zone is also local corrosion. These portions are protected by cathodic protection, sometimes combined with coatings. Flooded members should be closed to isolate the chamber from contacting with atmospheric oxygen. If necessary, particularly in the presence of corrosion promoting bacteria, cathodic protection and anti-fouling coatings can be used for these members. Moreover, it is also useful to raise the pH value of the water in flooded chambers.

C. Optimization of Inhibitor Mixture Formulation for Acid Pickling Baths in High-Strength Low-Alloy Steel

High-strength low-alloy steels have similar corrosion resistance with ordinary mild steels; however, very little is known about its corrosion behavior in hydrochloric acid. Corrosion inhibitors are commonly used in acid pickling baths in order to prevent metals dissolution after surface adherent products are taken. Industrial pickling process shows that the good use of corrosion inhibitor mixtures should make use of the synergistic effect. For example, Aoki et al. [44] put together three inhibitors: benzotriazole (BATH), *N*-phenylthiourea (*N*-PTU), and hexadecyltrimethylammonium chloride (HTAC) with different compositions in the hydrochloric acid (HCl) medium to measure HSLA steel surface protection degree from weight loss detection and anodic and cathodic potentiodynamic polarization curves of corrosion. The composition of the steel in weight percentage is listed as follows: C: 0.162; Mn: 0.94; Si: 0.38; P: 0.015; Cr: 0.52; Cu: 0.253; Ni: 0.015; S: 0.010; Fe: bal.

High-strength low-alloy specimens were prepared in the size of $18 \times 9 \times 4.5$ mm and then ground with SiC paper 320, 400 and 600 grit. After being washed and rinsed, these specimens are weighted before and after immersion in 16% w/w HCl solution naturally

Table 25 Experiment Design Matrix with Corrosion Rate and Protection Degree Results

Run	x_1	x_2	x_3	V_{corr} (mg cm^{-2} h^{-1})	θ
0	0	0	0	4.50	0.00
1	1	0	0	4.88	0.00
2	0	1	0	0.294	0.935
3	0	0	1	1.42	0.68
4	1/2	1/2	0	0.320	0.93
5	1/2	0	1/2	1.196	0.73
6	0	1/2	1/2	0.309	0.93
7	2/3	1/6	1/6	0.337	0.925
8	1/6	2/3	1/6	0.236	0.95
9	1/6	1/6	2/3	0.364	0.92
10	1/3	1/3	1/3	0.235	0.95
11	2/3	1/3	0	0.222	0.95
12	1/3	2/3	0	0.177	0.96
13	2/3	0	1/3	1.174	0.74
14	1/3	0	2/3	1.208	0.73
15	0	2/3	1/3	0.178	0.96
16	0	1/3	2/3	0.324	0.93

Source: Ref. 44.

aerated at 40°C for 30 min in the presence or absence of inhibitors. Corrosion rate for specimen was measured and then the protection degree θ was [44]:

$$\theta = (\upsilon_{corr,o} - \upsilon_{corr,i}) \tag{17}$$

where $\upsilon_{corr,o}$ is the corrosion rate in the absence and $\upsilon_{corr,i}$ is the corrosion rate in the presence of the inhibitor.

Sixteen experiments were run according to the simplex-lattice type, which yields the Sheffé polynomial expressions as the model. The experimental results are shown in Table 25 [44] where the independent variable x_i, the inhibitor concentration, was represented as in a linear relation in the 0–1 range. In this manner, $x_1 = 1$ corresponds to 10^{-3} M BATH maximum concentration, $x_2 = 1$ corresponds to 5×10^{-3} M *N*-PTU and $x_3 = 1$ corresponds to 10^{-3} M HTAC and $\sum x_i = 1$.

The regression analysis with cubic model fitting of the experimental results in Table 25 [44] was performed and the result was listed with only the statistical significant coefficients as follows:

$$\begin{aligned} Y = {} & 0.9454x_2 + 0.6874x_3 + 2.0238x_1x_2 + 1.68x_1x_3 + 0.5530x_2x_3 \\ & + 2.0706x_1x_2(x_1 - x_2) + 1.479x_1x_3(x_1 - x_3) \end{aligned}$$

Regression analysis predicted that in considering the effect of the individual inhibitor, the *N*-PTU led to the highest coverage degree followed by HTAC and BTAH. However, while considering the synergism, the cooperation of BATH and *N*-PTU showed the strongest effect followed by the *N*-PTU and HTAC cooperation. Also it predicted that the highest coverage degree must be obtained in a ternary mixture. After calculation of the maximum point of cubic model, the optimized formulation is $x_1 = 0.45$, $x_2 = 0.40$, and $x_3 = 0.15$.

High-strength low-alloy steel specimens for polarization test were prepared. The experimental curves were obtained, in the absence or presence of each one of the three inhibitors, from −900 mV×SCE to +200 mV×SCE with a static working electrode in a naturally aerated media. It follows from the three test results that *N*-PTU behaved most efficiently followed by HTAC and BTAH. All these obtained from polarization test were in good agreement with that of the weight loss measurements.

V. DESIGNING THE THERMO-MECHANICAL CONTROL PROCESS

Thermo-mechanical control process (TMCP), which includes straight controlled rolling and accelerated cooling after controlled rolling, is a significant technological developments in the steel industry. Starting from the fundamental research on controlled rolling in 1960s, the technology was then established in worldwide in plate mill in 1970s. After several decades of development, controlled rolling is characterized at present by low slab reheating temperatures and large reductions in the vicinity of the transformation temperature to obtain very fine austenite microstructure. Large reduction is also applied in the intercritical temperature range to increase the strength by work hardening of transformed ferrite. The concept of the process was defined by ASTM standard committee [45] as follows:

Thermo-mechanical control process has evolved from the "controlled rolling" process which has been known and used for a number of years. Thermo-mechanical control process is the new generation of controlled rolling. Thermo-mechanical control process produces fine grained steel by a combination of chemical composition and integrated controls of manufacturing process from slab reheating to post-rolling cooling, thereby achieving the specified mechanical properties in the required plate thickness. Thermo-mechanical control process requires accurate control of both steel temperature and rolling reduction.

A. Miscellaneous Aspects of Metallurgy, Engineering, and Application

The principles of TMCP design are shown in Fig. 16 [46], where the relationship among process, mechanism and resulting mechanical properties are drawn. Accelerated cooling of the recrystallized austenite can refine the ferrite grains to some extent because of the depression of Ar_3 temperature. However, the accelerated cooling of unrecrystallized austenite (after thermo-mechanical rolling) may also activate the nucleation of ferrite within grains of austenite, in addition to the ferrite activated by deformation bands. All these stated above contribute to the ferrite grain refinement. During accelerated cooling, moreover, pearlite can be replaced by an increased amount of finely disposed bainite, which contributes to further strengthening without sacrificing toughness. Also during accelerated cooling, Nb modifies the transformation behavior by its hardenability effect to produce an increased amount of bainite. Since this beneficial effect is remarkable, more addition of Nb is required in higher grade or heavier thickness steels to satisfy both mechanical properties and weldability requirements. The effect of microalloying and alloying elements in solution on strength can be empirically expressed as a carbon equivalent during accelerated cooling of 10°C/sec as follows [46]:

$$\mathrm{CE} = \mathrm{C} + 0.20\mathrm{Mn} + 0.03\mathrm{Cu} + 0.06\mathrm{Ni} + 0.14\mathrm{Cr} + 0.29\mathrm{Mo} + 1.03\mathrm{Nb} + 0.67\mathrm{V} \quad (18)$$

This equation has been widely used in the plate produced mill and results are shown in Fig. 17 [46] for the adjustment of water cooling rate to obtain minimum variation of strength.

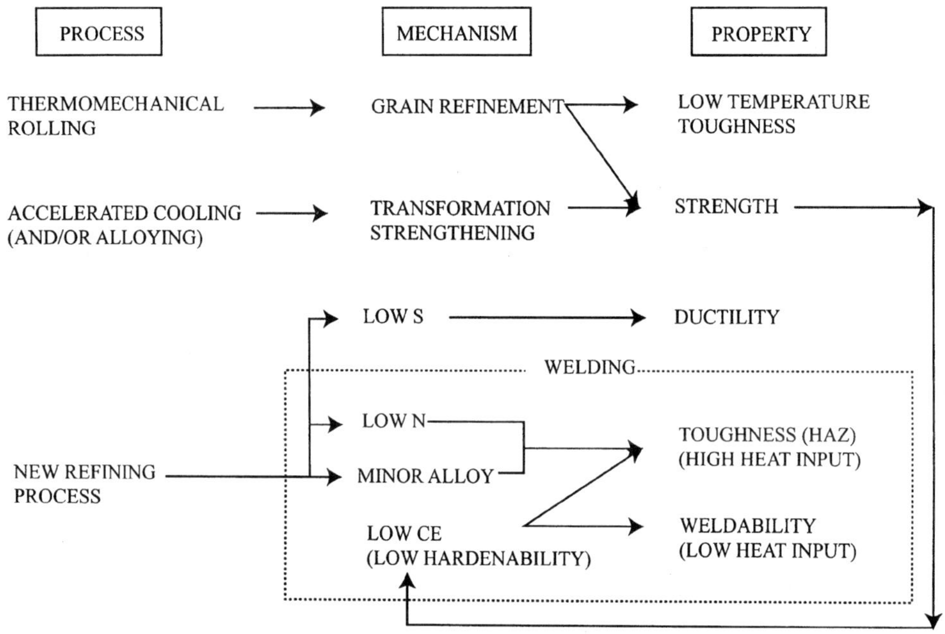

Figure 16 Relationship among new processing, mechanism, and resulting property. (From Ref. 47.)

Basically, rolling conditions give the same effect in accelerated cooled plate as well as in controlled rolled plate. For wide and heavy gauge plate, there exist several problems in accelerated cooling, the most serious are [46]:

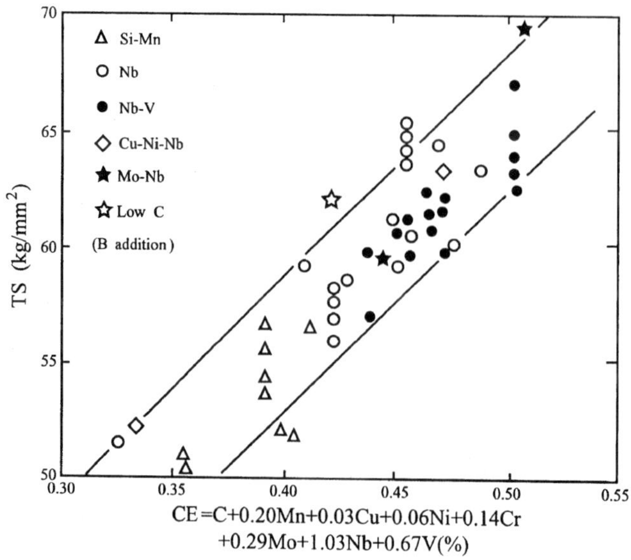

Figure 17 Relationship between tensile strength and carbon equivalent (during ACC)(From Ref. 46.)

(a) Cooling must be uniform at every location on the top and bottom surfaces, edge and center, and leading and ends.
(b) Any distortion and residual stress due to water cooling must be minimized.
(c) The cooling rate and cooling start and finish temperatures must be controlled accurately.
(d) The cooling system must be designed simply and compactly, and constructed at low cost and be maintained easily.

Great efforts have been made to solve these problems. A specification of accelerated cooling equipment of 18 companies was shown in Ref. [46]. Additionally, in order to obtain uniform cooling, the water for the top and bottom surfaces must be balanced at an optimum ratio to prevent plate deflection. The amount of water from center to edge must also be precisely controlled to minimize transverse temperature gradients and buckling. Cooling for leading and trailing ends must also be controlled to prevent excessive temperature drop. Most importantly, the interruption of cooling at ≈500°C as required for property control, would lead to amplification of temperature non-uniformity below the interruption temperature. Good rolling practice is the essential requirement to produce plates with satisfactory flatness because cooling only amplifies the existing deformation. Hydrogen is a harmful element which introduces hairline cracking during cooling and thus must be reduced.

The composition and mechanical properties of accelerated cooled type TMCP structural steels are shown in Table 26 [46], where the effects of alloying elements and carbon

Table 26 Chemistry and Mechanical Properties of Accelerated Cooled Type TMCP Structural Steels (Mechanical Properties: Transverse direction)

Grade	Thickness (mm)	C	Mn	S	Others	CE IIW	YS (MPa)	TS (MPa)	CVE at 0°C	J	FATT °C
AH36	25	0.15	0.97	0.010		0.31	383	515	0	162	−25
DH36	25	0.14	1.07	0.006		0.32	373	516	−20	197	−43
EH36	40	0.12	1.17	0.002		0.32	389	505	−40	264	−72
AH36*	25	0.16	0.65	0.009		0.27	354	507	−20	105	−24
EH36-060	40	0.08	1.50	0.001	Cu 0.20 Ni 0.34	0.37	430	509	−60	370	−119
	75	0.08	1.50	0.001	Cu 0.20 Ni 0.33	0.37	379	509	−60	322	−88
EH390	30	0.12	1.35	0.003	Nb 0.010	0.34	443	550	−40	161	−60
BS4360	50				Cu 0.25		466	564	−60	225	−110
−50D equi.	100	0.06	1.56	0.001	Ni 0.41 Nb 0.01	0.36	390	554	−60	166	−64
YS 355 for const.	50	0.13	1.34	0.002		0.35	388 (YR 75)	553	0	305	
	100	0.11	1.46	0.003	Nb 0.010	0.35	397 (YR 75)	532	0	306	

Note: * indicates data from Nippon Steel Technical Report, No. 22, Dec. 1983, pp 1–17; equi. and const. are the abbreviation of equivalent and construction, respectively.
Source: Ref. 46.

on the susceptibility to cold cracking during low heat input welding or on toughness in HAZ of high heat input welding were considered.

Accelerated cooling has not made full use of the hardenability of steel. For hardening and toughening, however, it utilizes the transformation strengthening due to the formation of bainitic microstructure and grain refinement of the microstructures. It is thus suitable for processing of steels with tensile strength of 500 MPa or sometimes up to 600 MPa. For the production of higher strength steel, direct quenching is quite suitable and is now being operated either in immersion-type cooling or roller-type quenching. The advantage of direct quenching is the full use of the hardenability of the steel, the conserving energy by deleting the reheat quenching, shortening the production process and reducing the amount of alloying elements to meet a given strength which results in the improvement of weldability. Direct quenching is applied to the heavy gauge HSLA structural steels with tensile strength of 600–1000 MPa.

B. Simulation of Thermo-Mechanical Control Processing in Laboratory Scale

The mechanical properties and the final microstructures of HSLA steels are largely dependent on the process parameters as [47]:

(a) reheating temperature (Treh),
(b) rolling schedules and finishing rolling temperature (FRT).
(c) accelerated cooling rate (ACC) and finish accelerated cooling temperature (FCT), and
(d) steel chemistry.

In the TMCP processing, the slab reheating temperature (Treh) has a strong influence on the strength, toughness and microstructure of microalloyed steel. The low Treh gives finer austenite grains, this in turn refines the final microstructure and improves the low temperature toughness of the steel. On the other hand, lower Treh reduces the

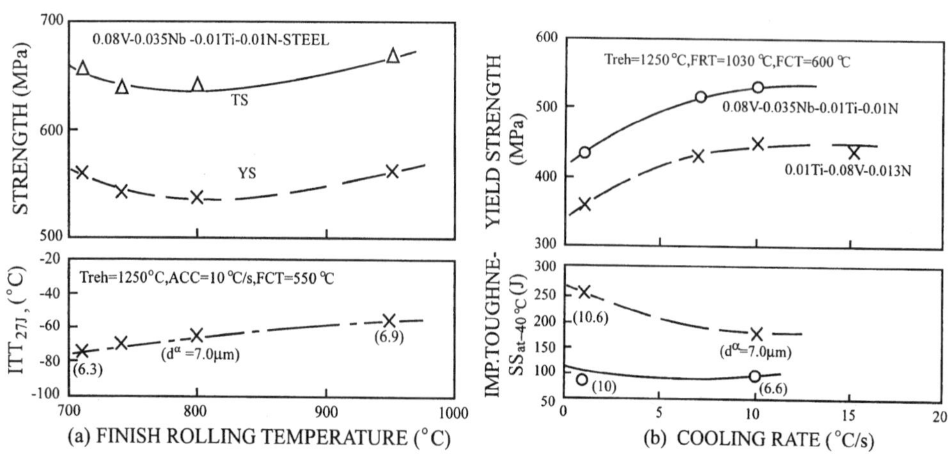

Figure 18 Effect of finish-rolling temperature (a) and cooling rate from FRT of 1030°C to FCT of 600°C (b) on the strength, impact toughness and ferrite grain size of Ti–V–Nb microalloyed steel (From Ref. 47.)

amount of dissolved microadditions in the austenite as well as the potential of precipitation hardening after cooling, the yield strength and tensile strength of the steel are therefore decreased.

The finish-rolling temperature (FRT) and the deformation in the final pass are important process parameters to affect the strength and toughness. The effect of FRT on the microstructure and mechanical properties for Ti–Nb–V microalloyed steel is shown in Fig. 18a [47], where d^{α} represents the diameter of ferrite. The best combination of low temperature toughness and tensile strength is obtained for FRT near to Ar_3.

The effect of ACC and FCT on the final microstructure, yield strength and toughness of Ti–V–(Nb) after recrystallization controlled rolling (RCR) is shown in Fig. 18b [47] where yield strength increases as the cooling rate increases. It can be shown in the same figure that impact toughness of RCR + ACC steel usually decreases as the cooling rate becomes higher despite the fact the ferrite grain size becomes smaller. This is caused by the precipitation of V(C,N) or (V,Nb)(C,N).

At low temperatures, dynamic recrystallization controlled rolling (DRCR) can take place during rolling and is an alternative softening process leading to more effective grain refinement than RCR and controlled rolling (CR). The criteria for DRCR to occur during TMCP processing of HSLA microalloyed steels is: low temperature range rolling, short intermediate pass time and sufficient accumulated strain. These should be executed below the no-recrystallization temperature (T_{NR}). The DRCR could be obtained during the finishing rolling stage of hot rolling, if enough strain has been accumulated and static recrystallization has not happened. Most importantly, the critical strain for dynamic recrystallization should be obtained before the strain-induced precipitation of

Table 27 Chemical Composition of the Steels (wt%)

Steel grade	C	Si	Mn	S	P	Al	V	Ti	N	Nb
Ti–V	0.19	0.36	1.70	0.008	0.014	0.027	0.15	0.012	0.0185	—
Nb–V	0.10	0.24	1.26	0.01	0.01	0.042	0.04	—	0.0085	0.027

Source: Ref. 48.

Table 28 Simulation

Pass no. —Type of mill	Equivalent strain per pass	Temperature, °C	Delay time between passes
1—piercer	1.60	1,230	54
2—MPM	0.45	1,085	1
3—MPM	0.40	1,072	1
4—MPM	0.30	1,060	0.5
5—MPM	0.20	1,050	0.5
6—MPM	0.18	1,020	75
Reheat furnace	—	1,000	5
7–22—SRM	0.10	1,000 to 860	0.5

Note: Soaking temperature is 1250°C for 15 min. Delay time between billet reheating furnace and piercing is 80 sec. Time in intermediate (tube) reheat furnace is 10 min. The strain rate is 2 sec^{-1} for each pass. The average cooling rate during finish rolling is ~8°C/sec.
Source: Ref. 48.

carbonitrides begins so as to ensure the initiation of dynamic recrystallization. If not, the strain accumulation continues, which results in the pancaking of austenite.

It is apparent that suitable design of finishing rolling schedule is more important for the application of DRCR. As an example, Pussegoda et al [48] described the design and prepared two kinds of test materials designated as Ti–V and Nb–V, respectively. The compositions of these two steels are listed in Table 27 [48].

Laboratory simulation of seamless tube piercing and rolling was applied and the schedule was divided into two temperature ranges. The piercing and multi-pipe mill (MPM) stages were carried in high temperatures (see in Table 28 [48], from pass 1 to 6) and the stretch reduction mill (SRM) deformation, which supplied sufficient strain to produce dynamic recrystallization, was applied at low temperatures for finishing.

The specimen was cooled to 950°C before being reheated to 1000°C to simulate the reheat furnace condition. The finishing schedule, i.e., SRM was applied at different temperatures to obtain different exit temperature, the lowest of which was 735°C being 20°C higher than Ar_3 of T–V steel. The cooling rates for SRM entry temperature at 1000°C and 840°C were 8°C and 6°C, respectively. The cooling rate after SRM in the $\gamma \rightarrow \alpha$ transformation period was about 3.5°C.

Results of simulation and experiment showed that for Ti–V steel, all the austenite at the exit of different SRM were dynamically recrystallized. The grain refinement caused by DRCR could be optimized by reducing the SRM entry temperature. By contrast, reducing the entry temperature from 1000°C to 910°C for Nb–V steel produced pancaking of austenite. This is because the strain-induced precipitates in Nb–V steel nucleated immediately after entry to SRM where the sufficient strain for dynamic recrystallization was not reached.

C. Industrial Thermo-Mechanical Control Processing

Changes in the mechanical properties and the microstructure of HS-350 (0.01 Ti–0.04 V wt%), W-420 (0.01 Ti–0.025 Nb wt%), and W-500 (0.01Ti–0.085V–0.04Nb wt%) microalloyed steels were studied in the industrial scale as a function of the finish-rolling temperature and plate thickness [47]. An example of the influence of these parameters on the

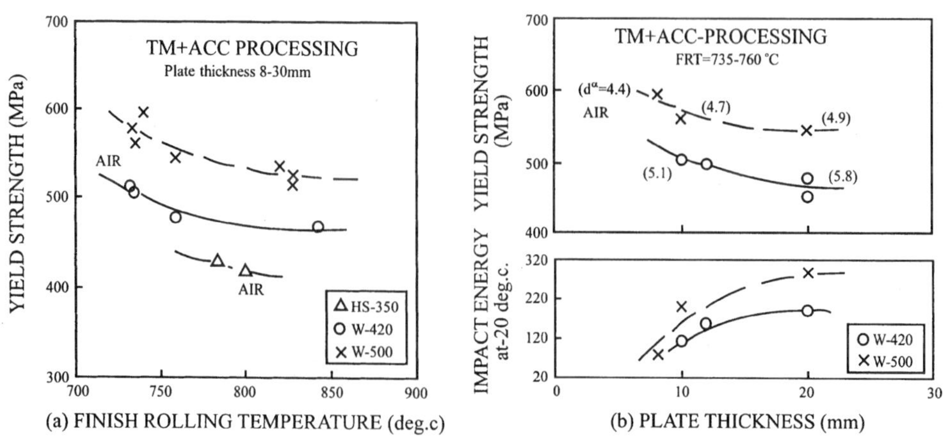

Figure 19 Mechanical properties and ferrite grain sizes of the commercial CR-processed HS-350, W-420 and W-500 steels as function of: (a) FRT, and (b) plate thickness. (From Ref. 47.)

yield strength, toughness, and ferrite grain size for these microalloyed steels is shown in Fig. 19 [47]. In Fig. 19a, the yield strength of CR-processed steels increases with the decreasing of FRT from 840°C to 730°C. But in Fig. 19b, the strength increases but toughness decreases with the decreasing of plate thickness with FRT in the interval of 735–760°C. Increasing strength is due to refinement of microstructure and precipitation, however, the low toughness is due to precipitates and the smaller Charpy-V specimen.

VI MICROSTRUCTURE DESIGN OF HIGH-STRENGTH LOW-ALLOY STEEL

A. Design of Microcomposite Austenite–Martensite Structure

Thomas [49] has studied the roles of C, Cr, Mn, Ni, Mo, Ti, and others individually on the microstructure-property relations in steels, especially their effect on martensite start temperature M_S. Data from Izumiyama et al [50] were collected for binary systems and are shown in Fig. 20. Since twinned carbon steels are far less tough than dislocated martensite steels, dislocation is an essential component for strength and toughness. The main factor controlling the substructures in martensite is composition, especially the carbon content and nitrogen. The amount of the elements must be regulated to maintain $M_s > 200°C$, that means, the upper limit of carbon (and/or nitrogen) should be set to about 0.35 wt% according to Fig. 20. Based on the data collected in Ref. [50] and equations of Andrews [51], M_s for steels of interest can be satisfactorily estimated. Thus a steel may be designed mainly on the estimation of M_s data to yield a microstructure with a rather wide range of suitable alloying elements.

It has been recognized that the retained interlath austenite films in the martensite packets promote increasing toughness to prevent crack propagation [49]. The microstructure which possesses optimum mechanical properties is therefore a microcomposite austenite–martensite structure where the thin austenite film is enriched with carbon and become stable. The maximum number of variants in the prior austenite is 4, while several packets of martensite of the same variant can exist. Through controlled rolling, these packets can

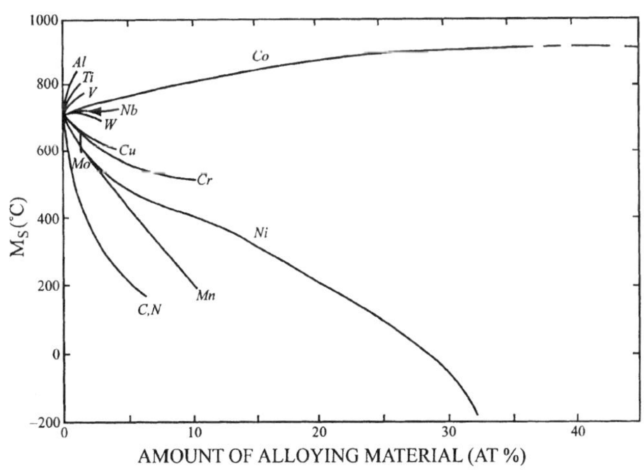

Figure 20 Martensite start temperatures M_s for binary iron-based alloys. (From Ref. 50.)

Table 29 Alloy Composition (wt%)

Alloy	C	Mn	Cr	Nb	Si	P	S
Non-Nb steel	0.26	1.20	1.99	—	0.14	0.012	0.015
With Nb steel	0.24	1.21	1.98	0.02	0.046	0.006	0.004

Source: Ref. 52.

be refined and the maximum grain refinement caused by austenite recrystallization can be obtained.

Relying on the microstructural basis of alloy design for HSLA steels, typical compositions of the recently developed alloys for long products and/or plates are provided in Table 29 [52]. Emphasis has been on steels containing 1–2% Cr and 1% Mn for retained austenite to develop the desired microcomposite microstructures by controlled hot rolling and ACC. The steels were rough rolled in 1000–950°C, finishing rolled at 900°C, water cooled in the rate bigger than 30°C/sec and then step quenched below M_s temperature and at last air cooled [52]. To obtain the desired autotempered packet martensite and to avoid interlath carbide precipitation or other unwanted austenite decomposition products during the quenching, high speed cooling is preferable. Actually, rapid quenching rates in plant operation are difficult to achieve and require enormous quantities of water. However, the addition of a small amount of Nb, 0.02 wt% to Fe–2Cr–1Mn–0.25C alloys, can successfully reduce the stringent quenching conditions by increasing hardenability and controlled hot rolling just above the recrystallization temperature to obtain the fine austenite grain. Thomas et al [52] controlled rolling by finishing just above the austenite recrystallization temperature, to produce a fine grained austenite which was then directly quenched to produce a fine microcomposite packet martensite–austenite structure. Actually at Dillinger–Huttenwerke (W. Germany), plates of Fe–2Cr–lMn–0.26C–0.26Nb were controlled rolled and quenched at 30°C/sec to between M_s and M_f to obtain on-line autotempering and high toughness (Charpy 90 J).

A further processing route has been investigated for rounds [52], where these low-alloy steels were controlled rolled, normalized and then spherodized to be cold formable. In this case, Nb additions may not be necessary [52]. Such processing may follow the process to produce products such as bolts, chains, etc. The final product was subsequently quenched and tempered to achieve the microcomposite structure as indicated above.

B. Designing of Acicular Ferrite Structure

The formation mechanism of phases during austenite decomposition in a titanium bearing HSLA steel has been discussed [53]. The γ-grain size plays an important role in deciding the transformation characteristics. If the size of the parent phase is large enough, it will induce the proeutectoid phase to form first; however, if there is sufficient nucleation potency introduced either by heavy deformation or the existence of inclusions in the transgranular regions, acicular ferrite will form [54]. Despite the difficulty in identification among polygonal and acicular ferrite, upper and lower bainite and martensite, Chatterjee and Mishra [53] still showed that the nucleation site density will determine the aspect ratio (length to thickness ratio) of acicular ferrite. The acicular ferrite forms at a low degree of undercooling or at slighter higher temperature of transformation induced by ACC. When the growth rate is slow, some amount of coherency might occur which leads to the formation of sidebranches, showing the special morphology of acicular ferrite. This phenomenon suggests that the aspect ratio of acicular ferrite should not be too high.

The proeutectoid ferrite phase is observed to be the finest after finishing rolling at 800°C. This was explained as the fact that γ-grains underwent maximum pancaking at that temperature and brought about maximum defects. In addition, this FRT might enhance the precipitation of carbonitrides rendering high-angle grain boundary migration more difficult. This would lead to the increased nucleation sites of ferrite at grain boundaries by way of protrusion, which in turn produced the finest structure.

In order to design and develop the acicular ferrite, Chatterjee and Mishra [53] further prepared the test material consisting of 0.06% C, 1.45% Mn, 0.010% S, 0.010% P, 0.02% Si, 0.01% Ti, and 0.02% Al in mass content, where titanium was chosen as the microalloying element since it was widely considered as the most effective in restraining γ-grain coarsening. The solubility of titanium in steel may be calculated with the following equations [55]:

$$\ln[\mathrm{Ti}][\mathrm{C}] = -7.000/\mathrm{T} + 2.75 \tag{19}$$

$$\ln[\mathrm{Ti}][\mathrm{N}] = -16.586/\mathrm{T} + 5.90 \tag{20}$$

thus, the solubility of titanium at 1200°C in the presence of 0.06% carbon is obtained as ≅2.32%, which is well above the concentration of titanium in the test steel. A similar result may also be obtained if a reliable value of nitrogen is assumed and substituted in the above equation, the calculated titanium concentration must be much higher than that in the steel. The specimen was then soaked at 1200°C for half an hour to dissolve all of the existing Ti-carbides and carbonitrides. After solution treatment, the specimen was rolled according to the design schedule shown in Fig. 21 where the FRT could be changed [53]. During rolling, the specimen thickness was reduced from 14 to 6 mm, yielding 57% cumulative deformation in three stages. The reduction in the first pass was 15%, the second 25% and the last 33%. In some cases the reduction varied between 15% and 25% at the finishing pass. The specimen was quenched into water after the last rolling step. Optical and TEM studies indicated that the acicular ferrite could be engendered by applying a minimum deformation of 25% at the finishing pass followed by

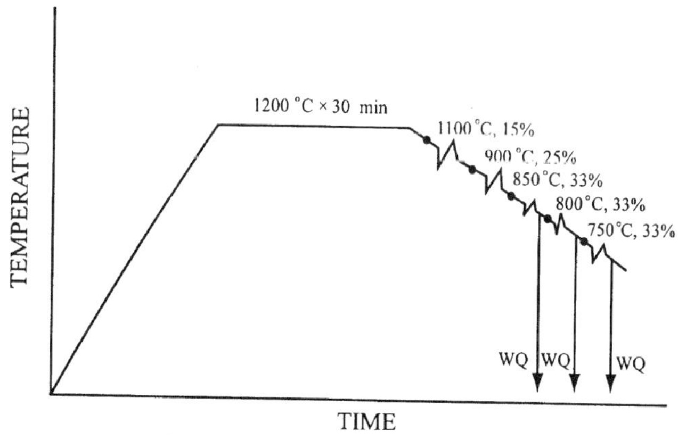

Figure 21 Schematic diagram representing three stage controlled rolling operation. (From Ref. 53.)

water quenching. Decreasing the FRT from 850°C, with deformation of 33% and water quenching, resulted in the refinement of polygonal ferrite, but the volume fraction of acicular ferrite could be increased progressively as the FRT was lowered. Evaluation of mechanical properties indicated a high yield strength (~600 MPa) and elongation rate (~25%) with the FRT of 750°C, and adequate toughness with an impact transition temperature of −50°C with the FRT of an 800°C.

C. Designing of Aging in High-Strength Low-Alloy Steel

Structural steels have been traditionally heat treated low-alloy steels. These steels derive strength from their carbon content. Carbon is a very cost-effective element in increasing strength in ferrite–pearlite and tempered martensite structures. However, carbon content (or CE) also leads to poor weldability, weldment toughness, and notch toughness. In order to simultaneously improve the quality of structural steels, a new family of HSLA steels with copper addition (HSLA-100) has been developed. This steel is designed with a reduction of carbon to improve the weldability and toughness in construction applications.

The steel in the WQ condition consists of a α'-phase (martensite supersaturated with copper). An example of the aging design was contributed by Mujahid et al. [56] who showed that the aging process in Fe–Cu alloys could be divided into the following stages:

(a) formation and growth of coherent bcc copper-rich clusters,
(b) transformation from bcc to fcc phase,
(c) apparent recovery and recrystallization of matrix, and
(d) austenite transformation.

Field ion microscopy in the aged Fe–Cu alloys showed that these initial clusters had bcc structure and contained significant amounts of iron [57]. In addition, they were coherent with the matrix and had a very small mismatch of lattice parameter. In stage (a) of aging, at about 450°C, the matrix still consisted of the lath structure with a dislocation density similar to that of the WQ condition. In stage (b), with the increase of aging temperature between 450°C and 600°C, martensite laths continuously recovered and precipitates coarsened. At this stage, spherical ε-Cu precipitates became resolvable in TEM because the diffraction contrast could be obtained. Further overaging of the steels occurred when the spherical ε-Cu precipitates changed their shape to rodlike particles.

At temperatures above 600°C, the lath structure was partially recovered and thus had a moderate dislocation density. At the same time, recrystallization of ferrite started in some areas. Formation of new austenite might occur in the steels when they were aged at or more than 640°C. The new austenite was observed to form within the prior-tempered-martensite matrix at the interlath as well as at high-angle boundaries. The growth of the new austenite took the direction both perpendicular and parallel to the grain boundaries. When temperature exceeded 665°C, the growth of austenite occurred at significant rates, with the principal growth direction paralleling to the martensitic ferrite laths or at the prior-austenite grain boundaries [58].

The transformation behavior of this newly formed austenite is quite important, since it has a strong and direct influence on the final properties. In the initial stages of its formation during aging, austenite becomes rich in solutes, especially those of nickel and copper by depleting the matrix, which makes this microstructure highly stable even during air cooling to room temperature. However, at higher aging temperatures, the growth of those newly formed austenite causes the dilution of solutes in the new austenite, which in turn

Table 30 Chemical Composition of the Plate Steel (wt%)

Steel	C	Mn	Ni	Cr	Cu	Mo	Nb	Si	Al	N	P	S
B	0.057	0.99	3.42	0.67	1.66	0.60	0.036	0.38	0.024	0.011	0.010	0.001
C	0.036	0.91	3.59	0.59	1.60	0.59	0.025	0.24	0.022	0.010	0.008	0.005

Source: Ref. 56.

transforms to martensite or bainite on cooling. A new type of HSLA steels was designed based upon the comprehensive knowledge of aging and heat-treatment effects and are listed in Table 30 [56].

Test material with the composition listed in Table 30 in the as control-rolled condition was heated to 900°C for 1 h for austenitizing and then water quenched. The test specimens were aged for 1 h in the temperature range of 450–730°C and then air cooled to room temperature. The mechanical properties of the tests were determined in all aged conditions. The yield strength and impact toughness are shown in Figs. 22–24. Fig. 22 shows aging curves plotted between yield strength and aging temperature of both steels. The aging behavior was characterized as four stages. Stage (a) was associated with the highest strength at 450°C. In stage (b), the strength continuously decreased in the temperature range of 450–665°C. There was a secondary strengthening around 708°C in stage (c). In stage (d), a decrease in strength above 708°C happened showing the austenite transformation began. Figs. 23–24 showed that the steels aged at 500°C exhibited very low resistance to brittle fracture. The toughness improved extraordinarily as aging temperature exceeded 600°C. The best combination of strength and toughness that can be obtained through comparison of Figs. 23 and 24.

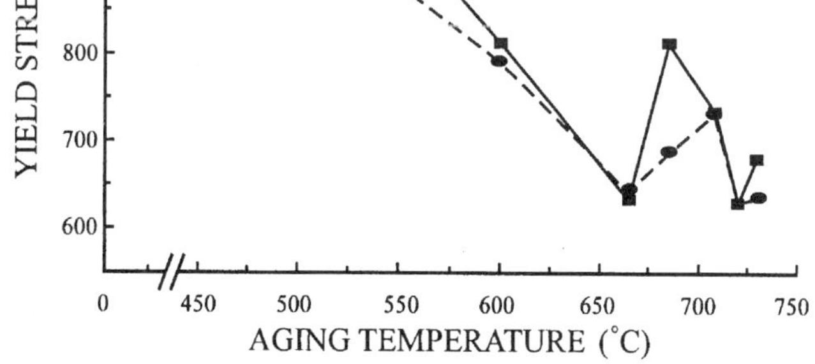

Figure 22 Variation of room-temperature strength with aging temperature of as-quenched HSLA-100 steels. (From Ref. 56.)

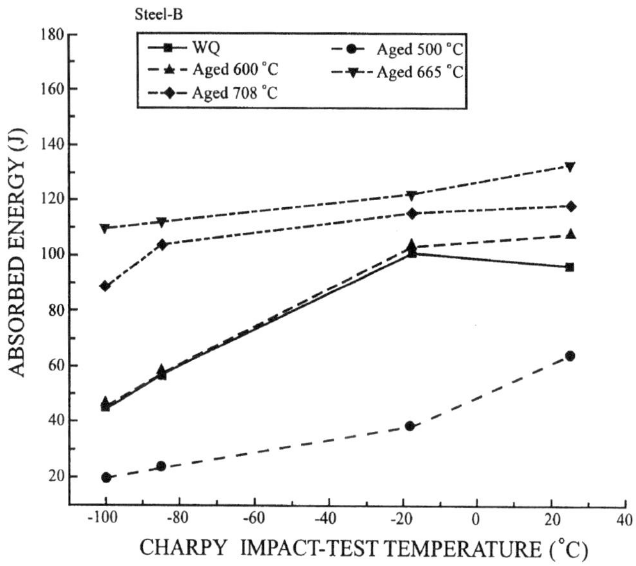

Figure 23 The influence of test temperature and aging condition on impact toughness of steel B. (From Ref. 56.)

VII. COMPUTER CALCULATION OF EQUILIBRIUM PHASES AND DYNAMIC PROCESS

A combination of basic thermodynamic principles with mathematical formulations to describe the various thermodynamic properties of phases has been performed for nearly

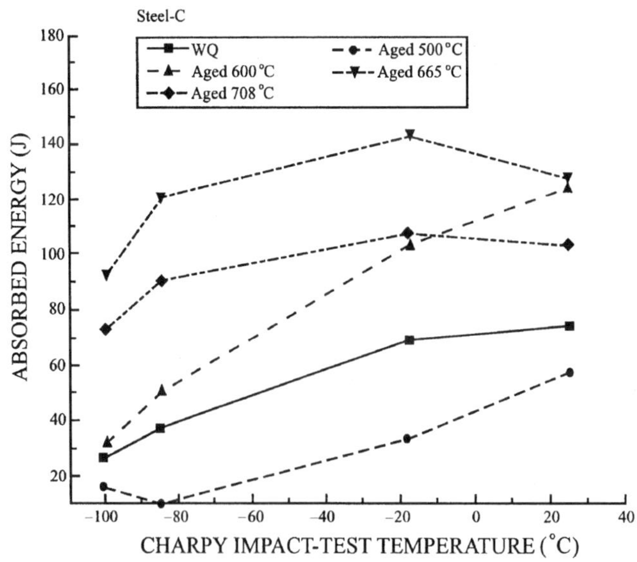

Figure 24 The influence of test temperature and aging condition on impact toughness of steel C. (From Ref. 56.)

50 years and models, formulations and calculation methods have continuously evolved. Among the pioneers, Meijering [59] was probably the first to complete the calculation of a real ternary system Ni–Cr–Cu, where he found that to make the calculation possible, the stability of a specified phase, e.g., f.c.c Cr must be deduced. He attempted to obtain this datum by the extrapolation from activity measurements, this idea directly led to the concept of lattice stability which became important later. Also early in 1950s, Kaufman and Cohen [60] initially put activity into use in steels to solve practical problems. In parallel to Kaufman and his co-workers in the USA, Mats Hillert made the use of thermodynamic calculations in a series of steel problems. Another fundamental approach, which could include the function of magnetism, was developed by Weiss and Tauer [61]. These determinations combined all the major workers and contributors in the field worldwide and formed the basic driving force for the eventual establishment of CALPHAD (CALculation of PHAse Diagrams) technique. As the knowledge of lattice stabilities of elements was undoubtedly the vital ingredient which was required to calculate a phase diagram, the discovery and applications of methods to deduce these parameters occurred next. Also required for the calculation of phase diagrams was a set of interaction parameters between elements or species in the system. The production of a whole set of the parameters was a time-consuming work; especially in iron-based alloys the magnetic effects should be added in many cases, which made the situation more complicated. Basic or commonly used models and parameters for the calculation of thermodynamic properties and phase diagram in multi-component steels will be introduced.

A. Gibbs Free Energy Formalism

The integral Gibbs energy of a pure species or stoichiometric compound phase is given by the following equation:

$$G(T, P) = H(T, P) - TS(T, P) \tag{21}$$

where $H(T, P)$ and $S(T, P)$ are the enthalpy and entropy as functions of temperature and pressure.

The Gibbs free energy for 1 mole solution phase in steels could be generally listed as follows:

$$G_{\text{m}} = G_{\text{m}}^{\text{ref}} - TS_{\text{m}}^{\text{ideal}} +^{\text{E}} G_{\text{m}} +^{\text{mg}} G_{\text{m}} \tag{22}$$

$G_{\text{m}}^{\text{ref}}$ is the Gibbs free energy of 1 mole phase in its reference state, which is not explicitly given in many calculation programs. However, the most common used reference state is the "stable element reference" (SER), given by a mechanical mixture of pure elements at 298.15K in their stable state. Using this reference state, G_{m} is alternatively the abbreviation of the difference:

$$G_{\text{m}} - \sum_{i=1}^{n} H_i^{\text{SER}}(298.15\text{K}) \tag{23}$$

$TS_{\text{m}}^{\text{ideal}}$ is the contribution of ideal mixing entropy, ${}^{\text{E}}G_{\text{m}}$ is the excess Gibbs free energy, ${}^{\text{mg}}G_{\text{m}}$ is due to the contribution of magnetic ordering. With the formalism of

Eq. (23), the molar Gibbs energy of pure element or species or stoichiometric compound phase can then be represented as

$$G_{\mathrm{m}} - H_{\mathrm{m}}^{\mathrm{SER}} = a + bT + cT\ln(T) + \sum_{2}^{n} d_n T^n \tag{24}$$

This datum, sometimes named as lattice stability parameter, can be found in the database contributed by Dinsdale [62].

B. Models of Thermodynamic Properties

1. Substitutional Model

A random substitutional model can be used for gas phase, solid solution or even simple metallic liquid assumes that components can mix randomly and occupy any spatial position of the phase. In gas or liquid phases, the crystallographic structure does not exist, but the positional occupation of components relies on random substitution of all the components rather than the preferential occupation of any special component.

a. Ideal Solution Model. The ideal substitutional solution is characterized by the random distribution of components in one lattice with zero interchange energy. The mixing Gibbs free energy is thus dependent on the configuration entropy, which can be listed as

$$G_{\mathrm{mix}}^{\mathrm{ideal}} = -TS_{\mathrm{mix}}^{\mathrm{ideal}} = RT\sum_{i} x_i \ln x_i \tag{25}$$

where R is the gas constant, and x_i the mole fraction of component i. The Gibbs energy of the ideal solution phase will be

$$G_{\mathrm{m}} = \sum_{i} x_i G_i^{\mathrm{ref}} + RT\sum_{i} x_i \ln x_i \tag{26}$$

where the first term is actually $G_{\mathrm{m}}^{\mathrm{ref}}$, with $G_{\mathrm{i}}^{\mathrm{ref}}$ defining the Gibbs energy of the phase containing the pure component i in steel.

b. Non-ideal Solution Model. The regular solution model is the simplest of the non-ideal models and basically considers the sign and magnitude of interaction between the components as constant and independent of the composition of the phase, which leads to the well-known Gibbs excess energy of mixing of regular solution:

$$^{\mathrm{E}}G_{\mathrm{m}} = \sum_{i}\sum_{j>i} x_i x_j \Omega_{ij} \tag{27}$$

where Ω is a temperature-dependent interaction parameter. When adding Eq. (27) into Eq. (26), one obtains Gibbs energy of one mole phase:

$$G_{\mathrm{m}} = \sum_{i} x_i G_i^{\mathrm{ref}} + RT\sum_{i} x_i \ln x_i + \sum_{i}\sum_{j>i} x_i x_j \Omega_{ij} \tag{28}$$

However, the assumption of composition-independent interaction is too simple for use with steels. In order to introduce complex composition dependency to Ω, general formulas in terms of a power series were suggested. The most common was based on the Redlish–Kister equation:

$$^{E}G_{m} = \sum_{i}\sum_{j>i} x_i x_j \Omega_{ij}^{\upsilon}(x_i - x_j)^{\upsilon} \tag{29}$$

where $\Omega^{\upsilon}{}_{ij}$ is a binary interaction parameter dependent on the value of υ. It is obvious that Eq. (29) becomes regular when υ equals to 0 and subregular when υ equals to 1. Then Eq. (28) is extended to become

$$G_{m} = \sum_{i} x_i G_i^{ref} + RT\sum_{i} x_i \ln x_i + \sum_{i}\sum_{j>i} x_i x_j \Omega_{ij}^{\upsilon}(x_i - x_j)^{\upsilon} \tag{30}$$

Equation (30) implies that the ternary interactions are small in comparison to those of the binaries. This may not be always the case and if so, there exists evidence for higher order interactions, a further term of $\Delta G = x_i x_j x_k L_{ijk}$ can be added to Eq. (30), where L_{ijk} is the excess ternary interaction parameter. When it is applied the calculation in multi-component steel systems, normally there is little evidence for the need of interaction parameters of any higher order than third. The prediction of the thermodynamic properties as well as the phase relations in steels is usually based on the assessment of binary and ternary terms.

c. The Extrapolation of the Gibbs Excess Energy from Lower Component Systems to Multi-component Steels. From Eq. (30), it is clear that the first and second term is already in the formalism of multi-component systems. Equation (30) can be valid in most of the cases only if the third term, the excess Gibbs free energy part, is modified and extrapolated into higher order systems. The following is the three main extrapolation methodologies, Muggianu equation (63), Kohler equation (64) and Toop equation (65), which are used widely in the calculation of the multi-component systems.

The predominant method was developed by Muggianu [63], which described the excess energy in a subregular model:

$$\begin{aligned} ^{E}G_{m} = {} & x_A x_B\{L_{AB}^{0} + L_{AB}^{1}(x_A - x_B)\} + x_B x_C\{L_{BC}^{0} + L_{BC}^{1}(x_B - x_C)\} \\ & + x_A x_C\{L_{AC}^{0} + L_{AC}^{1}(x_A - x_C)\} \end{aligned} \tag{31}$$

where L denotes the binary interaction parameter.

In the Kohler equation, the excess energy was described as [64]:

$$\begin{aligned} ^{E}G_{m} = {} & (x_A + x_B)^2 \frac{x_A}{x_A + x_B}\frac{x_B}{x_A + x_B}\left\{L_{AB}^{0} + L_{AB}^{1}\left(\frac{x_A - x_B}{x_A + x_B}\right)\right\} \\ & + (x_B + x_C)^2 \frac{x_B}{x_B + x_C}\frac{x_C}{x_B + x_C}\left\{L_{BC}^{0} + L_{BC}^{1}\left(\frac{x_B - x_C}{x_B + x_C}\right)\right\} \\ & + (x_A + x_C)^2 \frac{x_A}{x_A + x_C}\frac{x_C}{x_A + x_C}\left\{L_{AC}^{0} + L_{AC}^{1}\left(\frac{x_A - x_C}{x_A + x_C}\right)\right\} \end{aligned} \tag{32}$$

Toop listed the excess part of the Gibbs free energy as [65]:

$$\begin{aligned} ^{E}G_{m} = {} & x_A x_B\{L_{AB}^{0} + L_{AB}^{1}(x_A - x_B - x_C)\} + x_A x_C\{L_{AC}^{0} + L_{AC}^{1}(x_A - x_C - x_B)\} \\ & + x_B x_C\left\{L_{BC}^{0} + L_{BC}^{1}\left(x_B - x_C + \frac{x_B - x_C}{x_B + x_C}x_A\right)\right\} \end{aligned} \tag{33}$$

There exist some similarity in Muggianu and Kohler equations since these equations can be considered symmetrical in treating the components. The Toop equation was generally regarded as asymmetrical since it is considered one of the limiting binary that did not behave in the same way as the others did. In practice, the phase boundaries calculated by either the Muggianu and Kohler equations may provide comparable results. The Toop equation is not suitable for the calculation of metallic systems, although it be appropriate for some ionic liquid systems.

2. Sublattice Model

Early in 1970, Hillert and Staffansson [66] developed a two sublattices model. If both sublattices contained more than one element each, the simplest case can be represented by the formula $(A,B)_a(C,D)_c$, where a, c expressed the number of sites in each sublattice. If one envisages that $a=c=1$ for convenience, the Gibbs energy in reference state is chosen as:

$$G_m^{ref} = y_A^1 y_C^2 G_{AC}^O + y_B^1 y_C^2 G_{BC}^O + y_A^1 y_D^2 G_{AD}^O + y_B^1 y_D^2 G_{BD}^O \tag{34}$$

where $y_i^s = n_i^s/N^s$, y_i^s is the site fraction of component i in sublattice s, n_i^s the number of atoms of component i in sublattice s and N^s is the total number of sites on the sublattices.

The Gibbs excess energy can also be demonstrated with a two sublattice model with the formalism of $(A, B)_a(C, D)_c$. Since the interaction energies between A–C, A–D, B–C, and B–D have been included in reference part [66], mixing on the same sublattice is only necessary to be considered. With the regular solution format, it can be listed as

$$^{E}G_m = y_A^1 y_B^1 L_{A,B:X}^0 + y_C^2 y_D^2 L_{X:C,D}^0 \tag{35}$$

where $L_{A,B:X}{}^0$ and $L_{X:C,D}{}^0$ represent, respectively, the regular solution parameters between A–B and C–D irrespective of site occupation of the other sublattice. However, if these actions are included, a subregular model should be introduced:

$$^{E}G_m = y_A^1 y_B^1 y_C^2 L_{A,B:C}^0 + y_A^1 y_B^1 y_D^2 L_{A,B:D}^0 + y_C^2 y_D^2 y_A^1 L_{A:C,D}^0 + y_C^2 y_D^2 y_B^1 L_{B:C,D}^0 \tag{36}$$

Site fraction dependence for solution parameters can be introduced as

$$L_{A,B:C}^0 \Rightarrow \sum_{\upsilon} L_{A,B:C}^{\upsilon}(y_A^1 - y_B^1)^{\upsilon} \tag{37a}$$

$$L_{A,B:D}^0 \Rightarrow \sum_{\upsilon} L_{A,B:D}^{\upsilon}(y_A^1 - y_B^1)^{\upsilon} \tag{37b}$$

$$L_{A:C,D}^0 \Rightarrow \sum_{\upsilon} L_{A:C,D}^{\upsilon}(y_C^1 - y_D^1)^{\upsilon} \tag{37c}$$

$$L_{B:C,D}^0 \Rightarrow \sum_{\upsilon} L_{B:C,D}^{\upsilon}(y_A^1 - y_B^1)^{\upsilon} \tag{37d}$$

For the ideal mixing part, the Gibbs energy can be listed as

$$G_{mix}^{ideal} = -TS_{mix}^{ideal} = RT\sum_{S} N^S \sum_{i} y_i^S \ln y_i^S \tag{38}$$

After the extension by Sundman and Ågren [67], the sublattice model becomes suitable for a phase with an arbitrary number of sublattices, which allows the occupation of arbitrary number of components. To simplify the analytical expression of the integral Gibbs energy, a new component array I was introduced, with which the Gibbs free energy of one mole phase could be defined as [68]:

$$G_m = \sum_{I0} P_{I0}(Y) G_{I0}^O + RT \sum_S N^S \sum_i y_i^S \ln y_i^S + \sum_{Z>0} \sum_{IZ} P_{iz}(Y) L_{IZ} \tag{39}$$

where G_{I0}^o represents the Gibbs energy of the compound defined by I, $P_{IZ}(Y)$ the corresponding product of site fractions, L_{IZ} the interaction parameters, the subscript IZ there refers to the order of the component array. For example, $I1$ indicates the first order of a component array, where one sublattice contains two components, the remaining containing only one component with summation operated over all different $I1$.

a. Line Compounds. In some binary phases, there exists a well-defined stoichiometry with respect to the components. However, these phases can have partial or complete solubility of other components with preferential substitution for one of the two components. An example of this phase is $(Fe,Mn)_7C_3$. Such phases were named as line compounds and generally described as

$$(A, B, C)_u(Z)_v$$

where A, B and C are the components mixed in sublattice 1 and Z the component in sublattice 2 with fixed stoichiometry. The Gibbs energy of these phases can thus be described as

$$G_m = \sum_i y_i^1 G_{i:z}^0 + RTu \sum_i y_i^1 \ln y_i^1 + \sum_i \sum_{j>i} y_i^1 y_j^1 \sum_v L_{ij}^v (y_i^1 - y_j^1)^v \tag{40}$$

where I and j represent the components mixed in sublattice 1 while Z is the stoichiometric component in another sublattice. $G_{i:z}{}^0$ is the Gibbs energy of the compound while sublattice 1 is completely occupied by component I, and u the molar number of sites in sublattice 1. The excess energy is considered only with interactions between components in the same sublattice.

b. Interstitial Phases. When components such as C and N occupy the interstitial sites of ferrite or austenite lattices, the phase can be modeled as a two sublattice structure, with one occupied by substitutional elements as Fe, Mn, Cr, Ni, etc., and the other by the interstitial elements as C, N, and vacancy. The sublattices are occupied by the elements in the following format:

$$(\mathrm{Fe, Cr, Mn, Ni}, \ldots)_u(Va, \mathrm{C, N}, \ldots)_v$$

For example, the f.c.c.-Al structure of austenite phase in Fe–Cr–C system was commonly represented by the formula $(Cr, Fe)_1\,(C,Va)_1$ and Gibbs energy of this phase can be listed as

$$\begin{aligned} G_m = {} & y_{Cr}^1 y_{Va}^2 G_{Cr:Va}^0 + y_{Fe}^1 y_{Va}^2 G_{Fe:Va}^0 + y_{Cr}^1 y_C^2 G_{Cr:C}^0 \\ & + y_{Fe}^1 y_C^2 G_{Fe:C}^0 + RT(1 \cdot (y_{Cr}^1 \ln y_{Cr}^1 + y_{Fe}^1 \ln y_{Fe}^1) + 1 \cdot (y_C^2 \ln y_C^2 + y_{Va}^2 \ln y_{Va}^2)) \\ & + y_{Cr}^1 y_{Fe}^1 y_{Va}^2 \left(\sum_v L_{Cr,Fe:Va}^v (y_{Cr}^1 - y_{Fe}^1)^v \right) + y_{Cr}^1 y_{Fe}^1 y_C^2 \left(\sum_v L_{Cr,Fe:C}^v (y_{Cr}^1 - y_{Fe}^1)^v \right) \\ & + y_{Cr}^1 y_C^2 y_{Va}^2 \left(\sum_v L_{Cr:C,Va}^v (y_C^2 - y_{Va}^2)^v \right) + y_{Fe}^1 y_C^2 y_{Va}^2 \left(\sum_v L_{Fe:C,Va}^v (y_C^2 - y_{Va}^2)^v \right) \end{aligned} \tag{41}$$

where the first two terms represent the Gibbs energy of Cr and Fe in f.c.c.-Al structure, while the next two the Gibbs energy of CrC and FeC compounds. The last four terms denote the excess energy and the rest part of the equation denotes the ideal mixing entropy.

C. Database and Calculation Example of High-Strength Low-Alloy Steel

A large database named as Fe-DATA, which is compatible with that of Scientific Group Thermodata Europe (SGTE) and suitable for the designing steels, was contributed by Saunders and Miodownik [69]. Besides the high accuracy, one may obtain in the calculation of liquidus and solidus temperatures while installing the Fe-DATA, this database includes the following elements or components:

Fe, C, Co, Cr, Cu, Mg, Mn, Mo, N, Nb, Ni, Si, Ti, V, and W

and the phases listed as following:

Liquid, austenite, ferrite, cementite, $M_{23}C_6$, M_7C_3, M_3C_2, M_6C, WC, M(C,N), M_2(C,N), Fe_4N, Mg_3N_2, σ, μ, χ, Laves and R-phase.

As known, TRIP steel has received considerable interest recently because it can highly raise the strength and toughness of the steel simultaneously and make the drastic weight reduction of automobile structural parts. Some details of the TRIP steel have been introduced in Section 1.2.2. In order to design the steel in the light of the existing calculating facilities, Li et al. [70] estimated a large number of phase diagrams with assumed composition in Fe-Si–Mn–C system with Thermo-Calc software contributed by Royal Institute of Technology in Stockholm. As an example, compositions of seven calculated steels are listed in Table 31 [70]. During calculation, S and P were not put into the program since their content was too low to be regarded as alloying elements. Calculated phase diagrams of No. 2 and No. 3 steels are shown in Fig. 25, while the calculated volume percentage of austenite and its carbon content in each steel at 780°C are also listed in Table 31. The calculation result was validated by the following experiment:

The samples of No. 2 and No. 3 steels were heated to 820°C, respectively, and held for 5 min and then immediately quenched into water. Metallography examination showed that no ferrite remained in the No. 2 steel but there did exist a small quantity of ferrite in No. 3 steel; which correlated well with the calculated phase diagram in Fig. 25.

Table 31 Compositions of Assumed Steels (in wt%)

	C	P	S	Mn	Si	Aus.%	C%
No. 1 steel	0.17	0.006	0.004	1.85	1.48	40.90	0.417
No. 2 steel	0.185	0.0064	0.003	1.87	1.00	52.02	0.354
No. 3 steel	0.195	0.0035	0.006	1.58	1.60	42.31	0.463
No. 4 steel	0.20	0.009	0.0032	1.92	1.55	48.12	0.418
No. 5 steel	0.22	0.0068	0.0025	1.65	1.50	49.61	0.446
No. 6 steel	0.19	0.0064	0.0039	1.32	1.42	40.31	0.477
No. 7 steel	0.20	0.0046	0.0048	1.52	1.55	42.99	0.471

Note: Aus.% is in volume percent.
Source: Ref. 70.

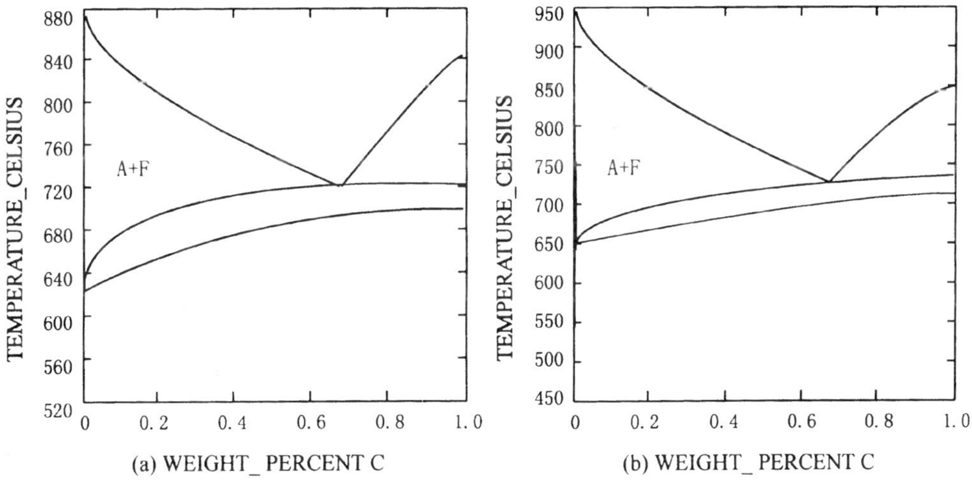

Figure 25 Phase Diagram of Fe–Si–Mn–C system: (a) no. 2 steel; (b) no. 3 steel. (From Ref. 70).

As described earlier the volume fraction of retained austenite and the carbon content in the phase considered are the key factors affecting the mechanical properties of TRIP steel. According to this rule, No. 2 steel was chosen as test material because it possessed the highest volume fraction of austenite. The carbon content in austenite was not high; however, it might be enriched during the bainite transformation period in the middle temperature range. No. 3 steel was also selected because it had rather a large amount of austenite, higher carbon weight percentage, and the lowest P content, Low content of P can improve the toughness of steel at low temperatures when a large amount of deformation is applied to induce phase transformation. Several different treatments were designed for these two steels and listed as follows:

(a) 760°C annealing 5 min + 400°C soaking 5 min,
(b) 780°C annealing 5 min + 370°C soaking 5 min,
(c) 780°C annealing 5 min + 400°C soaking 5 min,
(d) 780°C annealing 5 min + 430°C soaking 5 min,
(e) 780°C annealing 5 min + 460°C soaking 5 min,
(f) 800°C annealing 5 min + 400°C soaking 5 min,
(g) 820°C annealing 5 min + 400°C soaking 5 min, and
(h) 740°C annealing 5 min + 400°C soaking 5 min.

After heat treatment, mechanical properties of samples of No. 2 and No. 3 steels are measured and the results were shown in Tables 32 and 33 [70], where one can easily find that No. 2 steel after treatments a and b obtained the highest value of $\sigma_b \times \delta_5$, the best combination of strength and toughness.

D. Diffusion Dynamic Model and Estimation of Diffusion Quantity

The diffusion controlled phase transformation in steels was intensively studied. In every case, the local equilibrium at the moving interface between the interacting phases, e.g., α and γ was assumed to be established which seemed reasonable under most conditions.

Table 32 Detected σ_b (in MPa) and δ_5 of Steel No. 3 After Different Treatments

Treatment	a	b	c	d	e	f	g
$\sigma_b \times \delta_5$	17,643	17,325	19,866	19,120	18,314	18,979	19,179
σ_b	722.20	785.49	745.23	748.27	785.01	761.74	802.97
δ_5	24.43	22.05	26.66	25.55	23.33	24.90	23.88

Source: Ref. 70.

Table 33 Detected σ_b (in MPa) and δ_5 of Steel No. 2 After Different Treatments

Treatment	a	c	f	h
$\sigma_b \times \delta_5$	21,010	22,009	17,962	17,186
σ_b	738.50	848.08	862.26	640.38
δ_5	28.41	25.96	20.79	26.86

Source: Ref. 70.

The local equilibrium must obey the flux balance equation as follows [71]:

$$\upsilon^{\alpha} C_k^{\alpha} - \upsilon^{\gamma} C_k^{\gamma} = J_k^{\alpha} - J_k^{\gamma}, \qquad k = 1, 2, \ldots, n \tag{42}$$

where υ^{α} and υ^{α} are the interface migration rates, $C_k{}^{\alpha}$ and $C_k{}^{\gamma}$ are the concentrations of component k in the α and γ phases at the phase interface and J_k^{α} and J_k^{γ} correspond to the diffusional flux of component k in either side of the phase interface of the α and γ phases.

A frame of reference for the phase interface is then taken, through applying "number-fixed" with respect to the number of moles of component k, where [71]

$$\sum j_k^{\alpha} = \sum j_k^{\beta} = 0 \tag{43}$$

The summation is performed over the substitutional elements only.

For crystalline phases of iron-based alloys, the vacancy exchange mechanism of diffusion is predominant, diffusion occurs by the jumping of atoms into neighboring vacant sites in lattice. Thus, in the lattice fixed frame of reference, the diffusion flux of component k can be written as

$$\tilde{J}_k = -x_k \Omega_k \nabla \mu_k \tag{44}$$

where $\Omega_k = y_{\mathrm{Va}} M_{k\mathrm{Va}}$ (45)

k is substitutional. x_k is the mole fraction of k and $\nabla\mu_k$ is the gradient of chemical potential of species k. y_{Va} is the fraction of vacant lattice sites in the sublattice where k is dissolved, $M_{k\mathrm{Va}}$ is the mobility denoting the rate of exchange if there is a vacancy adjacent to a k atom. For the diffusivity in the number fixed frame of reference, it can be written as

$$D_{kj} = \sum_{i=1}^{n} (\delta_{ik} - x_k) x_i \Omega_i \frac{\partial \mu_i}{\partial x_j} V_{\mathrm{m}} \tag{46}$$

where δ_{ik} is Kronecker's symbol, when $i=k$, $\delta_{ik}=1$; $i=k$, $\delta_{ik}=0$. V_m is the molar volume. In practical calculations one usually chooses to eliminate one species since $\sum x_k=1$. Therefore the diffusivity becomes

$$D_{kj}^n = D_{kj} - D_{kn} \tag{47}$$

where n is taken as the dependent species. for diffusion couple experiments, one prefers to calculate $D_{kj}{}^n$ rather than D_{kj}.

The tracer diffusivity D_k^* is then directly related to the mobility Ω_k by means of the Einsten relation

$$D_k^* = RT\Omega_k \tag{48}$$

where Ω_k can be represented as a familiar formalism:

$$\Omega_k = \Omega_k^0 \exp(-Q_k/RT) \tag{49}$$

For steels containing interstitial solutes, it is better to introduce the U-fraction u_k, which is defined as

$$u_k = \frac{x_k}{\sum\limits_{i\in s} x_i} \tag{50}$$

where the symbol $i \in s$ denotes that the summation is taken over the substitutional species only. The choice of the volume fixed frame of reference in the diffusion calculation makes it necessary to use a composition variable, which is related to the volume. The definition of U-fraction meets with this idea and is based on the assumption that only the substitutional species contribute to the volume. Thus, Eq. (44) becomes

$$\tilde{J} = -u_k\Omega_k \nabla\mu_k \quad \text{(for } k \text{ is substitutional)} \tag{51}$$

$$\tilde{J} = -u_k y_{\mathrm{Va}}\Omega_k \nabla\mu_k \quad \text{(for } k \text{ is interstitial)} \tag{52}$$

where y_{Va} is the fraction of vacant interstitial sites.

For substitutional species, condition [43] exists.

The diffusivities in the number fixed frame of reference will be written as

$$D_{kj} = \sum_{i\in s}(\delta_{ik} - u_k)u_i\Omega_i\frac{\partial\mu_i}{\partial u_j}V_s + \sum_{i\notin S}\delta_{ik}u_i y_{\mathrm{Va}}\Omega_i\frac{\partial\mu_i}{\partial u_j}V_s \tag{53}$$

where V_s is the molar volume of the substitutional species. Also Eq. (47) becomes

$$D_{kj}^n = D_{kj} - D_{kn} \quad \text{(for } j \text{ is substitutional)} \tag{54}$$

$$D_{kj}^n = D_{kj} \quad \text{(for } j \text{ is interstitial)} \tag{55}$$

In steels, the occurrence of the transition between paramagnetic and ferromagnetic states affects thermodynamic and kinetic properties. In Ref. [71], Q_k and Ω_k^0 in Eq. (49) can be related to magnetic order in the following way:

$$\Omega_k^0 = \Omega_k^{0p}\exp(\Lambda\alpha\xi) \tag{56}$$

and

$$Q_k = Q_k^p(1 + \alpha\xi) \tag{57}$$

where $\Lambda \approx 6$, $\alpha \approx 0.3$ in the bcc structure. For fcc structure, the ferromagnetic effect on diffusion is usually neglected, i.e., $\alpha \approx 0$. ξ represents the state of the magnetic order at the temperature under consideration, being the value: $0 < \xi < 1$. Ω_k^{0p} and Q_k^{p} represent the frequency factor and the activation energy in the paramagnetic state.

The above model can give a satisfactory description of the diffusion of Cr, Ni, and Co in Fe. For the interstitial species, such as carbon, Ågren (Ref. [71]) suggested the following expression:

$$RT\Omega_C^{\alpha} = 2.10^{-6}\mathrm{e}^{-10,115/T}\exp\left\{0.5898\left[1+\frac{2}{\pi}\arctan\left(14.985-\frac{15309}{T}\right)\right]\right\}(\mathrm{m}^2\,\mathrm{sec}^{-1}) \tag{58}$$

In Thermo-Calc system, there is a software package, DICTRA, which possesses several main modules for solving the diffusion equation, estimating the local equilibrium at phase interface as well as setting up and solving flux balance equations. Powerful support also comes from the dynamic database, such as MOB making the package applicable to steels. Miscellaneous examples concerning elements diffusion and diffusion controlled phase transformation were supplied to make the access of the software easier. As an example, Fig. 26 shows the effect of the addition of alloying elements on the volume fraction of ferrite during continuous cooling. Calculation was performed in a Fe–0.2 wt% C steel with various alloy additions. The cooling rate is chosen as low as 0.1 K sec^{-1} to allow ferrite grow adequately.

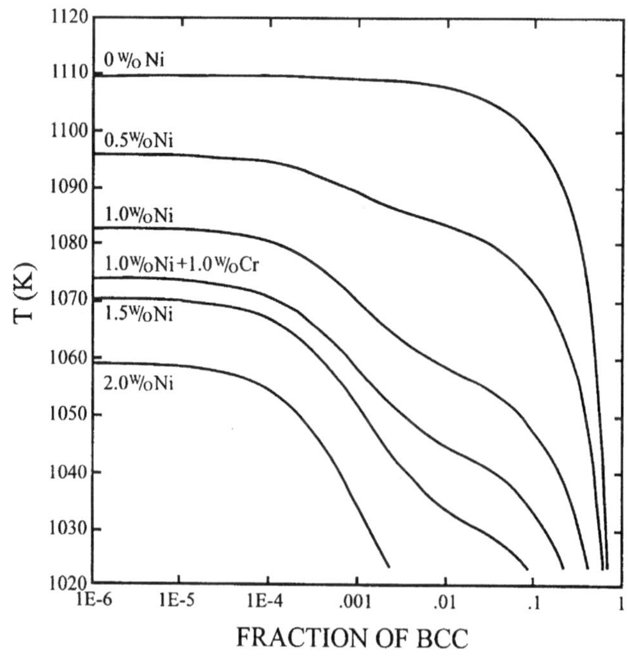

Figure 26 Simulation of ferrite growth during continuous cooling. Temperature vs. fraction of ferrite formed for Fe–0.2 wt% C alloys cooled at 0.1 K sec^{-1}. Labels indicate amount of alloying elements in addition to Fe and C. (From Ref. 71.)

REFERENCES

1. DeArdo, A.J. New developments in the alloy design of microalloyed and other modern HSLA steels. In *HSLA Steels: Processing, Properties and Applications*, Geoffrey, T., Zhang, S.H., Ed.; TMS: Pennsylvania, 1992; 21–32.
2. LeBon, A.B.; de Saint-Martin, L.N. Using laboratory simulations to improve rolling schedules and equipment *Microalloying 75*; Union Carbide Corporation: New York, 1977; 90–99.
3. Meyer, L.; Heisterkamp, F.; Mueschenborn W. Columbium, titanium and vanadium in normalized thermo-mechanically treated and cold-rolled steels *Microalloying 75*; Union Carbide Corporation: New York, 1977; 153–167.
4. Graville, B.A. Cold cracking in welds in HSLA steels. Proceedings of International Conference on Welding of HSLA Structureal Steels, ASM: Ohio, 1978; 85–101.
5. Haze, T.; Achara, S.; Mabuchi, H. Accelerated cooling of rolled steel. Proceedings of Seventh International Conference on Offshore Mechanics and Arctic Engineering, Texas, ASME: New York, 1988; 235–241.
6. Gräf, M.K.; Hillenbrand, H.-G.; Peters, P.A. New high-strength large diameter pipe steels *HSLA Steel: Metallurgy and Applications*; Metal Park: Ohio, 1986; 467–474.
7. Thomas, G. Design of strong, ductile, HSLA and dual phase steels Key Eng. Mater. 1993, *84–85*, 437–460.
8. Takechi, H. Recent developments in HSLA steel for automobile in Japan. Proceedings of fourth International Conference on HSLA Steels, Metallurgical Industry Press: Xi'an; 2000; 80–87.
9. Hanzaki, A.Z.; Hodgson, P.D.; Yue, S. Retained austenite characteristics in thermomechanically processed Si-Mn transformation-induced plasticity steels Metall. Mater. Trans. 1997, *A28*, 2405–2414.
10. Gong, Z.J.; Tan, S.K.; You, J.; Teng, X.H. Application of super deep-drawing cold-rolled IF steel sheet for autos. Proceedings of the International Conference on Advanced Automobile Materials, The Chinese Society For Metals: Beijing, 1997; 148–153.
11. Jiao, Y.M.; Dong, Y.H.; Yang, B.S.; Fu, Y.S.; Yin, Y.J. The bake hardening property of the China-developed bake hardening steel sheet BH340. Proceedings of the International Conference on Advanced Automobile Materials, The Chinese Society For Metals: Beijing, 1997; 127–131.
12. Engineer, S.; Huchtemann, B.; Schüler V. A review of the development and application of microalloyed medium-carbon steels In *Fundamentals of Microalloying Forging Steels*, Krauss, G.; Banerji, S.K., Eds.; TMS-AIME: Warrendale, 1987; 19–38.
13. Langeborg, R.; Sandberg, O.; Roberts, W.; Optimization of microalloyed ferrite-pearlite forging steels. In *Fundamentals of Microalloying Forging Steels*, Krauss, G.; Banerji, S.K., Eds.; TMS-AIME: Warrendale, 1987; 39–54.
14. Garcia, C.I.; Lis, A.K.; Deardo, A.J. Transformation behavior and properties of low-carbon multi-phase MA bar steels. In *Microalloyed Bar and Forging Steel*, Finn, M., Ed.; CIM-ISS: Warrendale, 1990; 25–42.
15. Szilva, W.A.; Grassl, K.J.; Weith, J.W.; Wright, P.H. Development of microalloy forging steels at chaparral steel In *Microalloyed Bar and Forging Steels*, Finn, M., Ed.; CIM-ISS: Warrendale, 1990; 227–236.
16. Ollilainen, V.; Kohoppa, J. Inform modern direct quenching steel In *Microalloyed Bar and Forging Steels*, Finn, M., Ed.; CIM-ISS: Warrendale, 1990; 237–248.
17. Garcia, C.I.; Lis, A.K.; DeArdo A.J. New microalloyed, multi-phase steel for high-strength forging applications. *Society of Automotive Engineers Transactions*; SAE: Pittsburgh, 1991; 143–145.
18. Bucher, J.H.; Held, J.F.; Butler, J.F. *Society of Automotive Engineers Transactions*, 790026; 1991.
19. Heitmann, W.E.; Bhattacharya, D. Comparison of the strength, ductility and machinability properties of cold draw medium carbon steels with those of as-rolled and cold draw medium

carbon steels. In *Microalloyed Bar and Forging Steels*, Finn, M., Ed.; CIM-ISS: Warrendale, 1990; 249–261.
20. Garcia, C.I.; Lis, A.K.; DeArdo A.J. High-strength cold forging applications of a new multiphase steel Society of Automotive Engineers Transactions; SAE, Pittsburgh, 1991, 145–149.
21. Poole, S.W.; Franklin J.E. High-strength structural and high-strength low allow steels 9th Metal Handbook; ASM: New York, 1978, 403–420.
22. Ma, S.Y.A.; Stannard, D.M.; Liimatainen, J.; Martikainen, H. State-of-the art cast joints technology for offshore use—foundry practices and design specifications. Tubular Structures: The Third International Symposium, Lappeenranta, Elsevier Applied Science: London, 1989; 118–128.
23. Sampath, K.; Green, R.S.; Civis, D.A.; Williams, B.E.; Konkol, P.J. Metallurgical model speeds development of GMA welding wire for HSLA steel Welding J. 1995, *74*, 69–76.
24. Fleming, D.A.; Bracarense, A.Q.; Liu, S.; Olson, D.L. Toward developing a SMA welding electrode for HSLA-100 grade steel Welding J. 1996, *75*, 171–183.
25. Wong, R.J.; Hayes, M.D. Arc welding consumables for HSLA steels with yield strengths of 80 ksi and above In *The Metallurgy, Welding, and Qualification of Microalloyed (HSLA) Steel Weldments*, Hickey, J.T.; Howden, D.G.; Randall, M.D., Eds.; American Welding Society: Houston, 1990; 450–489.
26. Denys, R.M. Implications of overmatching/undermatching of weld metal yield strength. In *The Metallurgy, Welding, and Qualification of Microalloyed (HSLA) Steel Weldments*, Hickey, J.T.; Howden, D.G.; Randall, M.D., Eds.; American Welding Society: Houston, 1990; 573–625.
27. Zhang, G.J.; Cai, H.B.; Cui, F.C. Low temperature fatigue property and its gray forecast of welded joint of HSLA steel Trans. China Welding Instit. 1993, *14* (1), 16–23.
28. Hulka, K.; Heisterkamp, F. Development trends in HSLA steels for welded constructions Mater. Sci. Forum 1998, Vols. *284–286*, 343–350.
29. Zhang, B.F.; Li, W.S.; Zeng, Y.J. Development of a high-toughness covered electrode for high-strength steels Trans. China Welding Instit. 1993, *14*, 227–232.
30. Steven, W.; Haynes, A.O. The temperature of formation of martensite and bainite in low-alloy steels J. Iron Steel Instit. 1956, *183*, 349–359.
31. Irvine, K.J.; Pickering, F.B. Low carbon bainitic steels J. Iron Steel Instit. 1957, *184*, 292–309.
32. Dearden, J.; O'Neill, H. A guide to the selection and welding of low-alloy structural steel Trans. Int. Welding 1940, *3*, 203–214.
33. Ito, Y.; Bessyo, K. Weldability formula for high-strength steel related to heat affected zone cracking Int. Instit. Welding Dec. 1968, 567–578.
34. Yuioka, N.; Suzuki, H.; Ohshita, S.; Saito, S. Determination of necessary preheating temperature in steel welding Welding J. 1983, *62*, 147s–153s.
35. Edwards, G.R.; Liu, S. Recent developments in HSLA welding. Proceedings on Advances in Welding Metallurgy, First US-Japan Symposium, San Francisco, American Welding Society: Miami, 1990; 215–293.
36. Donchenko, E.A.; Panasenko, L.K.; Larochkin, V.K. Some properties of welding slags Welding Prod. 1978, *6*, 48–50.
37. Braid, J.E.M.; Tyson, W.R. Fracture toughness considerations for arctic and cold marine structures. In *The Metallurgy, Welding, and Qualification of Microalloyed (HSLA) Steel Weldments*, Hickey, J.T.; Howden, D.G.; Randall, M.D., Eds.; Houston, American Welding Society: 1990; 427–447.
38. Pisarski, H.G.; Harrison, J.D. Fracture toughness considerations for offshore structures in UK waters. Proceedings of International Conferemce on Welding for Challenging Environments, Canada, 1985; New York, 1986, 131–144.
39. Yu, F.Z. *Anti-corrosion Ability of Metal Materials*; Science Pub.: Beijing, 1982; 128–129.
40. Gao, H.S.; Wang, Z.L. Alloy design of an atmospheric corrosion resistant cold rolled steel with 345 MPa yield strength In *HSLA Steels: Processing, Properties and Applications*, Geoffrey, T.; Zhang, S.H., Eds.; TMS: Pennsylvania, 1992; 341–346.

41. Gawne, D.T. *Strengthening Mechanism of HSLA Steel*; Metal Park: Ohio, 1986; 1–8.
42. Williams, J.G. Titanium microalloyed hot-rolled strip steels production, properties and applications. Proceedings of International Conference on Technology and Applications of HSLA Steel, Philadelphia, Metals Park: Ohio, 1983; 261–275.
43. Shiga, C. Steel for offshore platforms Steel Times International 1988, *12* (5), 38–43.
44. Aoki, I.V.; Guedes, I.C.; Taqueda, M.E.S. Polarization curve and experiment design as tools in the search of optimized inhibitors mixture formulation for HSLA steel in hydrochloric acid Mater. Sci. Forum 1998, Vols. *289–292*, 1237–1244.
45. ASTM Standard A841/A841M-88, pp 596–598.
46. Tsukada, K.; Watanabe, I. The progress of thermo-mechanical control process for HSLA plate steel Key Eng. Mater. 1993, *84–85*, 16–57.
47. Siwecki, T.; Engberg, G.; Cuibe, A. Thermo-mechanical controlled processes for high-strength and toughness in heavy plates and strips of HSLA steels. Proceedings of International Conference on Thermomechanical Processing of Steels and Other materials, Wollongong, Warrendale: PA, 1997; 7–11.
48. Pussegoda, L.N.; Yue, S.; Jonas, J.J.; Hunt, P.J. Development of dynamic recrystallization controlled rolled schedules during seamless tube rolling. In *HSLA Steels: Processing, Properties and Applications*, Geoffrey, T.; Zhang, S.H., Eds.; TMS: Pennsylvania, 1992; 159–163.
49. Thomas, G. Design and processing of strong-tough microcomposite steels Archives Metall. 1991, *36*, 147–167.
50. Izumiyama, M.; Tsuchiya, M.; Imai, Y. Effects of alloying element on supercooled A//3 J Jap. Inst. Met 1970, *34*, 291–295.
51. Andrews, K.W. Empirical formulae for the calculation of some transformation temperatures JISI 1965, *S203*, 721–727.
52. Thomas, G.; Kim, J.K.; Manojlovic, D.; Milovic, R. Metall. Trans. 1988, 19A, 399–408.
53. Chatterjee, S.; Mishra, N.S. Evolution of acicular ferrite microstructure in a titanium bearing HSLA steel Steel Res. 1994, *65* (4), 138–145.
54. Babu, S.S.; Bhadeshia, H.K.D.H. Mechanism of transition from bainite to acicular ferrite Mater. Trans JIM 1991, *32*, 679–688.
55. Kobayashi, H. Microstructure development in Ti bearing steel with simulated hot-rolling fracture ISIJ Inter. 1992, *32* (7), 873–881.
56. Mujahid, M.; Lis, A.K.; Garcia, C.I.; DeArdo, A.J. HSLA-100 steels: influence of aging heat treatment on microstructure and properties J. Mater. Eng. Performance 1998, *7* (2), 247–257.
57. Goodman, S.R.; Brenner, S.S.; Low, J.R. Fim-atom probe study of precipitation of copper from iron-1.4 At Pct copper. 1. Field-ion microscopy Metall. Trans. 1973, *4A* (10), 2363–2371.
58. Garcia, C.I.; DeArdo, A.J. Formation of austenite in 1.5 pct Mn steels Metall. Trans. 1981, *12A*, 521–530.
59. Meijering, J.L. Calculation of the nickel–chromium–copper phase diagram from binary data Acta Met. 1957, *5*, 257–264.
60. Kaufman, L.; Cohen, M. The martensitic transformation in the iron-nickel system. J Metals 1956, *10*, 1341–1393.
61. Weiss, R.J.; Tauer, K.J. Component of the thermodynamic function of iron Phys. Rev. 1956, *102*, 1490–1495.
62. Dinsdale, A.T. SGTE data for pure elements CALPHAD 1991, *15*, 317–426.
63. Muggianu, Y.M.; Gambino, M.; Bros, JP. Enthalpies de formation des alliages liquids bismuth-etain-gallium a 723K J. Chim. Phys. 1975, *22*, 83–88.
64. Kohler, F. Zur berechnung der thermodynamis chen daten enines tern ä ren aus den zurgehorigen bin ä ren systemen Monatsh Chem. 1960, *91*, 738–740.
65. Toop, G.W. Prediction ternary activities using binary data Trans. Met. Soc. AIME 1965, *233*, 850–855.
66. Hillert, M.; Staffansson, L.I. The regular solution model for stoichiometric phases and ionic melts Acta Chemica Scandinavica 1970, *24*, 3618–3626.

67. Sundman, B.; Ågren, J. A regular solution model for phases with several components and sublattices, suitable for computer applications J. Phys. Chem. Solids 1981, *42*, 297–301.
68. Saunders, N.; Miodownik, A.P. CALPHAD, 1st; Elsevier Science Ltd: New York, 1998, 95–106.
69. Saunders, N.; Miodownik, A.P. CALPHAD, 1st; Elsevier Science Ltd: New York, 1998; 332–333.
70. Li, L.; Si, W.; Fu, R.J.; Xu, L.P. Composition design and mechanical properties of cold rolled TRIP steel In *HSLA Steels '2000*, Liu, G.Q.; Wang, F.M.; Wang, Z.B.; Zhang, H.T., Eds.; Metallurgical Industry Press: Xian, 2000; 299–303.
71. Ågren, J. Computer simulations of diffusional reactions in complex steels ISIJ Inter. 1992, *32* (3), 291–296.

6

Designing with Microalloyed and Interstitial-Free Steels

David V. Edmonds
University of Leeds, Leeds, England

I. INTRODUCTION

Microalloyed and *interstitial-free steels* may be defined as steels which contain less than 0.1 wt% total addition of the elements niobium (columbium), titanium or vanadium, either singly or in combination, which control the microstructure during processing to produce significant improvement in the mechanical properties required for manufacturing or service. The microstructural control is produced by the formation of stable microalloy carbides and nitrides. In the case of the microalloyed steels, this produces a two-to-three fold increase in strength by grain refinement, possibly augmented by precipitation strengthening, whilst in the interstitial-free steels, it scavenges carbon and/or nitrogen from solid solution to improve formability. Thus, higher strength constructional and engineering steels or more formable sheet steels result.

II. HISTORICAL BACKGROUND

A. Microalloyed Steels

The introduction of microalloying to C–Mn constructional and engineering steels has probably been the most important steel development in modern times. A large number of conference proceedings have been published chronicling all aspects of this significant development, especially for constructional steels [1–41]. Although there is evidence in the literature on the use of niobium, vanadium, and titanium in steels in the early years of the 20th century [42,43], the early foundations for the technology were laid mainly during the 1950s, based upon niobium which became available at more competitive prices, and developed rapidly, both commercially and in understanding, during the 1960s [44–56]. A key driver was the need for increased strength in constructional steel plates and sections, initially mainly in connection with the recovery and transmission of oil and gas (e.g. Refs. [57,58]). Because offshore structures and linepipe required welding, the steels needed good weldability and additionally, high toughness. The use of microalloying to improve strength allowed reductions in carbon levels, thereby improving weldability.

A history of the emergence and development of microalloyed steels up to 1985 has been presented by Woodhead and Keown [42], which also explains the terminology used over the period. A more recent appreciation has also been given by Gladman [59]. Aluminum additions were being made to low-carbon steels to control interstitial nitrogen (by formation of AlN) and a beneficial side-effect of grain refinement was apparent (initially of the austenite during solution treatment, but this resulted in finer ferrite grains upon subsequent transformation during cooling). Aluminum has a high affinity for oxygen and so this practice could only be applied to fully killed (fully deoxidized) steels which gave low yield. Exploration of the addition of minor amounts of niobium, which has less oxygen affinity, followed, and similar effects were thus achieved for semi-killed steels, which gave greater yields. However, in addition to grain refinement, a precipitation strengthening effect was also obtained by NbC formation within the ferrite. The alternative use of vanadium and/or titanium additions followed and the practice of microalloying was established. The early aluminum-treated steels were never known as microalloyed steels, and this name seems to have been applied only to those steels containing minor additions of the strong carbide- and nitride-formers niobium, vanadium or titanium, which produce grain refinement and/or precipitation by the formation of microalloy carbides, nitrides, or carbo-nitrides. These steels are also often referred to as *high-strength low alloy* or *HSLA steels*, and this name, originally applied over a wider composition range to steels with mechanical properties improved over those of mild steel, appears to have become synonymous nowadays with *microalloy* steels, although not all HSLA steels are microalloyed [43].

The next step was to combine microalloying with control of the hot-rolling schedule, which was developed, also during the 1960s, to maximize the grain refinement effect, which became known as *controlled rolling* (e.g. Refs. [60,61]). *Accelerated cooling* (e.g. Ref. [62]) after rolling, by introducing water sprays, e.g. between the finishing rolls and the coilers, to refine the precipitation and enhance the precipitation strengthening component, came next in the early 1970s.

Considerable strength increases have thus been achieved whilst maintaining acceptable toughness, and at reduced carbon content and therefore improved weldability. Yield strengths of 400–500 MPa are readily achievable and the beneficial effect of grain refinement on reducing the impact transition temperature (ITT)* outweighs the increase due to solid solution and precipitation strengthening to maintain ITT below room temperature, and often lower than that of an equivalent normalized C–Mn steel. Although developed primarily for larger assemblies, e.g. linepipe, microalloyed steels have found wider application, e.g. in automotive parts where increased strength can be used to reduce section size, and hence weight, to give fuel economy.

The successful application of microalloying to low-carbon constructional steels was followed in the 1980s and 1990s by extension of the practice to higher-carbon steels (e.g. Refs. [63–77]). Thus, it has become possible to find examples of many former grades of C and C–Mn ferrite–pearlite steels to which minor additions of strong carbide-formers such as vanadium are specified, e.g., reinforcing bar steels at 0.25% C, direct hardening forging steels at 0.4% C and wire and rail steels at 0.6–0.9% C. In all of these, there is scope for austenite grain size control during austenitizing treatments, and precipitation strengthening. The latter was somewhat unexpected because many of the microstructures are predominantly pearlitic, but it has been discovered that carbo-nitride precipitation occurs within the pearlitic ferrite during austenite transformation so that beneficial strengthening is

*For each 15 MPa increase in yield strength by grain refinement, the ITT is lowered by 10°C.

achieved. In the case of medium carbon forging steels, the achievable strength levels are no more than those that could be obtained by other heat treatment routes, e.g. by quenching and tempering. However, direct hardening during cooling from the forging temperature gives considerable saving in the number of processing operations required, compared with quench and temper routes, and therefore significant cost benefits (e.g. Refs. [78–82]). The benefits of microalloyed steels can thus be summarized in terms of more efficient designs, reduced production or fabrication costs and/or improved performance [83].

B. Interstitial-Free Steels

The traditional cold-forming sheet steel was mild steel, at the lower end of the carbon concentration range to provide the highest ductility. Developments have paralleled those in microalloyed steels in terms of reducing the carbon content, and also in this case, the nitrogen content, both of which impair formability, whilst maintaining the potential for strengthening and compatibility with modern steelmaking, processing, and finishing and fabrication routes. Microalloying has played a significant role in these developments, but the function of the microalloying addition is fundamentally different from that described above. Essentially, it is to combine strongly with the carbon and/or nitrogen that would otherwise remain in solid solution and reduce formability. Niobium and titanium have been used in this way since the 1930s, to scavenge interstitial carbon and nitrogen in stainless steels to prevent their combination with chromium that could lead to weld decay [84], a procedure which has become known as "stabilization". They were similarly used to stabilize interstitial-free low-carbon steels soon after [85,86], but this approach really took off in the 1980s (e.g. Refs. [87,88]) when modern steelmaking allowed the production of steels with ultra-low-carbon content $<0.005\%$, virtually cementite-free and the microstructure, therefore, almost 100% ferritic. These steels display good deep drawing behavior and are suitable for continuous annealing and hot-dip galvanizing processes. Formability can still be limited by the carbon and/or nitrogen in solid solution and so additions of niobium and titanium were made to scavenge that remaining by forming stable NbC or TiN in ferrite, although more recent studies have suggested that sulfur can play a role via the formation of TiS and $Ti_4C_2S_2$ [89,90], thus enabling Ti also to scavenge carbon. These ultra-low-carbon (ULC)* steels have become known as *interstitial-free steels,* or *IF steels,* and are widely used in the more severe cold-press forming operations. Again, a number of recent conferences have been devoted to these significant developments [92–98].

Without the solid-solution strengthening effect of interstitial carbon, such steels are relatively soft (yield strength $<$200 MPa). Substitutional solid-solution hardening can be used to recover some of the strength—besides small additions of Mn and Si, phosphorus can be added, which is a potent substitutional solid-solution strengthener of iron (10 MPa per 0.01% P), giving rise to the so-called *rephosphorized steel* grades of autobody strip with yield strengths around 220–260 MPa. A related group of autobody steels is understabilized, with the free carbon and/or nitrogen level controlled to about 0.001%. During paint baking of the cold-formed component at around 170°C, the steel strain ages, i.e., the carbon and/or nitrogen, diffuses to the dislocations and pins them. This has the effect of

*ULC steels without interstitial-scavenging microalloying additions have also been finding new applications. An example is damping members in earthquake-proof constructions where their low yield strength but high elongation allows safe plastic deformation to provide energy absorption [91].

increasing the strength by some 50 MPa and thereby increasing the dent resistance. These are known, appropriately, as *bake-hardening steels* and have yield strengths around 250 MPa [99].

III. FUNDAMENTAL PRINCIPLES

A. Microalloy Carbide and Nitride Solubility

Microalloy precipitation in austenite can restrict grain growth and in ferrite can provide strengthening. In order to make efficient use of the microalloying elements, it is necessary to match their concentration with the processing conditions to control the dissolution and precipitation reactions of the carbides and/or nitrides. This requires an understanding of the solubilities of the microalloy carbides and nitrides.

If concentrations of microalloying element [M], and interstitial element [X] are dissolved in austenite, then for the reaction $M + X \leftrightarrow MX$, where X might be C or N, and MX a carbide or nitride phase, the equilibrium constant, k, may be written as

$$k = a_{[M]}a_{[X]}/a_{[MX]} \tag{1}$$

where $a_{[M]}$ is the activity of the dissolved microalloying element, $a_{[X]}$ is the activity of the dissolved interstitial element and $a_{[MX]}$ is the activity of the carbide or nitride phase formed.

By assuming activities equal to the concentrations of the dissolved elements, and equating the activity of the phase formed to unity, k can be written as

$$k = [M][X] \tag{2}$$

and is known as the *solubility product*. [M] and [X] can be expressed in weight percent.

The temperature dependence of k can be expressed by the integrated form of the Arrhenius relationship

$$\log k = A - B/T \tag{3}$$

where A and B are constants, dependent on the steel and microalloy concentration, and T is the temperature in K.

Thus, it is possible to express the solubilities of microalloy carbides and nitrides as in Fig. 1. Differences in solubility are readily apparent and determine the way in which the use of these microalloys has developed in controlling the steel microstructure in microalloyed steels. For example, titanium nitride stands out as being very stable in austenite, and thus fine coarsening-resistant titanium nitride particles have been favored as an austenite grain refiner, a practice which has become known as "titanium nitride technology" (e.g. Refs. [100,101]). Vanadium carbide, on the other hand, is very soluble in austenite, two to three orders of magnitude more so than the other carbides, and thus can be more easily taken into solution, to precipitate at a later stage of processing to give precipitation strengthening of ferrite. This behavior is used directly in normalized steels and in high-carbon steels.

More recent studies have been making use of advanced high-resolution microanalytical techniques to emphasize the role of microalloy carbo-nitrides, e.g. $Nb(C_xN_{1-x})$, or complex carbo-nitrides, e.g. $Nb_yV_{1-y}(C_xN_{1-x})$, [102,103] and non-equilibrium precipitation. In the latter case, a variation in composition within individual carbo-nitride particles (coring), and between particles in the same steel, is evidence of different non-equilibrium

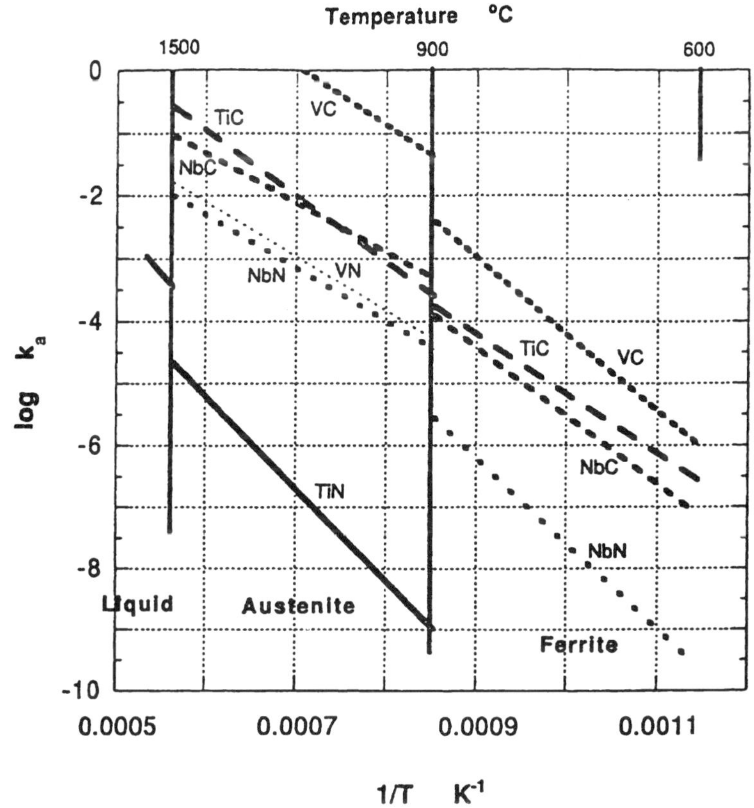

Figure 1 The solubility products of various microalloy carbides and nitrides. (From Ref. 59.)

chemistries in particles and matrix during precipitation. This results from their formation at different stages of the steel's processing history and relates to the instantaneous conditions prevailing, e.g. strain-induced precipitation in hot-worked austenite, interphase precipitation (see next section) and precipitation in ferrite (e.g. Refs. [104–110]). Thus, a spectrum of carbo-nitride particles could be expected to form in terms of chemistry, and size, distribution and location in the final microstructure.

B. Microstructural Characteristics

1. Microalloy Precipitation in Austenite

The dissolution and precipitation of microalloy carbides and nitrides in austenite is crucial to the development of microalloyed steels. Solubility data can be used to control dissolution and/or subsequent supersaturation and precipitation, and hence austenite grain refinement and precipitation strengthening. Precipitation kinetics can be represented by the customary C-curve for diffusionally controlled growth as indicated by Fig. 2. The kinetics of precipitation in undeformed austenite can be relatively slow, and thus when taken into solution the austenite composition could remain fairly constant during subsequent cooling; on the other hand, hot working can dramatically accelerate precipitation kinetics, reducing precipitation times dependent on the degree and rate of reduction [111–112], as illustrated in Fig. 2, by strain-induced precipitation on dislocations and

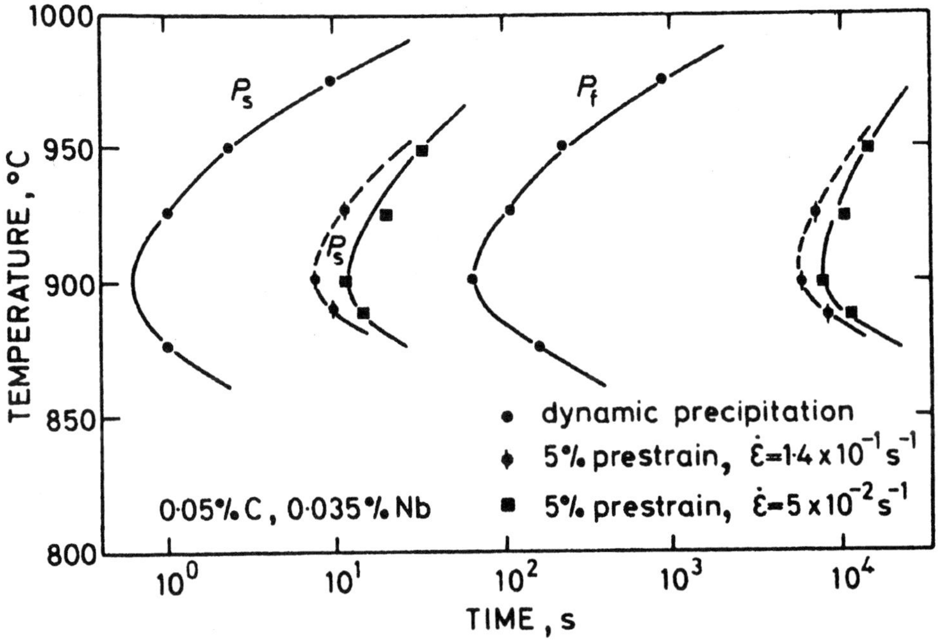

Figure 2 C-curve precipitation kinetics for niobium carbo-nitride in austenite. (From Ref. 111.)

sub-boundaries. Prestrain will increase the static precipitation kinetics, and dynamic precipitation during hot deformation will be accelerated even further [111].

The carbo-nitrides of interest, based upon titanium, niobium and vanadium, are face-centered cubic (fcc) and consequently have a simple cube–cube orientation relationship with the parent face-centered cubic austenite.

2. Microalloy Precipitation During the Austenite–Ferrite Transformation

Metallographic observations of microalloy carbides in ferrite frequently show them to have a morphology similar to that familiar in tempered secondary hardening steels, but to be dispersed non-randomly in rows, or sheets (Fig. 3). These rows have been shown to be parallel to the austenite/ferrite transformation interface, and delineate former positions of this interface. Thus, it is generally accepted that this characteristic dispersion of the microalloy carbides results from their precipitation at the transformation interface during austenite decomposition, a process that has been termed "interphase precipitation", and which has been of importance in understanding microstructure formation in microalloyed steels [54–56,113–127]. Further identifying characteristics resulting from this form of precipitation are that the precipitates have an orientation relationship with respect to the ferrite, and within one ferrite grain usually exhibit only a single variant of this relationship [56,127]. The Baker–Nutting orientation relationship [128] between fcc carbide and bcc ferrite is characteristically found:

$$\{1\ 0\ 0\}_{\text{carbide}}//\{1\ 0\ 0\}_{\text{ferrite}}$$
$$\langle 1\ 0\ 0\rangle_{\text{carbide}}//\langle 1\ 1\ 0\rangle_{\text{ferrite}}$$

Figure 3 Carbo-nitride precipitate dispersion characteristic of interphase precipitation observed in pro-eutectoid ferrite in microalloyed steels. (From Ref. 75.)

The row spacing is a function of transformation temperature, decreasing as the temperature is reduced, or cooling rate increased, which refinement can be used to increase yield strength [129–131]. The degree of refinement is subject to composition, e.g. Mn content, which can influence transformation temperature [132].

A further morphology is occasionally observed at concentrations of microalloying element generally higher than typically made to constructional microalloyed steels, and is unique to this form of microalloy precipitation at the austenite/ferrite interface. At high transformation temperatures and/or in steels with high hardenability, the microalloy carbides can appear as fibers (Fig. 4) [55–113]. Some early observations of this morphology were made in normalized regions of the microstructure of temperature-resistant power plant steels containing around 0.25 wt% V and also other strong carbide-forming elements not typically added to HSLA steels such as Mo and Cr [133].

Interphase precipitation to give microalloy precipitates dispersed in rows also occurs within the pearlitic ferrite of higher-carbon engineering steels microalloyed usually with vanadium because of its solubility behavior (Fig. 5). These precipitate dispersions in pearlitic ferrite display all of the characteristics cited above for that in proeutectoid ferrite [134], and can make a contribution to yield strength. There is also some evidence to suggest that the fibrous morphology could also form within the pearlitic ferrite at higher microalloy concentrations [135]. A more important additional microstructural observation, however, is that a precipitate-free-zone adjacent to the pearlitic cementite lamellae occurs [134–136], which is due to partitioning of vanadium to the cementite which is the more stable carbide at low vanadium concentrations [105]. This means that a minimum concentration of vanadium, $\geq$0.05 wt% approximately, needs to be added in order to provide the excess in solution to produce a strengthening precipitate dispersion in the pearlitic ferrite.

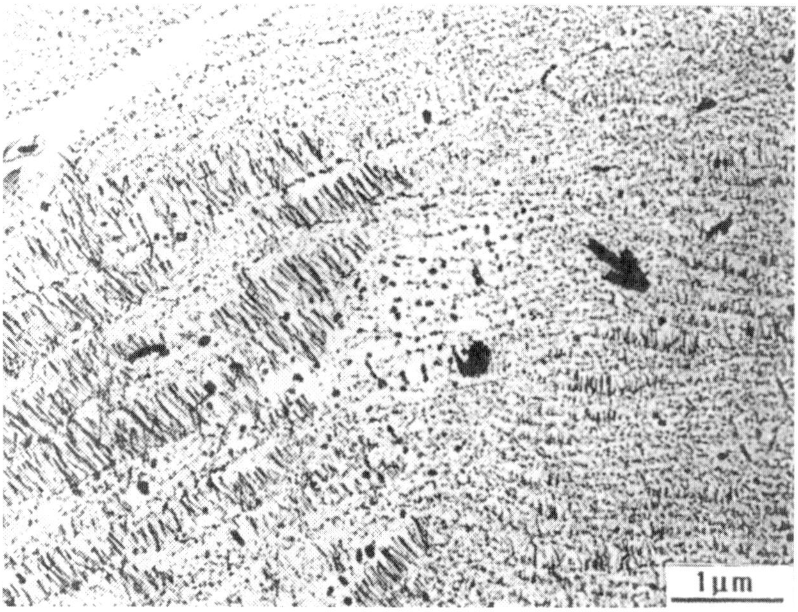

Figure 4 Fibrous morphology of microalloy precipitates in ferrite. (From Ref. 113.)

3. Microalloy Precipitation in Ferrite

If the microalloyed steel is cooled fast enough to suppress interphase precipitation and/or retain some supersaturation in ferrite, then further precipitation in ferrite can occur. This may take place during further cooling of the freshly formed ferrite over the higher temperature range required to allow substitutional diffusion [137–139], analogous to "auto-tempering", or during subsequent aging, analogous to "secondary hardening" [52–140]. Metallographic evidence comprises fine precipitates frequently observed between the rows of interphase precipitation or located upon dislocations and/or ferrite sub-grain boundaries. High-resolution microanalytical examination of these different forms of precipitation confirms dissimilar chemistry consistent with formation at different periods as discussed above.

4. Particle Coarsening

The coarsening of microalloy carbides and nitrides will be important to grain refinement in austenite and strengthening in ferrite. It has been suggested [105] that in the case of a simple carbo-nitride particle, it would not be unreasonable to treat coarsening according to Ostwald ripening by bulk diffusion described most simply by the Wagner equation in the form

$$r^3 - r_o^3 = 8\gamma\Omega SDt/9kT \tag{4}$$

where r is the mean particle radius after time t at temperature T, r_o the initial mean particle radius, S the equilibrium solubility, D the controlling diffusion coefficient, γ the interfacial energy, Ω the molar volume and k is Boltzmann's constant. The equilibrium microalloy concentration can be used for S and the substitutional diffusion coefficient of the microalloying element for D, which will be similar to iron and less than the interstitial diffusion

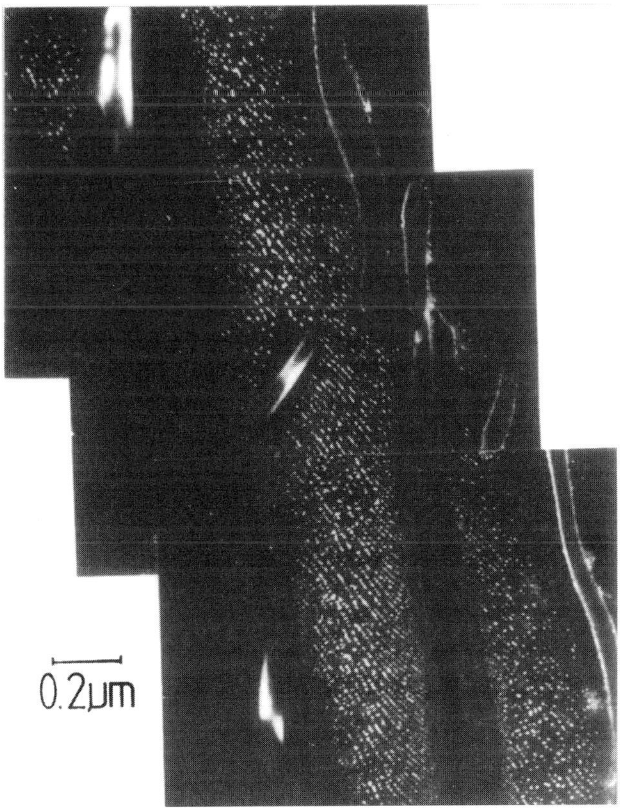

Figure 5 Interphase precipitation in pearlitic ferrite. (From Ref. 71.)

coefficients for carbon and nitrogen. This means that microalloying elements of low diffusivity which give carbo-nitrides with low solubility should have lower coarsening rates. For complex carbo-nitrides with non-equilibrium compositions, the coarsening behavior is less well understood and rates more difficult to predict [105].

5. Acicular Ferrite Microstructures

It is possible that microalloying has a part to play in the future development of tough HSLA steels that have an acicular microstructure. The term "acicular ferrite" is used to describe a particular microstructural morphology (Fig. 6), first recognized in steel weld metals, comprising fine interwoven ferrite laths or plates, and showing improved toughness over coarser transformation products, especially conventional bainite [142–143]. Its formation requires intragranular nucleation, and in weld metals, this has been associated with weld metal oxide inclusions. Attempts have been made to reproduce this behavior in plate steels, often suitably inoculated with oxygen, and titanium additions [144–146] have met with some success due to the apparent effectiveness of titanium oxides in nucleating intragranular ferrite (titanium oxide is the stable oxide if the steel is aluminum-free). Vanadium additions have also been reported to encourage the acicular ferrite microstructure [147–148], and explanations for this behavior are that the inclusions are "activated" by the precipitation of vanadium nitride upon them [149,150], which is then more effective

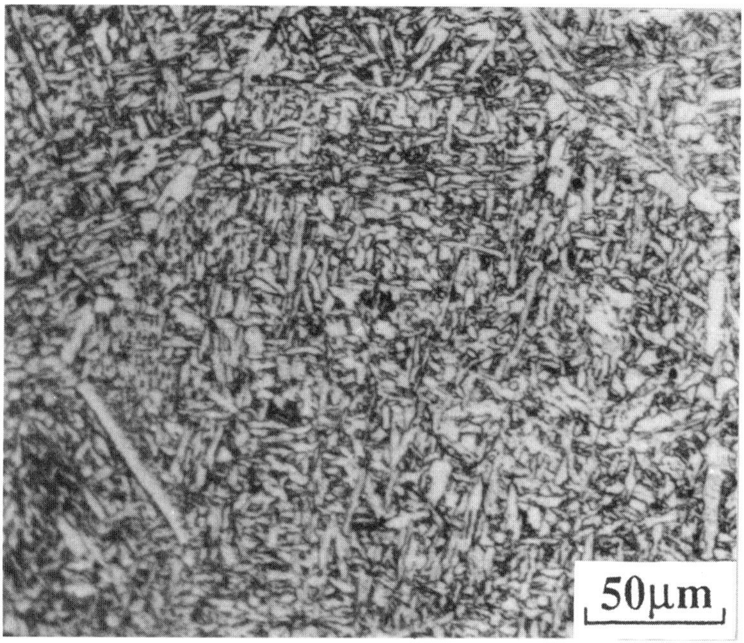

Figure 6 Characteristic acicular ferrite microstructure. (From Ref. 141.)

in nucleating ferrite, or that vanadium segregation enhances the effectiveness of suitable intragranular structural embryos, associated with inclusions or, perhaps, autocatalytic nucleation sites [141]. Exploitation of the acicular ferrite microstructure might be expected to develop where rolling to develop grain refinement is limited, or where uniform grain refinement is more difficult, e.g. in seamless tube rolling or in forging.

6. Ultra-Low-Carbon Ferrite Microstructures

A most significant development alluded to already has been the emergence over the last two decades of the availability of commercial steels with ultra-low-carbon content (<0.05 wt%) due to improvements in steelmaking practice. Moreover, if a microalloy addition is combined with the carbon, then the "free" carbon remaining in solution, which will determine the final transformation microstructure, is further reduced. This has had important consequences to the microstructure of both microalloyed constructional steels and interstitial-free steels. At such low-carbon levels, the microstructures developed on cooling from the austenite solution treatment temperatures are very different from those more familiar in steels with higher-carbon concentrations. Not only is cementite formation restricted, or virtually eliminated, but more importantly, the as-formed ferrite can adopt a variety of morphologies. These can range from recognizable equiaxed ferrite at slow cooling rates to recognizable martensite at high cooling rates, but in between, a number of less-easily recognized intermediate ferrite structures, referred to as Zwischenstufe or Z_w ferrite [151], can develop. Figure 7 illustrates some of these forms and their various descriptions [152]: polygonal (equiaxed), quasi-polygonal (irregular), "granular" ferrite or bainite, "acicular" bainite, lath martensite. A number of classificatory systems have been suggested to aid recognition of these various forms [153–158]. Although a measure of agreement is apparent in terms of identification, different nomenclatures have been

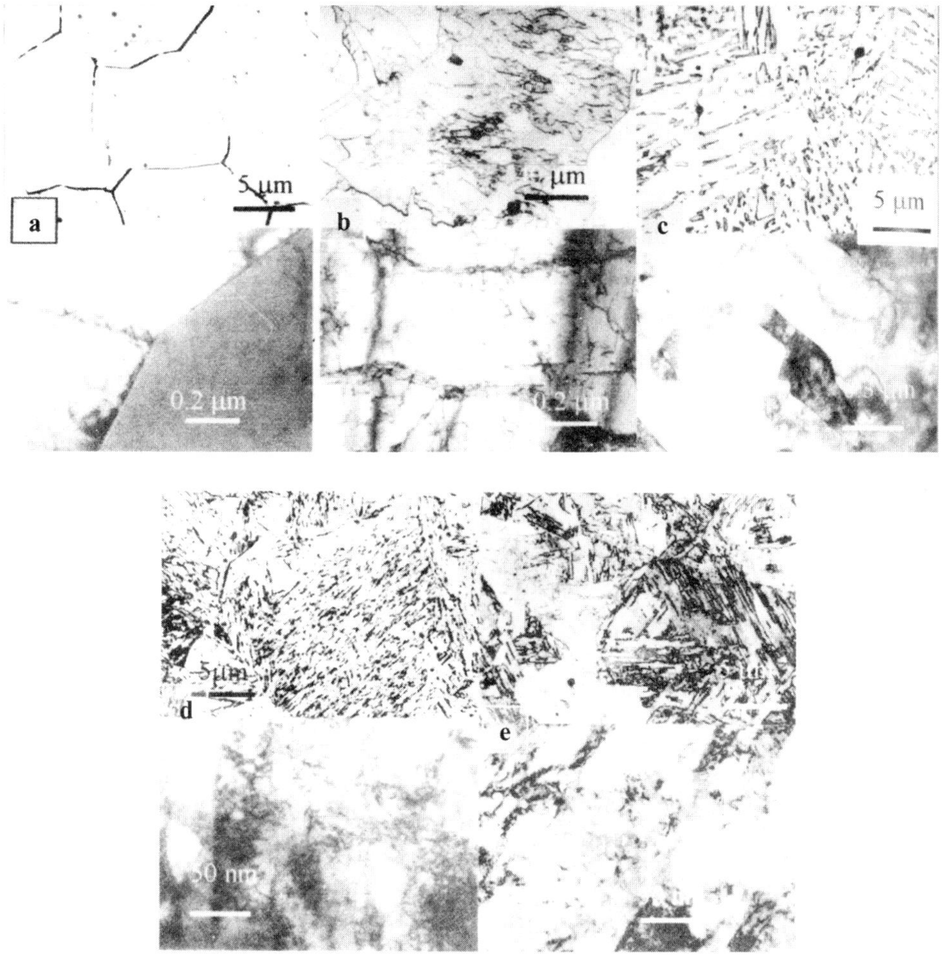

Figure 7 Ultra-low-carbon ferrite microstructures (comparative light and electron micrographs): (a) equiaxed ferrite, (b) quasi-polygonal ferrite, (c) granular bainitic ferrite, (d) acicular bainitic ferrite, and (e) lath martensite. (From Ref. 152.)

used: Table 1 suggests a simple list of the basic microstructural forms which may be related to the classical ferrite microstructures more familiar at higher-carbon content [164,165], but which highlights the main variations at ultra-low-carbon content that can then be related to the various more detailed descriptions in the literature.

C. Mechanical Properties

1. Strengthening Mechanisms

A number of quantitative relationships have been determined by statistical multiple regression analysis of experimental data to relate microstructural features to mechanical properties, especially in the case of yield strength, and have been collectively reviewed [166]. In general, most work has been carried out on ferrite–pearlite structures in carbon steels, but nonetheless, it is thought that this provides a basis for examining the microstructural

Table 1 A Simple Terminology and Nomenclature for the Ferritic Microstructures Observed in Ultra-Low-Carbon Steels (From Refs. 83 and 87.)

Microstructure	Description LM observation by light microscopy, TEM observation by transmission electron microscopy	Symbol (From Ref. 87.)
Polygonal (equiaxed) ferrite	LM: equiaxed grains, smooth boundaries, TEM: low dislocation concentration	α_p
Quasi-polygonal (massive) ferrite	LM: large irregular grains, sometimes elongated, irregular boundaries, intragranular veins or sub-boundaries, TEM: increased dislocation concentration but largely polygonized into sub-boundaries	α_q
Granular (bainitic) ferrite	LM: regions of coarse elongated structure, often only apparent because of aligned retained austenite (or other second phase), TEM: parallel lath structure, islands of retained austenite (possibly plus carbide andor martensite—the so-called MA constituent), high dislocation concentration	α_B
(Acicular) bainitic ferrite	LM: coarse *parallel* acicular grains, sheaves or packets, TEM: parallel lath structure, high dislocation concentration	$\alpha^{\circ}{}_B$
Lath (massive) martensite	LM: refined acicular lathsplatespackets, TEM: parallel lath structure, very high dislocation concentration	$\alpha'{}_m$

Notes: (i) Widmanstatten ferrite (α_W) has been excluded because there is little evidence, as yet, for its appearance, or its clear identification, at ultra-low-carbon content [159] It is possible that it may form but then suffer an immediate recovery or effective partial recrystallization [160,161]. (ii) The term quasi-polygonal ferrite is preferred to describe the irregular ferrite that is characteristic of ultra-low-carbon ferrite microstructures rather than massive ferrite which implies a formation mechanism. (iii) Granular ferrite, another form characteristic of ultra-low-carbon content, is often clearly similar to a low-carbon bainite when observed by transmission electron microscopy and so the term granular *bainitic* ferrite is probably more informative. (iv) The term *acicular* bainitic ferrite is used to differentiate this classical form of low-carbon bainite from the granular form. It is not to be confused with the term "acicular ferrite" used to describe the fine interlocking ferrite microstructures characteristic of intragranular inclusion-nucleated ferrite in steel weld metals. (v) There are no clear microstructural features to distinguish easily between acicular bainite and lath martensite at ultra-low-carbon content [162,163], although lath dimension has been used. The term "massive martensite" is used in early references, although lath martensite now seems to be preferred, but this serves to illustrate the long-standing nomenclatural difficulties.

effects in microalloyed steels [105,167]. The results may be illustrated by expanding the Hall–Petch equation [168,169], expressing the relationship between yield strength and grain size, for additional parameters.

According to the Hall–Petch equation:

$$\text{yield strength (MPa)}, \ \sigma_y = \sigma_i + k_y d^{-1/2} \tag{5}$$

where σ_i is the inherent resistance to dislocation motion of the pure lattice (for iron ~50 MPa), k_y is the coefficient related to spreading yielding from one grain to another, sometimes referred to as the dislocation locking or strengthening coefficient (for iron ~17 MPa mm$^{1/2}$), and d is the mean linear grain diameter (mm)(e.g. Refs. [59,166]). This leads to [170]:

$$\begin{aligned}\sigma_y &= \sigma_i + k_y d^{-1/2} \text{(the initial terms of the Hall–Petch equation)}\\ &+ \Sigma(k_i c_i) \text{(the solid-solution contribution where } k_i \text{ is the strengthening}\\ &\quad \text{coefficient of the } i\text{th solute and } c_i \text{ is the concentration of the } i\text{th solute)}\\ &+ 10.8 f^{1/2} x^{-1} \ln(x/6.125 \times 10^{-4}) \text{ (the precipitation strengthening contribution}\\ &\quad \text{where, } f \text{ is the volume fraction of particles and } x \text{ is the}\\ &\quad \text{particle size (mm))}\end{aligned} \tag{6}$$

When a significant volume fraction of pearlite is present in the microstructure, a simple modified approach may be adopted. The alternating lamellar distribution of the second phase in pearlite has been treated, for a *fully* pearlitic microstructure, in a similar manner to that resulting in the Hall–Petch equation, and thus it has been proposed [63,166] that

$$\sigma_y = 178 + 3.8 S^{-1/2} \tag{7}$$

where S is the interlamellar spacing (mm). Then, for a *mixed* ferrite–pearlite microstructure, a law of mixtures adequately represents the yield stress as a function of the relative mass fractions of ferrite and pearlite [63,166]:

$$\sigma_y = f_\alpha^n \sigma_\alpha + (1 - f_\alpha^n)\sigma_{\text{pearlite}} \tag{8}$$

where f_α is the mass fraction of ferrite, and σ_α and σ_{pearlite} are the yield stresses of the ferrite and pearlite, respectively. The index, n, is approximately equal to 1/3, and allows for a non-linear variation of σ_y with pearlite content. This analysis should be relevant to medium- and higher-carbon microalloyed steels; however, at low-carbon content of microalloyed HSLA steels, pearlite has little effect on yield strength [105].

2. Fracture Toughness

A multiple regression approach has also been used to describe the Charpy V-notch impact transition temperature, ITT, and for low-carbon structural steel containing up to about 0.2 wt% C, corresponding to pearlite content up to about 25–30%, the following equation has been determined [47,59,166]:

$$\text{ITT}(^\circ\text{C}) = -19 + 44(\%\text{Si}) + 700(\%\text{ free N})^{1/2} + 2.2(\%\text{pearlite}) - 11.5\text{d}^{-1/2} \tag{9}$$

It is immediately apparent that solid-solution elements and volume fraction of pearlite raise the impact transition temperature, whereas grain refinement has a beneficial effect

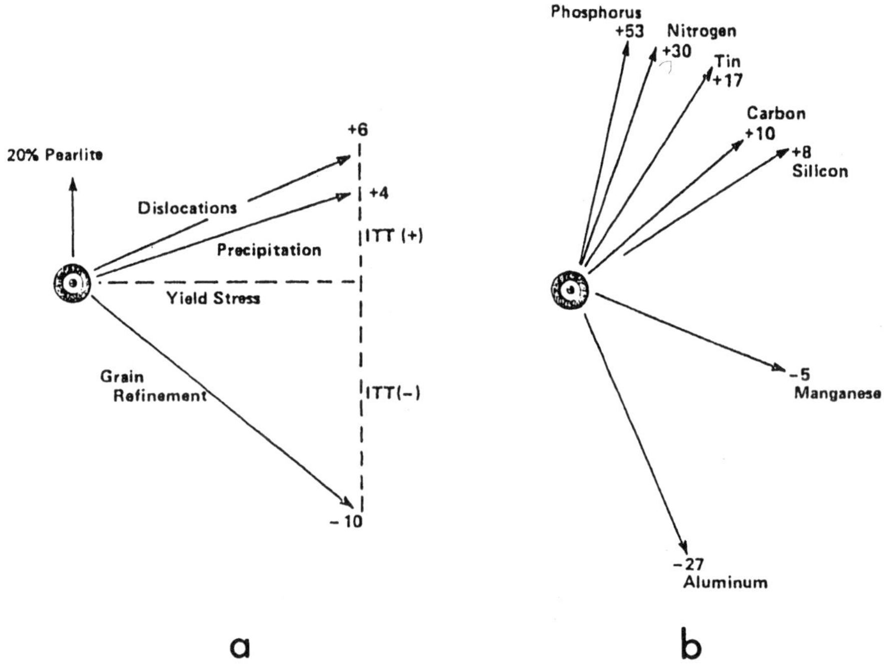

Figure 8 Diagrammatic representation of embrittlement vectors for (a) microstructural factors, and (b) solid-solution effects. (From Ref. 59.)

by lowering it. This can be elegantly demonstrated graphically through the use of "embrittlement vectors" [59], defined as the change in the impact transition temperature per unit increase in yield strength. Figure 8 illustrates this for various solid-solution elements, pearlite fraction and grain size, and also for the effects of dislocation strengthening and precipitation strengthening. This approach is useful to the design of a steel for optimum strength and impact resistance. However, an alloying element addition may have multiple effects, not just its effect on solid-solution strengthening. For example, it may change the transformation temperature and hence influence other key variables such as pearlite volume fraction, or combine with free nitrogen, not just remove the solid-solution effect of the nitrogen but perhaps form a grain-refining nitride precipitate dispersion. Thus, in considering the effect of an individual element, it is necessary to calculate the "resultant embrittlement vector" [59], taking into account its indirect effect through other variables, to determine the true picture. Additionally, another microstructural parameter important to toughness, the thickness of grain boundary carbides, t_{gb}, has sometimes been incorporated into the regression analysis [171], to give an additional term in the regression equation for ITT of the form: $+k t_{gb}^{1/2}$, where k is a constant.

3. Formability

Thin steel sheet suitable for cold pressing requires good formability and ductility. The limits to press forming are determined by the extent of uniform elongation, or resistance to plastic instability, and the ductile fracture behavior (e.g. Refs. [172,173]). Uniform elongation increases with increase in work hardening rate. Composition, grain size, and pearlite content of carbon steels can affect the flow stress, σ_f, and the work hardening rate, $d\sigma/d\varepsilon$, also dependent upon strain, ε, and this has been treated by regression analysis and reduced to relationships [166,174] such as, for example:

$$\text{At } \varepsilon = 0.2,\ \sigma_f\ (\text{MPa}) = 246 + 4.2\ (\text{mass\% pearlite}) + 15.0d^{-1/2} \quad (10)$$

$$d\sigma/d\varepsilon\ (\text{MPa}) = 385 + 1.4\ (\text{mass\% pearlite}) + 15.4d^{-1/2} \quad (11)$$

$$\text{At } \varepsilon = 0.8,\ \sigma_f\ (\text{MPa}) = 385 + 4.8\ (\text{mass\% pearlite}) + 27.7d^{-1/2} \quad (12)$$

$$d\sigma/d\varepsilon\ (\text{MPa}) = 161 + 23.1d^{-1/2} \quad (13)$$

It is noticeable that the grain size effect on σ_f increases at increased strain, and that the presence of pearlite increases the work hardening rate at low strains but has a lesser or negligible effect at high strains.

The uniform elongation, expressed as the true uniform strain, ε_u, is equivalent to the work hardening exponent, n, found if the flow curve is approximated to a simple power law relation such as the Ludwig equation:

$$\sigma_f = \sigma_y + k\varepsilon^n \quad (14)$$

thus

$$\varepsilon_u = n \quad (15)$$

where

$$d\sigma/d\varepsilon = n(\sigma/\varepsilon) \quad (16)$$

A steel with a high n value is better at distributing the strain during pressing, resulting in greater uniform strain and improved formability. However, the n value tends to decrease as the strength increases [175].

The following expression has been found for the uniform strain [174], ε_u:

$$\varepsilon_u = 0.27 - 0.016\ (\text{mass\% pearlite}) \quad (17)$$

which showed it to be independent of grain size, whereas it has been shown that the n value decreases with decreasing grain size [176], according to

$$n = 0.5(10 + d^{-1/2})^{-1} \quad (18)$$

The total strain at fracture, ε_T, is important if the forming operation is fracture limited and is given by [174]:

$$\varepsilon_T = 1.3 - 0.02\ (\text{mass\% pearlite}) + 0.015d^{-1/2} \quad (19)$$

In modern cold-forming steels the carbon content has been reduced to a level where pearlite is virtually eliminated from the microstructure and therefore the corresponding pearlite terms in these equations will be reduced to 0, and it is expected that the behavior will be controlled by the solid-solution elements and grain size.

The mechanisms of ductile fracture will be determined by the nucleation and growth of voids at high strains, which are dependent, in the absence of pearlite, upon the presence and type of any carbides and inclusions in the steel. A considerable amount of knowledge has been accumulated on the control of sulfide inclusions and desulfurization, and on deoxidation practices to produce cleaner steels, and modern steels developed for high formability have carbon content low enough to reduce the adverse effects of carbides on ductility.

An effective way of further improving formability in clean low-carbon steels is through control of crystallographic texture such that plastic anisotropy is developed in the sheet, resisting through-thickness thinning, and enhancing the uniform strain achievable [177] in a deep drawing operation. An increase of the {1 1 1} textural component in the plane of the sheet, at the expense of the cubic {1 0 0} component, is required (e.g. Ref. [178]). This is produced by a cold rolling and annealing process, which allows the desired preferred orientation to develop during recovery and recrystallization. The plastic anisotropy of cold-rolled and recrystallized low-carbon steel sheet can be simply represented by the mean ratio, $\bar{r}$, of width strain to thickness strain, determined by a simple tensile test [179]. $\bar{r}$ has also been shown by regression analysis to be a function of grain size, increasing with grain size, according to the expression [166,180]:

$$\bar{r} = 4.5 + 1.99 \log d \tag{20}$$

In accordance with the above, it has also been shown [166,181] that

$$\bar{r} = 0.8 + 0.6 \log\{\mathbf{I}_{(1\,1\,1)}/\mathbf{I}_{(1\,0\,0)}\} \tag{21}$$

where the **I** values represent the intensities of the planar textures.

The initial state of the steel prior to cold rolling is important. Low-carbon steels grain refined and strengthened by microalloy additions are, in general, not well suited to this routine. The carbo-nitride dispersions retard the recrystallization and grain growth process during annealing, and inhibit the development of preferred crystallographic orientations [59,182]. In general, precipitation strengthening increases the flow stress far more than the work hardening rate, thereby effectively reducing the maximum achievable uniform strain. Developments in high-strength formable steels have thus relied on the use of solid-solution strengtheners, particularly silicon and phosphorus, which give a relatively large increase in work hardening rate compared with the increase in flow stress.

Microalloy additions have, however, found the important alternative use in cold rolled and annealed formable steels as "getters" of interstitial carbon and nitrogen. The equilibrium solubility of carbon in ferrite at low temperatures is very low, of the order of 10^{-4} wt% at room temperature (e.g. Ref. [183]), and it has additionally been pointed out that the carbon level in ferrite, even in well-annealed steels, is likely to be a few orders of magnitude higher [59,184]. Added to the interstitial nitrogen in solid solution in the ferrite, it is likely that this concentration will be sufficient to give significant interstitial atom decoration of the dislocation population [183]. This will give dislocation pinning which results in discontinuous yielding behavior and the formation of Luders bands. In lightly deformed areas of cold-pressings, the Luders bands cause defects, termed "stretcher-strain" markings, on the surface. The problem can be overcome by subjecting the strip to a light rolling deformation of about 2% reduction that essentially deforms the surface layers, i.e., introduces mobile dislocations, such that in subsequent pressing, a more continuous yield is experienced and the surface stretcher-strain markings do not occur. However, in the presence of free carbon and nitrogen which can segregate to the dislocations again over a period of time, the discontinuous yield point can return, known as "strain aging". Moreover, the yield and tensile strengths may also increase, "strain age-hardening", and may be accompanied by an increase in the impact transition temperature, "strain age embrittlement" [185]. At free carbon levels above that required to give dislocation saturation, it has been suggested that carbide precipitation may also occur on the dislocations, providing strong locking and dislocation obstacles. The return of the yield point depends upon the kinetics of strain aging, which, in turn, is dependent upon the concentration of carbon and nitrogen in solid solution. This gives rise to the concept of a shelf-life for the steel, after which it will be unsuitable for press forming. Modern low-aging forming steels are thus microalloyed with strong carbide-formers that mop-up any free carbon and nitrogen that would otherwise be left in solid solution to cause strain aging, thereby increasing the shelf-life.

IV. THERMAL AND THERMOMECHANICAL TREATMENTS

A. Normalizing

Normalizing treatment, involving soaking at a constant temperature, generally aimed at around $Ac_3 + 30°C$, and air-cooling (e.g. Ref. [186]), provides a more consistent product than as-rolled. Microalloying can be employed during normalizing treatments to control the austenite grain size prior to its decomposition to ferrite during cooling. This is important because it results in a finer ferrite grain size. Austenite grain growth at the normalizing temperature can be inhibited by microalloy carbo-nitrides which act to pin the migrating austenite grain boundaries [187–189]. The concentration of the microalloy must be matched to the normalizing temperature through solubility data such that microalloy carbo-nitride remains undissolved, and resistant to coarsening, such that a sufficiently fine dispersion exists to pin and/or restrict austenite grain boundary mobility, and hence grain coarsening, during re-austenitization. Microalloying can be used in this way to produce a simple grain-refined steel; aluminum additions to produce aluminum nitride, or niobium to produce niobium carbo-nitride, are effective.

Excess microalloying addition taken into solution can reprecipitate during austenite decomposition upon cooling, mainly via interphase precipitation, to give a useful strengthening increment, if higher strength is required. In this instance, vanadium addition is commonly employed because of the higher solubility of vanadium carbo-nitride at the low austenitizing temperatures used that is required to achieve sufficient dissolved microalloy for subsequent interphase precipitation. Usually, the nitrogen concentration of the steel is enhanced to ensure that sufficient vanadium carbo-nitride remains undissolved at the normalizing temperature to inhibit grain growth. Alternatively, a combination of niobium and vanadium can be used.

B. Hot Working

An alternative to normalizing is control of the hot working process, be it rolling or forging, which can also give product consistency or improved properties. In the early days, most progress was made in developing a controlled rolling practice for plate steels, i.e. thickness down to about 15 mm. Essentially, this involved introducing delays into the rolling schedule so that significant rolling occurred at lower temperatures. The temperature and deformation during rolling were thus controlled to give the desired microstructure and mechanical properties.

The importance of microalloying is the extent to which it can be used to control evolution of the austenite grain structure by recovery, recrystallization and grain growth during the working operation. It is usual to carry out the deformation in several stages, and so several cycles of recovery and recrystallization, static or dynamic, may occur, and at successively lower temperatures. For example, during a multipass rolling schedule, the

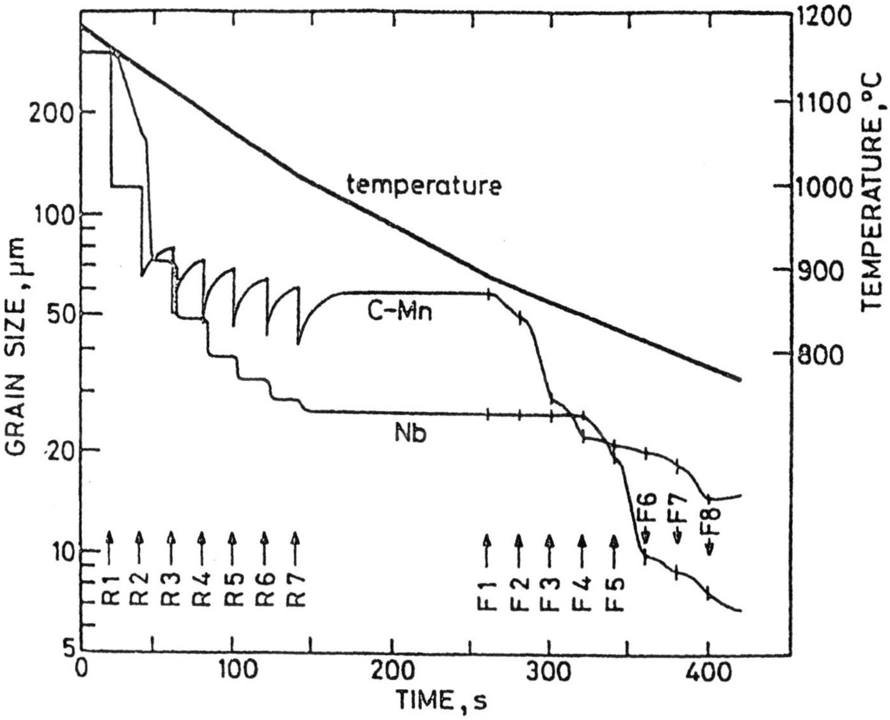

Figure 9 The calculated effect of hot-rolling schedule on grain size. (From Ref. 193.)

kinetics of the recrystallization and subsequent grain coarsening which can occur during the interpass time will be important. Microalloy addition in solution, undissolved carbo-nitride and strain-induced carbo-nitride precipitation can all have an effect on these microstructural processes.

The microalloy can have a number of effects during austenite deformation: Undissolved microalloy will influence the starting grain size at the commencement of hot working, in a manner identical to that described for normalized steels, although higher concentrations may be necessary due to the higher temperatures used for soaking prior to hot working. In addition, static or dynamic recovery and recrystallization processes will take place depending upon the exact working schedule, and microalloying additions can be used to control these. Dissolved microalloy will influence recovery and recrystallization, and strain-induced precipitation can inhibit recrystallization by pinning sub-grain boundaries (e.g. Refs. [190–192]). Grain coarsening can similarly be retarded by grain boundary pinning as for normalized steels. Niobium addition has the largest effect in retarding recrystallization and strain-induced precipitation of niobium carbo-nitride occurs in the common finish rolling temperature range 1000–800°C. Figure 9 shows the predicted effect of niobium microalloying in controlling grain size during controlled rolling as compared with a C–Mn steel, and is in good agreement with commercial rolling practice. Grain growth between deformation passes is suppressed in the niobium steel.

The evolution of the austenite structure depends upon whether working finishes above or below the effective recrystallization temperature. This temperature is best defined as that below which recrystallization becomes so sluggish as to be negligible. Figure 10 illustrates the effect of various microalloying elements on this recrystallization

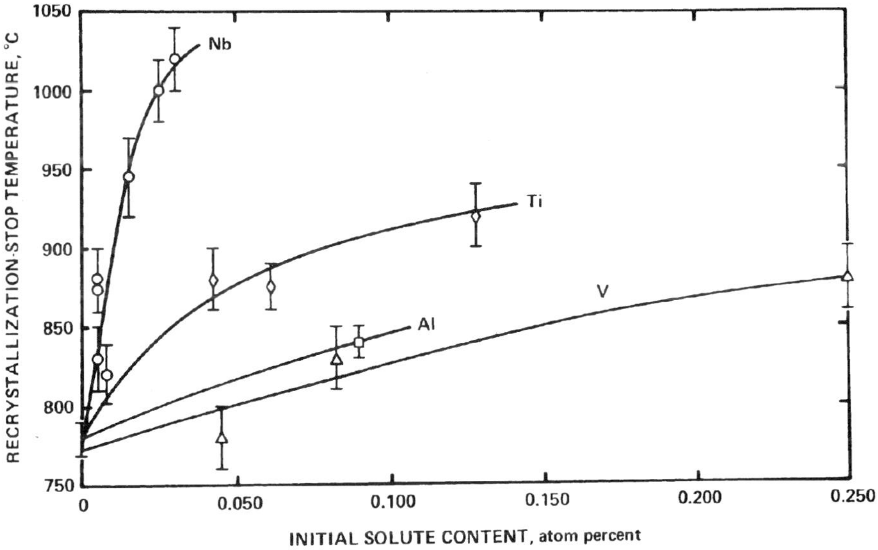

Figure 10 The effect of microalloying elements on austenite recrystallization temperature. (From Ref. 194.)

temperature. In the case of controlled rolling of microalloyed steels, grain refinement of the austenite structure by microalloys above the recrystallization temperature has come to be known as "recrystallization controlled rolling" (RCR), or stage 1 rolling, the behavior depicted by Fig. 9. If rolling continues beneath the effective recrystallization temperature, the process has been referred to as "deformation controlled rolling", or stage 2 rolling, and the austenite grain refinement is effectively mechanical. The complete inhibition of recrystallization and grain growth means that the austenite grain thickness, or cross-section, is significantly reduced normal to the major working direction. The large increase in grain boundary area per grain volume of the elongated austenite grains, and the effects of the deformation in enhancing the heterogeneous nucleation sites at the grain boundaries, ensures that a refined grain size results in the ferrite that forms during subsequent austenite decomposition. In addition to enhanced grain boundary nucleation, intragranular nucleation is also induced on deformation bands [195,196].

It is also possible to finish rolling at temperatures after austenite decomposition has commenced, in the two-phase austenite plus ferrite region, referred to as stage 3 rolling. In this case, the ferrite undergoes deformation, and if the temperature is low enough, typically $\leq 750^\circ$C, it may recover and recrystallize to a fine-grained ferrite with a sub-grain structure and a preferred orientation [197,198]. The presence of ferrite grains that have undergone this deformation can affect the final strength and toughness. Figure 11 illustrates the effects on strengthening of low finishing temperatures in the two-phase region. However, a disadvantage of this third stage rolling is that the final microstructure can develop laminar (cleavage) separations during impact testing. These may be associated with ductile fracture, but serve to reduce the ductile shelf energy. This could be of importance if high energy absorption is required to arrest a ductile crack propagating from a defect, e.g. in linepipe applications.

A summary of a possible thermal cycle for hot-rolled plate, involving all three stages of rolling, is shown schematically in Fig. 12.

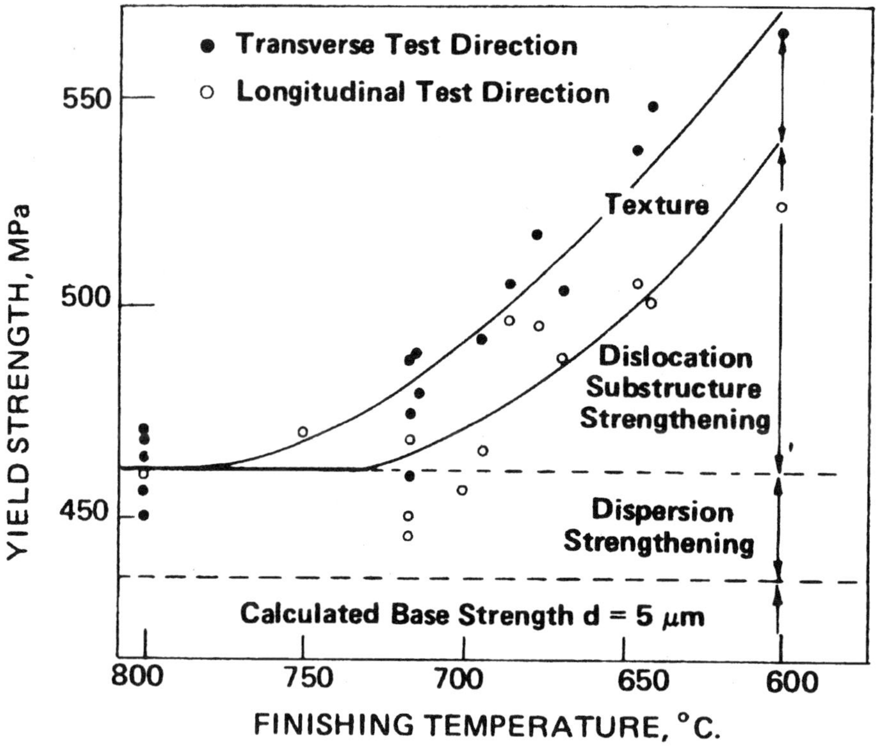

Figure 11 Contributions to strengthening by rolling into the two-phase region. (From Ref. 199.)

In the case of strip steels, hot rolled to thinner gauges than plate, the cooling cycle is different towards the final stage of the rolling schedule; the thinner strip cools rapidly as it exits the final rolling pass, but is then coiled and the thermal mass of the coil dramatically retards the subsequent cooling. The austenite decomposition occurs when the steel is in the coiled condition, which can better be approximated to isothermal heat treatment conditions. To ensure this and also to prevent the coils from sticking, the cooling of the strip can be accelerated before the coiler. The ferrite grain size and interphase precipitation are thus controlled by the coiling temperature. The effects of microalloy composition and hot-rolling and coiling variables on the yield strength of a titanium microalloyed steel are represented by a nomogram in Fig. 13. Such considerations illustrate the application of the principles discussed above to alloy design in the development of microalloyed steels.

V. MICROALLOYED MEDIUM- AND HIGH-CARBON STEELS

In the last two decades, many of the features of microalloying and controlled hot-working, combined with accelerated cooling, have been extended to higher-carbon steels with ferrite/pearlite microstructures. The main driver for this development, certainly in medium-carbon forging steels, has been the ability to achieve the required mechanical properties with less processing steps, i.e. after direct cooling from the working temperature, thereby reducing through-process costs. Thus, controlled cooling after forging medium-carbon grades, and controlled rolling and cooling of higher-carbon bar and rod stock, are now commonly

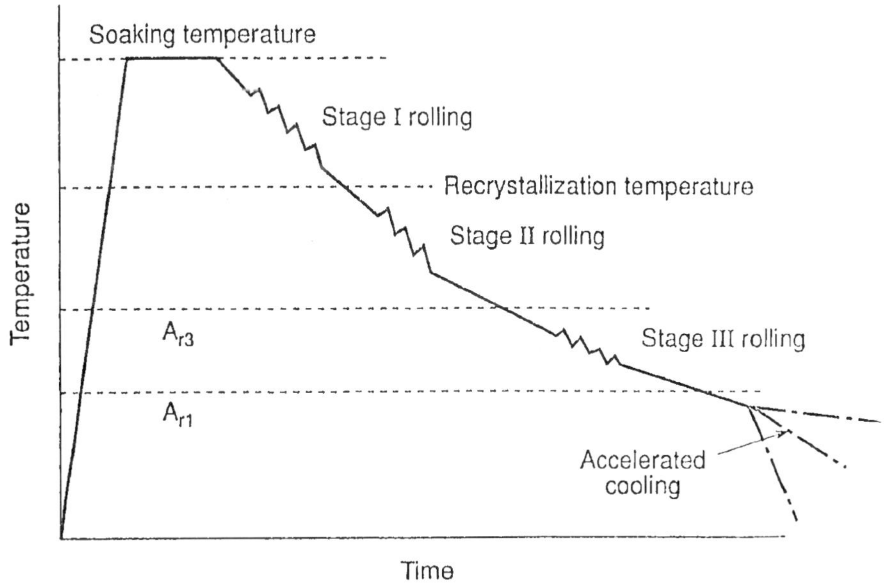

Figure 12 Schematic thermal cycle for hot rolling of plate. (From Ref. 59.)

practiced. Figure 14 illustrates the processing advantage of direct strengthening microalloyed forging steels over various alternative quench and temper heat treatment routes. The ferrite/pearlite structure can also display further advantage over heat-treated steels in terms of machinability [200].

The solubility characteristics of vanadium are the most suitable for providing the interphase precipitation strengthening. The microalloy carbide solubilities in higher-carbon austenites are shown in Table 2. In general, high soaking temperatures are used prior to hot working that can lead to grain coarsening. Figure 15 compares the effects of the various microalloys on austenite grain coarsening temperatures and highlights the importance of titanium, which in combination with controlled nitrogen content (titanium nitride technology), has been used with some success as a grain-refining addition. The practice of grain refinement, however, can be more complex in forgings or rail steels, e.g. due to the non-uniform deformations involved.

At higher carbon levels, the need to eliminate the expensive lead-patenting process of rod stock for wire drawing has led to air-patenting, but this produces a lower-strength, coarser pearlitic structure. Microalloying with vanadium can provide a compensatory strength increment. This results from interphase precipitation of microalloys in the pearlitic ferrite. An additional, but less well-known effect of a vanadium microalloying addition, is that it interferes with the pearlite reaction at the prior austenite grain boundaries during austenite decomposition; this feature can be used to increase the carbon content to hyper-eutectoid levels without producing extensive grain boundary cementite which would reduce drawability. In this way, a higher strength rod stock can be achieved which carries through to the fully drawn wire [202]. The wider exploitation of these potential microalloying benefits to commercial high-strength wire steels is anticipated.

The maximum strength that can be obtained in microalloyed steels with ferrite/pearlite microstructures is around 1000 MPa. However, with the introduction of accelerated cooling after forging and bar or rod rolling, it is possible to produce bainitic

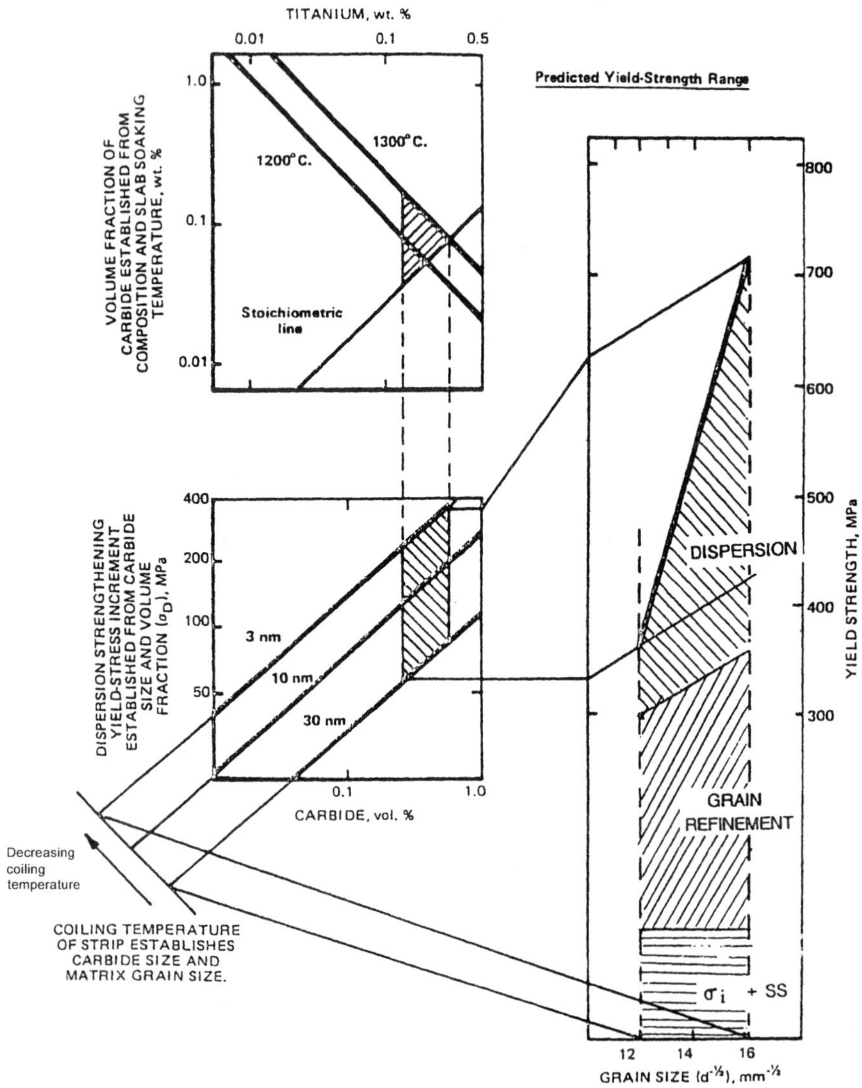

Figure 13 Nomogram predicting the effects of composition, and hot rolling and coiling conditions, on the yield strength of microalloyed steel strip. (From Refs. 59 and 199.)

Table 2 Solubility of Microalloy Carbides in High-Carbon Austenite at 1200°C (From Ref. 59.)

Carbide	Dissolved Microalloy at 0.4 wt C (in wt)	Dissolved Microalloy at 0.8 wt C (in wt)
Vanadium	>1.0	>1.0
Titanium	0.042	0.021
Niobium	0.027	0.015

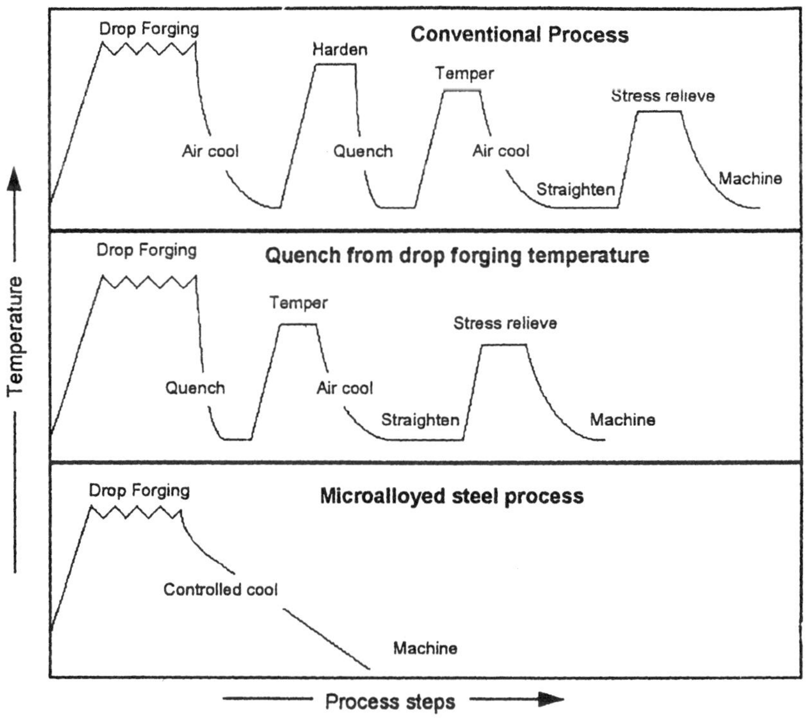

Figure 14 A comparison of the stages in alternative drop forging processes. (From Ref. 81.)

structures with a higher strength and toughness combination. Cooling to a bainitic structure gives less problems of distortion than quenching to martensite, and if no further heat treatment is required, it can also help in producing process cost savings. The microalloying addition can be retained to control austenite grain size during soaking and forging or rolling, and may also give some hardenability benefit during cooling. If the forging is subjected to further heat treatment, some precipitation strengthening could also result from secondary hardening [80,200,203–205].

VI. COLD-ROLLED AND ANNEALED STRIP STEELS

Hot-rolled strip can be produced down to a thickness of 2 mm but many cold-forming operations require thinner strip down to 0.3 mm and, additionally, better surface finish and dimensional control than can result from a hot strip mill. Consequently, thin strip for cold pressing is manufactured by cold rolling, which can produce the desired surface and dimensional control. The cold-rolled strip is then annealed to give a low strength product which is found to give the highest level of formability. If the coils are batch annealed, the differential cooling rates across the coil can lead to a variation in properties and so, in the modern mill process, a continuous anneal is preferred in which the coil is unwound and passed through an annealing furnace and re-coiled. This move away from batch annealing to continuous annealing, with attendant changes to the heat treatment cycle, has also influenced the development of cold-forming steels and is closely associated with the changes from batch annealed aluminum-killed deep drawing steels to modern continuous

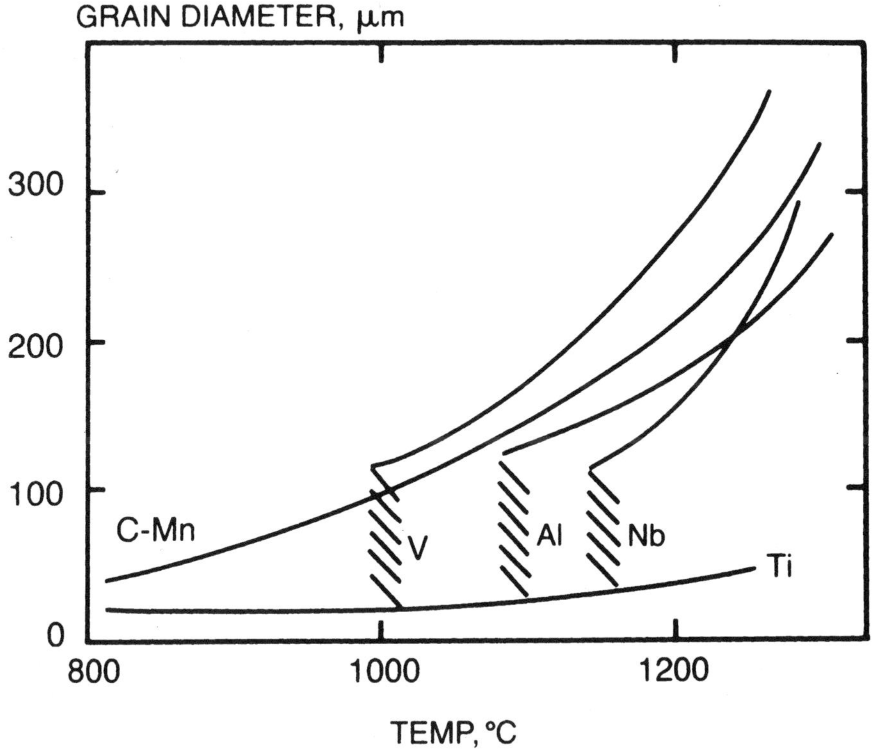

Figure 15 Comparative effect of microalloying elements on the grain coarsening temperature of austenite. (From Ref. 201.)

annealed interstitial-free steels [184,206]. Further in-line processes, such as pickling or ultrasonic cleaning, are being added to the continuous annealing line in order to effect greater economies in processing [207].

VII. INTERSTITIAL-FREE STEELS

These steels use microalloying to eliminate virtually all free carbon and nitrogen from solution in ferrite and thereby the discontinuous yield point and the formation of surface stretcher-strain markings during cold-press forming. The concentration of microalloy addition must exceed the stoichiometric ratio required to form the carbide and nitride. Current interstitial-free steels make use of modern steel refining methods to give a low starting interstitial content limited to about 0.003 wt% each of carbon and nitrogen, thus requiring the use of low levels of microalloy addition. The typical stoichiometric levels of Ti and Nb addition might thus be 0.010 and 0.020 wt%, respectively. The low levels of carbon in solution mean that these steels will transform to the irregular ferrite and ultra-low-carbon bainite morphologies mentioned above, dependent on the exact process routines, and it has also been pointed out that this can influence the final textures produced and hence the $\bar{r}$-values [208].

To increase the strength levels, solid-solution strengthening is employed. Re-phosphorized and/or re-siliconized high-strength grades are available. A typical

Table 3 Typical Composition of an Interstitial-Free Steel

Element	Concentration (wt)
C	0.003
N	0.003
Mn	0.1–0.2
Si	0.01–0.06
P	0.070
S	0.005
Ti	0.010
Nb	0.025
Al	0.035

composition profile is given in Table 3. If Mn is limited, then titanium sulfide (TiS) or carbo-sulfide ($Ti_4C_2S_2$) can form in preference to manganese sulfides [89,90,209,210]. The stabilization of the carbon could thus be more complicated than by direct carbide and nitride formation, and will influence the alloy design methodology [211]. This, coupled with the interactions between the interstitials and the microalloy additions at very low concentrations, and the segregation behavior of the interstitials, makes evolution of the microstructure complex. Indeed, there is continuing debate as to whether the precipitation of the interstitial compounds influences the development of the desired textures for enhanced formability, or whether this is simply due to the reduced interstitial concentration [212–216].

An alternative route to increasing strength is to employ the potential strain aging behavior in the interstitial-free steel. If, after cold pressing the steel in the continuous yielding condition, it is subjected to an anneal at a temperature which will allow free interstitial atoms to decorate the dislocations, then a useful strength increment is given to the formed component. This anneal can be the paint curing (or baking) process which is carried out at a temperature in the range 140–200°C. This strategy has been used to improve the dent resistance of the finished press formed sheet, and is known as "bake hardening".

VIII. SUMMARY

The importance of microalloying to the design of steels in the latter half of the 20th century, and into the present century, has been described, and although only a brief treatment of many of the important topics has been attempted, several references have been cited, many of which will lead the reader to more detailed information. The introduction of microalloying practice has been made in response to the challenges to improve one or more of strength, ductility, toughness, formability or weldability, in order to deliver improved performance, often combined with weight reduction and/or cost benefits. The resulting successful evolution from mild steel to the modern present-day microalloyed steel has also required concurrent improvements to the integral steelmaking and process control functions, requiring new or modified plant, in order to maximize the benefits attainable by microalloying. This period has provided both economic and environmental advantages and also preserved and expanded the market for steel against competition from alternative materials. It has thus been an exciting episode in steel metallurgy.

ACKNOWLEDGMENTS

This chapter is inspired by the recent treatment of microalloyed steels presented in *The Physical Metallurgy of Microalloyed Steels* by T. Gladman, published in 1997 by The Institute of Materials, London, and in which the reader can find many of the topics dealt with in greater detail. Furthermore, the author gratefully acknowledges many stimulating discussions with his colleagues, Emeritus Professor T. Gladman and Professor R.C. Cochrane, at the Department of Materials, University of Leeds.

REFERENCES

1. *Metallurgical Developments in Carbon Steels*, Harrogate, 1963; Special Report No. 81; Iron and Steel Institute: London, 1963.
2. *Strong Tough Structural Steels*, Scarborough, 1967; ISI Publication No. 104; Iron and Steel Institute: London, 1967.
3. Symposium on Low Alloy High Strength Steels, Nuremberg, 1970; The Metallurg Companies: Dusseldorf, 1970.
4. Gray, J.M., Ed., *Processing and Properties of Low-Carbon Steels*; The Metallurgical Society, AIME: Warrendale, PA, 1973.
5. *Controlled Processing of HSLA Steels*, York, 1976; British Steel Corporation: London, 1976.
6. *Microalloying '75*, Washington, DC, 1975; Union Carbide Corp.: New York, 1977.
7. *The Hot Deformation of Austenite*, Cincinnati, 1976; The Metallurgical Society, AIME: Warrendale, PA, 1977.
8. *Low Carbon Structural Steels for the Eighties*, Plymouth, 1977; Institution of Metallurgists: London, 1977.
9. *Welding of HSLA Structural Steels*; American Society for Metals: Metals Park, OH, 1978.
10. Abrams, H.; McNair, G.N.; Nail, D.A.; Solomon, H.D. *Optimisation of Processing, Properties and Service Performance Through Microstructural Control*; Micon '78; ASTM, STP 672: Philadelphia, 1979.
11. *Vanadium in High Strength Steels*, Publication No. 140; Vanadium International Technical Committee (Vanitec): London, 1979.
12. DeArdo, A.J.; Ratz, G.A.; Wray, P.J., Eds. *Formable HSLA and Dual Phase Steels*; The Metallurgical Society, AIME: Warrendale, PA, 1979.
13. Sellars, C.M.; Davies, G.J., Ed., *Hot Working and Forming Processes*; The Metals Society: London, 1980.
14. *Technology and Applications of Vanadium Steels*; Vanadium International Technical Committee (Vanitec): Krakow, London, 1980.
15. *Hot Deformation of Steels*; Iron and Steel Institute: Tokyo, Japan, 1981.
16. *Thermomechanical Processing of Microalloyed Austenite*; The Metallurgical Society, AIME: Warrendale, PA, 1982.
17. *Steels for Line Pipe and Pipeline Fittings*, Book 285; The Metals Society: London, 1982.
18. *HSLA Steels—Technology and Applications*, Philadelphia, 1983; American Society for Metals: Metals Park, OH, 1983.
19. *Advances in the Physical Metallurgy and Applications of Steels*; The Metals Society: London, 1983.
20. Dunne, D.P.; Chandra, T., Eds. *High Strength Low Alloy Steels*, Wollongong, 1984; The Metallurgical Society, AIME: Warrendale, PA, 1984.
21. Stuart, H.; Ed. *Niobium*, San Francisco; The Metallurgical Society, AIME: Warrendale, PA, 1984.
22. *Vanadium Structural Steels*; Vanadium International Technical Committee: London, and American Society for Metals: Metals Park, OH, 1985.

23. *Accelerated Cooling of Steel*, Pittsburgh; The Metallurgical Society, AIME: Warrendale, PA, 1986.
24. Gray, J.M.; Ko, T.; Zhang Shouhua; Wu Baorong; Xie Xishan, Eds. *HSLA Steels—Metallurgy and Applications*, Beijing, 1985; American Society for Metals: Metals Park, OH, 1986.
25. *Welding Metallurgy of Structural Steels*; The Metallurgical Society, AIME: Warrendale, PA, 1987.
26. *Processing, Microstructure and Properties of HSLA Steels*; The Metallurgical Society: Warrendale, PA, 1987.
27. *Accelerated Cooling of Rolled Steel*, Winnipeg; Pergamon Press: New York, 1988.
28. *Microalloying '88—Microalloyed HSLA Steels*, Chicago, 1988; American Society for Metals: Metals Park, OH, 1988.
29. Tamura, I. Ed. *Physical Metallurgy of Thermomechanical Processing of Steels and other Metals*; THERMEC-88; Iron and Steel Institute: Tokyo, Japan 1988.
30. *The Metallurgy, Welding and Qualification of Microalloyed (HSLA) Steel Weldments*; American Welding Society: Miami, FL, 1990.
31. *HSLA Steels '90*, Second International Conference on HSLA Steels, Beijing, 1990; ASM International: Metals Park, OH, 1991.
32. De Ardo, A.J., Ed. *Processing, Microstructure and Properties of Microalloyed and Other Modern High Strength Low Alloy Steels*, Pittsburgh, 1991; Iron and Steel Society, AIME: Warrendale, PA, 1992.
33. Tither, G.; Zhang, S. Eds. *High Strength Low Alloy Steels—Processing, Properties and Applications*; The Metallurgical Society, AIME: Warrendale, PA, 1992.
34. Asfahani, R.; Tither, G., Eds. Symposium on *Low-Carbon Steels for the 90's*, Pittsburgh, 1993; The Minerals, Metals and Materials Society, Warrendale, PA, 1993.
35. Liu, Guoxin; Stuart, H.; Zhang Hongtao; Li Chengji, Eds. *HSLA Steels '95*, Third International Conference on HSLA Steels, Beijing, 1995; Chinese Society for Metals: Beijing, 1995.
36. *Microalloying '95*, Pittsburgh, 1995; Iron and Steel Society: Warrendale, PA, 1995.
37. Chandra, T., Ed. *Thermomechanical Processing of Steels and Other Materials*, THERMEC '97, Wollongong, 1997; The Metallurgical Society, AIME: Warrendale, PA, 1997.
38. Asfahani, R., Ed. *Accelerated Cooling/Direct Quenching of Steels,* Indianapolis, 1997; American Society for Materials: Materials Park, OH, 1997.
39. Rodriguez-Ibabe, J.M.; Gutierrez, I.; Lopez, B. Eds. *Microalloying in Steels*; Trans Tech Publications, Inc.: Brandocein, 1998.
40. *Thermomechanical Processing of Steels*, London, 2000; Institute of Materials: London, 2000.
41. *HSLA Steels '00*, Fourth International Conference on HSLA Steels, Xi'an, 2000; Metallurgical Industry Press: Beijing, 2000.
42. Woodhead, J.H.; Keown, S.R. *HSLA Steels—Metallurgy and Applications*, Beijing, 1985; Gray, J.M., Ko, T., Zhang Shouhua, Wu Baorong, Xie Xishan, Eds.; American Society for Metals: Metals Park, OH, 1986; 15.
43. Morrison, W.B. *HSLA Steels '00,* Fourth International Conference on HSLA Steels, Xi'an, 2000; Liu, G., Wang, F., Wang, Z., Zhang, H., Eds.; Metallurgical Industry Press: Beijing, 2000; 11.
44. Grange, R.A.; Shortsleeve, F.J.; Hilty, D.C.; Binder, W.O.; Motock, G.T.; Offenhauer, C.M. *Boron, Calcium, Columbium and Zirconium in Iron and Steel*; John Wiley and Sons: New York, 1957.
45. Great Lakes Steel Corporation. Mechanical Engineer 1959, *81*, 53.
46. Beiser, C.A. ASM Preprint No. 138; 1959.
47. Pickering, F.B.; Gladman, T. *Metallurgical Developments in Carbon Steels*, Special Report No. 81; Iron and Steel Institute: London, 1963; 10.
48. Morrison, W.B.; Woodhead, J.H. J. Iron Steel Inst. 1963, *201*, 43.
49. Morrison, W.B. J. Iron Steel Inst. 1963, *201*, 317.

50. Leslie, W.C. *The Relation Between Structure and Mechanical Properties of Metals*; HMSO: London, 1963; 334.
51. Gray, J.M.; Webster, D.; Woodhead, J.H. J. Iron Steel Inst. 1965, *203*, 812.
52. Irvine, K.J.; Pickering, F.B.; Gladman, T. J. Iron Steel Inst. 1967, *205*, 161.
53. Gray, J.M.; Yeo, R.G.B. Trans ASM 1968, *64*, 255.
54. Davenport, A.T.; Berry, F.G.; Honeycombe, R.W.K. Metal Sci. J. 1968, *2*, 104.
55. Berry, F.G.; Davenport, A.T.; Honeycombe, R.W.K. *The Mechanisms of Phase Transformations in Crystalline Solids*; Institute of Metals: London, 1969; 288.
56. Davenport, A.T.; Honeycombe, R.W.K. Proc. R. Soc. A 1971, *322*, 191.
57. Clay, D.B.; McCutcheon, D.B. *The Contribution of Physical Metallurgy to Engineering Practice*, Rosenhain Centenary Conference, 1985; The Royal Society: London, 1976; 305.
58. van der Veen, J.H. *The Contribution of Physical Metallurgy to Engineering Practice*, Rosenhain Centenary Conference, 1985; The Royal Society: London, 1976; 319.
59. Gladman, T. *The Physical Metallurgy of Microalloyed Steels*; Institute of Materials: London, 1997.
60. Irani, J.J.; Burton, D.; Jones, J.D.; Rothwell, A.B. *Strong Tough Structural Steels*; ISI Publication No. 104. Iron and Steel Institute: London, 1967; 110.
61. Morrison, W.B.; Chapman, J.A. *The Contribution of Physical Metallurgy to Engineering Practice*, Rosenhain Centenary Conference, 1985; The Royal Society: London, 1976; 289.
62. Ouchi, C.; Tanaka, J.; Kozasu, I.; Tsukada, K. *Optimisation of Processing, Properties and Service Performance Through Microstructural Control*, Micon '78; Abrams, H., McNair, G.N., Nail, D.A., Solomon, H.D., Eds.; ASTM, STP 672: Philadelphia, 1979; 105.
63. Gladman, T.; McIvor, I.D.; Pickering, F.B. J. Iron Steel Inst. 1972, *210*, 916.
64. von den Steinen, A.; Engineer, S.; Horn, E.; Preis, G. Stahl und Eisen 1975, *95*, 209.
65. *Rail Steels—Developments, Processing and Use*; American Society for Testing and Materials, ASTM, STP 644: Philadelphia, PA, 1976.
66. Dunlop, G.L.; Carlsson, C.J.; Frimodig, G. Metall. Trans. A 1978, *9*, 261.
67. *Vanadium in Rail Steels*; Vanadium International Technical Committee (Vanitec): Chicago, London, 1979.
68. Laufer, E.E; Fegredo, D.M. Canadian Metall. Quart. 1983, *22*, 193.
69. Parsons, D.E.; Malis, T.F.; Boyd, J.D. *HSLA Steels—Technology and Applications*; American Society for Metals:Metals Park, OH, 1983; 1129.
70. Krauss, G., Banerjee, S.J., Eds. *Fundamentals of Microalloying Forging Steels*, Denver, 1986; The Metallurgical Society: Warrendale, PA, 1987.
71. Parsons, S.A.; Edmonds, D.V. Mater. Sci. Technol. 1987, *3*, 894.
72. *Information Day on Microalloyed Engineering Steels*; Commission of the European Communities, Dusseldorf, Verein Deutscher, Eisenhuttenleute, 1988.
73. Naylor, D.J. Ironmaking and Steelmaking 1989, *16*, 246.
74. *Microalloyed Bars and Forging Steels*, Hamilton, 1990; Canadian Institute of Mining and Metallurgy: Ottawa, 1990.
75. Khalid, F.A.; Edmonds, D.V. Mater. Sci. Technol. 1993, *9*, 384.
76. Wang, P.; Jerath, V.; Edmonds, D.V. *HSLA Steels '95*, Third International Conference on HSLA Steels, Beijing, 1995; Liu Guoxin, Stuart, H., Zhang Hongtao, Li Chengji, Eds.; Chinese Society for Metals: Beijing, 1995; 489.
77. Van Tyne, C.J.; Krauss, G.; Matlock, D.K., Eds. Proceedings of *Fundamentals and Applications of Microalloying Forging Steels*, Denver, 1996; The Minerals, Metals and Materials Society: Warrendale, PA, 1996.
78. Engineer, S.; Huchtemann, B.; Schuler, V. *Fundamentals of Microalloying Forging Steels*, Denver, 1986; Krauss, G., Banerjee, S.J., Eds.; The Metallurgical Society: Warrendale, PA, 1987; 19.
79. Naylor, D.J. *Microalloying '88—Microalloyed HSLA Steels*, Chicago, 1988; American Society for Metals: Metals Park, OH, 1988;43.

80. Cristinacce, M.; Reynolds, P.E. *Fundamentals and Applications of Microalloying Forging Steels*, Denver, 1996; Van Tyne, C.J., Krauss, G., Matlock, D.K., Eds.; The Minerals, Metals and Materials Society: Warrendale, PA, 1996; 29.
81. Engineer, S.; Huchtemann, B. *Fundamentals and Applications of Microalloying Forging Steels*, Denver, 1996; Van Tyne, C.J., Krauss, G., Matlock, D.K., Eds.; The Minerals, Metals and Materials Society: Warrendale, PA, 1996; 61.
82. Reynolds, P.E.; Reynolds, J.H. *Fundamentals and Applications of Microalloying Forging Steels*, Denver, 1996; Van Tyne, C.J., Krauss, G., Matlock, D.K., Eds.; The Minerals, Metals and Materials Society: Warrendale, PA, 1996; 79.
83. Hansen, S.S. *Microalloying '88—Microalloyed HSLA Steels*, Chicago, 1988; American Society for Metals: Metals Park, OH, 1988; 31.
84. Stuart, H. *Niobium*, San Francisco; Stuart, H., Ed.; The Metallurgical Society: Warrendale, PA, 1984; 1237.
85. Comstock, G.F.; Urban, S.F.; Cohen, S. *Titanium in Steel*; Pitman: New York, 1941.
86. Leslie, W.C.; Sober, R.J. Trans ASM Quarterly 1967, *60*, 99.
87. Ohashi, N.; Irie, T.; Satoh, S.; Hashimoto, O.; Takahashi, I. SAE Paper No. 810027; 1981.
88. Yamada, N.; Tokunaga, Y. Trans. Iron Steel Inst. Jpn. 1985, *25*, 640.
89. Wilshynski-Dresler, D.O.; Matlock, D.K.; Krauss, G. International Forum for Physical Metallurgy of IF Steels; Tokyo, Iron and Steel Institute: Japan, 1994.
90. Hua, M.; Garcia, C.I.; DeArdo, A.J. Metall. Trans. A 1997, *28*, 1769.
91. Takechi, H. *Processing, Microstructure and Properties of IF Steels*; Iron and Steel Society, AIME: Warrendale, PA, 2000; 1.
92. Pradhan, R. Ed. *Technology of Continuously Annealed Cold-Rolled Sheet Steel*; The Metallurgical Society, AIME: Warrendale, PA 1985.
93. Pradhan, R. Ed. *Metallurgy of Vacuum Degassed Products*; The Metallurgical Society, AIME:Warrendale, PA, 1990.
94. Collins L.E. Barager, D.L. Eds. *Interstitial-Free Steel Sheet: Processing, Fabrication and Properties*; The Canadian Institute of Mining, Metallurgy and Petroleum Montreal, Quebec, 1991.
95. Pradham, R.; Gupta, I. Eds. *Developments in the Annealing of Sheet Steels*; The Metallurgical Society, AIME: Warrendale, PA, 1992.
96. International Forum for Physical Metallurgy of IF Steels; Iron and Steel Institute: Tokyo, Japan, 1994.
97. 39th Mechanical Working and Steel Processing Conference, Indianapolis, 1997; Iron and Steel Society, AIME: Warrendale, PA, 1998; Vol. 35.
98. *Processing, Microstructure and Properties of IF Steels*; Iron and Steel Society, AIME: Warrendale, PA, 2000.
99. Hanai, N.; Takemoto, N.; Takunaga, Y.; Misuyama, Y. Trans. Iron Steel Inst. Japan. 1984, *24*, 17.
100. Pickering, F.B. 35th Mechanical Working and Steel Processing Conference; Iron and Steel Society, AIME: Warrendale, PA, 1994; Vol. 31, 477.
101. Baker , T.N. Ed. *Titanium Technology in Microalloyed Steels*; Institute of Materials: London, 1997.
102. Adrian, H. *Microalloyed Vanadium Steels*, Korchynsky, M.; Gorczyca, S.; Blicharski, M., Eds.; Akad. Gom-Gutz: Krakow, 1990; 105.
103. Speer, J.G.; Michael, J.R.; Hansen, S.S. Metall. Trans. A 1987, *18*, 211.
104. Adrian, H. Mater. Sci. Technol. 1992, *8*, 148.
105. Glwadman, T. In *Fundamentals and Applications of Microalloying Forging Steels, Denver, 1996*, Van Tyne, C.J.; Krauss, G.; Matlock, D.K., Eds.; The Minerals, Metals and Materials Society: Warrendale, PA, 1996; 3.
106. He, K.; Baker, T.N. In *Titanium Technology in Microalloyed Steels*, Baker, T.N., Ed.; Institute of Materials: London, 1997; 115.

107. Craven, A.J.; He, K.; Baker, T.N. Inst. Phys. Conf. Ser. No 153: Section 13; Butterworths: London, 1997; 605.
108. Subramanian, S.V.; Weatherley, G.C. In *Titanium Technology in Microalloyed Steels*, Baker, T.N., Ed.; Institute of Materials: London, 1997; 133.
109. Egner, D.J.; Cochrane, R.C.; Brydson, R. EMAG'99, Sheffield, 1999; Inst. Phys. Conf. Ser. No 161: Section 9; IOP Publishing Ltd.: London, 1999; 447.
110. Khalid, F.A.; Brydson, R.; Edmonds, D.V. EMAG '99, Sheffield, 1999, Inst. Phys. Conf. Ser. No 161: Section 9; IOP Publishing Ltd.: London, 1999; 511.
111. Jonas, J.J.; Weiss, I. Metal Sci. 1979, *13*, 238.
112. Hoogendorn, T.; Spanraft, M.J. *Microalloying '75*, Washington, DC, 1975; Union Carbide Corp.: New York, 1977; 75.
113. Edmonds, D.V. J. Iron Steel Inst. 1972, *210*, 363.
114. Suzuki, H.G.; Tanino, M. Trans. ISIJ 1972, *12*, 217.
115. Edmonds, D.V. Metall. Trans. 1973, *4*, 2527.
116. Heikkinen, V.K. Trans. ISIJ 1973, *13*, 79.
117. Heikkinen, V.K. Acta Metall. 1973, *21*, 709.
118. Campbell, K.; Honeycombe, R.W.K. Metal Sci. J. 1974, *8*, 197.
119. Honeycombe, R.W.K. Metall. Trans. A 1976, *7*, 915.
120. Aaronson, H.I.; Plichta, M.R.; Franti, G.W.; Russell, K.C. Metall. Trans. A 1978, *9*, 363.
121. Roberts, W.T.; Sandborg, A.; Siwecki, T. *Vanadium Steels*, Krakow, 1980; Vanadium International Technical Committee: London, 1980; D1.
122. Ricks, R.A.; Howell, P.R. Metal Sci. 1982, *16*, 317.
123. Ricks, R.A.; Howell, P.R. Acta Metall. 1983, *31*, 853.
124. Obara, T.; Shiflet, G.J.; Aaronson, H.I. Metall. Trans. A 1983, *14*, 1159.
125. Law, N.C.; Parsons, S.A.; Howell, P.R.; Edmonds, D.V. Mater. Sci. Technol. 1987, *3*, 642.
126. Furuhara, T.; Aaronson, H.I. Scripta Metall 1988, *22*, 1635.
127. Li, P.; Todd, J.A. Metall. Trans. A 1988, *19*, 2139.
128. Baker, R.G.; Nutting, J. *Precipitation Processes in Steels*, Special Report No. 64; Iron and Steel Institute: London, 1959; 1.
129. Batte, A.D.; Edmonds, D.V.; Honeycombe, R.W.K. *Strength of Metals and Alloys*, Asilomar, Pacific Grove, CA, 1970; American Society for Metals: Metals Park, OH, 1970; 585.
130. Freeman, S. *Effect of Second-Phase Particles on the Mechanical Properties of Steel*; Iron and Steel Inst.: London, 1971; 152.
131. Batte, A.D.; Honeycombe, R.W.K. Metal Sci. J. 1973, *7*, 160.
132. Batte, A.D.; Honeycombe, R.W.K. J. Iron Steel Inst. 1973, *211*, 284.
133. McCann, J.; Ridal, K.A. J. Iron Steel Inst. 1964, *202*, 441.
134. Edmonds, D.V. *Fundamentals and Applications of Microalloying Forging Steels*, Denver, 1996; Van Tyne, C.J., Krauss, G., Matlock, D.K., Eds.; The Minerals, Metals and Materials Society: Warrendale, PA, 1996, 111.
135. Wang, P.; Edmonds, D.V. 11th European Congress on Electron Microscopy, Dublin, 1996; Vol. II: Materials Science; Brussels: Committee of European Societies of Microscopy, 1998; 341.
136. Parsons, S.A. Ph.D. Dissertation, Cambridge University, Cambridge, UK, 1981.
137. Balliger, N.K.; Honeycombe, R.W.K. Metall. Trans. A 1980, *11*, 421.
138. Sakuma, T.; Honeycombe, R.W.K. Metal Sci. 1984, *18*, 449.
139. Sakuma, T.; Honeycombe, R.W.K. Mater. Sci. Technol. 1985, *1*, 351.
140. Bucher, J.H.; Grozier, J.D. J. Iron Steel Inst. 1966, *204*, 1253.
141. He, K.; Edmonds, D.V. Mater. Sci. Technol. 2001, *18*, 289.
142. Garland, J.G.; Kirkwood, P.R. Metal Constr 1975, *7*, 275.
143. Abson, D.J.; Dolby, R.E.; Hart, P.H.M. *Trends in Steels and Consumables for Welding*; TWI, Abington, Cambs., UK, 1979; 75.
144. Yamamoto, K.; Matsuda, S.; Haze, T.; Chijiiwa, R.; Mimura, H. *Residual and Unspecified Elements in Steel*; Melilli, A.S., Nisbett, E.G., Eds.; ASTM STP 1042, ASTM: Philadelphia, 1989; 266.

145. Terado, Y.; Chijiiwa, R.; Takehiro, H.; Kikuma, T.; Funato, K. In *HSLA Steels, Processing, Properties and Applications*, Tither, G.; Zhang, S., Eds.; The Metallurgical Society: Warrendale, PA, 1992; 519.
146. Gladman, T.; Senogles, D.J. In *Titanium Technology in Microalloyed Steels*, Baker, T.N., Ed.; Institute of Materials: London, 1997; 83.
147. Zhang, M.; Edmonds, D.V. *HSLA Steels '95*, Third International Conference on HSLA Steels, Beijing, 1995; Liu, G., Stuart, H., Zhang, H., Li, C., Eds.; Chinese Society for Metals: Beijing, 1995; 133.
148. Mitchell, P.S.; Hart, P.H.M.; Morrison, W.B. *Microalloying '95*, Pittsburgh, 1995; Iron and Steel Society: Warrendale, PA, 1995; 149.
149. Ishikawa, F.; Takahashi, T.; Ochi, T. Metall. Mater. Trans. A 1994, *25*, 929.
150. Yamamoto, K.; Yoshida, S.; Watanabe, K. US Patent 5,421,920, 1995.
151. Araki, T.; Enomoto, M.; Shibata, K. *HSLA Steels '90*, Second International Conference on HSLA Steels, Beijing; ASM International: Metals Park, OH, 1991; 93.
152. Rehman, A.; Edmonds, D.V. *HSLA Steels '00*, Fourth International Conference on HSLA Steels, Xi'an, 2000; Liu, G., Wang, F., Wang, Z., Zhang, H., Eds.; The Metallurgical Press: Beijing, 2000; 241.
153. Araki, T.; Enomoto, M.; Shibata, K. Mater. Trans. JIM 1991, *32*, 729.
154. Ohmori, Y.; Ohtani, H.; Kunitake, T. Trans. ISIJ 1971, *11*, 250.
155. *Guide to the Light Microscope Examination of Ferritic Steel Weld Metals*, IIW Doc., IX-1533–88/IXJ-123–87 Revision 2, 1988.
156. Bramfitt, B.L.; Speer, J.G. Metall. Trans. A 1990, *21*, 817.
157. Bainite Research Committee, T. Araki, Chairman. *Atlas for Bainitic Microstructures*; Iron and Steel Institute of Japan: Tokyo, 1992; Vol. 1.
158. Krauss, G.; Thompson, S.W. ISIJ Int. 1995, *35*, 937.
159. Araki, T.; Shibata, K. *HSLA Steels '95*, Third International Conference on HSLA Steels, Beijing, 1995; Liu, G., Stuart, H., Zhang, H., Li, C., Eds.; Chinese Society for Metals: Beijing, 1995; 13.
160. Bee, J.V.; Honeycombe, R.W.K. Metall. Trans. A 1978, *9*, 587.
161. Wilson, E.A. ISIJ Int. 1994, *34*, 615.
162. Ohmori, Y. *New Aspects of Microstructures in Modern Low Carbon High Strength Steels*; Iron and Steel Inst.: Tokyo, Japan 1994; 7.
163. Shibata, K.; Asakura, K. ISIJ Int. 1995, *35*, 982.
164. Dube, C.A.; Aaronson, H.I.; Mehl, R.F. Rev. Met. 1958, *55*, 201.
165. Aaronson, H.I. In *The Decomposition of Austenite by Diffusional Processes*, Zackay, V.F.; Aaronson, H.I., Eds.; Interscience: New York, 1962; 389.
166. Gladman, T.; Pickering, F.B. *Yield, Flow and Fracture of Polycrystals*, Baker, T.N., Ed.; Applied Science: London, 1983; 141.
167. Sawada, Y., et al., Proceedings of 35th Mechanical Working and Steel Processing Conference; Iron and Steel Society: Warrendale, PA, 1994; 263.
168. Hall, E.O. Proc. Phys. Soc., Ser. B 1951, *64*, 747.
169. Petch, N.J. J. Iron Steel Inst. 1953, *174*, 25.
170. Gladman, T.; Holmes, B.; McIvor, I.D. *Effect of Second Phase Particles on the Mechanical Properties of Steels*; Iron and Steel Institute: London, 1971; 68.
171. Morrison, W.B. *Controlled Processing of High Strength Low Alloy Steels*, British Steel Corporation: London, 1976, paper 1.
172. Hudd, R.C. *Low Carbon Structural Steels for the Eighties*; Institution of Metallurgists: London, 1977; IIIA-1.
173. Wilson, D.V. *Effect of Second Phase Particles on the Mechanical Properties of Steels*; Iron and Steel Institute: London, 1971; 28.
174. Gladman, T.; Holmes, B.; Pickering, F.B. J. Iron Steel Inst. 1970, *208*, 173.
175. Gensamer, M. Trans ASM 1946, *36*, 30.
176. Morrison, W.B. Trans ASM 1966, *59*, 824.

177. Abe, M. In *Materials Science and Technology*, Cahn, R.W.; Haasen, P.; Kramer, E.J., Eds.; VCH: Weinheim, 1992; 7, 285.
178. Held, J.F. Proceedings of Fourth Mechanical Working and Steel Processing Conference; Iron and Steel Society, AIME: Warrendale, PA, 1966, 3.
179. Lankford, W.T.; Snyder, S.D.; Bauscher, J.A. Trans ASM 1950, *42*, 1197.
180. Blickwede, D.J. Trans ASM 1968, *61*, 653.
181. Wilson, D.V. Metall. Rev. No 139 Metals Mater. 1969, *3*, 175.
182. Gladman, T. Mater. Sci. Technol. 1990, *6*, 1131.
183. Honeycombe, R.W.K. *Steels—Microstructure and Properties*; Edward Arnold: London, 1981.
184. Hudd, R.C. Metals Mater 1987, *3*, 71.
185. Hutchinson, W.B.; Ushioda, K Scand. J. Met. 1984, *13*, 269.
186. Roberts, J.E. *The Contribution of Physical Metallurgy to Engineering Practice*, Rosenhain Centenary Conference, 1985; The Royal Society: London, 1976; 277.
187. Gladman, T. Proc. Roy. Soc. A 1966, *294*, 298.
188. Gladman, T. *Recrystallisation and Grain Growth of Multi-Phase and Particle Containing Materials*, Hansen, N.; Jones, A.R.; Leffers, T., Eds.; RISO National Laboratories: Roskilde, Denmark, 1980; 183.
189. Gladman, T. *HSLA Steels '90*, Second International Conference on HSLA Steels, Beijing, 1990; ASM International, OH: Metals Park, 1991; 3.
190. Jonas, J.J.; Weiss, I. Metal Sci. 1979, *13*, 238.
191. Sellars, C.M. Recrystallisation and Grain Growth in Multi-phase Materials In Hansen, N.; Jones, A.R.; Leffers, T. Risø National Laboratory: Roskilde, 1980; 291.
192. DeArdo, A.J.; Gray, J.M.; Meyer, L. *Niobium*, Stuart, H., Ed.; The Metallurgical Society: New York, 1984; 685.
193. Sellars, C.M. *Hot Working and Forming Processes*, Sellars, C.M.; Davies, G.J., Eds.; The Metals Society: London, 1980; 3.
194. Cuddy, L.J. *Thermomechanical Processing of Microalloyed Austenite*, DeArdo, A.J.; Ratz, G.A.; Wray, P.J., Eds.; The Metallurgical Society, AIME: Warrendale, PA, 1982; 129.
195. Kosazu, I.; Ouchi, C.; Sampei, I.; Okita, T. *Microalloying '75*, Washington, DC, 1975; Korchynsky, M., Ed.; Union Carbide Corp.: New York, 1977; 120.
196. Roberts, W.; Lidefelt, H.; Sandberg, A. *Hot Working and Forming Processes*, Sellars, C.M.; Davies, G.J., Eds.; The Metals Society: London, 1980; 38.
197. Bramfitt, R.L.; Marder, A.R. Processing and Properties of Low Carbon Steel; The Metallurgical Society AIME: Warrendale, PA, 1967; 191.
198. Tanaka, T.; Tabata, N.; Hatomura, T.; Shiga, C. *Microalloying '75*, Washington, DC, 1975; Korchynsky, M., Ed.; Union Carbide Corp.: New York, 1977; 107.
199. Gladman, T.; Dulieu, D.; McIvor, I.D. Proceedings of Microalloying '75, Washington, DC, 1975; Union Carbide Corp.: New York, 1977; 25.
200. Ravenshorst, H. *Microalloying '95*, Pittsburgh, 1995; Iron and Steel Society: Warrendale, PA, 1995; 439.
201. Speich, G.R.; Cuddy, L.J.; Gordon, C.R.; DeArdo, A.J. In *Phase Transformations in Ferrous Alloys*, Marder, A.R.; Goldstein, J.I., Eds.; The Metallurgical Society, AIME: Warrendale, PA, 1984; 341.
202. Han, K.; Edmonds, D.V.; Smith, G.D.W. Metall. Trans. A 2001, *32*, 1313.
203. Heitmann, W.E.; Babu, P.B. *Fundamentals of Microalloying Forging Steels*, Denver, 1986; Krauss, G., Banerjee, S.J., Eds.; The Metallurgical Society: Warrendale, PA, 1987; 55.
204. Zhang, F.; Boyd, J.D. *Fundamentals and Applications of Microalloying Forging Steels*, Denver, 1996; Van Tyne, C.J., Krauss, G., Matlock, D.K., Eds.; The Minerals, Metals and Materials Society: Warrendale, PA, 1996; 127.
205. Takada, H.; Koyasu, Y. *Fundamentals and Applications of Microalloying Forging Steels*, Denver, 1996; Van Tyne, C.J., Krauss, G., Matlock, D.K., Eds.; The Minerals, Metals and Materials Society: Warrendale, PA, 1996; 143.

206. Goodenow, R.H. Trans ASM 1966, *59*, 804.
207. Fudaba, K.; Akisue, O.; Tokunaga, Y. *Ladle Steelmaking and Furnaces*; Canadian Inst. Mining and Metallurgy: Montreal, 1988; 290.
208. Matsumoto, T.; Hamanaka, S.; Yamada, T.; Tanaka, T. *New Aspects of Microstructures in Modern Low Carbon High Strength Steels*; Iron and Steel Inst.: Tokyo, Japan 1994; 35.
209. Takechi, H. Proceedings of HSLA Steels '95, Third International Conference on HSLA Steels, Beijing, 1995; Liu, G., Stuart, H., Zhang, H., Li, C., Eds.; Chinese Society for Metals: Beijing, 1995; 72.
210. Tither, G.; Stuart, H. Proceedings of HSLA Steels '95, Third International Conference on HSLA Steels, Beijing, 1995; Liu, G., Stuart, H., Zhang, H., Li, C., Eds.; Chinese Society for Metals: Beijing, 1995; 22.
211. DeArdo, A.J. New Aspects of Microstructures in Modern Low Carbon High Strength Steels; Iron and Steel Inst.: Tokyo, Japan 1994; 57.
212. Hughes, I.F.; Hudd, R.C. British Patent GB1 236,598, 1971.
213. Mould, P.; Gray, J.M. Metall. Trans. 1972, *3*, 3121.
214. Hutchison, W.B. International Metal Rev. 1984, *29*, 25.
215. Hutchison, W.B.; Nilsson, K.-I.; Hirsch, J. *Metallurgy of Vacuum Degassed Products*, Pradhan, R., Ed.; The Metallurgical Society, AIME: Warrendale, PA, 1990; 109.
216. Tither, G.; Garcia, C.I.; Hua, M.; DeArdo, A.J. International Forum for Physical Metallurgy of IF Steels; Iron and Steel Institute: Tokyo, Japan, 1994; 293.

7

Designing with Stainless Steel

Joseph Ki Leuk Lai
City University of Hong Kong, Kowloon, Hong Kong, China

I. TYPES OF STAINLESS STEELS

The term stainless steel is, strictly speaking, a misnomer. It refers to a class of steel with enhanced corrosion resistance. In fact, there is no stainless steel that can remain "stainless" in all aggressive environments. In some situations, such as stress corrosion in the presence of chloride ions, some stainless steels may be even more susceptible to corrosion than ordinary steels. To most metallurgists, stainless steels are iron-based alloys (i.e. $>50\%$ Fe) that contain at least about 11% chromium, which renders the material rust-free in ordinary atmospheres due to the formation of a protective chromium oxide film. In the literature, higher alloys containing less than 50% iron are sometimes also referred to as "stainless steels". Information on such alloys will be presented in a separate chapter of this handbook.

The discovery of stainless steel can be traced back to the early 1900s. During the period 1910–1919, patents on stainless steels were filed independently in Britain, USA, Germany, and France. Its first application was in the cutlery industry, followed by other applications in situations where enhanced corrosion resistance was required.

Nowadays, there are over 350 different grades of commercial stainless steels listed in the United Numbering System [1]. They can be classified into five categories according to their chemical composition and structure: ferritic, martensitic, austenitic, duplex, and precipitation hardenable.

A. Ferritic Stainless Steels

The chemical compositions of some selected ferritic stainless steels are listed in Table 1. The compositions given are not specifications. The figures represent typical values unless otherwise indicated. The information provided is by no means exhaustive. In fact, it is intended to supplement the information given in other reference books [1,2].

The characteristics of ferritic stainless steels are that:

- they have body-centered-cubic (bcc) crystal structure;
- they are ferromagnetic at room temperature;
- they cannot be quench hardened;

Table 1 Chemical Composition of Selected Ferritic Stainless Steels (wt%)

Type	C	Mn	P	S	Si	Cr	Ni	Al	N	Mo	Others
405	< 0.08	< 1.0	< 0.04	< 0.03	< 1.0	11.5–14.5	—	0.1–0.3	—	—	
405NA[a]	< 0.08	< 1.0	< 0.04	< 0.03	< 1.0	11.5–14.5	—	—	—	—	
408	< 0.08	< 1.0	< 0.045	< 0.045	< 1.0	11.5–13.0	< 0.8	—	—	—	Ti 12 × C-1.1
408L[a]	0.01	0.35	< 0.03	< 0.01	0.4	11.0	0.76	—	0.01	—	
409	< 0.08	< 1.0	< 0.045	< 0.045	< 1.0	10.5–11.75	< 0.5	—	—	—	Ti 6 × C-0.75
409L[a]	0.008	0.4	< 0.04	< 0.02	0.5	11.1	—	—	< 0.01	—	Ti 0.2
409HP[b]	0.02	< 1.0	< 0.045	< 0.03	< 1.0	11.0	< 0.5	—	—	—	Ti 0.25
AL409HP™[c]	0.01	0.3	0.023	0.001	0.41	11.0	0.25	—	0.012	—	Ti 0.17, Nb 0.15
AL409Cb™[c]	0.015	0.35	0.025	0.005	0.35	12.30	0.25	0.01	0.015	0.04	Ti 0.005, Nb 0.45
409Cb	< 0.06	< 1.0	< 0.045	< 0.04	< 1.0	10.5–11.75	< 0.5	—	—	—	Nb 10 × C-0.75
Carpenter 5F[d]	< 0.1	< 1.0	< 0.06	> 0.3	< 1.0	13.0–14.0	< 0.5				
430	< 0.12	< 1.0	< 0.04	< 0.03	< 1.0	16.0–18.0	—	—	—	—	
430L[a]	0.01	0.4	< 0.04	< 0.02	0.4	16.3	< 0.5	—	0.01	—	
430F	< 0.12	< 1.25	< 0.06	> 0.15	< 1.0	16.0–18.0	—	—	—	< 0.6	
430FSe	< 0.12	< 1.25	< 0.06	< 0.06	< 1.0	16.0–18.0	—	—	—	—	Se > 0.15
430F[d] Solenoid Quality	< 0.065	< 0.8	< 0.03	0.25–0.40	0.3–0.7	17.25–18.25	< 0.6	—	—	< 0.5	
430FR[d] Quality Solenoid	< 0.065	< 0.8	< 0.03	0.25–0.40	1.00–1.50	17.25–18.25	< 0.6	—	—	< 0.5	
AL433™[c]	0.01	0.30	0.21	0.001	0.39	20.0	0.25	—	0.019	—	Cu 0.5, Nb 0.54–0.8 Nb/(C+N)≥ 10.0
434	< 0.12	< 1.0	< 0.04	< 0.03	< 1.0	16.0–18.0	—	—	—	0.75–1.25	
436	< 0.12	< 1.0	< 0.04	< 0.03	< 1.0	16.0–18.0	—	—	—	0.75–1.25	5 × C − 0.7 Nb + Ta
AL436S™[c]	0.01	0.2	0.02	0.001	0.37	17.3	0.3	—	0.015	1.20	Ti 0.2, Ti/(C + N)≥ 8
Project 70® Stainless 182-FM[d]	< 0.08	1.25–2.5	< 0.04	0.15–0.4	< 1.0	17.5–19.5	—	—	—	1.5–2.5	
439	< 0.07	< 1.0	< 0.04	< 0.03	< 1.0	17.0–19.0	< 0.5	< 0.15	—	—	Ti 12 × C-1.1
AL439HP™[c]	0.012	0.45	0.02	< 0.001	0.55	17.50	0.23	—	0.013	—	Ti 0.4
441	0.02	0.33	0.025	0.002	0.39	18.0	0.32	0.05	0.021	—	Ti 0.27, Nb 0.69

AL441HP™[c]	0.009	0.35	0.0023	0.002	0.34	18.0	0.30	0.05	0.014	—	Ti 0.29, Nb 0.71
443	< 0.2	< 1.0	< 0.04	< 0.03	< 1.0	18.0–23.0	< 0.5	—	—	—	Cu 0.9–1.25
444	< 0.025	< 1.0	< 0.04	< 0.03	< 1.0	17.5–19.5	< 1.0	—	< 0.025	1.75–2.5	Nb + Ti [0.2 + 4(C + N)]–0.8
446	< 0.2	< 1.5	< 0.04	< 0.03	< 1.0	23.0–27.0	—	—	< 0.25	—	
Sandvik4C54[e]	< 0.2	0.8	< 0.03	< 0.015	0.5	26.5	—	—	0.2	—	
446MOD-1[b] (UNS S44660)	< 0.03	< 1.0	< 0.04	< 0.03	< 1.0	25.0–28.0	1–3.5		< 0.04	3.0–4.0	Nb + Ti [0.2 + 6(C + N)]–1.0
AL453™[c]	0.03	0.3	0.02	< 0.03	0.3	22.0	0.3	0.6			Ti 0.02, REM(Ce + La) < 0.1
AL468™[c]	0.009	0.4	0.024	0.001	0.55	18.25	0.22	0.03	0.016	—	Ti 0.1, Nb 0.25
Chrome Core® 8[d]	< 0.03	0.2–0.7	0.03	< 0.3	0.3–0.7	7.50–8.50			0.2–0.5		
Chrome Core 8-FM[d]	< 0.03	0.2–0.7	0.03	0.2–0.4	0.3–0.7	7.50–8.50			0.2–0.5		
Chrome Core 12[d]	< 0.03	0.2–0.7	0.03	< 0.3	0.3–0.7	11.5–12.0			0.2–0.5		
Chrome Core 12-FM[d]	< 0.03	0.2–0.7	0.03	0.2–0.4	0.3–0.7	11.5–12.0			0.2–0.5		

Note: The chemical compositions given in the table are not specifications. They are typical values unless otherwise indicated.

[a]Proprietary alloy, ATLAS Stainless Steels.
[b]Proprietary alloy, J&L Specialty Steel.
[c]Proprietary alloy, Allegheny Ludlum Corp.
[d]Proprietary alloy, Carpenter Specialty Alloys.
[e]Proprietary alloy, Sandvik Steel.

- they are resistant to chloride stress corrosion;
- they have moderate yield strength;
- some grades have poor impact resistance and weldability.

As the name indicates, ferritic stainless steels consist mainly of ferrite, which is a body-centered-cubic (bcc) phase and is ferromagnetic at room temperature. The development of ferritic stainless steels can be divided into three stages. The early types, so called "first generation ferritic stainless steels", have a relatively high carbon content (0.12–0.2 wt% max). Since carbon is an austenite stabilizer, other ferrite stabilizers such as chromium have to be kept sufficiently high to ensure a ferritic structure. With the advent of the argon–oxygen decarburization steel refining process in the early 1970s, a second generation of ferritic stainless steel was developed with maximum carbon content in the range of 0.01–0.08 wt%. This reduction in carbon content has improved weldability by reducing the risk of "sensitization" due to carbide precipitation. The third generation ferritic stainless steels, often called "superferritics", have higher chromium content (up to 30% max) for enhanced corrosion resistance.

From an engineering design point of view, corrosion resistance and mechanical properties are often the major factors for consideration in the selection of material for a particular application. In this respect, AISI type 405 lies at the lowest rung of the ladder amongst all commercial stainless steels in terms of these properties. It contains no nickel and its lowest allowable chromium content is 11.50 wt%, which is just sufficient for the material to be classified as a "stainless steel". The aluminum added (0.1–0.3 wt%) has the effect of increasing the chromium equivalent of the alloy and narrowing the gamma loop in the equilibrium phase diagram. Since an austenite field in the equilibrium phase diagram is a prerequisite for martensitic transformation, the aluminum addition helps render the material non-hardenable by heat treatment and improve electrical conductivity. This steel is weldable and is designed to be used in the as-welded condition. Typical applications include steam nozzles and partitions.

Type 405 stainless steel has a commercial variant, type 405NA, which is an aluminum-free version of the same grade. This proprietary alloy is produced by ATLAS Stainless Steels. Even without aluminum, this grade is still non-hardenable. With just sufficient chromium content to qualify for the "stainless status", this steel is intended to be used in mildly corrosive environments. In fact, even under relatively mild conditions, a light rust film may form on the surface, rendering this material unsuitable for applications where long-lasting surface brightness is required.

Ferritic stainless steels can suffer from intergranular corrosion in the heat-affected zone after welding. This phenomenon, called sensitization, is attributed to the precipitation of chromium carbides leading to chromium-depleted regions at grain boundaries. In type 408 stainless steel, titanium is added as a carbide stabilizer to prevent sensitization. Titanium is a stronger carbide former than chromium and titanium carbide is stable at much higher temperatures than their chromium counterpart. Thus, titanium carbide is formed in preference to chromium carbide and chromium-depleted regions are avoided.

Another way to lower the risk of sensitization is to reduce the carbon content of the alloy. Type 408L is a commercial variant of type 408 with low carbon content but without titanium. This proprietary alloy (ATLAS Stainless Steels), however, can be hardened by heat treatment because the presence of nickel can lead to the formation of a martensitic structure under certain circumstances. Strictly speaking, this is a ferritic/martensitic stainless steel rather than a ferritic stainless steel. Type 408L steel is used in the fabrication of coach and bus structural members. It is resistant to mild atmospheres and chemicals with the formation of a light rust film. It should not be used in applications where surface brightness is important.

Type 409 stainless steel is similar to type 408 in chemical composition except for slightly lower nickel, titanium, and chromium content. The lower nickel content enables this steel to have a ferritic structure under normal conditions. This steel is widely used in automotive exhaust systems and in thin section structural applications where good toughness and surface brightness are not needed. Because of its good weldability and formability, type 409 is one of the most popular grades amongst all stainless steels with an annual production tonnage almost equally that of AISI type 304.

With improvements in steel making practice, various steel producers have developed low carbon commercial variants of type 409 steel to further improve formability and weldability, e.g. ATLAS 409L, J&L 409HP and AL 409HP™. In AL 409HP™, both titanium and niobium are added as stabilizing elements. Nitrogen is also carefully controlled in both ATLAS 409L and AL 409HP™. The reductions in titanium and nitrogen content help minimize titanium nitride precipitation, improve surface properties and reduce tool and die wear during forming operations. Titanium is replaced by niobium (columbium) in type 409Cb and a proprietary alloy, AL 409Cb™. This has the effect of improving the brazability of the material without any deleterious effect on other properties compared to AL 409HP™.

Carpenter stainless No. 5F is a proprietary alloy which is essentially non-hardenable and can be classified as a ferritic stainless steel. With sulfur content specified at $>0.3\%$, this steel offers good machining properties and is used for applications involving intricate machining operations.

So far, the steel grades described have chromium levels close to the minimum required to qualify for "stainless" classification. For better corrosion resistance, type 430 is one of the most popular grades with chromium content in the range 16–18%. It is used extensively for trims on automobiles, cameras, and other products to enhance their appearance for marketability. There are many modifications of this alloy. Types 430F and 430Se have sulfur and selenium additions, respectively, to improve machinability. Type 430L is a low carbon commercial variant produced by ATLAS Stainless Steels. Types 430F and 430FR Solenoid Quality are special proprietary steels (Carpenter Specialty Alloys) with more precisely controlled composition to achieve consistent magnetic properties. They are specially developed for use in soft magnetic components operating in corrosive environments. The latter grade has a high silicon content to increase electrical resistivity and annealed hardness compared to type 430F Solenoid Quality.

The addition of molybdenum to type 430 stainless steel has led to the creation of a new grade—type 434. The molybdenum serves to improve corrosion resistance, particularly to road salt attack in automotive trim applications. Another grade derived from this family is type 436, which contains niobium as well as molybdenum to improve surface appearance of stretched and drawn parts. The addition of niobium helps prevent the phenomena of "ribbing" and "roping" on the surface of flat rolled products.

Type 436 has a commercial variant, AL 436S™ (Allegheny Ludlum Corp.), which contains titanium instead of niobium. It is used in applications such as automotive exhaust systems where good fabricability and resistance to chloride environments are required.

Project 70® Stainless 182-FM is a proprietary alloy produced by Carpenter Specialty Alloys. Its chromium, molybdenum, and manganese levels are higher than those of type 436, but it contains no titanium or niobium for carbide stabilization. Sulfur is added to improve machinability.

Increase in chromium content can provide improved resistance to pitting and crevice corrosion in chloride environments as well as resistance to high temperature oxidation. AL 433™ is a proprietary ferritic stainless steel (Allegheny Ludlum Corp.) which contains 20% chromium, about 2–4% higher than that of type 430 steel. Copper and niobium are added to further enhance properties. The latter element serves as a carbide stabilizer to prevent intergranular corrosion and to enhance high temperature strength.

Type 439 stainless steel is similar to type 430 in many ways. The former has a lower carbon content, and the addition of titanium for carbide stabilization. This grade of steel wad developed in the 1950s and later improved with the advent of the argon–oxygen decarburization process. It has good general corrosion resistance, particularly to chloride stress corrosion. Its composition is designed to retain good ductility after welding and to avoid intergranular corrosion, which makes it particularly suitable for applications such as heat exchanger tubing. An improved formability version of this steel, AL439HP™ (Allegheny Ludlum Corp.), has also been developed.

Type 441 stainless steel has chromium content at the 18% level but is dual stabilized with both niobium and titanium. Some of the niobium is present in solid solution during high temperature annealing. Upon cooling or holding at intermediate temperatures, fine particles of Laves phase (Fe_2Nb) are precipitated which markedly improve creep strength due to dispersion strengthening. This alloy has very good resistance to progressive scaling in both continuous and cyclic oxidizing conditions, with oxidation resistance comparable to that of AISI type 304 stainless steel.

For applications involving severe bending and forming operations, a proprietary grade, AL 441HP™ (Allegheny Ludlum Corp.), has been developed with tight control on interstitial elements such as carbon and nitrogen.

Type 443 is a first generation ferritic stainless steel with high carbon content. Chromium content is relatively high and copper is added to further improve corrosion resistance. It is, however, relatively weak in impact toughness, particularly in the welded condition.

Type 444 stainless steel has a high molybdenum content (1.75–2.5%). It is stabilized with niobium and titanium. Carbon and nitrogen are also kept at low levels. It is a general purpose stainless steel with corrosion resistance similar to type 316, but with much superior resistance to chloride stress corrosion cracking. In common with all ferritic stainless steels, however, it suffers from 475°C embrittlement.

The chromium level is increased to 23–27% in type 446 stainless steel in order to improve corrosion and oxidation resistance. There are a number of commercial variants of this grade. For example, Sandvik4C54 has a stricter control of phosphorous and sulfur. Type 446M0D-1 (UNS 44660) is a low carbon, high chromium, high molybdenum, and niobium and titanium stabilized version designed for applications in the chemical process industry. The excellent resistance to pitting, crevice and stress corrosion renders this material particularly suitable for heat exchanger applications.

The addition of certain rare earth metals (REM) can improve the high temperature oxidation resistance of ferritic stainless steels. AL453™ is a proprietary alloy produced by Allegheny Ludlum Corp. with cerium (Ce) and lanthanum (La) additions. These elements help stabilize the oxide scale and lower the risk of spallation.

One problem of titanium stabilized ferritic stainless steels is the susceptibility to surface defects, while in the niobium stabilized steels, there could be a risk of weld cracking. In AL468™ alloy (Allegheny Ludlum Corp.), titanium and niobium content is carefully controlled to achieve a balance that culminates in good weld ductility and resistance to intergranular corrosion. This dual stabilized alloy has a relatively low titanium content which also enhances brazing properties.

There are a number of proprietary iron–chromium alloys for magnetic applications that can be loosely classified as ferritic stainless steels. These are the Chrome Core 8, 8-FM, 12 and 12-FM alloys produced by Carpenter Specialty Alloys. Although the Chrome Core 8 and 8-FM alloys have chromium content below the 11% minimum required to qualify for "stainless" status, their corrosion resistance is not inferior to 18% chromium ferritic stainless steels in certain alcohol-base fuel environments. These alloys have been developed for applications in electromechanical devices requiring moderate corrosion resistance. They have the advantage over the 18% chromium ferritic stainless steels in that they have higher saturation magnetic induction.

B. Martensitic Stainless Steels

The chemical composition of martensitic stainless steels is very similar to that of their ferritic cousins except for the higher carbon content in the former. One effect of carbon is to expand the gamma loop in the iron–chromium phase diagram. Without the gamma loop, austenite cannot be formed and quench hardening to form martensite is not possible. Thus, the amount of austenite and ferrite formers must be carefully balanced in the chemical composition of the alloy in order to facilitate the formation of austenite at the heat treatment temperature. It is for this reason that the maximum chromium level in martensitic stainless steels is only about 18%, whereas in ferritic stainless steels, the maximum chromium content can be as high as 30%. Similar to their non-stainless counterparts, martensitic stainless steels are normally used in the quenched and tempered condition for better ductility and toughness. The chemical compositions of some selected martensitic stainless steels are listed in Table 2.

The characteristics of martensitic stainless steels are that:

- they have body-centered-tetragonal (bct) crystal structure;
- they are ferromagnetic at room temperature;
- they are quench hardenable;
- they are wear resistant;
- they have high yield strength and hardness;
- they have low room temperature impact strength;
- they are susceptible to hydrogen cracking, particularly in sulfide environments;
- they are susceptible to intergranular corrosion upon rapid cooling.

Type 410 is the basic grade of martensitic stainless steel. Most other martensitic stainless steels are modifications of type 410. It contains just sufficient chromium (11.5–13.5%) for "stainless" classification. The presence of up to 0.15% carbon leads to martensite formation even after air cooling from about 1000°C, resulting in high strength and hardness. This is one of the most popular grades of martensitic stainless steel, with applications ranging from cutlery to nuclear reactor control rod mechanisms.

Type 410S is a low carbon version of the basic alloy. The lower carbon content results in lower hardness but this disadvantage is offset by its better weldability.

Type 403 is similar to type 410 but has lower maximum chromium and silicon. This material has been developed for turbine blade applications where reasonable corrosion resistance, rigidity, and impact strength are required.

AL403™ and AL410™ are proprietary alloys produced by Allegheny Ludlum Corp. These proprietary grades contain up to 0.6% and 0.75% nickel, respectively. Nickel is an

Table 2 Chemical Compositions of Selected Martensitic Stainless Steels (wt%)

Type	C	Mn	P	S	Si	Cr	Ni	Mo	Others
403	< 0.15	< 1.0	< 0.04	< 0.03	< 0.5	11.5–13.0	—	—	
AL403™[a]	< 0.15	< 1.0	< 0.04	< 0.03	< 0.5	11.5–13.0	< 0.6	—	
410	< 0.15	< 1.0	< 0.04	< 0.03	< 1.0	11.5–13.5	—	—	
AL410™[a]	< 0.15	< 1.0	< 0.04	< 0.03	< 1.0	11.5–13.5	< 0.75	—	
410S (UNS S41008)	< 0.08	< 1.0	< 0.04	< 0.03	< 1.0	11.5–13.5	< 0.6	—	
414	< 0.15	< 1.0	< 0.04	< 0.03	< 1.0	11.5–13.5	1.25–2.5	—	
416	< 0.15	< 1.25	< 0.06	> 0.15	< 1.0	12.0–14.0	—	< 0.6	Mo is optional
416Se	< 0.15	< 1.25	< 0.06	0.06	< 1.0	12.0–14.0	—	—	Se > 1.0
420	> 0.15	< 1.0	< 0.04	< 0.03	< 1.0	12.0–14.0	—	—	
420F	> 0.15	< 1.25	< 0.06	> 0.15	< 1.0	12.0–14.0	—	< 0.6	Mo is optional
Sandvik4C27A[b]	0.22	1.6	< 0.03	0.18	0.6	13.0	0.8	1.2	
TrimRite[c]	0.15–0.30	< 1.0	< 0.04	< 0.03	< 1.0	13.5–15.0	0.25–1.0	0.4–1.0	
422	0.20–0.25	< 1.0	< 0.04	< 0.03	< 0.75	11.0–13.5	0.5–1.0	0.75–1.25	V 0.15–0.30, W 0.75–1.25
431	< 0.2	< 1.0	< 0.04	< 0.03	< 1.0	15.0–17.0	1.25–2.5		
440A	0.60–0.75	< 1.0	< 0.04	< 0.03	< 1.0	16.0–18.0	—	< 0.75	
440B	0.75–0.95	< 1.0	< 0.04	< 0.03	< 1.0	16.0–18.0	—	< 0.75	
440C	0.95–1.20	< 1.0	< 0.04	< 0.03	< 1.0	16.0–18.0	—	< 0.75	
440F	0.95–1.20	< 1.25	< 0.04	0.1–0.35	< 1.0	16.0–18.0	< 0.75	0.4–0.6	N < 0.08
440FSe	0.95–1.20	< 1.25	< 0.04	< 0.03	< 1.0	16.0–18.0	< 0.75	< 0.6	N < 0.08, Se > 0.15

Note: The chemical compositions given in the table are not specifications. They are typical values unless otherwise indicated.
[a]Proprietary alloy, Allegheny Ludlum Corp.
[b]Proprietary alloy, Sandvik Steel.
[c]Proprietary alloy, Carpenter Specialty Alloys.

austenite stabilizer which normally improves corrosion resistance and toughness. In type 414 stainless steel, nickel of 1.25–2.5% is added with a slight increase in tensile and impact strength compared with the basic type 410 grade.

For high strength, rigid components requiring intricate machining, such as shafts, axles, gears, and pinions etc., type 416 stainless steel with a minimum of 0.15% sulfur, is often used. Another free-machining version is type 416Se, which contains selenium. However, the improvement in machinability is accompanied by a countervailing effect of reduction in impact strength. Carpenter Specialty Alloys produces two commercial variants of type 416: Project 70® Stainless type 416 and 416BQ. The former has better machining characteristics than ordinary type 416, while the latter is capable of producing a minimum hardness of Rockwell C 40 when bright hardened.

While type 410 stainless steel and its variants have a *maximum* carbon content of 0.15%, this level of carbon has been set as the *minimum* for type 420. This increase in carbon content results in higher strength, hardness and wear resistance. Similar to type 410 steel, type 420 is difficult to weld. While it can be used at moderately high temperatures, it is not recommended for highly stressed components at subzero temperatures. Type 420 has been used for cutlery, dental and surgical instruments, nozzles, valve parts, and similar applications. It is also a popular steel for plastic moulds in view of its high strength, hardness, and corrosion resistance. A free-machining grade, type 420F with added sulfur, is commercially available for applications requiring intricate machining. However, the added sulfur has a slightly detrimental effect on notch toughness.

Sandvik 4C27A is a proprietary alloy similar to type 420 but with added molybdenum for enhanced corrosion resistance. This steel is a candidate material for the manufacture of components of precision instruments that require good machinability, high wear resistance, good toughness, and good corrosion resistance.

TrimRite is another commercial variant of type 420 stainless steel manufactured by Carpenter Specialty Alloys. The higher level of chromium, plus the presence of nickel and molybdenum, culminates in better corrosion resistance than the standard type 420 steel.

Although type 420 stainless steel can be used at moderately high temperatures, it is not a common material for applications involving significant departure from room temperature. Somewhat improved high temperature properties are found in type 422 steel. The additions of vanadium and tungsten result in the precipitation of fine vanadium and tungsten carbides and a concomitant improvement in elevated temperature strength. The strengthening effect of these carbides is further reinforced by solid-solution strengthening due to the addition of molybdenum. Nickel is also added to improve corrosion resistance and to balance the effect of the ferrite stabilizers such as vanadium, tungsten, and molybdenum.

Amongst all the common martensitic stainless steels, type 431 is one of the most corrosion resistant. The addition of 1.25–2.5% nickel and the relatively high chromium level (15–17%) are responsible for this improvement in corrosion resistance. It is used for highly stressed aircraft components such as fasteners, where a combination of good corrosion resistance, hardness, and toughness is required.

Types 440A, 440B, and 440C are a series of high-chromium high-carbon alloys. The carbon content increases from 0.6/0.75% in type 440A to 0.75/0.95% in type 440B and finally to 0.95/1.2% in type 440C with corresponding increase in hardness from Rockwell C 56 to 58 and 60, respectively. There is no stainless steel with a higher hardness than type 440C. These steels are used for applications such as cutlery and ball bearings. They are not suitable for high temperature applications. Types 440F and 440FSe are free-machining grades with sulfur and selenium additions, respectively. These grades are used when high hardness and machinability are major considerations. In view of their brittleness, type 440 steels are not suitable for applications such as gas cylinders.

C. Austenitic Stainless Steels

Austenitic stainless steels differ from ferritic and martensitic stainless steels in that they remain fully austenitic at room temperature due to the addition of austenite promoting elements such as nickel, and in some instances, manganese. It is by far the most widely used amongst all categories of stainless steels.

The characteristics of austenitic stainless steels are that:

- they have face-centered-cubic (fcc) crystal structure;
- they are non-ferromagnetic at room temperature;
- they cannot be quench hardened;
- they have good ductility, formability, and toughness;
- they can be hardened by cold work;
- they are susceptible to chloride stress corrosion.

Austenitic stainless steels using nickel as the major austenite stabilizer are grouped into the 300 series under the AISI designations. Of over the 350 different grades of commercial stainless steels listed in the UNS system [1], over 40% fall into this category. The other category of austenitic stainless steels, i.e. the 200 series, consists of grades containing manganese to replace part of the nickel as the austenite stabilizer.

1. Type 300 Series Austenitic Stainless Steels

The first member of this series, type 301, has 16–18% chromium, 6–8% nickel, and a carbon content of up to 0.15% (Table 3). This steel can be substantially work hardened by cold work, which also promotes the formation of martensite. While the steel is non-ferromagnetic at room temperature in the annealed condition, it becomes ferromagnetic in the cold-worked condition due to the formation of martensite. It is supplied in the annealed as well as different levels of work-hardened conditions to meet the mechanical and magnetic properties requirements in a variety of applications, including automobile wheel covers, stainless steel sinks, springs, etc.

Type 302 is the basic 18/8 stainless steel grade which is frequently used as a general purpose corrosion-resistant material. The higher nickel content compared to type 301 steel (8–10% instead of 6–8%) reduces the susceptibility to martensite formation during cold working. It is extremely tough and ductile in the annealed condition. This makes it ideally suited to the manufacture of components that require deep drawing, bending, forming or upsetting. The tendency to work harden is further reduced in a specially designed proprietary alloy, Carpenter Stainless Custom Flo 302HQ, which contains 3–4% copper and less than 0.03% carbon to minimize tool wear and cracking during cold heading operations. If good machining characteristics are also required, another proprietary grade, 302HQ-FM, has been developed by Carpenter Specialty Alloys. However, the high level of sulfur in the latter alloy renders it unsuitable for applications involving welding. Type 302B is another modification of type 302 involving an increase in silicon content to improve high temperature oxidation resistance.

Since austenitic stainless steels have a high work-hardening rate and ductility, they are often not easy to machine and they have a tendency to generate stringy and tangling

Table 3 Chemical Compositions of Selected Type 300 Series Austenitic Stainless Steels (wt%)

Type	C	Mn	P	S	Si	Cr	Ni	Others
301	< 0.15	< 2.0	< 0.045	< 0.03	< 1.0	16.0–18.0	6.0–8.0	
302	< 0.15	< 2.0	< 0.045	< 0.03	< 1.0	17.0–19.0	8.0–10.0	
302HQ[a]	< 0.03	< 2.0	< 0.045	< 0.03	< 1.0	17.0–19.0	8.0–10.0	Cu 3.0–4.0
302HQ-FM[a]	< 0.06	< 2.0	< 0.040	< 0.14	< 1.0	16.0–19.0	9.0–11.0	Cu 1.3–2.4
302B (UNS S30215)	< 0.15	< 2.0	< 0.045	< 0.03	2.0–3.0	17.0–19.0	8.0–10.0	
303	< 0.15	< 2.0	< 0.2	> 0.15	< 1.0	17.0–19.0	8.0–10.0	Mo < 0.6 (optional)
303Se	< 0.15	< 2.0	< 0.2	< 0.06	< 1.0	17.0–19.0	8.0–10.0	Se > 0.15
303MA	< 0.15	< 2.0	< 0.05	0.11–0.16	< 1.0	17.0–19.0	8.0–10.0	Al 0.6–1.0, Mo 0.4–0.6
304	< 0.08	< 2.0	< 0.045	< 0.03	< 1.0	18.0–20.0	8.0–10.5	
304L	< 0.03	< 2.0	< 0.045	< 0.03	< 1.0	18.0–20.0	8.0–12.0	
304N	< 0.08	< 2.0	< 0.045	< 0.03	< 1.0	18.0–20.0	8.0–10.5	N 0.1–0.16
Micro-Melt NeutroSorb Plus[a]	< 0.08	< 2.0	< 0.045	< 0.03	< 0.75	18.0–20.0	12.0–15.0	N < 0.1, B < 2.25
305	< 0.12	< 2.0	< 0.045	< 0.03	< 1.0	17.0–19.0	10.0–13.0	
309	< 0.2	< 2.0	< 0.045	< 0.03	< 1.0	22.0–24.0	12.0–15.0	
309S	< 0.08	< 2.0	< 0.045	< 0.03	< 1.0	22.0–24.0	12.0–15.0	
310	< 0.25	< 2.0	< 0.045	< 0.03	< 1.5	24.0–26.0	19.0–22.0	
310S	< 0.08	< 2.0	< 0.045	< 0.03	< 1.5	24.0–26.0	19.0–22.0	
316	< 0.08	< 2.0	< 0.045	< 0.03	< 1.0	16.0–18.0	10.0–14.0	Mo 2.0–3.0
316L	< 0.03	< 2.0	< 0.045	< 0.03	< 1.0	16.0–18.0	10.0–14.0	Mo 2.0–3.0
316N	< 0.08	< 2.0	< 0.045	< 0.03	< 1.0	16.0–18.0	10.0–14.0	Mo 2.0–3.0, N 0.1–0.16
316LN	< 0.03	< 2.0	< 0.045	< 0.03	< 1.0	16.0–18.0	10.0–14.0	Mo 2.0–3.0, N 0.1–0.16
Carpenter Stainless Custom Flo 316HQ[a]	< 0.03	< 2.0	< 0.03	< 0.015	< 1.0	16.0–18.25	10.0–14.0	Mo 2.0–3.0, Cu 3.0–4.0
317	< 0.08	< 2.0	< 0.045	< 0.03	< 1.0	18.0–20.0	11.0–15.0	Mo 3.0–4.0
317L	< 0.03	< 2.0	< 0.045	< 0.03	< 1.0	18.0–20.0	11.0–15.0	Mo 3.0–4.0
321	< 0.08	< 2.0	< 0.045	< 0.03	< 1.0	17.0–19.0	9.0–12.0	Ti > 5 × C
347	< 0.08	< 2.0	< 0.045	< 0.03	< 1.0	17.0–19.0	9.0–13.0	Nb > 10 × C
348	< 008	< 2.0	< 0.045	< 0.03	< 1.0	17.0–19.0	9.0–13.0	Nb > 10 × C, Co < 0.2,Ta < 0.1
334	< 0.08	< 1.0	< 0.03	< 0.015	< 1.0	18.0–20.0	19.0–21.0	Al 0.15–0.6, Ti 0.15–0.6
384	< 0.08	< 2.0	< 0.045	< 0.03	< 1.0	15.0–17.0	17.0–19.0	

Note: The chemical compositions given in the table are not specifications. They are typical values unless otherwise indicated.
[a]Proprietary alloy, Carpenter Specialty Alloys.

chips. Type 303 is the first chrome-nickel stainless steel developed with free-machining characteristics. There are a number of proprietary alloys based on this composition with varying degrees of improvement in machining speeds, e.g. Carpenter Project 70 & 7000 types 303 and 303DQ. The good machining properties are achieved with the addition of at least 0.15% sulfur, which has the countervailing effects of lowering formability and rendering the material susceptible to hot cracking during welding. The detrimental effect on cold formability can be somewhat ameliorated by replacing part of the sulfur with selenium (type 303Se) or aluminum (type 303MA).

With the advent of the argon–oxygen decarburization process, the carbon content in steels could be reduced without severe cost penalty. Type 304 is essentially a low carbon version of type 302 with corresponding improvements in corrosion resistance and fabricability. This is the most popular grade of stainless steel in terms of production volume. It is so versatile that the list of applications is endless. For applications involving welding a low carbon version, type 304L is recommended. The low carbon content ($<0.03\%$) minimizes the precipitation of chromium carbides and reduces the risk of intergranular corrosion in service.

Interstitial elements such as nitrogen and boron are known to have a significant effect on the strength and high temperature ductility of 18/8 stainless steels [3,4]. In type 304N, nitrogen of between 0.1% and 0.16% is added to improve strength without any significant countervailing deleterious effects. Other varieties include type 304LN, which has a low carbon content but with nitrogen addition to make up for the loss in strength.

While traces of boron at the ppm level would increase the long term ductility of austenitic stainless steels [3,4], a large amount of boron addition would lead to reductions in tensile ductility, impact toughness, and corrosion resistance with concomitant increases in hardness and tensile and yield strength. For nuclear reactor applications, the addition of large amount of boron has the advantage of improving the neutron absorption capability of the material. Special alloys, such as the Carpenter Micro-Melt® NeutroSorb Plus® Alloys, have been developed for this specific application.

Allegheny Ludlum Corp. has developed a special three-ply stainless/carbon steel composite for use in conventional cookware applications. In AL304DA, a titanium and niobium stabilized low carbon ferritic steel center is sandwiched between two layers of type 304 stainless steel cladding. The 304 steel cladding is bonded to the low carbon ferritic steel core by hot rolling. This composite has the good corrosion resistance and food compatibility of type 304 stainless steel, as well as the good heat transfer properties of ferritic steel. This unique combination of properties renders it particularly suitable for cookware and food processing applications.

Severe cold work can lead to martensite formation in the 300 series stainless steels described so far, leading to a tendency to work harden during cold-forming operations. Increasing nickel content can inhibit martensite formation, culminating in lower work-hardening rates and better cold formability. In type 305 stainless steel, the nickel content is raised to 10–13% and this material is particularly suitable for applications involving severe cold work operations, such as cold headed bolts and screws.

In type 309 stainless steel, the chromium and nickel contents are substantially increased to 22–24% and 12–15%, respectively, for better heat-resisting properties such as creep strength and oxidation resistance. However, the carbon content of up to 0.2% in this material is rather high and may lead to intergranular corrosion during welding applications. In this respect, type 309S, with a maximum carbon content of 0.8%, offers better as-welded corrosion resistance.

The chromium and nickel content is further increased to 24–26% and 19–22%, respectively, in type 310 stainless steel. This alloy has excellent corrosion resistance and elevated temperature strength, and can resist oxidation in continuous service at temperatures up to 1150°C. For welding applications, the lower carbon version, type 310S, is recommended if intergranular corrosion is to be avoided.

Type 316 stainless steel differs from the austenitic stainless steels described so far in that it contains molybdenum to enhance corrosion resistance, particularly resistance to pitting and crevice corrosion. It is used in a wide variety of applications and has popularity approaching that of type 304 steel. Similar to type 304 steel, low carbon (316L) and nitrogen added (316N, 316LN) versions are commercially available. The low carbon version is less susceptible to intergranular corrosion in welding applications, while the nitrogen added will enhance strength.

For enhanced resistance to sulphurous acid compounds, a proprietary alloy, Carpenter Stainless Custom Flo 316HQ, has been developed. It differs from ordinary 316 stainless steel in that between 3% and 4% of copper is added to this alloy. It has low tensile strength and low work-hardening rate during cold forming, which make it suitable for applications such as cold headed nuts, bolts, screws, fittings, etc. This material has been used in equipment for handling various chemicals such as bleaches and dyestuffs.

Type 317 stainless steel is similar to type 316 but with higher molybdenum, chromium and nickel content. The increase in the content of these elements improves the pitting and general corrosion resistance of the alloy. It is generally used in the chemical process industries, particularly in situations where good resistance to sulfurous acids compounds is required. For applications involving welding, a low carbon version, 317L, is available to minimize the risk of intergranular corrosion.

In order to avoid chromium depletion due to chromium carbide precipitation, strong carbide formers are sometimes added to remove supersaturated carbon from solid solution. In type 321 stainless steel, titanium is added as a carbide stabilizer. This has the benefit of enabling welded joints to be used in the as-welded condition without post-weld heat treatment because susceptibility to sensitization in the form of intergranular corrosion is minimized.

In type 347 stainless steel, niobium is used as the carbide stabilizer instead of titanium. For both type 321 and 347 steels, the mechanical properties at room temperature are not significantly affected by titanium or niobium additions. However, the elevated temperature creep strength is improved by these additions due to the dispersion strengthening effect of the finely dispersed titanium or niobium carbides. Type 347 is slightly more corrosion resistant than 321 in strongly oxidizing environments. Moreover, thermal aging at temperatures between 450°C and 800°C lowers the overall corrosion resistance of type 321 to a greater extent than type 347.

Some variants of type 347 steel contain tantalum as well as niobium as carbide stabilizers (e.g. UNS 34751, UNS 34781, UNS 34788). For nuclear applications involving strong ionizing radiations type 348 steel, which is similar to type 347 but with restricted cobalt and tantalum content, is recommended.

Type 334 is an aluminum and titanium bearing austenitic stainless steel with higher chromium and nickel content than the standard 18/8 grade. At 18–20% chromium, this alloy can maintain a protective chromium oxide scale at elevated temperatures. In addition, the relatively high nickel content (19–21%) helps improve the oxidation resistance of the alloy. In view of its good formability, weldability, and corrosion resistance, type 334 is a good candidate material for the sheath of resistance-heated electric heating elements.

Type 384 is similar to type 334 but without the aluminum and titanium. Its chromium and nickel content is also slightly lower than that of type 334. This alloy has a low work-hardening rate and is particularly suitable for applications involving severe forming operations.

2. Type 200 Series Austenitic Stainless Steels

Type 200 series austenitic stainless steels are characterized by replacing part of the nickel with manganese as an austenite-forming element. As a rule of thumb, approximately 2% of manganese is used to replace every 1% of nickel. Manganese alone, even if added in large amounts, cannot guarantee a fully austenitic structure in alloys with more than 15% chromium. Other austenite stabilizers such as carbon, nitrogen, copper, and cobalt are often needed to ensure a completely austenitic structure. The addition of manganese also has the concomitant effect of increasing the solubility of nitrogen in the austenite phase. Thus, nitrogen can be added in larger quantities of type 200 series stainless steels than to the type 300 series counterparts. Nitrogen is known to inhibit the precipitation of chromium and molybdenum-rich phases, decrease the grain size in recrystallized material and enhance the resistance to acid attack through the formation of ammonia at the metal surface. Nitrogen-containing steels also tend to have a superior combination of strength and toughness relative to conventional steels (Table 4).

Type 201 and its low carbon version, type 201L, are amongst the most popular of the type 200 series austenitic stainless steels. This grade was developed as a substitute for its 300 series counterpart, type 301. The partial replacement of the expensive nickel with the cheaper manganese results in significant cost reductions. In terms of properties, it is similar to type 310 in many respects. However, at temperatures above about 800°C, the oxidation resistance of type 201 is worse than that of type 301 and is not recommended for service above this temperature, particularly when rapid cooling is involved. For cryogenic applications such as the containment of liquefied gases, type 201LN has been specially developed to provide sufficient austenite stability to inhibit martensite formation at low temperatures.

In the same family of type 201 steel are proprietary alloys such as Gall-Tough® and Gall-Tough® PLUS produced by Carpenter Specialty Alloys. These wear resistant steels contain high silicon content, manganese, and nitrogen. The Gall-Tough® PLUS alloy also contains between 0.5% and 2.5% of molybdenum. These alloys are designed for moving components under aggressive environments and high stresses without the ameliorating effects of lubricants.

In parallel with type 302 steel, which has higher nickel than type 301 to reduce the susceptibility to martensite formation during cold work, type 202 has higher manganese than type 201 for austenite stability. In the annealed condition, type 202 is stronger and slightly more ductile than its 300 series counterpart. In general, type 202 can replace type 302 in most applications.

For applications involving intricate machining, type 203EZ, with 0.18–0.35% sulfur, is recommended.

Type 204 steel, also known as Nitronic 30 or AL30™, has higher strength than its 300 series counterpart, type 304. It also has better stress corrosion resistance. Type 204-Cu is a proprietary alloy made by Carpenter Specialty Alloys. The copper addition helps reduce the work-hardening rate to the same level as type 304. Its cold-forming characteristics are similar to type 304 and are better than most 200 series stainless steels.

Type 205 is a grade where nickel is almost completely replaced by austenite stabilizers such as manganese, carbon, and nitrogen. This steel has a low work-hardening rate and is particularly suitable for applications involving heavy cold work.

Table 4 Chemical Composition of Selected Type 200 Series Austenitic Stainless Steels (wt%)

Type	C	Mn	P	S	Si	Cr	Ni	N	Others
201	< 0.15	5.5–7.5	< 0.06	< 0.03	< 1.0	16.0–18.0	3.5–5.5	< 0.25	
201L	< 0.03	5.5–7.5	< 0.06	< 0.03	< 0.75	16.0–18.0	3.5–5.5	< 0.25	
201LN	< 0.03	6.4–7.5	< 0.045	< 0.015	< 0.75	16.0–17.5	4.0–5.0	0.1–0.25	Cu < 1.0
Gall-Tough®[a]	< 0.15	4.0–6.0	< 0.04	< 0.04	3.0–4.0	15.0–18.0	4.0–6.0	0.08–0.2	
Gall-Tough Plus™[a]	< 0.15	4.0–8.0	< 0.04	< 0.04	2.5–4.5	16.5–21.0	6.0–10.0	0.05–0.25	Mo 0.5–2.5
202	< 0.15	7.5–10.0	< 0.06	< 0.03	< 1.0	17.0–19.0	4.0–6.0	< 0.25	
203EZ	< 0.08	5.0–6.5	< 0.04	0.18–0.35	< 1.0	16.0–18.0	5.0–6.5	—	Cu 1.75–2.25, Mo < 0.5
204 (AL30™)[b]	< 0.03	7.0–9.0	< 0.04	< 0.03	< 1.0	15.0–17.0	1.5–3.0	0.15–0.3	
204-Cu[a]	< 0.15	6.5–9.0	< 0.06	< 0.03	< 1.0	15.5–17.5	1.5–3.5	0.05–0.25	Cu 2.0–4.0
205	< 0.25	14.0–15.5	< 0.06	< 0.03	< 1.0	16.5–18.0	1.0–1.75	< 0.4	
22Cr–13Ni–5Mn (UNS S20910)	< 0.06	4.0–6.0	< 004	< 0.03	< 1.0	20.5–23.5	11.5–13.5	0.2–0.4	Nb 0.1–0.3, Mo 1.5–3.0, V 0.1–0.3
SCF 19[a](UNS S21000)	< 0.1	4.0–7.0	< 0.03	< 0.03	< 0.6	18.0–23.0	16.0–20.0	> 0.15	Mo 4.0–6.0, Cu < 2.0
Tenelon (UNS S21400)	< 0.12	14.5–16.0	< 0.045	< 0.03	< 1.0	17.0–18.5	< 0.75	< 0.35	
Cryogenic Tenelon (UNS S21460)	< 0.12	14.0–16.0	< 0.06	< 0.03	< 1.0	17.0–19.0	5.0–6.0	< 0.5	
Esshete 1250 (UNS S215)	0.06–0.15	5.5–7.0	< 0.04	< 0.03	0.2–1.2	14.0–16.0	9.0–11.0	—	B 0.003–0.009, V 0.15–0.4 Nb 0.75–1.25, Mo 0.8–1.2
216	< 0.08	7.5–9.0	< 0.045	< 0.03	< 1.0	17.5–22.0	5.0–7.0	0.25–0.5	Mo 2.0–3.0
216L	< 0.03	7.5–9.0	< 0.045	< 0.03	< 1.0	17.5–22.0	5.0–7.0	0.25–0.5	Mo 2.0–3.0
2lCr–6Ni–9Mn (UNS S21900)	< 0.08	8.0–10.0	< 0.06	< 0.03	< 1.0	19.0–21.5	5.5–7.5	0.15–0.4	
21Cr–6Ni–9Mn LC)(UNS S21904)	< 0.04	8.0–10.0	< 0.06	< 0.03	< 1.0	19.0–21.5	5.5–7.5	0.15–0.4	
241	< 0.15	11.0–14.0	< 0.06	< 0.03	< 1.0	16.5–19.5	0.5–2.5	0.2–0.45	
18-18 PLUS (UNS S28200)	< 0.15	17.0–19.0	< 0.045	< 0.03	< 1.0	17.0–19.0	—	0.4–0.6	Cu 0.5–1.5,Mo 0.5–1.5
15-15 LC[a]	0.04	15.0–19.0	< 0.05	< 0.05	< 1.0	16.0–21.0	< 3.0	0.2–0.8	Mo 0.5–3.0

Note: The chemical compositions given in the table are not specifications. They are typical values unless otherwise indicated.
[a]Proprietary alloy, Carpenter Specialty Alloys.
[b]Proprietary alloy, Allegheny Ludlum Corp.

There are a number of commercial alloys which correspond to UNS S20910:22Cr–13Ni–5Mn, Nitrogen 50, AL50™, XM19, Aquamet 22. It has better strength and corrosion resistance than type 316 steel. This is a versatile alloy which has been used in the petrochemical, pulp and paper, and food processing industries.

The proprietary alloy, SCF 19®, patented by Carpenter Specialty Alloys, has higher nickel (16.0–20.0%) and molybdenum (4.0–6.0%) content than type 209. Nitrogen is also specified at more than 0.15%. Its characteristics are resistance to stress corrosion cracking, high strength and low magnetic permeability.

Nickel is reduced to less than 0.75% in the commercial alloy "Tenelon", which is very similar to type 205 but with lower carbon and nickel content. For low temperature applications, nickel is raised back to 5.0–6.0% in order to improve low temperature toughness in another commercial alloy "Cryogenic Tenelon". Tenelon and Cryogenic Tenelon have been given the UNS designations S21400 and S21460, respectively.

Esshete 1250 is a high creep strength alloy which has been considered for use as power plant boiler tubes. Enhanced creep properties are provided by additions of boron, niobium, and vanadium. It has been given the UNS designation S21500.

Types 216 and 216L are the equivalents of types 316 and 316L. In terms of the carbon and molybdenum levels, they are similar. However, the presence of manganese in the 200 series grades increases the solubility of nitrogen. The nitrogen content is 0.25–5.0% in type 216 compared with 0.16% in type 316N. The former also has higher upper limit in chromium—22% in type 216 compared with 18% in type 316. The nitrogen helps to promote resistance to pitting and crevice corrosion.

Type 219 stainless steel, also known as 21Cr–6Ni–9Mn, Nitronic 40, AL40™, or XM-18, has a good combination of strength and corrosion properties. It is resistant to corrosion by molten lead oxide and high temperature oxidation in air. It has a high work-hardening rate and remains non-magnetic even after severe cold work. There is a low carbon version of this grade, 21–6-9LC, with UNS designation S21904.

Type 241 has a chemical composition similar to type 205 but with lower carbon and manganese content. It has higher yield and tensile strengths than type 304 stainless steel and has been used to replace type 304 in applications where the strength or magnetic permeability of the latter, after cold working, is unsuitable.

18-18 PLUS (UNS S28200) is a nickel-free austenitic stainless steel. As mentioned before, as a rule of thumb, 2% of manganese is equivalent to about 1% of nickel as far as the replacement of nickel by manganese is concerned. Thus, 18-18 PLUS is equivalent to 18% Cr 9Ni, such as types 304 or 316. In fact, the yield strength of 18-18 PLUS is about double that of type 304, with corrosion resistance intermediate between types 304 and 316. It is a gall-resistant material and does not form martensite even after severe cold work. Despite its high strength, it can be cold formed, but the work-hardening rate is considerable. It is also less machinable than type 316.

15-15 LC® Modified Stainless is a proprietary alloy produced by Carpenter Specialty Alloys. The relatively low carbon content of this material reduces its susceptibility to intergranular corrosion. It is a candidate material for non-magnetic drill collars, stabilizers, and MWD housings in the oil and gas industries.

D. Precipitation-Hardening Stainless Steels

We have described three categories of stainless steels so far. Ferritic and austenitic stainless steels are not hardenable by heat treatment. Martensitic stainless steels derive their hardness from the martensite formed as a result of quenching from the austenite phase. Precipitation-hardening stainless steels, or PH stainless steels, are chromium–nickel steels which can be hardened by thermal aging treatments. These steels usually contain small amounts of elements such as copper, niobium, titanium or aluminum which form precipitates at temperature between 450°C and 800°C. Thermal aging results in the precipitation of very fine intermetallics, carbides, and/or phosphides. These precipitates are obstacles to dislocation movement, making the material more resistant to deformation.

Austenitic stainless steels, which are not hardenable by thermal aging treatments, can be strengthened by cold work. However, if the cold-worked material is subsequently exposed to elevated temperatures, such as in the heat-affected zones of a welded joint, the strength derived from prior cold work may be lost due to recovery processes. Precipitation-hardening stainless steels do not suffer from this disadvantage. The thermal aging treatment may be carried out after welding and fabrication of the component to achieve the desired properties. The strength of the PH steels will only deteriorate if the service temperatures are so high that coarsening of the precipitates occurs.

The characteristics of PH stainless steels are that:

- they can be hardened by aging treatment;
- they have moderate corrosion resistance;
- they have good ductility and toughness;
- they have high yield (515–1415 MPa) and tensile (860–1520 MPa) strengths;
- they are susceptible to hydrogen embrittlement.

There are three types of PH stainless steels: martensitic, semi-austenitic, and austenitic.

The chemical composition of the martensitic PH stainless steels is such that the M_S (martensite start) temperature is well above room temperature. Thus, when the steel is heated to the austenite phase field at about 1000°C and then rapidly cooled, a fully martensitic structure is formed. The steel is then aged at 480–620°C to produce precipitation hardening (Table 5).

The addition of elements such as chromium, nickel, carbon, and nitrogen lowers the M_S temperature and reduces the tendency to form martensite upon cooling. In semi-austenitic PH stainless steels, rapid cooling from the austenite phase field to room temperature results in a metastable austenitic structure. The austenite is very ductile and the steel is easily formed into shape in this condition. After forming, the metastable austenite is transformed into martensite by further cooling to about −73°C, or by tempering at about 760°C to reduce the carbon in solution followed by cooling to room temperature. The martensite produced by cooling to subzero temperatures contains more carbon, and is of a higher strength. In both cases, the steel is then aged at 450–579°C to produce precipitation hardening.

In austenitic PH stainless steels, the content of austenite stabilizing elements is sufficiently high to maintain a fully stable austenitic structure at room temperature. Such steels are aged at 700–800°C to produce fine intermetallic compounds for strengthening.

1. Martensitic Grade

Stainless W (UNS S17600 or AISI 635) is the first commercial PH stainless steel introduced in 1946. The carbon is precipitated as titanium carbide in this material and the

Table 5 Chemical Compositions of Selected Precipitation-Hardening Stainless Steels (wt%)

Type	C	Mn	P	S	Si	Cr	Ni	Al	Others
Martensitic									
17-4PH (UNS S17400)(AISI 630)	< 0.07	< 1.0	< 0.04	< 0.03	< 1.0	15.0–17.5	3.0–5.0	—	Nb 0.15–0.45, Cu 3.0–5.0
Stainless W (UNS S17600)(AISI 635)	< 0.08	< 1.0	< 0.04	< 0.03	< 1.0	15.0–17.5	6.0–7.5	< 0.4	Ti 0.4–1.2
15-5PH (UNS S15500)	< 0.07	< 1.0	< 0.04	< 0.03	< 1.0	14.0–15.5	3.0–5.5	—	Nb 0.15–0.45, Cu 2.5–4.5
13-8Mo (UNS S13800)	< 0.05	< 0.2	< 0.01	< 0.008	< 0.1	12.25–13.25	7.5–8.5	0.9–1.35	Mo 2.0–2.5, N < 0.01
Custom 450[a]4 (UNS S45000)	< 0.05	< 1.0	< 0.03	< 0.03	< 1.0	14.0–16.0	5.0–7.0	—	Nb > 8 × C, Cu 1.25–1.75, Mo 0.5–1.0
Custom 455[a] (UNS S45500)	< 0.05	< 0.5	< 0.04	< 0.03	< 0.5	11.0–12.5	7.5–9.5	—	Nb + Ta 0.1–0.5, Cu 1.5–2.5, Mo < 0.5, Ti 0.8–1.4
Custom 465[a]	< 0.02	< 0.25	< 0.015	< 0.01	< 0.25	11.0–12.5	10.5–12.5	—	Mo 0.75–1.25, Ti 1.5–1.8
Semi-austenitic									
PH15-7Mo (UNS S15700)	< 0.09	< 1.0	< 0.04	< 0.03	< 1.0	14.0–16.0	6.5–7.75	0.75–1.5	Mo 2.0–3.0
Pyromet® Alloy 350 (UNS S35000) (AISI 633)[a]	0.07–0.11	0.5–1.25	< 0.04	< 0.03	< 0.5	16.0–17.0	4.0–5.0	—	Mo 2.5–3.25, N 0.07–0.13
Pyromet® Alloy 355 (UNS S35500) (AISI 634)[a]	0.1–0.15	0.5–1.25	< 0.04	< 0.03	< 0.5	15.0–16.0	4.0–5.0	—	Mo 2.5–3.25, N 0.07–0.13
Austenitic									
A286 (UNS S66286)	< 0.08	< 2.0	< 0.04	< 0.03	< 1.0	13.5–16.0	24.0–27.0	< 0.35	B 0.001–0.01, Mo 1.0–1.5 Ti 1.9–2.35, V 0.1–0.5

Note: The chemical compositions given in the table are not specifications. They are typical values unless otherwise indicated.
[a]Proprietary alloy, Carpenter Specialty Alloys.

as-quenched hardness is lower than that of comparable martensitic stainless steels such as type 410. Thus, the material can be easily worked in this condition before aging treatment is carried out to achieve the final hardness. A yield strength of 1240 MPa can be achieved after aging treatment.

Type 17-4PH is a martensitic PH stainless steel offering high strength and hardness. It has better general corrosion resistance than the 400 series martensitic stainless steels and in many situations, its corrosion resistance is close to that of type 304. The presence of 3–5% copper is responsible for the formation of a very fine precipitate-hardening phase. In common with most PH stainless steel, it is supplied in several heat treatment conditions. Condition A is the solution treated condition with no aging. Solution treatment is carried out at 1035°C for 30 min, followed by oil cooling for sections under 75 mm, and rapid air cooling for sections larger than 75 mm. The final structure is completely martensitic. Material in this condition is not normally used because of its susceptibility to stress corrosion. Condition H is the aged–hardened condition. The letter H is followed by the age-hardening temperature in °F. For example, H900 indicates aging at 900°F (482°C). The ageing temperature varies from 900°F to 1150°F (480–620°C). Higher ageing temperature can improve impact toughness, but this is achieved at the expense of reduction in yield strength. There is also some reverse transformation of martensite to austenite at higher aging temperatures. The age-hardening treatment also leads to a slight dimensional contraction in the component.

15-5PH has composition specification similar to that of 17-4PH. In fact, there is some overlap in their composition specifications. The slightly lower chromium and higher nickel in 15-5PH are designed to eliminate the presence of traces of delta ferrite, with concomitant improvement in transverse toughness, ductility, and formability.

For heavy section components, improvement in toughness and ductility in both longitudinal and transverse directions is desirable. In PH13-8Mo steel, tight control of elements such as sulfur, silicon and manganese is achieved through specialized melting techniques such as vacuum induction.

Custom 450 (Carpenter Specialty Alloys) has corrosion resistance comparable to type 304 but three times its yield strength. This age-hardening martensitic stainless steel is often used in situations where the strength of type 304 is not sufficient. It has good welding and brazing properties and no preheating is required before welding.

Another proprietary alloy developed by Carpenter Specialty Alloys is Custom 455. It has a low work-hardening rate and can be extensively cold formed in the annealed condition. Very high yield and tensile strengths can be achieved after a one-step age-hardening treatment.

Custom 465™ (Carpenter Specialty Alloys) also has very high yield and tensile strengths in the peak aged (H900) condition. It is specifically developed for applications requiring excellent notch tensile strength and fracture toughness.

2. Semi-Austenitic Grade

PH15-7Mo is a semi-austenitic precipitation-hardening stainless steel with chemical composition adjusted so that quenching from the solution treatment temperature produces metastable austenite. In this condition, the yield strength of the material is not a lot higher than that of ordinary austenitic stainless steels, and it can be readily cold formed in the same way. In common with other PH steels, this steel is supplied in a variety of heat treatment conditions. After solution treatment, a "conditioning" treatment is applied to turn the metastable austenite into martensite. This involves heat treatment at 955°C to reduce carbon from solution and destabilize the austenite, or by subzero cooling. Subsequently, aging at 510°C produces NiAl intermetallic phases which provide significant increases in strength and hardness.

Pyromet® Alloy 350 (AISI 633, UNS S35000) is similar to PH15-7Mo. The notable differences are lower nickel content and addition of nitrogen in the former. This alloy is used in applications requiring high strength at moderate temperatures. Hardening is achieved by subzero cooling and tempering (SCT) or by double aging (DA). The precipitate responsible for strengthening is chromium nitride. Subzero cooling has the advantage that the resulting material is of a higher strength and is less susceptible to intergranular corrosion. For applications requiring higher ductility and better machinability, an equalizing and overtempering treatment is applied. Pyromet® Alloy 355 (AISI 634, UNS S35500) is a similar alloy with slightly lower chromium and higher carbon content.

3. Austenitic Grade

The austenitic grade is not as popular as the other two grades because the increase in strength due to precipitate hardening is relatively modest. A286 (UNS S66286) is a typical example of austenitic PH stainless steel. The high nickel content ensures a fully austenitic structure after solution treating and quenching. Strengthening is due to the precipitation of the intermetallic phase $Ni_3(Al,Ti)$. However, even when fully hardened, its yield strength is only about 650 MPa, which is substantially lower than that of the martensitic or semi-austenitic grades. Its main advantages are its excellent corrosion resistance due to its high alloy content, and its suitability for applications involving high magnetic fields because of the absence of magnetic phases.

E. Duplex Stainless Steels

Duplex stainless steels contain both ferrite and austenite in significant quantities at room temperature. The first commercial duplex stainless steel was produced in 1929 by Avesta Jernverk. The steel, designated as 453E, had 25% Cr and 5% Ni. The shortage of nickel during the Korean War (1950–1951) stimulated the development of duplex stainless steels, which contained less nickel than the austenitic grades. However, while the duplex grades were known to have advantages such as higher strength and resistance to stress corrosion cracking relative to the fully austenitic grades, they were inferior in toughness, ductility, and ease of fabrication. In this first generation of duplex stainless steels, the problem was due to the continuous network of ferrite, particularly in the heat-affected zones of welded joints. As such, the world production of duplex stainless steels was much less than the austenitic grades and they tended to be used in very specific circumstances (Table 6).

The development of duplex stainless steels was given another boost in the 1970s when there was another nickel shortage. Moreover, the boom in the offshore oil industry led to a demand for steels with good resistance to the aggressive marine environments. The advent of vacuum and argon-oxygen decarburization processes in steel making allowed tighter control of residual elements, in particular carbon, nitrogen, sulfur, and oxygen, resulting in better formability and less likelihood of cracking during hot working. The addition of nitrogen marked the development of the so-called second generation duplex stainless steels. The nitrogen helps promote austenite stability and break up the ferrite network. This has enabled these steels to be used in the as-welded condition without post-weld heat treatment.

Table 6 Chemical Compositions of Selected Duplex Stainless Steels (wt%)

Type	C	Mn	P	S	Si	Cr	Ni	N	Mo	Cu	Others
Alloy lean											
UNS S31500 (3RE60)[a]	< 0.03	1.2–2.0	< 0.03	< 0.03	1.4–2.0	18.0–19.0	4.25–5.25	—	2.5–3.0	—	
UNS S32304 (2304)	< 0.03	< 2.5	< 0.04	< 0.04	< 1.0	21.5–24.5	3.0–5.5	0.05–2.0	—	0.05–0.6	
UNS S32404 (UR50)	< 0.04	< 2.0	< 0.03	< 0.01	< 1.0	20.5–22.5	5.5–8.5	< 0.2	2.0–3.0	1.0–2.0	
Standard 22% Cr											
UNS S31803 (2205)	< 0.03	< 2.0	< 0.03	< 0.02	< 1.0	21.0–23.0	4.5–6.5	0.08–0.2	2.5–3.5	—	
UNS S35505 (2205, high N)	< 0.03	< 2.0	< 0.03	< 0.02	< 1.0	22.0–23.0	4.5–6.5	0.14–0.2	3.0–3.5	—	
High alloy											
UNS S31200 (44LN)	< 0.03	< 2.0	< 0.045	< 0.03	< 1.0	24.0–26.0	5.5–6.5	0.14–0.2	1.2–2.0	—	
UNS S31260 (DP3)	< 0.03	< 1.0	< 0.03	< 0.03	< 0.75	24.0–26.0	5.5–7.5	0.1–0.3	2.5–3.5	0.2–0.8	W 0.1–0.5
UNS S32550 (Ferralium 255)	< 0.04	< 1.5	< 0.04	< 0.03	< 1.0	24.0–27.0	4.5–6.5	0.1–0.25	2.9–3.9	1.5–2.0	
UNS S32900 (329)	< 0.08	< 1.0	< 0.04	< 0.03	< 0.75	23.0–28.0	2.5–5.0	—	1.0–2.0	—	
7-Mo®[b]	< 0.1	< 2.0	< 0.04	< 0.03	< 1.0	25.0–30.0	3.0–6.0	—	1.0–2.0	—	
7-Mo PLUS® (UNS S32950)[b]	< 0.03	< 2.0	< 0.035	< 0.01	< 0.6	26.0–29.0	3.5–5.2	0.15–0.35	1.0–2.5	—	
Superduplex											
UNS S32520 (Uranus 52N+)(SD40)	< 0.03	< 1.5	< 0.035	< 0.02	< 0.8	24.0–26.0	5.5–8.0	0.2–0.35	3.0–5.0	0.5–3.0	
UNS S32750 (Uranus 47N+)(SAF2507)	< 0.03	< 1.2	< 0.035	< 0.02	< 0.8	24.0–26.0	6.0–8.0	0.24–0.32	3.0–5.0		
UNS S32760 (Zeron 100)(FALC100)	< 0.03	< 1.0	< 0.03	< 0.01	< 1.0	24.0–26.0	6.0–8.0	0.2–0.3	3.0–4.0	0.5–1.0	W 0.5–1.0
UNS S39274 (DP3W)	< 0.03	< 1.0	< 0.03	< 0.02	< 0.8	24.0–26.0	6.0–8.0	0.24–0.32	2.5–3.5	0.2–0.8	W 1.5–2.5
UNS S39277 (AF918)(25.7NCu)	< 0.025	< 1.0	< 0.025	< 0.002	< 0.8	24.0–26.0	6.5–8.0	0.23–0.33	3.0–4.0	1.2–2.0	W 0.8–1.2

Note: The chemical compositions given in the table are not specifications. They are typical values unless otherwise indicated.
[a]Proprietary alloy, Sandvik Steel.
[b]Proprietary alloy, Carpenter Specialty Alloys.

Modern development of duplex stainless steels is focused on more highly alloyed materials—superduplex grades. These materials have very good pitting and crevice corrosion resistance. In fact, duplex stainless steels are categorized by a pitting resistance equivalent number (PRE_N):

$$\{PRE_N\} = \%\{Cr\} + 3.3 \times (\%\{Mo\}) + 16 \times (\%\{N\})$$

Superduplex grades have PRE_N number equal to or greater than 40. They were specially developed to meet the hostile marine conditions in the offshore oil industry.

The characteristics of duplex stainless steels are that:

- they have structure consisting of significant amounts of bcc ferrite and fcc austenite;
- they are ferromagnetic at room temperature due to the ferrite;
- they have about twice the yield strength of common austenitic steels;
- they have low nickel content (2.5–8%);
- they often contain nitrogen, molybdenum, and copper additions;
- they are susceptible to embrittlement at temperatures above 450°C, particularly for the higher alloyed grades;
- they have good resistance to pitting, crevice, and stress corrosion.

Commercial duplex stainless steels are sometimes classified by their alloy content. The system used by Gunn [5] divides the alloys into four groups in order of increasing alloy content and corrosion resistance: "alloy lean", "standard 22% Cr", "high alloy", and "superduplex".

1. Alloy Lean Group

UNS S31500 (Sandvik 3RE60) is the duplex stainless steel grade with the lowest chromium content (18–19%). Nickel is also relatively low at 4.25–5.25%. It has good resistance to stress corrosion and has given satisfactory service in the past. However, in recent years, there is a tendency for it to be replaced by the newer nitrogen-strengthened grades.

UNS S32304 is a nitrogen-strengthened duplex stainless steel belonging to the "alloy lean" group. It is sometimes referred to by its chromium and nickel content as 2304. It is supplied by various steel manufacturers such as Avesta 2304, CLI UR35N, and Sandvik SAF2304. It is used in domestic water heaters where the risk of pitting corrosion is not high.

Uranus 50 (UNS S32404) is one of the earliest duplex stainless steels developed in the 1950s. It contains about 25–30% ferrite and has been used in the oil refining, food processing, pulp and paper and pharmaceutical industries. Despite efforts to control the content of residual elements, the material is susceptible to cracking.

2. Standard 22%Cr Group

This group is represented by UNS S31803 which is supplied by a large number of manufacturers under various trade names: Allegheny Ludlum AL2205™, Avesta 2205, Bohler A903, CLI UR45N, Knupp Falc223, Mannesmann AF22, Nippon Kokan NKCr22, Sandvik SAF2205, Sumitomo SM22Cr, Thyssen Remanit 4462, Valourec VS22, British Steel Hyresist22/5. It is used as piping and tubing in general corrosion and chloride stress corrosion environments. UNS S32205 is a high nitrogen version of UNS S31803.

3. High Alloy Group

This category includes alloys with chromium content in the region of 25%.

Type 329 (UNS S32900) is a first generation duplex stainless steel which has a relatively high carbon content and contains no nitrogen. It tends to form continuous network of ferrite which is detrimental to ductility.

The ductility is somewhat improved in the second generation UNS S31200 which has a lower carbon content and contains nitrogen. It is manufactured under a variety of trade names (UHB44LN, UR47N, VLX547). This material is used as a pressure vessel steel and has been shown to have good resistance to intergranular corrosion [6]. A further improved version of the alloy, UNS S31260 (DP3), has increased molybdenum, nitrogen, and nickel content relative to S31200 as well as copper and tungsten additions.

The grade UNS S32550 is supplied under the trade names Ferralium 255 (Haynes International) or URS2N (Creusot-Loire Industrie). This alloy was developed in the 1970s, originally as castings for pumps and valves in offshore oil rigs and for applications involving exposure to sulfuric acid.

In 7-Mo® (Carpenter Specialty Alloys), the chromium content is increased to between 25% and 30%. At this high level of ferrite stabilizers, the structure consists of a ferritic matrix with islands of austenite. This alloy can be hardened by heat treatment but the toughness is poor in this condition. The process involves heating to 732°C followed by furnace cooling. Because of its high chromium content, this alloy is very corrosion resistant and is a candidate material for use in extremely corrosive environments. However, in view of the poor toughness after heating, this steel is not considered to be weldable.

7-Mo PLUS® (UNS S32950) is an improved version of 7-Mo®. Nitrogen is added while carbon, sulfur, and silicon are reduced. At this composition, the ratio of ferrite to austenite is about 55:45. It has better impact toughness compared with 7-Mo® and can be welded using standard electric arc techniques.

4. Superduplex Group

The alloys in this group are rich in chromium, molybdenum, and nitrogen and are characterized by a PRE_N greater than 40. These steels were originally developed to meet the requirement for extremely high resistance to pitting and crevice corrosion in offshore oil rigs. Their current applications range from pipes for handling seawater to human body implants. UNS S32520 (Uranus 52N+) and UNS S32750 (2507) are essentially similar in composition with the former having copper addition. UNS S32760 (Zeron 100), UNS S39274 (DP3W), and UNS S39277 (AF918) are similar to UNS S32520 but with tungsten addition.

II. DESIGN CONSIDERATIONS

The design of engineering components involves the selection of suitable material to meet functional requirements under the anticipated service conditions, as well as considerations of factors such as fabrication, cost, and availability. The reason why stainless steels are used instead of other materials is often because of their good combination of corrosion resistance and mechanical properties [8–11]. For most applications, the "best all-rounder" is the basic 18/8 type (i.e. 304). One simple approach to the design of components using stainless steels is to ask the question: is type 304 steel adequate for the purpose? If not, and if higher strength was required, or if stress corrosion was a problem, or if cost was to be minimized, one could then proceed to find other alloys with further improvements in the properties considered important.

A. Physical Properties

In the majority of cases, the physical properties of various grades of stainless steels are similar and do not influence the selection of material. Notable exceptions are magnetic properties and thermal conductivity, where there are significant differences between the ferritic and austenitic grades.

1. Density

There is not much variation in the density of stainless steels. The density is 8000 kg/m^3 for the 300 series and 7800 kg/m^3 for the other grades. These values are about the same as those of carbon or low alloy steels.

2. Elastic Modulus

The room temperature elastic modulus of most stainless steels is usually between 190 and 200 GPa. However, in heavily cold-worked condition, some may have modulus as low as 150 GPa. This is significantly lower than that of plain carbon or low alloy steels, which have an elastic modulus of around 207 GPa. Thus, if stainless steels are used to replace carbon steels, a slightly larger cross-section is required to maintain the same level of deflection. Depending on the application, this could be an important factor to consider in engineering design.

3. Thermal Conductivity

The thermal conductivity of the austenitic grades is about 14.6W/m K while that for the ferritic and martensitic grades is about 25W/m K at room temperature. Heat transfer is often very much affected by surface conditions such as corrosion and scale accumulation. The ability of stainless steels to retain a relatively clean surface under most conditions is one major factor to consider in engineering design requiring heat transfer properties to remain constant during long-term service.

4. Thermal Expansion

The coefficient of thermal expansion for non-austenitic grades is about the same as carbon steels (~10.5 μm/m °C) while that for the austenitic grades is about 17 μm/m °C. Thus, the austenitic grades are more susceptible to thermal distortion and this must be allowed for in design.

5. Magnetic Properties

The austenitic grades in the fully annealed condition are paramagnetic at room temperature. However, certain low nickel grades form martensite in the cold-worked condition and the increase in magnetic permeability can be significant. Since the magnetic permeability of austenitic stainless steels is generally very low (1.003–1.005 at 16 kA/m), they are used in applications involving magnetic detection and measuring where a non-magnetic material is required.

The ferritic grades are ferromagnetic at room temperature. Their magnetic properties are dependent on the heat treatment condition. In general, annealing treatments leading to lower hardness result in higher magnetic permeability and lower coercive force. They are soft magnetic materials with maximum magnetic permeability and coercive force up to 2500 and 550 A/m, respectively. In terms of soft magnetic properties, they are not as good as the high nickel alloys such as Supermalloy but they can be a cost-effective alternative to these materials.

The duplex grades containing ferrite and austenite are intermediate between the austenitic and ferritic grades in terms of magnetic permeability. However, upon exposure to elevated temperatures, the ferrite will transform into carbides, intermetallic phases and austenite, culminating in a reduction in permeability.

All martensitic and most PH grades are hard ferromagnetic materials at room temperature. The magnetic properties vary from grade to grade and are markedly influenced by the heat treatment conditions. In general, the higher the hardness, the lower is the magnetic permeability and the higher is the coercive force. As a rough guide, the magnetic permeability of these grades is about an order of magnitude lower than that of the ferritic grades, whereas the coercive force is about an order of magnitude higher.

6. Electrical Resistivity

The electrical resistivity of stainless steels does not vary significantly from grade to grade. For most grades, the electrical resistivity is around 700 nΩ m. Addition of elements such as aluminum can significantly increase resistivity. For example, the iron–chromium–aluminum alloys for electrical resistance heating elements can have resistivity of around 1400nΩ m.

7. Melting Range

The melting range of stainless steels (solidus to liquidus) does not vary significantly from grade to grade and is around 1400°C for most grades. The 400 series grades in general have a slightly higher melting range than the other grades.

8. Mechanical Properties

The room temperature mechanical properties of various stainless steels described so far are listed in Tables 7–12. The properties of type 304 stainless steel provide a benchmark for comparison.

Apart from a few exceptions, ferritic stainless steels generally have 0.2% yield strength in the range 250–350 MPa and ultimate tensile strength in the range 450–550 MPa. The yield strength is comparable to type 304 and is on the low side compared with other stainless steels. The ultimate tensile strength of 450–550MPa is the lowest among all the various grades of stainless steels. Moreover, even in the second generation ferritic stainless steels with controlled carbon and nitrogen, the ductility and toughness of ferritic stainless steels are lower than those of type 304. The ferritic grades are also more susceptible to phenomena such as 475°C embrittlement and sigma phase embrittlement. On the other hand, they are better than type 304 in their resistance to chloride stress corrosion and hydrogen embrittlement.

Martensitic stainless steels are usually sold in the annealed "soft" condition for ease of forming and machining by the customer. In the annealed condition, the yield strength and hardness are comparable to, or slightly higher than, type 304. In the hardened condi-

Table 7 Typical Mechanical Properties of Selected Ferritic Stainless Steels

Type	0.2% Yield strength (MPa)	Ultimate tensile strength (MPa)	Elongation (%)	Hardness (Rockwell B)
405	262	449	31	69
405NA[a]	276	449	28	75
408L[a]	297	484	30	78
409	260	425	32	70
409L[a]	242	415	32	70
409HP[b]	240	400	36	68
AL409HP™[c]	241	434	33	71
AL409Cb™[c]	225	420	35	70
Carpenter 5F[d]	517	621	15	78
430	310	517	30	82
430L[a]	311	518	28	80
430F	379	551	25	86
430FSe	379	551	25	86
430F[d] Solenoid Quality	310	517	20	82
430FR[d] Solenoid Quality	345	538	30	86
AL433™[c]	323	438	31	79
434	365	531	23	83
436	365	531	23	83
AL436S™[c]	297	456	31	77
Project 70® Stainless 182-FM[d]	345	517	30	88
439	280	482	30	80
AL439HP™[c]	290	455	34	78
441	331	510	29	84
AL441HP™[c]	310	496	31	82
443	345	551	25	86
444	276	483	30	83
446	345	551	25	86
446M0D-1[b] (UNS S44660)	587	677	20	98
AL453™[c]	310	470	35	78
AL468™[c]	290	463	32	75
Chrome Core® 8[d]	228	421	40	73
Chrome Core 8-FM[d]	228	421	40	73
Chrome Core 12[d]	228	421	40	73
Chrome Core 12-FM[d]	228	421	40	73

[a]Proprietary alloy, ATLAS Stainless Steels.
[b]Proprietary alloy, J&L Specialty Steel.
[c]Proprietary alloy, Allegheny Ludlum Corp.
[d]Proprietary alloy, Carpenter Specialty Alloys.
Source: Manufactures' data sheets and Refs. 2 and 7.

tion, the 0.2% yield stress is often above 1000MPa and can be as high as ~1900MPa. Ductility is low and the material may suffer from temper and hydrogen embrittlement. In the high carbon grade such as type 440C, the elongation is as low as 2%. This makes martensitic stainless steels difficult to handle and they are used in applications where high hardness is the overriding factor in design.

Table 8 Typical Mechanical Properties of Selected Martensitic Stainless Steels

Type	Condition	0.2% Yield strength (MPa)	Ultimate tensile strength (MPa)	Elongation (%)	Hardness (Rockwell)
403,410	Annealed	276	517	30	B82
414	Annealed	655	827	17	C22
414	[1]	1,034	1,379	15	C43
414	[2]	1,000	1,310	15	C41
414	[4]	827	1,000	20	C34
416, 416Se	[1]	1,000	1,310	15	C41
416, 416Se	[2]	965	1,241	15	C39
416, 416Se	[4]	793	1,000	20	C31
420, 420F	Annealed	345	655	25	B92
420, 420F	[7]	1,345	1,586	8	C50
Sandvik 4C27A[a]	Annealed	450	<800	25	—
TrimRite (bar)[b]	Annealed	372	607	28	B88
422	Annealed	586	793	22	B98
422	[3]	1,282	1,627	10	—
422	[4]	1,145	1,476	13	C42
422	[5]	862	1,000	14	C32
431	Annealed	655	862	20	C24
431	[6]	1,069	1,413	15	C43
431	[7]	1,034	1,345	15	C41
431	[8]	896	1,034	18	C34
440A	Annealed	414	724	20	B95
440A	[7]	1,655	1,793	5	C51
440B	Annealed	428	738	18	B96
440B	[7]	1,862	1,930	3	C55
440C,440F	Annealed	448	758	13	B97
440C,440F	[7]	1,896	1,965	2	C57

[a]Proprietary alloy, Sandvik Steel.
[b]Proprietary alloy, Carpenter Specialty Alloys.
[1] Hardened at 982°C, tempered at 205°C.
[2] Hardened at 982°C, tempered at 316°C.
[3] Hardened at 982°C, tempered at 427°C.
[4]Hardened at 982°C, tempered at 538°C.
[5] Hardened at 982°C, tempered at 649°C.
[6] Hardened at 1038°C, tempered at 205°C.
[7] Hardened at 1038°C, tempered at 316°C.
[8] Hardened at 1038°C, tempered at 538°C.
Source: Manufacturers' data sheets and Refs. 2 and 7.

All the 300 series austenitic stainless steels are similar in mechanical properties. The 0.2% yield stress is low but the ultimate tensile strength is higher than that of the ferritic stainless steels. They have excellent ductility and toughness, and can be hardened by cold work. The 200 series stainless steels resemble the 300 series in mechanical properties with slightly higher yield strength.

The martensitic and semi-austenitic PH stainless steels have yield and ultimate tensile strengths comparable to the martensitic stainless steels. This advantage of high strength and hardness is achieved with less reduction in ductility and toughness. The austenitic

Table 9 Typical Mechanical Properties of Selected 300 Series Austenitic Stainless Steels

Type	0.2% Yield strength (MPa)	Ultimate tensile strength (MPa)	Elongation (%)	Hardness (Rockwell B)
301	276	758	60	85
302	276	612	50	85
302HQ[a]	214	496	69	—
302HQ-FM[a]	221	510	57	—
302B(UNS S30215)	276	655	55	85
303, 303Se	241	620	50	76
304	290	579	55	80
304L	269	558	55	79
304N	331	620	50	85
305	262	586	50	80
309	310	620	45	85
309S	310	620	45	85
310	310	655	45	85
310S	310	655	45	85
316	290	579	50	79
316L	220	517	50	79
316N	331	620	48	85
316LN	279	573	52	80
317	276	620	45	85
317L	262	593	55	85
321	241	620	45	80
347	276	655	45	85
348	276	655	45	85
334	241	572	40	82
384	241	517	55	70

[a]Proprietary alloy, Carpenter Specialty Alloys.
Source: Manufacturers' data sheets and Refs. 2 and 7.

PH stainless steels are more ductile but not as strong. The high strength, with reasonable ductility and toughness, renders the PH alloys particular useful in applications involving highly stressed component.

The duplex stainless steels have yield and ultimate tensile strengths nearly twice that of type 304. They are more resistant to chloride stress corrosion but are more susceptible to 475°C and sigma phase embrittlement.

B. Corrosion Properties

The reason why stainless steels are chosen for a specific application is often due to their excellent corrosion resistance. However, as a general rule, the richer the alloy content, the better is the corrosion resistance, and the higher is the cost. Thus, the degree of corrosion resistance one wishes to attain is dependent on how much one is prepared to spend. In general, the martensitic grades have the lowest corrosion resistance among the various types of stainless steels because of their rather low chromium and nickel content. Their high carbon content also has a negative effect on corrosion resistance due to precipitation of carbides.

Table 10 Typical Mechanical Properties of Selected 200 Series Austenitic Stainless Steels

Type	0.2% Yield strength (MPa)	Ultimate tensile trength (MPa)	Elongation (%)	Hardness (Rockwell B)
201	379	758	55	90
201L	310	744	57	—
201LN	320	685	51	—
Gall-Tough®[a]	365	1,000	59	93
Gall-Tough Plus™[a]	476	779	69	—
202	379	689	55	90
203EZ	276	551	58	90
204-Cu[a]	290	635	75	—
205	482	862	45	98
22Cr–13Ni–5Mn (UNS S20910)	414	827	50	98
SCF 19[a] (UNS S21000)	434	827	52	93
Tenelon (UNS S21400)	448	827	40	95
Cryogenic Tenelon (UNS S21460)	380 min	725 min	40 min	—
216, 216L	379	689	45	92
21Cr–6Ni–9Mn (UNS S21900)	345 min	620 min	45 min	—
21Cr–6Ni–9Mn LC)(UNS S21904)	345 min	620 min	45 min	—
241	380 min	690 min	30 min	—
18–18 PLUS (UNS S28200)	476	827	65	95
15–15 LC[a]	434	758	50	—

[a]Proprietary alloy, Carpenter Specialty Alloys.
min = minimum value.
Source: Manufacturers' data sheets and Refs. 2 and 7.

1. Effect of Composition

Alloying elements in stainless steels often have significant effects on corrosion properties. Such effects can be direct, such as enhancing the formation of the chromium oxide protective film, or indirect, such as the removal of carbon from solid solution to avoid sensitization (e.g. Ti, Nb, etc.).

The formation of a protective chromium oxide film at chromium level greater than about 11% was mentioned at the beginning of this chapter. Generally, the stability of this film increases with increasing chromium content. However, the chromium content of stainless steels does not normally exceed 30% because of the possible adverse effects of chromium on other properties, such as susceptibility to embrittlement, etc.

Nickel is an austenite stabilizer which will normally enhance mechanical properties, fabrication characteristics, and corrosion resistance. However, stainless steels with about 9% nickel (e.g. type 304) are susceptible to stress corrosion cracking (SCC). At higher levels of nickel, the resistance to SCC or other forms of corrosion is improved.

Manganese is also an austenite stabilizer and its effects are similar to nickel. It is used to replace part of the nickel in type 200 series stainless steels. However, there is a tendency for manganese to combine with sulfur to form manganese sulfide particles. These particles could act as potential sites for pitting corrosion.

Table 11 Typical Mechanical Properties of selected Precipitation-Hardening Stainless Steels

Type	Condition	0.2% Yield strength (MPa)	Ultimate tensile strength (MPa)	Elongation (%)	Hardness (Rockwell)
Martensitic					
17-4PH (UNS S17400)(AISI 630)	A	758	1,034	10	C33
17-4PH (UNS S17400)(AISI 630)	PH	1,227	1,379	12	C44
Stainless W (UNS S17600)(AISI 635)	A	517	827	7	C30
Stainless W (UNS S17600)(AISI 635)	PH	1,241	1,344	7	C46
15-5PH (UNS S15500)	A	586	862	10	C27
15-5PH (UNS S15500)	PH	1,275	1,379	14	C44
13-8Mo(UNS S13800)	A	586	896	12	C28
13-8Mo(UNS S13800)	PH	1,379	1,551	13	C48
Custom 450[a] (UNS S45000)	A	814	979	13	C28
Custom 450[a] (UNS S45000)	PH	1,296	1,351	14	C42.5
Custom 455[a] (UNS S45500)	A	793	965	12	C31
Custom 455[a] (UNS S45500)	PH	1,516	1,585	10	C48
Custom 465[a]	A	772	951	20	C29.5
Custom 465[a]	PH	1,703	1,779	14	C50
Semi-austenitic					
PH15-7Mo (UNS S15700)	A	379	896	35	B88
PH15-7Mo (UNS S15700)	[1]	1,551	1,655	6	C48
PH15-7Mo (UNS S15700)	[2]	1,379	1,448	7	C44
Pyromet® Alloy 350 (UNS S35000)(AISI 633)[a]	A	414	1,103	30	B95
Pyromet® Alloy 350 (UNS S35000)(AISI 633)[a]	[3]	1,117	1,365	15	C48
Pyromet® Alloy 350 (UNS S35000)(AISI 633)[a]	[4]	1,034	1,124	22	C38
Pyromet® Alloy 350 (UNS S35000)(AISI 633)[a]	[5]	979	1,179	12	C40
Pyromet® Alloy 355 (UNS S35500)(AISI 634)[a]	[3]	1,255	1,489	19	C48
Pyromet® Alloy 355 (UNS S35500)(AISI 634)[a]	[4]	1,179	1,276	19	C40
Austenitic					
A286 (UNS S66286)	A	241	620	45	B81
A286 (UNS S66286)	[6]	689	1006	25	C34

[a]Proprietary alloy, Carpenter Specialty Alloys.
A = annealed.
PH = Precipitation hardened (maximum value).
[1] Cooled to −73°C then aged at 510°C.
[2] Tempered at 760° then aged at 566°C.
[3] Conditioned at 932°C for 90 min. followed by rapid cooling, cooled to −73°C for 3 hr, then tempered at 454°C for at least 3 hr.
[4] Conditioned at 932°C 90 min. followed by rapid cooling, cooled to −73°C for 3 hr, then tempered at 538°C for at least 3 hr.
[5] Double age treatment: hold for 3 hr at 732/760°, air cool to room temperature, heat to 440/468°C, hold 2–3hr, air cool.
[6] Aged at 718°C.
Source: Manufactures' data sheets and Refs. 2 and 7.

Table 12 Typical Machanical Properties of Selected Duplex Stainless Steels

Type	0.2% Yield strength (MPa)	Ultimate tensile strength (MPa)	Elongation (%)	Hardness (Rockwell)
Alloy lean				
UNS S31500 (3RE60)[a]	482	717	48	B92
UNS S32304 (2304)	400 min	600 min	25 min	B97 max
UNS S32404 (UR50)	386	600	25	C8
Standard 22% Cr				
UNS S31803 (2205)	520	760	27	—
High alloy				
UNS S31200 (44LN)	450	800	25	—
UNS S32550 (Ferralium 255)	676	869	30	B100
UNS S32900 (329)	551	724	25	B98
7-Mo®[b]	552	724	25	B99
7-Mo PLUS® (UNS S32950)[b]	586	800	31	B99.5
Superduplex				
UNS S32750 (Uranus 47N+)(SAF2507)	550 min	800 min	25	C28
UNS S32760 (Zeron 100)(FALC100)	550 min	750 min	25	C28
UNS S39274 (DP3W)	568	862	42	C25

[a]Proprietary alloy, Carpenter Specialty Alloys.
[b]Proprietary alloy, Sandvik Steel.
min = minimum value.
Source: Manufactures' data sheets and Refs. 2 and 7.

Molybdenum is a ferrite stabilizer and has the effect of enhancing the pitting and crevice corrosion resistance of stainless steels. For example, type 316 steel with about 2.5% Mo has superior pitting resistance over the Mo-free type 304.

Carbon has a deleterious effect on corrosion resistance. It combines with chromium to form chromium carbides, resulting in chromium-depleted regions around these carbide particles. This problem is particularly acute at the heat-affected zones of welded structures, where the precipitation of chromium carbides at grain boundaries causes chromium-depleted zones which are susceptible to intergranular corrosion. The solution lies in reducing the carbon content (e.g. type 304L) or adding a strong carbide former such as titanium or niobium to remove the carbon from solution (e.g. types 321 and 347). However, elements such as titanium and niobium can be detrimental to corrosion properties at high concentrations.

Nitrogen is also an austenite stabilizer which has the concomitant effects of improving the pitting resistance and strength of austenitic stainless steels. It does not cause sensitization to these steels as carbon does and is sometimes added to the low carbon grades to make up for the loss of strength due to carbon reduction. Nitrogen promotes passivity and is regarded as generally beneficial to corrosion resistance.

Elements such as copper, cobalt, lead, phosphorus, tantalum, tin, tungsten, and zirconium are present in solid solution and have no significant effect on corrosion resistance. Elements such as silver, rhenium, and vanadium are beneficial to corrosion resistance, whereas elements such as selenium, sulfur, tellurium form second phases which may have negative effects on corrosion resistance. Boron is different in that when a small amount ($<0.007\%$) is present in solution, it is beneficial, but in larger quantities, the precipitation of borides would be detrimental to corrosion resistance. Aluminum, sulfur, and silicon are also detrimental to corrosion resistance at high concentrations.

2. General Corrosion

For most applications involving exposure to industrial atmospheres and acid media, the austenitic type 304 stainless steel is adequate. Surfaces which were brightly polished and reasonably well maintained would stay bright and rust-free under most ambient conditions. If surface brightness is not required, the leanest ferritic grade, type 405, may be the cheapest option. For more severe environments, more highly alloyed grades should be used (e.g. type 316, 310, nickel-based alloys, etc.).

3. Pitting Corrosion

The pitting resistance of stainless steels is very much affected by the elements chromium, molybdenum, and nitrogen. As mentioned previously under the section on duplex stainless steels, the pitting resistance is categorized by a pitting resistance equivalent number:

$$\{PRE_N\} = \%\{Cr\} + 3.3 \times (\%\{Mo\}) + 16 \times (\%\{N\})$$

For stainless steels containing tungsten, a modified pitting resistance equivalent number is used:

$$\{PRE_W\} = \%\{Cr\} + 3.3 \times (\%\{Mo\} + 0.5 \times \%\{W\}) + 16 \times (\%\{N\})$$

Type 316 is preferred over type 304 if pitting corrosion is considered to be a problem. The addition of molybdenum in the former is responsible for the improved resistance to pitting. However, even type 316 will not withstand aggressive media with high chloride content (>1000 ppm) under relatively stagnant conditions.

The ferritic stainless steels are also susceptible to pitting corrosion and grades with at least 23% chromium and 2% molybdenum should be used to reduce the susceptibility to pitting corrosion.

Duplex staianless steels, particularly those with high pitting resistance equivalent numbers, are superior to the 300 series grades in terms of pitting corrosion resistance.

The pitting resistance is impaired by the presence of inclusion particles or precipitation of carbides during the heating cycles of welding and heat treatment.

4. Intergranular Corrosion

The precipitation of chromium carbides at grain boundaries in stainless steels removes chromium from solution and creates chromium-depleted regions adjacent to the grain boundaries. This often occurs at the heat-affected zones of welded joints. The chromium-depleted boundaries are prone to intergranular corrosion and the phenomenon is called "sensitization". The austenitic grades are particularly susceptible to this effect and the remedy lies in lowering the carbon content to below 0.3%, or by the addition of strong carbide formers such as titanium and niobium. Duplex stainless steels are not prone to sensitization because the chromium carbides are precipitated predominantly around the ferrite which is relatively chromium rich.

5. Stress Corrosion

The austenitic grades are prone to stress corrosion, particularly in chloride-containing solutions. Concentrated caustic solutions can also cause stress corrosion in austenitic stainless steels. The ferritic and the duplex grades are generally resistant to stress corrosion. The PH grades are susceptible to stress corrosion cracking, and the degree of susceptibility increases with strength.

6. Crevice Corrosion

Crevice corrosion occurs at wet tightly fitting surfaces where a concentration gradient exists. The crevice is depleted in oxygen and enriched in metal ions. An oxygen concentration cell is established culminating in an anodic condition within the crevice. In general, the behavior of stainless steels in crevice corrosion situations is similar to that in pitting corrosion. Elements such as chromium, molybdenum, and nickel are beneficial to crevice corrosion resistance. Precipitation-hardening and martensitic stainless steels are usually not recommended for applications involving crevice corrosion.

7. Galvanic Corrosion

Galvanic corrosion occurs when two different metals are immersed in an electrolyte. Since the corrosion potentials of the two metals are different, one will act as the anode and the other as the cathode. A current will flow and the metal at the anode will go into solution as ions.

Stainless steels are protected by a chromium oxide passive film which gives them a noble position in the galvanic series. However, when the film is damaged either mechanically or chemically, they become active. This is a matter of some importance in design because under some conditions, such as chloride solutions, galvanic corrosion may occur in stainless steels when coupled with noble metals such as copper or nickel, whereas no galvanic corrosion occurs in freshwater.

8. Erosion Corrosion

Generally, erosion–corrosion resistance is dependent on hardness, the harder the alloy, the more resistant it is against erosion corrosion. In most situations, the austenitic stainless steels can perform adequately. However, ferritic stainless steels with high chromium and molybdenum content are better in this respect. Materials hardened by cold work also have their resistance enhanced. On the other hand, although the PH and martensitic grades have high hardness and are therefore resistant to erosion corrosion, there are countervailing considerations of poorer performance in pitting and other forms of corrosion.

C. Ease of Fabrication

We shall once again use type 304 as a benchmark. If ease of fabrication is a major consideration in the selection of material for a particular application, type 304's advantages lie in its excellent forming and welding properties. With carbon limited to less than 0.03% in type 304L, the risk of intergranular corrosion in the heat-affected zone of welded joints is reduced. The filer metal used for welding austenitic stainless steels usually contains a relatively high content of ferrite stabilizing elements to produce between 2% and 12% of ferrite in the weld metal. The presence of ferrite reduces the risk of hot cracking during welding.

The ferritic grades are generally considered to be less easy to weld than the austenitic grades. With the advent of the argon–oxygen decarburization process, interstitial elements such as carbon and nitrogen are better controlled, culminating in better weldability. However, there is still a risk of unfavorable grain growth in the heat-affected zone if the heat input is not properly controlled.

In general, stainless steels are more difficult to machine than other metals. Because of their high toughness and ductility, they tend to produce long and stringy chips during machining. These chips have a tendency to build up at the edge of the tool, culminating in reduced tool life and poor surface finish. Project 70® stainless 304 is a proprietary alloy produced by Carpenter Specialty Alloys with improved machinability over ordinary type 304. The Project 70® alloys are specially produced alloys with fine adjustments to chemical composition and processing procedure within the specification to provide optimum machining performance. If further improvement on machinability is required, type 303 stainless steel and its associated variations (see Sec. 1.3.1) should be considered.

D. High Temperature Properties

Design for high temperature service is usually more demanding than that for room temperature because above about 480°C for stainless steels time-dependent properties such as creep become important. Usually the allowable design stresses at temperatures below 480°C are based on tensile properties (0.2% yield strength or ultimate tensile strength) and are not time dependent. At temperatures above 480°C when creep effects become important, the component in question is usually designed for a period of service known as the "design life". For example, high temperature components in power generating plants are usually designed for a 30-year service life.

Our benchmark material, type 304 stainless steel, is better than the carbon and low alloy steels in terms of creep resistance and is a suitable material for many high temperature applications. For higher creep strength, type 316, with 2.5% molybdenum addition, should be considered. The molybdenum in type 316 provides enhanced high temperature strength through solid-solution strengthening. The titanium or niobium stabilized grades, types 321 and 347, are also stronger in creep than type 304 due to the precipitation-strengthening effect of titanium or niobium carbide.

Interstitial elements such as carbon and nitrogen are beneficial to creep resistance (3,4). However, the trend nowadays is to use the low carbon grades in order to minimize the risk of sensitization. Thus, the use of nitrogen-strengthened grades is becoming more and more popular.

Creep rupture ductility is also an important property to consider in design. Although quantitative data on elongations at rupture are not directly used in design, these values provide useful information on making judgment on a material's ability to withstand stress concentrations. Rupture ductility also has a major influence on creep fatigue interactions. The effect of composition on creep rupture ductility of type 316 stainless steel has been thoroughly analyzed [3,4]. Boron, even present in small quantities, can be rather beneficial to long-term creep ductility.

Long-term exposure of stainless steels to elevated temperatures will lead to the precipitation of carbides and intermetallic phases such as sigma, chi, etc. [12–14]. Such microstructural changes can have major effects on properties, particularly the room temperature impact strength. In some cases, thermal aging can lead to a 90% reduction in impact strength.

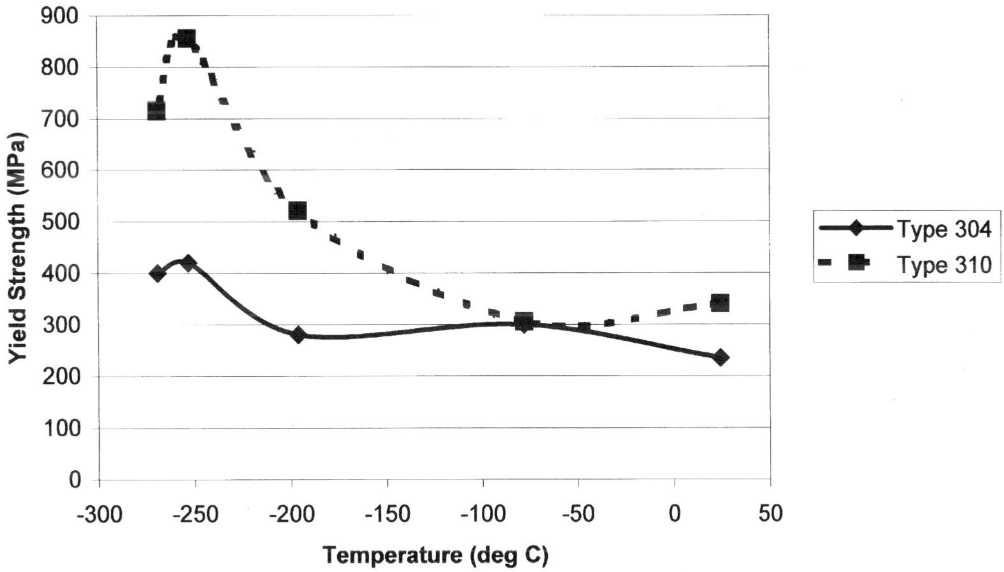

Figure 1 Yield strength vs temperature. (From Ref. 2.)

E. Low Temperature Properties

When designing engineering structures for subzero temperature service, the major factor to consider is brittle fracture. Steels with a body-centered-cubic crystal structure will undergo a phenomenon called ductile-to-brittle transition when cooled to low temperatures. The loss in impact strength can be very drastic, and care must be exercised in the proper selection of material and the minimization of stress concentrations.

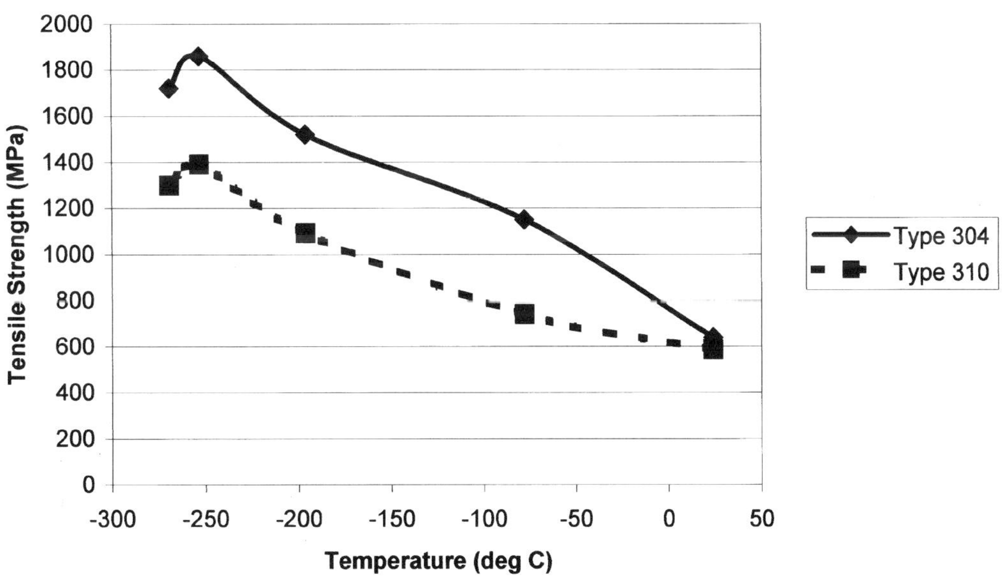

Figure 2 Tensile strength vs temperature. (From Ref. 2.)

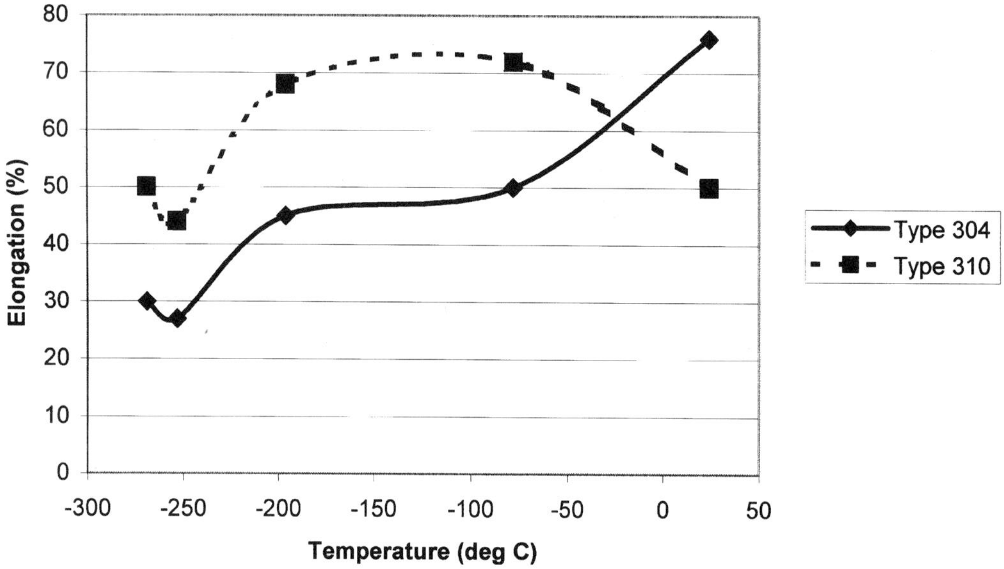

Figure 3 Tensile elongation vs temperature. (From Ref. 2.)

The austenitic stainless steels usually suffer from a less drastic ductile-to-brittle transition when cooled to cryogenic temperatures because of their face-centered-cubic rather than body-centered-cubic crystal structure. Consequently, they are commonly used for low temperature applications such as containment for liquefied gases. Our benchmark material, type 304 stainless steel, is suitable for low temperature applications and has been extensively used as such. However, it can form martensite when cold worked, particularly at low temperatures. In some instances, type 310 stainless steel is preferred because of its stability against transformation to martensite.

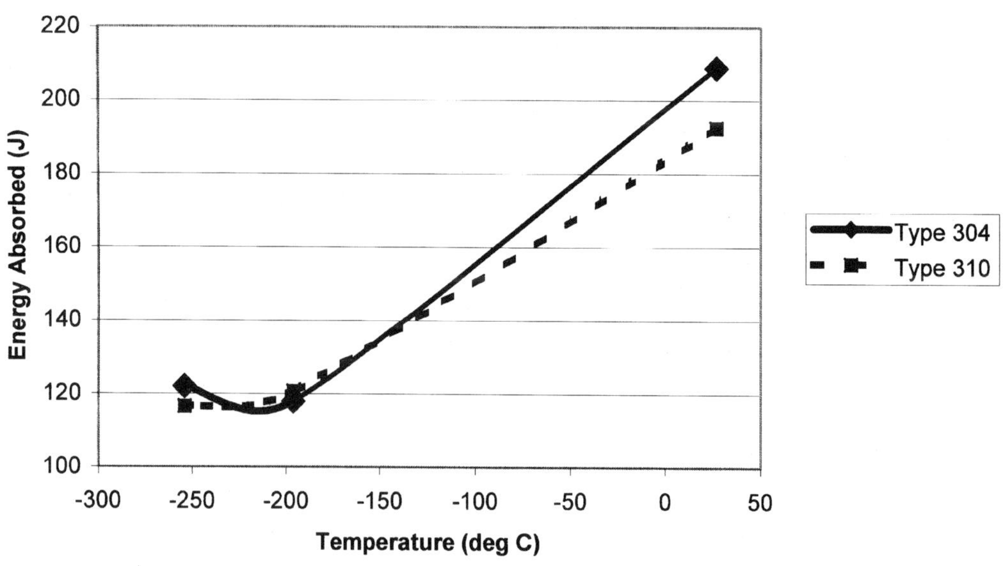

Figure 4 Transverse Charpy V-notch impact strength vs temperature. (From Ref. 11.)

Figure 1 shows that the yield strength of type 310 stainless steel increases with decreasing temperature, whereas that for type 304 is not severely affected even when cooled to liquid helium temperatures. On the other hand, the tensile strength of both materials increases as the temperature is reduced, with the increase for type 304 higher than that of type 310 (Fig. 2).

The tensile ductility of both type 304 and 310 is good even at cryogenic temperatures. The tensile elongation at fracture for type 304 shows a steady decrease as the temperature is reduced, whereas for type 310 the elongation vs. temperature curve shows a broad maximum between −50°C and −200°C (Fig. 3).

The Charpy V-notch impact strength of both steels is reduced by about 50% when cooled from room temperature to −254°C (Fig. 4). There is not a lot of difference in behavior between types 304 and 310 in this respect.

REFERENCES

1. Redmond, J.D. *Metals and Alloys in the Unified Numbering System*, 8th Ed.; Society of Automotive Engineers: Warrendale, 1999.
2. Davies, J.R. Ed.; Stainless steels. In *ASM Speciality Handbook*, ASM International: Materials Park 1994.
3. Lai, J.K.L. A set of master curves for the creep ductility of type 316 stainless steel. J. Nuclear Materials 1979, *82*, 123–128.
4. Lai, J.K.L. Optimizing the stress rupture properties of AISI type 316 stainless steel with the aid of computers. J. Nuclear Materials 1981, *99*, 148–151.
5. Gunn, R.N *Duplex Stainless Steels, Microstructure, Properties and Applications*; Woodhead Publishing Ltd.: Cambridge, 1997.
6. Dirscherl, R.; Barth, S. ASM Symposium on Duplex Stainless Steels, ASM International: Materials 1983, 503–533.
7. Sedriks, A.J. In *Corrosion of Stainless Steels*, 2nd Ed.; John Wiley & Sons: New York, 1996.
8. Dillon, C.P. *Corrosion Resistance of Stainless Steels*; Marcel Dekker: New York, 1995.
9. Beddoes, J.; Parr, A.P. *Introduction to Stainless teels*; ASM International: Materials Park, 1999.
10. Lula, R.A. *Stainless Steels*; ASM International: Materials Park, 1986.
11. Peckner, D.P.; Bernstein, I.M. *Handbook of Stainless Steels*; McGraw Hill: New York, 1977.
12. Lai, J.K.L. A study of precipitation in AISI type 316 stainless steel. Mater. Sci. Engineering, 1983, *58*, 195–209.
13. Lai, J.K.L. A review of precipitation behaviour in AISI type 316 stainless steel. Mater. Sci. Engineering, 1983, *61* (2), 101–109.
14. Lai, J.K.L. Precipitate phases in type 321 stainless steel. Mater. Sci. Technol. 1985, *1*, 97–100.

8

Cast Iron Design: Processes, Alloys, and Properties

Magnus Wessén and Ingvar L. Svensson
Jöngköping University, Jöngköping, Sweden

I. INTRODUCTION

Cast iron, especially ductile iron, is being used more and more in most countries of the world due to its excellent mechanical properties, castability, and good price. Cast iron alloys can be given a wide range of properties by changing the alloy composition, inoculation, and treatment, heat treatment, or cooling conditions. On a weight basis, most castings are made in gray cast iron. Some examples of gray and ductile iron products are shown in Fig. 1 [1]. The main factors for the high usage are good casting properties, low price, good cutability, and unique properties like damping and good tribology. Lamellar gray cast iron has very good damping capacity. This property is used in many components where damping of sound and vibration is important. Gray iron, also called lamellar graphite iron, is an iron–carbon–silicon alloy with different alloying elements.

To understand the microstructure formations and to model them and other important phenomena, both the solidification and the solid-state transformations must be considered. These phase transformations are, to a large extent, affected by nucleation and growth kinetics which are dependent on, for example, handling of the melt, charge material, melting method, metal treatment, inoculation, pouring method, casting process and mold material, cooling power of the mold, and other factors. Many of these important material and process factors and combinations of them are still not fully understood today. Consequently, it is necessary that a foundry producing cast iron components has very strict process control in order to avoid unpredictable problems in production. When this is the case, it will also be possible to use simulation tools for predicting solidification sequence, microstructures, mechanical properties, as well as formation of defects.

II. CAST IRON MATERIALS

Cast iron is a group of materials with a wide variety of microstructures and properties. The type of cast iron is often related to the morphology of graphite. Its properties can be further improved by changing the chemical composition (alloying).

Class of Cast Iron	Morphology of Graphite Phase
Gray cast iron	Lamellar graphite eutectic
Compacted graphite cast iron	Short chubby graphite flakes ("worms")
Nodular cast iron (ductile iron)	Round graphite spheres
White cast iron	No graphite. Contains eutectic cementite

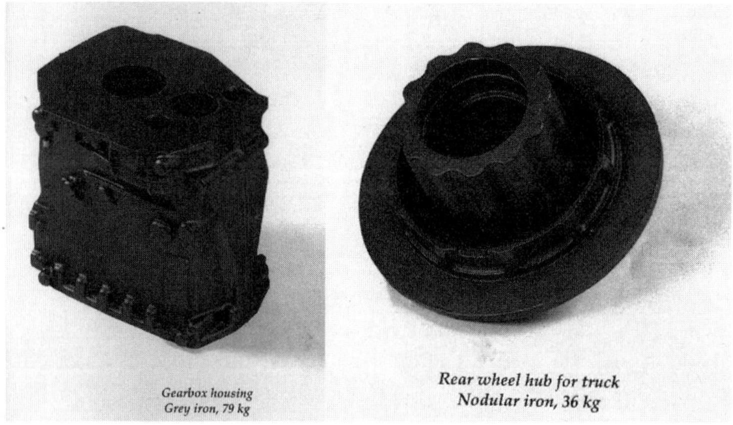

Figure 1 Examples of products made in gray and ductile cast iron. (From Ref. 1.)

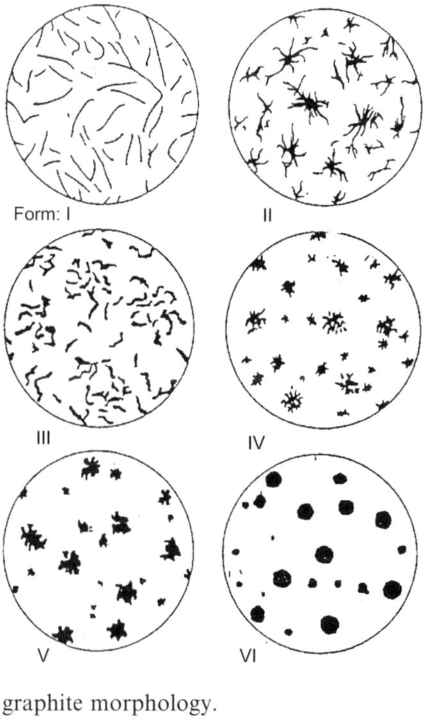

Figure 2 Classification of graphite morphology.

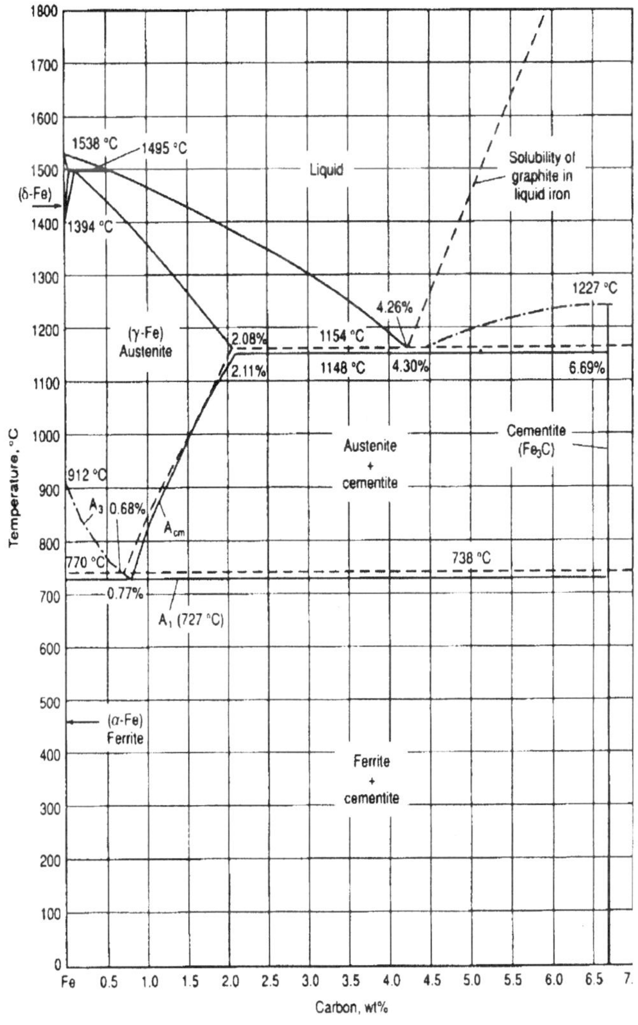

Figure 3 Phase diagram, Fe–C.

Figure 2 shows different types of graphite shapes. Some common abbreviations for the different classes are shown.

The growth rate of the solidifying phases in the melt depends on kinetic- and thermodynamic factors, of which the slowest controls the growth. The growth rate of the eutectic can, in most cases, be related to the super cooling relative to the phase diagram (Fig. 3). The basis for understanding the solidification of cast irons is the phase diagram. The three most important elements in cast irons are iron (Fe), carbon (C), and silicon (Si). The compositions for cast irons are normally around the eutectic composition.

Carbon is mainly precipitated as lamellar graphite in gray iron, and the graphite/austenite eutectic grows in a spherical way. The graphite precipitation at solidification depends on cooling conditions and the number of growing cells.

Additions of inoculants facilitate nucleation of graphite. Inoculants based on Fe SiCa-alloys, where about 0.2–0.5% is added in the pouring stream or in the ladle, are most frequently used. The lifetime of inoculants is short and its potential to nucleate graphite decreases with time, the so-called fading. Therefore, in order to have a good effect, inoculation should be done as late as possible.

One important limit for graphitic cast irons is set by the section thickness of the casting. In thin sections, the cooling rate becomes high which may cause gray solidification to suppress and the white eutectic, consisting of cementite and austenite, to form. Additional factors which increase the risk of white solidification include bad inoculation, low silicon and carbon content and high content of cementite stabilizers such as Cr and Mo. The white eutectic, also called ledeburite, has an unwanted high hardness in graphitic cast irons.

The liquid cast iron melt shrinks during cooling until the eutectic reaction starts and the graphite precipitation turns the shrinking melt into expansion (approx. 1% volume expansion in total).

This gives a low demand for feeders, at least in the lower strength classes, which is one major reason for graphitic cast irons to result in good economy.

(a)

(b)

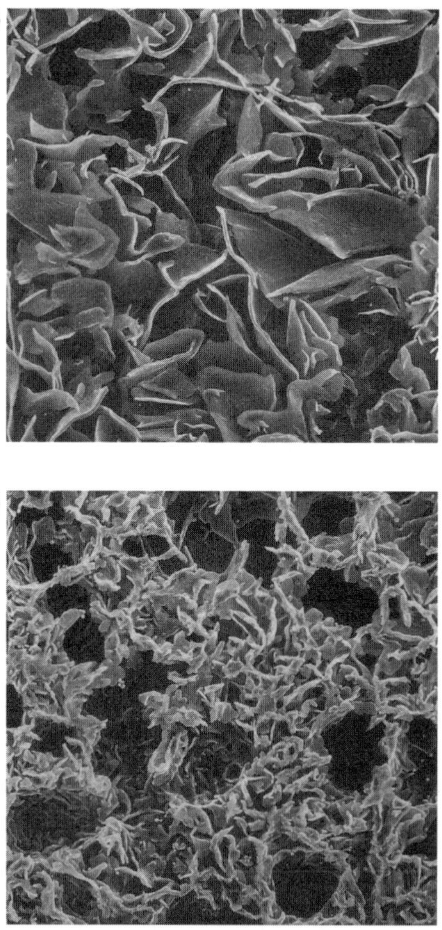

Figure 4 (a) Gray iron with graphite A-type. (From Ref. 2.) (b) Gray iron with graphite D-type. (From Ref. 2.)

A. Gray Cast Iron

Figure 4a and b shows how the graphite morphology looks like in a gray iron of A-type and D-type, respectively. The properties of gray cast irons are to a large extent controlled by the chemical- and phase composition. The chemical composition can be expressed in terms of a carbon equivalent, given by the expression

$$C_{eqv} = \%C + \frac{\%Si}{4} + \frac{\%P}{2} \tag{1}$$

In general, mechanical properties are improved when the carbon equivalent is decreased. Cast irons can be heat treated in different ways. The most common is stress relaxation heat treatment, but cast iron can be hardened, annealed, and soft annealed. Lamellar gray cast iron has very good damping capacity. This property is used in many components where damping of sound and vibration is important.

B. Compacted Graphite Cast Iron

An intermediate form of graphite growth is the compacted graphite, sometimes also called vermicular graphite, which gives a structure and properties between that of lamellar and nodular cast irons.

By controlling the modification of the graphite, the morphology of graphite can be more compacted, resulting in improved properties. Figure 5 [2] shows the graphite morphology in a compacted graphite iron.

Usually, compacted graphite irons also contain some fraction of nodular graphite. The main factors which increase the amount of nodular graphite are a high cooling rate, increased inoculation and higher Mg or Ce content.

The combination of high strength, some ductility, and good thermal conductivity makes compacted graphite irons attractive for a number of engineering applications, particularly for castings to withstand thermal cycling, such as ingot molds, cylinder heads, engine blocks, and braking components.

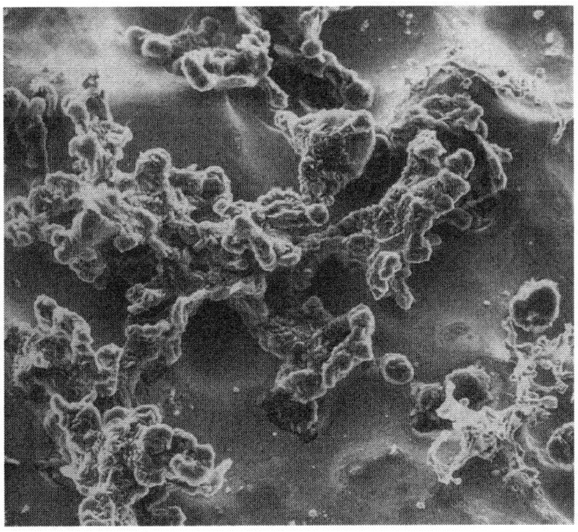

Figure 5 Graphite worms in compact graphite cast iron. (From Ref. 2.)

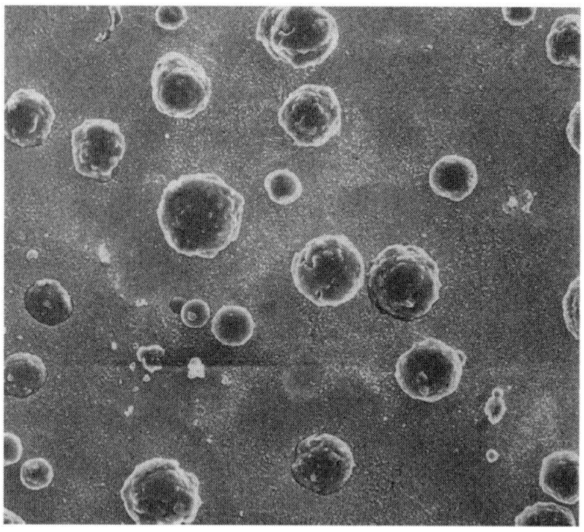

Figure 6 Graphite nodules in ductile cast iron. (From Ref. 2.)

C. Ductile Iron

Ductile iron, often also called nodular cast iron, was developed around 1950. The graphite precipitates as spheres or nodules (Fig. 6), without sharp edges, as in lamellar cast iron. To obtain nodule-shaped graphite, magnesium is normally used to change the growth mechanisms. Magnesium controls the sulfur and oxygen level in the melt. Mixing with other deoxidants and changing the level of the addition can control the shape of graphite.

Nodular cast iron has a good castability, though not as good as gray cast iron. By having a good inoculation, thin sections can be cast without white solidification taking place. However, ductile iron has a higher tendency for white solidification than lamellar cast iron. This is due to the solidification mechanisms, with the diffusion of carbon through the austenite layer. This relatively slow process makes ductile irons solidify at higher super-cooling as compared to gray cast irons, where solidification mainly is controlled by carbon diffusion in the liquid.

When increasing the pearlite content, a higher tensile strength and hardness is obtained, but the ductility will be decreased. The structure of the ductile iron matrix, and therefore also the strength, is normally controlled by the manganese and/or copper content, which promotes the formation of pearlite. The cooling rate during solidification and solid-state transformations, as well as the number of graphite nodules, are other important factors for the mechanical properties.

The round morphology of the graphite considerably improves the mechanical properties, with an increased ductility, tensile strength, and Young's modulus (E). The cutting properties are nearly as good as gray iron, but the heat conductivity and damping capacity is lower. Ductile iron is often an alternative to rolled, forged and cast steel in components.

D. Austempered Ductile Iron

"ADI" [3] is an acronym for austempered ductile iron. The acicular microstructural constituent in steel was discovered long ago and was called "bainite". Austempered steel has found many applications because of its good mechanical properties.

There is a big difference between austempered conventional steels and austempered ductile iron. Austempered ductile iron contains silicon, which is a very strong graphitizing

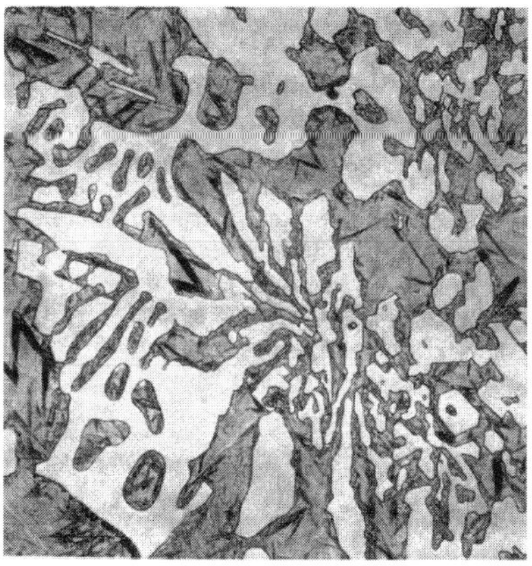

Figure 7 White cast iron. (From Ref. 2.)

element, and it retards carbide formation during austempering. Consequently, the metallic matrix of ADI normally consists of ferrite and austenite. In an optical microscope, this microstructure resembles bainite, but, of course, it is not bainite because bainite consists of ferrite and carbides. The austenite amounts of ADI can be very high, between 20% and 50%, depending on the austempering treatment.

The size of the component influences the microstructure of ductile irons, in general, and thus also of ADI. In thick-walled castings, the nodules and cells become larger because of the lower cooling rate after casting, causing stronger microsegregation, which is harmful from the point of view of mechanical properties.

E. Malleable Cast Iron

Like gray iron and nodular cast iron, malleable cast iron contains graphite. The major difference is that, in malleable cast irons, the graphite is obtained after a heat treatment, where the metastable cementite formed at solidification decomposes to graphite. The shape of the graphite is normally that of irregular nodules or balls. Malleable cast irons can be obtained in ferritic malleable, pearlitic malleable and martensitic malleable.

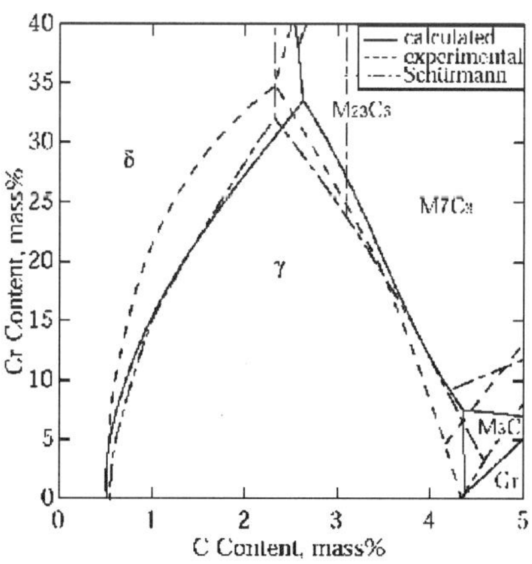

Figure 8 Phase diagram Fe–C–Cr. (From Ref. 68.)

F. White Cast Iron

White cast irons are normally used for abrasion-resistant applications such as pump wheels and machinery for crushing and grinding.

Most white irons are alloyed with chromium to promote white solidification and nickel, molybdenum, copper to avoid pearlite in the structure.

In the higher alloyed white cast irons, chromium-carbides (M_7C_3) are formed. The composition range [2] can be 2–3.6% C, max 2.0 Mn, max 2.0 Si, 2–5% Ni, 1.4–25% Cr, 1–3% Mo, max 1.2% Cu (Figs. 7 and 8).

III. MELTING AND PROCESS PARAMETERS

A. Melting

The base for a good cast iron starts in the melting shop. For successful production of cast iron castings, the entire production should be selected carefully. It starts with the type of melting furnace, charge materials, treatment of the melt, inoculation of the melt, pouring system, mold and core practice, and quality system.

The main methods of melting cast irons are *cupola* and *induction furnaces*. For gray iron, the cupola furnace is good, but for compacted graphite and nodular cast iron, an induction furnace is preferred. In the cupola furnace, the charge contains foundry scrap or used gray iron products (e.g. 50–60%), steel scrap (20–30%), pig iron (15%), and alloying elements. The charge and gray iron composition will vary among the foundries around the world. Typical for iron produced in a cupola furnace is that the sulfur content is higher, which promotes the solidification of a eutectic with lamellar graphite. Induction furnaces are designed to induce energy into the metal. The charge materials for induction furnaces are more sensitive to the composition of the charge material. In induction furnace melted material almost no pig iron is used. The normal charge materials are steel scrap and foundry returns, with carbon and ferrosilicon added to match the desired composition. The melts from induction furnaces are suitable for vermicular and ductile irons but gray irons can also be made.

B. Magnesium Treatment of Cast Iron

Magnesium used can be added as *pure metallic* or in an *alloy*. The reaction between the metallic magnesium and the liquid iron is violent and the magnesium vaporizes quickly. Specially covered containers have to be used. The magnesium can be added by using ferrosilicon with 5–10% Mg with a recovery of 50–60%. To improve the indoor environment, *wire feeding* can be added to the melt to treat the metal by magnesium material.

The amount of residual magnesium to obtain different graphite shapes is shown in Fig. 9 [4].

The addition of magnesium can be done in several ways, and Fig. 10 [2] shows the sandwich method. Due to the high vapor pressure of magnesium and the high reactivity with oxygen, the treatment must be done a short time before pouring.

C. Inoculation of Graphite

The inoculation and cooling rate are very important factors for the solidification behavior of cast irons. Inoculation controls the number of growing eutectic cells and therefore also

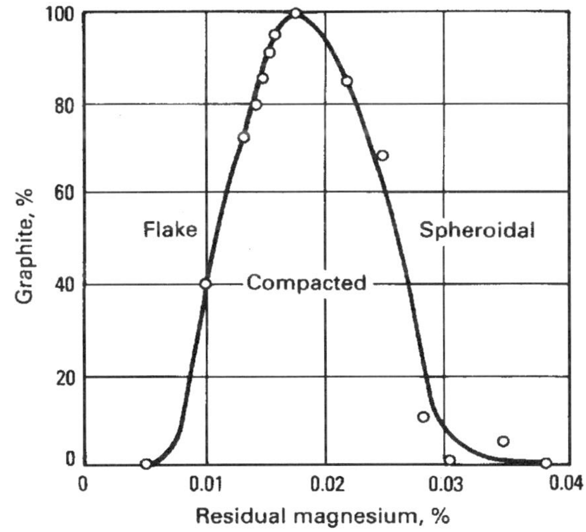

Figure 9 Graphite shape and residual magnesium. (From Ref. 4.)

the growth temperature. If the cooling rate is high and there is a high alloy content of Cr or Mn for example, the metastable temperature can be passed and a white solidification structure can be formed.

Skaland et al. [5] have proposed that the majority of the inclusions in ductile iron is a primary or secondary product of the magnesium treatment (e.g. MgS, CaS, $MgO–SiO_2$, and $2MgO–SiO_2$). After inoculation with (X,Al)-containing ferrosilicon (X denotes calcium, strontium, or barium), hexagonal silicate phases of the $XO–SiO_2$ or the

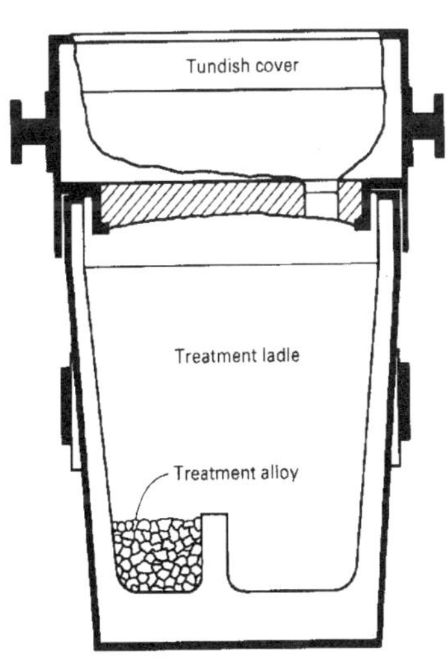

Figure 10 Magnesium treatment of nodular cast iron with sandwich method. (From Ref. 2.)

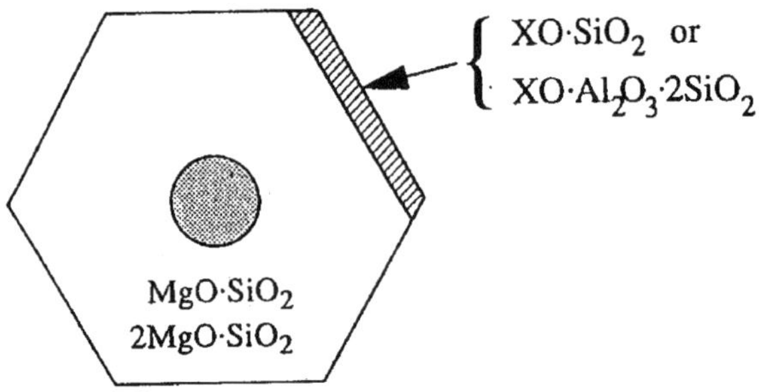

Figure 11 Proposed modification of the surface chemistry of a nucleus made by using a ferrosilicon inoculants containing Ca, Sr or Ba, X denotes Ca, Sr or Ba. (From Ref. 5.)

$XO–Al_2O_3–2SiO_2$ type form at the surface of the oxide inclusions, probably through an exchange reaction with MgO (Fig. 11). The presence of these phases will enhance the nucleation potency of the inclusions with respect to graphite. In particular, the [0 0 0 1] basal planes of the crystals are favorable sites for graphite nucleation, since these facets allow for the development of coherent/semicoherent low-energy interfaces between the substrate and the nucleus. In contrast, the fading of inoculation can be explained by a general coarsening of the inclusion population with time, which reduces the total number of catalyst particles for graphite in the melt.

Okada and Miyake [6] proposed that the fading effect of inoculants containing calcium occurs due to a decrease of the residual Ca during handling of the melt.

The residual Ca forms CaC_2 with the Ca in the melt. CaC_2 is proposed to nucleate graphite.

The choice of inoculants used in cast iron production today is probably one of the most important parameters in obtaining the right quality of a casting. The inoculants selected determine the conditions for the nodule count, the structure and the microshrinkage tendency of the nodular cast iron. Surveying suppliers of inoculants for iron foundries, one often gains the impression that specific inoculants from a certain supplier are much better than all the other inoculants on the market. If that were the case, everybody would use one particular inoculant. However, the production set-up and conditions are, in principle, not the same for all the foundries worldwide. Consequently, each foundry must find the best inoculants for their use.

In principle, there are five types of inoculants in the market today. One type is the pure *calcium–aluminum*-based inoculants with a silicon content slightly above 70%. Another type is the *zirconium*-based inoculants. Sometimes, manganese is added to this type of inoculants in order to facilitate its dissolution into the melt. A third type is the *barium*-based inoculants, which is often used in ladle inoculation due to its relatively long inoculating effect. The *strontium*-based inoculants are a newer development for in-stream inoculation. Finally, there is the *rare-earth*-based inoculants that have recently been developed. The latest generation of inoculants for nodular cast iron contains bismuth and rare-earth material, and has been found to increase the nodule count significantly, primarily due to the bismuth content.

Directly after the Mg-treatment, the melt contains a certain number of potential nuclei, and therefore even an uninoculated casting solidifies into a gray structure. Due to fading and coalescence, the amount of nuclei decreases rapidly after the Mg-treatment,

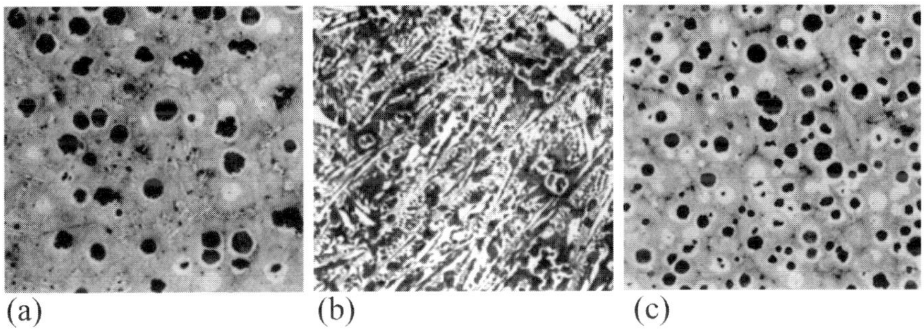

Figure 12 Inoculation of graphite in ductile iron at different stages and times. (a) Un-inoculated, directly after Mg-treatment, (b) Un-inoculated, 20 minutes after Mg-treatment, and (c) Inoculated, 20 minutes after Mg-treatment. (From Refs. 7,8.)

and therefore old and uninoculated melt produces a mainly white structure. By adding a suitable amount of inoculants, usually about 0.05–0.3%, solidification again can proceed according to the stable phase diagram, producing a gray structure. In Fig. 12, this sequence is shown for a ductile iron melt cast in standard quick cups [7,8].

D. Pouring Temperature

For most casting alloys and components, the cast iron is poured when in the temperature interval between 1315°C and 1430°C. Higher temperatures and long holding times of the melt increase the risk of decreasing the nucleation potential for the graphite, and white solidification can occur. The casting must also be compensated for the shrinkage of the liquid, which is around 1–1.6% shrinkage per 100°C temperature above solidification temperature.

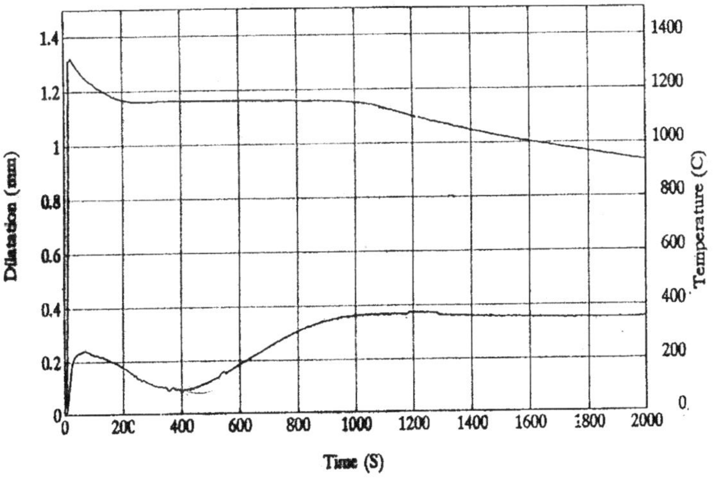

Figure 13 Upper curve temperatures and lower curve outward dilatation curve in a green sand mold, casting of 50 mm plate, measured by thermocouple and dilatation transducer. (From Ref. 9.)

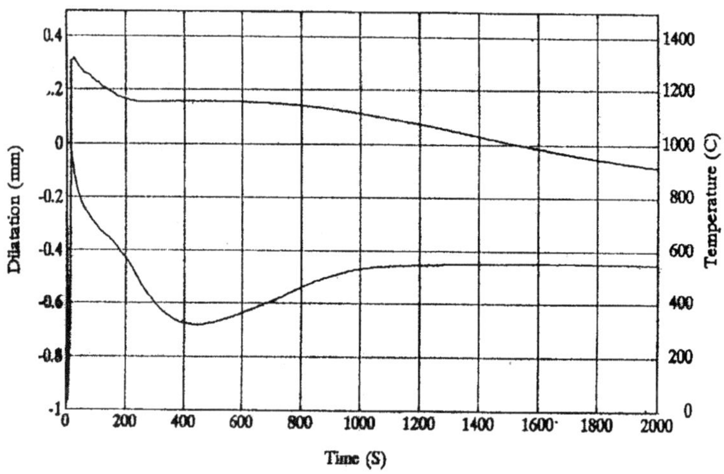

Figure 14 Upper curve temperatures and lower curve outward dilatation curve in a chemical bounded mold (furan), casting of a 50 mm plate, measured by thermocouple and dilatation transducer. (From Ref. 9.)

IV. MOLD WALL MOVEMENT

Mold stability has an influence on the defect formation in most cast irons. If the mold is stable enough, the pressure within the solidifying material will prevent the formation of shrinkage porosities. The dilatation of a ductile iron 50 mm plate during solidification, cast in a greensand mold and furan bonded mold, respectively, are shown in Figs. 13 and 14 [9]. For example, it be seen that the casting at the end of solidification in the green sand mold has increased in size by about 0.35 mm, while in the furan bonded mold, the casting has *decreased* in size by about 0.45 mm.

V. STRUCTURE FORMATION DURING SOLIDIFICATION

A. Primary Austenite Precipitation

Primary austenite precipitation involves two stages. The first part is the austenite growth in the fresh melt, forming either columnar or equiaxed dendrites. When the dendrites impinge, the second stage starts. The network is formed and from then on coarsening will be the growth mechanism. The growth of austenite during this period is slow and follows the phase diagram. The segregation of the alloying elements can be calculated by the lever rule or by a modified Scheil model including back diffusion, depending on the diffusion rate of the atoms in austenite. As the primary and off-eutectic austenite precipitation is calculated, the super cooling is influenced and also the nucleation rate of graphite, due to less super cooling. Many researchers have treated modeling of the dendritic growth in a more general way, which can be applied to the austenitic growth in cast iron. Mampaey [10] showed that austenite growth calculations could be based on the tip velocity for solid fractions lower than 0.335 and on dendrite arm coarsening by diffusion at higher fractions of solid.

B. Growth of Eutectics

Cast iron alloys can produce different microstructures: flake-like graphite, short flakes of graphite (undercooled graphite), compacted graphite, nodular graphite, and cementite. All these microstructures can be achieved from the same melt by small adjustments of the chemical composition or by changing the cooling rate. From a modeling point of view, cast iron is therefore a very interesting material to investigate.

The driving force for the reaction describes the growth rate of the phases. The driving force is coupled to the phase diagram, and by assuming local equilibrium at the phase boundaries, the chemical composition can be taken into account in the growth rate equation. Phase diagrams, solubility lines, phase equilibria, partition coefficients, and other alloy-dependent parameters can be calculated from thermodynamic data and calculation programs as, e.g. Thermo-Calc software [11].

Considering the shape of graphite in a cast iron, a complex growth can be observed. It can be expected that several different mechanisms are active at the same time, which changes the shape from flakes to spheres. The crystal structure of graphite [12] is a hexagonal sheet structure (Fig. 15). To explain the observed microstructures of graphite, it is proposed that the growth rate is different in the two main crystallographic planes, the basal plane and the prism plane (Fig. 16) [12]. Selection of the dominant growth plane is thought to be made by small amounts of oxygen and sulfur.

The D-type of flake graphite is a fully coupled eutectic growth, whereas graphite is the leading phase in A-type growth. The change of growth morphology can be explained

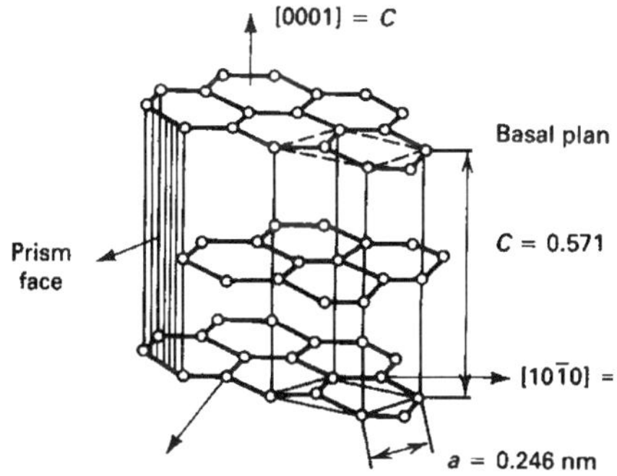

Figure 15 The crystallographic structure of graphite. (From Ref. 12.)

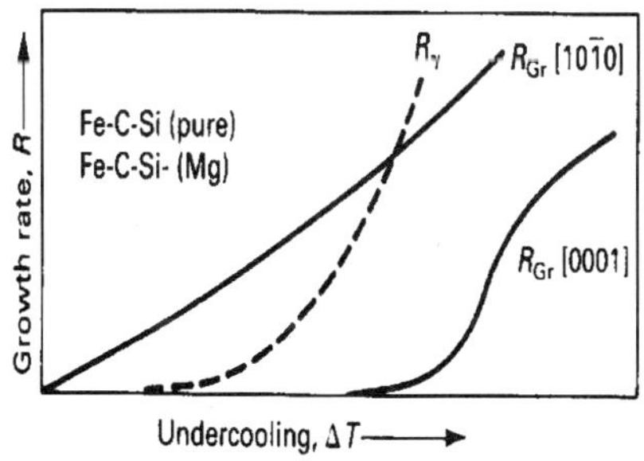

Figure 16 Growth rate of the different graphite planes and austenite vs. super cooling. (From Ref. 12.)

by considering the super cooling [13–16]. The growth of graphite can be modeled by using an interfacial kinetic constant, describing the rate of carbon to enter the graphite interface from the melt [13]. Interface kinetics has also been found to be of great importance during the eutectoid transformation [17].

In a simplified way, taking into account the controlling kinetic rate parameters, the eutectic lamellar growth of flake graphite can be written as

$$v = k\Delta T^n \tag{2}$$

where k is an alloy-dependent parameter and n is a constant taking the value 2 for normal coupled eutectic growth.

The growth rate of the eutectic cells in nodular cast iron is often calculated from a model derived by Wetterfall et al. [18]. This model is used when a shell of austenite surrounds the graphite:

$$\frac{\mathrm{d}R_{\mathrm{gr}}}{\mathrm{d}t} = \frac{V_{\mathrm{m}}^{\mathrm{gr}}}{V_{\mathrm{m}}^{\gamma}} D_{\mathrm{c}}^{\gamma} \frac{R_{\gamma}(X^{\gamma/\mathrm{L}} - X^{\gamma/\mathrm{gr}})}{R_{\mathrm{gr}}(R_{\gamma} - R_{\mathrm{gr}})(X^{\mathrm{gr}} - X^{\gamma/gr})} \tag{3}$$

where $\mathrm{d}R_{\mathrm{gr}}/\mathrm{d}t$ is the growth rate of graphite, $V_{\mathrm{m}}^{\mathrm{gr}}$ and V_{m}^{γ} are the molar volume of graphite and austenite, respectively, D_{c}^{γ} is the diffusivity of carbon in austenite, R_{gr} and R_{γ} are the radii of graphite and austenite, respectively, $X^{\gamma/\mathrm{L}}$ and $X^{\gamma/\mathrm{gr}}$ are the molar fraction of carbon in austenite at the liquid and graphite interfaces, respectively, and X^{gr} is the molar fraction of carbon in graphite (= 1). A simplified and approximate growth rate (m/sec) can be obtained from the following equation:

$$\frac{\mathrm{d}R_{\mathrm{gr}}}{\mathrm{d}t} = 2.87 \cdot 10^{-11} \frac{\Delta T}{R_{\mathrm{gr}}} \tag{4}$$

In compacted graphite irons, the graphite precipitation and its morphology reveal that the graphite growth can be divided into two types depending on the process of formation [19]. In one type, growth proceeds from a flake, whereas in the other, it is initiated from spheroids (Fig. 17).

It has been observed that the growth direction of CGI often changes from the a- to the c-direction, or vice versa. Because of the influence of modifying elements, the CG is

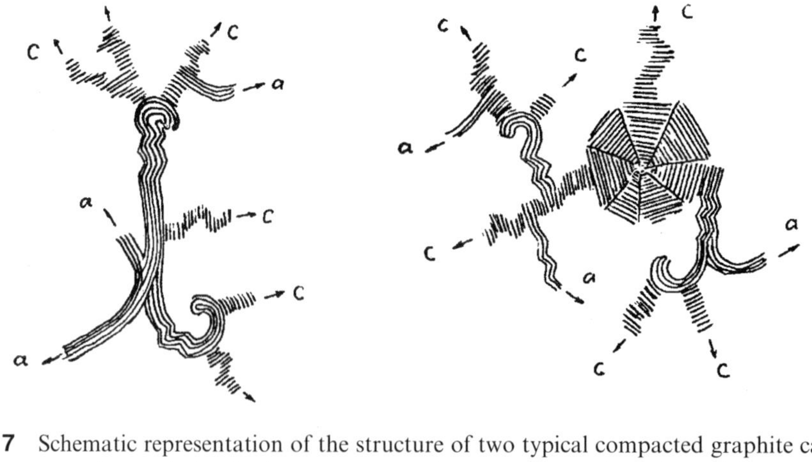

Figure 17 Schematic representation of the structure of two typical compacted graphite cast irons. Left: growing out from a flake. Right: growing out from a sphere. (From Ref. 19.)

twisted severely, changing its growth direction continually and also as it undergoes frequent branching. Growth of the branches may be in the *a*- and/or *c*-direction, and their tips may be round and/or sharp. In all cases, there exists a large fraction of thick flake graphite in the compacted graphite. The observation is that an austenite shell surrounds the growing tip of graphite. A combined model has been proposed by Fredriksson and Svensson [16]. Growth model of compacted graphite was modeled by using two different growth laws on a cylindrical shape. The growth of the radius will be due to diffusion of carbon through the solid in a similar way as in nodular cast iron (Eq. (3)). Growth in the a-direction was described in Eq. (2) (flake graphite).

According to an early discovery by Hillert and Steinhauser [20], white solidification structures are normally formed in two consecutive steps. In one step, the primary Fe_3C plates are formed, and, in the other, a normal eutectic structure is formed. The white eutectic is modeled [16] by assuming the formation of a plate of Fe_3C and the thickness is given by the phase diagram and the number of plates. When the plate has stopped growing, the white eutectic can grow perpendicular to the first plate. From then on, the white structure can be calculated by using coupled eutectic growth models similar to the gray structure. The growth constant is considerably higher in the white solidification.

The transition between gray to white and white to gray structures is of particular interest. The transition from gray to white normally needs higher cooling growth rates, and the opposite transition needs lower rates. Magnin and Kurz [21] found that the growth rates in the two cases are given by

$$V_{\text{gr}\to\text{white}} = \left(\frac{\Delta T_{\text{e}} + \Delta T_{\text{n}}^{\text{c}}}{K}\right)^2 \tag{5}$$

$$V_{\text{white}\to\text{gr}} = \left(\frac{\Delta T_{\text{n}}^{\text{c}}}{K}\right)^2 \tag{6}$$

where V is the growth velocity (m/sec), ΔT_{e} is the difference between the stable and metastable eutectic temperatures, K is the growth constant (C $\text{m}^{-1/2}$ $\text{sec}^{1/2}$), and $\Delta T_{\text{n}}^{\text{c}}$ is the super cooling necessary to nucleate cementite. However, in some cases, it can be expected that both gray and white structures grow at the same time. Mostly white structure dominates due to the higher growth rate.

C. Deterministic Modeling

The models presented above result in a fixed density of growing cells in each computational volume, and for simulation purposes, all cells have the same size. This is called deterministic modeling. To obtain a realistic transformation rate, the solidification rate has to be corrected when the cells or phases impinge. The transformed fraction is usually calculated by the Johnson–Mehl equation:

$$f_{\text{s}} = 1 - \exp\left(-\frac{4\pi R^3 N}{3}\right) \tag{7}$$

where f_{s} is the fraction transformed, R is radius of solid phase, and N is the number of growing cells per unit volume.

The deterministic modeling approach gives a good picture of the microstructure distribution in a casting. This type of modeling has in most cases proved to give good results when it comes to microstructure formation, porosity formation as well as mechanical

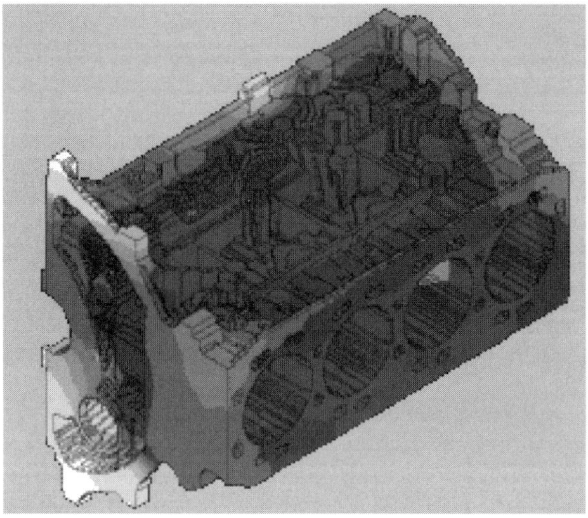

Figure 18 Simulated hardness in an engine block using kinetic growth laws for solidification and solid-state transformations together with a deterministic modeling approach for gray iron. (From Ref. 22.)

properties. An example of calculated hardness in an engine block is shown in Fig. 18 [22]. In this simulation, both solidification and the solid-state transformations have been modeled using kinetic growth laws.

D. Probabilistic Modeling

Probabilistic modeling [23,24] of microstructure formation in solidification processes is a rather new technique to simulate structure formation and phase morphology in a photograph-like way. Different principles of modeling have been developed, and one is based upon minimization of the interfacial energy [25].

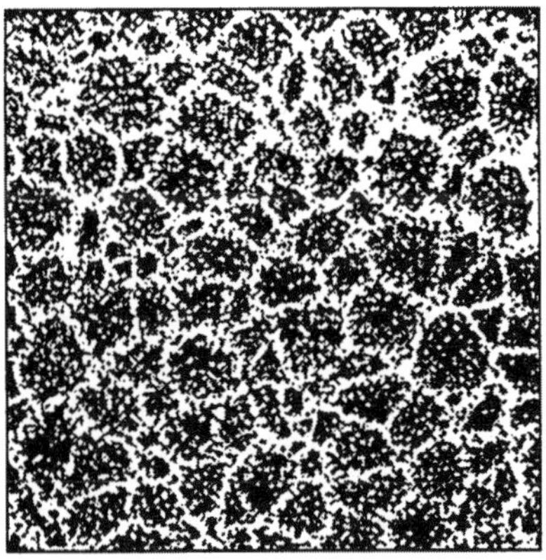

Figure 19 Microstructure of eutectic cast iron from a cylinder with a diameter of 20 mm. (From Ref. 26.)

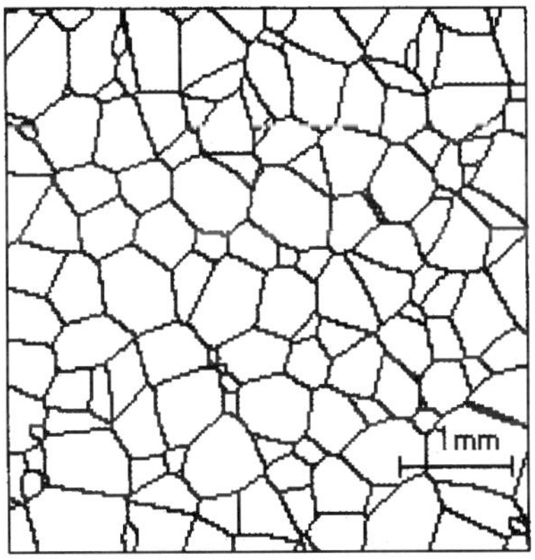

Figure 20 Simulation of cast structure of a cylinder with a diameter of 20 mm. (From Ref. 26.)

Mai et al. [26] used modified micro and macro models to visualize the eutectic cell size in the case of eutectic gray iron. Figure 19 shows the experimentally evaluated metallographic structure and Fig. 20 shows the corresponding simulated structure.

Furthermore, the probabilistic approach uses nucleation laws and kinetic growth laws, as in the deterministic approach, but the laws are applied separately on each growing cell or phase. The simulation will then calculate the growth kinetics of each phase and its crystallographic direction. The nucleation sites are randomly selected. If the phase is nucleated at the wall of the mold and nucleation is suppressed in the liquid, columnar growth will occur. The grains have also been allowed to move within the liquid [27] according to different convection models, such as no movement, random movement of the grains associated with turbulent effects, inhomogeneous fluid flow velocity, or sedimentation. In order to trace the position of the grain interface, each grain surface is subdivided into a given number of elementary surfaces. Grain impingement, volume fraction of solid, the effective solid–liquid interface, the average number of nearest neighbors, the grain size distribution, and stereological relationships may all be calculated by the means of such models.

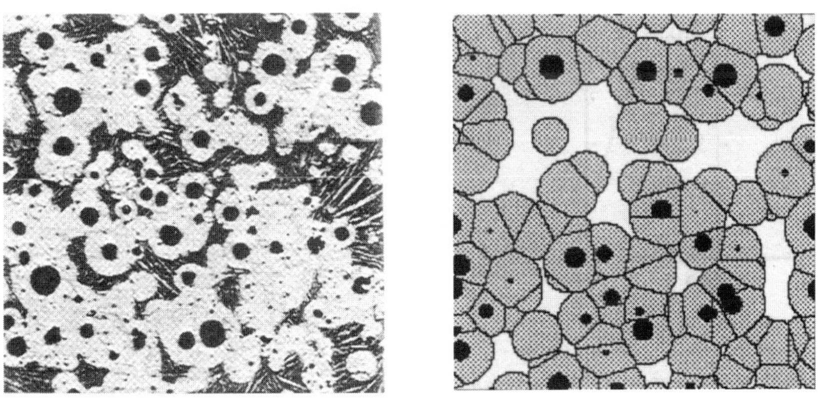

Figure 21 Microstructure in a ductile iron. Metallographic structure (left) and simulated (right). (From Ref. 28.)

For ductile iron, this approach has been applied by Charbon [28]. An example from Charbon's simulations can be seen in Fig. 21, which shows the structure during solidification at a solid fraction of 0.5.

However, today, it is not realistic to perform this kind of simulation on large real castings due to the enormous amount of data, which must to be stored. Further, when comparing the results from a stochastic simulation with those of a deterministic simulation, it can be concluded that the latter method gives satisfactory results when it comes to nodule count as well as cooling curves.

VI. VOLUME CHANGES AT SOLIDIFICATION AND POROSITY FORMATION

A. Volume Changes During Solidification

Cast iron, which solidifies with graphite eutecticum, gives a volume increase. The volume changes in cast irons start with changes in the volume of the liquid. This liquid shrinkage is normally compensated by the feeders or by the ingate system, during filling and before the ingates solidifies. The primary austenite shrinks during solidification, and often rather large porosities, coming from this early shrinkage, are found in the casting (Fig. 22) [2].

During the eutectic reaction, austenite and graphite are precipitated at the same time. The carbon is solved interstitially in the melt. When graphite is formed, the graphite structure is rather open with layers of carbon hexagon-rings. The density of graphite is around one-third of the melt and austenite. After the austenite has formed, the graphite precipitation can compensate for the volume loss. In gray iron and some types of compact graphite irons, the melt is in contact with that during solidification of the eutectic, which secures the internal feeding.

In ductile irons, the graphite precipitation starts directly in the melt, which compensates some of the shrinkage occurring at primary austenite precipitation or during austenite precipitation resulting from divorced eutectic growth. When an austenite shell surrounds the nodules, compensation is more difficult. The outward movements get stronger on the mold and, if the mold is too weak, the total volume increases and loss of material occurs. As a rigid network has been formed, the volume of the melt due to segregation

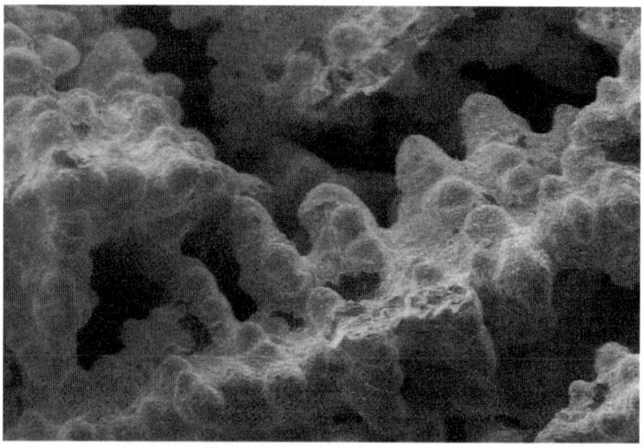

Figure 22 Interdendritic shrinkage porosity formed at austenite precipitation. (From Ref. 2.)

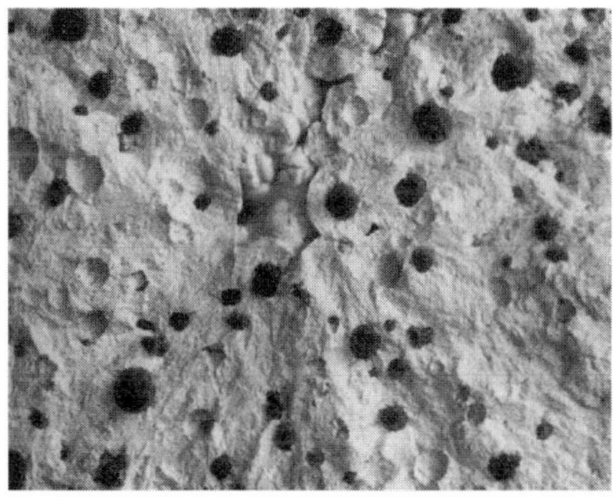

Figure 23 Microshrinkage in ductile iron. (From Ref. 2.)

Figure 24 Microshrinkage in ductile iron. (From Ref. 2.)

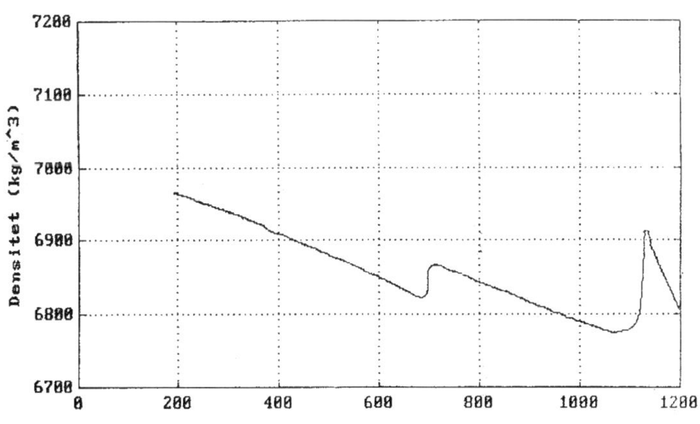

Figure 25 Density of a ductile iron as a function of temperature. (From Ref. 9.)

can be of larger density, and microshrinkage can occur between the austenite-covered nodules (Figs. 23 and 24).

A typical density curve as function of temperature for a ductile iron is shown in Fig. 25 [9].

B. Modeling of Molar Volumes

In simulation of various metallurgical processes, it is of interest to follow the volume changes taking place. This chapter discusses a way to calculate the molar volumes and how they can be used in solidification calculations in the system Fe–C–Si.

The volume changes on solidification can be divided into four different regions:

- melt,
- primary precipitation of austenite or graphite,
- eutectic reaction, and
- solidstate.

The modeling is made of phases occurring during solidification.

The Fe–C–Si system is of special interest for cast iron castings. The complete route of cast iron solidification is still not fully understood, and the modeling of the volume changes needs to be improved to accurately predict shrinkage and eutectic expansion. In this paper, modeling of the volume is done by using temperature-dependent molar volumes of the atoms and using a sublattice model to describe the influence of carbon as an interstitially dissolved element.

1. Substitutionally Dissolved Elements

To express the molar volumes in simulation or calculations, it is practical to use a mathematical expression similar to the one defined to describe the Gibbs energy. For a phase, it is common to apply the substitutional solution model and in line with that model, one can choose the following expression for the molar volume

$$V_{\mathrm{m}} = \sum x_i{}^0V_i + \sum_i \sum_j x_i x_j I_{ij} \tag{8}$$

${}^0V_{\mathrm{i}}$ represents the volume of the pure components i, x_i molar fraction of atom i and I_{ij}, parameters representing the deviation from ideal behavior [29] (Vegard's law). Higher-order parameters can be used but the information is usually too uncertain to allow these to be evaluated. The ${}^0\mathrm{V_i}$ quantities depend on the temperature and an exponential law is used in this work to describe the temperature dependence

$${}^0V_i(T) = {}^0V_{i,\mathrm{ref}} \cdot \exp\left(\alpha_1 T^2 + \alpha_2 T + \alpha_3 + \alpha_4/T\right) \tag{9}$$

2. Interstitially Dissolved Elements

Interstitially dissolved atoms, such as carbon, nitrogen and most gaseous elements, cannot be treated using the regular mixing law.

An alternative way to describe the influence of carbon in an fcc-lattice is to use a sublattice model. The model used here was described by Sundman and Ågren [30]. This model places iron and other larger atoms in one lattice and interstitials in a second lattice. If no interstitial atoms are dissolved, the second lattice contains only vacancies (*Va*). The molar volume can be described as follows. The colon (:) in the equation below separates species entering the first and second sublattices, respectively.

$$V_m = y_i y_C {}^0V_{i:C} + y_j y_C {}^0V_{j:C} + y_i y_{Va} {}^0V_i + y_j y_{Va} {}^0V_j + \sum_i \sum_j \sum_{Va} y_i y_j y_{Va} L_{i,j,Va} \quad (10)$$

where i and j substitute for each other on the first sub-lattice and C is an interstitial element. The y:s are site fractions, given by

$$y_i = \frac{x_i}{1 - x_C}, \quad y_j = \frac{x_j}{1 - x_C} \quad (11)$$

$$y_C = \frac{x_C}{1 - x_C} \quad (12)$$

C. Density Calculation from Molar Volumes

The density, ρ, can be calculated by using the molar weight M and the molar volume of the pure element or alloy

$$\rho = \frac{M}{V_m} \quad (13)$$

The density of a multiphase material can be calculated by a linear relationship involving the volume fraction of the phases and their respective densities:

$$\rho = \sum f_V \rho_i \quad (14)$$

D. Assessment of Molar Volumes and Densities of Phases Included in the Iron-Rich Fe–C–Si System

1. Molar Volume of Graphite

The molar volume of graphite has been assessed as a function of temperature by Gustafsson [31].

2. Molar Volume of fcc-iron with Carbon

In iron alloys containing carbon, the carbon dissolves interstitially between the iron atoms. The crystal structure has the same number of vacancies as the lattice. In saturated iron–carbon alloys, the maximum molar fraction of carbon is thus 0.5, although this situation is in practice impossible. The remaining unoccupied sites are vacant (Va). The molar volume of austenitic pure iron was assessed by Guillermet and Gustafsson [32], and the graphs of the molar volumes are shown in a report on modeling of density in cast irons [33].

The lattice parameters of the fcc-phase of Fe containing carbon have been measured by Ridley and Stuart [34]. The experiments have been made with more than 6 at% C and the temperature dependence of the lattice parameter has been obtained. The lattice data of Ridley and Stuart have been used to assess the molar volume of the austenite. Their measurements are extrapolated to x_C=0.5. The data show that the experimental measurements lie on straight lines, which can be extrapolated. The influence of carbon on the volume of the austenite phase is very important and has great influence. In this work, the lattice parameters have been adjusted to fit dilatometer experiments [35,36]. At higher carbon content, as in cast irons, the molar volumes are less than the extrapolated ones obtained from the experiments of Ridley and Stuart [34].

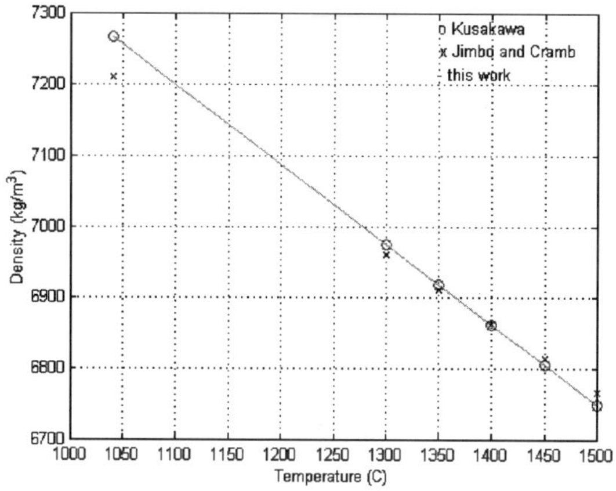

Figure 26 Liquid density of 3.65% C and 2.65% Si, used to evaluate the liquid density model. (From Ref. 41.)

3. Molar Volume of fcc-silicon

Pure silicon (cubic, diamond-crystal structure) can exist in several lattice-structures; however, it does not exist in the fcc-crystal structure. The density of cubic silicon at various temperature has been collected by Tatsumi and Ohsaki [37].

4. Molar Volume of Liquid Fe

The volume of the pure Fe-liquid has been assessed by Guillermet and Gustafsson [32]. The results of this study are adopted in the present work.

5. Molar Volume of Liquid Fe Containing Carbon

Many investigators have measured the density of liquid iron–carbon alloys, with a large scatter in the results. Olsson [38] has made measurements and compared them with literature results. Jimbo and Cramb [39] reviewed data in the literature for the density of liquid Fe–C alloys and reported also their own measurements. Kusakawa et al. [40] made careful experiments on liquid cast iron using ductile and gray iron melts. The data of Jimbo and Cramb [39] and Kusakawa et al.[40] have been used in assessment of ${}^{0}V^{\ell}_{\mathrm{Fe:C}}$, by using eqns. (9) and (10). The experimental density results from Kusakawa et al. [40] and Jimbo and Cramb [39] and the evaluated sub-lattice model transformed to density are shown in Fig. 26.

6. Molar Volume of Liquid Silicon in an Iron Melt

Olsson [38] assessed the molar volume of liquid silicon in an iron melt.

E. Coupling of Solidification Modeling and Molar Volume Modeling

Micromodeling of the solidification of gray and ductile iron coupled with molar volume calculations of the phases is expected to give the volumes and densities observed by experiments. Kagawa et al. [35] have made measurements of volume change and the graphite expansion during cooling and solidification. These experiments are very difficult to perform, and for this reason, the results in the literature show significant scatter. Combining

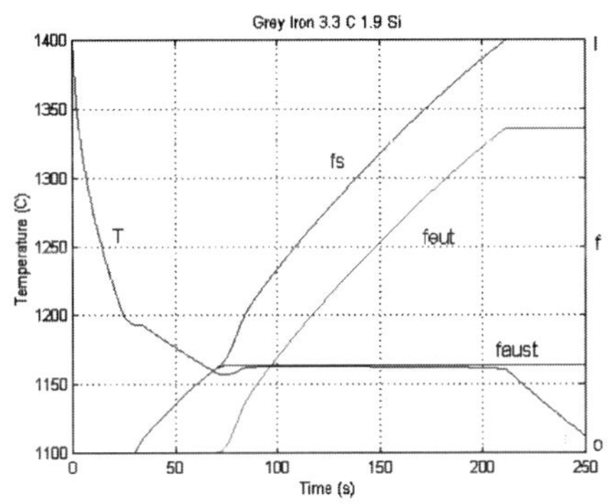

Figure 27 Calculated cooling curve and fraction of solid, austenite and eutectic. Gray cast iron 3.3% and 1.9% Si. (From Ref. 41.)

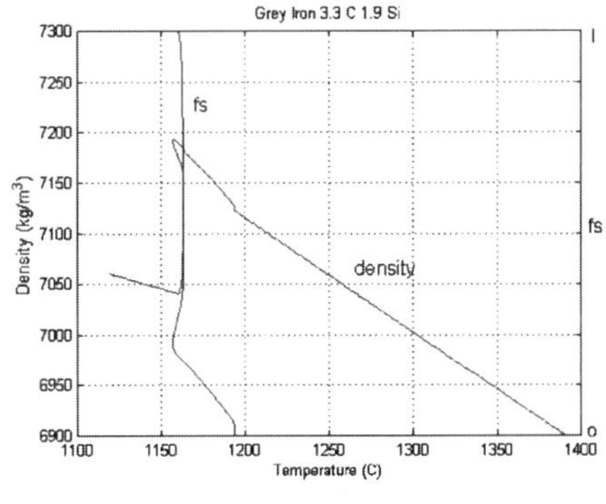

Figure 28 Fraction solid and density curve as function of temperature. (From Ref. 41.)

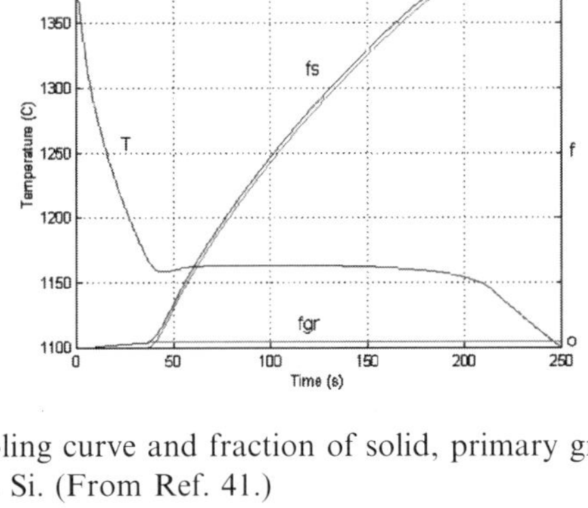

Figure 29 Calculated cooling curve and fraction of solid, primary graphite and eutectic. Ductile cast iron 3.65% and 2.65% Si. (From Ref. 41.)

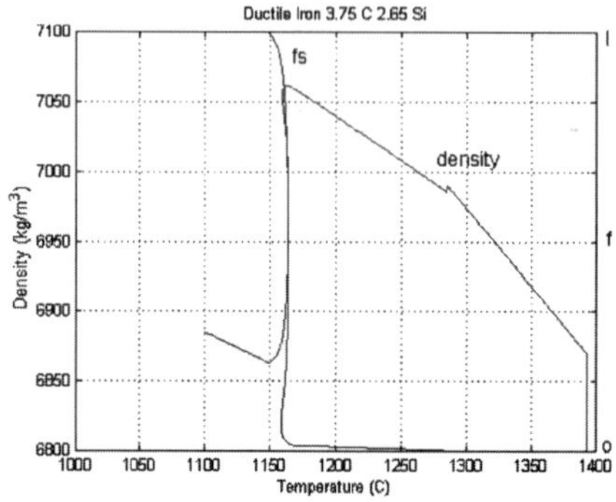

Figure 30 Fraction solid and density curve as function of temperature. Ductile cast iron 3.65% and 2.65% Si. (From Ref. 41.)

the results of solidification modeling and molar volume calculations gives the results for a gray iron casting containing 3.3% C and 1.9% Si, shown in Figs. 27 and 28 [41]. The modeling of the volume changes in gray iron alloy shows lower expansion when compared to results reported by Kagawa et al. [35]. The density curve in Fig. 28 [41] shows the shrinkage of the liquid down to 1200°C. The austenite precipitation gives an increase in the slope until the graphite eutectic has been formed at roughly 1160°C. The eutectic density decreases the total density until all the melts have solidified. The calculated fraction of precipitated graphite in the case of gray iron was fvgr = 6.8%, also according to the phase diagram and values in the literature [40].

Ductile iron, 3.65% C and 2.65% Si, i.e. the same composition as used by Kusakawa et al. [40], gave the result illustrated in Figs. 29 and 30 [41]. This eutectic ductile iron starts with primary precipitation of graphite, which can be seen as the change of the slope of the liquid density curve in Fig. 30. The primary graphite precipitation compensates some of the shrinkage of the liquid before the start of solidification. Thermal analysis of the precipitation of primary graphite cannot be seen on the cooling curve (Fig. 29), and the fraction of graphite is a few percent by volume, as shown in Fig. 31. Using the same database for the ductile iron as for the gray iron should give a larger expansion, since ductile irons normally have higher carbon content. The total volume expansion in the

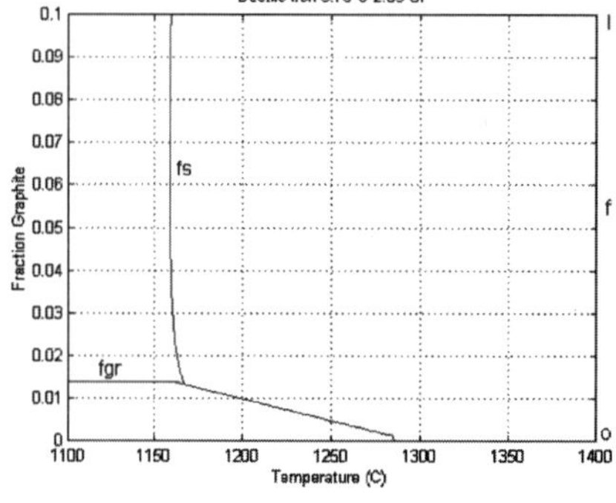

Figure 31 Fraction precipitated primary graphite and as function of temperature. Ductile cast iron 3.65% and 2.65% Si. (From Ref. 41.)

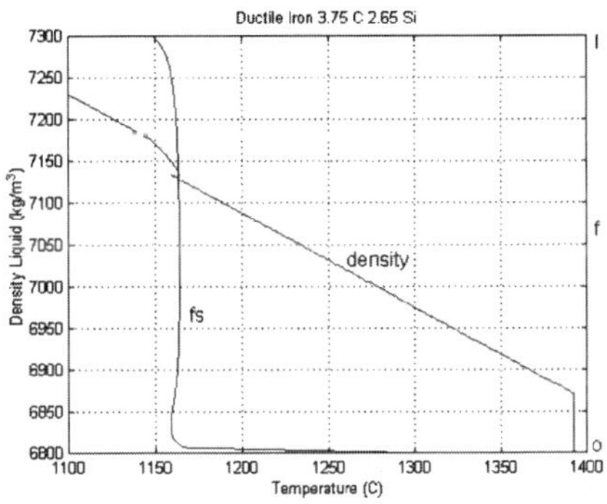

Figure 32 Density of the liquid during cooling, solidification and influence of temperature. Ductile cast iron 3.65% and 2.65% Si. (From Ref. 41.)

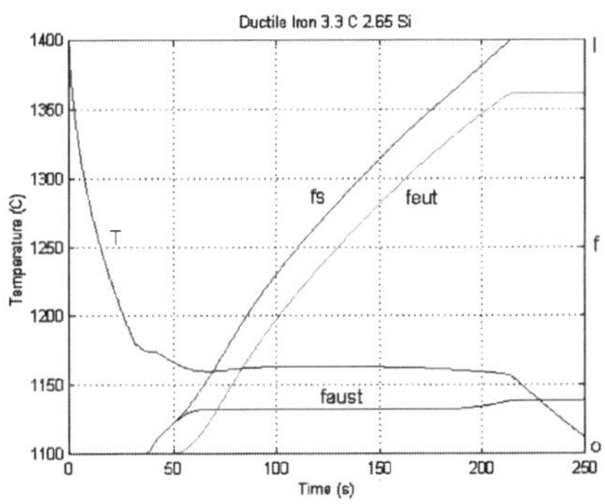

Figure 33 Calculated cooling curve and fraction of solid, austenite and eutecticum. Ductile cast iron 3.3% and 2.65% Si. (From Ref. 41.)

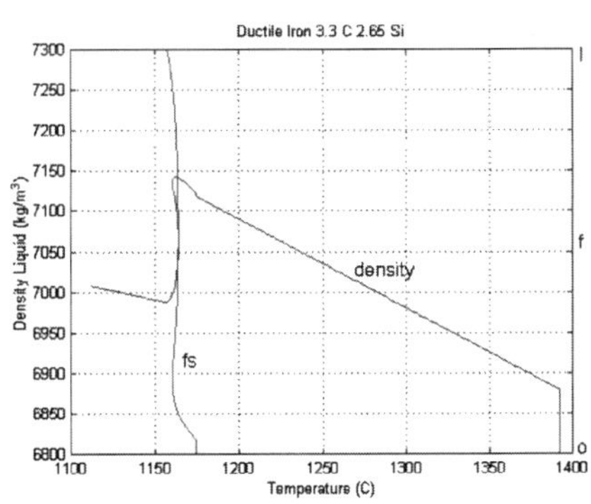

Figure 34 Fraction solid and density curve as function of temperature. Ductile cast iron 3.3% and 2.65% Si. (From Ref. 41.)

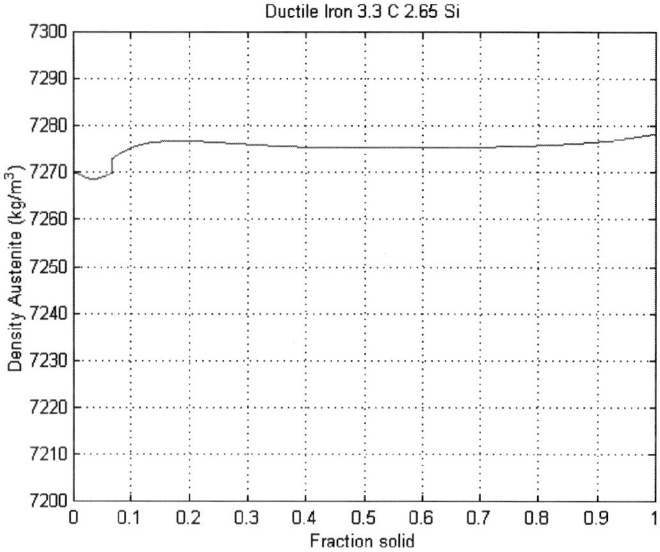

Figure 35 Density of the austenite during solidification. Ductile cast iron 3.3% and 2.65% Si. (From Ref. 41.)

simulated ductile iron is $\Delta V/V$=2.8%, which is larger than the results obtained by Kagawa et al. [35], who observed $\Delta V/V$=1.1–1.4%.

It is interesting to follow the density changes of the liquid and the different phases. Figure 32 shows the density change of the liquid during cooling. When the eutectic growth occurs, the liquid has about the same carbon content. The inverse segregation of silicon increases the density of the melt, as can be seen in Fig. 37 [41].

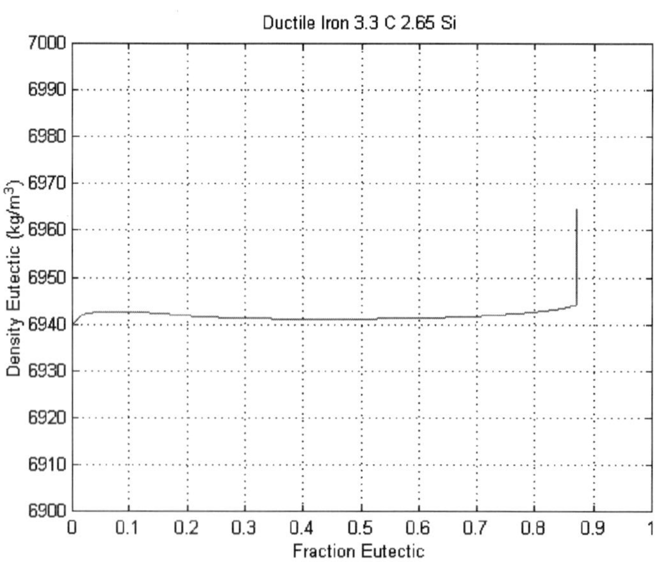

Figure 36 Density of the eutectic during solidification. Ductile cast iron 3.3% and 2.65% Si. (From Ref. 41.)

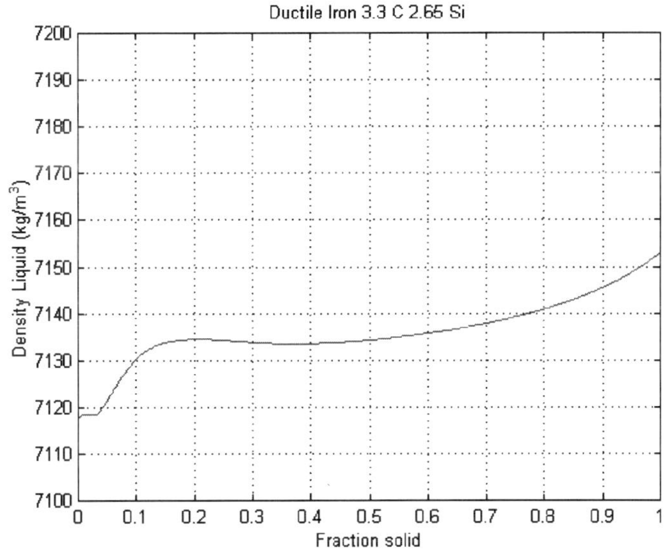

Figure 37 Density of the liquid during solidification. Ductile cast iron 3.3% and 2.65% Si. (From Ref. 41.)

In alloys which are hypoeutectic (Figs. 33 and 34), the solidification starts with primary austenite dendrites, which increases the total density. The density of the austenite during the solidification is shown in Fig. 35. The calculation shows that the density change of the austenite is mostly a temperature effect rather than a dependence on the chemical composition.

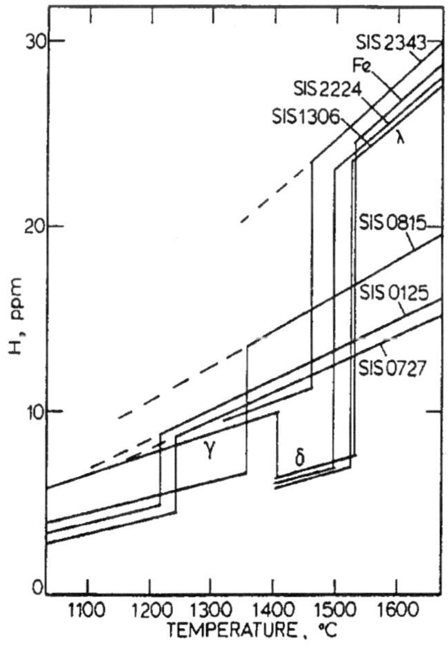

Figure 38 Hydrogen solubility in some gray iron and steels. (From Ref. 42.)

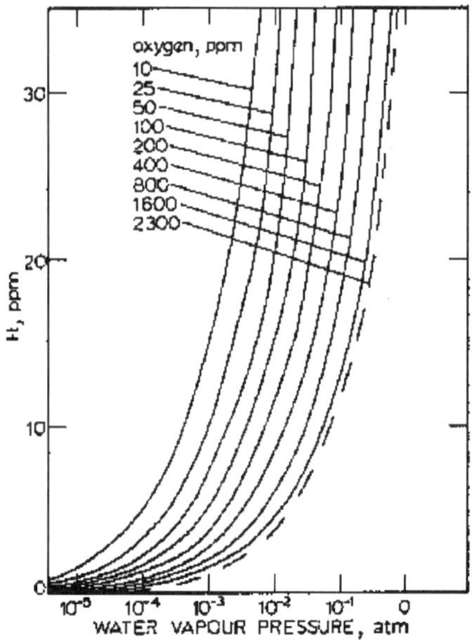

Figure 39 Equilibrium solubility of hydrogen concentration at different solved oxygen and water pressure. (From Ref. 43.)

The eutectic density is rather constant and can be seen in Fig. 36. The largest change is found in the density of the liquid during solidification. The density increases from 7118 to 7154 kg/m^3. The increase in density is largest at the end of the solidification and can be seen in Fig. 37.

F. Gas Solubility and Gas Pressure

Cast irons dissolve gases, which can create problems of porosity formation at solidification. The liquid metal has an internal strength, which has to be exceeded to form porosity.

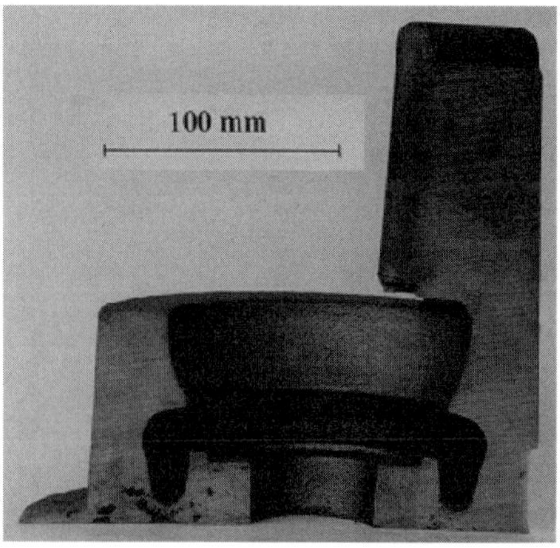

Figure 40 Porosities formed in ductile iron by external pressure.

Figure 41 Porosities in the casting.

The nucleation in a pure melt is thus very difficult, but in cast irons, there are normally slags, oxide- and carbon films, which can be sites for nucleation of gas pores.

The cast iron melt is exposed to air and, by this exposure, the melt will dissolve nitrogen. Air contains water and it will react with the metal and increase the hydrogen content. The hydrogen solubility at one atmosphere hydrogen gas pressure is shown in Fig. 38 for some iron-base materials [42]. The reason for porosity formation is that the solubility of the gas is much lower in the solid state than in the liquid. The casting of iron carbon alloys in more or less water-saturated green sand molds increases the dissolution of hydrogen in the melt. The oxygen content in the liquid will influence the possible hydrogen pressure at the surface and the possible concentration of hydrogen. Figure 39 shows the influence of the deoxidation level on the hydrogen content in cast iron melts [43].

1. Porosity Formation by External Gas Pressure

In a metal which shrinks at solidification, gravity feeding can compensate the shrinkage. The feeder is placed at a higher level than the area to be fed. It is also important that the feeder is in contact with the atmosphere, in order to have a transport of liquid from the feeder to the shrinking area. If a thin metal shell is formed at the surface of the feeder, the direct contact is broken between the atmosphere and interior. If at the same time, the melt has contact with the atmosphere at a point lower than the shrinking area, external pressure can transport the metal upwards. In the example shown in Figs. 40 and 41, the sand core in the ductile iron casting has been heated, and the external gas pressure has transported the melt to the shrinking areas, including the feeder. The result is that the feeder is completely free from porosities, while the casting contains an excessive amount of porosities.

VII. SOLID-STATE TRANSFORMATIONS AND MATRIX MICROSTRUCTURES

The mechanical properties of a cast iron material are to a large extent affected by the phase transformations taking place in the solid state, usually also referred to as the eutectoid transformations. After solidification, the cast iron structure mainly consists of graphite and/or carbides dispersed in an austenitic matrix. The decomposition of austenite into different phases and structures is very much affected by cooling conditions, alloying elements and the graphite morphology. Consequently, a great number of matrix microstructures can be obtained in different grades of cast irons. The following phases or structures may form as a result of the austenite decomposition: ferrite (+graphite), pearlite, martensite as well as different kind of bainitic structures.

Martensitic cast irons can be produced by heat treatment, i.e. austenitizing at around 900°C followed by rapid cooling (oil quenching or air blasting). The resulting material is very hard and brittle which makes these cast irons suitable in abrasive applications. The ductility may increase somewhat by a tempering heat treatment. However, the use of these materials is rather limited.

During the last 20 years, there has been a rather large interest in austempered ductile irons (ADI) for use in, e.g. automotive applications. The microstructure in these irons is produced by a heat treatment process in three steps:

- Austenitizing in the temperature range 825–925°C.
- Rapid cooling to the austempering temperature (230–400°C).
- Isothermal holding at the austempering temperature.

During the austempering process, plate-like ferrite (acicular ferrite) and metastable carbon-rich austenite is formed, and possibly also some carbides depending on the transformation temperature and time. The ADI-material can obtain a very attractive combination of high strength and high ductility. The high cost related to the heat treatment process is probably the main reason for the rather limited production of ADI-castings.

In the majority of the cast irons produced, the matrix microstructure consists of ferrite and/or pearlite. The relative amount of ferrite and pearlite is one major factor in controlling the mechanical properties in the different grades of cast irons. Therefore,

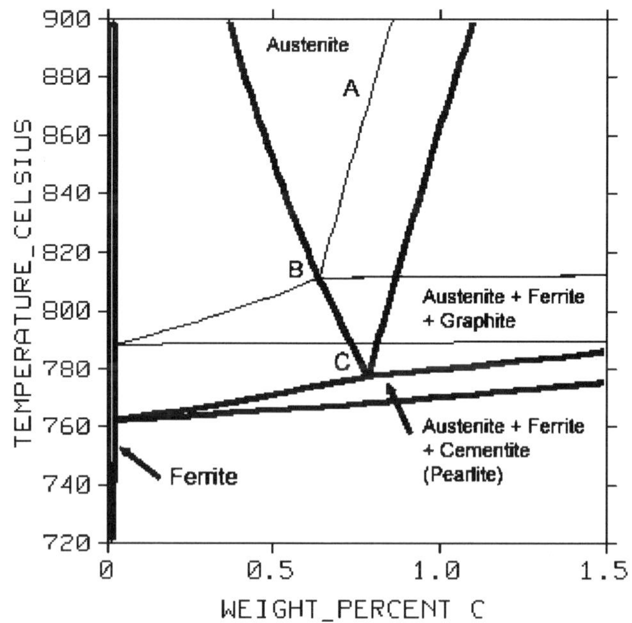

Figure 42 Isopleth at 2.5% Si in the Fe–C–Si phase diagram. Bold lines: metastable system, and thin lines: stable system.

the main focus in this section will be to describe the formation of ferrite and pearlite, and the mechanisms that control the rate of the related phase transformations.

Firstly, it is instructive to take a look at the Fe–C–Si phase diagram. In Fig. 42, an isopleth of this diagram at 2.5% Si is shown. The diagram has been calculated using the Thermo-Calc software [11]. In the figure, both the stable diagram (containing the graphite phase) as well as the metastable diagram (containing the cementite phase) are shown. The transformations in solid state can proceed either according to the stable or the metastable phase diagrams. In the former case, austenite decomposes into ferrite and graphite. If the alloy transforms according to the metastable diagram, the austenite transforms into ferrite and cementite. Ferrite and cementite grow in a lamellar way to form the pearlite structure. Consequently, it is the existence of two competing phase diagrams which is the reason for the varying amounts of ferrite and pearlite in gray cast irons.

During cooling after solidification, in cast irons containing graphite (gray, compacted and ductile iron), the austenite composition follows the austenite/graphite phase boundary (line A in Fig. 42). Due to the decreasing solubility of carbon in the austenite during cooling, the graphite content increases from about 5 vol% directly after solidification to about 10 vol% before the onset of the solid-state transformations. When the austenite reaches the stable eutectoid composition (point B), ferrite becomes thermodynamically stable and has the possibility to nucleate. Nucleation of graphite in the solid state is very unlikely during normal cooling conditions. Therefore, ferrite will nucleate on an existing graphite surface, which then acts as a natural sink for the carbon diffusing away from the prior austenite.

For pearlite to form, it is necessary that the austenite reaches point C, where the cementite becomes thermodynamically stable, which is at a considerably lower temperature than point B. Consequently, the difference in temperature between point B and C is a very significant factor for the possibilities for ferrite to form. A low cooling rate also favors ferrite formation at the expense of pearlite, thereby providing more time for ferrite growth before pearlite can nucleate. Since ferrite only can form if carbon can diffuse away to existing graphite, it is also natural that a fine graphite structure promotes ferrite formation, since it results in shorter diffusion distances for carbon.

A. Gray Cast Iron

If the austenite was allowed to transform according to equilibrium conditions, it should result in formation of ferrite and graphite. However, it is well known that gray iron for most practical purposes has mainly pearlitic matrixes, and consequently, transformation proceeds under non-equilibrium conditions. In Fig. 43, a typical gray iron structure without any free ferrite is shown. The sample has been etched in a picric acid solution to reveal the solidification structure [44]. The pearlite in the light areas has formed from austenite which grew dendritically, while the more brownish pearlite has grown from austenite formed during the eutectic reaction. Using normal etching techniques, it is not possible to make this differentiation. Some microporosities as well as microcarbides can also be seen in the figure.

Factors which promote formation of free ferrite (as opposed to ferrite contained in the pearlite) are slow cooling rate through the eutectoid transformation range, a fine graphitic structure and a high content of silicon [45]. Most other alloying elements act as pearlite promoters, i.e. tin, antimony, copper, manganese, nickel and chromium. When adding pearlite promoting elements to ensure a fully pearlitic structure in heavy sections, one

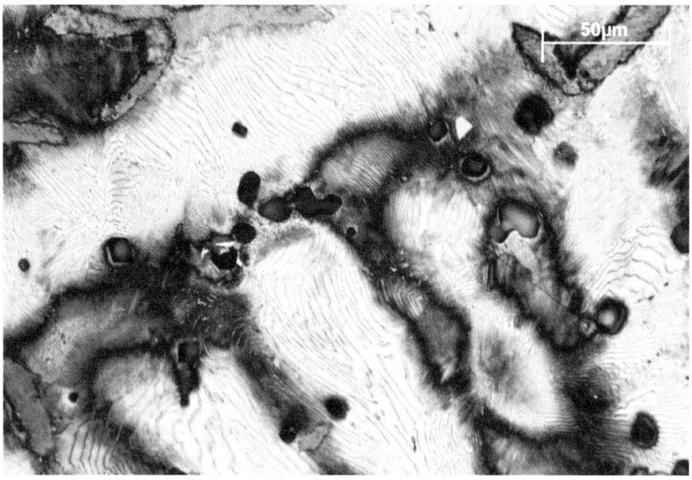

Figure 43 Etched gray iron, showing the pearlite and prior austenite dendrites (white area). (From Ref. 44.)

should bear in mind that most of these elements also increase chilling tendency during solidification, which therefore could lead to chill in thin sections.

Once a fully pearlitic structure has been obtained, the mechanical properties can be further improved by adding alloying elements which have a refining effect on the pearlite. Such elements include vanadium, molybdenum and chromium, all of which have a very strong pearlite refining effect, and copper and nickel, which have a somewhat weaker effect. Additionally, an increased cooling rate also results in a refinement of the pearlite.

Catalina et al. [46] have suggested a model for ferrite nucleation in gray iron, which assumes that ferrite nucleates at the edges of the graphite flakes. The number of ferrite grains, N_α, was found to depend on the eutectic fraction in the microstructure f_e, and the average graphite lamellar spacing according to the relation

$$N_\alpha = \frac{379.5 \cdot 10^6 f_e}{L_{Gr}} (\text{grains/m}^3) \tag{15}$$

The relation was derived from experiments where the CE was in the range 3.9–4.1. No information was given about the amount of alloying elements. From the equation, it is clear that the tendency for ferrite formation in gray iron becomes larger for a fine graphitic structure. It is also well known that the risk of ferrite formation is considerably larger if undercooled graphite (type D) exists.

It is important to emphasize that the formation of ferrite is not purely controlled by diffusion of carbon to the graphite. If this were the case, it could be expected that more ferrite should form in gray irons than in, i.e., ductile irons, due to the fact the free distances between graphite lamellas are usually considerably smaller than they are between graphite nodules. This is contrary to what is generally found. Instead, the planes of the graphite exposed to the austenite are the most important factor. Carbon can easily be incorporated on the graphite prism planes, the planes which are exposed to the austenite in ductile iron. However, the basal planes, which are those exposed on the surfaces of graphite flakes, do not easily accept addition of carbon atoms [47,48].

Pearlite nucleation and growth in cast irons can for most purposes be considered to take place in a way similar as it does in steels, since the reaction does not involve the graphite. At low undercoolings, as is the case in most castings, pearlite nucleates in the austenite grain boundaries and then grows in a spherical manner [49]. In a binary Fe–C alloy, the growth of pearlite is mainly controlled by a complex interplay between diffusion of carbon and by the interface mobility. If alloying elements are present, as is always the case in cast irons, it is rather the partitioning of these elements between the ferrite and the cementite which controls the growth rate. Based on the experimental results of Al Salman et al. [50] (Fe–C–2% Si steel), Lacaze and Gerval [51] described the growth rate of a pearlite colony as

$$\frac{\mathrm{d}R_p}{\mathrm{d}t} = 1.63 \cdot 10^{-5}(\Delta T_\mathrm{p})^3(\mu\mathrm{m/sec}) \tag{16}$$

which later was modified to [52]

$$\frac{\mathrm{d}R_\mathrm{p}}{\mathrm{d}t} = 7 \cdot 10^{-5}(\Delta T_\mathrm{p})^3 \exp(-20\,\Delta T_\mathrm{p}/T)(\mu\mathrm{m/sec}) \tag{17}$$

where R_p is the radius of the pearlite colony and ΔT_p is the undercooling with respect to the equilibrium pearlite formation temperature, given by

$$T_\mathrm{p} = 727 + 21.6w_\mathrm{Si} + 0.023(w_\mathrm{Si})^2 - 21w_\mathrm{Cu} - 25w_\mathrm{Mn} + 8w_\mathrm{Mo} + 13w_\mathrm{Cr} \tag{18}$$

For the mechanical properties of gray iron, the pearlite lamellar spacing has a major influence. Alloying elements and growth temperature have a strong effect on the lamellar spacing. The relationship between growth rate and lamellar spacing of the pearlite is given by Hillert [53]

$$\frac{\mathrm{d}R_\mathrm{p}}{\mathrm{d}t}(\lambda_\mathrm{p})^2 = \mathrm{Const.} \tag{19}$$

However, the complex alloy systems in cast irons, and the lack of precise information about the influence of alloying elements on the growth rate, causes reliance upon empirical or semiempirical relations for the pearlite lamellar spacing. Often, alloy content is related directly to the mechanical properties by multiplication factors, or regression equations.

B. Compacted Graphite Iron

The matrix in compacted graphite iron can vary from mostly ferritic to fully pearlitic, even though most applications seem to require a mostly pearlitic matrix to fulfill strength requirements. Therefore, pearlite promoters such as copper and tin are usually added to ensure a predominantly pearlitic matrix. The amount of pearlite promoting elements needed to prevent ferrite formation is considerably higher for compacted irons as compared to gray irons. Pearlite refining elements can also be used in the same way as for gray iron. An example of a compacted graphite iron microstructure is shown in Fig. 44a.

C. Ductile Iron

Ductile iron can for most purposes be considered as a steel with dispersed graphite nodules in the matrix. The graphite content normally ranges from about 9 to 12 volume%.

(a)

(b)

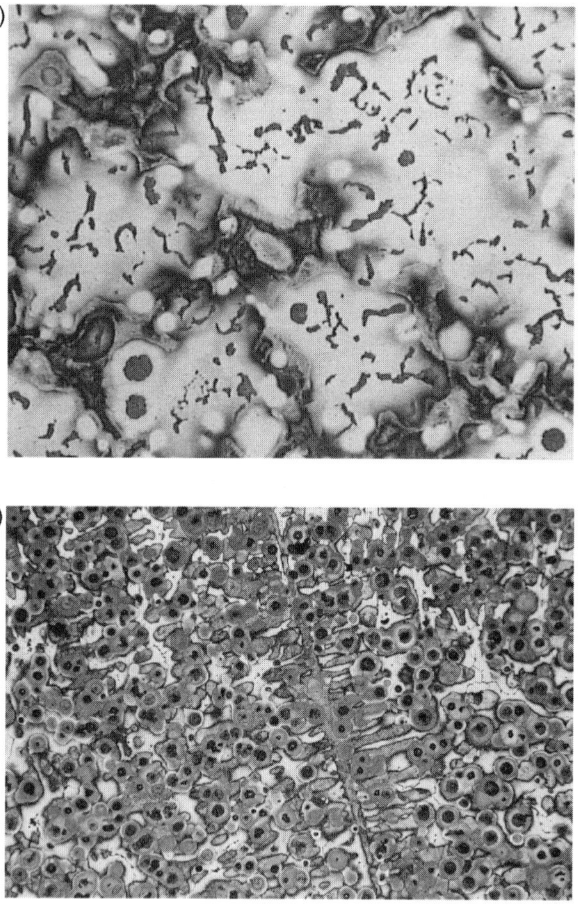

Figure 44 (a) Microstructure in compacted graphite iron. (b) Microstructure in ductile iron.

In Figs. 44b and 45 a typical ductile iron microstructures are shown. It is seen that the graphite nodules (black) in most cases are surrounded by a ferrite shell (white), the so-called bull's eye ferrite, and that the remaining matrix consists of pearlite (gray). Since ferrite has a lower strength and hardness than the pearlite, the mechanical properties of a nodular iron casting are strongly dependent on the relative amounts of ferrite and pearlite. In addition, the amount of alloying elements which strengthen the structure by solid solution hardening, and the interlamellar spacing of the pearlite both have a strong influence on the mechanical properties.

A casting often has a large variation in wall thickness. This will lead to a situation where the cooling conditions will vary during both solidification and solid-state transformations in different parts of the casting. As a consequence, the structure, i.e. the nodule count, ferrite/pearlite ratio and pearlite interlamellar spacing, will not be uniform throughout the casting which will therefore have a variation in mechanical properties.

The main factors which will result in an increased ferrite content in ductile iron are:

1. High nodule count, which is found in thin sections where a high cooling rate is obtained or by using a efficient inoculant. This creates shorter diffusion distances during the ferrite growth and therefore also an increased transformation rate [54].
2. Low cooling rate through the eutectoid transformation range, thus providing time for ferrite growth before pearlite nucleates. The growth rate of pearlite is considerably higher than that of ferrite since it is only dependent on short range diffusion corresponding to the interlamellar spacing. Therefore, there will be only a small increase in ferrite content after the nucleation of pearlite.
3. Addition of alloying elements which increase the difference between the stable and the metastable eutectoid temperatures, or which facilitate the nucleation of ferrite or obstruct the nucleation of pearlite (e.g. Si).
4. Addition of alloying elements which increase the transformation temperatures and therefore also increase the growth rate of the ferrite (e.g. Si).
5. Minimizing the amount of elements which accumulate at the graphite surface thereby creating an interfacial resistance for the carbon flux to the nodules (e.g. Cu, Sb, and Sn).

The very strong effect of nodule count on the solid-state transformations is illustrated in Fig. 45, showing the structure in two quick-cup castings cast under identical conditions, but using two different inoculants [7,55]. On the left-hand figure, the matrix ferrite content is 14% as opposed to 71% on the right-hand figure. Consequently, inoculation of ductile irons plays a major role not only for the solidification, but also for the solid-state transformations and consequently also for the mechanical properties.

The alloying element most commonly used in ductile iron for controlling the pearlite content is copper. Tin is an even stronger pearlite promoter, but its use is limited by the tendency to increase brittleness due to the formation of intercellular carbides. Manganese is a weak pearlite promoter. In Fig. 46 [7], the matrix ferrite content formed in 50 mm plates cast under identical conditions is plotted vs. additions of Mn, Cu, and Sn. The base alloy contained in these experiments relatively has silicon content (about 4%).

Considerable efforts have been made to model the eutectoid transformations in ductile irons. The reason for this is of course that there is a huge interest in predicting the final structure and the mechanical properties by means of simulation.

Pearlite grows more or less in the same way as it does in steel or in gray cast irons. Due to the short diffusion distances, corresponding to the pearlite lamellar spacing, the growth

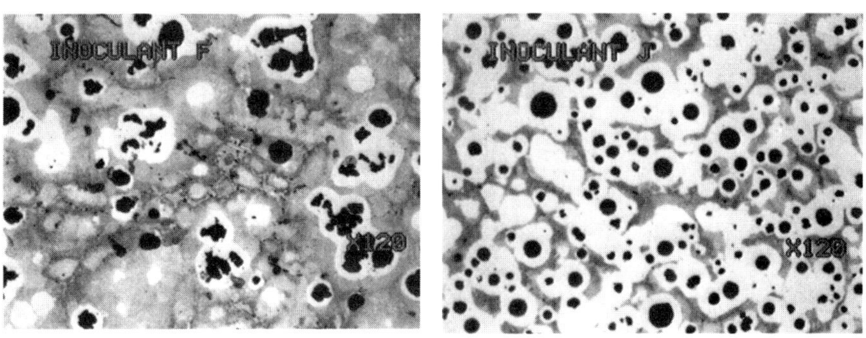

Figure 45 Influence of nodule count on ferrite formation in ductile irons.

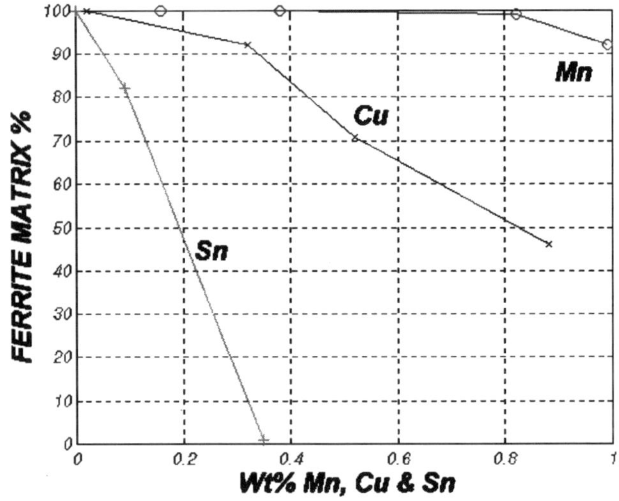

Figure 46 Effect of alloy additions on ferrite formation in a ductile iron 50 mm plate. (From Ref. 7.)

rate of pearlite is considerably higher than that of the ferrite. Ferrite growth depends on long range diffusion corresponding to the thickness of the ferrite shell. Consequently, only a small amount of ferrite will form after the nucleation of pearlite. Therefore, the main focus has been to describe how ferrite grows around the graphite nodules, and how this growth is affected by alloying elements and the distribution of graphite nodules.

In Fig. 47, the sequence of ferrite and pearlite growth in ductile iron is shown. The pictures are taken from samples quenched at different temperatures during the transformation [7]. In Fig. 47a, ferrite has just nucleated on one graphite nodule, while the other nodule is almost completely surrounded by a ferrite shell. The rest of the matrix consists of untransformed austenite (transformed into martensite at quenching). In Fig. 47b, ferrite has continued to grow radially around the nodules. Note the ferrite nuclei on the small nodule in the lower right corner. This nodule has probably nucleated at a later stage of solidification, and the austenite around this nodule has probably a higher alloy content (Cu, Mn, Sn, etc.), which has retarded the ferrite nucleation. Figure 47c shows a pearlite colony which has nucleated between two ferrite shells, probably at an austenite grain boundary. Note that the ferrite growth is efficiently stopped by the pearlite, which means that there will only be a very small increase in ferrite content after the moment of pearlite

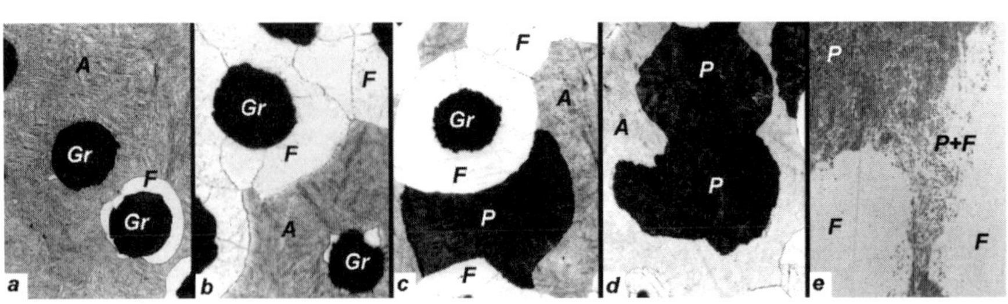

Figure 47 Sequence of ferrite and pearlite growth in ductile iron. (From Ref. 48.) A = Austenite, Gr = Graphite, F = Ferrite, and P = Pearlite.

nucleation. Figure 47d shows two pearlite colonies which have impinged. In Fig. 47e, it is shown that the metastable pearlite can decompose into the stable phases ferrite and graphite during normal cooling. In this case, carbon has to diffuse away to any nearby graphite nodule (not shown in the picture).

The first approach to combine a model for the eutectoid transformations with a simple heat transfer calculation was made by Stefanescu and Kanetkar in 1985 [56]. They used a model for ferrite based on the assumptions that the transformation is controlled by diffusion of carbon through the ferrite shell and that it occurs at steady state. The growth rate of pearlite was assumed to be equal to that in steel. This ferrite growth model was later revised and extended [17]. This model was able to account for differences in alloying and trace elements. From experimental observations, it was concluded that a pure diffusion model was not sufficient to describe the transformation rate of austenite into ferrite and graphite as previously proposed in the literature. By introducing an interfacial mass transfer resistance at the graphite/ferrite interface, to control the incorporation rate of carbon atoms on the graphite nodules, it was possible to reproduce the ferrite growth experimentally observed by numerical simulation. The ferrite growth was divided into three stages, which are described briefly below.

First stage: At low supercoolings below the upper stable eutectoid temperature, growth proceeds by carbon diffusion into the austenite. The ferrite grows in the plane of the graphite surface until the graphite is totally surrounded by a thin ferrite shell ($S^{\alpha}\sim 6\,\mu\text{m}$), see Fig. 47a. The growth rate of the ferrite nuclei in the plane of the graphite surface is given by

$$\frac{dy}{dt} = \frac{D_c^{\gamma}\left(C_c^{\gamma/\alpha} - C_0^{\gamma}\right)}{S^{\alpha}\left(C_c^{\gamma/\alpha} - C_c^{\alpha/\gamma}\right)} \tag{20}$$

where D_c^{γ} is the diffusion rate of carbon in austenite and C_c are the solubilities of carbon calculated from the phase diagram.

Second stage: The rate controlling mechanism during the second stage is assumed to be related to an interface reaction at the graphite surface, corresponding to the incorporation of carbon atoms. It was found that this mechanism controlled ferrite growth up to a transformed austenite fraction of ~0.8. The radial growth rate of the ferrite shell can be calculated from

$$\frac{dl^{\alpha}}{dt} = \frac{\rho^{\alpha}\left({}^{i}C_c^{\alpha} - C_c^{\alpha/gr}\right)}{\rho^{\gamma}\left(C_c^{\gamma/\alpha} - C_c^{\alpha/\gamma}\right)}\left(\frac{r_g}{r_g + l_{\alpha}}\right)^2 \exp\left(\frac{4\pi\left(r_g + l_{\alpha}\right)^3 N_v}{3}\right)\mu\ (\text{m/sec}) \tag{21}$$

where ρ is the density, r_g is the graphite radius, and N_v is the volumetric nodule count. The interface coefficient, μ, having the dimension of velocity, was evaluated from experiments and is given by the following expression [7]:

$$\mu = 9.49 \cdot 10^{-5}\left(r_g\right)^{-0.65}\exp\left(-9775/T(K) + \%\text{Cu}\left(0.070 - 7.72 \cdot 10^4 r_g\right)\right) \tag{22}$$

Third stage: During ferrite growth, the diffusion distances for carbon through the ferrite shell increase continuously. As a consequence, the diffusion rate of carbon will control ferrite growth at later stages of the transformation. There is no sharp transition between the second and third stages. Instead, the increased diffusion distances will continuously cause

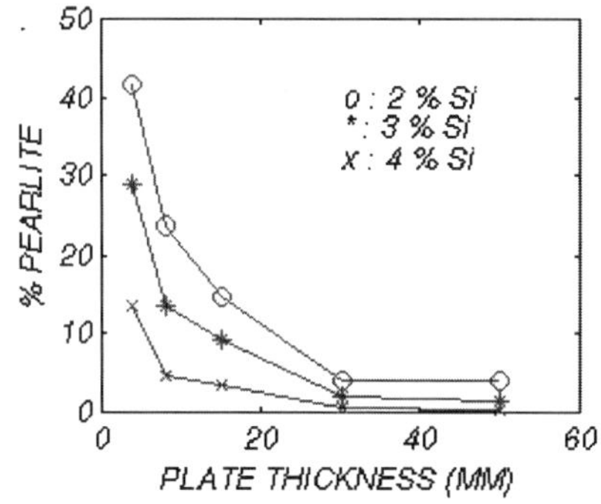

Figure 48 Pearlite content vs. plate thickness. (From Ref. 57.)

the transformation to become more diffusion controlled. The growth rate of the ferrite shell in the diffusion controlled stage is given by

$$\frac{dl^{\alpha}}{dt} = \frac{\rho^{\alpha}\left(C_{c}^{\alpha/\gamma} - C_{c}^{\alpha/\mathrm{gr}}\right)}{\rho^{\gamma}\left(C_{c}^{\gamma/\alpha} - C_{c}^{\alpha/\gamma}\right)} \frac{r_{g} D_{c}^{\alpha}}{l^{\alpha}(r_{g} + l^{\alpha})} \tag{23}$$

From Figs. 48–53, some examples from simulations using the models described above are shown [57]. Some of the relations used for the mechanical properties are discussed in the next section. Figures 48 and 49 illustrate the effect of silicon content and cooling rate (plate thickness) on the pearlite content and hardness in a low alloyed ductile iron (0.35% Mn and 0% Cu).

From Figs. 50–53, the importance of inoculation (nodule count), together with the effect of alloying elements, on the pearlite content and hardness is shown. These calculations were performed for the cooling conditions obtained in sand cast 15 mm plate [57]. Figure 50 illustrates clearly how the pearlite content increases when inoculation is insufficient (decreased nodule count). Naturally, this is accompanied with an increase in hardness, as shown in Fig. 51. The same calculations were performed for identical alloy compositions, except for the copper content which was increased to 0.7%, see Figs. 52 and 53. Note that the hardness for a nodule count of 400 nodules/m^2 is about the same for the three silicon content. The effect on hardness by the ferritizing effect of silicon seems to be exactly counterbalanced by the solid solution hardening effect of silicon.

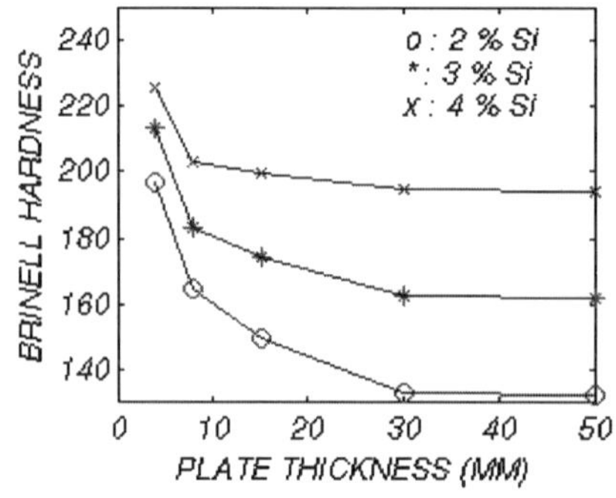

Figure 49 Hardness content vs. plate thickness. (From Ref. 57.)

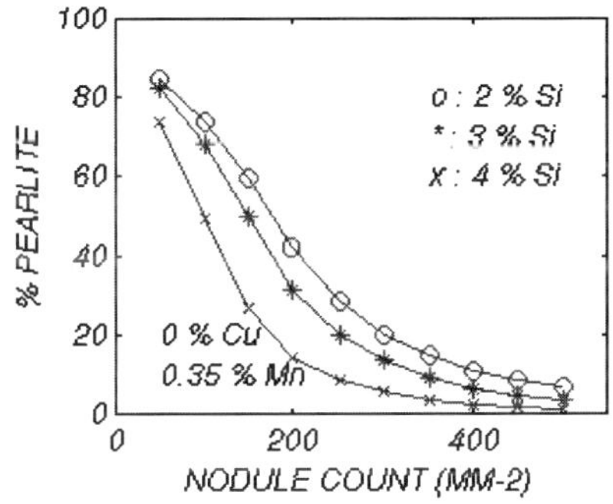

Figure 50 Pearlite content vs. nodule count for a base alloy content of 0% Cu and 0.35% Mn. (From Ref. 57.)

In Fig. 54, the results from a simulation of a plate casting of a ductile iron containing 4% Si and 0.88% Cu are shown. Experimentally measured values of the pearlite content in some locations are also shown.

VIII. MECHANICAL PROPERTIES

The mechanical properties of cast irons are determined by the microstructure, i.e. the amount, size and finesse of different phases. Consequently, cast irons can be considered as quite complex composite materials. The main reasons for the extremely wide range of properties which can be obtained in these materials are the existence of two competing phase diagrams; the stable and the metastable ones, and the variety of morphologies in which graphite may precipitate during solidification. Some general factors which affect the mechanical properties are listed in Table 1.

A. Gray Cast Iron

In gray cast irons, a wide variety of properties can be obtained by changing the carbon equivalent. When lowering the carbon equivalent, the amount of primary austenite dendrites increases. Thereby, the coherency of the dendritic network increases which gives

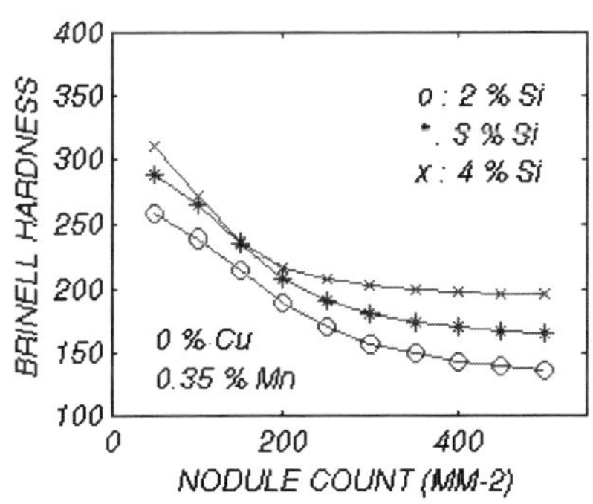

Figure 51 Hardness content vs. nodule count for a base alloy content of 0% Cu and 0.35% Mn. (From Ref. 57.)

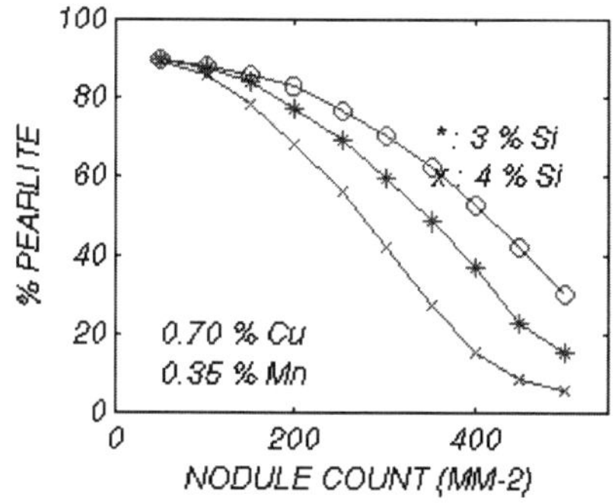

Figure 52 Pearlite content vs. nodule count for a base alloy content of 0.7% Cu and 0.35% M. (From Ref. 57.)

the material more steel-like properties. In Fig. 55, the relationships between hardness, tensile strength, modulus of elasticity, section thickness and carbon equivalent are shown [2].

As aforementioned, most commercial gray irons are predominantly pearlitic. Therefore, at a fixed carbon equivalent, the strength of a gray cast iron can very easily be increased by adding alloying elements which refines the pearlite, as shown in Fig. 56 [58]. It can also be expected that the alloying elements in this case have contributed to an increased strength by solid solution hardening of the ferrite contained in the pearlite.

The graphite lamellas act as stress risers, which is the main reason for the generally low ductility of gray irons. Therefore, there is a rather strong influence from cell size (or flake length) on the mechanical properties, which is shown in Fig. 57 [59]. It should be mentioned that this figure also reflects the effect of pearlite lamellar spacing, since different flake lengths were produced using different test bar sizes. At smaller bar sizes (smaller flake lengths), the pearlite lamellar spacing will decrease.

The complex interaction between the graphite morphology, the austenite dendrites, and the matrix structure, makes it rather difficult to describe the separate influence of each factor on the properties. A number of approaches have been used to predict the properties, but most of them are based on some kind of regression analysis or multiplication factors. Yang et al.[60] developed the following equation for the influence of alloying elements and cooling rate on hardness:

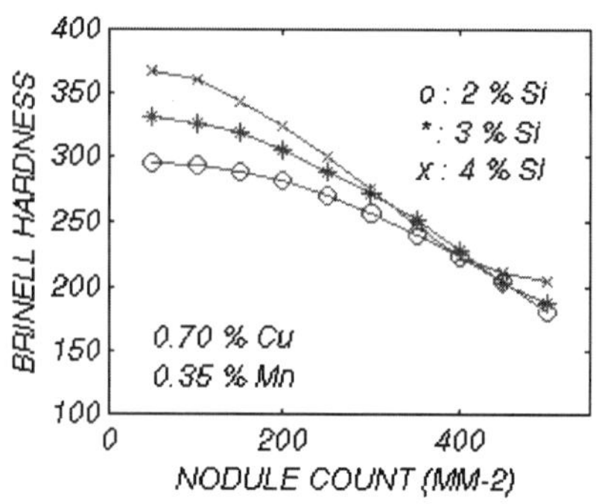

Figure 53 Hardness content vs. nodule count for a base alloy content of 0.7% Cu and 0.35% Mn. (From Ref. 57.)

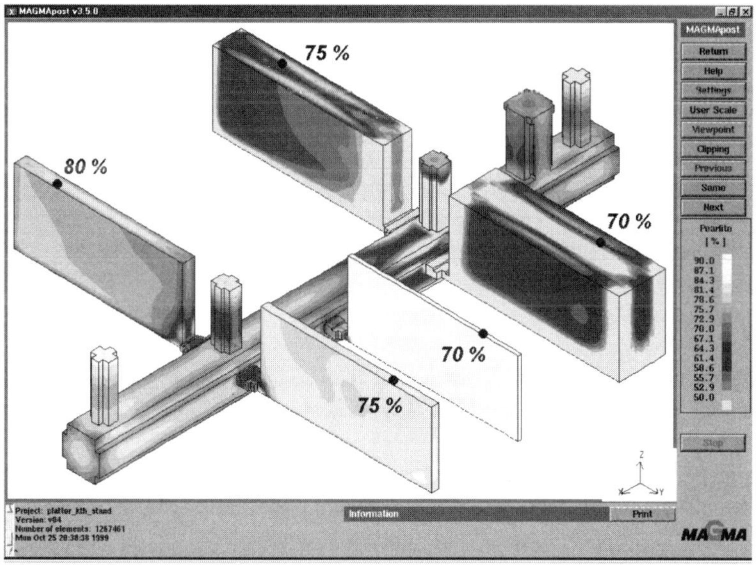

Figure 54 Simulated pearlite content in a plate casting of ductile iron containing 4.0% Si and 0.88% Cu. Experimental values in some locations are also shown.

$$HB = 106.7 + 111.1\,(\%\text{Cr}) + 15.8\,(\%\text{Cr})^2 + 150.8\,(\%\text{V}) - 9.6\,(\%\text{V})^2 - 93.7\,(\%\text{Mo}) + 167.4\,(\%\text{Mo})^2 + 20\,(\%\text{Cu}) - 10.6\,(\%\text{Ni}) + 74.1\,v_{900} - 15.3\,(v_{900})^2 \tag{24}$$

where v_{900} is the cooling rate at 900°C given in °C/sec. They also found a linear relation between tensile strength and hardness given by

$$TS = -3.3 + 1.4335^{*}HB \tag{25}$$

Catalina et al. [46] instead assumed that the mechanical properties can be calculated from the rule of mixtures of different phases:

$$HB = 100f_{\text{gr}} + HB_{\alpha}f_{\alpha} + HB_{\text{p}}f_{\text{p}} \tag{26}$$

Table 1 Factors Affecting Mechanical Properties

Factor	Effect
Graphite shape	The smoother the size, the better the properties. Mainly controlled by the content of some elements (such as Mg, Ce, S, O) and the cooling rate
Alloying elements	Control solidification mode (gray/white), amount of phases, and solid solution hardening of phases
Cooling rate	The higher the cooling rate, the finer the phases (and structures; particularly the pearlite)
Defects	The influence of defects such as oxide films, slag inclusions, and porosities cannot be neglected in cast irons

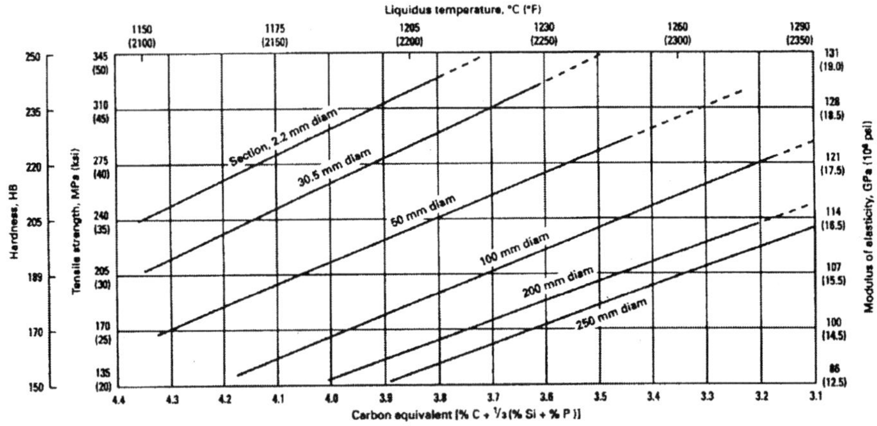

Figure 55 Relation between hardness, tensile strength, modulus of elasticity, section thickness, and carbon equivalent in gray irons. (From Ref. 2.)

$$HB_\alpha = 54 + 37\,(\%\text{Si}) \tag{27}$$

$$HB_\text{p} = 110 + 87(L_\text{p})^{-0.5} \tag{28}$$

$$TS = 80 + \left(2.25 + 1.98(L_\text{p})^{-0.5}\right)(L_\text{max} \cdot 10^{-3})^{-0.05} \tag{29}$$

where L_p is the pearlite lamellar spacing, and L_max is the maximum graphite length (both given in μm).

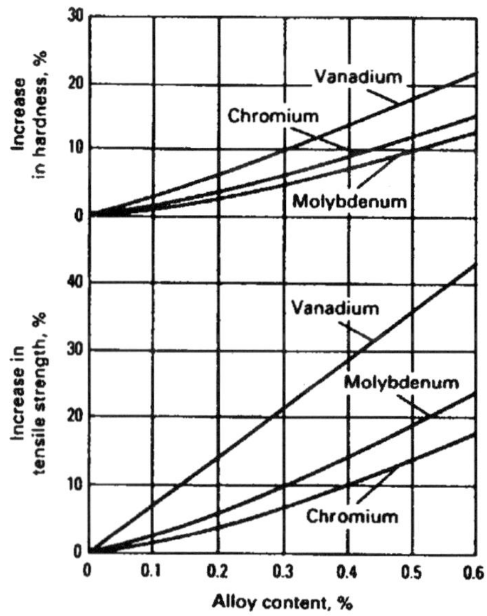

Figure 56 Effects of some pearlite refining alloying elements on strength and hardness of pearlitic gray irons. (From Ref. 58.)

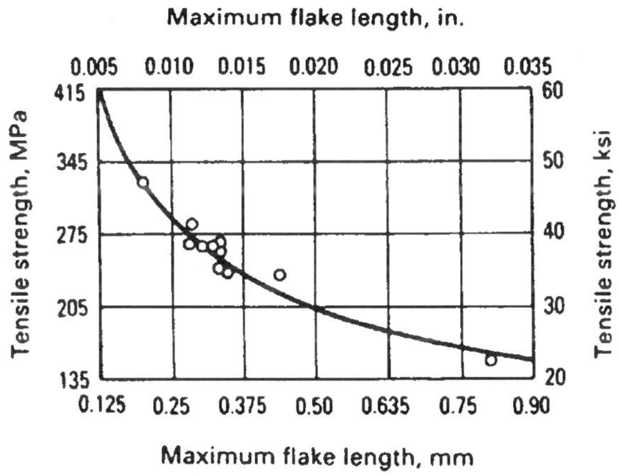

Figure 57 Effect of flake length on tensile strength in gray irons. (From Ref. 59.)

B. Compacted Graphite Iron

Due to the more rounded form and shorter flakes of the graphite in compacted graphite irons as compared to gray cast irons, considerably higher mechanical properties can be obtained. Compacted graphite irons usually contain a certain number of graphite nodules, where the nodularity increases with an increasing cooling rate and magnesium content. The relation between nodularity and elastic modulus is shown in Fig. 58 for compacted graphite irons having a pearlite content between 85% and 100% [61].

Since compacted irons can contain a large variation in pearlite content depending on cooling conditions and alloying elements, this also has to be considered. Ferrite has a considerably lower strength and hardness than pearlite. However, as the pearlite content increases, the ductility will decrease. The effect of pearlite content on tensile strength is shown in Fig. 59 [61].

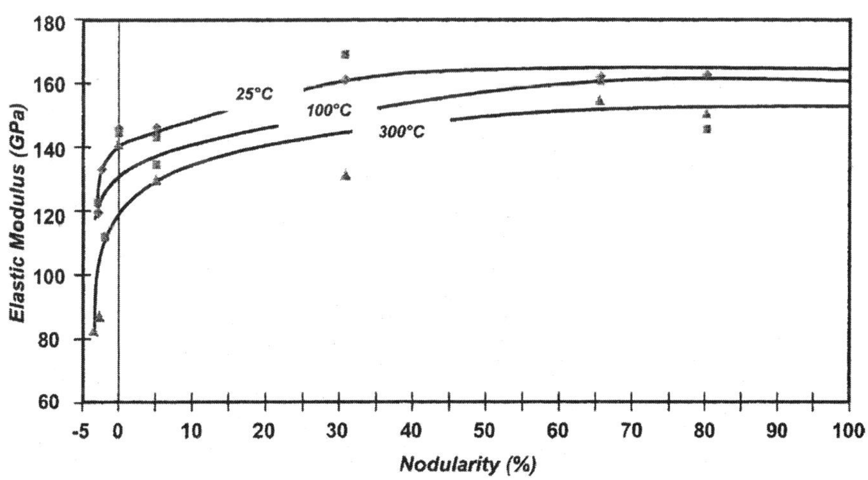

Figure 58 Relation between nodularity and elastic modulus in an 85–100% pearlitic compacted iron. (From Ref. 61.)

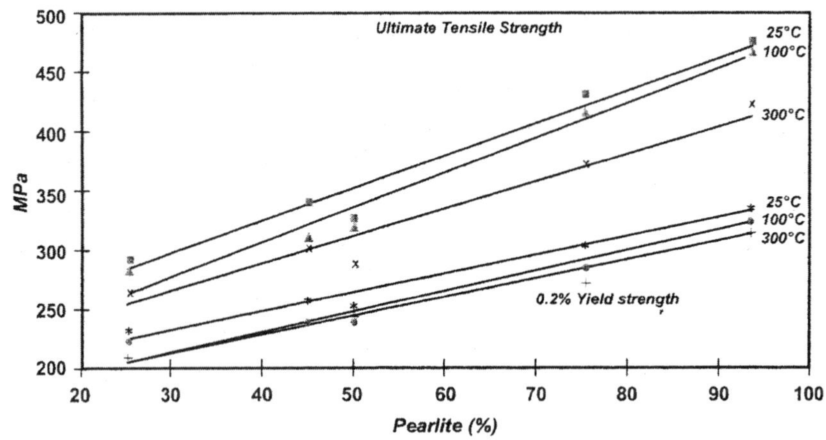

Figure 59 Relation between pearlite content and tensile strength in compacted irons. (From Ref. 61.)

C. Ductile Iron

Depending on the relative amounts of ferrite and pearlite, a ductile iron casting can be given a wide range of properties. Generally, ferritic and pearlitic grades have the features given in Table 2 [62].

It should be mentioned that even if the two grades both have a good machinability due to the presence of graphite, a variation in ferrite/pearlite ratio will often cause difficulty in machining, since a different set of cutting parameters are needed for each grade.

Ductile iron castings can obtain more or less steel-like properties due to the roundness of the graphite. The different strength classes are obtained primarily by varying the matrix, i.e. the relative amounts of ferrite and pearlite. Figure 60 illustrates how the pearlite fraction influences the hardness in a ductile iron with different silicon content [63]. In addition to the strong ferrite promoting effect, silicon also contributes by solid solution hardening of the ferrite.

The pearlite lamellar spacing also has a significant effect on hardness and strength. This relation is shown in Fig. 61 [64]. The measurements, which were made on samples taken from plates with thickness 4, 8, 15, 30 and 50 mm, show the hardness of the pure pearlite, i.e. excluding the graphite and any free ferrite, if present.

In ductile irons, the round shape of the graphite makes it reasonable, and more justifiable than in the case of gray irons, to assume that the mechanical properties can be calculated from the rule of mixtures. This approach was used by Lundbäck et al. in

Table 2 General Properties of Ferritic and Pearlitic Grades of Ductile Iron [62]

Ferritic grade	Pearlitic grade
Soft	Relatively hard
High ductility	Moderate ductility
Relatively low strength	High strength
Poor wear resistance	Good wear resistance
Relatively good thermal conductivity	Somewhat reduced thermal conductivity
Very good machinability	Good machinability

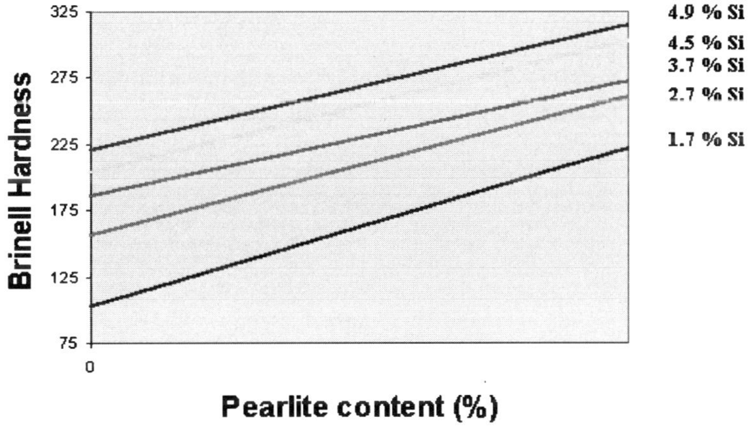

Figure 60 Relation between hardness, pearlite content and silicon content in ductile irons. (From Ref. 63.)

1988 [65], who made a full coupling between microstructure and properties. The equations were later revised to include the solid solution hardening of the ferrite and pearlite [7,66]. Basically, the following relations can be used to predict the hardness of ductile iron castings:

$$HB = HB_{gr} f_{gr} + HB_{\alpha} f_{\alpha} + HB_{p} f_{p} \tag{30}$$

$$HB_{\alpha} = 54 + 37\,(\%\mathrm{Si}) + 14\,(\%\mathrm{Cu}) \tag{31}$$

$$HB_{p} = 167 + 31\,(\%\mathrm{Si}) \tag{32}$$

No significant effect of the nodule size on the hardness has been found. However, Dahlberg [67] has made attempts to model the ductile iron properties with micromechanical models. The mechanical and short crack properties of the cast iron composite were

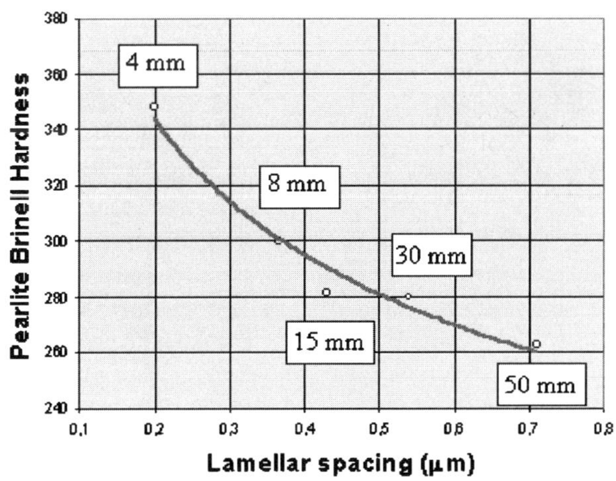

Figure 61 Relation pearlite lamellar spacing and pearlite hardness. (From Ref. 64.)

analyzed using finite element models with symmetrical patterns of holes or voids, representing the graphite nodules. An approximate expression for the stress intensity factor (K) has been derived for a crack emanating from a nodule in a nodule pattern. Introducing a fatigue limit condition based on the assumption of the existence of a short crack fatigue threshold, simulations were performed with variations in crack size, nodule size and spacing. The fatigue results were essentially qualitative, but were considered capable of capturing some important trends since the qualitative agreement with experiments was good.

D. White Iron

Cementite, a major phase in white irons, is extremely hard and brittle. Therefore, these irons are only used in applications requiring a good wear resistance. Depending on the alloy content, hardness values in the range from 350 to 800 HB can be obtained.

REFERENCES

1. Arvika Gjuteri AB Product information.
2. *ASM Specialty Handbook Cast Iron*, 1996.
3. Vuorinen, J. What is ADI-history and state of the art. In International ADI Seminar, Helsinki, 1991.
4. Dunki, W. Cash iron with compacted graphite—production and properties. Escher Wyss Mittellungen 1980, *53*, 215.
5. Skaland, T.; Grong, O.; Grong, T. Metall Trans A 1993, *24A*, 2321–2345.
6. Okada, A.; Miyake, H. Inoculation mechanism of grey cast iron. Technol. Reports Kansai Univ. Jpn. 1994, *36*, 85–95.
7. Wessén, M. Ph.D. Thesis, Royal Institute of Technology, Sweden, 1997, ISRN KTH/MG-AVH-004-SE: ISSN 1104–7127: TRITA-MG.
8. Aagaard, R. Ph.D. Thesis, Technical University of Denmark, 1998, Publication no. TM 97.42.
9. Svensson, I.L. Unpublished work.
10. Mampaey, F. Proceedings of 62nd World Foundry Congress, Philadelphia, USA, 1996, Paper 4.
11. Sundman, B.; Jansson, B.; Andersson, J.O. CALPHAD 1985, *9*, 153–190.
12. Minkoff, I. *The Physical Metallurgy of Cast Iron*; John Wiley & Sons: 1983.
13. Fredriksson, H. Study of transition from undercooled to flake graphite in cat iron. Proceedings of Second International Symposium on the Metallurgy of Cast Iron, 1976, 277–293.
14. Thorgrimsson, J.T. Thesis, Royal Institute of Technology, Stockholm, Sweden, 1986.
15. Fredriksson, H. Coupled zone in gray cast iron. Metall. Trans. A 1975, *6A*, 1658–1660.
16. Fredriksson, H.; Svensson, I.L. *Metall. Soc.* 1988, 153–162.
17. Wessén, M. Svensson, I.L. Modeling of ferrite growth in nodular cast iron. Metall. Trans A 1996, *27A*, 2209–2220.
18. Wetterfall, S.E.; Fredriksson, H.; Hillert, M. Solidification Process of Nodular Cast Iron. J. Iron Steel Institute 1972, *5*, 323–333.
19. Den, X.; Zhu, P.; Liu, Q. Proceedings of Structure and Formation of Vermicular Graphite. Elsevier Science Publishing Company, 1985, 141–150.
20. Hillert, M.; Steinhauser, H. Jernkontorets annaler 1960, *144*, 520.
21. Magnin, P.; Kurz, W. Proceedings of Transition From Grey to White and White to Grey in Fe–C–X Eutectic Alloys; Elsevier Science Publishing Company, New York, 1985; 263–272.
22. MAGMA Giesseritechnologi GmbH, Eisenwerk Bruhl, Germany.
23. Brown, S.G.; Spittle, R. Proceedings of Modelling of Casting, Welding and Advanced Solidification Processes, TMS, Warrendale, 1991, 395.
24. Rappaz, M.; Gandin. Probabilistic modeleing of microstructure formation in solidification processes. Acta Metall. Mater. 1993, *41*(2), 345–360.
25. Kolotov, Y.A.L.; Vashukov, I.A. Changes in the density of grey cast irons during slow solidifications. Izv. Akad. Nauk SSSR Met (May–June 1978) No.3, 70–75.

26. Mai, R.; Leube, B.; Schule, E. Modeling of solidification of cast iron alloys. Giesseeriforschung 1995, *47* (1), 1–5.
27. Charbon, C.; Jacot, A.; Rappaz, M. Three-dimensional probabilistic modelling of equiaxed eutectic solidification. Mater. Sci. Eng. A 1993, *173A* (1–2), 143–148.
28. Charbon, C. Ph.D. Thesis, Ècole Polytechnique Fédérale de Lausanne, Switzerland, 1995.
29. Friedel Deviatons from Vegard's Law. J. Phys. Mag. 1954, *46*, 514.
30. Sundman, B.; Ågren, J. The Sublattice Model: J. Phys. Chem. Solids 1981, *42*, 297–301.
31. Gustafsson, P. Carbon 1984, *24*, 169–176.
32. Guillermet, A.F.; Gustafsson, P. An assessment of the thermodynamic properties and the (p, T) phase diagram of iron. High Temperature High Pressure 1985, *16*, 591–610.
33. Svensson, I.L.; Dugic, I. Int. J. Cast Metals Res. 1999, *11* (6).
34. Ridley, N.; Stuart, H. Partial molar volumes from high temperature lattice parameters of iron–carbon austenites. Metal Sci. J. 1970, *4*, 219–222.
35. Kagawa, A.; Nakamura, H.; Kiguchi, S.; Osada, M. Volumetric change in freexing cast irons. Trans. Jpn. Foundrymen's Soc. 1995, *18*, 18–23.
36. Svensson, I.L. Unpublished work.
37. Tatsumi, Y.; Ohsaki, H. EMIS DataRev. June 1987.
38. Olsson, A. The influence of carbon, silicon and molybdenum on the density of liquid iron. Scand. J. Metallurgy 1981, *10*, 263–271.
39. Jimbo, I.; Cramb, A.W. The density of liquid iron–carbon alloys. Met Trans B 1993, *24B*, 5–10.
40. Kusakawa, T.; Kim, S.Y.; Kondo, T. Volume change of molten spheroidal graphite cast iron. Report of the Castings Research Laboratory, Waseda University, 1972; Vol. 23, 23–32.
41. Svensson, I.L.; Dioszegi, A. On modelling of volume related defect formation in cast irons. Proceedings of Modeling of Casting, Welding and Advanced Solidification Processes IX; Sahm, P.R.; Hansen, P.N.; Conley, J.G., Eds.; Shaker Verlag GmbH, Germany, 2000, 102–109.
42. Svensson, I.L. Proceedings of Solidification Technology in Foundry and Casthouse; The Metals Society Conference, University of Warwick, 1980.
43. Svensson, I.L.; Fredriksson, H. Proceedings of Solidification Technology in Foundry and Casthouse; The Metals Society Conference, University of Warwick, 1980.
44. Diószegi, A.; Svensson, I.L.; Millberg, A. Proceedings Conference of The Science of Casting and Solidification, Bratsov, Romania, May 28–31, 2001.
45. Janowak, J.F.; Gundlach, R.B. A modern approach to alloying gray iron. AFS Trans. 1982, *90*, 847–863.
46. Catalina, A.; Guo, X.; Stefanescu, D.M.; Chuzoy, L.; Pershing, M.A.; Biltgren, G.L. Proceedings of MCWASP VIII; Thomas, B.G.; Beckerman, C., Eds., TMS, 1998, 455–462.
47. Voight, R.C.; Loper , C.R. Jr. Matrix structure development in ductile cast irons. AFS Trans 1989, *97*, 595–600.
48. Pan, E.N.; Loper , C.R. Jr. Matrix development in graphite cast irons. AFS Trans 1986, *94*, 545–556.
49. Cahn, J.W.; Hagel, W.C. *Decomposition of Austenite by Diffusional Processes*; Zackey, V.F.; Aaronsen, H.I., Eds.; Interscience Pub., NY, 1962, 131–192.
50. Al Salman, A.; Lorimer, G.W.; Ridley, N. Partitioning of silicon during pearlite growth in a eutectoid steel. Acta Metall. 1979, *27*, 1391–1400.
51. Lacaze, J.; Gerval, V. Modelling of the eutectoid reaction in spheroidal graphite Fe–C–Si alloys. ISIJ Int. 1998, *38*, 714–722.
52. Lacaze, J. Pearlite growth in cast irons: a review of literature data. Int. J. Cast Metals Res. 1999, *11*, 431–436.
53. Hillert, M. The role of interfacial energy during solid state phase transformation. Jernkontorets Ann 1957, *141*, 757–765.
54. Askeland, D.R.; Gupta, S.S. AFS Trans 1975, *83*, 313–320.
55. Aagaard, R. Ph.D. Thesis, Technical University of Denmark, 1998, Publication no. TM 97.42.
56. Stefanescu, D.M.; Kanetkar, C.S. Computer modelling of the solidification of eutectic alloys. TMS Conference Proceedings of Computer Simulation of Microstructural Evolution; Srolovitz, D.J., Ed.; 1985, 171–188.

57. Svensson, I.L.; Wessén, M. Simulation of mechanical properties in cast iron castings. Proceedings from CIATF Technical Forum 99, Düsseldorf, June 10–11, 1999, Paper 16, 116–123.
58. Powell, J. AIME Electric Furnace Conference Proceedings; Iron and Steel Society: 1986; Vol. 44, 215–231.
59. Bates, C.E. Alloy element effects on gray iron properties II. AFS Trans 1986, *94*, 889–912.
60. Yang, Y.; Louvo, A.; Rantala, T. 57th World Foundry Congress, Osaka, Japan, 1990, Paper 21.
61. Dawson, S. Sintercast Ltd, Internal Publication, United Kingdom, 1999.
62. Karsay, S.I. QIT Corporation, Canada, 1971.
63. Svensson, I.L. Proceedings of 57th World Foundry Congress, Osaka, Japan, Sept 1990; Paper no. 3.
64. M. Wessén, Unpublished work.
65. Lundbäck, E.; Svensson, I.L.; Persson, P.E. Modelling of structure formation and relation to mechanical properties of nodular cast iron. Proceedings from Modelling of Metal Forming Processes; Chenot, J.L.; Onate, E., Eds; Kluwer Academic Publishers, 1988, 37–46.
66. Svensson, I.L. 57th World Foundry Congress, Osaka, Japan, 1990, Paper 2.
67. Dahlberg, M. Micromechanical modelling of nodular cast iron, a composite material. Cast Metals (UK) 1997, *9*(6), 319–330.
68. Yamamoto, K; Ogi, K Thermodynamic evaluation of the solidification sequence of high chromium white cast irons. Conference Proceedings, The Science of Casting and Solidification, Brasov, Romania, 2001, 1–13.

9

Designing with Aluminum Alloys

Nack J. Kim
Pohang University of Science and Technology, Pohang, Korea

Since the discovery of age-hardening capability of Al–Cu alloys by Wilm in 1906, aluminum alloys have become important structural materials especially for aircraft and aerospace vehicles. This has been due to characteristics such as light weight, good corrosion resistance, good strength and ductility, to name a few. It also has good electrical and thermal conductivities. Aluminum can be cast and worked into almost any form and can be given a wide variety of surface finishes. It is anticipated that the importance of aluminum alloys will grow due to increasing demands on the energy conservation and environmental protection.

I. ALLOYING OF ALUMINUM

Pure aluminum has very low strength and cannot be used for structural applications. When alloyed with other elements, however, it gains strength by various strengthening mechanisms. Aluminum can be alloyed with most metallic elements, but only some have enough solid solubility to be used as major alloying elements. Of more importance are copper, magnesium, silicon, and zinc (Table 1). However, a considerable number of other elements have pronounced effect on improving the properties of aluminum alloys. These include chromium, manganese, and zirconium, which are used primarily to control grain structure. Maximum solid solubility in aluminum alloys usually occurs at eutectic, peritectic, or monotectic temperature. Solid solubility decreases with decreasing temperature. Such change of solid solubility with temperature is the basis of age hardening (Sec. 2.3).

A. Alloy Designations

1. Wrought Alloys

There are various systems describing the compositions and temper conditions of wrought aluminum alloys. Among these systems, the International Alloy Designation System (IADS) is widely accepted by most countries and it is used to describe alloys in this book.

Alloy is identified by a four-digit number in IADS. The first digit indicates the alloy group on the basis of major alloying element(s) as shown in Table 2. The second digit indicates alloy modifications. The original alloy is indicated when the second digit is 0. Other

Table 1 Invariant Reactions and Maximum Solubilities in Binary Aluminum Alloys

Elements	Temperature[a] (°C)	Liquid solubility wt%	Liquid solubility at.%	Solid solubility wt%	Solid solubility at.%
Ag	566	72.0	60.9	55.6	23.8
Ca	620	7.6	5.25	< 0.1	< 0.05
Cd	649[b]	6.7	1.69	0.47	0.11
Co	657	1.0	0.46	< 0.02	< 0.01
Cr	661[c]	0.41	0.21	0.77	0.40
Cu	548	33.15	17.39	5.67	2.48
Fe	655	18.7	9.1	0.052	0.025
Li	600	9.9	30.0	4.0	13.9
Mg	450	35.0	37.34	14.9	16.26
Mn	658	1.95	0.97	1.82	0.90
Ni	640	6.12	2.91	0.05	0.023
Si	577	12.6	12.16	1.65	1.59
Sn	228	99.5	97.83	< 0.01	< 0.002
Y	645	7.7	2.47	< 0.1	< 0.03
Zn	380	95.0	88.7	82.8	66.4
Zr	660[c]	0.11	0.033	0.28	0.085

[a]Eutectic reactions unless designated otherwise.
[b]Monotectic reactions.
[c]Peritectic reactions.

Table 2 Wrought Aluminum Alloy Designation System

Designation	Major alloying elements
1XXX	99.00% aluminum minimum
2XXX	Cu
3XXX	Mn
4XXX	Si
5XXX	Mg
6XXX	Mg and Si
7XXX	Zn
8XXX	Other than above
9XXX	Unused

integers indicate alloy modifications. Therefore, the alloys with the same four-digit number except the second digit usually have similarity in composition. The last two digits are used to identify different aluminum alloys in the group and have no other special meaning. The only exception is the 1XXX group, where the last two digits designate the minimum percent of aluminum in excess of 99.00. A prefix X is used to denote an alloy, which is still considered experimental.

2. Casting Alloys

Aluminum casting alloys are usually identified by the three-digit registration system of the Aluminum Association. The first digit indicates the alloy group on the basis of major

Table 3 Cast Aluminum Alloy Designation System

Designation	Major alloying elements
1XX	99.00% aluminum minimum
2XX	Cu
3XX	Si with added Cu and/or Mg
4XX	Si
5XX	Mg
7XX	Zn
8XX	Sn

alloying element(s) as shown in Table 3. No commercial alloys are currently available in the 6XX and 9XX series. The second two digits have different meanings depending on the alloy systems. In the 1XX group, the second two digits indicate the minimum percentage of aluminum, e.g., 150 indicates a composition containing a minimum of 99.50% aluminum. In the 2XX–9XX alloy groups, the second two digits are used to identify different aluminum alloys in the group and have no special meaning. Aluminum casting alloys are often represented by the four-digit system, with the last digit at the right of the decimal point. This last digit indicates the product form, 0 being castings and 1 and 2 being ingots.

Prefix letter is also included when there is a modification to an original alloy, or to the normal impurity limits. X is assigned for experimental alloys.

II. HEAT TREATMENT OF ALUMINUM ALLOYS

A. Homogenization

The initial thermal treatment applied to ingots prior to secondary operation such as hot working is homogenization. Homogenization has one or more purposes depending on the alloy and product. One of the main purposes of homogenization is to reduce the effects of harmful microstructural features existing in cast structures. The cast structure is a cored dendritic structure with solute content increasing from center to edge with an interdendritic distribution of second phase particles or eutectic. Since homogenization involves diffusion of alloying elements, the time required depends on the grain size or dendrite arm spacing and the rates of diffusion of the alloying elements. Solution of interdendritic second phase particles also occurs during homogenization. However, in commercial high-strength aluminum alloys, it is not always possible to completely dissolve the interdendritic second phase particles. These are usually intermetallic compounds of the impurity elements, e.g., iron and silicon, with aluminum and/or the alloying elements. Therefore, it is necessary to minimize the size and number of these particles by reducing the iron and silicon content.

In other cases, homogenization is designed to induce formation of high temperature dispersoids in fine scale. Transition metals may supersaturate in aluminum during solidification. During homogenization, these elements form intermetallic compounds such as $Al_{20}Cu_2Mn_3$, $Al_{12}Mg_2Cr_3$, Al_6Mn, and Al_3Zr. These intermetallic compounds effectively control grain structure during subsequent fabrication and heat treatment by preventing recrystallization and/or grain growth. Homogenization temperature must be kept below the melting point of the constituent with the lowest melting point.

B. Solution Treatment

The main purpose of solution treatment is to obtain complete solution of most of the alloying elements for the subsequent aging treatment. Therefore, it should be carried out at a temperature within the single solid phase region. As in the case of homogenization, however, temperature must be limited to a safe level below the solidus temperature to avoid overheating and partial melting. It is especially true for the alloys with high solute content.

The grain size of hot- or cold-worked products can be significantly affected by solution treatment. Precautions should be taken in solution treatment of hot- or cold-worked products to prevent the growth of coarse, recrystallized grains. Particularly with products that have varying amount of working at different locations, care should be taken not to have unnecessarily high temperatures and excessively long solution treatment times. Otherwise, the products would have a heterogeneous grain structure.

After solution treatment, aluminum alloy components must be fast cooled or quenched, usually to room temperature. The objective of quenching is to preserve the solid solution formed at the solution treatment temperature in preparation for subsequent aging. In general, the best combinations of strength and toughness are associated with the most rapid quenching rates. However, rapid quenching results in the distortion of thinner products such as sheet and the introduction of residual stresses into thicker products that are normally compressive at the surface and tensile in the core.

The effects of residual stress from quenching become serious when the products have an irregular shape or are machined. Machining disturbs the balance of the residual stresses, resulting in distortion of products. Also, the level of residual stresses may approach the yield stress in some high-strength alloys that can cause premature failure during service.

Several methods are available either to minimize the residual stresses generated during quenching or to relieve them after quenching. The most common methods used for stress relieving include stretching, roller leveling, and flattening in a press. The aging treatment also allows some relaxation of stresses. High residual stresses can be reduced by using slower cooling rates. This is particularly important for the irregularly shaped parts such as forging since the once formed residual stresses cannot be removed by subsequent working. Hence, the quenching of forgings is commonly conducted using hot or boiling water. In some cases, however, quench sensitive alloys may suffer a loss in mechanical properties when the slower quenching rates are used. Microstructural change which can occur by slow cooling is the precipitation of coarse particles which lowers the age-hardening response. Also, there can be a segregation of solutes such as copper to grain boundaries, which reduces toughness and increases the susceptibility to intergranular corrosion.

C. Aging Treatment

Aging treatment is the final stage in the development of properties in the heat-treatable aluminum alloys. Most of the heat-treatable alloys show age-hardening behavior at room temperature (natural aging) after quenching. But age hardening is greatly accentuated by heating at elevated temperatures (artificial aging), which are usually in the range of 100–190°C. The age-hardening behavior of aluminum alloys varies depending on the alloy compositions. Metallurgical changes that occur during aging are discussed below.

Table 4 Precipitation Sequences in Selected Aluminum Alloys

Alloy	Precipitation sequence
Al–Cu	GP (disc)→ θ″ (disc)→ θ′ (plate)→ θ (Al_2Cu)
Al–Mg	GP (sphere)→ β′ (plate)→ β (Mg_5Al_8)
Al–Cu–Mg	GP (rod)→ S′ (lath)→ S (Al_2CuMg) (lath)
Al–Mg–Si	GP (rod)→ β″ (needle)→ β′ (rod)→ β (Mg_2Si) (plate)
Al–Zn–Mg	GP (sphere)→ η′ (sphere or plate)→ η ($MgZn_2$)
	GP → T′ → T
Al–Li–Mg	δ′ → Al_2LiMg (rod)
Al–Li–Cu	δ′ → δ
	T_1 (plate)
	θ″ → θ′

1. Precipitation

The decomposition process of the supersaturated solid solution (SSSS) is very complex and the resulting precipitation sequence varies depending on the respective alloy systems. Probable precipitation sequences in typical aluminum alloys are shown in Table 4. In general, Guinier–Preston (GP) zones and one or more metastable transition phases may be formed prior to the formation of equilibrium phase. GP zones are ordered, solute-rich clusters of atoms which are homogeneously nucleated in the matrix. They are about one or two atomic layers thick and fully coherent with the matrix. The equilibrium shape of a GP zone is a sphere such as those found in Al–Ag and Al–Zn systems (Fig. 1a). However, in Al–Cu system, GP zones have disc shape perpendicular to the elastically soft ⟨1 0 0⟩ directions in the aluminum matrix (Fig. 1b). This arises because the atomic size difference is large enough to make the strain energy more important in determining the equilibrium shape of the zone.

The formation of GP zones is usually followed by the precipitation of transition phases. The crystal structures of the transition phases are usually intermediate between those of the matrix and the equilibrium phase and are partially coherent with the matrix. For example, in Al–Cu system, there are two transition phases, θ and θ (Fig. 2). θ has a tetragonal unit cell in which the copper and aluminum atoms are ordered on (0 0 1) planes. Since distortion in the [0 0 1] direction is very small, θ forms a fully coherent interface with the aluminum matrix. θ is also tetragonal but has a large misfit in the [0 0 1] direction. Therefore, θ is partially coherent with the aluminum matrix.

Nucleation of transition phases occurs at either the sites of GP zones or lattice defects, such as dislocations depending on the alloy systems. S (Al_2CuMg) and T_1 (Al_2CuLi) phases are typical examples of transition phases which precipitate heterogeneously on dislocations. Therefore, 2XXX alloys and Al–Li alloys are usually cold worked prior to aging to promote the formation of such phases in the microstructure. Maximum hardness occurs when a critical combination of coherent and semi-coherent particles is present. Formation of the equilibrium phase occurs with complete loss of coherency with the aluminum matrix and appreciable coarsening.

2. Precipitate-Free Zones at Grain Boundaries

It is a well-known fact that the grain boundaries are the main sinks for excess vacancies and solute atoms. This has important effects on the distribution of precipitates that form

Figure 1 Morphologies of GP zones; (a) spherical shape (Al–Mg alloy), and (b) disc shape (Al–Cu alloy). (Courtesy of J.K. Oh.)

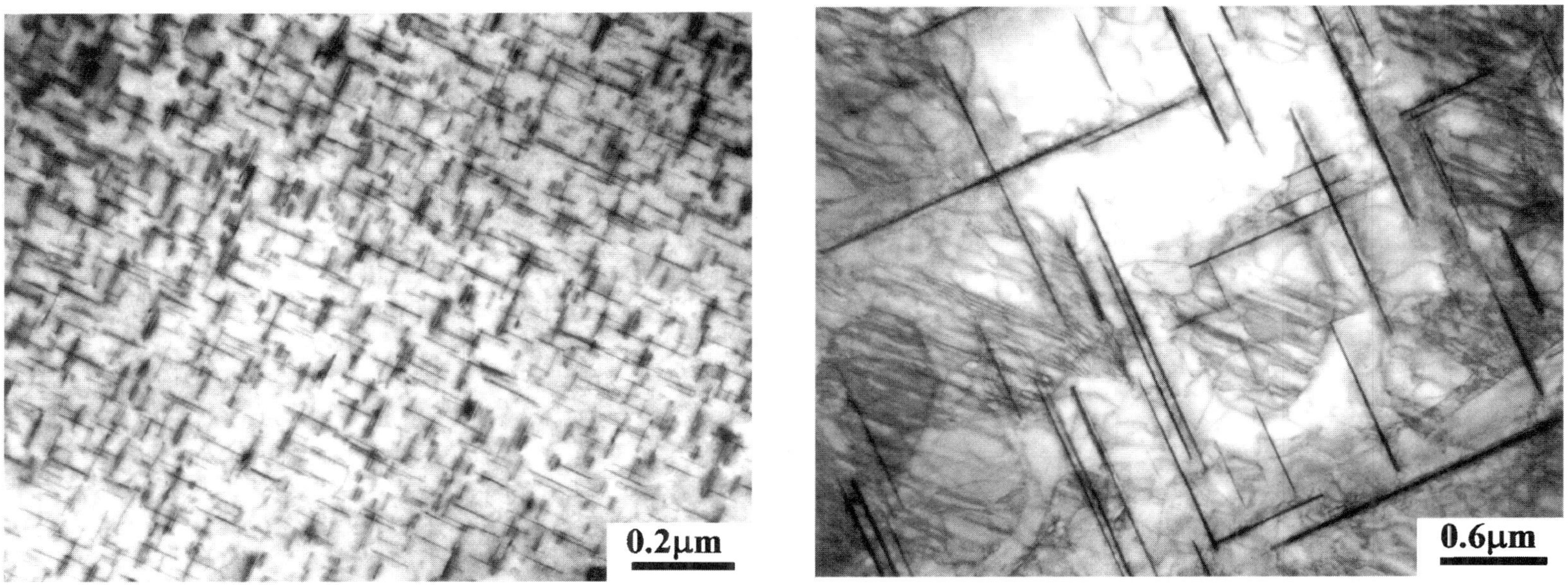

(a) (b)

Figure 2 Transition phases in Al–Cu alloy; (a) θ'', and (b) θ'. (Courtesy of J.K. Oh.)

in the vicinity of grain boundaries upon subsequent aging. Nucleation and growth of precipitates at grain boundary deplete the solute near the grain boundary and a precipitate-free zone (PFZ) results upon aging as shown in Fig. 3a. Also, since the vacancy concentration near the grain boundary is lower than that of the critical vacancy supersaturation required for the nucleation to occur, a PFZ is formed near the grain boundary upon aging (Fig. 3b). In this case, the width of the PFZ is determined by the vacancy concentration near the grain boundary, which is in turn controlled by heat-treatment conditions. High solution treatment temperatures and fast quench rates produce narrow PFZs by reducing the width of vacancy concentration profile. Low temperature aging also produces narrow PFZs since the driving force for precipitation is high and the critical vacancy supersaturation is low. Similar PFZs can form at dislocations and inclusions.

All age-hardenable aluminum alloys show PFZs along grain boundaries except when the alloys are aged at temperatures below the GP zone solvus since GP zones can form homogeneously in some aluminum alloys. PFZs, in general, have a deleterious effect on the properties of aluminum alloys. Since they are softer than the surrounding matrix, the strain is localized along grain boundaries under applied stress and a premature failure can occur. It is also known that PFZs increase the susceptibility to stress-corrosion cracking.

III. PROPERTIES OF ALUMINUM ALLOYS

A. Hardening (Strengthening) Mechanisms

1. Precipitation and Dispersion Hardening

The strength of an age-hardenable alloy is controlled by the interaction between dislocations and precipitates. In general, an increase in strength is synonymous with an increased difficulty of moving dislocations. Either a dislocation must cut through the precipitate particles in its path, or it must move between them. In the early stage of aging, the precipitates are coherent with the matrix and can be sheared by dislocations. Shearing of precipitate particles increases the precipitate–matrix interfacial areas as shown in Fig. 4. High applied stress is needed for this to occur and the critical resolved shear stress is proportional to the precipitate particle size.

Although this so-called chemical strengthening significantly contributes to the strength of high-strength aluminum alloys, the shearing of precipitate particles may cause a degradation of other mechanical properties such as ductility and toughness. Once the precipitate particles are sheared, dislocations continue to pass through the precipitate particles on the sheared planes and hence deformation becomes localized on only a few slip planes developing intense slip bands as shown in Fig. 5. Such developed intense planar slip bands allow the pile-up of dislocations at grain boundaries, resulting in a premature failure along grain boundaries.

As the precipitate particles grow in size, they lose their coherency with the matrix and would not be sheared by dislocations. In this case, the precipitate particles are by-passed by moving dislocations which bow out between them (Fig. 6). Rejoining of dislocations results in the formation of loops of dislocations around precipitate particles. Strength is then inversely proportional to the distance between precipitate particles. This so-called Orowan mechanism is operative in overaged alloys and other dispersion strengthened alloys. The maximum strength is usually observed at the transition point when deformation behavior changes from shearing to by-passing of precipitate particles.

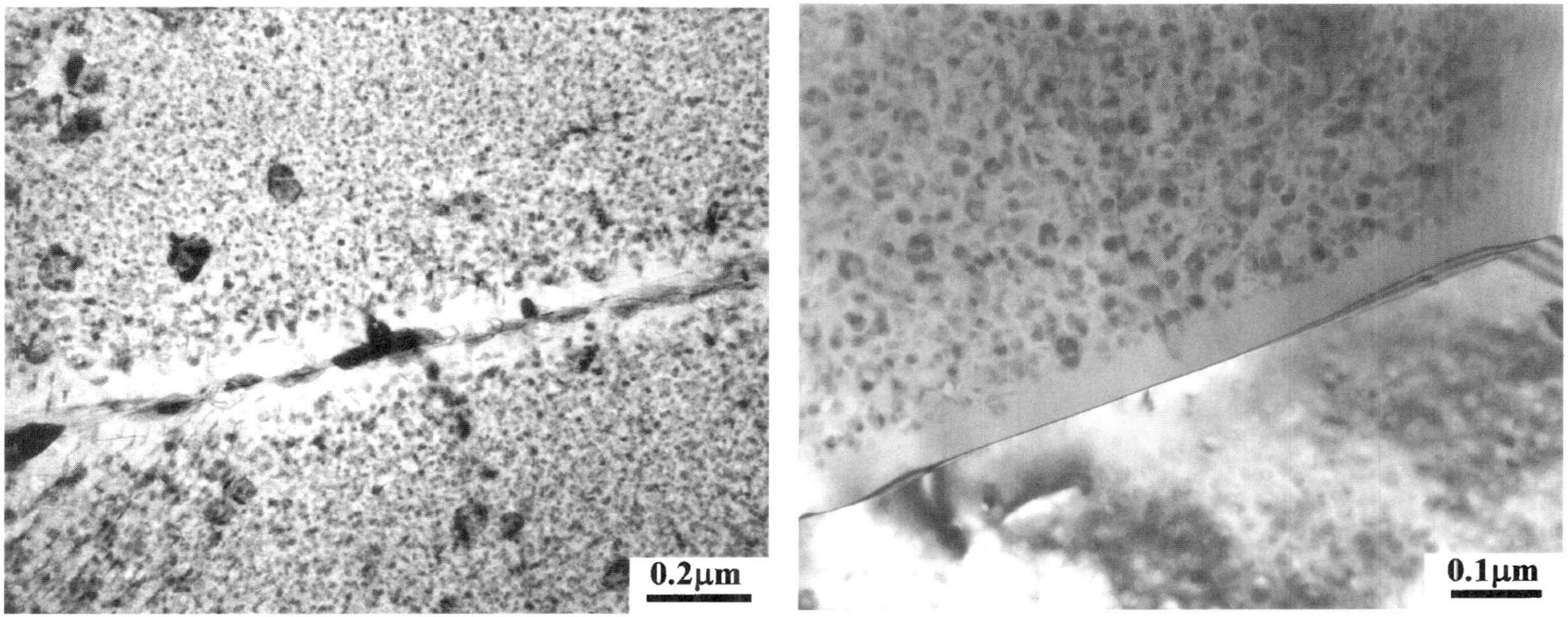

(a) (b)

Figure 3 Precipitate-free zones; (a) formed by the depletion of solute atoms around coarse grain boundary precipitates (courtesy of J.K. Oh), and (b) formed by the depletion of vacancies near grain boundaries.

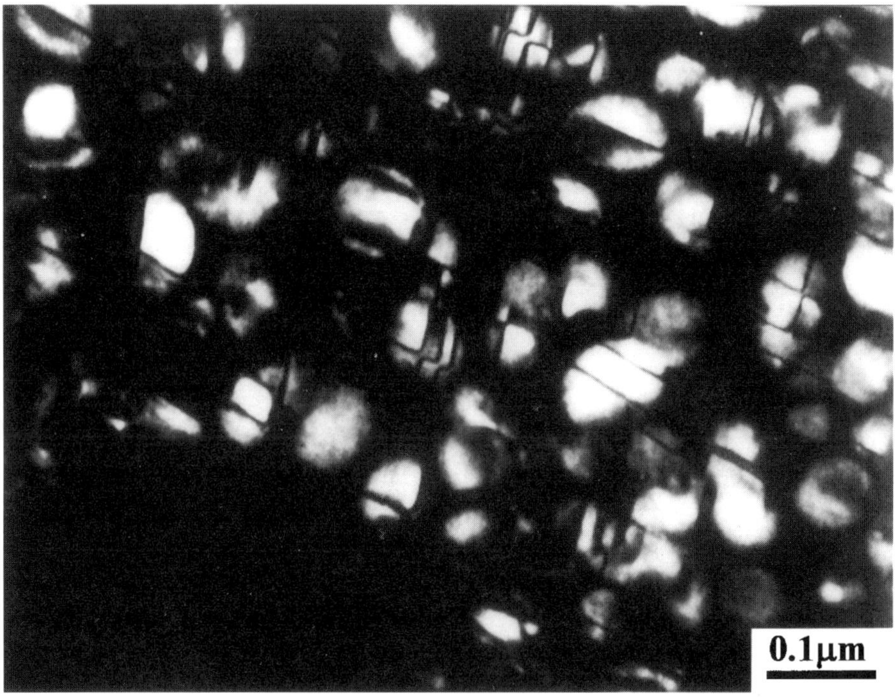

Figure 4 Shearing of δ′ precipitates by moving dislocations.

The yield strength of the alloy is low but the rate of work hardening is high, and plastic deformation tends to be spread more uniformly throughout the grains.

2. Work Hardening

Work, or strain hardening, is a natural consequence of most working and forming operations and is used extensively to increase the strength of the non-heat-treatable aluminum alloys. For heat-treatable aluminum alloys, work hardening may supplement the strength achieved by precipitation hardening. Work-hardened products can be restored to a fully soft, ductile condition by annealing.

During deformation, the dislocation content increases when dislocation generation and multiplication occur faster than annihilation by dynamic recovery. Dislocation tangles, cells, and subgrain walls are formed, all of which decrease the mean free slip distance and increase the strength.

Usually, there is an initial rapid increase in yield strength of non-heat-treatable alloys by cold working, after which the increase is more gradual and roughly equals the change in tensile strength. The heat-treatable alloys in both the annealed and T4 temper conditions show a similar behavior. The work-hardened alloys, however, have much reduced ductility and formability as compared to the non-work-hardened alloys. Exceptions are 3003 and 3004 alloys, which exhibit better drawing properties in the cold worked rather than in the annealed condition. Good drawing properties of these alloys are obtained by the control of sheet texture by cold rolling and these alloys are used for making thin-walled beverage cans.

Figure 5 Occurrence of planar slip in Al–Li alloys due to the shearing of δ′ precipitates.

In the case of fully hardened heat-treatable alloys, the increase in strength by cold working after aging is comparatively small, except at very high strains, and is often limited by the poor workability of alloys in this condition. The principal use of this practice is for some extruded and drawn products such as wire, rod, and tube that are cold drawn after heat treatment to increase strength and improve surface finish.

The work-hardening characteristics of aluminum alloys vary considerably with temperature. Work hardening is much greater at cryogenic temperatures than at room temperature. The gain in strength by working at −196°C can be as much as 40% although there is a significant reduction in ductility. At elevated temperatures, the work-hardening characteristics vary both with temperature and strain rate. Work hardening decreases progressively as the working temperature increases until a temperature is reached above which no effective strain hardening occurs due to dynamic recovery and recrystallization. Dynamic recovery results in the formation of a subgrain structure that is similar to the structure resulting from heating a previously cold-worked alloy. Subgrain structure also increases the strength of aluminum alloys to some extent.

3. Textures

During hot or cold working, considerable numbers of the deformed grains approach certain orientations, forming crystallographic textures. Such texture occurs because

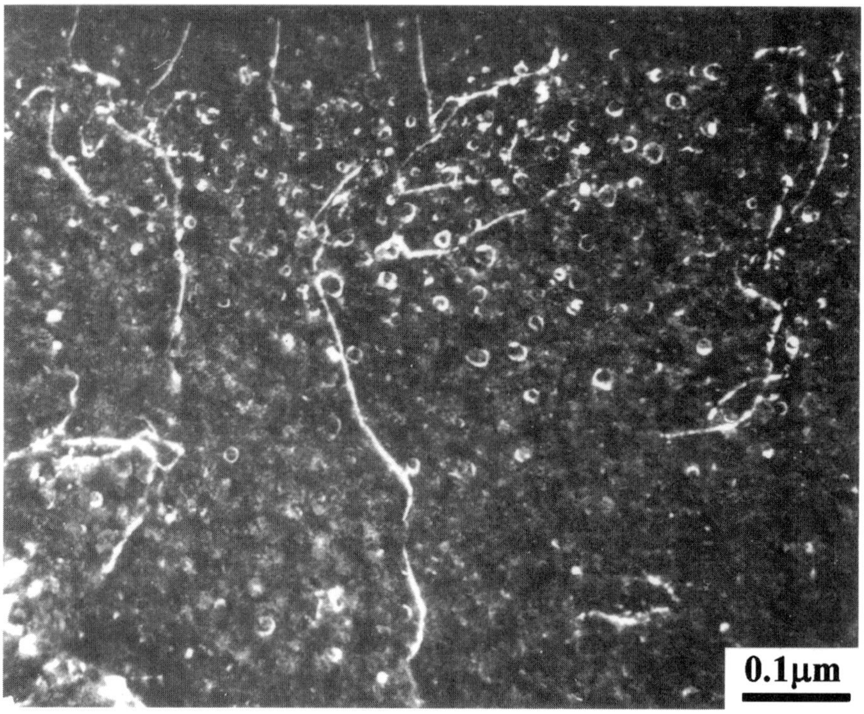

Figure 6 By-passing of dislocations around shear-resistant dispersoids.

deformation, or slip, in aluminum is confined to the {1 1 1} planes in the ⟨1 1 0⟩ directions. Large amounts of deformation at ambient temperatures lead to some strengthening through the development of textures, the nature of which depends in part on the mode of working. In rolled sheet, the deformation texture may be described as a mixture of the three ideal textures (1 1 0) $\langle \bar{1}\ 1\ 2 \rangle$, (1 1 2) $\langle 1\ 1\ \bar{1} \rangle$ and (1 2 3) $\langle \bar{2}\ 1\ 1 \rangle$. Wire, rod, and bar usually have a "fiber" texture in which the ⟨1 1 0⟩ direction is parallel to the axis of the product, with a random orientation of crystal directions perpendicular to the axis.

When cold-worked aluminum alloys are recrystallized by annealing, new grains form with orientations that differ from the principal components of the above-described deformation texture. In rolled sheet, there is a strong tendency for the new grains to form with a cube plane {1 0 0} parallel to the surface and a cube edge parallel to the rolling direction, i.e. (1 0 0)[0 0 1] texture. The texture and final grain size in recrystallized products are determined by the amount of cold work, annealing conditions (i.e., rate of heating, annealing temperature and time), composition and the size and distribution of intermetallic compounds which tend to restrict grain growth.

A consequence of the textures, particularly deformation texture, is the production of some directionality in properties. Texture hardening causes moderate increases in both yield and tensile strength in the direction of working. Forming characteristics of sheet may also be improved through an increase in the *R*-value for sheet. However, the texture may present a problem for the formation of ears during deep drawing of sheet. Earing describes the phenomenon of small undulations that may appear on the top of drawn cups. Earing may be minimized by careful control of rolling and annealing schedules, e.g., a sheet is sometimes cross-rolled so that textures are less clearly defined.

B. Toughness

The application of high-strength aluminum alloys in the aerospace industry has resulted in increased performance in the areas of fracture and fatigue. Minimum fracture toughness requirements become more stringent and, in the high-strength alloys, it is necessary to place a ceiling on the level of yield strength that can be safely employed by the designer. It is generally understood that the fracture of brittle constituent particles or the decohesion at the particle/matrix interface is responsible for the reduction of fracture toughness of commercial aluminum alloys (Fig. 7). Consequently, the main approach to improve the toughness of high–strength aluminum alloys has been the control of the levels of the impurity elements such as iron and silicon. It has been shown that the reduction of the combined levels of these elements from 1.0% to 0.5% doubles the plane strain fracture toughness of some Al–Cu–Mg alloys. As a consequence of this, a range of high toughness versions of older alloy compositions is now in commercial use in which the levels of impurities have been reduced (Table 7).

The amount of the submicron dispersoid forming elements should be held to the minimum required for control of grain structure, other mechanical properties, or

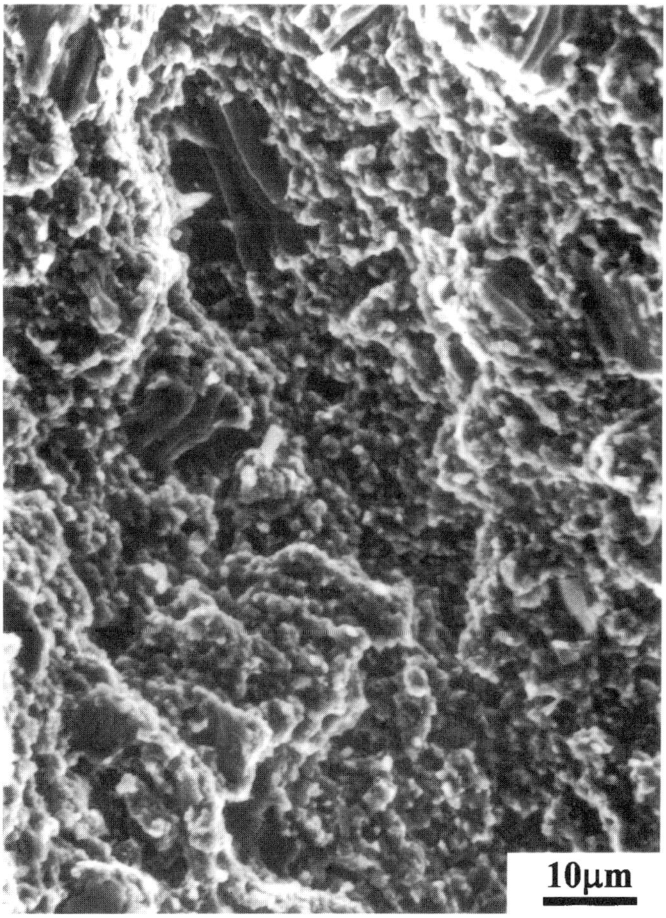

Figure 7 Cleavage fracture of Fe containing intermetallic particles.

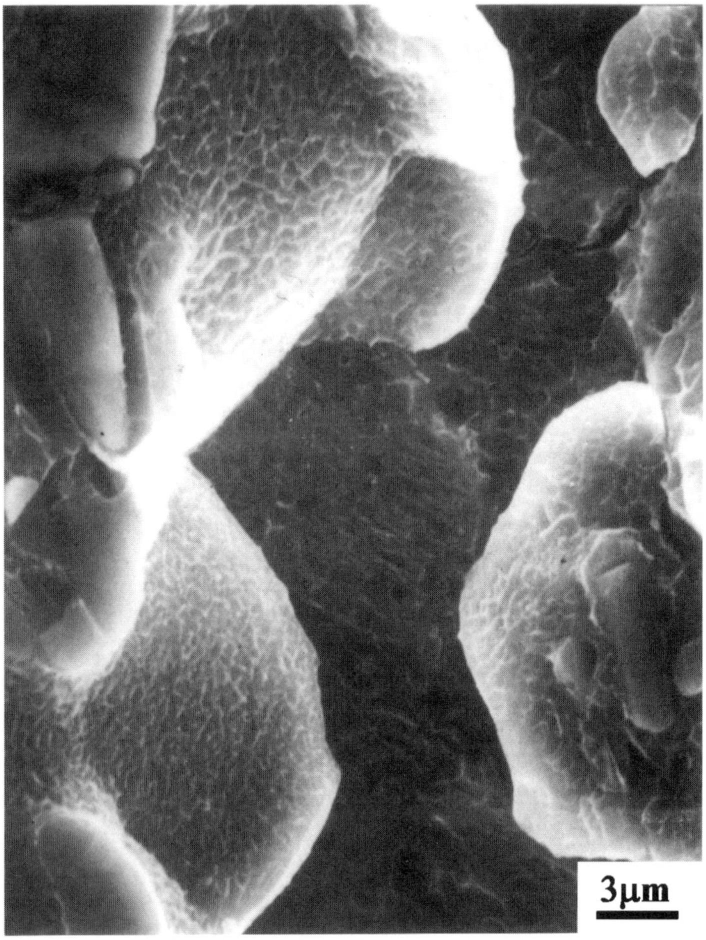

Figure 8 Intergranular fracture caused by strain localization along grain boundaries (Al–Li alloy).

resistance to stress-corrosion cracking. It has been shown that the alloys containing zirconium are more resistant to fracture than those containing manganese or chromium since zirconium forms relatively small (~20 nm in diameter) Al_3Zr particles.

The effect of the fine precipitates developed by age hardening is more complex. Usually, toughness is greatest in the underaged condition and decreases as aging proceeds to peak strength. Toughness can increase or decrease as the alloys are overaged, depending on the microstructure. When the overaged microstructure of an alloy consists of coarse precipitate particles or PFZs along grain boundaries, the alloy fails by intergranular fracture having poor toughness (Fig. 8). Lithium containing alloys are the typical example. On the other hand, in other alloys, improvement in toughness may occur on overaging due to a reduction in yield strength.

C. Fatigue

To characterize the fatigue behavior of a material, the resistance to both initiation and growth of fatigue cracks must be evaluated. The resistance to fatigue crack initiation is expressed as a fatigue strength (stress) for a given number of cycle. Unlike steels, aluminum alloys do not exhibit the sharply defined fatigue limit in S–N tests. When the fatigue crack growth (FCG) is of interest, the fatigue properties are expressed by the crack growth rate (da/dN) as a function of stress intensity range (ΔK). This type of FCG test is of prime importance for alloys used in aerospace applications.

It is generally known that improving the tensile strength increases the fatigue strength of aluminum alloys. However, the improvement in fatigue strength of aluminum alloys incurred by increase in tensile strength is lower than those of steels. The effects of large constituent particles and dispersoid particles on the fatigue strength of high-strength aluminum alloys are dependent on the type of fatigue test or stress regime chosen for evaluation. It has been shown that commercial-purity alloys have higher fatigue strength than equivalent high-purity compositions. The presence of inclusions and intermetallic compounds in commercial-purity alloys disperse slip during fatigue deformation, thereby delaying the crack initiation. However, the fatigue performance characteristics are reversed if FCG is measured. In this condition, the FCG is faster in the commercial-purity alloy because voids nucleate more readily at large constituent particles and dispersoid particles which are within the plastic zone of the advancing crack.

The effect of the fine precipitates developed by age hardening is also dependent on the type of fatigue test. During deformation of aluminum alloys, shearing of precipitates can occur resulting in a localization of strain in a limited number of slip bands. The occurrence of this so-called planar slip decreases the fatigue strength of aluminum alloys. On the other hand, the resistance to FCG can significantly be increased by the occurrence of planar slip. Planar slip results in a crack path along crystallographic planes and the resulting tortuous crack path induces crack closure effects associated with rough fracture surfaces and fretting debris (Fig. 9).

D. Corrosion

Aluminum alloys rapidly form a natural, protective film of alumina on their surface, making them quite corrosion resistant. Moreover, if the surface of aluminum is scratched sufficiently to remove the oxide film, a new film quickly re-forms in most environments. However, the oxide film can be dissolved in some chemicals, notably strong acids and alkalis. When the film is removed, the metal corrodes rapidly by uniform dissolution.

The mechanism of the corrosion process is usually associated with a local cell action between the solid solution of the matrix and the intermetallic second phase with a different electrode potential. Therefore, the corrosion behavior of aluminum alloys is affected by their microstructure. For example, the susceptibility to localized corrosion or stress-corrosion cracking can be reduced by appropriate treatments such as T7X tempers for 7XXX alloys or T8X tempers for 2XXX alloys.

Corrosion resistance of aluminum alloys can be enhanced by the thickening of oxide films. This can be achieved by various chemical and electrochemical treatments such as conversion coating and anodizing. Both conversion and anodic coatings can be dyed to give attractive colors and the latter process is widely applied to architectural products. The use of Alclad products is another way to improve the corrosion resistance. The core alloy is cathodically protected by the cladding alloy due to their different corrosion potentials. The design of an aluminum structure can also have a large effect on its corrosion behavior. The design of joints and the presence of other metals are important factors.

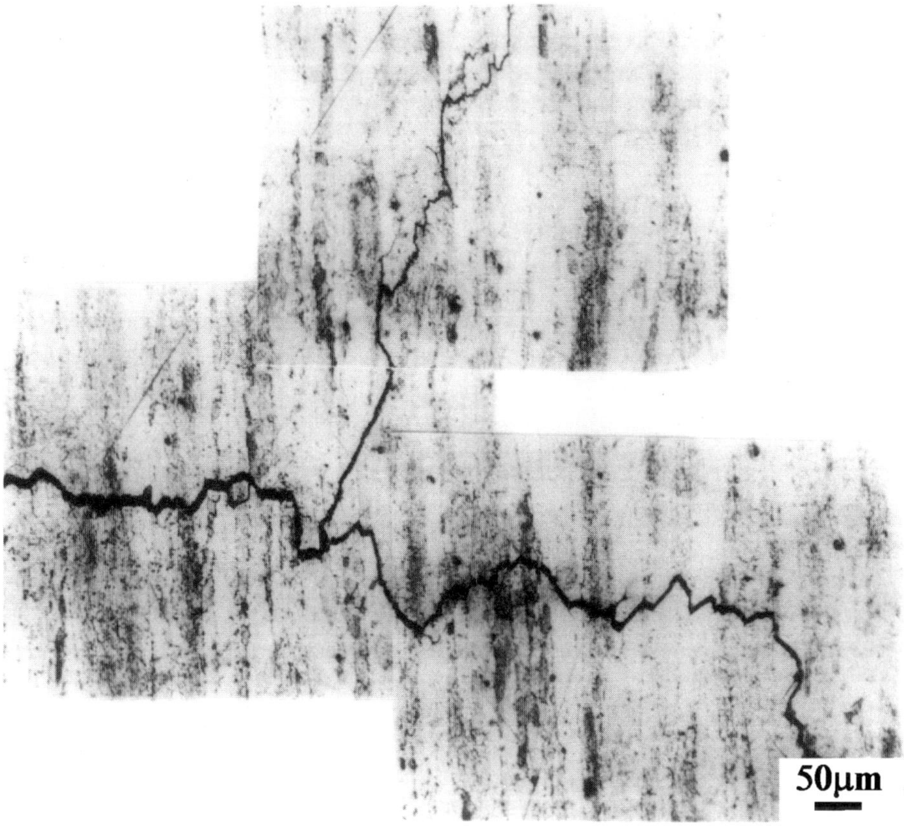

Figure 9 Tortuous fatigue crack path in 8090 Al–Li alloy. (Courtesy of C.S. Lee.)

IV. WROUGHT ALUMINUM ALLOYS

Most of aluminum alloys are used for wrought products, e.g., rolled plate, foil, extrusions, tube, rod, bar, and wire. Working operations and thermal treatments transform the cast ingot structure into a wrought structure. Working characteristics and subsequently developed properties are greatly affected by the compositions of the alloys.

A. Temper Designations

In addition to the alloy designation mentioned in Sec. 1.1 (Table 2), IADS has temper designation system to describe the thermal/mechanical history of alloys. Temper condition is identified by suffix letter and digits to the alloy number.

Basic Temper Designations

- F: As-fabricated.
- O: Annealed (wrought products only). Temper with the lowest strength and highest ductility.
- H: Cold worked (strain hardened).
- T: Heat treated to produce stable tempers other than F or O.

Strain Hardened Subdivisions

H1: Strain hardened only. The degree of strain hardening is indicated by the second digit and varies with quarter hard (H12) to full hard (H18).

H2: Strain hardened and partially annealed. Tempers ranging from quarter hard to full hard obtained by partial annealing of cold-worked materials with strengths initially greater than desired.

H3: Strain hardened and stabilized. Tempers for age softening Al–Mg alloys that are strain hardened and heated at a low temperature to increase ductility and stabilize mechanical properties.

Heat-Treated Subdivisions

T1: Naturally aged.

T3: Solution treated, cold worked, and naturally aged to a substantially stable condition.

T4: Solution treated and naturally aged to a substantially stable condition.

T5: Cooled from an elevated temperature shaping process and then artificially aged.

T6: Solution treated and then artificially aged.

T7: Solution treated and stabilized.

T8: Solution treated, cold worked, and then artificially aged.

B. Non-Heat-Treatable Alloys

Non-heat-treatable alloys include the various grades of pure aluminum and other alloys in which strength is developed largely by solid solution hardening and by strain hardening from the annealed temper. The major alloying elements include magnesium, manganese, and silicon. Chromium, copper, iron, and zinc are also added as minor elements (Table 5). It is worth mentioning that some of the above elements are often regarded as impurities in the other aluminum alloys. These elements form dispersoids with aluminum and other elements that provide dispersion strengthening. Non-heat-treatable alloys are in the 1XXX, 3XXX, 4XXX, and 5XXX series, with a few alloys in the 7XXX and 8XXX series. Typical mechanical properties of non-heat-treatable alloys are given in Table 6 along with their commercial forms and typical applications.

1. Pure Aluminum (1XXX Series)

There are various grades of aluminum: super-purity (SP) aluminum (99.99%) and commercial-purity (CP) aluminum containing up to 1% of impurities or minor additions. The structure of CP aluminum consists of small volume fraction of iron and/or silicon containing insoluble constituents in aluminum matrix. Strength of CP aluminum comes mostly from strain hardening. The maximum strength level occurs in alloy 1100, which is the "standard" CP aluminum alloy. Fully hardened alloy 1100 has a tensile strength of 165 MPa, with a yield strength of 150 MPa and an elongation of only 5%. Wrought products of CP aluminum are used in virtually all fields, with major consumption in the packaging, chemical, electrical, and architectural fields.

2. Al–Mn and Al–Mn–Mg Alloys (3XXX Series)

The only binary Al–Mn alloy of commercial interest is 3003. This alloy has about 1.2% manganese, which strengthened the aluminum by solid solution hardening and dispersion

Table 5 Compositions of Selected Non-Heat-Treatable Wrought Aluminum Alloys

Alloy	Si	Fe	Cu	Mn	Mg	Zn	Cr	Ti	Other
1100	1.0+	Fe	0.05–0.2	0.05	–	0.10	–	–	Al min 99.0
1145	0.05+	Fe	0.05	0.05	0.05	0.05		0.03	Al min 99.45
1199	0.006	0.006	0.006	0.002	0.006	0.006		0.002	Al min 99.99
3003	0.6	0.7	0.05–0.20	1.0–1.5		0.10			
3004	0.30	0.7	0.25	1.0–1.5	0.8–1.3	0.25			
3005	0.6	0.7	0.30	1.0–1.5	0.20–0.6	0.25	0.10		
3105	0.6	0.7	0.30	0.3–0.8	0.20–0.8	0.40	0.20	0.10	
5005	0.30	0.7	0.20	0.20	0.50–1.1	0.25	0.10		
5050	0.40	0.7	0.20	0.10	1.1–1.8	0.25	0.10		
5052	0.25	0.40	0.10	0.10	2.2–2.8	0.10	0.15–0.35		
5082	0.20	0.35	0.15	0.15	4.0–5.0	0.25	0.15	0.10	
5083	0.40	0.40	0.10	0.40–1.0	4.0–4.9	0.25	0.05–0.25	0.15	
5086	0.40	0.50	0.10	0.20–0.7	3.5–4.5	0.25	0.05–0.25	0.15	
5454	0.25	0.40	0.10	0.50–1.0	2.4–3.0	0.25	0.05–0.20	0.20	
5456	0.25	0.40	0.10	0.50–1.0	4.7–5.5	0.25	0.05–0.20	0.20	
8001	0.17	0.45–0.7	0.15	–	–	–	0.05	–	0.09–1.3 Ni
8009[a]	1.7	8.5							1.3V
8010	0.40	0.35–0.7	0.10–0.30	0.10–0.8	0.10–0.50	0.40	0.20	0.10	
8011	0.5–0.9	0.6–1.0	0.10	0.10	0.05	0.10	0.05	0.08	
8019[a]		8.3							4.0 Ce
8081	0.7	0.7	0.7–1.3	0.10	–	0.05	–	0.10	18–22 Sn
8280	1.0–2.0	0.7	0.7–1.3	0.10	–	0.05	–	–	5.5–7.0 Sn 0.20–0.7 Ni

[a]Alloys prepared by powder metallurgy.

hardening of Al_6Mn particles. Addition of magnesium to Al–Mn alloy considerably improves the tensile properties by solid solution hardening. For example, alloy 3004 has a tensile strength of about 180 MPa in the annealed condition as compared to 110 MPa of alloy 3003. Manganese content in the 3XXX series of alloys is limited well below the maximum solid solubility. Iron as an impurity reduces the solubility and the coarse Al_6Mn particles may form with harmful effect on ductility.

Both alloys 3003 and 3004 have excellent corrosion resistance and moderate strength combined with high ductility. These alloys are used for general purposes where moderate strength and good weldability are required and find the applications previously mentioned for pure aluminum. In particular, alloy 3004 is mostly used for manufacturing beverage cans. Several lower strength alloys such as 3005 and 3105 have desirable combinations of strength, formability, and corrosion resistance for applications in building and specialty products areas.

3. Al–Mg Alloys (5XXX Series)

The 5XXX series alloys contain about 0.8–5% magnesium as a major alloying element. Although magnesium has a greatly decreasing solid solubility with decreasing temperature in aluminum (Table 1), Al–Mg alloys do not show appreciable age hardening at concentrations less than 7% magnesium. However, magnesium provides substantial solid solution hardening and increases work-hardening capability. For example, the strength values of annealed Al–1Mg alloy are 40 MPa yield strength and 125 MPa tensile strength,

Table 6 Typical Mechanical Properties and Applications of Selected Non-Heat-Treatable Wrought Aluminum Alloys

AA number	Temper	Yield strength (MPa)[a]	Tensile strength (MPa)	Elongation (% in 50 mm)	Fatigue limit (MPa)[b]	Typical applications
1100	O	35	90	35	35	Sheet, fin stock, wire, spun hollow-ware
	H14	115	125	9	50	
	H18	150	165	5	60	
1145	O	35	75	40		Foil, fin stock, semi-rigid containers
	H18	115	145	5		
1199	O	10	45	50		Electrical and electronic foil
	H18	110	115	5		
3003	O	25	110	30	50	Chemical equipment, cooking utensils, rigid containers, sheet, plate
	H14	145	150	8	60	
	H18	185	200	4	70	
3004	O	70	180	20	95	Sheet, plate, rigid containers
	H34	200	240	9	105	
	H38	250	280	5	110	
	H19	285	295	2		
3005	O	55	130	25		Building products, high-strength foil
	H14	165	180	7		
	H18	225	240	4		
3104	H19	260	290	4		Sheet, plate, rigid containers
3105	O	55	120	24		Building products, general sheet material
	H25	160	180	8		
	H18	195	215	3		
5005	O	40	125	30		General sheet material, appliances, utensils, electrical conductors
	H18	195	200	4		
	H34	140	160	8		
	H38	185	200	5		
5050	O	55	145	24	85	Sheet and tube, rip tops for cans
	H34	165	190	8	90	
	H38	200	220	6	95	
5052	O	90	195	25	110	Sheet, plate, hydraulic tubes, appliances, marine uses
	H34	215	260	10	125	
	H38	255	270	7	140	
5056	O	150	290	35	140	Cable sheathing, zippers, screen wire
	H18	405	435	10	150	
	H38	345	415	15	150	
5083	O	145	290	22		Welded structure, pressure vessels, marine uses, transportation equipment, cryogenics
	H116	230	315	16	160	
	H343	280	360	8		

(Continued)

Table 6 (Continued)

AA number	Temper	Yield strength (MPa)[a]	Tensile strength (MPa)	Elongation (% in 50 mm)	Fatigue limit (MPa)[b]	Typical applications
5086	O	115	260	22		Welded structure, pressure vessels, marine uses, transportation equipment, cryogenics
	H34	255	325	10		
	H112	130	270	14		
	H116	205	290	12		
5154	O	115	240	27	115	Welded structure, rigid containers
	H34	230	290	13	130	
	H38	270	330	10	145	
	H112	115	240	25	115	
5252	O	85	180	23		Auto and appliance trim
	H25	170	235	11		
	H28	240	285	5		
5454	O	120	250	22		Welded structure, pressure vessels, marine uses, transportation equipment, cryogenics
	H34	240	305	10		
	H111	180	260	14		
	H112	125	250	18		
5456	O	160	310	24		High-strength welded structure, storage tanks, pressure vessels, marine uses
	H24	280	370	12		
	H112	165	310	22		
	H116	255	350	16		
5457	O	50	130	22		Anodized auto and appliance trim
	H25	160	180	12		
	H28	185	205	6		
5657	O	40	110	25		Anodized auto and appliance trim
	H25	140	160	12		
	H28	165	195	7		
8001	O	40	110	30		Sheet, tubing for water-cooled nuclear reactors
	H18	185	200	4		
8009[c]	F	405	536	9		Elevated temperature
8019[c]	F	338	427	5		Elevated temperature

[a]Yield strength, 0.2% offset.
[b]Based on 500 million cycles using an R.R. Moore-type rotating-beam machine.
[c]Alloys prepared by powder metallurgy.

which increase to 160 MPa yield strength and 310 MPa tensile strength by increasing magnesium content to 6%. Elongation over this range of magnesium is relatively high and exceeds 25%.

Alloys containing up to 3% magnesium are structurally stable at room and elevated temperatures, but structural instability may occur in alloys with higher magnesium content. The main cause of structural instability is the precipitation of $\beta(Mg_5Al_8)$ phase, along grain boundaries or along slip bands. Although the precipitation of β phase is slow at ambient temperatures, it is accelerated in the presence of severe strain hardening. Such structure is undesirable since it makes the alloys susceptible to intergranular attack and stress-corrosion cracking in corrosive conditions. Stress relieving at higher temperature (i.e., above solvus) is effective in preventing the formation of grain boundary precipitates, although it decreases the strength. Al–Mg alloys also have a tendency to soften with time (age softening) at ambient temperatures after strain hardening. Age softening occurs due to localized recovery within the deformed grains. This may be avoided by using H3 temper consisting of cold working to a level slightly greater than desired followed by stabilization at a temperature of 120–150°C.

Except a few general purpose commercial binary Al–Mg alloys, most of the 5XXX series alloys contain small additions of dispersion hardening elements such as chromium and manganese. These provide the increased tensile properties and raise the recrystallization temperatures.

Al–Mg alloys combine a wide range of strength, good forming, and welding characteristics, and high resistance to general corrosion. Probably, the most important property is the good welding response of the higher strength alloys under commercial welding processes. Weld strengths equal the annealed strengths of the various alloys, and the welds exhibit good ductility. The high strength and good welding qualities make them widely used in transportation and structural fields including boats, truck bodies and large carriage tanks. Good low temperature properties also make them valuable in applications requiring good cryogenic properties.

4. Miscellaneous Alloys

This group covers a rather broad range of alloys which contain alloying elements not normally used in other series of aluminum alloys. Alloy 8001 containing nickel (1.1 wt%) and iron (0.6 wt%) has good resistance to corrosive attack in water at high temperatures and pressures and is used in nuclear energy applications. Its mechanical properties resemble those of 3003. Alloy 8011 (Al–0.75Fe–0.7Si) is used for bottle caps because of its good deep drawing qualities.

Tin containing alloys such as 8081 (Al–20Sn–1Cu) and 8280 (Al–6.2Sn–1Cu–1.5Si–0.45Ni) have excellent anti-friction properties, making them ideal for applications as bearings in automobiles. Alloy 8009 (Al–8.5Fe–1.3V–1.7Si) is a rapid solidification/powder metallurgy (RS/PM) processed alloy and contains iron in amounts much larger than its maximum solid solubility (0.05%). The alloy has good elevated temperature properties due to the presence of thermally stable $Al_{12}(Fe,V)_3Si$ dispersoids (Fig. 10). General characteristics of PM alloys are described in Sec. IV.D. One of the recently developed 6XXX alloys can be classified as non-heat treatable.

C. Heat-Treatable Alloys

Heat-treatable alloys develop their strength by age hardening and are one of the most important structural materials for aerospace applications. They are covered by the three series of alloys: 2XXX (Al–Cu, Al–Cu–Mg), 6XXX (Al–Mg–Si), and 7XXX (Al–Zn–Mg, Al–Zn–Mg–Cu). Lithium containing alloys are covered in separate section. They can be classified into two groups: those that have medium strength and are readily weldable (Al–Mg–Si and Al–Zn–Mg), and the high-strength alloys that have been developed primarily for aircraft construction (Al–Cu, Al–Cu–Mg, and Al–Zn–Mg–Cu), most of which have very limited weldability. Compositions of representative commercial alloys are shown in Table 7. Typical mechanical properties are given in Table 8 along with their commercial forms and typical applications.

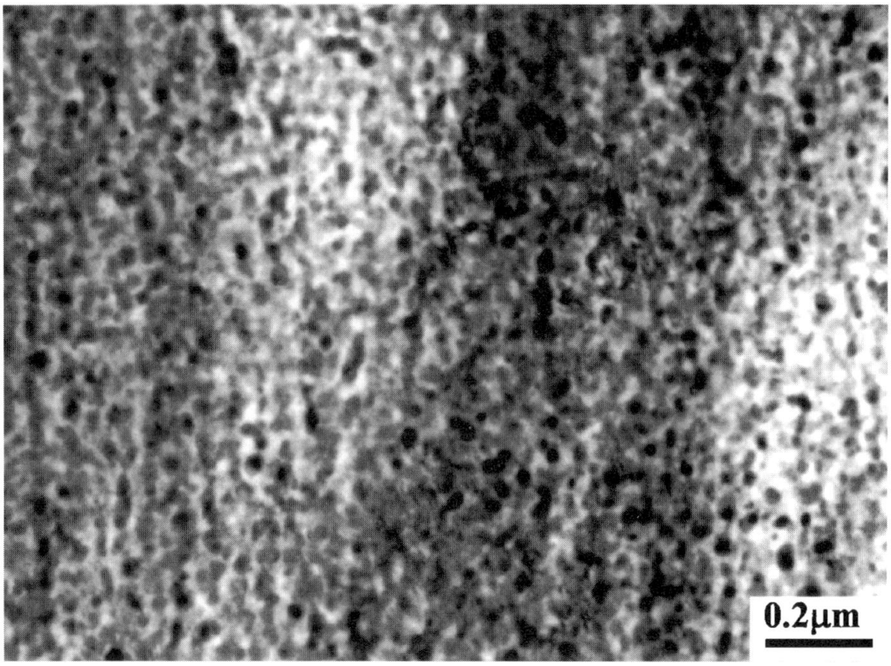

Figure 10 Fine dispersions of thermally stable $Al_{12}(Fe,V)_3Si$ particles in RS/PM 8009 alloy.

1. Al–Cu Alloys (2XXX Series)

The important wrought Al–Cu alloys are alloys 2011, 2025, and 2219. Alloy 2011 (Al–5.5Cu) contains small amounts of bismuth and lead that form discrete particles in the microstructure. It has good cutting characteristics and produces fine, easily broken chips during machining. Since its introduction in 1934, it has been the basic aluminum screw-machine alloy. The bismuth and lead, however, lower the corrosion resistance of the Al–Cu alloys; hence, the T8 temper is used to obtain a higher resistance to corrosion for applications involving corrosive atmosphere.

Alloy 2025 (Al–4.5Cu) is the first Al–Cu alloy developed in North America and is still used extensively for forgings. But this alloy has been replaced in many applications by alloy 2219 (Al–6.3Cu), which has a more useful combination of properties and is also available as sheets, plates, and extrusions. Alloy 2219 has a wide range of strength, good corrosion resistance and excellent elevated temperature properties. It also has good weldability and has been used for fuel tanks for storing liquefied gases such as hydrogen and oxygen that serve as propellants for missiles and space vehicles. Alloy 2219 is one of the alloys which show enhanced response to age hardening by strain hardening prior to artificial aging (T8 temper). Alloy 2219 in the T8 condition has yield strength more than 30% higher than in the T6 condition.

To meet requirements for increased tensile properties, several modified versions of 2219 have been developed. Alloy 2021 contains micro-addition of 0.15% cadmium and 0.05% tin, which refine the size of the semi-coherent θ precipitates during aging. It has a yield strength of 435 MPa, tensile strength of 505 MPa and elongation of 9% with no sacrifice of weldability or toughness at low temperatures. Another alloy is 2519 which contains a small amount of magnesium. One noticeable experimental alloy is the Al–6.3Cu–

Table 7 Compositions of Selected Heat-Treatable Wrought Aluminum Alloys

Alloy	Si	Fe	Cu	Mn	Mg	Zn	Cr	Ti	Other
2011	0.40	0.70	5.0–6.0			0.30			0.2–0.6 Bi 0.2–0.6 Pb
2014	0.50–1.20	0.70	3.9–5.0	0.40–1.2	0.20–0.80	0.25	0.10	0.15	0.2 Zr + Ti
2017	0.20–0.80	0.70	3.5–4.5	0.40–1.0	0.40–0.80	0.25	0.10	0.15	0.2 Zr + Ti
2021	0.20	0.30	5.8–6.8	0.20–0.40	0.02	0.10		0.02–0.10	0.10–0.25 Zr 0.05–0.20 Cd
2024	0.50	0.50	3.8–4.9	0.30–0.9	1.2–1.8	0.25	0.10	0.15	0.20 Zr + Ti
2025	0.50–1.2	1.0	3.9–5.0	0.40–1.2	0.05	0.25	0.10	0.15	
2036	0.50	0.50	2.2–3.0	0.10–0.40	0.30–0.6	0.25	0.10	0.15	
2048	0.15	0.20	2.8–3.8	0.20–0.6	1.2–1.8	0.25		0.10	
2090			2.4–3.0		0.00–0.25				0.08–0.15 Zr 1.9–2.6 Li
2091			1.8–2.5		1.1–1.9				0.04–0.16 Zr 1.7–2.3 Li
2124	0.20	0.30	3.8–4.0	0.30–0.9	1.2–1.8	0.25	0.10	0.15	0.20 Zr + Ti
2219	0.20	0.30	5.8–6.8	0.20–0.40	0.02	0.10		0.02–0.10	0.05–0.15 V 0.10–0.25 Zr
2519	0.25	0.30	5.3–6.4	0.10–0.50	0.05–0.40	0.10		0.10	0.05–0.15 V 0.10–0.25 Zr
2618	0.10–0.25	0.9–1.3	1.9–2.7		1.3–1.8	0.10		0.04–0.10	0.9–1.2 Ni
6009	0.6–1.6	0.50	0.15–0.6	0.2–0.8	0.40–0.8	0.25	0.10	0.10	
6010	0.8–1.2	0.50	0.15–0.6	0.2–0.8	0.6–1.0	0.25	0.10	0.10	
6013	0.6–1.0	0.50	0.6–1.1	0.2–0.8	0.8–1.2	0.25	0.10	0.10	
6017	0.55–0.7	0.15–0.30	0.05–0.20	0.10	0.45–0.60	0.05	0.10	0.05	
6061	0.40–0.8	0.70	0.15–0.40	0.15	0.8–1.2	0.25	0.04–0.35	0.15	
6063	0.20–0.6	0.35	0.10	0.10	0.45–0.9	0.10	0.10	0.10	
6151	0.6–1.2	1.0	0.35	0.20	0.45–0.8	0.25	0.15–0.35	0.15	
6262	0.40–0.8	0.70	0.15–0.40	0.15	0.8–1.2	0.25	0.04–0.14	0.15	0.40–0.7 Bi 0.40–0.7 Pb

(Continued)

Table 7 (Continued)

Alloy	Si	Fe	Cu	Mn	Mg	Zn	Cr	Ti	Other
6351	0.7–1.3	0.50	0.10	0.40–0.8	0.20		0.20		
6463	0.20–0.6	0.15	0.20	0.05	0.45–0.9	0.05			
7001	0.35	0.04	1.6–2.6	0.20	2.6–3.4	6.8–8.0	0.18–0.35	0.20	
7004	0.25	0.35	0.05	0.20–0.7	1.0–2.0	3.8–4.6	0.05	0.05	0.10–0.20 Zr
7005	0.35	0.40	0.10	0.20–0.7	1.0–1.8	4.0–5.0	0.06–0.20	0.01–0.06	0.08–0.20 Zr
7009	0.20	0.20	0.6–1.3	0.10	2.1–2.9	5.5–6.5	0.10–0.25	0.20	0.25–0.40 Ag
7010	0.10	0.15	1.5–2.0	0.30	2.2–2.7	5.7–6.7	0.05		0.11–0.17 Zr
7016	0.10	0.12	0.45–1.0	0.03	0.8–1.4	4.0–5.0		0.03	
7017	0.35	0.45	0.20	0.05–0.50	2.0–3.0	4.0–5.2	0.35	0.15	0.10–0.25 Zr 0.15 min Mn + Cr
7039	0.30	0.40	0.10	0.10–0.40	2.3–3.3	3.5–4.5	0.15–0.25	0.10	
7049	0.25	0.35	1.2–1.9	0.20	2.0–2.9	7.2–8.20	0.10–0.22	0.10	
7050	0.12	0.15	2.0–2.6	0.10	1.9–2.6	5.7–6.7	0.04	0.06	0.08–0.15 Zr
7075	0.40	0.50	1.2–2.0	0.30	2.1–2.9	5.1–6.1	0.18–0.28	0.20	0.25 Zr + Ti
7079	0.30	0.40	0.40–0.8	0.10–0.30	2.9–3.7	3.8–4.8	0.10–0.25	0.10	
7090[a]	0.12	0.15	0.6–1.3		2.0–3.0	7.3–8.7			1.0–1.9 Co 0.20–0.50 O
7091[a]	0.12	0.15	1.1–1.8		2.0–3.0	5.8–7.1			0.20–0.6 Co 0.20–0.50 O
7178	0.40	0.50	1.6–2.4	0.30	2.4–3.1	6.3–7.3	0.18–0.35	0.20	
7475	0.10	0.12	1.2–1.9	0.06	1.9–2.6	5.2–6.2	0.18–0.25	0.06	
8090			1.0–1.6		0.6–1.3				0.04–0.16 Zr 2.2–2.7 Li

[a]Alloys prepared by powder metallurgy techniques.

Table 8 Typical Mechanical Properties and Applications of Selected Heat-Treatable Wrought Aluminum Alloys

AA number	Temper	Yield strength (MPa)[a]	Tensile strength (MPa)	Elongation (% in 50 mm)	Fatigue limit (MPa)[b]	Typical applications
2011	T3	295	380	15	125	Screw machine parts
	T6	295	390	17		
	T8	310	405	12	125	
2014	O	95	185	18	90	Aircraft structures, truck frames
	T4, T451	290	425	20	140	
	T6, T651	410	480	13	125	
2017	O	70	180	22	90	Screw machine products, fittings
	T4, T451	275	425	22	125	
2020	T6	530	580	7		Aircraft structures
2024	O	75	185	20	90	Aircraft structures, truck wheels, screw-machine products
	T3	345	485	18	140	
	T361	395	495	13	125	
	T4, T451	325	470	20	140	
	T6	395	475	10		
	T81, T851	450	480	6	125	
	T861	490	515	6	125	
2025	T6	255	400	19	125	Forgings, aircraft propellers
2036	T4	195	340	24		Automotive body panels
2048	T85	440	480	10		Aircraft structures
2090	T83	496	552	6.3		Aircraft structures
2124	T351	325	470	20	140	Aircraft structures
	T8	440	490	8		
2218	T72	255	330	11		Jet engine impellers, compressor rings, engine heads and pistons
2219	O	70	170	18		Structure use at elevated temperatures, high-strength weldments
	T42	185	360	20		
	T31, T351	250	360	17		
	T37	315	395	11		
	T62	290	415	10	105	
	T81, T851	350	455	10	105	
	T87	395	475	10	105	
2618	T61	330	435	10		Aircraft engine, structure use at elevated temperatures
6009	T4	125	230	25		Auto body sheet
6010	T4	170	290	24		Auto body sheet
6053	O	55	110	35	55	Wire and rods for rivets
	T6	220	255	13	90	
6061	O	55	125	25	60	Welded structures, heavy-duty structure
	T4, T451	145	240	22	90	
	T6, T651	275	315	12	95	
	T91	395	405	12	95	
	T913	455	460	10		

(Continued)

Table 8 (Continued)

AA number	Temper	Yield strength (MPa)[a]	Tensile strength (MPa)	Elongation (% in 50 mm)	Fatigue limit (MPa)[b]	Typical applications
6063	O	50	90		55	Railings, architectural extrusions, pipes, truck flooring
	T1	90	150	20	70	
	T4	90	170	22		
	T5	145	185	12	70	
	T6	215	240	12	70	
	T83	240	255	9		
	T831	185	205	10		
	T832	270	290	12		
6066	O	85	150	18		Forging and extrusions for welded structure
	T4, T451	205	360	18		
	T6, T651	360	395	12	110	
6070	O	70	145	20	60	Pipelines
	T6	350	380	10	95	
6101	T6	195	220	15		High-strength bus conductors
6151	T6	295	330	17	85	Medium-strength forgings
6201	T81		330	6[c]	105	Electrical conductor wire
6262	T9	380	400	10	90	Screw machine products
6463	O	50	90		55	Architectural and trim extrusions
	T1	90	150	20	70	
	T5	145	185	12	70	
	T6	215	240	12	70	
7001	O	150	255	14		High-strength aircraft structure, missile structure
	T6, T651	625	675	9	150	
	T75	495	580	12		
7004	T6	340	400	12		Medium-strength welded structure
7005	O	80	195	20		
	W	205	345	20		
	T53	345	395	15		
	T6	290	350	13	150	
7009	T6	470	535	12		Aircraft structures
7010	T6	485	545	12		
7039	T61	345	415	13		Medium-strength welded structures
7049	T73	470	530	11		Hydraulic fittings, high-strength aircraft structures
7050	T736	510	550	11		Hydraulic fittings, high-strength aircraft structures
	T74, T7452	450	510	13		
7075	O	105	230	17	115	Hydraulic fittings, high-strength aircraft structures
	T6, Y651	500	570	11	160	
	T73, T735X	430	500	13	150	
	T76	470	540	12		
7090[d]	T7E71	580	620	9		Aircraft structures
7091[d]	T7E69	545	590	11		

(Continued)

Table 8 (Continued)

AA number	Temper	Yield strength (MPa)[a]	Tensile strength (MPa)	Elongation (% in 50 mm)	Fatigue limit (MPa)[b]	Typical applications
7175	T736, T74SX	485	550	12		Hydraulic fittings, high-strength aircraft structures
	T7351	435	505	13		
7178	O	105	230	15		Aircraft structures
	T6, T651	540	610	10	150	
	T76, T7651	505	570	9		
7475	T651	560	590	12		Hydraulic fittings, high-strength aircraft structures
	T7351	435	505	14		
8090	T8	436	503	5.0		Aircraft structures

[a]Yield strength, 0.2% offset.

[b]Based on 500 million cycles using an R.R. Moore-type rotating-beam machine.

[c]Elongation, % in 250 mm.

[d]Alloys prepared by powder metallurgy techniques.

0.45Mg–0.3Ag–0.3Mn–0.15Zr. Additions of small amounts of silver and magnesium to Al–Cu alloy result in the precipitation of the Ω phase instead of the θ and θ phases. Ω phase is very close to the equilibrium phase θ (Al_2Cu) that forms in binary Al–Cu alloys, but is quite stable at temperatures up to 200°C. In the T6 condition, the alloy has a yield stress as high as 520 MPa at room temperature, which compares with a typical value of 290 MPa for 2219. It also has superior creep properties to 2219.

2. Al–Cu–Mg Alloys (2XXX Series)

Al–Cu–Mg alloys have originated from the first precipitation hardenable alloy developed by Wilm in 1906. This alloy, called Duralumin (Al–3.5Cu–0.5Mg–0.5Mn), was used for the frames for Zeppelin airships. Alloy 2017 is a modified version of Duralumin and is now used mainly for rivets. Al–Cu–Mg alloys are now widely used for aircraft construction.

Alloy 2014 (Al–4.4Cu–0.5Mg–0.9Si–0.8Mn) contains silicon which make the alloy more responsive to artificial aging and therefore higher strengths than 2017. This alloy is used as forgings or extrusions for aircraft. Another alloy, 2024, in which the magnesium content is raised to 1.5% and the silicon content is reduced to impurity levels, has much enhanced response to hardening by natural aging at room temperature and is frequently used in T3 or T4 tempers. The strengthening phase in 2024 is S (Al_2CuMg) which preferentially nucleates on dislocations. Thus, a homogeneous distribution of dislocation produces a very fine and uniform distribution of S. Hence, the alloy 2024 has a high response to deformation prior to aging. The fuselage skin and the lower wing skins of most commercial passenger airplanes are made of 2024-T3 and it is used almost exclusively for the smaller general aviation aircraft. However, 2024 shows poor resistance to corrosion and stress-corrosion cracking so that its sheet and plate are normally roll-clad with pure aluminum or Al–1Zn.

In general, for equivalent yield strengths, the 2XXX series alloys have lower fracture toughness than the 7XXX series alloys (Al–Zn–Mg–Cu). This is mainly due to the larger sizes of intermetallic compounds in the 2XXX series alloys. Alloys 2124 and 2224 have been developed with reduced levels of iron and silicon impurities. They contain much reduced amounts of insoluble intermetallic compounds in the microstructure and have improved fracture toughness and ductility over 2024.

Although 2XXX series alloys have good combinations of mechanical properties which are needed for aircraft applications, they are prone to softening at around 120°C. Development of supersonic aircraft such as Concorde stimulated a need for a sheet alloy with improved creep strength on prolonged exposure at high temperatures. The alloy 2618 (Al–2.2Cu–1.5Mg–1Ni–1Fe) was developed to meet this need. In the alloy 2618, iron and nickel are important alloying elements, because they form Al_9FeNi phase which causes dispersion strengthening and assists in stabilizing the microstructure at elevated temperatures. Copper and magnesium contribute to strengthening through precipitation of S phase.

Lower strength Al–Cu–Mg alloys are being investigated as possible sheet materials for automotive applications. One example is 2036 which has nominal copper and magnesium content of 2.5% and 0.45%, respectively.

3. Al–Mg–Si Alloys (6XXX Series)

The 6XXX alloys contain additions of magnesium and silicon which, in the aged condition, precipitate Mg_2Si as a strengthening phase. In most cases, the magnesium and silicon are present in the alloy in amounts to nominally combine to Mg_2Si (Mg:Si 1.73:1). But silicon in excess of that needed to form Mg_2Si is also used. The excess silicon provides an appreciable increase in strength over that obtained from a specific quantity of Mg_2Si by refining the size of the Mg_2Si particles and also precipitating as silicon. Manganese or chromium is added to most 6XXX series alloys because of their effects on promoting fine grain size and inhibiting recrystallization during solution treatment. The 6XXX alloys are widely used as medium-strength structural alloys which have the additional advantages of good weldability, corrosion resistance, and immunity to stress-corrosion cracking. The 6XXX alloys are used for the majority of extrusions, with smaller quantities being available as sheet and plate.

Alloy 6061 is the most well-known and widely used general purpose structural alloy in 6XXX series. It has a balanced composition of Al–1Mg–0.6Si (1.5% Mg_2Si) with 0.25% copper and 0.2% chromium. Copper is added to improve the strength and chromium is added to offset any adverse effect copper may have on corrosion resistance. Alloy 6053 is another alloy which has balanced composition (2% Mg_2Si). Moderate strength alloy 6063 was developed for the applications where strength requirement was less stringent. This alloy has about 1% Mg_2Si and can be readily extruded. The advantage of this alloy is that it can be quenched adequately during the extrusion operation as it comes from the press. Thus, a separate solution treatment is not needed. Alloy 6063 has excellent finishing qualities, ideal for application for architectural and decorative finishes. In alloy 6463, the iron content is kept to a low level ($<0.15\%$) so that the alloy responds well to chemical brightening and anodizing for use as automotive trim.

Examples of the alloys which have excess silicon are the alloys 6066, 6070, 6151, and 6351. The presence of this excess silicon promotes an additional response to age hardening by both refining the size of the Mg_2Si particles and precipitating as silicon. However, these alloys have reduced elongation and toughness compared to the alloy 6061. The presence of chromium and/or manganese helps to counter this effect by promoting fine grain size and inhibiting recrystallization during solution treatment.

The alloy 6013 has a higher copper content than the other 6XXX alloys and is considerably stronger than 6061. It has excellent stretch forming characteristics and its properties in the T6 condition are comparable to 2024-T3, making it a potential replacement alloy for 2024-T3 sheet. However, alloy 6013 shows slight susceptibility to intergranular corrosion which is attributed to the presence of copper containing precipitates in grain boundaries.

Alloy 6262 contains lead and bismuth, which give the alloy improved machining characteristics. Alloy 6262-T9 has higher strength with somewhat lower machinability than the aluminum–copper alloy 2011. Alloy 6262 has much higher resistance to stress-corrosion cracking than 2011 and is preferred for more highly stressed fittings.

Recently, a new type of non-heat-treatable 6024 alloy has been developed which has all the characteristics of conventional heat-treatable 6XXX alloys. This alloy, having a nominal composition of Al–0.75Mg–1Si–1.4Mn–0.2Zr–0.3Cr, has tensile properties equivalent to those of 6061-T6 in as-fabricated condition: a yield strength of 200 MPa, tensile strength of 350 MPa and elongation of 22%. A unique characteristic of this alloy comes from the presence of a rather large volume fraction of Al_6Mn dispersoids which provide dispersion strengthening. The alloy is subjected to faster solidification rate than normal to have a fine distribution of Al_6Mn dispersoids (Fig. 11). The size and distribution of Al_6Mn dispersoids are optimized during homogenization. Since aging treatment is not required for the alloy to obtain required properties, it is ideal for applications for long structural components of which the heat treatment is often not possible.

4. Al–Zn–Mg(–Cu) Alloys (7XXX Series)

These alloys have very pronounced age-hardening response and therefore have been the primary material for aircraft applications. 7XXX alloys mostly contain 4–8 wt% zinc

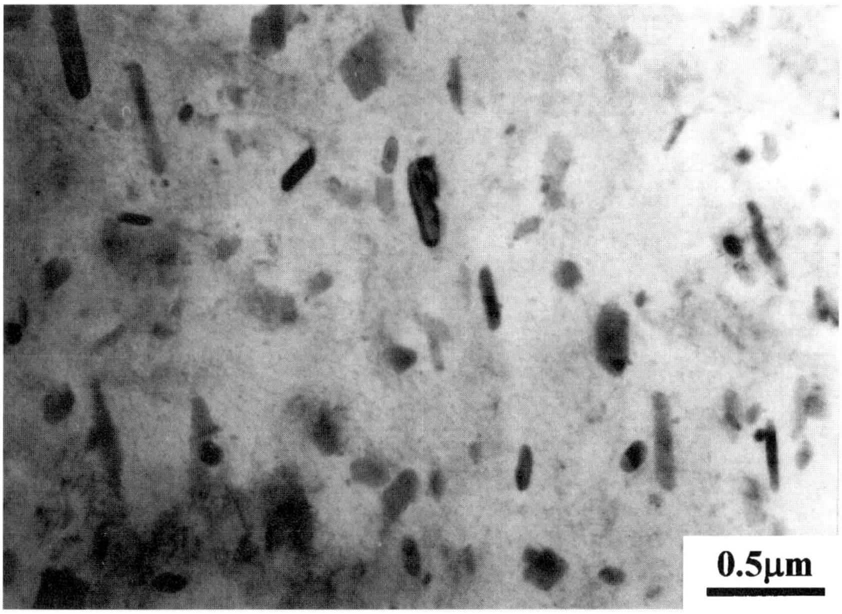

Figure 11 Al_6Mn dispersoids in non-heat-treatable Al–0.75Mg–1Si–1.4Mn–0.2Zr–0.3Cr alloy. (Courtesy of S.W. Nam.)

and 1–3 wt% magnesium. Al–Zn–Mg alloys without copper have intermediate strength and are weldable. Copper additions of 1–2 wt% increase the strengths of Al–Zn–Mg alloys. 7XXX alloys are strengthened by the precipitation of GP zones and/or semicoherent η' ($MgZn_2$) phase. One unique characteristic of the precipitation behavior of Al–Zn–Mg alloys is that η' precipitates can be nucleated from pre-existing GP zones. Also, deformation prior to aging has no effect on the precipitation behavior of these alloys. Copper additions do not change the basic precipitation mechanism of Al–Zn–Mg alloys. However, copper in GP zones increases its stability so that GP zones can exist at higher temperatures than in comparable Al–Zn–Mg alloys.

Although these alloys have excellent tensile properties and good fabricating characteristics, they have somewhat high susceptibility to stress-corrosion cracking. Improvements in resistance to stress-corrosion cracking have come by control of both compositions and heat-treatment procedures.

In general, overaging of 7XXX alloys results in a large decrease in tensile properties but with a significant increase in resistance to stress-corrosion cracking. The T73 temper was developed to improve the resistance to stress-corrosion cracking with a small sacrifice in tensile properties. The T73 temper is a duplex aging treatment consisting of the first aging step at low temperature (e.g. 120°C) followed by an overaging step at a higher temperature (160–170°C). A fine dispersion of the η' precipitates could be obtained through nucleation from pre-existing GP zones. Thick sections of 7075-T73 have high resistance to stress-corrosion cracking, albeit at the expense of strength. 7075-T73 is used for critical aircraft components such as very large die forging. Another duplex ageing treatment, designated T76, has been applied successfully to 7XXX alloys to increase resistance to exfoliation (layer) corrosion.

Aluminum alloys are usually cold water quenched to exploit their full potential for age hardening. However, it can introduce high levels of residual stress that contribute to stress-corrosion cracking. Using the slower quench rates (e.g. air cooling) from the solution treatment temperature minimizes residual stresses and decreases differences in electrode potentials throughout the microstructure, resulting in an improvement to resistance to stress-corrosion cracking. However, slow quench cannot be utilized to the chromium or manganese bearing alloys which have high quench sensitivity (Sec. 2.2). As a consequence, zirconium replaces chromium and manganese for the purpose of inhibiting recrystallization during solution treatment. Such alloys (e.g. 7010 and 7050) are not quench-sensitive and thus are used for thick sections in military aircrafts. 7050 also has somewhat higher copper content than 7075 in order to increase the strengthening associated with the second stage of the T73 treatment. The slightly modified version of 7050, 7150, is used as an improved performance upper wing skin alloy for several types of passenger aircrafts.

To combine the high strength of T6 with the resistance to stress-corrosion cracking equal to T73, another heat treatment has been developed. This is known as retrogression and re-aging (RRA) which involves the following stages for 7075: (1) conventional T6 treatment (solution treatment at 465°C, cold water quench, age 24 hr at 120°C), (2) short heat treatment (e.g. 5 min) at a temperature in the range 200–280°C and water quench, (3) re-aging at 120°C for 24 hr. The RRA treatment results in a microstructure having a matrix similar to that obtained for a T6 temper, combined with grain boundary regions characteristic of a T73 temper in that precipitates are larger and more widely separated. Reduced susceptibility to stress-corrosion cracking has been attributed to this latter change. However, the RRA treatment has been difficult to apply to thick sections because of the short time interval initially proposed for stage (2). More recently, the time and temperature conditions for this stage of the process have been optimized (e.g. 1 hr at 200°C) and the RRA treatment has been given the commercial temper designation T77. It has been applied to the new highly alloyed composition 7055 (Al–8Zn–2.05Mg–2.3Cu–0.16Zr) that has been used for structural members of the wings of the new Boeing 777 passenger aircraft.

Other compositional changes have been made to improve the strength and toughness. One typical example is alloy 7475. 7475 has a much reduced level of iron and silicon impurities (combined maximum of 0.22%) compared to 7075 (0.90%) (Table 7). 7475 also has the manganese content reduced from 0.30% maximum to 0.06% maximum. Hence, the size and number of deleterious intermetallic compounds are much reduced in 7475, resulting in an improved fracture toughness over 7075 at same temper condition and tensile strength. This alloy is widely used in aircraft in the form of thick plate and forgings.

Compared to high-strength 7XXX alloys, medium-strength Al–Zn–Mg alloys have the advantage of being readily weldable. Unlike the other weldable aluminum alloys, these alloys age harden significantly at room temperature. Also, solution treatment temperature for the alloys can be as low as 350°C and the strength is relatively insensitive to cooling rates from high temperatures. Therefore, the alloys retain most of their strengths after welding. Most of these alloys contain smaller amounts (0.1–0.3%) of one or more of the elements chromium, manganese, and zirconium. These elements are added mainly to retain the dynamically recovered structure and prevent excess growth of recrystallized grains during fabrication. Weldable Al–Zn–Mg alloys find their applications in various areas including truck bodies, railroad cars, and portable lightweight military bridges.

5. Al–Li Alloys

Probably one of the most important aluminum alloys developed in the last decades is the Al–Li alloy. By adding lithium, a new class of aluminum alloys has been developed which offers significant weight savings over conventional aluminum alloys. Lithium decreases the density (3% for each 1% added) and increases the stiffness (6% for each 1% added) of aluminum. Al–Li alloys also have unusually high resistance to fatigue crack growth. Because of these unique features, some of Al–Li alloys have now attained commercial status for use in aircraft structures.

Age hardening of binary Al–Li alloys produces the precipitation of an ordered, metastable phase δ' (Al_3Li) that is coherent and has a particularly small misfit with the matrix. Small amounts of zirconium are also added in all variants of Al–Li alloys to control recrystallization and grain size. One other effect of this element is that nucleation of some of the δ' precipitates occurs on Al_3Zr dispersoids (see arrow, Fig. 12).

Due to the coherency, the δ' particles are sheared by moving dislocations giving rise to planar slip (Fig. 5). The resulting planar slip bands lead to strain localization at grain boundaries, giving rise to cracking along grain boundaries. Therefore, the aged binary Al–Li alloys suffer from low ductility and toughness and further alloying elements are needed for the formation of additional precipitates to homogenize deformation and prevent strain localization.

The addition of copper leads to the enhanced precipitation of δ' and co-precipitation of θ' (Al_2Cu) and T_1(Al_2CuLi) phases. T_1 precipitates are the predominant copper bearing strengthening precipitate in the commercial Al–Li–Cu alloys such as 2090. T_1 nucleates heterogeneously on dislocations and low angle boundaries so that 2090 is normally cold worked before aging (T8 temper) to promote a more uniform distribution. The presence of T_1 along grain boundaries mitigates strain localization. There can also be a formation

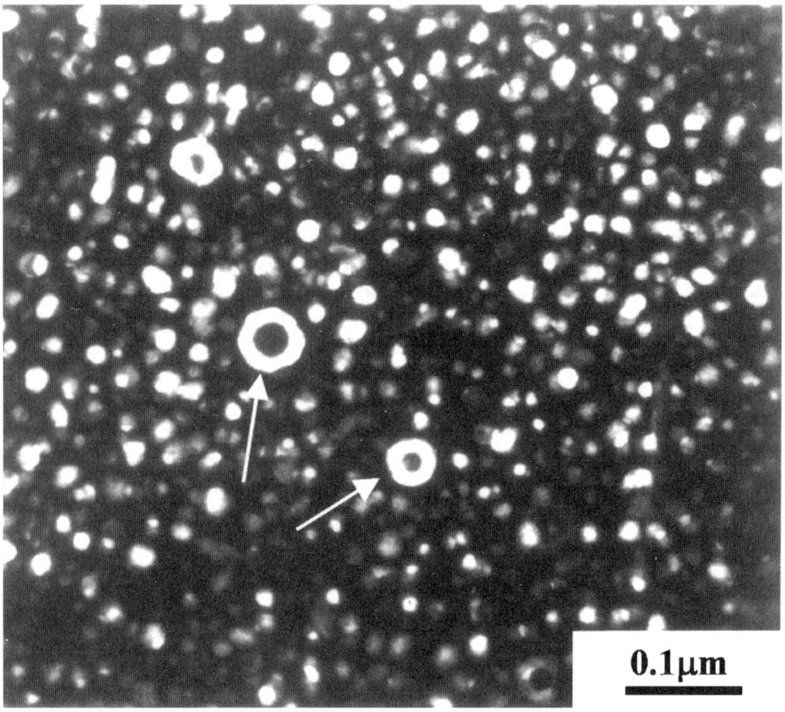

Figure 12 δ′ precipitates in an Al–Li–Cu–Mg–Zr alloy. The arrows indicate Al_3Zr dispersoids. (Courtesy of C.S. Lee.)

of quasicrystalline $T_2(Al_6CuLi_3)$ phase. T_2 phase nucleates predominantly at high angle grain boundaries, reducing ductility and toughness.

Addition of manganese to binary Al–Li alloys results in the precipitation of the incoherent Al_2LiMg phase. However, this phase precipitates preferentially at grain boundaries and does not reduce planar slip behavior. It has an adverse effect on ductility, toughness, and corrosion behavior due to coarse grain boundary precipitates and associated PFZs. However, the addition of manganese to ternary Al–Li–Cu alloys introduces the co-precipitation of $S'(Al_2CuMg)$ phase in addition to T_1 and δ phases. The S phase nucleates heterogeneously on matrix dislocations and low angle grain boundaries so that Al–Li–Cu–Mg alloys are normally cold worked prior to aging (T8 temper) to promote a more uniform distribution. S′ phase forms as laths that are not readily sheared by dislocations, thereby promoting more homogeneous slip. The relative amounts of δ′, T_1, and S′ phases in Al–Li–Cu–Mg alloys depend on the relative concentrations of three alloying elements. High copper and lithium content relative to magnesium results in the dominant precipitation of T_1 phase. Preferential precipitation of S occurs when alloys have higher copper and magnesium content than lithium.

Another type of Al–Li alloy is available, which is weldable and has high strength at cryogenic temperatures. This alloy, 2095, is based on Al–Li–Cu–Mg alloy and contains small amounts of silver. Addition of silver promotes the precipitation of a fine dispersion of T_1 phase along with S′ and θ′ in the peak aged condition. It has a minimum yield strength of around 600 MPa in T6 condition, which is double that of 2219-T6. It also maintains its high strength and high toughness at the cryogenic temperatures associated with storage of liquified gases. It is currently considered for applications for the welded fuel tanks of space vehicles in place of the existing alloys 2229 and 2014.

The original objective in developing the Al–Li alloys was as direct substitutes for conventional aircraft alloys. While a number of applications were pursued in aircraft and launch vehicles, these alloys had shortcomings that limited their applications. Alloys 2090, 8090, and 2091 have the solute content near the maximum that could be dissolved in solid solution, resulting in pronounced quench sensitivity. Except for thin gauges, embrittling equilibrium phases nucleated during quench and coarsened during aging. Moreover, prolonged exposure of stressed alloys to elevated temperatures ($> 50°C$) can lead to relatively high rates of creep crack growth. Alloy 2297 (Al–2.8Cu–1.3Li–0.3Mn–0.1Zr) contains lower lithium content than the other Al–Li alloys and has a good combination of strength and toughness in all orientations, thermally stable properties and good corrosion resistance. This alloy is currently used as replacements for fighter aircraft structural parts which have limited fatigue life due to severe flight spectra loading.

D. Powder Metallurgy Aluminum Alloys

Conventional powder metallurgy (PM) processing of aluminum alloys involves the production of powders which are then cold pressed and sintered at elevated temperatures. Such processing has the advantage of near net shape production, thereby enabling the costs associated with machining to be minimized. But there are only a few aluminum alloys with moderate properties that permit a sintering despite the oxidation of powder surface. This has led to the development of several novel processes such as rapid solidification, spray forming, and mechanical alloying. These new processes have enabled new, and sometimes unconventional, alloys to be produced, thereby extending the range of properties available from powder compacts. They have certain advantages when compared to the ingot metallurgy processing, such as finer and more homogeneous microstructure, extension of solid solubility, and formation of metastable phases, with the possibility of better mechanical properties and corrosion resistance.

Rapid solidification (RS) processing of aluminum alloys involves solidification at very high rate (up to 10^{5}°C/sec) from the melt to produce a powder or ribbon. Powders are produced by gas atomization in which molten alloy is sprayed through a nozzle into a stream of high velocity gas such as nitrogen or argon. Ribbons are made by melt spinning in which the molten alloy is forced into an orifice onto a water-cooled rotating wheel. Ribbons are pulverized into flakes for subsequent consolidation. Mechanical alloying is a dry, high energy ball milling process that makes it possible to produce dispersions of oxides, carbides or nitrides in fine powders. Consolidation of powders or flakes is usually done by canning, degassing and hot pressing followed by extrusion or forging. Steps of canning, degassing, and hot pressing can be replaced by vacuum hot pressing. The major obstacle is the presence of oxide layers along the surface of powder particles, which may prevent inter-particle bonding during consolidation with consequent deleterious effect on mechanical properties (Fig. 13). Spray forming is basically the same as atomization; however, in this case, the resulting spray of droplets is deposited on a rotating collector plate to produce a compact perform. Preforms, which may be 98–99% of the theoretical density, can be extruded or forged to obtain a full density. The advantage of spray forming over atomization is that some of the steps involved in normal powder processing are eliminated. Also, particulates (e.g. SiC) may be injected into the gas stream to produce metal matrix composites directly.

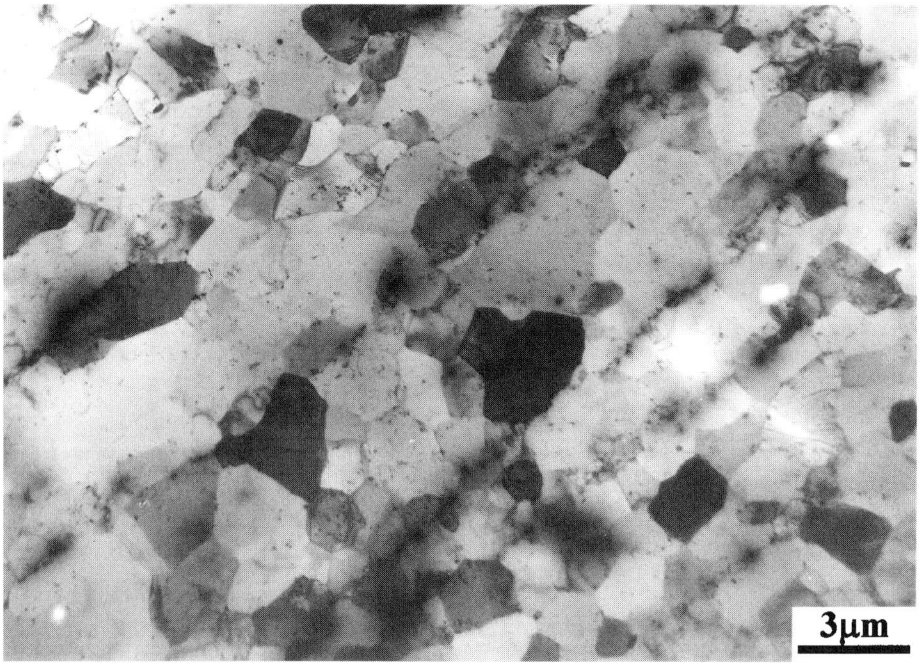

Figure 13 Oxide layers along the prior powder particle boundaries of RS/PM Al–Li alloy.

Although the refinement of microstructure can be obtained by PM processing, the gain in mechanical properties by utilizing rapidly solidified powders without alloy modification is only marginal. The real advantage of PM processing is that novel alloys can be developed that cannot be fabricated by conventional ingot metallurgy. Alloying aluminum with oxygen, carbon, nitrogen, transition metals, rare earth metals, and lithium has been made.

Better combinations of strength and resistance to both corrosion and stress-corrosion cracking have been obtained by the additions of cobalt, zirconium or nickel to an Al–Zn–Mg–Cu alloy which produces a fine dispersion of the intermetallic phases in the PM product. Commercial alloys such as 7090 and 7091 (Table 7) are now available.

Iron, which is regarded as the most deleterious impurity in aluminum alloys, has been utilized for the development of PM aluminum alloys for elevated temperature applications. There are several variants of these so-called high temperature aluminum alloys, including 8009 (Al–8.5Fe–2.4V–1.3Si) and 8019 (Al–8.3Fe–4.0Ce) alloys. Good elevated temperature properties of these alloys are achieved by the formation of a large volume fraction of thermally stable dispersoids. For 8009, additions of vanadium and silicon to Al–8.5Fe alloy result in the fine dispersion of thermally stable $Al_{12}(Fe,V)_3Si$ dispersoids (Fig. 8). Sintered aluminum powders (SAP) were one of the first aluminum alloys which utilize oxide dispersion effect of mechanical alloying. SAP has superior creep resistance at temperatures above about 200–250°C compared to conventional aluminum alloys, but has insufficient strength at low temperatures. It is currently used for fuel element cans in nuclear fission reactors which use organic coolants.

Powder metallurgy processing has also been used for the development of several low density Al–Li alloys. These alloys incorporate higher lithium content than ingot alloys and

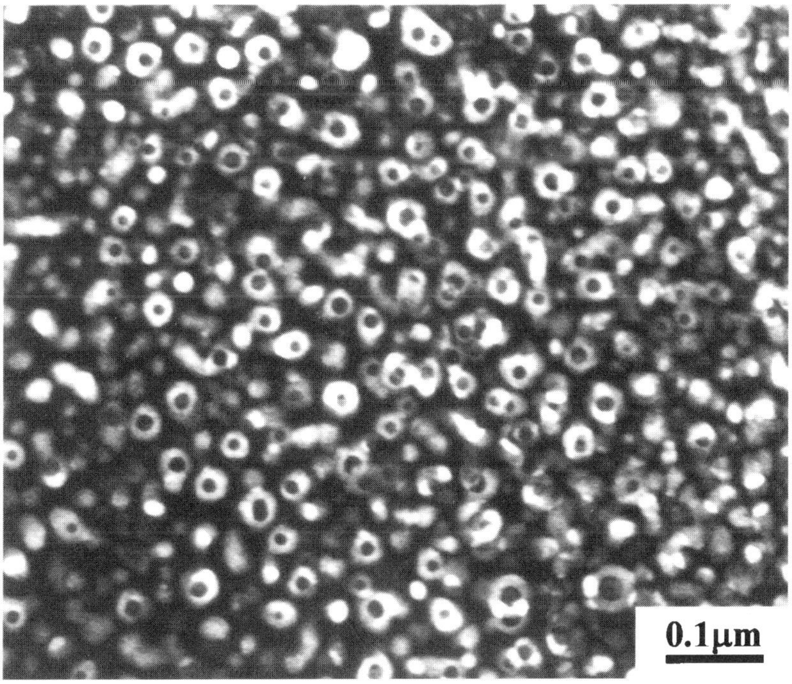

Figure 14 Al_3Zr dispersoids in RS/PM Al–Li–Zr alloy.

this offers further potential for reducing density and increasing stiffness. The lithium content in these alloys is high, however, the amounts of copper and/or magnesium should be reduced so as not to exceed the solid solubility limit. Consequently, the alloys do not have enough amounts of T_1 and/or S′ to disperse slip. The formation of a large volume fraction of Al_3Zr dispersoids, which is possible only via PM approach, has been shown to be quite effective in dispersing slip (Fig. 14).

Powder metallurgy techniques also enable the creation of materials with improved wear properties. An example is the hypereutectic Al–Si alloys and mechanical properties are again much greater than those of the equivalent conventionally cast alloy due mainly to the finer dispersion of the primary silicon phase (Fig. 15). Another example is Al–17Si–6Fe–(1–4.5)Cu–0.5Mg alloy which is currently used as piston materials of two-stroke snowmobile (Yamaha) and Lysholm compressor rotor (Mazda Motor Co., Ltd.).

Powder metallurgy has been successful in the development of new aluminum alloys that exhibit superior properties for certain applications. However, it is necessary to appreciate that the PM alloys carry a significant cost premium so that potential applications are likely to be confined to specialized products, where an increased price is justified by the demand for improved properties.

V. CAST ALUMINUM ALLOYS

Aluminum alloys can be cast in virtually all molds, including sand molds, permanent molds, and high pressure dies. The aluminum melt can be fed either by gravity or by applying pressure. Solidification is much more rapid when the alloys are cast in metal molds

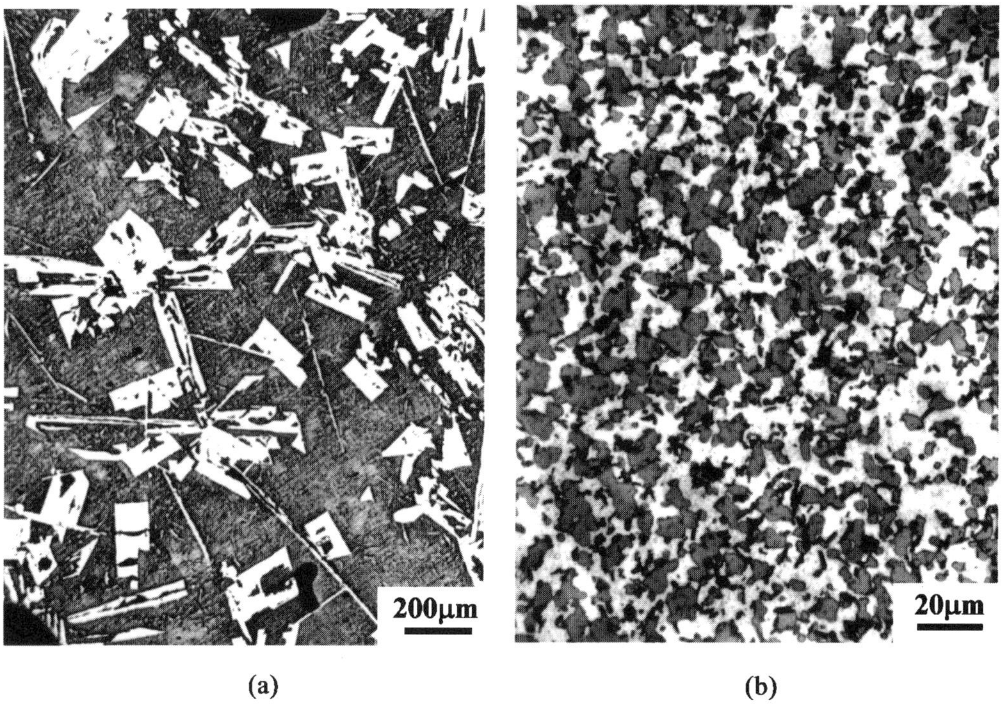

Figure 15 Effect of solidification rate on refining the size of primary particles in hypereutectic Al–Si alloys; (a) ingot cast, and (b) spray cast. (Courtesy of W.J. Park.)

and, therefore, a given alloy will be fine grained as a permanent mold casting. Aluminum alloys have several advantages for casting; relatively low melting temperatures, good fluidity, negligible solubility for all gases except hydrogen, and a good surface finish. However, aluminum castings have the problem of the relatively high level of shrinkage (3.5–8.5%) that occurs during solidification. Usually, the mold is designed to accommodate the occurrence of shrinkage and achieve dimensional accuracy. Selection of alloys for particular casting processes depends primarily on composition which, in turn, controls characteristics such as solidification range, fluidity, susceptibility to hot cracking, etc. Compositions of representative commercial alloys are shown in Table 9. Typical mechanical properties are given in Table 10 along with their typical applications.

Recently, several novel casting processes have been developed which offer several advantages over conventional casting processes; semi-solid processing and squeeze casting. In semi-solid processing, the dendrites are broken up by vigorously agitating the melt during solidification, increasing the fluidity. Each broken piece of dendrite becomes a separate crystal and a very fine grain size can be achieved. Semi-solid processing has the advantage of much lower capital investment and operating costs when compared with conventional casting methods. Squeeze casting process involves compressing the liquid metal in a hydraulic press during solidification. The high pressure promotes intimate contact between casting and mold walls, resulting in an increase in solidification rate. Both semi-solid processing and squeeze casting can be used for the production of composite materials.

Table 9 Compositions of Selected Aluminum Casting Alloys

AA number	Casting process	Si	Fe	Cu	Mn	Mg	Cr	Ni	Zn	Ti	Other
150.1	Ingot	[a]	[a]	0.10					0.05		
201.0	S	0.10	0.15	4.0–5.2	0.20–0.50	0.15–0.55				0.15–0.35	0.4–1.0 Ag
208.0	S	2.5–3.5	1.2	3.5–4.5	0.50	0.10		0.35	1.0	0.25	
213.0	P	1.0–3.0	1.2	6.0–8.0	0.6	0.10		0.35	2.5	0.25	
	S, P	4.0–6.0	0.8	2.0–4.0	0.20–0.6	0.15		0.30	0.50	0.20	
222.0	S, P	< 2.0	< 1.5	9.0–11.0		0.15–3.5		< 0.8			
238.0	P	3.5–4.5	1.5	9.0–11.0	0.6	0.15–3.5		1.0	1.5	0.25	
242.0	S, P	0.7	1.0	3.5–4.5	0.35	0.15–0.35	0.25	1.7–2.3	0.35	0.35	
295.0	S	0.7–1.5	1.0	4.0–5.0	0.35	0.03			0.35	0.25	
308.0	P	5.0–6.0	1.0	4.0–5.0	0.50	0.10			1.0	0.25	
319.0	S, P	5.5–6.5	1.0	3.0–4.0	0.50	0.10		0.35	1.0	0.25	
A332.0	P	11.0–13.0	1.2	0.50–1.5	0.35	0.7–1.3		2.0–3.0	0.35	0.25	
355.0	S, P	4.5–5.5	0.6[b]	1.0–1.5	0.50	0.40–0.60	0.25		0.35	0.25	
C355.0	S, P	4.5–5.5	< 0.20	1.0–1.5		0.40–0.60			< 0.10		
356.0	S, P	6.5–7.5	0.6	0.25	0.35	0.20–0.40			0.35	0.25	
A356.0	S, P	6.5–7.5	0.20	0.20	0.10	0.20–0.40			0.10	0.20	
357.0	S, P	6.5–7.5	< 0.15	< 0.05		0.45–0.65			< 0.05		
A357.0	S, P	6.5–7.5	< 0.20	< 0.20		0.45–0.65			< 0.10		0.05 Be
359.0	S, P	8.5–9.5	< 0.20	< 0.20		0.50–0.70					
360.0	D	9.0–10.0	2.0	0.6	0.35	0.40–0.6		0.50	0.50		
364.0	D	8.0–9.0	< 1.5	< 0.20		0.20–0.40	0.35		< 0.15		0.03 Be
380.0	D	7.5–9.5	2.0	3.0–4.0	0.50	0.10		0.50	3.0		
A380.0	D	7.5–9.5	1.3	3.0–4.0	0.50	0.10		0.50	3.0		
390.0	D	16.0–18.0	1.3	4.0–5.0	0.10	0.45–0.65			0.10		
	S, P, D	10.0–13.0	0.6	0.10	0.50	0.10		0.10	0.10	0.20	
A390.0	S, P	16.0–18.0	< 0.50	4.0–5.0	< 0.10	01.2.45–0.65			< 0.10		

(Continued)

Table 9 (Continued)

AA number	Casting process	Si	Fe	Cu	Mn	Mg	Cr	Ni	Zn	Ti	Other
392.0	D	18.0–20.0	< 1.5	0.60	0.40	1.0			< 0.50		
413.0	D	11.0–13.0	2.0	1.0	0.35	0.10		0.50	0.50		
A413.0	D	11.0–13.0	< 1.3	< 1.0		< 0.10					
443.0	S	4.5–6.5	0.8	0.6	0.50	0.05	0.25		0.50	0.25	
514.0	S	0.35	0.50	0.15	0.35	3.5–4.5			0.15	0.25	
518.0	D	0.35	1.8	0.25	0.35	7.5–8.5		0.15	0.15		
520.0	S	0.25	0.30	0.25	0.15	9.5–10.6			0.15	0.25	
535.0	S	0.15	0.15	0.05	0.1–0.25	6.2–7.5				0.10–0.35	
705.0	S, P	0.20	0.8	0.20	0.4–0.6	1.4–1.8	0.20–0.40		2.7–3.3	0.25	
707.0	S, P	0.20	0.60	0.20	0.4–0.6	1.4–1.8	0.20–0.40		4.0–4.5	0.25	
712.0	P	0.15	0.50	0.25	0.10	0.50–0.65	0.4–0.6		5.0–6.5	0.10–0.25	
713.0	S, P	0.25	1.18	0.40–1.0	0.6	0.20–0.50	0.35	0.15	7.0–8.0	0.25	
850.0	S, P	0.70	0.70	0.7–1.3	0.10	0.10		0.7–1.3		0.20	5.5–7.0 Sn

S, sand casting; P, permanent mould (gravity die) casting; D, pressure die casting.
[a]Ratio Fe:Si minimum of 2:1.

Table 10 Typical Mechanical Properties and Applications of Selected Casting Alloys

AA number	Casting process	Temper	Yield strength (MPa)[a]	Tensile strength (MPa)	Elongation (% in 50 mm)	Fatigue limit (MPa)[b]	Typical applications
201.0	S	T43	255	414	17		Aircraft parts
		T6	345	415	5.0		
		T7	414	467	5.5	97	
	P	T43	255	414	17.0		
		T6	379	448	8.0		
		T7	414	469	5.0	97	
208.0	S	F	97	145	2.5	76	General purpose sand castings, manifolds and valve
		T533	105	185	1.5		
213.0	S	F	103	165	1.5	62	Washing machine agitators, cylinder heads, timing gear
	P	F	165	207	1.5	66	
		T533	185	220	0.5		
	S, P	T21	95	175	3.0		
		T6	230	295	2.0		
222.0	S	O	138	186	1.0	65	Pistons, air-cooled cylinder heads
		T61	276	283	< 0.5	59	
		T62	331	421	4		
	P	T52	214	241	1.0		
		T551	241	255	< 0.5	59	
		T65	248	331	< 0.5	62	
242.0	S	F	214	217	0.5		Air-cooled cylinder heads, pistons
		O	124	186	1.0	55	
		T571	207	221	0.5	76	
		T77	159	207	2.0	72	
	P	T571	234	276	1.0	72	
		T6	230	295	1.0		
		T61	290	324	0.5	66	
295.0	S	T4	110	221	8.5	48	General purpose castings requiring shock resistance
		T6	165	250	5.0	52	
		T62	221	283	2.0	55	
	P	T61	195	260	4.0		
308.0	P	F	110	193	2.0	90	General purpose permanent mold castings
319.0	S	F	124	186	2.0	69	General purpose alloy, cylinder heads and engine parts
		T21	125	185	1.0		
		T5	179	207	1.5	76	
		T6	164	250	2.0	76	
	P	F	131	234	2.5		
		T21	125	200	2.0		
		T6	186	276	3.0		
	S, P	T61	240	260	0.5		
332.0	P	T5	193	248	1.0		Automotive and diesel engine pistons
		T65	295	325	0.5		

(Continued)

Table 10 (Continued)

AA number	Casting process	Temper	Yield strength (MPa)[a]	Tensile strength (MPa)	Elongation (% in 50 mm)	Fatigue limit (MPa)[b]	Typical applications
333.0	P	F	131	234	2.0	100	General purpose alloy, engine parts
		T5	172	234	1.0	83	
		T6	207	290	1.5	103	
		T7	193	255	2.0	83	
355.0	S	F	83	159	3.0		General purpose alloy used where high strength and pressure tightness are required
		T4	125	210	3.0		
		T51	159	193	1.5	55	
		T6	172	241	3.0	62	
		T61	241	269	1.0	65	
		T7	250	264	0.5	69	
		T71	200	241	1.5	69	
	P	T4	140	245	6.0		
		T6	235	280	1.0		
C355.0	S	T6	200	269	5.0		Similar to 355, but stronger and more ductile
356.0	S	F	124	164	6.0		Intricate casings requiring good strength and ductility
		T51	138	172	2.0	55	
		T6	205	230	4.0	59	
		T7	207	234	2.0	62	
		T71	145	193	3.5	59	
	P	F	124	179	5.0		
		T51	138	186	2.0		
		T6	225	240	4.0	90	
		T7	165	221	6.0	76	
A356.0	S	F	83	159	6.0		Similar to 356, but stronger and more ductile
		T51	124	179	3.0		
		T6	207	278	6.0		
		T71	138	207	3.0		
	P	T61	207	283	10.0	90	
357.0	S	F	90	172	5.0		Highly stressed casings requiring corrosion resistance
		T51	117	179	3.0		
		T6	296	345	2.0		
		T7	234	278	3.0		
	P	F	103	193	6.0		
		T51	145	200	4.0		
		T6	296	359	5.0	90	
A357.0	S	T6	248	317	3.0	83	Aircraft and missile parts
	P	T6	290	359	5.0	103	
359.0	P	T62	290	345	5.5	110	High-strength aircraft and missile parts
360.0	S	F1	65	185	8.0		General purpose castings
		T5	110	185	2.0		
		T6	215	255	–		

(Continued)

Table 10 (Continued)

AA number	Casting process	Temper	Yield strength (MPa)[a]	Tensile strength (MPa)	Elongation (% in 50 mm)	Fatigue limit (MPa)[b]	Typical applications
	P	T5	130	245	2.5		
		T6	265	310	1.0		
		F1	90	205	9.0		
	D	F	172	324	3.0	131	
		F1	130	250	2.5		
380.0	D	F	165	331	3.0	145	General purpose castings
A380.0	D	F	159	324	4.0	138	
413.0	D	F	145	296	2.5	131	Large, intricate casings with thin sections
		F1	140	265	2.0		
A413.0	D	F	110	241	3.5	131	Large, intricate casings with thin sections
443.0	S	F	55	131	8.0	55	Castings requiring high resistance to corrosion and shock
		F1	65	130	5.0		
	P	F	62	159	10.0	55	
		F1	70	160	6.0		
	D	F	110	228	9.0	117	
514.0	S	F	83	172	9.0	48	
		F1	80	170	5.0		
	P	F1	80	230	10.0		
518.0	D	F	186	310	8.0	138	
		F1	130	260	10.0		
520.0	S	T1	175	320	15.0		
		T4	179	331	16.0	55	
535.0	S	F	145	275	13.0		
A535.0	S	F	124	250	9.0		
705.0	S	T1	130	240	9.0		
707.0	S	T1	185	255	3.0		
710.0	S	F	172	241	5.0	55	
711.0	P	F	124	241	8.0	76	
850.0	S	T5	76	138	8.0		
	P	T5	76	159	12.0	62	

[a]Yield strength, 0.2% offset.
[b]Based on 500 million cycles using an R.R. Moore-type rotating-beam machine.

A. Temper Designations

The temper designations for castings are the same as those used for wrought products (Sec. 4.1). However, this does not apply for ingots. As with wrought aluminum alloys, casting alloys include both heat-treatable and non-heat-treatable alloys. Non-heat-treatable alloys are usually used in as-cast condition, identified by a suffix F or by omission of any suffix. Solution treatment (T4 temper), however, is often applied to improve the ductility of some alloys such as aluminum–silicon alloys. Both heat-treatable and non-heat-treatable alloys may also be thermally stress relieved (O temper). For the heat-treatable alloys, the O, T4, T5, T6, and T7 tempers are normally applied to improve mechanical properties.

Pressure die castings are not normally solution treated since heating to the high temperature causes blistering due to the expansion of air entrapped during the casting process. Also, the distortion of castings may occur as residual stresses are relieved. However, recent improvements in die casting process have resulted in a quality which is compatible with solution treatment and have permitted the use of T6 and T7 tempers.

B. Al–SI Alloys

Aluminum casting alloys with silicon as the major alloying addition are the most important commercial casting alloys because of their superior casting characteristics imparted by the presence of relatively large volumes of the Al–Si eutectic. These alloys also have the advantage of high corrosion resistance, good weldability, and the low thermal expansion coefficient. Machining may present difficulties because of the presence of hard silicon particles in the microstructure. However, using tungsten carbide tools with appropriate coolants and lubricants or polycrystalline diamond tools results in good quality machining.

Microstructure of binary Al–Si casting alloys is characterized by the presence of a eutectic, consisting of plates or needles of silicon in a continuous aluminum matrix. When slowly solidified, the structure becomes very coarse with large plates or needles of silicon. Alloys having this coarse eutectic exhibit low ductility because of the brittle nature of the large silicon plates. Fast cooling, as it occurs during permanent mold casting, greatly refines the microstructure and the silicon phase assumes a fibrous form with the result that both ductility and tensile strength are much improved. Die casting produces an even faster rate of cooling with a much more refined microstructure. The eutectic may also be refined by alloy modification.

The modification of Al–Si alloys is commonly achieved by the addition of a small amount of sodium (0.001–0.003%), either as metallic sodium or as sodium salts before casting. It is usually necessary to add several times the amount of sodium required since sodium is lost continuously during holding of the melt. However, overmodification promotes the formation of coarse silicon particles. Another problem involved with the modification by sodium is that the effects of modification are lost if Al–Si castings are re-melted which prevent foundries being supplied with pre-modified ingots.

Modification can also be achieved by the addition of other elements such as strontium or antimony. Although strontium is also lost during holding of the melt, its loss is much less and the degree of modification increases with holding time for several hours. The modified microstructure is retained after re-melting and the overmodification is also less of a problem with strontium. Another advantage of strontium additions is that they suppress formation of primary silicon in hypereutectic alloys, resulting in improvements in ductility and toughness. Either Al–Sr or Al–Si–Sr master alloys can be used in adding strontium. Although the use of antimony has not been generally accepted as a result of hygiene and toxicological concerns, the effect of antimony on modification is essentially permanent in comparison with the transient effects of sodium or strontium.

Binary Al–Si alloys are used for sand and permanent mold castings for which strength is not a prime consideration. Alloy 443 (5.3% Si) may be used with all casting processes for parts in which good ductility, good corrosion resistance, and pressure tightness are important. Die cast alloy 413 (12% Si) has good corrosion resistance and has superior castability and pressure tightness compared to alloy 443. Alloy A444 (7% Si) has high ductility when cast in permanent mode and heat treated to T4 condition and is used when impact resistance is a prime consideration, e.g., domestic cookware, pump casings, and certain automobile castings, including water-cooled manifolds.

Strength of Al–Si alloys can be significantly increased by the addition of other alloying elements such as copper and magnesium. Copper increases strength and improves machinability, at the expense of reduced ductility and corrosion resistance. The higher silicon alloys are used for pressure die castings, whereas alloys with lower silicon and higher copper are used for sand and permanent mold castings. In general, Al–Si–Cu alloys are used for many of the applications listed for the binary alloys but where higher strength is needed. Many castings of Al–Si–Cu alloys are supplied only in the as-cast temper. However, the strength and machinability of some of these castings can be improved by T5 heat treatment. One example is the use of alloy 319 (Al–6Si–3.5Cu) for die cast (permanent mold) automotive cylinder heads in place of cast iron. As with the wrought alloys, some compositions contain minor additions of elements such as bismuth and lead which improve machining characteristics.

Large quantities of sand and permanent mold castings are made from Al–Si–Mg alloys which are strengthened by the precipitation of Mg_2Si. Typical alloys are 356 (Al–7Si–0.3Mg) and 357 (Al–7Si–0.5Mg) alloys. These alloys have excellent casting characteristics, weldability, pressure tightness, and corrosion resistance. They are heat treatable to provide various combinations tensile and physical properties. These alloys find particular use for aircraft and automotive applications, one recent example being their increasing use for lighter-weight, cast motor car wheels. In some cases, the properties of these alloys are quite comparable to those of high-strength wrought alloys. This suggests that the possible replacement of some wrought components by these relatively cheaper castings may be a trend in the future.

Additions of both copper and magnesium to Al–Si alloys result in higher strength than the ternary alloys, with some sacrifice in ductility and corrosion resistance. One example is A390 (Al–17Si–4Cu–0.55Mg) which has been used for sand and permanent mold castings of all-aluminum alloy automotive cylinder blocks. The alloy has high wear resistance due to a large volume fraction of fine primary silicon particles. Refinement of primary silicon particles is achieved by the addition of 0.01–0.03% phosphorus. More complex compositions are available where special properties are required. A336 (Al–12Si–1Cu–1Mg–2Ni) contains nickel which forms stable intermetallic compounds. This alloy has improved elevated temperature properties, a low thermal expansion coefficient, and improved wear resistance. This alloy is used as the piston alloy for internal combustion diesel engines and higher output engines.

C. Al–Cu Alloys

Although Al–Cu casting alloys were the first aluminum casting alloys to be extensively used, most of them have now been replaced by other alloys. The main reasons for the replacement of the Al–Cu alloys are that they have poor casting characteristics, poor corrosion resistance, and high specific gravities.

One of the most important characteristics of the Al–Cu casting alloys is that they have excellent elevated temperature properties. Alloy 238 (Al–10Cu–3Si–0.3Mg) is used to cast soleplates for electrical hand irons and has the required high hardness at elevated temperatures. Alloys 240, 242 (Al–4Cu–2Ni–1.5Mg), and 243 also have high strength and hardness at elevated temperatures and are used for diesel engine pistons and air-cooled cylinder heads for aircraft engines. Good elevated temperature properties of these alloys come from the combination of precipitation hardening together with dispersion hardening by intermetallic compounds.

The more recently developed 201 alloy has a particularly high response to age hardening, providing tensile properties significantly higher than those of any previous aluminum casting alloys. This alloy has the nominal composition Al–4.7Cu–0.7Ag–0.3Mg and is being used to cast premium quality aerospace parts. The high response to aging appears to arise due to the effect of silver on modifying the precipitation behavior expected for the Al–Cu–Mg system. Tensile properties of 345 MPa yield strength and 415 MPa tensile strength with a minimum elongation of 5% are guaranteed for the T6 tempered premium quality castings. These tensile properties are quite comparable to the properties of the high-strength wrought alloys. Like the other Al–Cu casting alloys, this alloy is also susceptible to stress-corrosion cracking in the T6 condition but resistance can be greatly improved by a T73 temper.

D. Al–Mg Alloys

Al–Mg alloys are characterized by high resistance to corrosion, good machinability, and attractive appearance when anodized. Controlled melting and pouring practices are needed since the presence of magnesium increases the tendency towards oxidation in the molten state. Most show little or no response to heat treatment.

The magnesium content of the binary alloys ranges from 4% to 10%. Most are sand cast although compositions with 7% and 8% magnesium have limited application for permanent mold and pressure die castings. The tensile properties and casting characteristics can be modified by the addition of alloying elements such as silicon, zinc, and manganese. However, these are not of major importance for the Al–Mg casting alloys because these alloys are used when corrosion resistance and decorative appearance are the essential requirements.

Al–Mg casting alloys are basically non-heat treatable. An exception is 520-T4 alloy (Al–10%Mg) which has an outstanding combination of strength, ductility, and high impact resistance. Slow quenching after solution-treatment is needed to provide optimum resistance to stress-corrosion cracking. The alloy tends to be unstable, even at warm ambient temperature, leading to the loss of ductility and resistance to stress-corrosion cracking after a period of time. Consequently, other alloys such as Al–Si–Mg tend to be preferred unless the higher strength and ductility of Al–10Mg are mandatory.

E. Al–Zn–Mg Alloys

The unique characteristic of Al–Zn–Mg alloys is their high melting points, making them suitable for castings which are to be assembled by brazing. The alloys also have good machinability, dimensional stability, and resistance to corrosion. The alloys are normally sand cast because the use of permanent molds tends to cause hot cracking.

Al–Zn–Mg alloys in the as-cast condition show natural precipitation hardening behavior and their properties change rapidly by aging at room temperature. These alloys are ideal for castings with shapes that are difficult to solution treat and quench. The alloys are not recommended for use at elevated temperatures because overaging causes rapid softening.

ACKNOWLEDGMENT

The author is grateful to Dr. J.K. Oh, Prof. S. Lee, Dr. S. Ahn and Prof. D.H. Kim for their support during the preparation of the manuscript.

FURTHER READING

Kelly, A.; Nicholson, R.B. Precipitation hardening. Progress. Mater. Sci. 1963, *10*, 149.

Godard, H.P.; Jepson, W.B.; Bothwell, M.R.; Kane, K.L. *Corrosion of Light Metals*; Wiley: New York, 1967.

Van Horn, K.R. Ed. *Aluminum*; American Society for Metals: Metals Park, 1967.

Mondolfo, L.F. *Aluminum Alloys: Structure and Properties*; Butterworths: London, 1976.

Sanders, T.H., Jr.; Staley, J.T. *Fatigue and Microstructure*; American Society for Metals: Metals Park, 1979.

Martin, J.W. *Micromechanisms in Particle-Hardened Alloys*; Cambridge University Press: Cambridge, 1980.

McQueen, H. The experimental roots of thermomechanical treatments for aluminum alloys. J. Met. 1980, *32*(2), 17.

Pampillo, C.A.; Bilani, H.; Erabury, D.E. Eds. *Aluminum Transformation, Technology and Applications*; American Society for Metals: Metals Park, 1980.

Altenpohl, D. *Aluminum Viewed from Within*; Aluminum-Verlag: Düsseldorf, 1982.

Froes, F.H.; Pickens, J.R. Powder metallurgy of light alloys for demanding applications. J. Met. 1984, *36*, 14.

Hatch, J.E. Ed. *Aluminum: Properties and Physical Metallurgy*; American Society for Metals: Metals Park, 1984.

Sanders, T.H., Jr.; Starke, E.A., Jr. Eds. *Aluminum Alloys—Their Physical and Mechanical Properties*; Engineering Materials Advisory Services: Warley, 1986.

Sheppard, T. Ed. *Aluminum'86*; Institute of Metals: London, 1986.

Froes, F.H.; Cull, R.A. Eds. *Space Age Metals Technology*; SAMPE: Covina, 1988.

Vasudevan, A.K.; Doherty, R.D. Eds. *Treatise on Materials Science and Technology, Aluminum Alloys—Contemporary Research and Applications*; Academic Press: New York, 1989; Vol. 31.

Polmear, I.J. *Light Alloys, Metallurgy of the Light Metals*, 2nd Ed.; Arnold: London, 1989.

Lee, E.W.; Chia, E.H.; Kim, N.J. Eds. *Light-Weight Alloys for Aerospace Applications*; TMS: Warrendale, 1989.

Starke, E.A. Jr.; Sanders, T.H. Jr. Eds. *Aluminum-Lithium V*; MCE: Birmingham, 1989.

Grimes, R. Aluminum–lithium alloys. *Encyclopedia of Materials Science and Engineering*, Cahn, R.W., Ed.; Pergamon: London, 1990; Supplementary Vol. 1,

Sekhar, J.A.; Dantzig, J., Eds. *Nature and Properties of Semi-Solid Materials*; The Minerals, Metals and Materials Society: Warrendale, 1991.

Lee, E.W.; Kim, N.J. Eds. *Light-Weight Alloys for Aerospace Applications II*; TMS: Warrendale, 1991.

Hirano, K.; Oikawa, H.; Ikeda, K. Eds. *Science and Engineering of Light Metals*; Japan Institute of Light Metals: Tokyo, 1991.

Arnberg, L.; Lohne, O.; Nes, E.; Ryum, N., Eds. Third International Conference on Aluminum Alloys: Their Physical and Mechanical Properties, NTH/SINTEF: Norway, 1992.

Davis, J.R. Ed. *ASM Specialty Handbook on Aluminum Alloys*; ASM International: Materials Park, 1993.

Bickert, C., Ed. Proceedings of International Symposium on Light Metals; Metallurgical Society of CIM: Quebec, 1993.

Anonymous. *Tempers for Aluminum Alloy Products*; Aluminum Association: New York, 1993.

Sanders, T.H., Jr.; Starke, E.A., Jr., Eds. Fourth International Conference on Aluminum Alloys—Their Physical and Mechanical Properties; Georgia Institute of Technology: Atlanta, 1994.

Anonymous. *Wrought Aluminum Alloys: Registration Board of International Alloy Designations and Chemical Composition Limits for Wrought Aluminum and Wrought Aluminum Alloys*; Aluminum Association: New York, 1993.

Peters, M.; Winkler, P.-J. Eds. *Aluminum–Lithium Alloys VI*; DGM Informationsgesellshaft: Germany, 1993.

Kirkwood, D.H. Semisolid processing. Int. Mater. Rev. 1994, *39*, 173.

Kim, N.J. Ed. *Light Materials for Transportation Systems*; Center for Advanced Aerospace Materials, POSTECH: Pohang, 1993.

Lee, E.W.; Kim, N.J. Eds. *Light-Weight Alloys for Aerospace Applications III*; TMS: Warrendale, 1995.

Lee, E.W.; Frazier, W.E.; Kim, N.J.; Jata, K. Eds. *Light-Weight Alloys for Aerospace Applications IV*; TMS: Warrendale, 1997.

Jata, K. Ed. *Light-Weight Alloys for Aerospace Applications VI*; TMS: Warrendale, 2001.

Kim, N.J.; Lee, C.S.; Eylon, D. Eds. *Light Materials for Transportation Systems II*; Center for Advanced Aerospace Materials, POSTECH: Pohang, 2001.

10
Designing with Magnesium: Alloys, Properties, and Casting Processes

Trevor B. Abbott and Mark A. Easton
Monash University, Clayton, Victoria, Australia

Carlos H. Cáceres
The University of Queensland, Brisbane, Queensland, Australia

I. INTRODUCTION

Magnesium alloys have been in use for a period of time similar to that of aluminum alloys; magnesium was isolated before aluminum and magnesium alloys were used significantly during the Second World War. However, consumption of magnesium is a small fraction of that for aluminum. In recent years there has been a rapid growth in the application of magnesium alloys [1–4], with growth in pressure die cast applications exceeding 18% per annum over the last decade. Magnesium is particularly attractive in applications where weight is critical, as it is the lightest of all structural metals with a density of 1.74 g/cm^3 compared to 2.70 g/cm^3 for aluminum and 7.87 g/cm^3 for iron (pure metals at 25°C) [5]. The automotive industry has provided a major impetus for growth in magnesium consumption in recent years. Magnesium and other light weight materials provide a means of counteracting the tendency for vehicle weight increases from the incorporation of new features. Overall vehicle weight reduction benefits both cost and performance through improved fuel economy, performance (handling, acceleration, and braking), and durability. Table 1 gives a weight basis comparison of properties between casting alloys of aluminum and magnesium.

As with other common metals, magnesium can be used in wrought and cast forms. However, the majority is consumed in pressure die castings, while the amount used for wrought applications is less than 5% of that for gravity and pressure die castings. This is due in part to the excellent casting properties of magnesium, which enables the production of complex components with thin walls.

This chapter will focus on cast magnesium alloys, including pressure die casting, gravity die casting, and sand casting, and is divided into five main sections:

- overview of magnesium alloys,
- solidification structure,
- mechanical behavior at room temperature,

Table 1 Yield Strength (YS, MPa), Specific Modulus E/ρ, Specific Strength, YS/ρ, of Aluminum Alloy A356 and Magnesium Alloy AZ91. W_E and W_Y are the Relative Weights of Panels of Given Width and Specified Stiffness or Strength, Respectively, Loaded in Bending. $E_{Al} = 70$ GPa, $E_{Mg} = 44$ GPa, $\rho_{Al} = 2.7$ g/cm^3; $\rho_{Mg} = 1.8$ g/cm^3

Alloy	Condition	YS	E/ρ	YS/ρ	W_E [a]	W_Y [a]
A356	Sand cast, T6	275[b]	25.9	102	1	1
AZ91	Sand cast, T6	155[c]	24.3	86	0.78	0.89
AZ91	Die cast, 2 mm thick	160[d]	24.3	88	0.78	0.88

[a]Calculated using the methods of Ref. 127.
[b]Ref. 125.
[c]Ref. 126.
[d]Ref. 53.

- elevated temperature properties, and
- pressure die casting process.

II. OVERVIEW OF MAGNESIUM ALLOYS

A. Alloy Designation

The American Society for Testing and Materials designation system (ASTM-B275) is the most commonly adopted practice worldwide and is used throughout this chapter. This system uses a three part code with the first part consisting of two letters indicating the major elements, the second part represents the compositions of these elements rounded to whole numbers and the third part is a letter code to differentiate alloys which would otherwise have the same designation. For example, AZ91D contains approximately 9 wt% Al, 1 wt% Zn. The "D" indicates that it is the fourth AZ91 alloy to be registered. The convention for the use of the letters to designate elements is given in Table 2.

B. Mg–Al System

The most widely used magnesium alloys are based on the Mg–Al system. Aluminum has the advantages of being relatively inexpensive and it improves strength and castability. Aluminum is also a low-density element causing only marginal increases to the density of magnesium alloys.

The Mg–Al phase diagram is shown in Fig. 1. Commercial alloys fall in the range of 2–9 wt% Al. At the lower end of this composition range, casting becomes difficult, while beyond 9 wt% Al, the alloys become increasingly brittle due to the high fractions of β-$Mg_{17}Al_{12}$ intermetallic phase [6]. The solidification of magnesium alloys is discussed further in Sec. III.

Increased aluminum content improves the castability and tensile strength, but decreases the ductility and impact strength [7,8]. Figure 2 shows the effects of aluminum content on tensile properties while Fig. 3 shows the effect on impact strength. The factors influencing mechanical properties are discussed in greater detail in Sec. IV. Alloys with higher aluminum content such as AZ91 are often selected for their excellent casting behavior that allows large complex parts with thin walls to be cast. Alloys with lower aluminum content such as AM20, AM50, and AM60 are chosen where ductility and impact energy absorption characteristics are more critical (see Sec. IV.D).

Table 2 Element Code Letters for the Magnesium Alloy Designation System (ASTM-B275)

Code	Element
A	Aluminum
B	Bismuth
C	Copper
D	Cadmium
E	Rare earth elements
F	Iron
H	Thorium
K	Zirconium
L	Lithium
M	Manganese
N	Nickel
Q	Silver
R	Chromium
S	Silicon
T	Tin
W	Yttrium
Y	Antimony
Z	Zinc

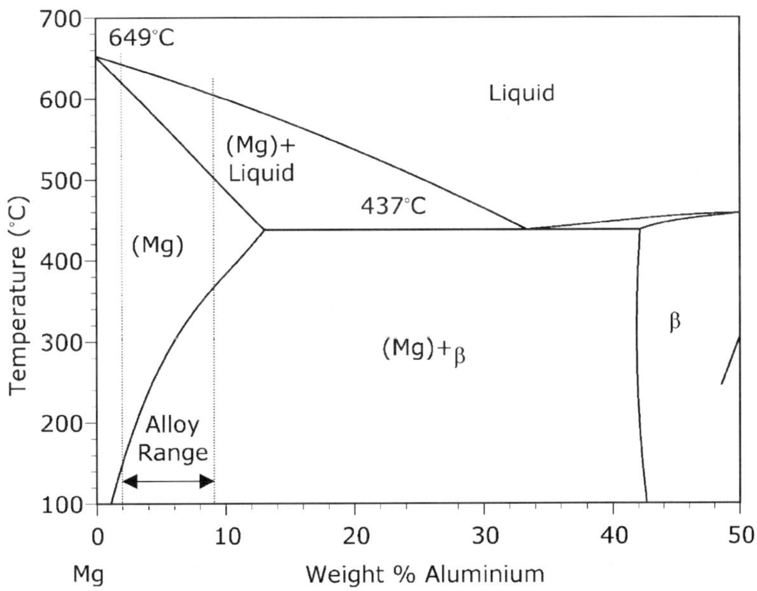

Figure 1 The Mg–Al phase diagram showing the composition range of common commercial magnesium pressure die casting alloys. (From Ref. 128.)

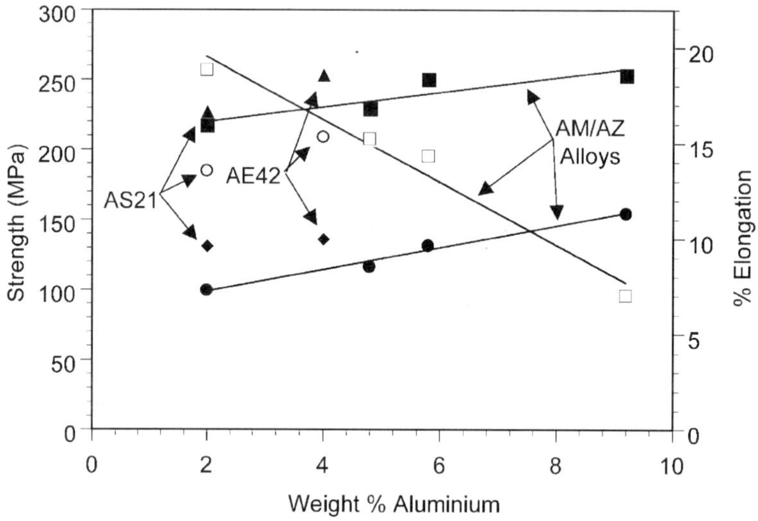

Figure 2 Variation of room temperature tensile properties with aluminum content for die cast AM series alloys and AZ91. Also plotted are data for AS21 and AE42. Solid symbols are tensile and yield strength, open symbols are % elongation. (From Ref. 7.)

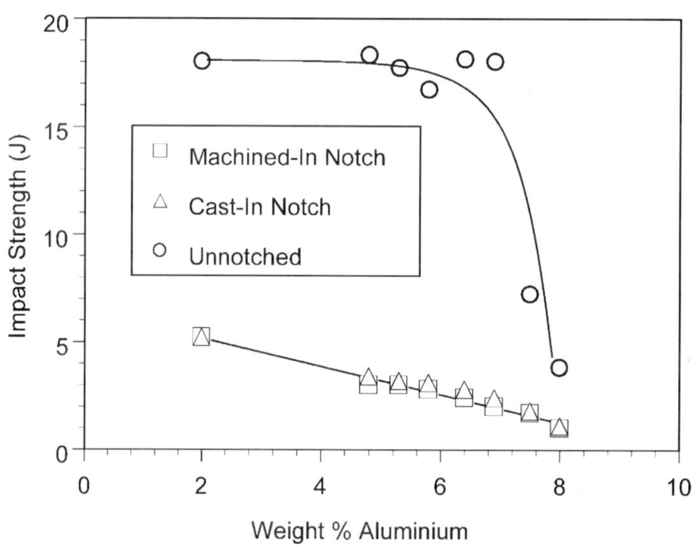

Figure 3 The variation of impact strength of AM series alloys with aluminum content. (From Ref. 8.)

C. Effects of Impurities and Alloying Elements

Certain impurity elements, notably iron, copper, and nickel, have a detrimental effect on corrosion resistance. In alloys where corrosion is critical, these elements must be kept to low levels [9]. For example, the specification for AZ91D sets maxima of 0.005 wt% Fe, 0.002 wt% Ni, and 0.030 wt% Cu. Manganese has been shown to be beneficial as it reduces the adverse effects of iron [9]. Consequently, most magnesium alloys have manganese additions, typically in the range of 0.2–0.4 wt%.

After manganese, the next most commonly added element is zinc. Zinc additions to pressure die cast Mg–Al based alloys, such as AZ91, are typically 0.5–1 wt%. The added zinc improves fluidity and provides some increase to hardness and strength. Pressure die cast alloys with zinc level above about 2 wt% tend to suffer hot cracking [10]; however, it has been reported that there are alloys with higher zinc content that are less prone to hot cracking [10]. This upper boundary for hot cracking varies with aluminum content and is approximated by the following expression (determined from Ref. [10]):

$$\text{wt\% Zn} = 12-3(\text{wt\% Al}) + 0.36(\text{wt\% Al})^2 - 0.015(\text{wt\% Al})^3 \quad (\text{for Al}: 0-9\,\text{wt\%})$$

Despite this, very few pressure die cast alloys have compositions within this upper range.

Alloys for other processes, such as sand and permanent mold castings, forgings, extrusions, and sheet, can have relatively high zinc levels up to about 7 wt%.

Silicon is added to pressure die cast alloys for improved elevated temperature properties with the common alloys AS21 and AS41 being originally developed for use in engine components. The intermetallic compound Mg_2Si has a melting point of 1085°C and forms a eutectic with magnesium at 1.5 wt% Si. These alloys are suitable only for pressure die casting where the fast cooling rates serve to refine the Mg_2Si particles. In sand castings, the Mg_2Si phase is coarse and makes the alloy brittle.

Rare earth elements are used in many alloys to improve elevated temperature creep properties. The AE42 alloy contains 2 wt% mischmetal which is typically 40–45 wt% cerium, 20–25 wt% lanthanum and 15–20 wt% neodymium, with smaller amounts of other rare earths. In alloys containing aluminum, the creep resistance is attributed to the presence of $Al_{11}RE_3$ precipitates at grain boundaries [11]. For sand and permanent mold castings, there are a number of creep-resistant alloys that contain rare earths and are free of aluminum. The characteristics of these alloys are discussed in more detail later in this chapter.

The pressure die casting process tends to produce a fine grain size due to the rapid cooling rates. In gravity and sand castings, where the cooling rates are slower, zirconium is often used to refine the grain size. The typical zirconium levels are about 0.7 wt% and are used only for aluminum free alloys. Zirconium and aluminum combine to form an intermetallic phase, which interferes with the refining effect.

III. SOLIDIFICATION STRUCTURE

The growth of primary α-magnesium has been investigated for AZ91 alloy using the Bridgman solidification techniques [12]. Two dendrite stem growth directions can occur; however, one of these occurs only under conditions of high temperature gradients and low growth velocities. Under conditions relevant to castings, dendrites grow in a $\langle 1\,1\,\bar{2}\,0\rangle$ direction within the basal plane. Side arms also within the basal plane grow readily along other $\langle 1\,1\,\bar{2}\,0\rangle$ directions but side arm growth out of the basal plane is relatively slow, occurring along $\langle 2\,2\,\bar{4}\,5\rangle$ directions. Consequently, the grains form in a platelike manner parallel to the basal plane. It is not clear whether the platelike growth tendency described above is actually observed in castings or whether it plays any significant role in influencing the cast microstructure.

The rapid cooling rates typical of the pressure die casting process result in solidification under non-equilibrium conditions best predicted using the Gulliver–Scheil

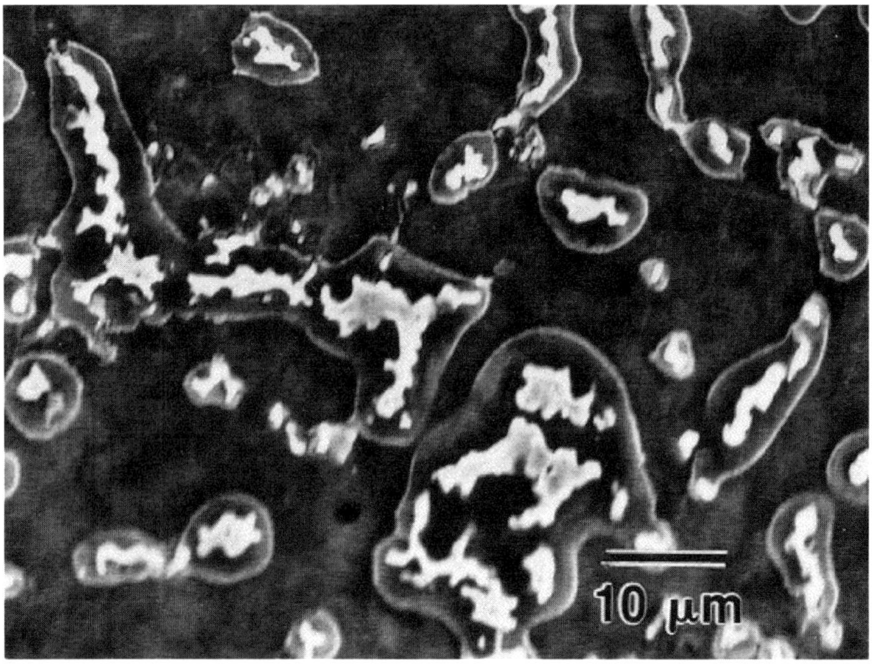

Figure 4 As-cast microstructure of pressure die cast AZ91D. The area immediately surrounding the β-$Mg_{17}Al_{12}$ (bright contrast) is α-Mg solid solution of higher aluminum content. SEM secondary electron image. (From Ref. 72.)

equation. Using this equation, Han et al. [13] have shown that AZ91 contains up to 16% of Mg–$Mg_{17}Al_{12}$ eutectic while even AS21 may contain 2% eutectic. Aluminum tends to be highly segregated within the α-Mg phase and reaches the solubility limit of approximately 12% towards the grain boundaries, as confirmed by measurements of Dargusch et al. [14]. An example of the microstructure of pressure die cast AZ91 is shown in Fig. 4.

The level of zinc has been shown [15] to affect the morphology of the Mg–$Mg_{17}Al_{12}$ eutectic. Alloys of Mg–9 wt% Al form a eutectic with a partially divorced structure, while additions of zinc tend to promote a fully divorced structure [15]. A zinc content of 1.6 wt% is sufficient to complete the transition to a fully divorced eutectic for cooling rates of 80°C/sec, while at lower cooling rates, the transition is less complete at this zinc level.

IV. MECHANICAL BEHAVIOR AT ROOM TEMPERATURE

Sand and permanent mold castings can be heat treated in the same way as aluminum alloys, with some improvement in the mechanical properties in comparison with the as-cast material. In contrast, pressure die castings normally are not heat treated and thus the as-cast microstructure largely determines the final material properties, particularly yield strength and ductility.

This section describes the room temperature mechanical properties of many of the existing magnesium alloys for both pressure die casting and gravity casting processes.

More detailed property information, specifically related to the pressure die casting process, is also given in Sec. VI.

A. Ductility and Strength: Microstructural Aspects

Polycrystalline magnesium shows low ductility. This is the main result of the limited number of slip systems of the HCP structure. Plastic deformation in polycrystalline Mg occurs first by $(0001)\langle 11\bar{2}0\rangle$ basal slip [16,17], which provides only two independent components to the strain tensor. Prismatic and pyramidal slip is observed as well in regions of stress concentration [18,19] such as grain corners or on a very limited scale, e.g., dislocations piled up at particles double cross-slip to another basal plane [20,21]. However, since these slip systems share a common $[2\bar{1}\bar{1}0]\mathbf{a}$ type Burger's vector, none of them can accommodate tensile strain in the **c**-direction [16].

Grain boundary sliding occurs rather readily in pure magnesium at room temperature [18], and it seems to relax localized stress concentrations [19]. Although its contribution to the total deformation in large grain size materials such as sand castings can be expected to be minimal, it may be more significant in pressure die castings, due to their small grain size, and particularly at high service temperatures (see Sec. V).

Twinning occurs readily in magnesium and it has been associated with microcracking during tensile deformation [22,23] and fatigue [24] and with strain-dependent anelastic behavior [19,25].

Work on single crystals of dilute alloys [17,26–29] has shown that many solute species, including aluminum and zinc, increase the critical resolved shear stress for basal slip by an amount that is proportional to $c^{2/3}$, where c is the atomic concentration. Solid solution hardening is not isotropic in magnesium since the same solute species lower the critical resolved shear stress for prismatic slip, presumably by decreasing the Peierls force, in what seems to be a general solid solution effect [17,27]. As a consequence, solute additions may have opposing effects on the yield strength and strain hardening behavior.

Easier prismatic slip with solute in solution has a very important side effect on the behavior of the alloys in comparison with polycrystalline pure magnesium: it limits the stress concentrations at the head of slip bands, increasing the ductility [17].

The flow stress is very dependent on the Schmidt factor due to the anisotropy of the HCP structure and thus, in polycrystalline material, some grains are soft while others are hard with respect to the tensile axis [16,30]. Consequently, magnesium is plastically inhomogeneous and this results in a long-range stress with a wavelength of the order of the grain size in the deformed material [25]. This picture is consistent with results from studies that show the yield strengths of both pure magnesium [19] and AZ91 alloy to vary with grain size [31–35]. In comparison with cubic crystals, a much stronger dependence of the flow stress on grain size is expected for HCP magnesium due to the limited number of slip systems [36]. See Sec. IV.B.3.

Aging of the solution-treated Mg–Al alloys produces continuous (i.e., general) precipitation in the interior of grains and, in alloys with a high degree of supersaturation [37], cellular or discontinuous precipitates at the grain boundaries. The relative proportions of either type of precipitate depend on the aging temperature [38,40], discontinuous precipitation being more prevalent at the intermediate aging temperatures (150–200°C) normally used in current industrial practice. On aging, plates of the equilibrium phase β-$Mg_{17}Al_{12}$ form from the supersaturated solid solution with no intermediate phases or GP zones. Although these precipitates are not sheared by dislocations, Orowan strengthening is limited by the large [25,39] interparticle spacing

(0.2–1 μm). Moreover, the most frequent habit plane of the precipitate plates is the basal plane and the aspect ratio of the precipitates is such that they do not interfere strongly with basal slip [21,39]. The alloys are, therefore, characterized by a mild response to aging [21,39]. Continuous precipitation seems to affect the deformation mode. Clark [21,39] reported a decrease in the amount of $\{10\bar{1}2\}$ twinning in aged material and an increase in prismatic slip near the $Mg_{17}Al_{12}$ plates which gave rise to the formation of tangles of dislocations and accompanying forest hardening.

Precipitates [39] in aged Mg–Zn alloys are long thin rods of a transition lattice of the MgZn phase that forms perpendicular to the basal plane. However, the limited solubility of zinc in magnesium and the relatively large size of the precipitates prevent efficient blocking of dislocations and the aging response of Mg–Zn alloys is also weak.

Despite relatively mild precipitation hardening in Mg–Al alloys, the large solubility of aluminum at high temperatures allows for a large volume fraction of precipitates to form after aging at low temperatures. Thus, up to 15–20% volume fraction of β-$Mg_{17}Al_{12}$ particles can be expected to form [39,41] in alloy AZ91–T6. This large volume fraction of particles produces a measurable dispersion strengthening [25] and has also some effects on the elastic modulus of the material [42]. See Sec. IV.B.2. For a detailed discussion of the strengthening mechanisms, see Ref. [43].

Yielding is very gradual and plastic deformation that can be detected at very low stresses in magnesium and its alloys. However, since gross plastic deformation [44] does not occur until prismatic slip is activated, at a much higher stress, a stage of high strain hardening rate develops right after the onset of plastic strain in the basal planes. This makes the determination of the yield strength using the usual 0.2% offset method more difficult [16]. An additional complication arises from the early activation of twinning, which tends to alter the elastic slope of the stress–strain curve. An alternative approach is to define the yield strength as the intersection between the flow curve and a reference line drawn at an offset strain of 0.2% with the slope of the nominal value of the elastic modulus as determined by ultrasound [8,45,46]. See Sec. IV.B.1.

The mechanical properties of magnesium gravity casting alloys are influenced by process-related variables such as porosity [47], solidification rate [31], and second phase particles formed during the solidification as well as by the post-casting heat treatment [48]. For example, Couture and Meier [31] studied the effect of section thickness on the mechanical properties of chilled sand castings of AZ80, AZ91, and AZ92 alloys in T4 and T6 condition. The strength and ductility increased with decreasing section thickness and this was associated with the faster cooling rates of the thinner sections, which, in turn, refined the grain size and reduced the porosity.

Both aluminum and zinc are known to have a grain refining effect in gravity cast Mg–Al, Mg–Al–Zn and Mg–Zn alloys, increasing the ductility [49–51], although this may be more than offset by the formation of brittle intermetallic phases at higher aluminum and zinc levels [43,49–52]. Mg–Zn alloys are also amenable to strong grain refinement by inoculation with zirconium [3,49,51]. Unfortunately, zirconium tends to combine with aluminum and thus its grain refining effects are lost in AM and AZ series alloys. This grain-refinement effect has led to the development of several series of alloys [3] that combine the grain refining ability of zirconium with zinc and other solutes that confer room and high temperature strength.

Data from the literature will now be presented that illustrate some of these effects in greater detail. Whenever possible, gravity cast and pressure die cast materials are compared on a common base. The aim is to help the engineer to select the magnesium alloys that are best suited for a particular application.

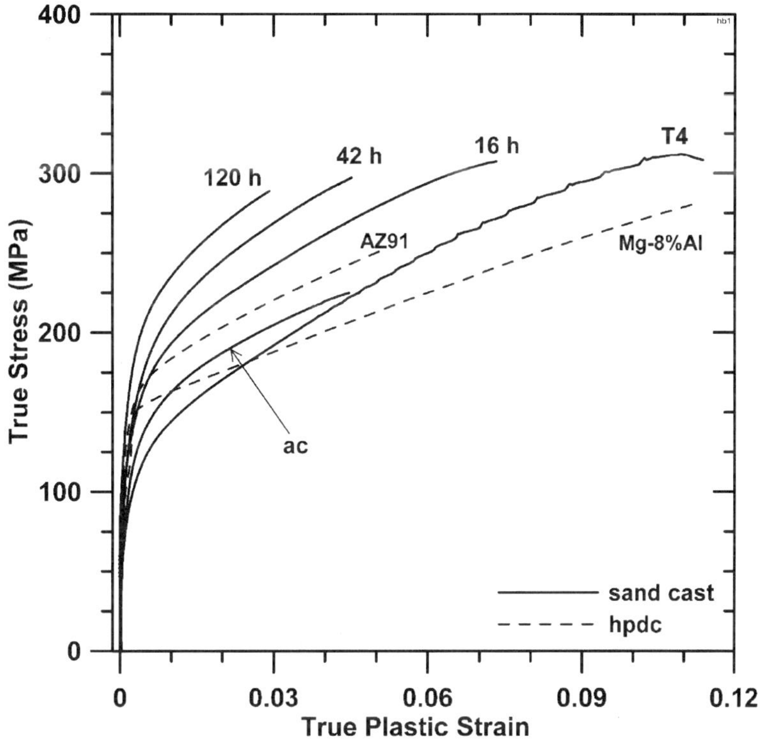

Figure 5 The solid lines are flow curves of sand-cast AZ91 alloy, aged [56] as cast (ac), as quenched (T4), and aged for different times (hour) at 165°C. The dashed lines represent pressure die castings and have been replotted from Ref. 53 (AZ91) and Ref. 129 (Mg–8% Al), respectively.

B. Yield Strength and Strain Hardening Behavior

Figures 5, 6, 7 show several examples of stress–strain curves of gravity and pressure die cast alloys. Figure 5 shows the effect of solution heat treatment and aging of AZ91 on the flow behavior. Aging increases the yield strength but has little effect on the tensile strength. Note that the departure from linear elastic deformation occurs at very low stress levels even in the aged samples, indicating that the elastic limit is not greatly increased by the aging. Figure 5 also includes two flow curves of pressure die cast alloys to show the differences in strain hardening behavior.

Figure 6 shows the different effects of aluminum and zinc additions. At comparable at.% content, zinc increases the yield strength significantly more than aluminum, but at the same time, the strain hardening rate of the alloy containing aluminum is larger, a result of the trade off between hardening of the basal plane and softening of the prismatic planes. Both effects are stronger in Mg–Zn alloys [17,26–28].

Figure 7 shows the flow curves of a wide range of pressure die cast alloys. In general, pressure die cast materials show a relatively well-defined yield strength and a near linear strain hardening. In comparison with sand-cast alloys, pressure die cast materials exhibit lower strain hardening than the sand-cast counterparts as shown in Fig. 5. These differences can be ascribed to the finer grain size which increases the yield strength of pressure die castings, and the lack of small-scale precipitation in the pressure die cast material which results in low strain hardening rate [43].

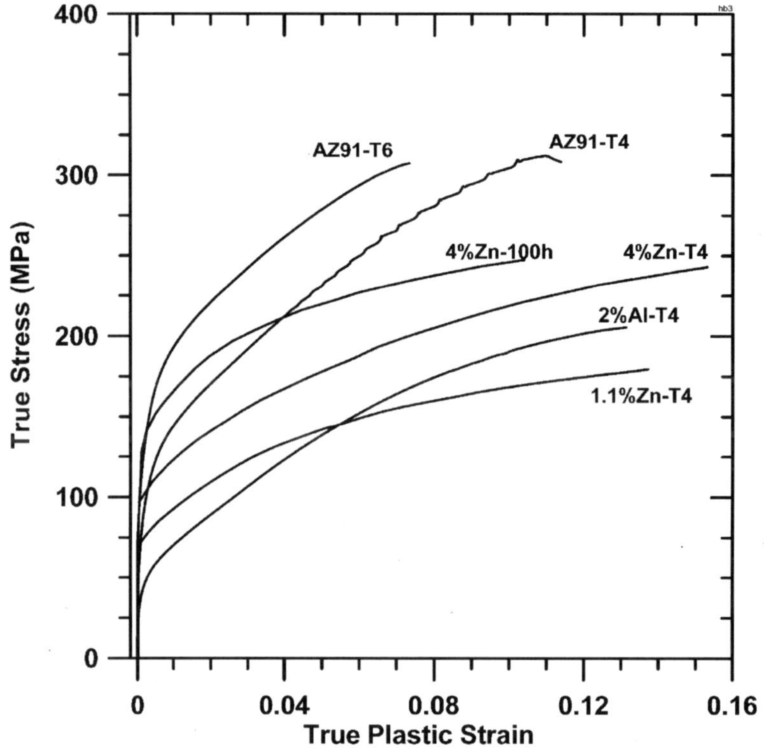

Figure 6 Flow curves of an Mg–4% Zn alloy in T4 temper and aged for 100 hr at 150°C [130], an Mg–1.1% Zn alloy [17] an AZ91–T6 alloy [131] and a binary Mg–2% Al–T4 alloy [56].

1. The Measurement of Yield Strength

Two methods of measuring the (0.2%) proof stress are commonly applied to magnesium alloys. These are illustrated in Fig. 8. One is based on the apparent elastic modulus of the tensile specimen determined as the slope of the first portion of the flow curve, and using this value to locate the 0.2% plastic strain. This is the method normally applied by the software of most tensile testing machines, and is called here method "S". The other method, suggested by Aune et al. [8] and Carbonneau et al. [45], assumes a nominal value for the elastic modulus of, for instance, 44 GPa (method N) as determined by dynamic methods. The results obtained by the two methods have been compared on several occasions, and both produce similar average values [43,45], although the dispersion of values determined by the N method can be slightly larger.

The reasons for the larger scatter of values from the method N are shown in Fig. 8, which illustrates the possible situations encountered with samples whose apparent elastic moduli differ from the "nominal" value of 44 GPa. The intersection of the flow curves with the solid line drawn at an offset strain of 0.002 defines the yield strength according to the method N. The intersection with the small segments, drawn also at an offset strain of 0.002 but with the apparent E-values determined as the slope of the flow curve, defines the proof stress as measured by the method S. It is seen that the method S tends to "compensate" for the departure from the theoretical value, under-reading the yield strength for high apparent E and over-reading it for lower apparent E. The net result is a narrower span of yield strength values in comparison with the method N.

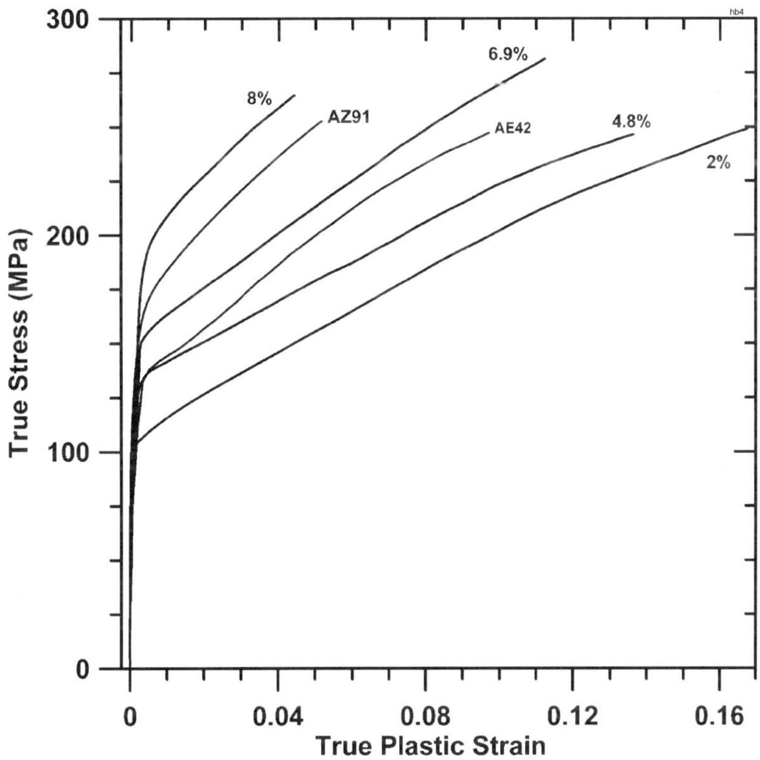

Figure 7 True stress–true plastic strain flow curves of a variety of die cast alloys (AM alloys replotted from Ref. 129, the Al content is indicated as %). AZ91 alloy from Ref. 132. and alloy AE42 from Ref. 133.

2. Elastic Modulus

Experimental values of the elastic modulus of magnesium alloys, determined with ultrasound methods, lie between 39 and 46 GPa, with lower values for pressure die cast material, as shown by Fig. 9. Possible reasons for the large scatter of values are differences in porosity, especially in pressure die cast material, anisotropy in the case of large grained sand-cast material, varying content of second phase particles, and solid solution effects. The most significant parameters are second phase content and porosity. A detailed discussion regarding alloy AZ91 is found elsewhere [42].

The volume fraction of second phases in Mg–Al may differ for a number of reasons. In as-cast material, the volume fraction may approach 5% (pressure die castings [53]) or 7% (sand castings [42]). The elastic modulus of the β-phase particles [43,54] (58 GPa) is only slightly larger than that of magnesium, but at 20% volume fraction, it is enough to increase E by approximately 2 GPa. Application of the rule of mixtures [55] is illustrated in Fig. 9. Controlled aging experiments on a single bar of sand cast AZ91 seem to conform to the rule of mixtures with increasing β-phase content.

A theoretical estimate of the effect of porosity, which in pressure die castings may vary between 1% and 5%, on the E-value (dashed line) is also shown in Fig. 9. It is seen that the combined effect of porosity and volume fraction of second phases, although not completely systematic, may account for a large share of the scatter of data.

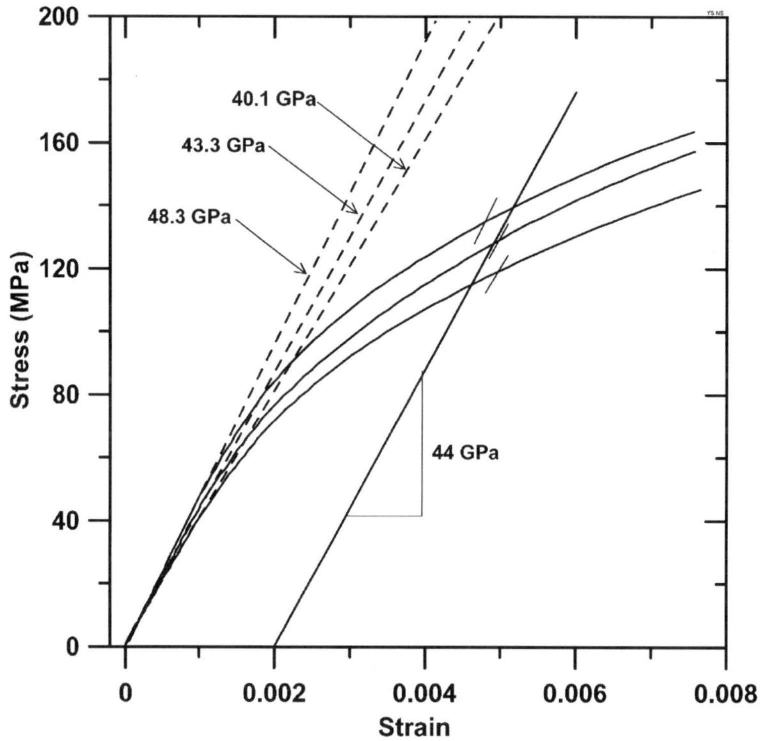

Figure 8 Experimental flow curves illustrating the two methods of measuring the proof stress. The dashed lines correspond to the modulus as determined by the testing machine's software (method S). The solid line represents a nominal elastic modulus of 44 GPa (method N). The short segments are parallel to the corresponding dashed lines. Notice that the method N results in a wider span of proof strength values for the three stress–strain curves.

3. Grain Size and Yield Strength

Figure 10 compares the grain size and yield stress of the material in T4 and T6 tempers. Although the σ_o and k-values vary widely [43], the individual datasets generally conform to the Hall–Petch relationship,

$$\sigma = \sigma_o + kd^{-1/2} \tag{1}$$

where σ_o and k are material parameters and d is the grain size. For cast AZ91 alloy, an average $k = 370$ MPa $\mu m^{-1/2}$ has been suggested [43].

4. Solid Solution and Aging Effects on Yield Strength and Hardness

Figure 11 shows the effect on the yield strength of zinc and aluminum in solid solution in Mg–Zn and Mg–Al alloys (T4 temper). Zinc has approximately two to three times the effect of aluminum on an at.% basis. Figure 12 shows similar data regarding the effect on hardness. It must be borne in mind that since both aluminum and zinc act as grain refiners, the observed strengthening effects are actually a combination of solid solution and grain size strengthening. See Ref. [56] for a quantitative discussion of both effects in Mg–Al alloys.

Figure 13 shows that while in general terms the hardness of alloy AZ91 scales linearly with strength in aged sand-cast material, in pressure die cast materials, the hardness is

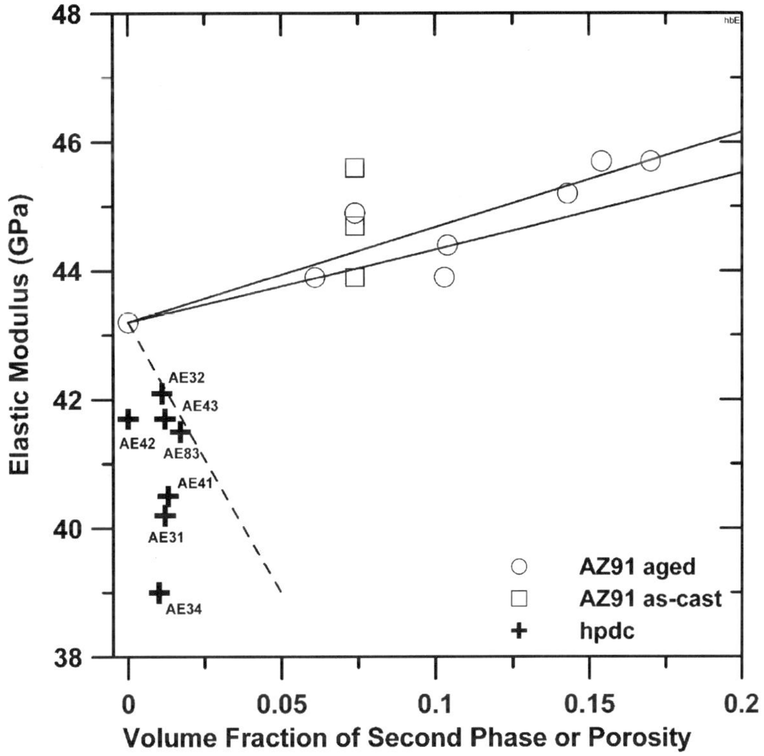

Figure 9 Elastic modulus as a function of the volume fraction of second phase for sand-cast AZ91 alloy and as a function of the volume fraction of porosity for several die cast alloys. The solid lines are upper and lower bounds (rule of mixtures [55]). The dashed line represents the effect of porosity on the elastic modulus according to Mackenzie's equation [134]. The circles are aged AZ91 from Ref. 42. Squares are as-cast AZ91 from Ref. 45 and the crosses, from Ref. 135, are AE alloys.

consistently less than what can be expected from the yield strength, possibly due to a sub-surface porosity layer.

C. Strength–Ductility Behavior

In the following sections, strength–ductility charts [57] are used to assess mechanical performance of sand casting and pressure die casting alloys. The method to develop the charts is described in Appendix A.

1. Sand-Cast Mg–Al Alloys

Data showing the effect of heat treatment on the strength and ductility of sand-cast AZ91 and AM100 alloys have been plotted in Fig. 14. With reference to alloy AZ91, it is seen that the solution heat treatment (T4 temper) increases the strength and ductility of the alloy, but subsequent to aging, while increasing the yield strength, decreases the ductility at constant tensile strength.

The chart of Fig. 15 shows the effect of aluminum content on the strength and ductility of binary Mg–Al alloys (data from Lagowski and Meier [52]). At low aluminum content, the ductility increases monotonically up to ≈4 wt% Al. At higher

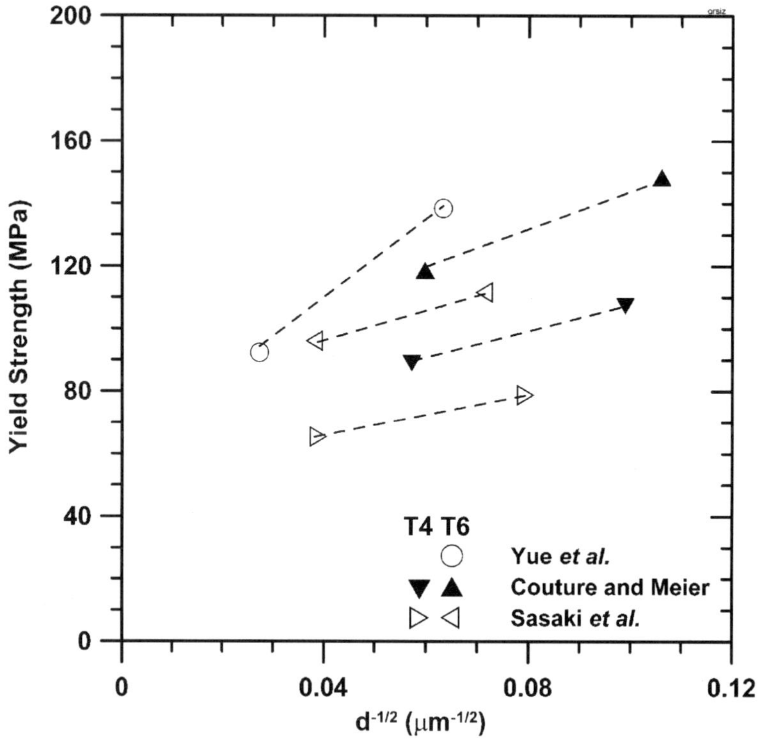

Figure 10 Yield strength as a function of the grain size of sand-cast alloy AZ91 in T4 and T6 temper. The dashed lines conform to Eq. (1), with k-values ranging between 324 and 1253 MPa $\mu m^{-1/2}$. (From Refs. 31, 33, and 34.)

concentrations of aluminum, the strength keeps increasing but the ductility decreases, especially in the as-cast material, due to the formation of brittle $Mg_{17}Al_{12}$ particles. The latter is especially important for aluminum content above 7 wt%. The shaded region shows the effect of changing the chemical composition of alloy AZ91–T6 within the nominal limits [52].

Figure 16 compares plates and bars of AZ91 and AZ63 alloys, cast with and without chills. In the case of unchilled bars, the loss of ductility and strength is due to the deleterious influence of increased shrinkage [31]. In the case of chilled plates of AZ91, the variation in ductility was mostly due to undissolved $Mg_{17}Al_{12}$ particles in the heat-treated plates [43]. The AZ63 alloy is particularly prone to porosity even in a chilled casting [58], as illustrated by the decreased ductility of a 50 mm thick plate as the distance from the chill end is increased. Note that the individual sets of data points in Fig. 16 tend to lie approximately on a single flow line, which means that although the shrinkage defects decrease both the strength and ductility, they do so without significantly affecting the yield strength or the strain hardening rate.

2. Sand-Cast Mg–Zn Alloys

The dashed lines in Fig. 17 illustrate the effect of zinc content on Mg–Zn–0.04% Ti alloys in T4 and T6 tempers. The ductility of the alloy in the T4 temper increases with the zinc content, up to about 7.5%. For higher zinc content, the ductility decreases due to the

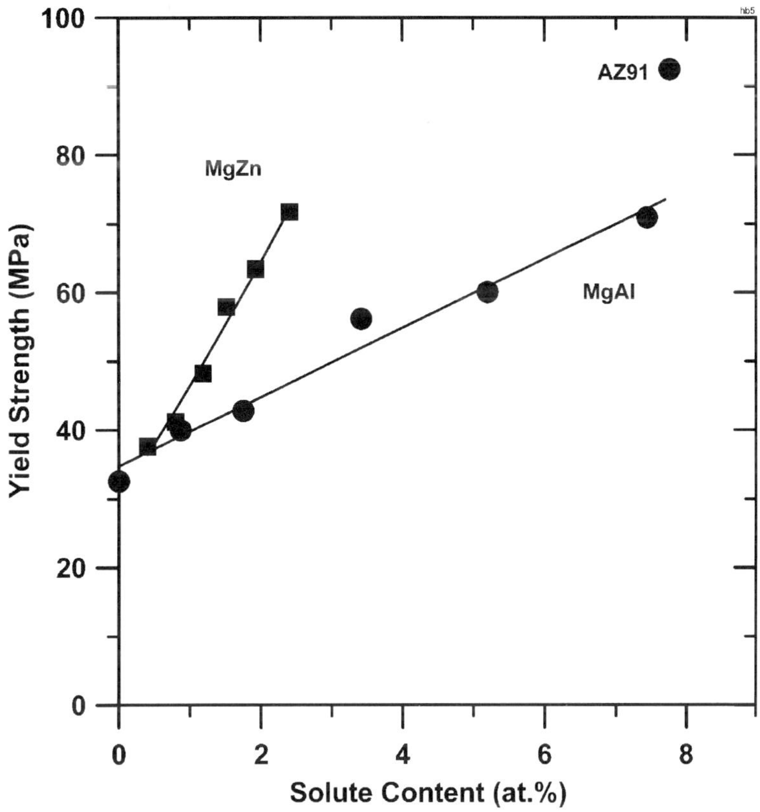

Figure 11 Solid solution effects on binary Mg–Zn and Mg–Al alloys (alloys in T4 temper). Data replotted from Ref. 136 (Mg–Zn) and Ref. 56 (Mg–Al).

lower temperature of solution heat treatment imposed by the solidus line in the Mg–Zn phase diagram, which results in the incomplete dissolution of the Mg_7Zn_3 particles. Upon aging to T6, the strength substantially increases, particularly for the higher zinc content.

Figure 17 also illustrates the effect of grain refining by titanium and zirconium on Mg–Zn alloys in T6 condition. The solid line represents binary Mg–Zn alloys (i.e., without grain refiners) and the dot–dash line alloys with 0.7% Zr, both with zinc content of up to 10%, while the dashed line represents alloys grain refined with 0.04% Ti. Titanium has a limited grain refining effect, which slightly increases the strength and ductility in comparison with the unrefined alloys. The strong grain refining effect of zirconium greatly increases both the strength and ductility of the alloys.

The beneficial effect of zirconium additions is the basis of the current high strength commercial alloys based on the Mg–Zn system. For comparison, data points of the high strength ZE63 (Mg–Zn–Rare Earth), ZQ71 (Mg–Zn–Ag–Zr) and ZK61 (Mg–Zn–Zr) alloys are also included in Fig. 17. A range of other zirconium refined alloys have been developed [3] containing rare earths (EZ33, ZE41), silver (QE22), silver and copper (EQ21), thorium (HZ32) and yttrium (WE54, WE43), aimed at increased strength at both room and elevated temperatures (see Sec. V.B). Figure 17 compares the respective performances.

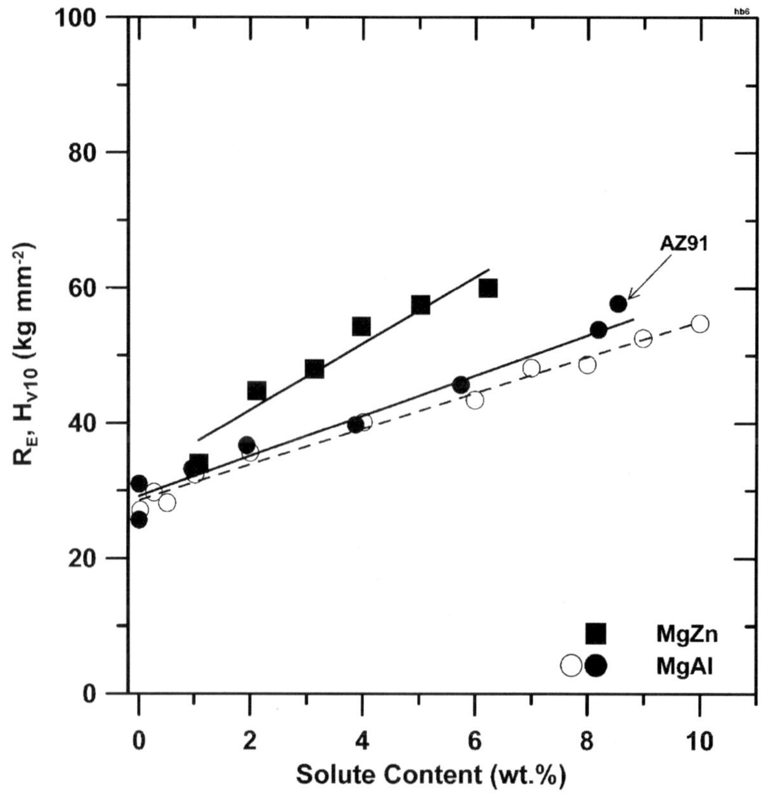

Figure 12 Solid solution effects on hardness on binary Mg–Zn and Mg–Al alloys in T4 temper. Hardness numbers for Mg–Zn alloys are [136] Rockwell E, while for Mg–Al alloys are Vickers. Data replotted from Ref. 41 (open circles) and Ref. 56 (solid circles).

3. Pressure Die Casting Alloys

Figure 18 is used to compare the performance of a number of commercial pressure die casting alloys. Generally, the room temperature properties of pressure die cast alloys fall in a more narrow range than for gravity cast alloys. Note with reference to the AM alloys that the alloys with higher aluminum content, 7.5% and 8%, are notably less ductile than the rest. This loss in ductility at high aluminum content can be related to differences in volume fraction of β-$Mg_{17}Al_{12}$ particles present in the as-cast structure [53,59].

In Fig. 19, the effect of different cross-section thicknesses on the strength and ductility of a range of alloys is compared [53,60–62]. In general, increased section thickness results in decreased strength and ductility, although reports of effects in the opposite direction also exist (Fig. 37). This suggests that pressure die casting parameters, other than section thickness, may have a dominant effect on the ductility and tensile strength [63], most likely porosity [62,64], and non-metallic inclusion content [5] (see also Sec. VI.B).

The properties of vacuum die cast AM alloys are compared in Fig. 20 with those of normally vented pressure die cast specimens. The normally vented specimens (crosses) outperformed the vacuum cast specimens for all aluminum content in these sets of data, which shows, again, that the final properties of pressure die cast material depend on a multiplicity of factors.

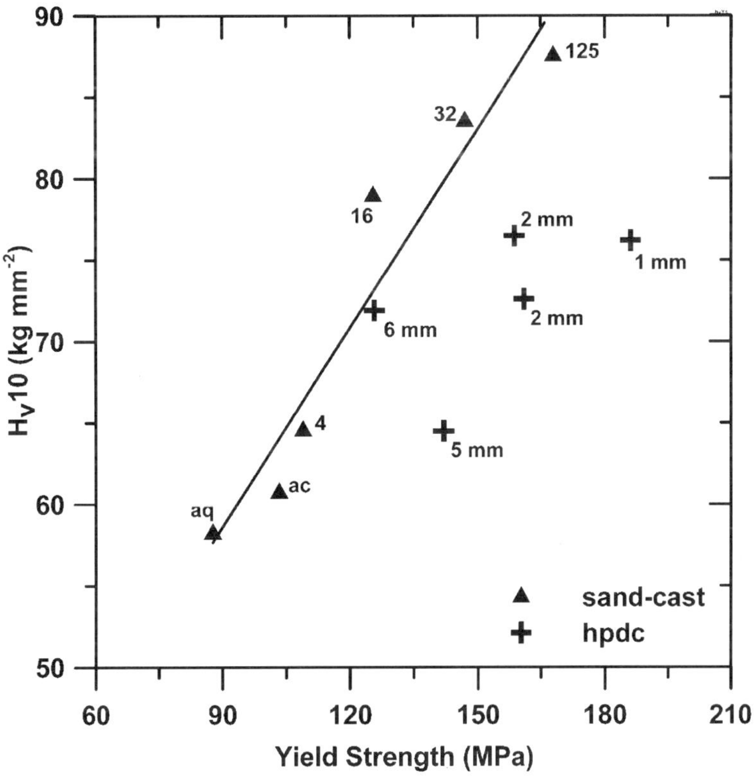

Figure 13 Hardness as a function of yield strength for sand and die cast AZ91 alloy. The line of best fit applies to aged AZ91 data points only. The numbers next to the triangles indicate the aging time in hour at 165°C; ac, aq, indicate as-cast and as-quenched, respectively. Numbers in mm indicate the wall thickness of die castings. Data for sand-cast AZ91 from Ref. 43. Die casting data from Refs. 53 and 132.

Figure 20 also shows the effect of porosity [65] on the strength and ductility of AM50, AZ91, AS41, and AE42 alloys. The arrow indicates the effect of increased area fraction of porosity and oxide films as measured on the fracture surface of thin walled (2–3.5 mm) AM60 castings [64]. It is noteworthy that the sets of data points showing the effect of porosity approximately follow a single flow line, which indicates that although increasing levels of porosity generally decrease the strength and ductility, the strain hardening rate and the yield strength [65] of the material are not strongly affected, as already concluded for sand castings from Fig. 16. In alloy AM20, a decrease in strength and ductility was observed when the gate velocity was increased from 50 to 80 m/sec on specimens 12 mm thick, but an opposite and much smaller effect was observed in alloy AM50 [63].

Figure 21 compares the effect of low temperature aging for alloys AZ91 and several AM and AE alloys. Low temperature aging generally increases the yield strength and decreases the ductility, although not all alloys conform to this behavior [59]. In the case of alloys AZ91 and AM60, low temperature aging increases the volume fraction of β-$Mg_{17}Al_{12}$ particles from about 1.9% and 5% in the as-cast condition to about 7% and 17%, respectively, this being the main reason for the loss of ductility as aging progresses [53].

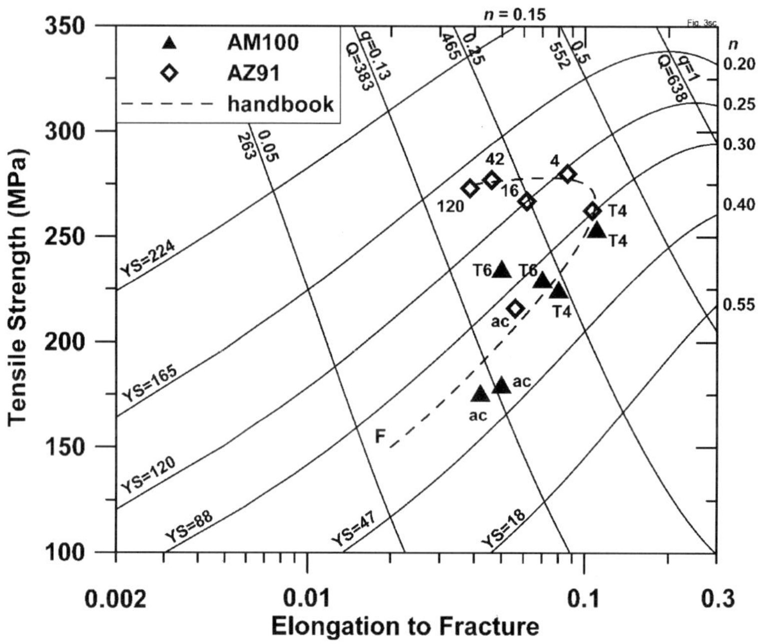

Figure 14 The effect of heat treatment on the ductility and strength of sand-cast AZ91 [43] and AM100 alloys [48]. Labels: ac = as cast, T4 = as quenched; T6 = peak aged. The numbers next to the diamonds indicate aging time (hour) at 165°C. The dashed line indicates the range of values specified by Avedesian and Baker [5] for AZ91 alloy, as cast (F), T4 (maximum ductility) and T6 (maximum strength).

D. Tensile Behavior and Energy Absorption

The experimental data in Fig. 22 show that the binary Mg–Al alloys with aluminum content of the order of 4–7% and the lower strength varieties of sand-cast alloy AZ91 offer the best energy absorption characteristics. The effect of deforming at high strain rate to simulate impact loading [66] is also shown in Fig. 22 for alloy AZ91 in different tempers. Figure 23, in turn, shows that most commercial pressure die cast alloys, within a factor of 2, absorb the same amount of energy regardless of their strength. In the case of AM alloys (crosses), the energy absorption is independent of the aluminum content except for the two higher concentrations (7.5% and 8%), for which the energy absorption falls by a factor of about 2 and 4, respectively. Actual impact strength data collected by Aune et al. [8] on unnotched bars of the same alloys showed that indeed this is the case (Fig. 3).

V. ELEVATED TEMPERATURE PROPERTIES

Previous sections have focused primarily on room temperature mechanical properties. The behavior of magnesium alloys at elevated temperatures is a significant issue for some current and proposed magnesium applications such as automotive transmission housings and engines. The properties that are an issue for these applications are creep deformation behavior and elevated temperature mechanical properties.

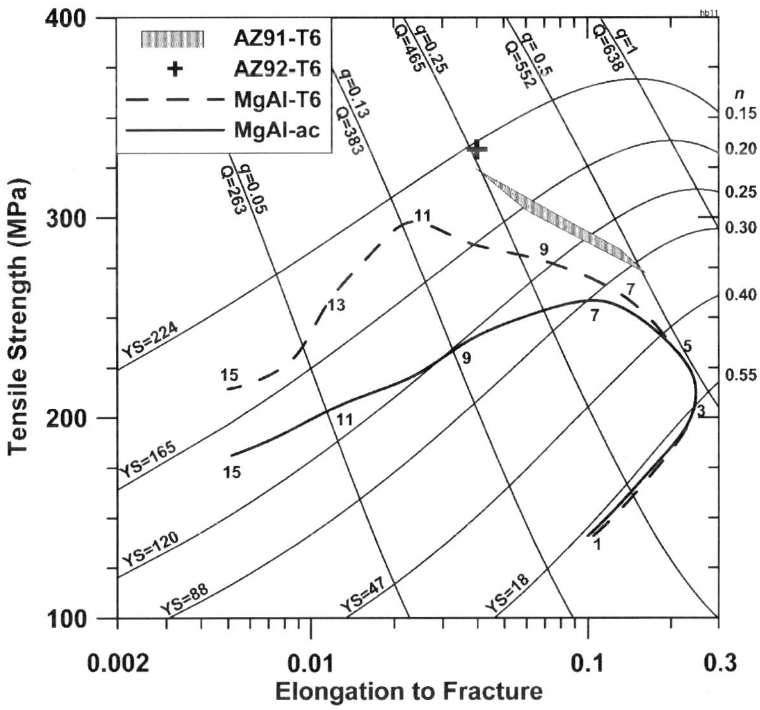

Figure 15 The effect of aluminum content on strength and ductility of sand-cast Mg–Al alloys. The solid and dashed lines represent binary Mg–Al alloys in the as-cast (ac) and T6 conditions, respectively. Labels indicate the Al content (%). The shaded region indicates the effect of varying the Al and Zn content within the nominal minimum and maximum content for alloy AZ91–T6. Experimental data from Ref. 52.

Creep deformation generally becomes significant above 0.3Tm, which for magnesium alloys is approximately room temperature, and consequently, magnesium alloys are prone to excessive creep deformation at slightly elevated temperatures ($>100°C$). At service temperatures common in automotive engine components ($\sim 150°C$), the common pressure die cast magnesium alloys, for example AM60 and AZ91, undergo significant deformation and are consequently unsuitable for such applications. Whilst there has been some success in developing creep-resistant sand-cast magnesium alloys using relatively expensive alloying elements, such as rare earth elements, a low cost creep-resistant alloy is still elusive despite some significant recent advances (see Sec. V.C).

The following section reviews the current magnesium alloy systems and attempts to relate the creep performance of the alloy with its microstructure for both sand cast and pressure die cast materials. More recently developed experimental and semi-commercial alloys are also discussed.

A. Creep Properties of Pressure Die Cast Magnesium Alloys

1. Mg–Al(–Zn). Alloys

The alloys AM60 and AZ91 exhibit relatively high yield strengths even at temperatures up to 150°C. The yield strengths of these alloys are generally higher, or at least similar to

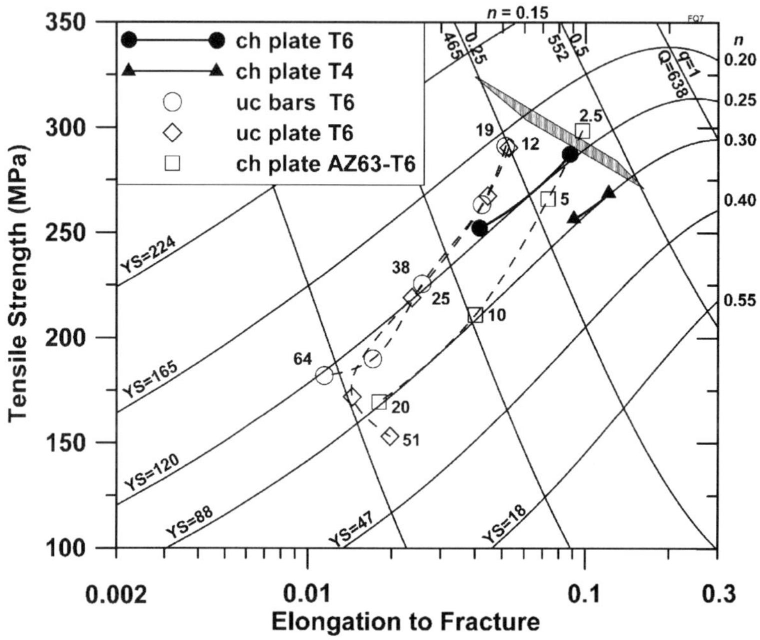

Figure 16 The effect of chills and section thickness on the strength and ductility of chilled (ch) and unchilled (uc) plates and bars of AZ91 and AZ63 alloys. The labels indicate the diameter or thickness (mm, open circles and diamonds), or the distance from the chill end (cm, squares). The solid symbols are from Ref. 43, the squares from Ref. 58 and the circles and diamonds from Ref. 31. The shaded area is alloy AZ91 as in Fig. 15.

those of AS21 and AE42 as shown in Fig. 24. However, the creep performances of AM60 and AZ91 are markedly inferior to AS21 and AE42 as shown in Figs. 25 and 26. Whilst adding aluminum to magnesium increases the room temperature yield strength (Fig. 2), the creep resistance decreases (Fig. 25).

Most current interpretations attribute the excessive creep deformation of the Mg–Al based pressure die casting alloys to the discontinuous precipitation of the equilibrium $Mg_{17}Al_{12}$ phase at grain boundaries [11,14,67,68] increasing the propensity for grain boundary sliding. It has been suggested that softening of particles of $Mg_{17}Al_{12}$ [69] during the creep process leads to excessive grain boundary sliding. However, $Mg_{17}Al_{12}$ retains its hardness up to 275°C [70] rendering this explanation unlikely [11]. In such interpretations, grain boundary sliding has been regarded as a key factor contributing to excessive creep deformation of the alloys [11,14,67,68,71]. Only moderate improvements in creep resistance are obtained in alloys containing reduced volume fractions of discontinuous precipitates of the $Mg_{17}Al_{12}$ phase.

2. Mg–Al–Si Alloys

AS21 and AS41 were used by Volkswagen in the late 1960s and early 1970s, in parts such as the crank case and transmission housing [68]. AS41 has superior creep resistance compared with AZ91, but as engine performance increased, the higher temperature meant that the improved creep resistance of AS21 was required for these components. However, AS21 is more difficult to cast compared to AS41.

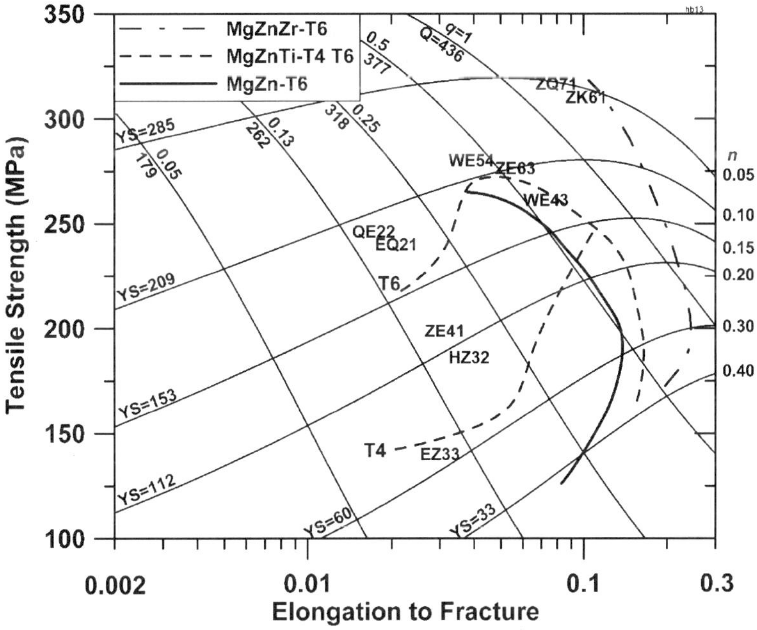

Figure 17 A strength–ductility chart for Mg–Zn alloys. The solid line represents binary Mg–Zn alloys for several zinc content. The dashed lines are Mg–Zn alloys grain refined with 0.04 wt% Ti and the dot–dash line represents Mg–Zn grain refined with 0.7 wt% Zr. Data for several commercial zirconium containing alloys (as indicated by the position of the labels) are included for comparison. From Refs. 136 (Mg–Zn–Ti), 50 (Mg–Zn, Mg–Zn–Zr), 137 (ZQ71), 5 (ZE63 and ZK61) and 3 (EZ, EQ, QE, HZ and WE alloys).

Comparisons of creep behavior of AS and corresponding AM pressure die cast alloys are provided in Figs. 25 and 26. It is apparent that the creep resistance of ternary AS alloys is better than AM alloys with similar aluminum content. The improved performance of the AS alloys is due to a combination of relatively low aluminum content and the presence of silicon. Dargusch [72] found that the main difference in microstructure between AM20 and AS21 was the presence of the high melting point Mg_2Si phase. An example of the microstructure of AS21 is shown in Fig. 27. The improved creep properties are likely due to the presence of Mg_2Si dispersoids on the grain boundaries.

3. Mg–Al–Rare Earth Alloys

Under similar testing conditions, the rare earth containing alloy AE42 has a creep resistance that is superior to that of AS alloys (Figs. 25 and 26). Rare earths have a stronger affinity with aluminum than magnesium and form an $Al_{11}RE_3$ phase, with a branched morphology that straddles the grain boundaries [73]. An example of the microstructure of AE42 is shown in Fig. 28. An Al_2RE phase can also form with the amount relative to $Al_{11}RE_3$ being dependent upon the proportions of the individual rare earths [74].

Above 150°C the creep properties of AE42 decrease significantly due to a transformation of $Al_{11}RE_3$ to Al_2RE and/or the appearance of $Mg_{17}Al_{12}$ during aging [74]. This suggests that the improved creep resistance of AE42 is due to the presence of the $Al_{11}RE_3$ grain boundary phase, which seems to be particularly successful in hindering grain boundary sliding.

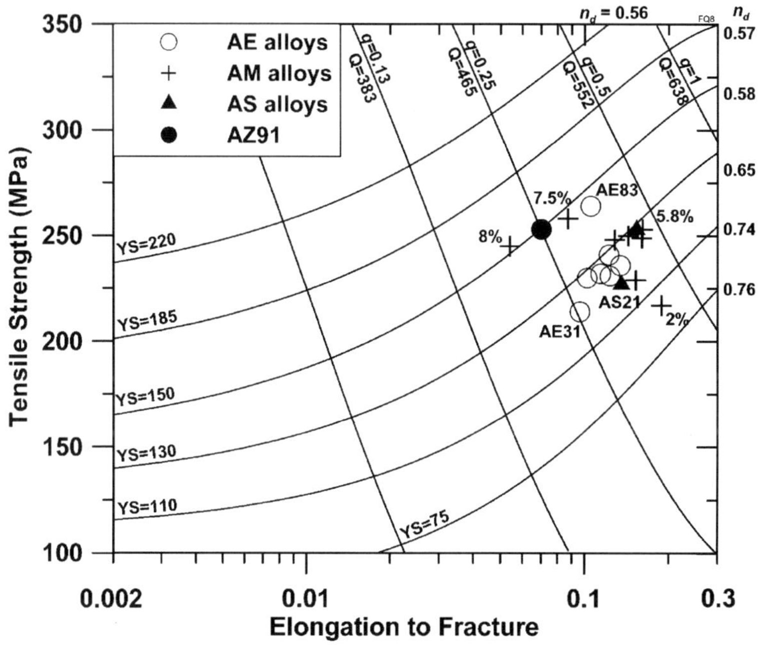

Figure 18 A strength–ductility chart for die castings. The crosses represent AM alloys with Al content between 2% and 8%. The solid circle is alloy AZ91 while the open circles are, in increasing order of strength, alloys AE31, 32, 42, 41, 34, 43 and 83, respectively. The solid triangles are AS21 and AS41 alloys. Data points from Refs. 129, 135, and 138 obtained with cylindrical specimens, 6.4 mm in diameter.

B. Properties of Sand-Cast Creep-Resistant Alloys

Higher strength and creep-resistant magnesium alloys have been developed for sand-cast components. The alloying additions are usually expensive elements, such as thorium, rare earths, and silver, and the alloys are difficult to pressure die cast. Consequently, these alloys are only used for specialty components. The most creep-resistant magnesium alloy is thorium containing HZ22 [75]. However, due to the radioactivity of thorium, this alloy is rarely used.

Rare earth additions to magnesium alloys improve the creep resistance significantly, with neodymium being particularly effective [76]. However, the room temperature tensile properties of magnesium–rare earth alloys are relatively poor [51]. The earlier alloys developed contained zinc additions, e.g., ZE41 and EZ33, to improve the strength, castability, and melt handling of these alloys. Higher zinc, lower rare earth alloy has higher strength and improved castability, while the lower zinc and high rare earth containing alloy has improved creep resistance [51]. Zirconium additions are made to refine the grain size and improve the tensile properties of these alloys.

Silver-containing alloys, such as QE22, were found to have substantially improved elevated temperature tensile properties (Fig. 29). The alloy QE22 can be used in service in the range 200–250°C in the T6 condition.

While the silver-containing alloys perform well at elevated temperatures, the yttrium-containing alloys WE54 and WE43 are generally superior. Consequently, these alloys have

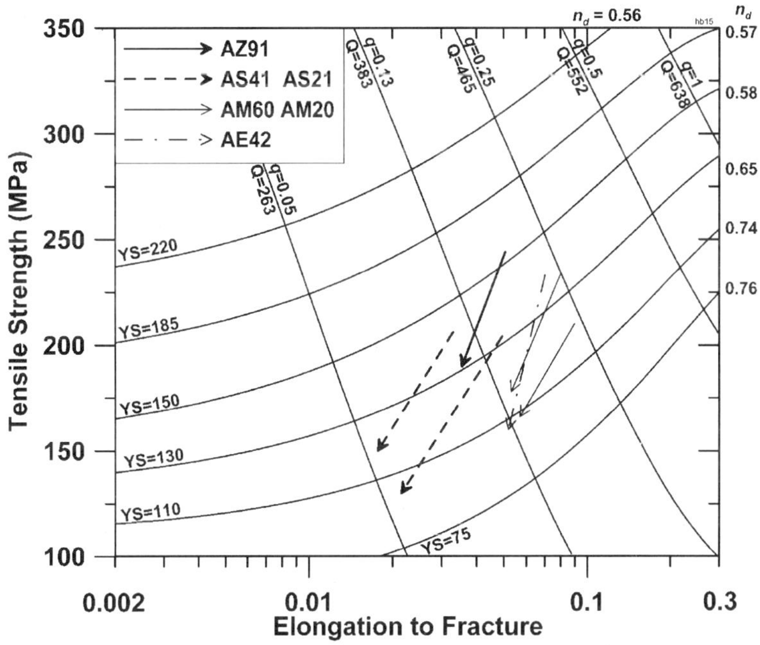

Figure 19 The effect of cross-section thickness on the strength and ductility of several commercial die casting alloys. The arrows indicate the trend line for increased thickness from 2.5 to 10 mm. (From Refs. 62 and 139.)

become the most commonly used creep-resistant magnesium alloys in high value applications. Of these two alloys, WE54 has higher strength, while WE43 has better ductility, slightly improved castability, and improved long-term stability [77]. WE54 is used in aircraft and racing car engine components, while WE43 has been used in helicopter transmission components.

The WE alloys show substantial precipitation hardening and it has been found that the strengthening phase contains magnesium, neodymium, and yttrium [78] and forms on prismatic planes providing impediment to dislocation motion on the basal planes.

C. New Alloy Systems for Magnesium Creep-Resistant Alloys

Several new alloys are currently under development or have been recently commercialized. This section discusses these recent developments.

1. Mg–Zn–Al and Mg–Zn–Al–Ca Alloys

Alloys based on the Mg–Zn–Al alloy system, with zinc as the major alloying element, have been examined as potential creep-resistant alloys from as early as the 1970s [10,79]. While pressure die cast Mg–Zn alloys with 2–12 wt% Zn are prone to hot cracking [10], the addition of aluminum reduces the tendency for hot cracking allowing for zinc content of 8 wt% or less (see Sec. II.C). The potential advantage of these alloys is that the major second phase present in the microstructure is not β-$Mg_{17}Al_{12}$. Consequently, discontinuous precipitation should be less prevalent leading to superior creep properties relative to Mg–Al based alloys [80–82].

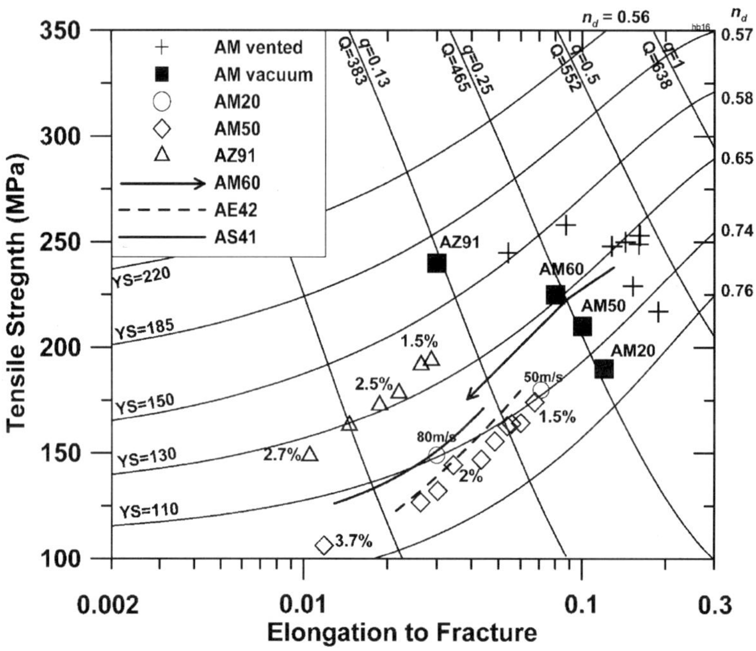

Figure 20 Comparison of properties of normally vented (crosses) and vacuum cast (squares, from Ref. 140) AM alloys, and the effect of porosity content and gate velocity on a range of die casting alloys. Crosses as in Fig. 18. The arrow [64], triangles, diamonds, and the solid and dashed lines [65] show the effect of increasing porosity in different alloys. The volumetric porosity content is indicated in %. Open circles show the effect of gate velocity (see text) [63].

The potential for improved creep properties in alloys based on the Mg–Zn–Al system is economically attractive and has received considerable attention [80,83–85]. The improved creep behavior of ZA85 relative to AZ91 and AM60 is shown in Fig. 26. Figure 26 also shows ZA85 to be generally inferior to AS21 and AE42 and much research has focused on the addition of extra elements to further improve the creep performance. Calcium additions have been investigated [84] including the composition 8% Zn, 5% Al, 0.03–0.12% Ca (plus other variations on zinc and aluminum content) which is the subject of a recent patent application [86]. This alloy composition is sometimes referred to as ZAC alloy, where, contrary to ASTM guidelines, C stands for calcium rather than copper. The creep resistance of this alloy is reported [86] to be better than AE42, although this is at a low stress (35 MPa) and moderate temperature (150°C). The alloy is also reported to have good castability and corrosion resistance [86].

2. New Mg–Al Based Alloys

Significant advances have been made in developing creep-resistant alloys based on Mg–Al binary alloys, particularly those with additions of calcium and/or strontium. Mg–Al–Sr alloys are claimed to have creep resistance as good as, or exceeding that, of AE42 [88,88], while Mg–Al–Ca alloys have been developed with creep resistances significantly superior to AE42 and approaching that of aluminum alloy, A380 [89]. The performance of the Mg–Al–Ca alloys is further improved by the addition of strontium and/or silicon. An alloy designated AMC522 (5 wt% Al, 2 wt% rare earths and 2 wt% Ca) has been used

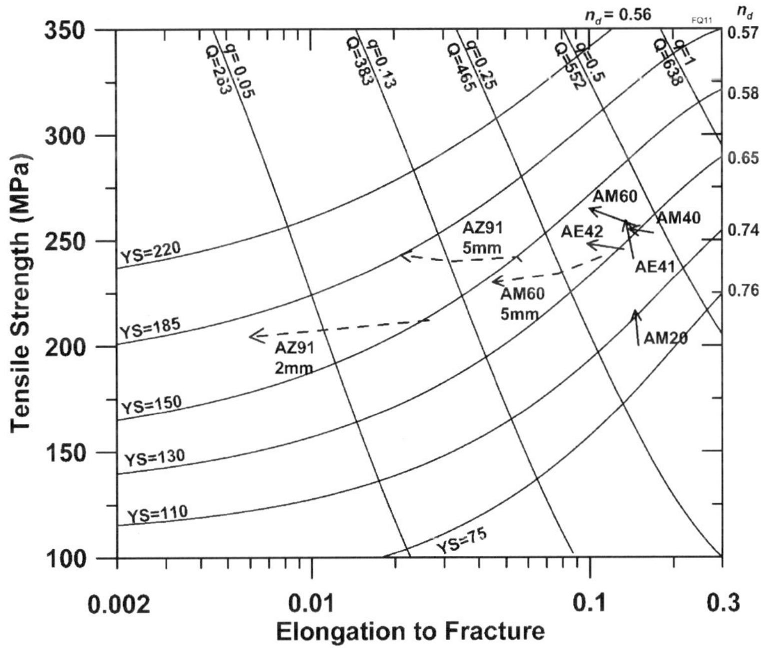

Figure 21 The arrows show the effect of low temperature aging on a range of die casting alloys. The dashed arrows are AM60 and AZ91 alloys, thickness as labeled, aged for up to 5000 hr at 120°C [53]. The solid arrows identify AM and AE alloys aged for up to 2000 hr at 107°C (round bars, 6.4 mm diameter) [141].

for the manufacture of oil pans [90]. This alloy has creep properties almost equal to the aluminum alloy, A384. It should be noted that limited data are available for these alloys and therefore further testing is required to fully verify their performance.

The above alloys contain approximately 5 wt% Al which takes advantage of the castability and strengthening provided by aluminum additions to magnesium. However, the addition of ternary and quaternary elements alters the microstructure significantly. The effects of rare earth additions on the microstructure of Mg–Al alloys are discussed in Sec. V.A.3.

Calcium additions, from minor additions up to 6%, to existing Mg–Al alloys have been shown to improve creep properties considerably [51,91]. Calcium also improves melt stability [92] and the oxidation of alloys at elevated temperatures [93]. However, susceptibility to hot cracking and die sticking has hindered the development of calcium-containing alloys, although additions of approximately 2% and above may overcome this.

The reason for the improved creep resistance from calcium additions has not yet been determined, although calcium does significantly alter the microstructure. In AZ91, 0.1 at.% Ca decreases the amount of discontinuous precipitation considerably [94]. The microstructure of the Mg–5 wt% Al–(1.9–3) wt% Ca alloys exhibits a relatively continuous distribution of the eutectic along the grain boundaries. The eutectic consisted of magnesium and a phase that was probably $(Al,Mg)_2Ca$. Both lamellar and divorced eutectic structures are observed. No $Mg_{17}Al_{12}$ was observed [95].

The effects of strontium additions on the microstructures of magnesium alloys containing 5 wt% Al alloys have been examined [87]. The microstructure of a 1.2 wt% Sr alloy

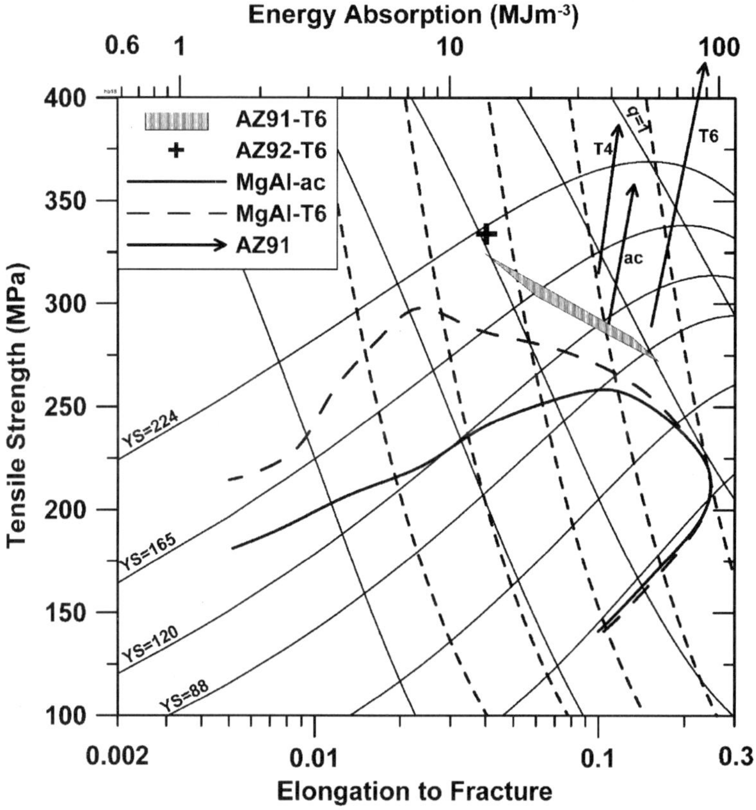

Figure 22 The chart of Fig. 15 reproduced with superimposed contours of constant absorption energy W (dashed lines), calculated with Eq. (for $W = 55$; 27.5; 13.8; 6.9 and 3.4 MJ m^{-3}, respectively, as explained in Appendix A. The arrows indicate the effect of high strain rate [66] on the strength–ductility of AZ91 alloy, for three tempers. The top x-axis provides a scale for the energy contours.

contained a lamellar eutectic around the grain boundaries, while the 1.8 wt% Sr alloy contained a divorced eutectic phase and lamellar eutectic. The minor phases were ternary magnesium, aluminum, and strontium compounds.

Calcium, strontium, and rare earths are aluminum scavengers and form aluminum containing intermetallics. The improved creep resistance of these alloys may be due to:

- the presence of a continuous grain boundary network of intermetallics that limits grain boundary sliding and other grain boundary processes,
- other phases forming in place of $Mg_{17}Al_{12}$, or
- lower matrix aluminum content limiting discontinuous precipitation.

Further work is required to determine which mechanisms are more significant.

3. New Sand-Cast Alloys

There has been some recent work on magnesium alloys containing scandium [77,96–100]. The Mg–Sc system was chosen for investigation because scandium has good solu-

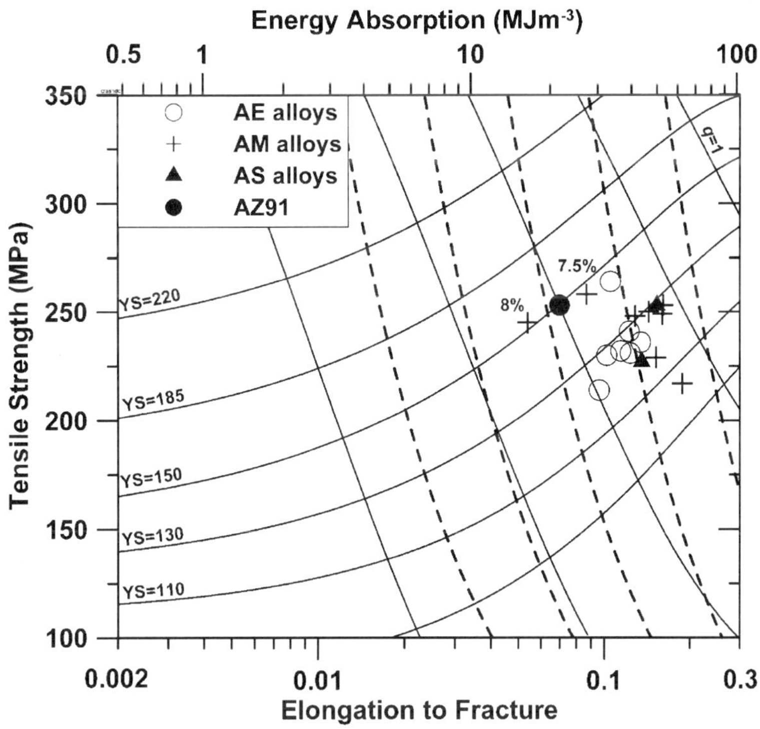

Figure 23 The chart of Fig. 18 reproduced with superimposed contours of constant energy absorption (dashed lines) calculated as for Fig. 22. The top *x*-axis provides a scale for the energy contours.

bility in magnesium at elevated temperatures and low solubility at lower temperatures, making the Mg–Sc system potentially age-hardenable [100]. Scandium also has a low density (3 g/cm^3), low diffusivity in magnesium and has potential for property improvements by alloying with additional elements [97]. Despite the age-hardening potential implied by the phase diagram, Mg–Sc systems have a poor age-hardening response as MgSc precipitates are coherent with the matrix [100]. However, the addition of manganese causes Mn_2Sc precipitates to form with significantly improved age-hardening behavior. The Mg–6 wt% Sc–1 wt% Mn alloy had a significantly lower strength than WE54 and QE22 (Fig. 29); however, increasing the scandium content to 15% increased the strength to that of WE54. Both the Mg–6 wt% Sc–1 wt% Mn and the Mg–15 wt% Sc–1 wt% Mn alloys in the T5 condition were found to have creep properties significantly superior to WE43 in the T6 condition. At 350°C and 30 Mpa, the minimum creep rates for the Mg–Sc–Mn alloys were of the order of 10^{-8} sec^{-1}, while for WE54, it was 10^{-6} sec^{-1} [97].

Because scandium is very expensive, alloys containing less scandium have been investigated. Alloys containing 1 wt% Sc or less with the rare earth elements, gadolinium, cerium, and yttrium have also proved to have superior creep properties than WE54 and WE43 [99,100] (Fig. 30). However, cheaper alternatives need to be developed if these alloys are to be used outside the aerospace industry.

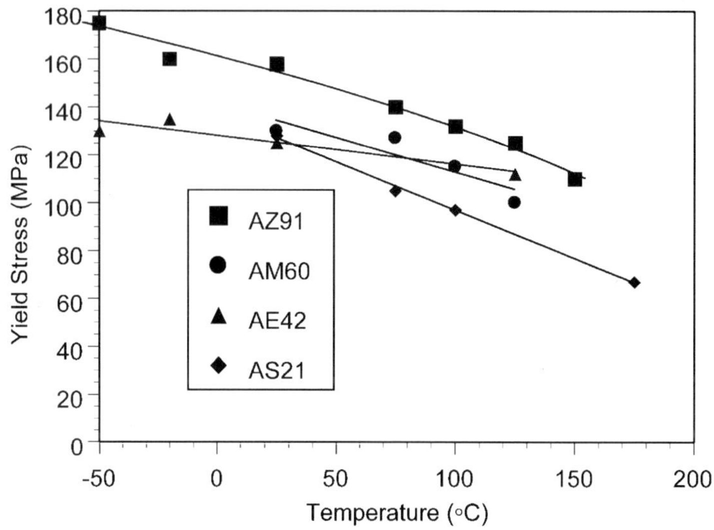

Figure 24 The variation of yield strength with temperature for four die cast magnesium alloys. (From Ref. 7.)

D. Effect of Processing on Creep Properties

While most magnesium alloy components are manufactured by pressure die casting, there is an increasing interest in other methods of producing magnesium components. Processing conditions can affect the microstructure substantially, which can cause significant variations in creep properties, even for the same alloy. Figure 31 shows the data for AZ91D produced using pressure die casting, thixoforming, and ingot casting. The coarse grained microstructure found in ingot has an order of magnitude improvement in minimum creep rate over pressure die cast material. The thixoformed material behaves similar to the ingot at lower stresses (below 80 MPa), while at higher stresses (above 90 MPa), the thixoformed material behaves similar to the pressure die cast material.

The difference in the creep properties of the alloys appears to be due to microstructural differences. The pressure die cast specimens had a grain size of 15 μm [102,102], while the ingot had a mixture of columnar grains and large equiaxed grains 300 μm in size [103]. The thixoformed material contained a bi-modal distribution of grains, with the larger grains being 300 μm in size and making up 60–80% of the microstructure, with fine (6μm) grains in between. The other difference in the microstructures is that the pressure die cast material contains a greater density of β-$Mg_{17}Al_{12}$ phase than the ingot casting.

The creep mechanism in fine grained materials seems to be grain boundary controlled whether it is due to grain boundary sliding and migration, grain boundary diffusion or a combination of these processes. It appears that the thixoformed material's behavior is dominated by coarse grain mechanisms at low stresses at 150°C, while at higher stresses, the fine grain mechanisms dominate [101].

However, a coarse structure is not always beneficial to the creep properties of alloys. Slower cooling rates with AS21 and AE42 lead to the presence of coarse intermetallics that are the cause of poor properties when sand cast or permanent mold cast.

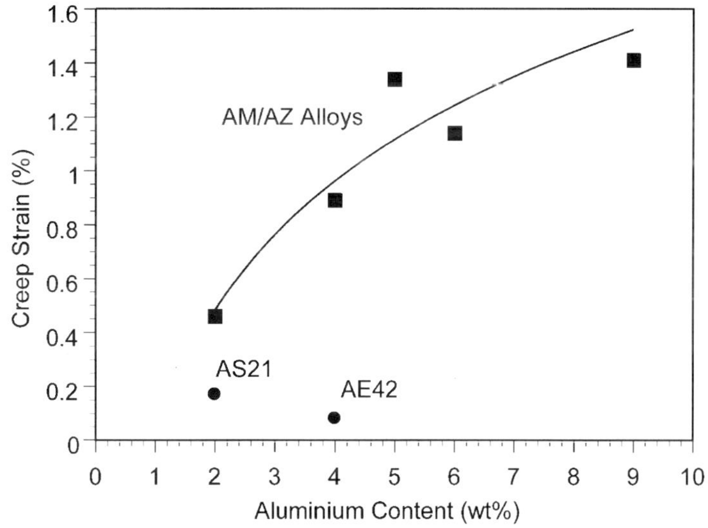

Figure 25 Variation of creep strain at 150°C, 50 MPa and 100 hr for die cast AM series alloys and AZ91. Also plotted are AS21 and AE42. (From Ref. 72.)

VI. PRESSURE DIE CASTING PROCESS

The vast majority of magnesium alloy component production is by the pressure die casting process. The process uses a permanent mold clamped together by hydraulic forces that is filled rapidly (5–100 msec) by injecting molten metal through a narrow gate. There are two basic types of pressure die casting processes. In the cold chamber process, shown in Fig. 32, a measured amount of molten metal is transferred to a shot sleeve and is then injected into the cavity using a piston. The hot chamber process, shown in Fig. 33, has a gooseneck immersed directly into a crucible holding the molten metal. Both processes are suitable for magnesium casting; however, the cold chamber process is favored in operations already equipped to cast aluminum.

Magnesium has a number of advantages that make it particularly suited to pressure die casting:

- Alloys with higher aluminum content, particularly AZ91, have high fluidity that allows complex thin wall parts to be cast. A wall thickness of less than 1 mm can be achieved.
- Magnesium has low specific and latent heats per unit of volume, which leads to short cycle times and reduced die wear.
- Iron has a relatively low solubility in magnesium reducing the tendency for die sticking and increasing die life.
- Magnesium has excellent machining characteristics allowing for faster cutting speeds and longer tool life.

While the intrinsic characteristics of alloys are important determinants of properties, the pressure die casting process itself also substantially influences properties. Rapid injection of molten metal and the high cooling rates in high pressure die casting result in a fine microstructure and other distinctive characteristics, such as defect bands

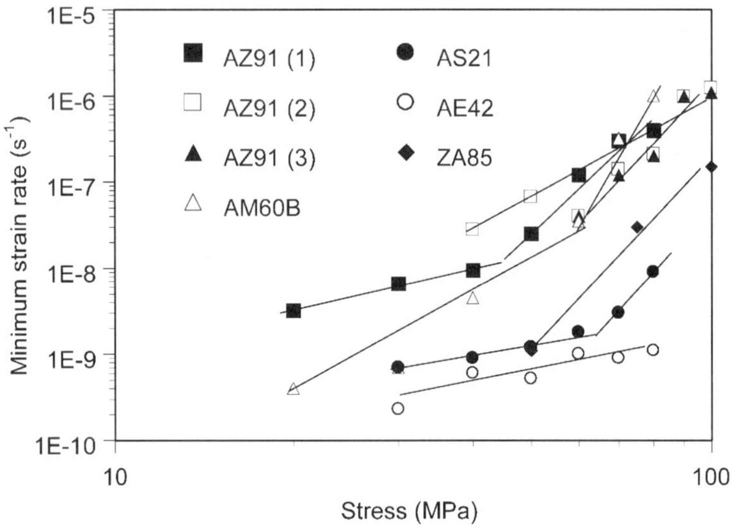

Figure 26 Minimum creep rate of die cast Mg alloys for a variety of stresses at 150°C. From Ref. 72 (AZ91 [1], AS21, AE42), Ref. 102 (AZ91 [2]), Ref. 101 (AZ91 [3]), Ref. 142 (AM60), and Ref. 85 (ZA85).

discussed below. Microstructures, and hence properties, are amenable to some control through casting parameters such as:

- melt temperature,
- shot velocity (low and high speed stages),

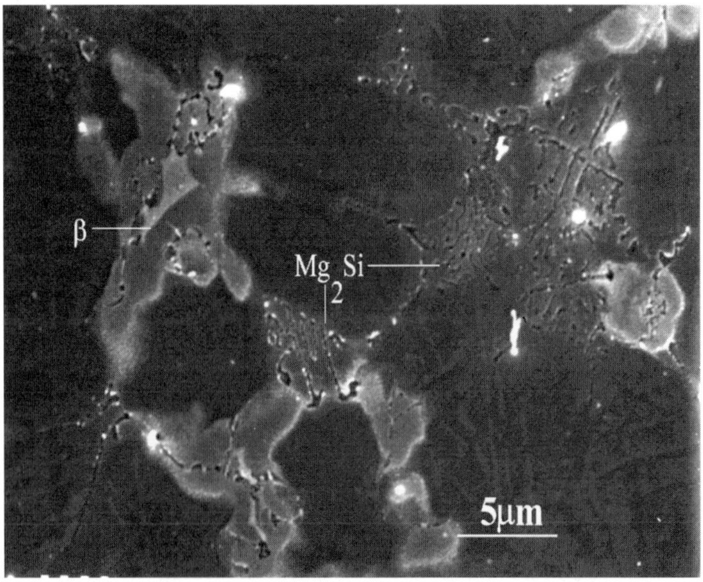

Figure 27 Microstructure of pressure die cast AS21 showing both β phase and Mg_2Si particles along with a region of supersaturated α-Mg. SEM secondary electron image. (From Ref. 72.)

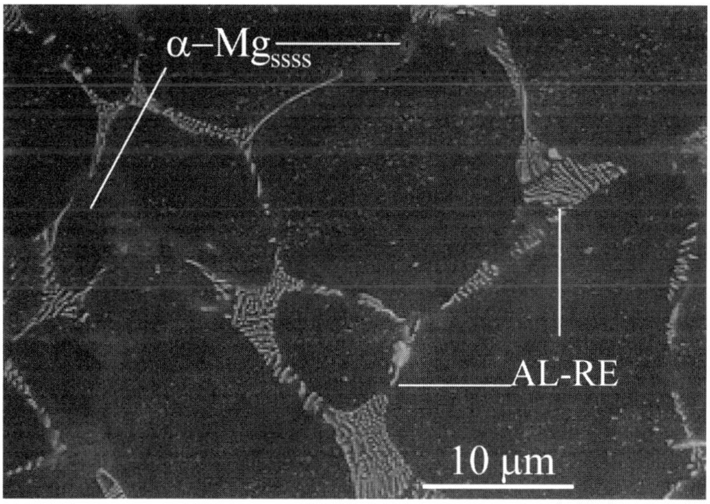

Figure 28 Microstructure of die cast AE42. SEM secondary electron image. (From Ref. 72.)

- intensification pressure,
- die temperature,
- section thickness, and
- distance from ingate.

Process parameters influence the microstructure, which in turn, influences the mechanical properties. In this section, the microstructural features of pressure die cast magnesium will be discussed followed by a review of process parameter–microstructure–property relationships.

A. Defects and Microstructural Features

A number of casting defects can occur which are usually unacceptable and lead to component rejection:

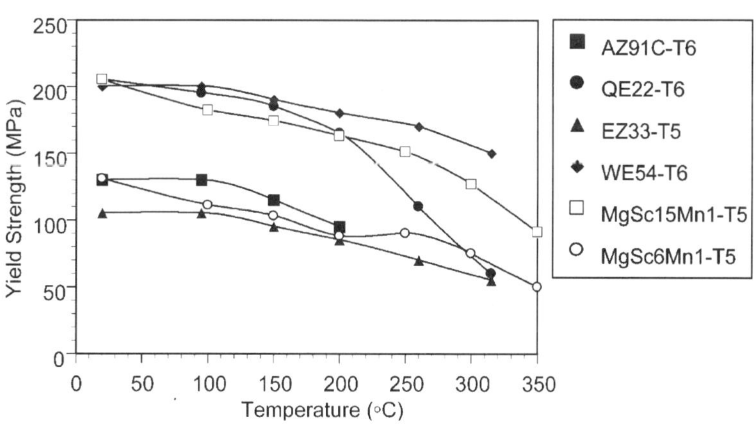

Figure 29 The variation of yield strength with temperature for commercial sand-cast magnesium based alloys, AZ91, QE22, EZ33 and WE54 [5] and the experimental Mg–Sc–Mn alloys [99].

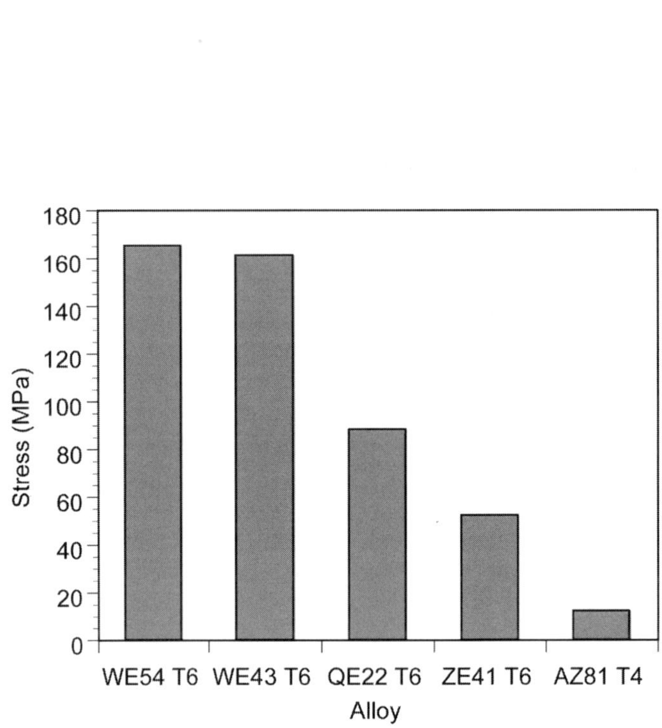

Figure 30 The creep strength of sand-cast alloys for 0.2% strain at 100 hr and 200°C. (From Ref. 100.)

- Incomplete filling or misruns can occur if the metal or die temperature is too low, if the metal has excessive oxide content, if the design and position of ingates and vents is inappropriate or if the injection speed is too low.
- Cold shuts occur when two flow fronts meet and fail to join. Low metal temperature is a contributing factor as are design factors influencing the flow pattern.

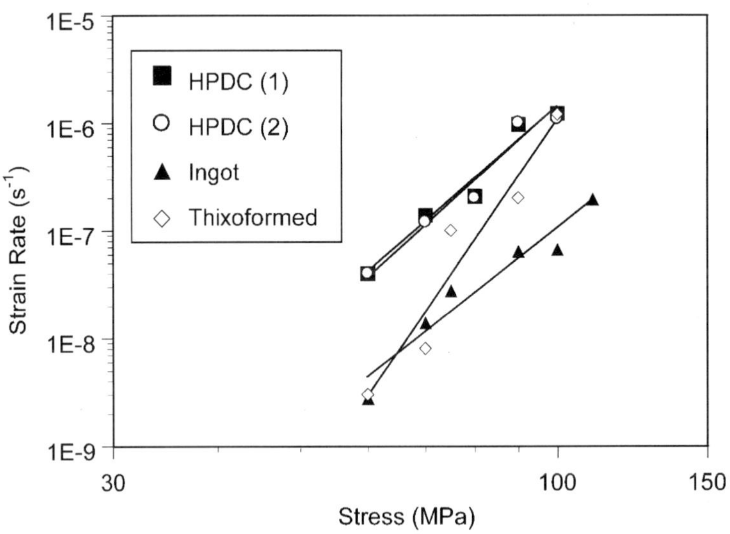

Figure 31 Plot of minimum creep rate against applied stress of AZ91 formed by HPDC, thixoforming and ingot casting in the as-cast condition. (From Refs. 101–103.)

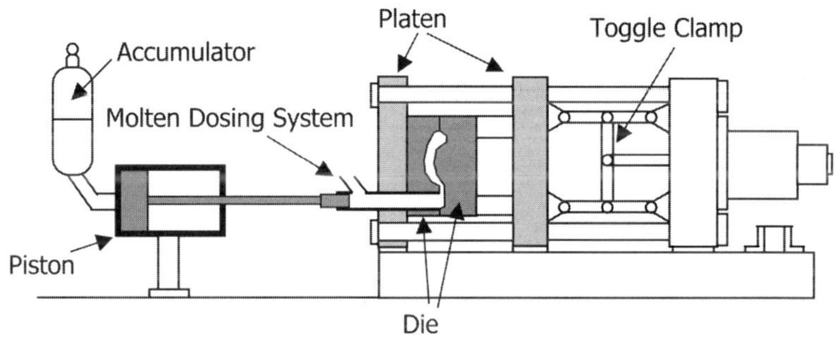

Figure 32 Schematic illustration of the cold chamber pressure die casting process.

- Hot cracking occurs in circumstances where the casting experiences substantial stresses while solidifying and cooling. The tendency for hot cracking is strongly influenced by the alloy solidification characteristics.
- Distortions can arise during cooling of the component following ejection from the die.
- Swirls and flow lines may also form on the surface. These can be an issue for applications where the casting surface is exposed.

In addition to the above defects, there are microstructural features that occur in most castings and do not necessarily lead to rejection:

- porosity,
- large grains formed in the shot sleeve,
- skin,
- defect bands, and
- non-metallic inclusions.

Porosity can arise from dissolved gases, entrapped air or from solidification shrinkage and is typically in the range of 1–5%. The amount of gas porosity is influenced by

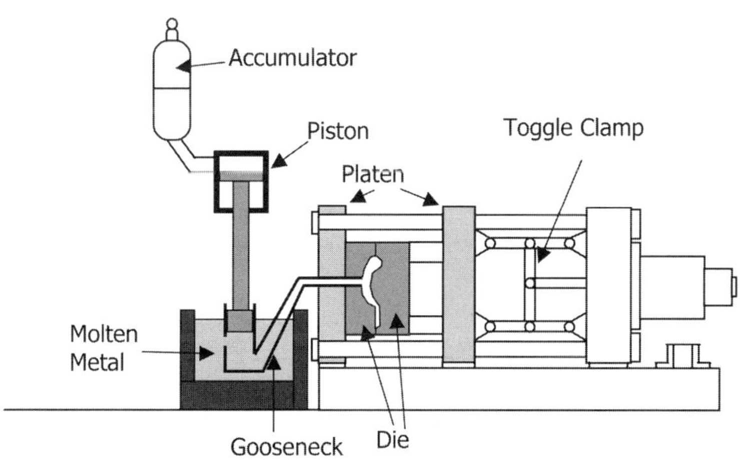

Figure 33 Schematic illustration of the hot chamber pressure die casting process.

metal temperature and the turbulence of filling, while the shrinkage porosity can be affected by casting designs that alter feeding behavior and intensification pressure. Porosity can occur in a dispersed manner or be concentrated in bands as discussed below.

In cold chamber machines, it is common for some solidification to take place in the shot sleeve prior to the commencement of the shot. The solid phase formed in this way is able to grow considerably larger than the material solidified during die filling, giving rise to a bimodal distribution of primary magnesium grains. The grains formed in the shot sleeve are referred to as externally solidified grains. The amount of these grains varies depending upon the metal temperature and other factors, and can range from 0 up to about 20% [104].

Another common feature of pressure die cast magnesium is the presence of a surface layer or skin. The skin is characterized by a higher volume fraction of second phases, $Mg_{17}Al_{12}$ in the case of Mg–Al alloys and, often, a finer microstructure [105]. The formation of the skin suggests the segregation of solute elements to the surface that may occur due to the semi-solid state of the metal as it flows in the cavity. The depletion of solid phase adjacent to surfaces is a common phenomenon for flowing suspensions [106]. Alternatively, the higher $Mg_{17}Al_{12}$ content may be a consequence of a faster cooling rate near the surface without enrichment in solute.

Possibly, the most visually apparent microstructural feature of magnesium pressure die castings is the bands of defects which, in most cases, form parallel to the casting surface. The make up of these bands seems to vary considerably and may consist of cracks, porosity (shrinkage and/or gas porosity), and/or segregation. Examples of defect bands are shown in Fig. 34. The bands may be single or multiple and can sometimes intersect with the surface. According to a mechanism proposed by Dahle and StJohn [107], the solid phase, within the semi-solid mixture flowing in the die cavity, migrates away from the walls and towards the center. While this effect raises the solid fraction towards the center, heat extraction through the die results in increased solid fraction near the surface. As a consequence, there is an intermediate zone with relatively high liquid fraction. As solidification and filling proceeds, the center and surface regions begin to develop some mechanical strength resulting in a localization of shearing in the high liquid fraction region. This mechanism should result in the central region having a lower solute content than the outer region, which has been reported to be the case in at least one instance [104].

Mao [108] observed bands of pores that appeared spherical suggesting they were due to gas porosity. The porosity was thought to have formed due to a sudden drop in pressure, possibly as a result of a small back movement of the plunger tip during intensification. This mechanism would appear only to explain instances of gas porosity and not shrinkage porosity or segregation.

Rodrigo and Ahuja [109] investigated the effects of casting parameters on the formation of defect bands. Although they did not propose a formation mechanism, they did assess the validity of the above mechanisms in light of their experimental findings. The model put forward by Dahle and StJohn [107] was disputed by Rodrigo and Ahuja [109] on the basis that the presence or absence of externally solidified grains did not influence the appearance of the pore bands and that when externally solidified grains were present they often appeared on either side of the defect bands. However, the theory of Dahle and StJohn [110] also applies to the situation where externally solidified grains are not present so it is not clear whether this is in fact contradictory evidence. The theory put forward by Mao [108] was also disputed by Rodrigo and Ahuja [109] as the observed porosity was due to shrinkage rather than gas.

A patent application by Murray and Cope [111] suggests an alternative mechanism for the formation of these bands. Although they do not explicitly refer to bands, they claim, from microstructural examination, that in the runner and in sections of the casting where flow is uni-directional, flow occurs through a cylindrical section much smaller than the physical cross-section. The size of the cylindrical section appeared to be determined by the flow velocity, which typically fell in the range of 140–165 m/sec. It was claimed that initially solidification occurred in the runner, reducing the effective cross-section and increasing the velocity. Once a velocity of about 150 m/sec was attained, viscous heating prevented further solidification. It was claimed that because viscosity rises rapidly for solid percentages above 50%, the viscous heating process would stabilize the solid fraction of material entering the cavity. As this phenomenon was also observed in sections of the cavity where flow was uni-directional, this suggests that the bands of defects occur where solidification was temporarily arrested by viscous heating. Rodrigo and Ahuja [109] also claim that defect bands are more defined in regions where flow is largely unidirectional.

Defect bands occur in a range of forms, such as cracks, bands of porosity (gas and/or shrinkage), and segregation. All of them have superficial similarity; however, it is not clear whether they form by the same process or whether multiple mechanisms are involved. Further investigations are necessary before their formation mechanism is fully understood.

Non-metallic inclusions may also be regarded as microstructural features. They differ from the features described above in that they do not form within the die but rather are oxide particles that were present in the molten metal. Control of molten metal handling practices is the key to controlling their influence on properties.

B. Effects of Porosity and Non-metallic Inclusions on Properties

The effects of porosity on mechanical properties of pressure die castings have been discussed earlier in Sec. IV.C.3 and are shown in Fig. 35 for AZ91 and AM50 alloys. The level of porosity does not have a significant effect on the shape of the stress–strain curve but instead affects the point along the curve where failure occurs. Consequently, increasing levels of porosity reduce the tensile strength and elongation while yield strength is relatively unaffected [63,65]. Variations in porosity levels account for some of the property variations with section thickness as porosity levels often increase with section thickness [112,113].

The presence of non-metallic inclusions, such as oxides produced during melt handling, act in a similar manner to porosity reducing strength and ductility but having little effect on yield strength [114] (Fig. 36).

C. Effect of Casting Design on Tensile Properties

A number of factors relating to the design or geometry of a casting may influence mechanical properties. The factors considered here are section thickness and the effect of distance, and section changes with distance, from the ingate.

The effect of section thickness has been studied for most common alloys with data from a range of sources for alloy AZ91D shown in Fig. 37. There have been a number of different studies [61,62,112,113,115] on the effects of section thickness on the properties of AZ91D. Some studies report an increase in yield and tensile strength with decreasing section thickness [112,115]. However, this is not always the case [113] and when results from different studies are combined, a wide range of values are observed as shown in

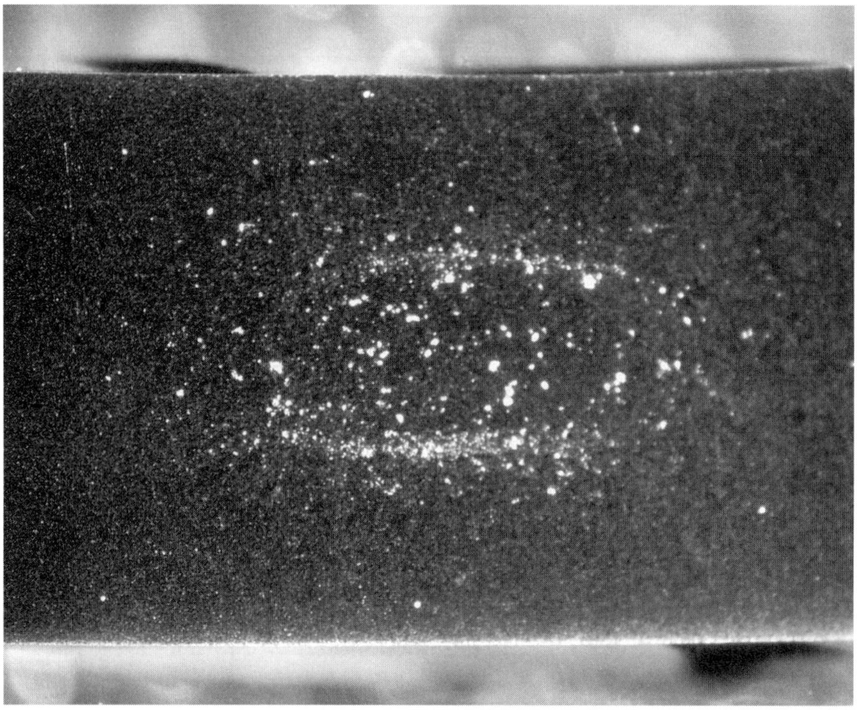

(a)

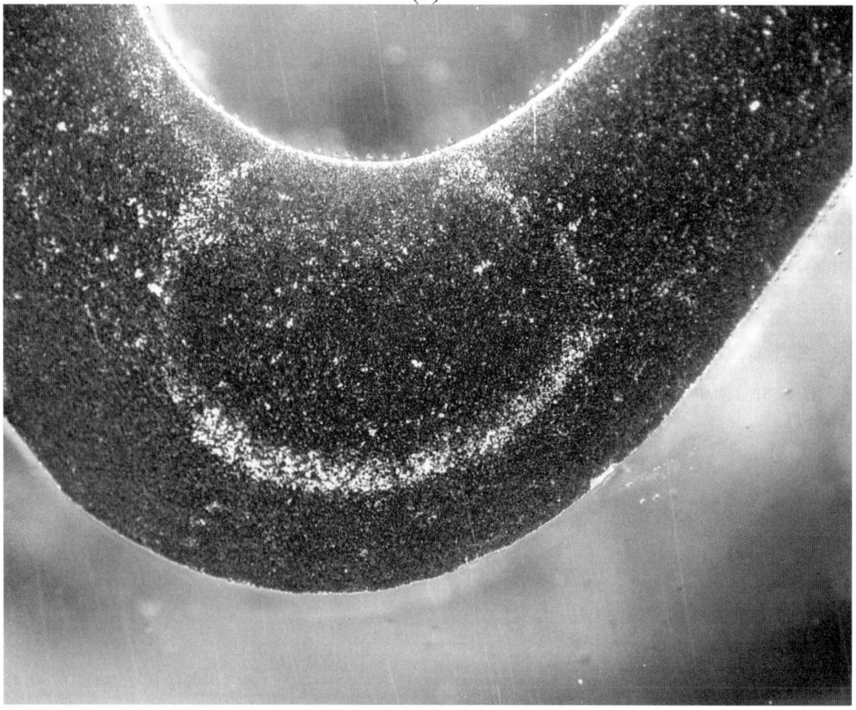

(b)

Figure 34 Example of defect bands (a) in a 2 mm thick cast tensile bar, and (b) in a commercial casting (3mm thick).

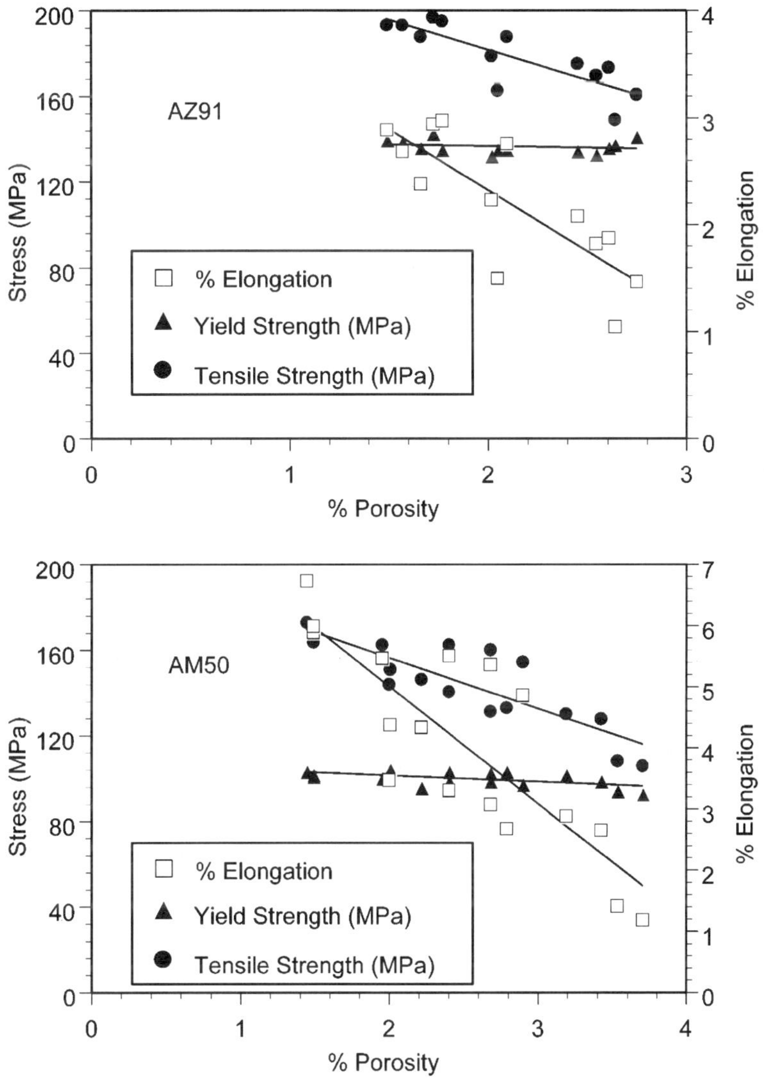

Figure 35 Relationship between porosity level and mechanical properties for 12 mm thick pressure die cast tensile samples. Alloys are AZ91D (top) and AM50 (bottom). (From Ref. 65.)

Fig. 37. In the case of tensile strength and elongation, the combined results do not show any clear trend. In the case of yield strength, the values are confined to a more narrow range and an increase in thinner sections is still evident in the combined data (for AZ91).

A number of studies attempt to relate the formation of a surface layer to mechanical properties [61,105,113]. Metallographic examinations of castings often reveal a surface layer with finer microstructural features and, in the case of AZ91, at least, a higher volume fraction of second phase particles [61]. Other characteristics reported for the surface layers include a higher hardness and a lower fraction of voids. The higher hardness contributes to increased yield strengths in thinner sections if the surface layer thickness is partially independent of section thickness [116]. Other reports, however, suggest that surface layer thickness can vary in proportion to the section thickness [105].

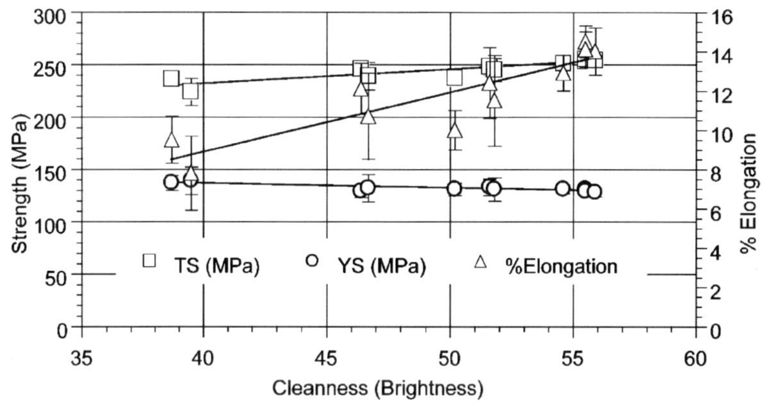

Figure 36 Effect of cleanness (an inverse measure of the quantity of non-metallic inclusions) on mechanical properties of pressure die cast alloy AZ91. (From Ref. 114.)

The wide spread of tensile strength and elongation values suggests that casting quality issues have a more significant effect on these properties than section thickness. Variations in casting quality (amount of porosity and casting defects) are suspected when the shape of stress–strain curves (and hence yield stress) remains largely unchanged, but the point along the curve where failure occurs changes (and hence tensile strength and elongation vary together).

Round bars tend to have better properties with yield and tensile strength and elongation values at the high end of the range for flat samples (Fig. 37).

In summary, section thickness may be of some value in predicting yield strength of AZ91 but not for tensile strength or elongation. One investigation [113] has taken this assessment further and found that even after the effects of porosity are corrected for, section thickness is still not a good predictor of these properties.

As the distance from the ingate increases, the ability to feed solidification shrinkage decreases, so porosity levels will tend to increase as shown in Fig. 38 [63]. This effect is exacerbated if the cross-sectional area increases with distance from the ingate. In cast to size test bars, the section increases at the grip sections relative to the central test region. Consequently, the grip section closest to the overflow tends to have relatively high levels of porosity.

D. Effects of Casting Parameters on Microstructure, Porosity, and Properties

Studies of the effects of a number of casting parameters have been reported [61,63,109,117–119]. The parameters include metal injection rate or velocity, applied solidification pressure, die temperature, metal temperature, venting gate area relative to ingate, cavity fill time, and ram position at the start of the high speed shot.

The metal injection rates or velocities reported in the literature are not always directly comparable, despite all the papers examined presenting their results in units of velocity (m/sec or equivalent). In some instances, the velocity of the ram is quoted, while in others, the velocity through the ingate is used. The casting cycle usually has a low velocity stage and a high velocity stage and the effects of both have been reported. In one

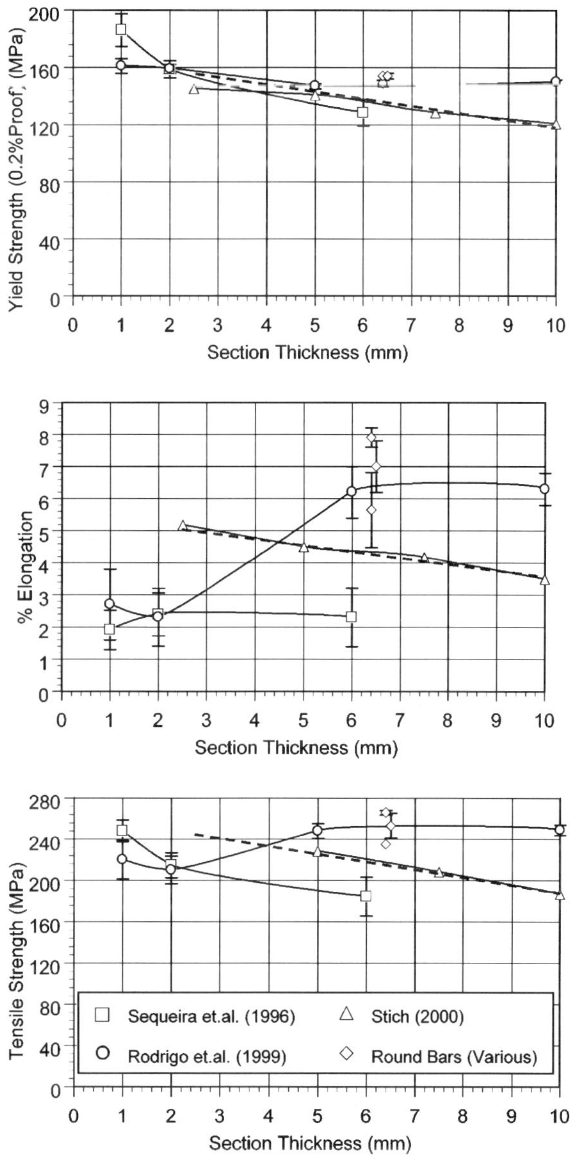

Figure 37 Effects of section thickness on tensile properties of pressure die cast AZ91 alloy from various sources. Results from Ref. [112] (squares) and Ref. [113] (circles) are from cast test bars. The dashed lines are from Ref. [62] and represent results from tensile specimens cut from a stepped casting. The results presented by Stich and Haldenwanger [115] appear to be from the same source as Schinelbacher and Rösch [62]. The results from round cast tensile bars are from Refs. 7, 112, and 143.

paper, an intermediate velocity is reported [119]. Other papers refer to an injection rate [117,118].

The different casting parameters appear to interact strongly and so it is difficult to separate out the effects of individual parameters. Furthermore, their influence will depend on cavity size and shape and on the runner and gating design. The metal injection rate or

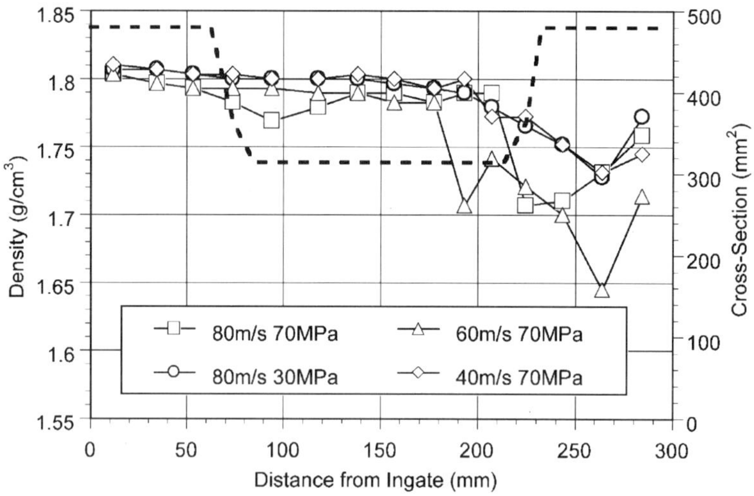

Figure 38 Effect of distance from ingate and section size on porosity in cast test bars of alloy AZ91D for various gate velocities and intensification pressures. Although the results indicate an increase in porosity with distance from ingate, they are influenced by the increase in section size (shown by the dashed line) associated with the grip sections. (From Ref. 63.)

velocity sometimes increases [61,117] and sometimes reduces [63,117–119] porosity (Figs. 39 and 40). In one report, where porosity increased with velocity, the effect was attributed to increased gas entrapment [61] (Fig. 39). It is likely that, in cases where increased metal injection rate reduces porosity, solidification shrinkage induced porosity is more significant.

Increases in applied pressure during solidification usually reduce porosity; however, one report shows the reverse for AZ91 [63] (Fig. 40).

The effects of die and metal temperature are also variable, with one report suggesting that porosity levels reach a maximum level at certain temperatures [118] (Fig. 41). Another report shows that porosity decreases with increasing metal temperatures [119].

An investigation into the relative effects of different parameters [119] found injection velocity to be the most significant factor for porosity but not significant for strain at fracture. The only parameter which showed a significant effect on both porosity and strain at fracture was metal temperature.

Other discrepancies between the effects on porosity and properties have been reported. In one study [61] where porosity continued to increase with injection velocity, the mechanical properties initially increased and then decreased (Fig. 39). The poor properties at low injection velocities (and low porosity levels) were attributed to poor surface finish.

Reports on the effects of casting parameters on other microstructural features are scarce. One report [109] investigated the effects on defect bands. The intensity of this band seems to be affected in a manner opposite to total porosity. The intensity of the defect band increased with applied pressure and die temperature, and decreased with slow shot speed (Fig. 42).

The effects of changes to defect bands on mechanical properties have been investigated to a limited extent. An investigation [120] using as-cast test bars found little effect of casting parameters (shot profile and melt temperature) on defect bands or mechanical properties. The aluminum content was shown to strongly influence defect bands with the

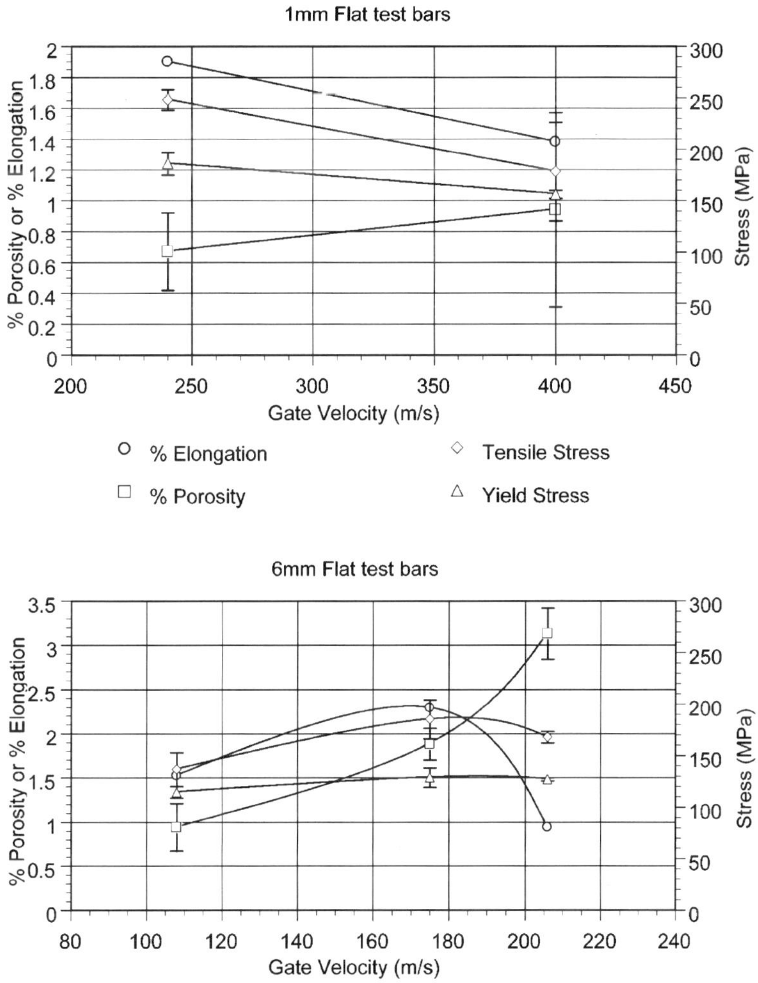

Figure 39 Effect of gate velocity on porosity and tensile properties of AZ91. (From Ref. 61.)

bands becoming less well defined as the aluminum content was increased from AM20 levels through AM50 and AM60 to AZ91 levels.

For AM50 alloy, a slight effect of the distance of the defect band from the surface on fracture elongation was noted. The spread of elongation values was large with a range of 6–27% observed. In order to attain the highest elongation values, it was necessary for the defect band to be near to the center of the section (Fig. 43), but this alone did not guarantee high ductility. It was noted that material between the defect band and the surface was sound while the material contained within the band had distributed porosity. The increased proportion of sound material would account for the improved ductility results for samples with defect bands closer to the center.

E. Overview of Variability in Pressure Die Casting Properties

The literature data demonstrate that a considerable variation in properties occurs for a given pressure die cast magnesium alloy. For example, the tensile strength of AZ91D can

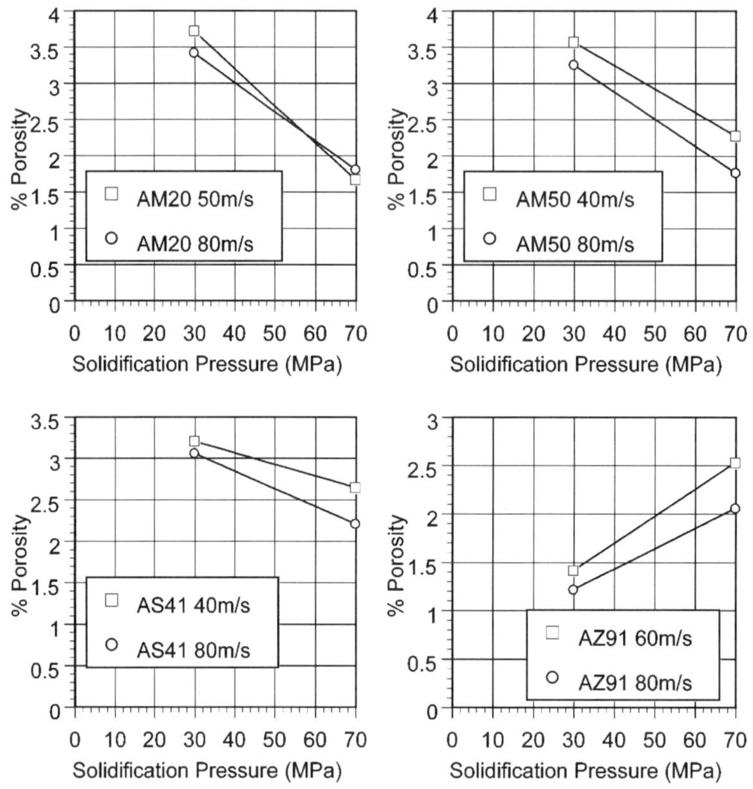

Figure 40 Effects of gate velocity, solidification pressure and alloy on porosity. (From Ref. 63.)

vary over a range of over 50 MPa (Fig. 37). While the literature demonstrates that a range of property values can occur, it only partly quantifies the factors contributing to this range. This is partially due to the limited number of investigations, and partly due to the various factors interacting in a complex manner making them difficult to separate.

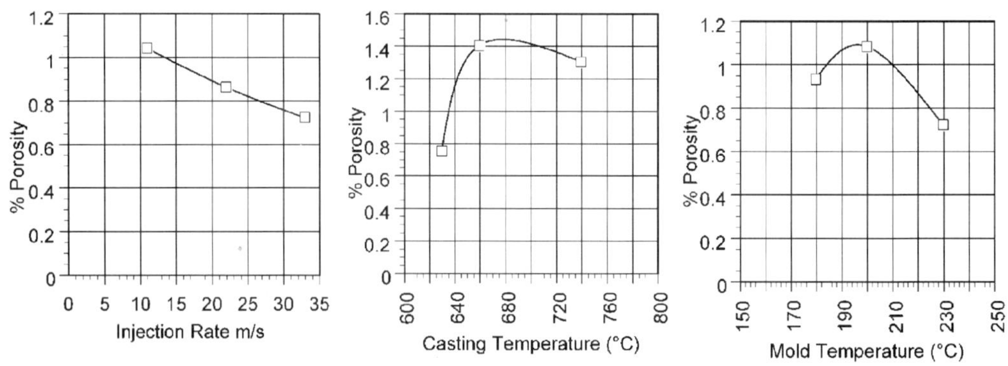

Figure 41 Effects of injection rate, casting temperature, and mold temperature on porosity in pressure die cast AZ91 alloy. (From Ref. 118.)

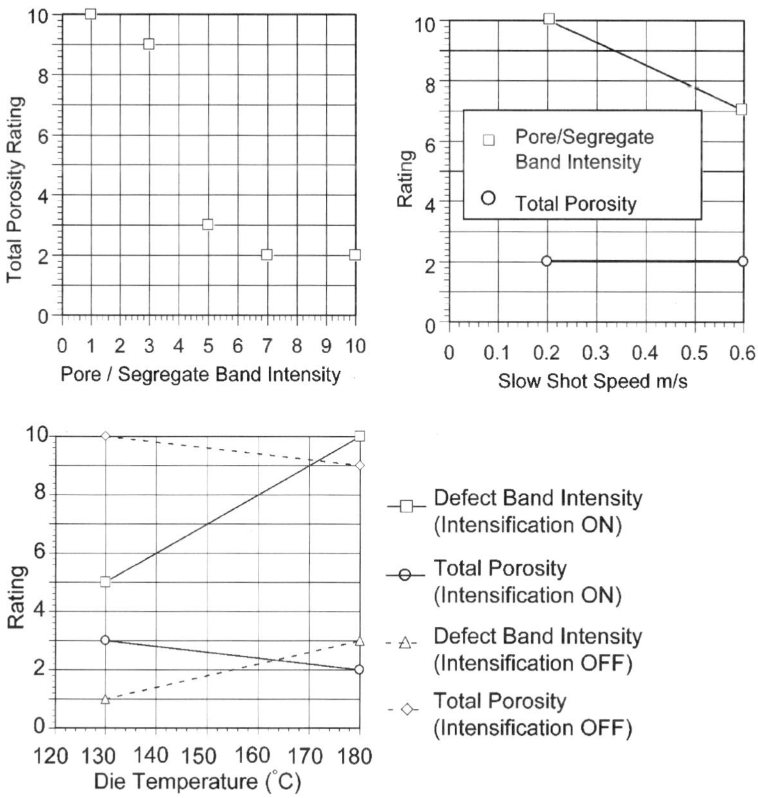

Figure 42 Effects of casting parameters on the distribution of porosity (whether it is concentrated in defect bands or randomly distributed). A higher rating value indicates either a higher total porosity level or a greater proportion of the porosity within defects bands as indicated by the legend. (From Ref. 109.)

Some of the discernible parameter–property relationships include:

- Tensile strength and elongation tend to decrease with increasing porosity and oxide inclusion levels.
- Tensile yield strength increases with decreasing section thickness for thicknesses less than 3 mm.
- Porosity levels tend to increase with distance from the ingate, particularly if there is an increase in section size.

Defect bands, which are highly prominent in metallographic examination, do not appear to have a strong effect on tensile properties. However, this assessment may give a misleading impression of their role in actual magnesium, components as in tensile bars, the bands are oriented parallel to the stress direction which would minimize their effects. In actual components, this is not always the case, as the example in Fig. 34 shows. In situations where they intersect the surface, their effects on mechanical properties may be more pronounced.

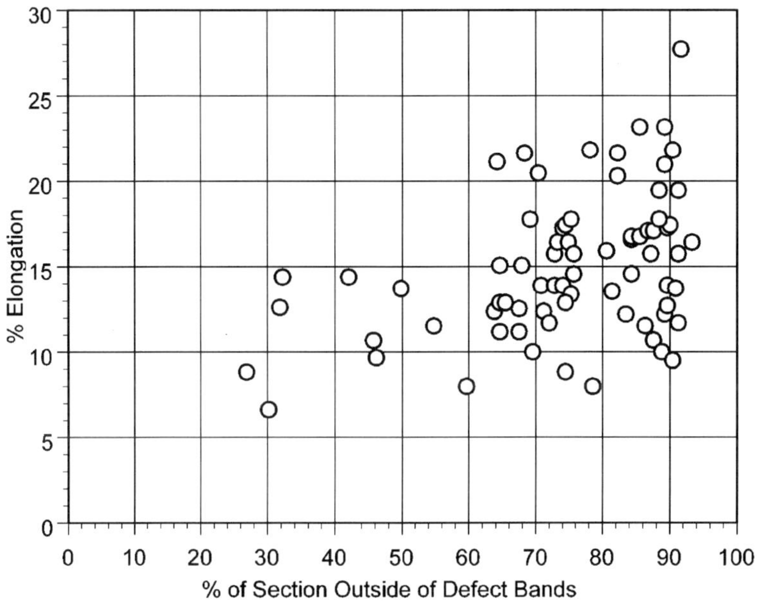

Figure 43 Effect of defect band location on elongation for as-cast tensile bars. (From Ref. 120.)

VII. APPENDIX A

A. Strength–Ductility Charts for Mg-Based Alloys

Strength–ductility charts, first developed by Drouzy et al. [121] for aluminum based casting alloys, are an efficient tool to compare on a common base the performance of different alloys or to assess the effect of changes to the alloy processing or chemical composition. The charts can be created using a simple model [122] based on the assumption that the deformation curves of the material can be described with a power law relationship:

$$\sigma = K\varepsilon^n \tag{A1}$$

where σ is the true flow stress, K is the alloy's strength coefficient, ε is the true plastic strain and n is the strain hardening exponent. The engineering stress–strain curve can be approximated from Eq. (A1) as

$$P \cong Ks^n e^{-s} \tag{A2}$$

where P and s are the engineering values of the stress and the plastic strain. The flow lines in Fig. A1 (solid lines, identified by the n-value) have been generated with Eq. (A2).

Application of the Considère criterion [123] to Eq. (A1) indicates the onset of necking at strain $\varepsilon = n$ (i.e., the end of the uniform deformation). For samples that fail at or before the onset of necking, it is useful to define the relative ductility parameter, q:

$$q = \frac{s_f}{n} \tag{A3}$$

where s_f is the elongation to fracture. Solving Eq. (A3) for n and substituting in Eq. (A2)

$$P = Ks^{s_f/q}e^{-s_f} \tag{A4}$$

Equation (A4) has been used to generate the iso-q contour lines in Fig. A1, which are identified by the q-value. A specimen that fails at the onset of necking is represented by a datum point on the line $q = 1$. Points representing specimens whose tensile ductility is 50%, 25%, 10% of the maximum uniform strain lie on the lines $q = 0.5, 0.25, 0.10$, etc., respectively.

Drouzy et al. [121] used their charts to define a "Quality Index" parameter, Q, that combines the tensile strength and the ductility. In terms of the analytical model, the Q-value can be expressed as [124]

$$Q = K((qn)^n e^{-qn} + 0.4 \log(100qn)) \tag{A5}$$

Figure A1 is a strength–ductility chart for Mg–Al alloys, calculated for $K = 570$ MPa and given n-values. The material's (0.2%) proof stress is the y-intercept of the flow curves and is indicated on each curve. The Q-values have been calculated with (25). The chart indicates that the highest Q-value attainable with Mg–Al alloys is $Q = 638$ MPa, corresponding to tensile specimens deforming up to the onset of necking, indicated by the line $q = 1$.

Figures 14–16 and 22 include charts similar to Fig.A1. The chart of Fig. 17 corresponds to MgZn alloys. For these alloys, a lower K-value (390 MPa) was used due to their lower strain hardening rate in comparison with Mg–Al alloys [57]. The maximum Q-value for MgZn alloys is also lower, 436 MPa, in comparison with Mg–Al alloys. The charts of Figs. 18–21 and 23, which correspond to pressure die casting alloys, have been generated with $K = 570$ MPa as for Fig. A1, but the flow lines follow the eqn. [57]:

$$P = (\mathrm{YS} + Ks^{n_d})e^{-s} \tag{A6}$$

which provides a better fit at low strains to the flow curves of Fig. 7. The fit parameter YS is slightly less than the actual proof stress of the alloys whose data points lie on the flow lines.

B. Energy Absorption

To a first order approximation, the amount of energy per unit volume, W, absorbed in deforming the material up to the fracture strain, ε_f, can be found by integrating Eq. (A1) with respect to the strain, which leads to

$$W = \frac{K\varepsilon_f^{n+1}}{n+1} \tag{A7}$$

In terms of engineering tensile stress, TS, and elongation to fracture, s_f, Eq. (A7) takes the form [59]:

$$W = \frac{\mathrm{TS}s_f e^{s_f}}{n+1} \cong 0.8\mathrm{TS}s_f e^{s_f} \tag{A8}$$

By solving Eq. (A8) for TS, contours of constant energy absorption can be superimposed on the quality index charts of both pressure die castings and sand castings. This was done in Figs. 22 and 23.

ACKNOWLEDGMENTS

The authors wish to thank their colleagues in the CRC for Cast Metals Manufacturing for valuable discussions and their input into the manuscript, particularly J.F. Nie for his input

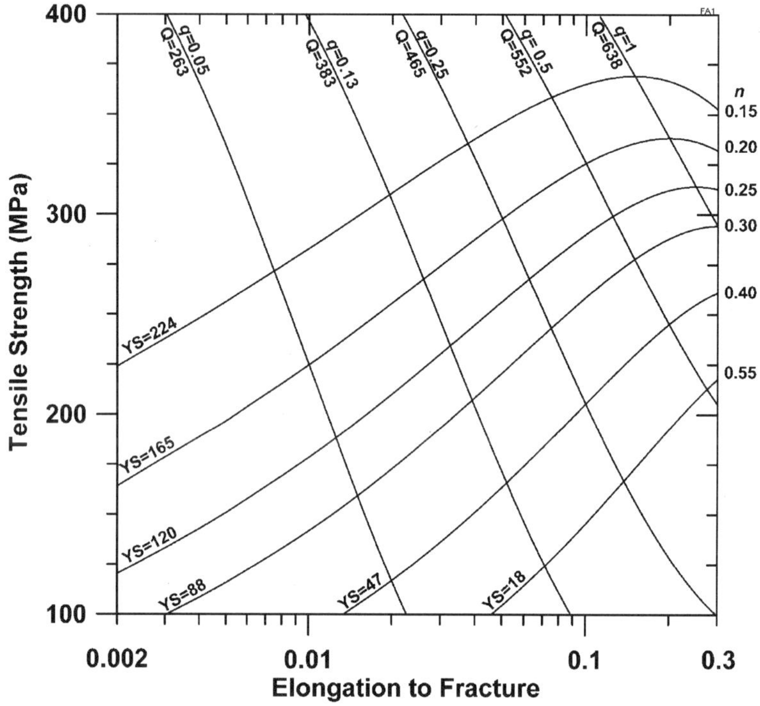

Figure A1 A strength–ductility chart for Mg–Al alloys [59]. The lines identified by the yield strength value are flow lines generated using Eq. (A4), with $K = 570$ MPa and given n-values. The constant relative ductility lines, identified by the q and Q values, have been generated with Eq. (A4). The line $q = 1$ identifies the onset of necking on the flow lines.

into the section on elevated temperature properties. The CRC for Cast Metals Manufacturing was established under, and supported in part by the Australian Government's Cooperative Research Centers Scheme.

REFERENCES

1. Brungs, D. Light weight design with light metal castings. Mater. Design 1997, *18*, 285–291.
2. Cole, G.S. Magnesium in the automotive industry. Die Casting Management 1995, October, 36–39.
3. Duffy, L.B. The development of magnesium sand casting alloys. Foundry Trade J. 1991, May 3, 319–321.
4. *Magnesium and magnesium alloys (ASM Specialty Handbook),* M. M. Avedesian and H. Baker (Eds.), ASM International, Materials Park, Ohio, 1999, pp. 55–65.
5. Avedesian, M.M.; Baker, H. Eds. ASM Specialty Handbook Magnesium and Magnesium Alloys; ASM International: Materials ParkOH, USA 1999.
6. Westengen, H. Magnesium alloys for structural applications; recent advances. Journal Physique IV 1993, 491–501.
7. Aune, T.K.; Westengen, H. Magnesium die casting properties. Automotive Engineering 1995, *103* (8), 87–92.
8. Aune, T.K.; Westengen, H.; Ruden, T. Mechanical properties of energy absorbing magnesium Alloys. SAE Technical Paper 930418 1993, 51–57.

9. Makar, G.L.; Kruger, J. Corrosion of magnesium. International Mater. Rev. 1993, *38*, 138–153.
10. Foerster, G. Improved magnesium die casting alloys. In Eighth SDCE International Die Casting Exposition and Congress, 1975; Detroit; 1–6.
11. Humble, P. Towards a creep resistant magnesium alloy. Mater. Forum 1997, *21*, 45–56.
12. Pettersen, K.; Lohne, O.; Ryum, N. Dendritic solidification of magnesium alloy AZ91. Metall. Trans. A 1990, *21A*, 221–230.
13. Han, Q., et al. Solidification behavior of commercial magnesium alloys. In *Magnesium Technology 2001*; The Minerals, Metals and Materials Society: Warrendale, PA 2001.
14. Dargusch, M.; Dunlop, G.L.; Pettersen, K. Elevated temperature creep and microstructure of die cast Mg–Al alloys. In *Magnesium Alloys and Their Applications*; Verksoff-Informationsgesellschaft:Wolfsburg, Germany, 1998.
15. Nave, M.D.; Dahle, A.K.; StJohn, D.H. The role of zinc in the eutectic solidification of magnesium–aluminum–zinc alloys. In *Magnesium Technology 2000*; The Minerals, Metals and Materials Society: Warrendale, PA, 2000.
16. Partridge, P.G. The crystallography and deformation modes of hexagonal close-packed metals. Metallurgical Rev. 1967, *12*, 169–194.
17. Akhtar, A.; Teghtsoonian, E. Solid solution hardening in magnesium alloys. Trans. Japan Inst. Metals 1968, *9 (Supplement)*, 692–697.
18. Hauser F.E., et al. Deformation mechanisms in polycrystalline aggregates of magnesium. Trans. American Society Metals 1955, *47*, 102–134.
19. Hauser, F.E.; Landon, P.R.; Dorn, J.E. Fracture of magnesium alloys at low temperature. Trans. Metallurgical Society AIME 1956, *206*, 589–593.
20. Hirsch, P.B.; Lally, J.S. The deformation of magnesium single crystals. Philosophical Magazine 1965, *12*, 595–648.
21. Clark, J.B. Age hardening mechanisms in two magnesium-base alloys. Trans. Japan Inst. Metals 1968, *9 (Supplement)*, 354–355.
22. Reed-Hill, R.E.; Robertson, W.D. The crystallographic characteristics of fracture in magnesium single crystals. Acta Metallurgica 1957, *5*, 728–737.
23. Reed-Hill, R.E.; Robertson, W.D. Additional modes of deformation twinning in magnesium. Acta Metallurgica 1957, *5*, 717–727.
24. Partridge, P.G. Cyclic twinning in fatigued close-packed hexagonal metals. Philosophical Magazine 1965, *12*, 1043–1054.
25. Gharghouri, M.A., et al. Study of the mechanical properties of Mg–7%at Al by in-situ neutron diffraction. Philosophical Magazine 1999, *79*, 1671–1696.
26. Akhtar, A.; Teghtsoonian, E. Solid solution strengthening of magnesium single crystals—I. Alloying behaviour in basal slip. Acta Metallurgica 1969, *17*, 1339–1349.
27. Akhtar, A.; Teghtsoonian, E. Solid solution strengthening of magnesium single crystals—II. The effect of solute on the ease of prismatic slip. Acta Metallurgica 1969, *17*, 1351–1356.
28. Akhtar, A.; Teghtsoonian, E. Substitutional solution hardening of magnesium single crystals. Philosophical Magazine 1972, *25*, 897–916.
29. Lukác, P. Solid solution hardening in Mg–Cd single crystals. Phys. Stat. Sol. (a) 1992, *131*, 377–390.
30. Levine, E.D.; Sheely, W.F.; Nash, R.R. Solid-solution strengthening of Magnesium single crystals. at room temperature. Trans. Metallurgical Society AIME 1959, *215*, 521–526.
31. Couture, A.; Meier, J.W. The effect of wall thickness on tensile properties of Mg–Al–Zn alloy castings. Trans. American Foundrymen's Society 1966, *74*, 164–173.
32. Nussbaum, G., et al. Strengthening mechanisms in the rapidly solidified AZ91 magnesium alloy. Scripta Metallurgica 1989, *23*, 1079–1084.
33. Yue, T.M.; Ha, H.U.; Musson, N.J. Grain size effects on the mechanical properties of some squeeze cast light alloys J. Mater. Sci. 1995, *30*, 2277–2283.
34. Sasaki, H.; Adachi, M.; Sakamoto, T.; Takimoto, A. In *IMA-53* (International Magnesium Association, 1996), p. 86.

35. Mabuchi, M.; Yamada, Y.; Shimojima, K.; Wen, C.E.; Chino, Y.; Nakamura, M.; Asahina, T.; Iwasaki, Aizawa, T.; Higashi, K. *The grain size dependence of strength in the extruded AZ91. Mg alloy*, K. U. Kainer, Magnesium alloys and their applications, Munich, Wiley-CH, New York, 2000, pp. 280–284.
36. Embury, J.D. Strengthening by dislocation substructures. In *Strengthening Methods in Crystals*, Kelly, A.; Nicholson, R.B., Eds.; Elsevier Materials Science Series Elsevier: London, 1972; 331–402.
37. Smith, C.S. Microstructure. Trans. American Society Metals 1953, *45*, 533–575.
38. Fox, F.A.; Lardner, E. An investigation of the effects of precipitation treatment of binary magnesium–aluminum alloys. J. Inst. Metals 1943, *69*, 373–398.
39. Clark, J.B. Age hardening in a Mg–9wt.% Al alloy. Acta Metallurgica 1968, *16*, 141–152.
40. Duly, D.; Simon, J.P.; Brechet, Y. On the competition between continuous and discontinuous precipitations in binary Mg–Al alloys. Acta Metallurgica Materialia 1995, *43*, 101–106.
41. Celotto, S., et al. A study of the kinetics of precipitation in binary Mg–Al alloys and AZ91E using nuclear magnetic resonance. In *Third International Magnesium Conference*, Lorimer, G.W., Ed.; The Institute of Materials: London, 1997; 391–402.
42. Sumitomo, T.; Cáceres, C.H.; Veidt, M. Elastic modulus effects in Mg–Al alloys. J. Light Metals, 2002, *2*, 49–56.
43. Cáceres, C.H.; Griffiths, J.R.; Davidson C.J.; Newton, C.L. Effects of solidification rate and ageing on the mechanical properties of AZ91 alloy. Mater. Sci. Engng. A, 2002, *325*, 344–355.
44. Smallman, R.E. *Modern physical metallurgy*; Butterworths: London, 1985.
45. Carbonneau, Y., et al. On the development of a new approach for the determination of yield strength in Mg-based alloys. Light Metal Age 1998, (October), 50–53.
46. Zhang, Z.; Couture, A.; Luo, A. An investigation of the properties of Mg–Zn–Al alloys. Scripta Materialia 1998, *39*, 45–53.
47. Fox, F.A. The properties of some magnesium–aluminum–zinc casting alloys and the incidence of microporosity. J. Inst. Metals 1945, *71*, 415–439.
48. Westengen, H.; Holta, O. Low pressure permanent mould casting of magnesium—recent developments. SAE Technical Paper 880509 1988.
49. Roberts, S.C. *Magnesium and its Alloys*; John Wiley and Sons: New York, 1960.
50. Lagowski, B.; Meier, J.W. Characteristics of sand-cast magnesium–zinc alloys. Trans. American Foundrymen's Society 1964, *72*, 561–574.
51. Emley, E.F. *Principles of Magnesium Technology*; Pergamon Press: London, 1966.
52. Lagowski, B.; Meier, J.W. Recent trends in Mg–Al–Zn casting alloys. Trans. American Foundrymen's Society 1968, *76*, 133–141.
53. Bowles, A.L., et al. The effect of low temperature ageing on the tensile properties of high-pressure die-cast Mg–Al alloys. In *Magnesium Technology 2000*, Kaplan, H.I.; Hryn, J.; Clow, B., Eds.; The Minerals, Metals and Materials Society (TMS): Warrendale, PA, 2000; 295–300.
54. Gharghouri, M.A. Elastic constants of $Al_{12}Mg_{17}$ intermetallic. unpublished, 2001.
55. Hull, D. *An Introduction to Composite Materials*; Cambridge University Press: Cambridge, 1981.
56. Cáceres, C.H.; Rovera, D.M. Solid solution strengthening in concentrated Mg-Al alloys. J. Light Metals, 2001, *1/3*, 151–156.
57. Cáceres, C.H. Quality Index charts for permanent-mould and sand-cast Mg-based alloys. AFS Trans. 2001, *109*, 1–8.
58. Green, R.D. Porosity-free magnesium alloy castings thermal requirements. Trans. American Foundrymen's Society 1960, *68*, 245–252.
59. Cáceres, C.H. "Material properties and process assessment in Mg-Al casting alloys using strength-ductility charts", Int. J. Cast Metals Res., 2001, *14*, 185–197.
60. Suman, C. The effect of direct aging on mechanical properties and corrosion resistance of diecast magnesium alloys AZ91D and AM60B. In *SAE Technical Paper 900794*; SAE:Warrendale, PA, 1990; 849–859.

61. Sequeira W., et al. Effect of section thickness and gate velocity on the microstructure and mechanical properties of high pressure die cast magnesium alloy AZ91D. Automotive Alloys; The Minerals, Metals and Materials Society: 1997; 169–183.
62. Schindelbacher, G.; Rösch, R. Mechanical properties of magnesium die casting alloys at elevated temperatures and microstructure in dependence of wall thickness. In *Magnesium Alloys and Their Applications*, ; Mordike, B.L.; Kainer, K.U., Eds.; Werkstoff-Informationsgesellschaft mbH: Frankfurt, 1998; 247–252.
63. El-Mahallawy, N.A., et al. On the influence of process variables on the thermal conditions and properties of high pressure die-cast magnesium alloys. J. Materials Processing Technol. 1998, *73 (1–3)*, 125–138.
64. Gokhale, A.M.; Patel, G.R. Origins of variability in the mechanical properties of AM60 magnesium alloy castings. In *Magnesium Technology 2001*, Hryn, J., Ed.; The Minerals, Metals and Materials Society (TMS): Warrendale, PA, 2001; 195–199.
65. Liu, Z.; Chen, L.; Zhao, H.; Wang, Y.; Wang, Z.; Klein, F. The influence of porosity of magnesium die casting alloys on its mechanical properties. *Metall* 2000, *54*, 122–125.
66. Kanahashi, H., et al. Experimental study for the improvement of crashworthiness in AZ91 magnesium foam controlling its microstructure. Mater. Sci. Engineering A 2001, *308*, 283–287.
67. Polmear, I.J. 3rd Ed. *Light Alloys: Metallurgy of the Light Metals*; John Wiley and Sons: New York, 1995.
68. Luo, A.; Pekgüleryüz, M.Ö. Review cast magnesium alloys for elevated temperature applications. J. Mater. Sci. 1994, *29*, 5259–5271.
69. Waltrip, J.S. Fresh look at some old Mg die casting alloys for elevated temperature applications. In 47th Annual World of Magnesium Conference; International Magnesium Association:Cannes, France, 1990.
70. Fukuchi, M.; Watanabe, K.J. Japan Inst. Light Metals 1975, *39*, 493.
71. Raynor, G.V. *The Physical Metallurgy of Magnesium and its Alloys*; Pergamon Press: UK, 1959
72. Dargusch, M. The role of microstructure in the creep of die cast magnesium alloys. Ph.D. Dissertation, Department of Mining, Minerals and Materials Engineering, University of Queensland: Brisbane, 1999.
73. Pettersen, G., et al. Microstructure of a pressure die cast magnesium–4wt.percent aluminum alloy modified with rare earth additions. Mater. Sci. Engineering A 1996, *207 (1)*, 115–120.
74. Powell, B.R., et al. The relationship between microstructure and creep behaviour in AE42 magnesium die casting alloy. *Magnesium Technology 2001*; The Minerals, Metals and Materials Society: New Orleans, 2001.
75. Mordike, B.L.; Ebert, T. Magnesium properties—applications—potential. Mater. Sci. Engineering A 2001, *302*, 37–45.
76. Unsworth, W. The role of rare earth elements in the development of magnesium base alloys. International J. Mater. Product Technol. 1989, *4* (4), 359–378.
77. King, J.F. Development of practical high temperature magnesium casting alloys Magnesium Alloys and their Applications; Deutsche Gesellschaft fur Materialkunde e.V:Munich, Germany, 2000.
78. Nie, J.F.; Muddle, B.C. Characterisation of strengthening precipitate phases in a Mg–Y–Nd alloy. Acta Materialia 2000, *48* (8), 1691–1703.
79. Foerster, G.; Statham, C.D. Magnesium Alloy for Die Casting. US Patent 3,892,565, NL Industries: United States, 1975.
80. Zhang, Z., et al. Solidification microstructure of ZA102, ZA104 and ZA106 magnesium alloys and its effect on creep deformation. Canadian Metallurgical Quarterly 2000, *39 (4)*, 503–512.
81. Zhang, Z., et al. Mechanical properties, microstructure and solidification characteristics of Mg–Zn–Al alloys Light Metals 200 Métaux Légers 2000, 273–284.

82. Anyanmu, I.A., et al. Heat resistance of Mg–Zn–Al–Ca alloy castings. Mater. Sci. Forum 2000, *350–351*, 73–78.
83. Zhang, Z.; Couture, A.; Luo, A. An investigation of the properties of Mg–Zn–Al alloys. Scripta Materialia 1998, *39 (1)*, 45–53.
84. Zhang, Z.; Tremblay, R.; Dubé, D. The Mg–Zn–Al alloys and the influence of calcium on their creep properties. Magnesium Technol. 2001, *2001*, 147–151.
85. Vogel, M., et al. Creep of Mg–Zn–Al alloys. In *Magnesium Alloys and their Applications*; Wiley-VCH: Munich, 2000.
86. Luo, A.; Shinoda, T. Magnesium Alloy having Superior Elevated-Temperature Properties and Die Castability. US Patent 5,855,697; Imra America, Inc: 1999.
87. Pekgüleryüz, M.Ö.; Baril, E. Development of creep resistant Mg–Al–Sr alloys. Magnesium Technol. 2001, *2001*, 119–125.
88. Argo, D., et al. Diecastability and properties of Mg–Al–Sr based alloys. Magnesium Technol. 2001, *2001*, 131–136.
89. Luo, A.; Powell, B.R. Tensile and compressive creep of magnesium–aluminum–calcium based alloys. Magnesium Technol. 2001, *2001*, 137–44.
90. Koike, S., et al. Development of lightweight oil pans made of a heat-resistant magnesium alloy for hybrid engines. SAE Paper no. 2000–01–1117, 2000.
91. Ninomiya, R.; Ojiro, T.; Kubota, K. Improved heat resistance of Mg–Al alloys by the Ca addition. Acta Metallurgica et Materialia 1995, *43 (2)*, 669–674.
92. Sakamoto, M.; Akiyama, S.; Ogi, K. Characterisation of the oxidation surface layer on the non-combustible Ca-bearing Mg melts. In Fourth Asian Foundry Congress, 1997; 467–476.
93. You, B.-S.; Park, W.-W.; Chung, I.-S. The effect of calcium additions on the oxidation behaviour in magnesium alloys. Scripta Materialia 2000, *42*, 1089–1094.
94. Bettles, C.J.; Humble, P.; Nie, J.F. The effect of trace additions on the ageing behaviour of AZ91E. Proceedings of the Third International Magnesium Conference; : 1996; 403–417.
95. Ozturk, K.; Zhong, Y.; Liu, Z.-K. Computational thermodynamics and experimental investigation of Mg–Al–Ca alloys. Magnesium Technol. 2001, *2001*, 113–118.
96. Pisch, A.; Gröbner, J.; Schmid-Fetzer, R. Application of computational thermochemistry to Al and Mg alloy processing with Sc additions. Mater. Sci. Engineering A 2000, *289*, 123–129.
97. von Buch, F., et al. Development of Mg–Sc–Mn alloys. Mater. Sci. Engineering A 1999, *263*, 1–7.
98. von Buch, F.; Mordike, B.L. Microstructure, mechanical properties and creep resistance of binary and more complex magnesium scandium alloys. Magnesium Alloys Applications 2000, 145–150.
99. Mordike, B.L.; von Buch, F. Development of high temperature creep resistant alloys. Magnesium Alloys Applications 2000, 35–40.
100. Gröbner, J.; Kevorkov, D.; Schid-Fetzer, R. Magnesium alloy development guided by thermodynamic calculations. Magnesium Technol. 2001, *2001*, 105–112.
101. Spigarelli, S., et al. Process and microstructural effects on the creep properties of the AZ91 magnesium alloy. In Magnesium 2000—Proceedings of the Second Israeli International Conference on Magnesium Science and Technology, 2000; 293–300.
102. Regev, M.; Aghion, E.; Rosen, A. Creep studies of AZ91D pressure die casting. Mater. Sci. Engineering A 1997, *234*, 123–126.
103. Regev, M., et al. Creep studies of coarse-grained AZ91D magnesium castings. Mater. Sci. Engineering A 1998, *252* (1), 6–16.
104. Dahle, A.K., et al. Optimisation of the quality of high pressure die cast magnesium alloys. Automotive Alloys 1999, .
105. Mao, H., et al. Microstructural characteristics of die cast AZ91D and AM60 magnesium alloys. SAE Paper no. 999–01–0928, 1999.
106. McCormack, P.D.; Crane, L. *Physical Fluid Dynamics*; Academic Press: New York, 1973; 407.

107. Dahle, A.K.; StJohn, D.H. Rheological behaviour of the Mushy zone and its effect on the formation of casting defects during solidification. Acta Materialia 1999, *47 (1)*, 31–41.
108. Mao, H. Microstructural features of cold chamber die cast AM60B. Masters Thesis, Ohio State University, Columbus, OH, 1999.
109. Rodrigo, P.D.D.; Ahuja, V. Effect of casting parameters on the formation of 'pore/segrega-segregation bands' in magnesium die castings. In Magnesium 2000—Proceedings of the Second Israeli Conference on Magnesium Science and Technology, 2000; 97–104.
110. Dahle, A.K.; StJohn, D.H. The origin of banded defects in high pressure die cast magnesium alloys. In NADCA Conference Cleveland, 1999.
111. Murray, M.T.; Cope, M.A. Magnesium pressure casting. International Patent WO 99/28065, 1999.
112. Sequeira, W.; Dunlop, G.L.; Murray, M.T. Effect of section thickness and microstructure on the mechanical properties of high pressure die cast magnesium alloy AZ91D. In Proceedings of the Third International Magnesium Conference, 1996; 63–73.
113. Rodrigo, P.D.D., et al. Effects of section size and microstructural features on the mechanical properties of die cast AZ91D and AM60B magnesium alloy test bars. SAE Paper no. 1999–01–0927, 1999.
114. Haerle, A.G., et al. The effect of non-metallic inclusions on the properties of die cast magnesium. SAE paper no. 970331, 1997.
115. Stich, A.; Haldenwanger, H.G. Dimensioning strategy for high-stress cast magnesium. In Magnesium 2000—Proceedings of the Second Israeli International Conference on Magnesium Science and Technology, 2000; 27–34.
116. VanFleteren, R. Magnesium for automotive applications. Advanced Mater. Processes 1996, *5*, 34–35.
117. Gutman, E.M., et al. Influence of porosity and casting conditions on creep of die-cast Mg alloy. J. Mater. Sci. Lett. 1998, *17*, 1787–1789.
118. Gutman, E.M., et al. Influence of technological parameters of permanent mold casting and die casting on creep and strength of Mg alloy Az91d. Mater. Sci. Engineering A—Structural Materials Properties Microstructure and Processing 1997, *234*, 880–883.
119. Renaud, J.; Beaulieau, É.; Fairchild, S. Improvement of strain at fracture and porosity levels in magnesium die castings using the taguchi design of experiments and analysis. SAE Paper no. 970327, 1997.
120. Sannes, S.; Westengen, H. *The Influence of Process Conditions on the Microstructure and Mechanical Properties of Magnesium Die Castings*, Magnesium Alloys and Their Applications, Wolfsburg, Germany, Werkstoff-Informations GmbH, 1998, 223–228.
121. Drouzy, M.; Jacob, S.; Richard, M. Interpretation of tensile results by means of quality index and probable yield strength. AFS International Cast Metals J. 1980, *5*, 43–50.
122. Cáceres, C.H. A rationale for the Quality Index of Al–Si–Mg casting alloys. International J. Cast Metals Res. 1998, *10*, 293–299.
123. Dieter, G.E. 3rd *Mechanical Metallurgy*; McGraw-Hill: New York, 1986.
124. Cáceres, C.H. A phenomenological approach to the Quality Index. International J. Cast Metals Res. 2000, *12*, 367–375.
125. Cáceres, C.H.; Davidson, C.J.; Griffiths, J.R. The deformation and fracture behaviour of an Al–Si–Mg casting alloy. Mater. Sci. Engng. A, 1995, *A197*, 171–179.
126. Bag, A.; Zhou, W. Tensile and fatigue behaviour of AZ91D magnesium alloy J. Mater. Sci. Lett. 2001, *20*, 457–459.
127. Ashby, M.F. In 2ndMaterials Selection in Mechanical Design. Butterworth-Heinemann: Oxford, 1999.
128. Okamoto, H. Al–Mg (Aluminum–Magnesium). J. Phase Equilibria 1998, *19 (December)*, 598.
129. Aune, T.; Westengen, H.; Ruden, T. Mechanical properties of energy absorbing magnesium alloys. SAE Technical Paper 930418; SAE: Warrendale, PA, 1993; 51–57.
130. Mima, G.; Tanaka, Y. Mechanism of precipitation hardening of magnesium–zinc alloys Trans. Japan Inst. Metals 1971, *12*, 323–328.

131. Bancroft, C.L.; Caceres, C.H.; Griffiths, J.R. On the relation between hardness and yield strength in aged sand cast AZ91 alloy. In *Magnesium Technology 2000*, Kaplan, H.I.; Hryn, J.; Clow, B., Eds.; The Minerals, Metals and Materials Society: Warrendale, PA, 2000; 291–294.
132. Sequeira, W.P., et al. Effect of section thickness and gate velocity on the microstructure and mechanical properties of high pressure die cast magnesium alloy AZ91D. TMS Symposium on Automotive Alloys; The Minerals Metals and Materials Society (TMS):Warrendale, PA, 1997; 169–183.
133. Sakkinen, D.J. Physical metallurgy of magnesium die casting alloys. SAE Technical Paper 940779; SAE: Warrendale, PA, 1994; 558–569.
134. Mackenzie, J.K. The elastic constants of a solid containing spherical holes. Proc. Phys. Soc. B 1950, *63*, 2–11.
135. Aune, T.; Westengen, H.; Ruden, T. The effect of varying aluminum and rare-earth content on the mechanical properties of die cast magnesium alloys. SAE Technical Paper 940777; SAE:Warrendale, PA, 1994; 553–557.
136. Lagowski, B.; Meier, J.W. Properties of sand-cast magnesium alloys. Part III. Effect of titanium additions to magnesium–zinc alloys. Mines Branch Research, Department of Mines and Technical Suveys, reports R56, 1960 (28 pages).
137. Lagowski, B.; Meier, J.W. Further development of Mg–Zn–Ag–Zr casting alloys. Trans. American Foundrymen's Society 1965, *73*, 246–254.
138. Aune, T.; Westengen, H. Property update on magnesium die casting alloys. SAE Technical Paper 950424; SAE:Warrendale, PA, 1995; 332–342.
139. Stich, A.; Haldenwanger, H.G. *Dimensioning strategy for high-stress cast magnesium components*, Aghion, E. and Eliezer, D. Magnesium 2000 (2nd Israeli Intl. Conf. on Magnesium Science and Technology, Dead Sea), MRI (Beer-heva), 2000, pp. 27–34.
140. Siedersleben, M. Production of magnesium structural parts in vacuum die casting. In *Magnesium Alloys and Their Applications*, Mordike, B.L.; Kainer, K.U., Eds.; Werkstoff-Informationsgesellschaft mbH: Frankfurt, 1998; 477–482.
141. Basner, T.G.; Evans, M.; Sakkinen, D.J. The effect of extended time aging on the mechanical properties of vertical vacuum cast aluminum–manganese and aluminum–rare earth magnesium alloys. SAE Technical Paper 930419; SAE International: Warrendale, PA, 1993; 59–64.
142. Agnew, S.R.; Liu, K.C.; Kenik, E.A.; Viswanathan, S. *Tensile and Compressive Creep Behaviour of Die Cast Magnesium Alloy AM60B*, Magnesium Technology 2000, Nashville, USA, The Minerals, Metals and Materials Society (TMS) Warrendale, PA, 2000, pp. 285–290.
143. Aune, T.K.; Westengen, H. Mechanical properties of pressure die cast Mg-alloys. Magnesium Alloys Applications 1992, 221–228.

11

Designing with Titanium Alloys

Michelle L. McCann and John Fanning
TIMET Henderson Technical Laboratory, Henderson, Nevada, U.S.A.

I. INTRODUCTION

Titanium offers an excellent combination of mechanical properties, high strength-to-weight ratio and corrosion resistance. These features, coupled with availability of product forms and ease of fabrication, have led to extensive use of titanium and its alloys in the chemical process industry, the aerospace industry, and numerous other industries. Titanium is now a standard material of construction.

Successful utilization requires careful consideration of titanium's unique characteristics at the design stage as well as during fabrication. Factors such as titanium's high strength-to-weight ratio, low elastic modulus, corrosion and erosion resistance, its tendency toward galling, and its reactivity at high temperatures must be considered in order to optimize designs in titanium.

II. DESIGNING WITH TITANIUM

The successful design of a titanium component begins with consideration of the environment to which the part is to be exposed. The corrodents present and maximum operating temperature (under upset conditions, possibly) will dictate which titanium alloy should be selected. The physical and mechanical properties of the alloy selected may, in turn, dictate some design features. For example, the ductility of an alloy limits the minimum bend radius, which is feasible for sheet, plate, or tubing. The excellent corrosion resistance of titanium often permits a zero corrosion allowance to be specified. This chapter will cover design parameters for commercially pure (CP) titanium (ASTM Grades 1, 2, 3, 4, 7, and 11), Alpha–Beta Titanium Alloy Ti–6Al–4V (ASTM Grade 5) and Beta Titanium Alloys Ti–15V–3Cr–3Sn–3Al (Ti-15-3), Ti–10V–2Fe–3Al, and Beta 21S.

The contents of this chapter were derived from TIMET internal documents in conjunction with the references noted.

III. MATERIAL PROPERTIES

"The material properties of titanium and its alloys are determined mainly by their alloy content and heat treatment, both of which are influential in determining the allotropic forms in which this material will be bound. Under equilibrium conditions, pure titanium has a hexagonal close packed (alpha) structure up to 1620°F (882°C), above which it transforms to a body center cubic (beta) structure. The inherent properties of these two structures are quite different. Through alloying and heat treatment one or the other or a combination of these two structures can be made to exist at service temperatures, and the properties of the material vary accordingly." [1]

IV. COMMERCIALLY PURE TITANIUM

Commercially pure (CP) titanium is used for applications that require moderate strength combined with good formability and excellent corrosion resistance. CP titanium has been a mill product since the 1950s and was developed largely because of aerospace demands for a material lighter than steel and more heat resistant than aluminum [2].

CP titanium is available in several different grades, which are identified by varying amounts of trace elements such as carbon, hydrogen, iron, nitrogen, and oxygen. The strength and ductility of the material are greatly affected by small amounts of these trace elements (primarily oxygen). Therefore, the CP grades are generally distinguished by mechanical properties instead of chemical properties. Palladium (Ti–0.2Pd), ruthenium (Ti–Ru) or nickel–molybdenum (Ti–0.3Mo–0.8Ni) may also be added to CP to enhance the corrosion resistant and/or strength properties of the titanium in certain environments [2]. The chemical composition for the different CP grades is shown in Table 1.

ASTM Grade 1 has the highest purity, lowest strength, and best room-temperature ductility and formability of the four CP grades. Grade 1 should be used where maximum formability is required such as explosive bonding and plate-type heat exchangers. It exhibits excellent corrosion resistance in highly oxidizing to mildly reducing environments, including chlorides.

ASTM Grade 2 is the workhorse for industrial applications, having a guaranteed minimum 0.2% yield strength of 40 ksi (275 MPa) and good ductility and formability. Grade 2 is used extensively for seawater piping, reactor vessels, and heat exchangers.

Table 1 Chemical Composition for ASTM Commercially Pure Titanium Grades

ASTM Grade	Oxygen (wt %)	Nitrogen (wt %)	Carbon (wt %)	Iron (wt %)	Hydrogen[a] (wt %)	Residual elements (wt %)	Titanium (wt %)
1	0.18	0.03	0.08	0.20	0.015	0.40	Remainder
2	0.25	0.03	0.08	0.30	0.015	0.40	Remainder
3	0.35	0.05	0.08	0.30	0.015	0.40	Remainder
4	0.40	0.05	0.08	0.50	0.015	0.40	Remainder

[a]Hydrogen content depends on product form.

It is also ASME approved, has good impact properties at low temperatures, and has excellent resistance to erosion and corrosion by seawater and marine atmospheres.

ASTM Grade 3 is a general purpose grade of CP that has excellent corrosion resistance in highly oxidizing to mildly reducing environments, including chlorides, and an excellent strength-to-weight ratio. Grade 3 offers the highest ASME allowable design stresses of any CP grade of titanium.

ASTM Grade 4 has the highest strength of the four CP grades in addition to good ductility and moderate formability. The benefits of the good strength-to-weight ratio of Grade 4 are retained at moderate temperatures.

All the four ASTM grades are typically used in continuous service up to 800°F (426°C) and in intermittent service up to 1000°F (538°C). They have good impact properties at low temperatures. In addition, all four grades can be readily welded, machined, cold worked, hot worked, and cast.

A. Product Forms Available/ASTM Specifications

The titanium product forms available are given in Table 2.

B. Physical Properties

The density for all the CP grades is 0.163 lb in.$^{-3}$ (4.51 g cm^{-3}). The elastic modulus at room temperature [68–78°F (20–25°C)] is 15.2–17.4 Msi (105–120 GPa) and all the CP titanium grades are nonmagnetic. All other physical properties for the CP grades are given in Table 3.

C. Mechanical Properties

Mechanical properties for the CP grades are given in Table 4. Typical elevated tensile properties are given in Table 5.

D. Fatigue and Tensile Properties

Fatigue and tensile properties for the CP grades are given in Table 6.

Table 2 Commercially Pure and Modified Titanium Product Forms

	ASTM Grade 1	ASTM Grade 2	ASTM Grade 3	ASTM Grade 4	ASTM Grade 7	ASTM Grade 11
Strip, sheet, plate	X	X	X	X	X	X
Welded pipe	X	X	X	X	X	X
Welded tubing	X	X	X	X	X	X
Bars and billet	X	X	X	X	X	X
Weld fittings	X	X	X	X	X	X
Castings	X	X	X	X	X	
Forgings	X	X	X	X	X	X

Table 3 Physical Properties for Commercially Pure Titanium Grades

	Beta transus	Thermal conductivity	Electrical resistivity	Mean coefficient of thermal expansion		
				68–212°F (20–100°C)	68–572°F (20–300°C)	68–932°F (20–500°C)
ASTM	1640°F	12.70 Btu $hr^{-1} ft^{-1} °F^{-1}$	18 μΩ in.	4.8×10^{-6} in. $in.^{-1} °F^{-1}$	5.3×10^{-6} in. $in.^{-1} °F^{-1}$	5.4×10^{-6} in. $in.^{-1} °F^{-1}$
Grade 1	890°C	21.97 W $m^{-1} K^{-1}$	0.45 μΩ m	8.6×10^{-6} m $m^{-1} °C^{-1}$	9.5×10^{-6} m $m^{-1} °C^{-1}$	9.7×10^{-6} m $m^{-1} °C^{-1}$
ASTM	1680°F	12.60 Btu $hr^{-1} ft^{-1} °F^{-1}$	21 μΩ in.	4.8×10^{-6} in. $in.^{-1} °F^{-1}$	5.3×10^{-6} in. $in.^{-1} °F^{-1}$	5.4×10^{-6} in. $in.^{-1} °F^{-1}$
Grade 2	915°C	21.79 W $m^{-1} K^{-1}$	0.53 μΩ m	8.6×10^{-6} m $m^{-1} °C^{-1}$	9.5×10^{-6} m $m^{-1} °C^{-1}$	9.7×10^{-6} m $m^{-1} °C^{-1}$
ASTM	1690°F	12.60 Btu $hr^{-1} ft^{-1} °F^{-1}$	21 μΩ in.	4.8×10^{-6} in. $in.^{-1} °F^{-1}$	5.3×10^{-6} in. $in.^{-1} °F^{-1}$	5.4×10^{-6} in. $in.^{-1} °F^{-1}$
Grade 3	920°C	21.79 W $m^{-1} K^{-1}$	0.54 μΩ m	8.6×10^{-6} m $m^{-1} °C^{-1}$	9.5×10^{-6} m $m^{-1} °C^{-1}$	9.7×10^{-6} m $m^{-1} °C^{-1}$
ASTM	1740°F	9.80 Btu $hr^{-1} ft^{-1} °F^{-1}$	24 μΩ in.	4.8×10^{-6} in. $in.^{-1} °F^{-1}$	5.3×10^{-6} in. $in.^{-1} °F^{-1}$	5.4×10^{-6} in. $in.^{-1} °F^{-1}$
Grade 4	950°C	16.95 W $m^{-1} K^{-1}$	0.60 μΩ m	8.6×10^{-6} m $m^{-1} °C^{-1}$	9.5×10^{-6} m $m^{-1} °C^{-1}$	9.7×10^{-6} m $m^{-1} °C^{-1}$

Table 4 Typical Mechanical Properties for Commercially Pure Titanium Grades

	UTS ksi (MPa)	0.2%YS ksi (MPa)	Elongation (%)	Reduction in area (%)	Bend radius
ASTM Grade 1	50 (345)	35 (220)	35	70	2.0T
ASTM Grade 2	70 (485)	50 (345)	28	57	2.5T
ASTM Grade 3	85 (585)	65 (450)	25	48	2.5T
ASTM Grade 4	99 (680)	81 (560)	23	46	3.0T

E. Microstructure

The microstructures for CP grades and modified CP grades (7, 11 and 12) are typically 100% α-crystal structures at room temperature. However, as levels of trace elements (primarily iron) increase, small amounts of spherical β are observed at the grain boundaries [2]. A typical microstructure for CP is shown in Fig. 1.

F. Manufacturing Considerations

CP titanium is supplied in the annealed condition and can be formed at elevated [300°F (149°C)–900°F (482°C)] or room temperatures. Property degradation can be experienced after forming. However, re-annealing the as-received material will restore the properties [1].

CP titanium is fully annealed by heat treating at 1000°F (538°C)–1300°F (704°C) for 10–30 min. It is stress relieved by heating at 900°F (482°C)–1000°F (538°C) for 30 min. Commercially pure titanium cannot be hardened by heat treatment [1].

G. Environmental

Titanium has an unusually high affinity for oxygen, nitrogen, and hydrogen at temperatures above 1050°F (565°C), which results in embrittlement of the material. Therefore, usage of the material should be limited to temperatures below 1050°F (565°C). Additional chemical reactivity between titanium and selected environments, such as methyl alcohol, chloride salt solutions, hydrogen, and liquid metal, can take place at lower temperatures [1].

V. Ti-6Al-4V (ASTM Grade 5)

Ti-6Al-4V (Ti-6-4) is the most widely used of all the titanium alloys. Introduced in 1954, this workhorse of the industry has a broad spectrum of good-to-excellent properties. As such, Ti-6-4 may rightfully be considered the general purpose titanium alloy.

Ti-6-4 is available as sheet, plate, billet, extrusion, castings, bar, and ingot. Ti-6-4 responds to heat treatment and typically its full ultimate strength of 160 ksi (1100 MPa) can be developed in sections up to 1 in. (25 mm) thick. Thicker sections can be heat treated but to lesser strength. Ti-6-4 is characterized as an alpha-rich alpha–beta titanium alloy.

With a density of 0.160 lb/in.^{3} (4.43 g/cm^3), Ti-6-4 possesses high structural efficiency.

Fatigue properties are excellent. Crack initiation is not affected by water or by salt below 450°F (230°C). Crack propagation resistance is also excellent under static or

Table 5 Typical Elevated Tensile Properties for Commercially Pure Titanium Grades

	ASTM Grade 1			ASTM Grade 2			ASTM Grade 3			ASTM Grade 4		
Test Temperature	0.2% YS ksi (MPa)	UTS ksi (MPa)	% E	0.2% YS ksi (MPa)	UTS ksi (MPa)	% E	0.2% YS ksi (MPa)	UTS ksi (MPa)	% E	0.2% YS ksi (MPa)	UTS ksi (MPa)	% E
212°F (100°C)	23 (156)	41 (284)	38	37 (256)	56 (383)	31	41 (283)	64 (438)	31	53 (364)	79 (537)	27
392°F (200°C)	12 (85)	32 (219)	40	28 (192)	41 (280)	37	28 (195)	47 (327)	36	33 (231)	55 (381)	33
572°F (300°C)	7 (51)	24 (163)	48	18 (127)	33 (229)	43	19 (129)	36 (250)	39	22 (149)	41 (284)	35
752°F (400°C)	5 (36)	19 (130)	51	13 (92)	27 (186)	38	15 (101)	28 (195)	33	15 (101)	33 (227)	33
842°F (500°C)	4 (31)	17 (117)	49	11 (74)	26 (178)	34	13 (92)	25 (174)	—	13 (92)	30 (209)	—

Table 6 Fatigue and Tensile Properties for Commercially Pure Titanium Grades

	ASTM Grade 1			ASTM Grade 2			ASTM Grade 3			ASTM Grade 4		
Condition	UTS ksi (MPa)	Fatigue limit 10^7 cycles ksi (MPa)	Fatigue ratio	UTS ksi (MPa)	Fatigue limit 10^7 cycles ksi (MPa)	Fatigue ratio	UTS ksi (MPa)	Fatigue limit 10^7 cycles ksi (MPa)	Fatigue ratio	UTS ksi (MPa)	Fatigue limit 10^7 cycles ksi (MPa)	Fatigue ratio
Smooth $K_t=1$	51 (354)	±28 (193)	0.55	60 (417)	±34 (232)	0.56	85 (589)	±40 (278)	0.47	98 (674)	±55 (376)	0.56
Notched $K_t=3$	51 (354)	±18 (123)	0.35	60 (417)	±22 (154)	0.37	85 (589)	±18 (123)	0.21	—	—	—

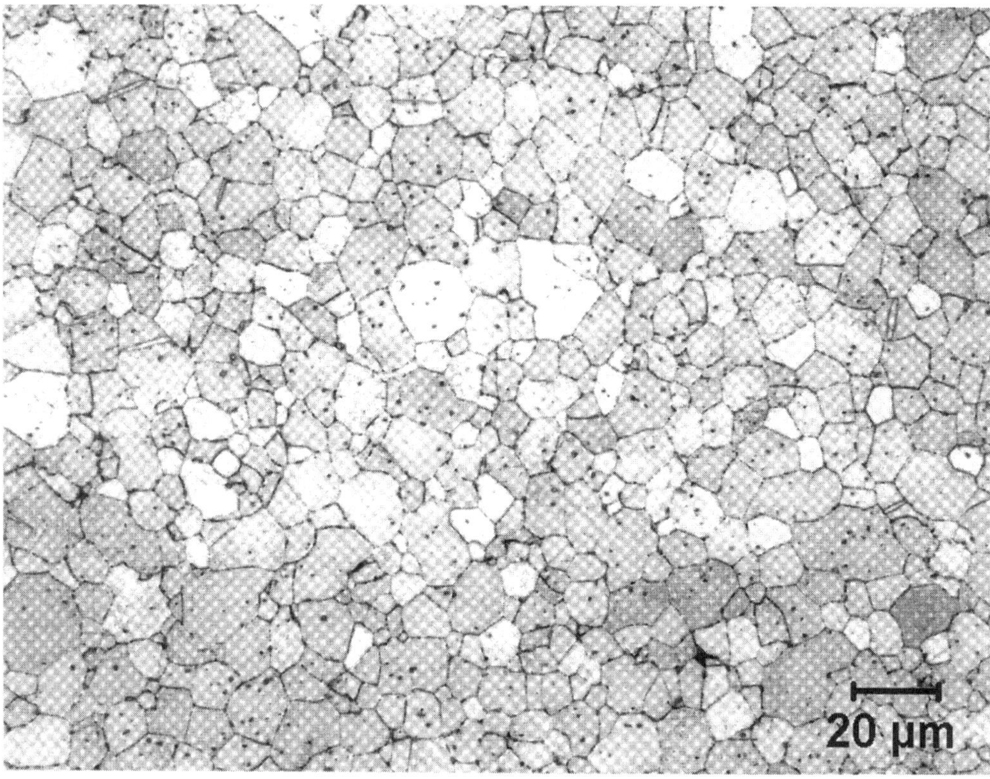

Figure 1 Typical microstructure of commercially pure Grade 2 Titanium. Equiaxed α grains; with iron stabilized spherical β.

dynamic load spectra. Salt water can affect crack propagation, the degree depending on interstitial content, and load spectra. Crack initiation and propagation are both affected by heat treatment.

Ti-6-4 is typically used at temperatures from −350°F to 750°F (−210°C to 400°C). Applications outside this temperature range are possible in certain situations.

The alloy is weldable. Complex shapes can be made via hot forming and machines similar to an austenitic stainless steel.

A. Chemical Composition

The properties available in Ti-6-4 are influenced significantly by composition. Two basic levels of interstitial content are available, standard Ti-6-4 and Ti-6-4 ELI. ELI is an acronym for Extra Low Interstitial, meaning primarily low oxygen. Typical compositions are given in Table 7 as exemplified by AMS and Military Specifications.

B. Product Forms Available/ASTM Specifications

> "Ti-6-4 is available in all mill product forms as well as casting and powder metallurgy forms. It can be used in either the annealed or solution-treated plus aged (STA) conditions. For maximum toughness, Ti-6-4 should be used in the

Table 7 Chemical Composition for Ti-6-4

	Standard wt % (AMS 4911D)		ELI wt % (AMS 4907 C)	
	Minimum	Maximum	Minimum	Maximum
Aluminum	5.50	6.75	5.50	6.50
Vanadium	3.50	4.50	3.50	4.50
Iron	—	0.30	—	0.25
Oxygen	—	0.20	—	0.13
Carbon	—	0.08	—	0.08
Nitrogen	—	0.05	—	0.05
Hydrogen	—	0.015	—	0.0125
Yttrium	—	0.005	—	—
Residual elements, each	—	0.10	—	0.10
Residual elements, total	—	0.40	—	0.30
Titanium	Remainder		Remainder	

mill annealed, duplex-annealed, or beta annealed conditions whereas for maximum strength, the STA condition is used. The full strength potential of this alloy is not available in sections greater than 1 inch" [1].

C. Physical Properties

Most of the physical properties of titanium and its alloys depend on the direction in which they are taken. This feature arises for two reasons: (1) both alpha and beta crystals are anisotropic such that most physical properties will have values that depend on the direction in which they are measured in the crystal, and (2) both alpha and beta in Ti-6-4 tend to be textured; that is, the crystallite axes tend to lie along preferred directions with respect to the direction of metal flow set up during processing. In general, any physical property that is not a simple scalar quantity will show at least some anisotropy in Ti-6-4. In the following, those physical properties that depend on test direction will be so indicated.

1. Density

The density of Ti-6-4 is 0.160 lb/in.^{3} (4.43 g/cm^3). This density is only 56% of that for steel.

2. Thermal Expansion

Figure 2 summarizes the reported data. The thermal expansion of Ti-6-4 is about half of that of austenitic stainless steel and about one-third that of aluminum. The following equation may be used to approximate $\Delta L/L_o$ in percent. $\Delta L/L_o$ (%)$= -0.220 + 5.992 \times 10^{-4}T + 5.807 \times 10^{-7}T^2 - 1.994 \times 10^{-10}T^3$. The attending error bar is approximately $\pm$ 0.025%. T is in K.

3. Thermal Conductivity

Values reported are shown in Fig. 3 as function of temperature. These values are similar to those for austenitic stainless steel.

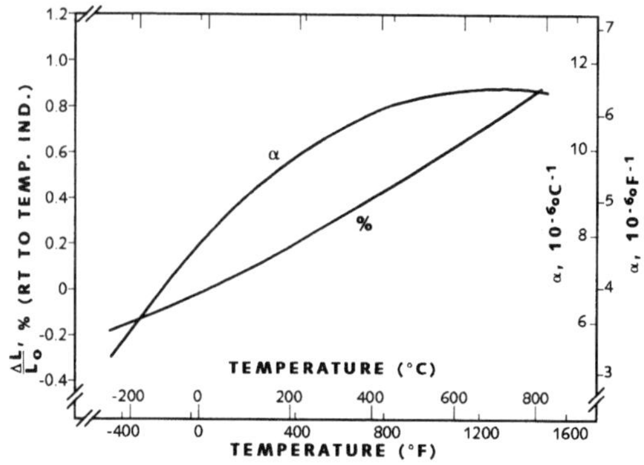

Figure 2 Effect of temperature on thermal expansion of *TIMETAL* 6-4. (From Ref. 3.)

4. Thermal Diffusivity

This quantity also varies with direction and the literature data are scattered. The trend line with temperature, along with two sigma values attending the data plotted, is given in Fig. 4.

5. Specific Heat

Figure 5 illustrates some determinations of specific heat. The variations between investigations here are due to compositional variance or experimental error or both.

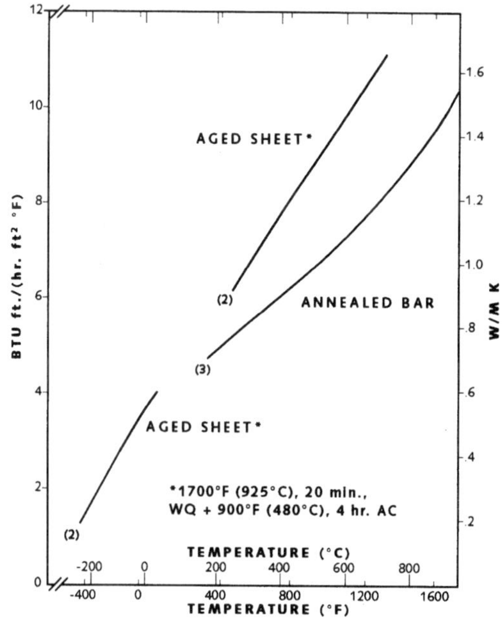

Figure 3 Thermal conductivity of *TIMETAL* 6-4 (numbers in parentheses indicate references). (From Ref. 4.)

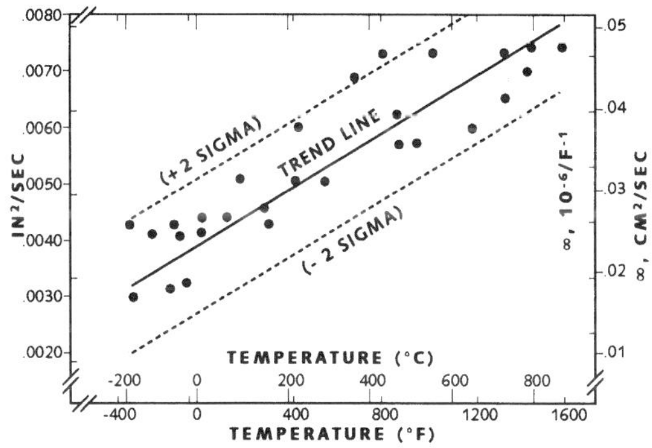

Figure 4 Thermal diffusivity for *TIMETAL* 6-4. (From Ref. 5.)

6. Electrical Resistivity

The resistivity of Ti-6-4 is shown in Fig. 6. Resistivity depends on measurement direction. Therefore, the trend shown would be expected to have an associated scatter band ascribable to texture variation.

7. Emittance

Total emittance and spectral emittance for oxidized and polished surfaces are shown in Figs. 7 and 8. Emittance is both temperature and wavelength sensitive. Values given are in reference to a black body rated at unity.

8. Magnetic Permeability

At 20 oersteds, the permeability of Ti-6-4 is given by one source [8] as 1.00005. The alloy is nonmagnetic. Permeability is direction dependent.

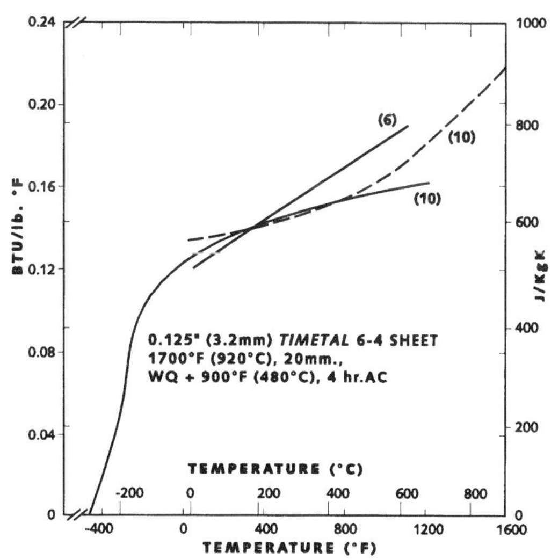

Figure 5 Specific heat. (Numbers in parentheses indicate references).

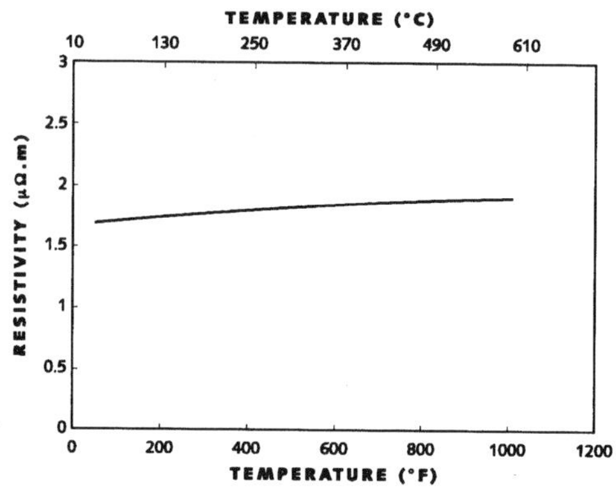

Figure 6 Electrical resistivity. (From Ref. 6.)

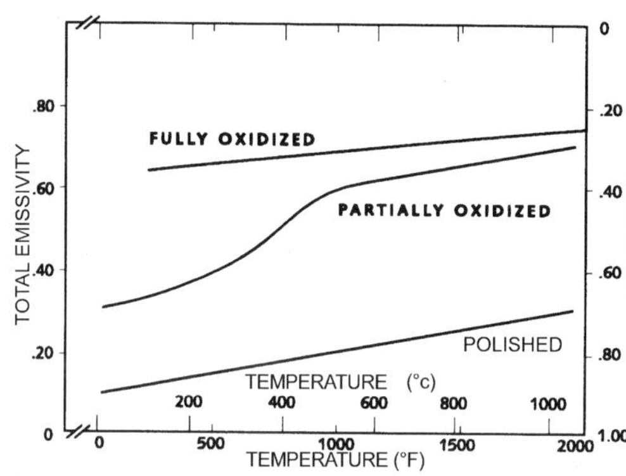

Figure 7 Emissivity of *TIMETAL* 6-4 as dependent upon temperature and surface condition. (From Ref. 7.)

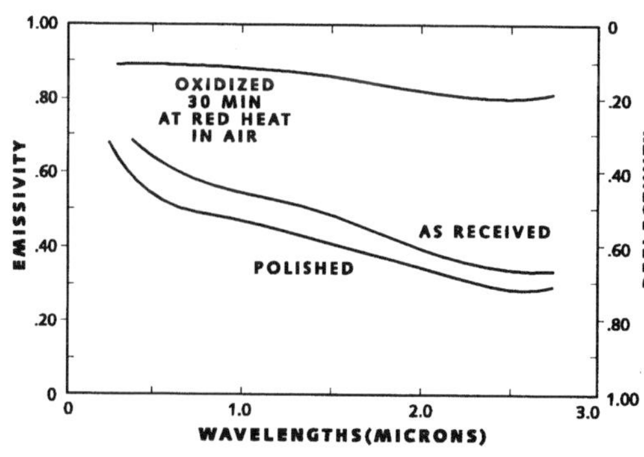

Figure 8 Spectral emissivity of *TIMETAL* 6-4 as function of surface condition. (From Ref. 7.)

9. Melting Range

The melting range of Ti-6-4 is 2965–3000°F (1630–1650°C). This compares with 3047°F (1675°C), the melting point of CP.

10. Phase Change

The temperature at which the alloy transforms completely to the β phase (T_β) depends on the composition. The standard grade transforms at approximately 1825 ± 25°F (996 ± 14°C), whereas Ti-6-4 ELI transforms at approximately 1805 ± 25°F (985 ± 14°C). Note that these are the transformation temperatures for equilibrium conditions. In cases of coarse or blocky alpha microstructures, longer solution times, up to 1 hr or more, may be required for metallographic evaluation of the transformant. The coarse alpha, requiring a variable but finite time to dissolve, may lead one to conclude a higher transformation temperature by metallography or dilatometry than would be found under conditions of thermodynamic equilibrium. One should note also, that under conditions of thermodynamic equilibrium, beta phase is present at room temperature.

D. Mechanical Properties

Mechanical properties for Ti-6-4 and Ti-6-4 ELI are given in Table 8.

E. Design Properties

This section serves several purposes. One is to acquaint the potential user with Ti-6-4. Another is to provide enough in-depth information to avoid certain recognized pitfalls in the process of designing to new highs in efficiency. Toward the first end, typical data are presented for several properties. Some potential pitfalls are discussed in the course of this development. Finally, there is a discussion of how several properties depend on the underlying metallurgy.

1. Modulus and Related Quantities

Tensile modulus depends strongly on crystallographic texture in both alpha and beta titanium. It also depends upon heat treatment. The temperature effect on an annealed sheet is illustrated in Fig. 9. Also shown are the effects from thermal exposure.

Texture-dependent directional effects in sheet are presented in Fig. 10. Compressive moduli are shown in Fig. 11, where scatter again is largely due to texture and heat-treatment effects.

The shear modulus at room temperature also depends on test direction. One reference gives 6.2 × 106 psi (43 × 103 MPa) at room temperature [13]. Another reference [11] gives 6.66 × 106 psi (45.9 × 103 MPa) for the shear modulus. The variation may be due to texture.

Poisson's ratio depends on material texture and measurement directions. Ten observations at TIMET, using a two-element rosette strain gauge, gave a mean value of 0.342 with a range of observations from 0.287 to 0.391 [14]. A further reference gives a single value of 0.31 [13].

2. Internal Friction and Modulus Defects

At low amplitudes, the internal friction Q^{-1} has been found in one investigation to be independent of frequency from 17 kHz to 10 MHz [15]. Internal friction at low amplitudes

Table 8 Mechanical Properties of *TIMETAL* 6-4 and *TIMETAL* 6-4 ELI

TIMETAL 6-4 Product (in.)	Condition	Specification	Dir.	Temperature °F (°C)	UTS ksi (MPa)	0.2% YS ksi (MPa)	Elongation (%)	RA (%)
0.025–1.00 Sheet and Plate	Annealed[a]	ASTM B265	L & LT	68 (20)	130 (895)	120 (828)	10[b]	—
< 3.00 Round or thick	Annealed	ASTM B348	L	68 (20)	130 (895)	120 (828)	10	25
< 4.00 Round or thick	Annealed	Mil-T-9047G[c]	All	68 (20)	130 (895)	120 (828)	10	25
> 4.00–6.00	Annealed	Mil-T-9047G[c]	All	68 (20)	130 (895)	120 (828)	10	20
< 0.500	STD[d]	Mil-T-9047G		68 (20)	165 (1,137)	155 (1,068)	10	20
> 0.500–1.00	STD[d]	Mil-T-9047G		68 (20)	160 (1,103)	150 (1,034)	10	20
> 1.00–1.500	STD[d]	Mil-T-9047G		68 (20)	155 (1,068)	145 (999)	10	20
> 1.500–2.00	rd, sq, hex	Mil-T-9047G		68 (20)	150 (1,034)	140 (965)	10	20
> 2.00–3.00	rd, sq, hex	Mil-T-9047G		68 (20)	140 (965)	130 (896)	10	20
TIMETAL 6-4 ELI								
0.025–1.00 Sheet and plate	ST	ASTM B265	L & LT	68 (20)	120 (828)	110 (759)	10	—
< 3.00 Round or thick	Annealed	ASTM B348	L	68 (20)	120 (828)	110 (759)	10	25
1.75 Round or thick	Annealed[d]	ASTM F136	L	68 (20)	125 (860)	115 (795)	10	25[e]
1.75–2.50	Annealed[d]	ASTM F136	L & LT	68 (20)	120 (825)	110 (760)	8	20
2.50–4.00	Annealed[d]	ASTM F136	L, LT & ST	68 (20)	120 (825)	110 (760)	8	15

[a] 1300°F/1 hr/AC or slower.
[b] For materials less than 0.025″ thick, the elongation must be negotiated with the manufacturer.
[c] Partial list of tensile requirements from the specifications.
[d] Solution treat and age (1650–1775°F)/2 – 120 min/WQ + 900 – 1275°F/2–8 hr/AC.
[e] Only for 0.187–<1.75″ thick material.

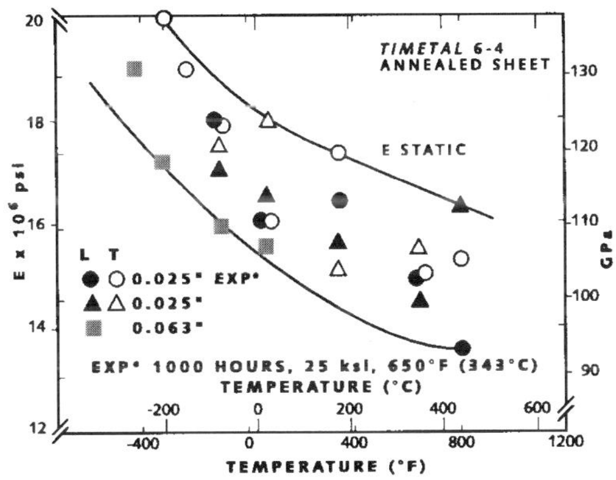

Figure 9 Modulus of elasticity at low and elevated temperatures for annealed Sheet. (From Refs. 9–11.)

is also relatively independent of temperature (within about 30%) from −452°F to 621°F (−269°C to 327°C). There may be an impurity-type peak at −135°F (−93°C). The internal friction Q^{-1} of annealed Ti-6-4 has a low value of 4×10^{-3}. High amplitude measurements show that the internal friction is constant up to longitudinal strains of approximately 4×10^{-3}. Instability sets in at higher strains. Figure 12 illustrates the internal friction Q^{-1} and the modulus defect $\Delta S/S$ as a function of maximum longitudinal stress [15].

The authors reported neither the texture nor test direction with respect to the material processing sequence. However, since internal friction is direction dependent in textured Ti-6-4, these results may not be typical of all the products.

3. Tensile and Notch Tensile Properties

Table 9 shows typical tensile and notch tensile properties of 0.25″ plate as they depend on alloy oxygen content and test temperature. Note the excellent efficiency in the presence of a $K_t = 6.7$ notch. For the annealed condition, Ti-6-4 retains its excellent ductility to liquid

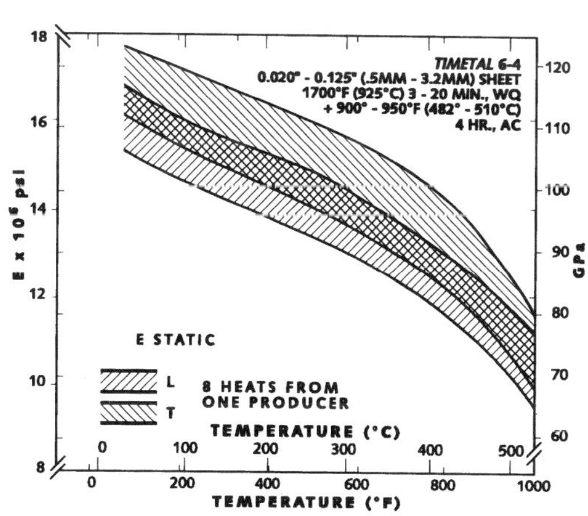

Figure 10 Spread of elastic modulus at room and elevated temperature for eight heats of aged sheet. (From Ref. 12.)

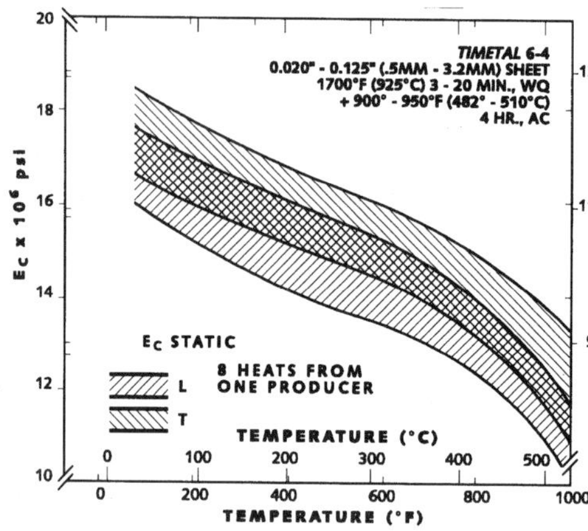

Figure 11 Spread of elastic compressive modulus at room temperature and elevated temperature for eight heats of aged sheet. (From Ref. 12.)

nitrogen temperatures. Both tensile and notch tensile properties of Ti-6-4 may be directional if the hot work done below the transformation temperature has been unidirectional. When such processing cannot be avoided tensile strength in the transverse direction is typically high.

Figures 13 and 14 show the effects of alloyed oxygen and temperature on the tensile and notch tensile properties of Ti-6-4 sheet. Ti-6-4 ELI is to be recommended for service below $-320°F$ ($-196°C$).

Caution should be exercised when designing pressure vessels that contain hydrogen. High hydrogen pressure can lead to embrittlement [16].

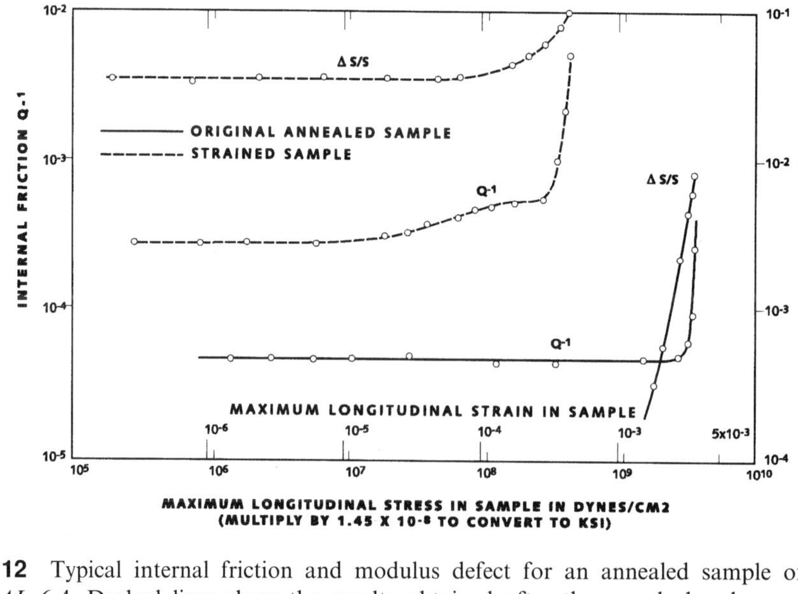

Figure 12 Typical internal friction and modulus defect for an annealed sample of the alloy *TIMETAL* 6-4. Dashed lines show the results obtained after the sample has become unstable. (From Ref. 15.)

Table 9 Typical Tensile and Notch Tensile Properties of Ti-6-4, $\frac{1}{4}''$ (6.4 mm) Plate at Various Temperatures and Two Oxygen Levels

Oxygen content (%)	Test temperature °F	°C	UTS[a] ksi	MPa	YS 0.2% ksi	MPa	El %	RA %	NTS[b] K_t=6.7 ksi	MPa	NTS/UTS
					Annealed Condition						
0.08	−320	−196	211	1,455	201	1,385	19	44	273	1,880	1.29
0.08	−110	79	159	1,095	152	1,050	20	38	222	1,530	1.40
0.08	RT	RT	134	925	126	870	21	44	201	1,385	1.50
0.16	−320	−196	230	1,585	216	1,490	21	36	245	1,690	1.07
0.16	−110	−79	177	1,220	168	1,160	24	35	220	1,515	1.24
0.16	RT	RT	151	1,040	143	985	21	42	209	1,440	1.38
					STA Condtions[c]						
0.08	−320	−196	251	1,730	236	1,625	2	8	270	1,860	1.08
0.08	−110	−79	194	1,340	183	1,260	16	46	240	1,655	1.24
0.08	RT	RT	166	1,145	153	1,055	18	57	226	1,560	1.36
0.08	340	171	142	980	123	850	18	66	214	1,475	1.51
0.08	400	204	123	850	100	690	17	69	—	—	—
0.16	−320	−196	269	1,855	251	1,730	4	6	246	1,695	0.91
0.16	−110	−79	212	1,460	197	1,360	12	29	230	1,585	1.08
0.16	RT	RT	182	1,255	169	1,165	14	45	222	1,530	1.22
0.16	340	171	156	1,075	134	925	17	56	225	1,550	1.44
0.16	800	427	132	910	102	705	16	67	—	—	—

[a]Smooth tensiles were 1/8″ dia. (3.2 mm) × 112″ (12.7 mm)gauge length.
[b]Notched tensiles were 0.133″ (3.4 mm) notch diameter.
[c]1725°F (940°C), $\frac{1}{2}$ hr WQ 1000°F (540°C), 4 hr AC.

In general, the effect of temperature on strength will trend as shown in Figs. 15 and 16. The slightly flatter region between about 400°F and 800°F (205°C and 425°C) is thought to be caused in part by dynamic strain aging. Temperature affects bearing and shear values in an analogous way.

4. Charpy Impact Energy Absorption

Impact resistance of Ti-6-4 depends inversely on strength and alloy interstitial content as shown in Fig. 17. The alloy exhibits good Charpy V-notch impact energy absorption. Note the absence of sharp transition behavior. Charpy impact energy absorption is a directional property.

5. Tangent Moduli

Typical tangent moduli are given in Fig. 18 for Ti-6-4 annealed bar at several temperatures. Again, while other product forms may differ somewhat in detail, Fig. 18 is illustrative of trends with temperature.

6. Creep and Stress Rupture Properties

Typical creep and stress rupture properties on bar are presented in Fig. 19. A note of caution: if one needs to extrapolate short time creep or rupture data to a long time, it is

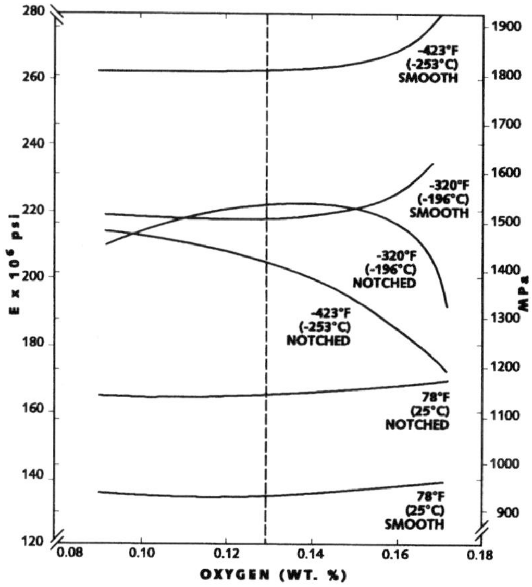

Figure 13 Effect of oxygen content on the room and cryogenic smooth and notched (KT=6.3) tensile properties of *TIMETAL* 6-4 sheet.

generally unwise to employ any of the usual stress–time–temperature parameters such as that of Larson–Miller. It is a better practice to establish the strain–time law for the stress and temperature of interest and develop design values by statistical means. One important reason for this is that creep mechanisms change with temperature, strain rate and possibly with strain and texture. Another reason is that the strain–time laws for creep can be non-linear; the commonly observed steady state region may never appear. If one does employ a

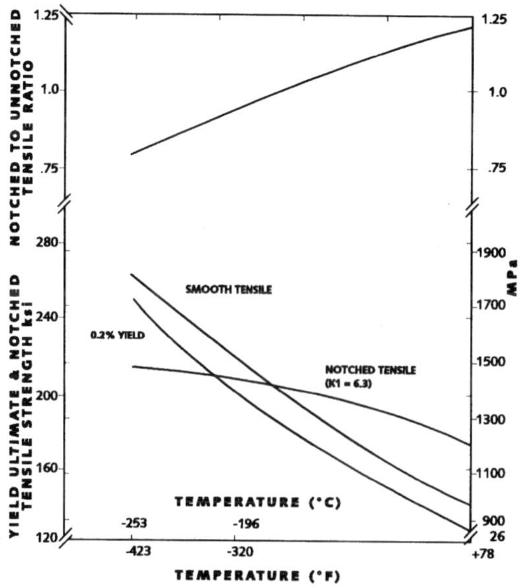

Figure 14 Effects of temperature on the cryogenic behavior of *TIMETAL* 6-4 ELI sheet.

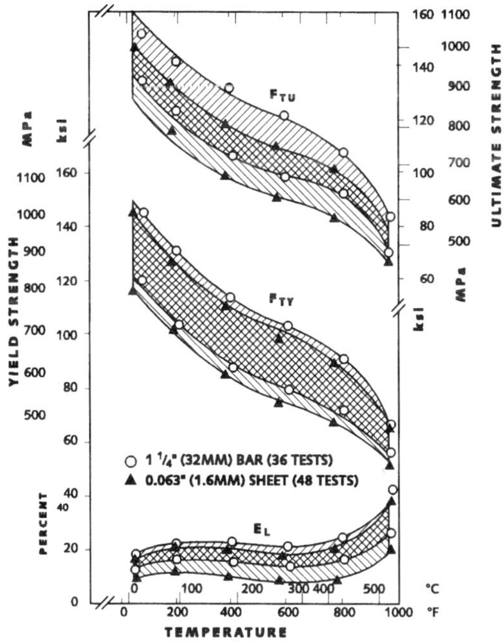

Figure 15 Spread of tensile test data at room and elevated temperature for annealed sheet and bar. (From Refs. 8, 17.)

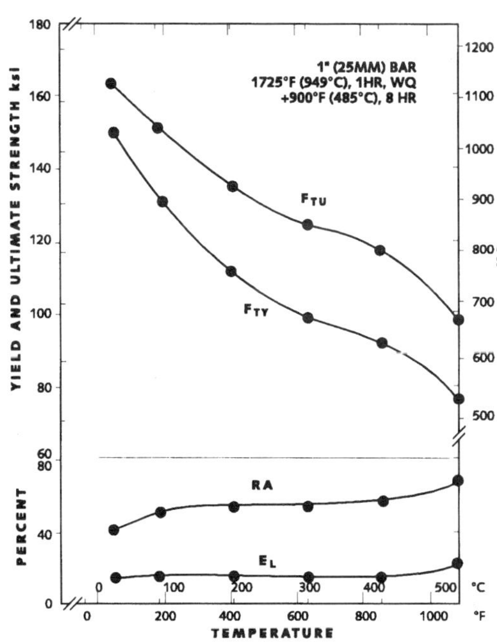

Figure 16 Effect of test temperature on tensile properties of aged bar. (From Ref. 18.)

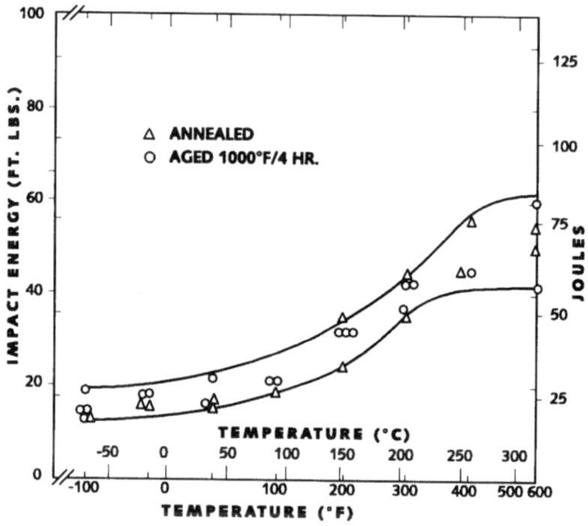

Figure 17 Effect of temperature on the charpy V-Notched impact energy absorption of annealed and STA sheet and bar.

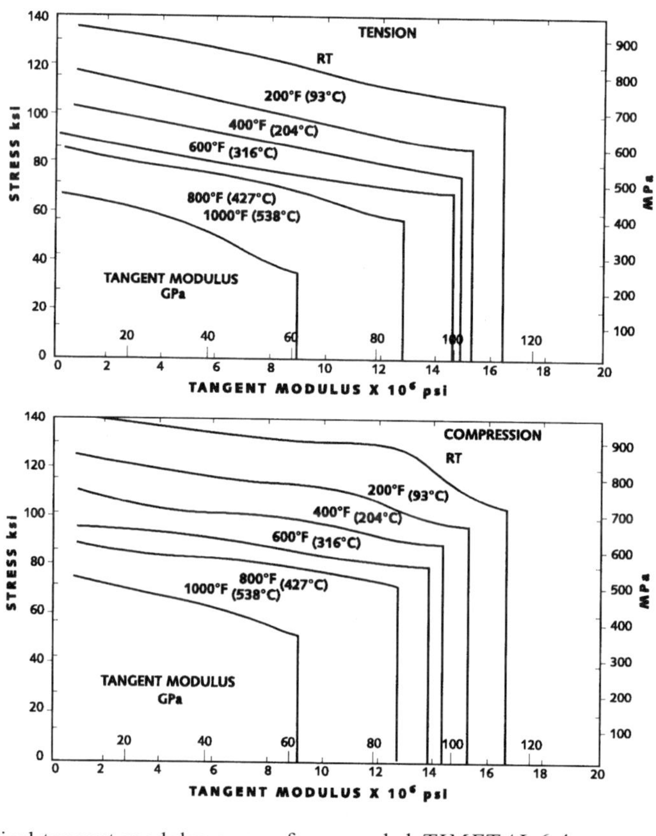

Figure 18 Typical tangent modulus curves for annealed *TIMETAL* 6-4.

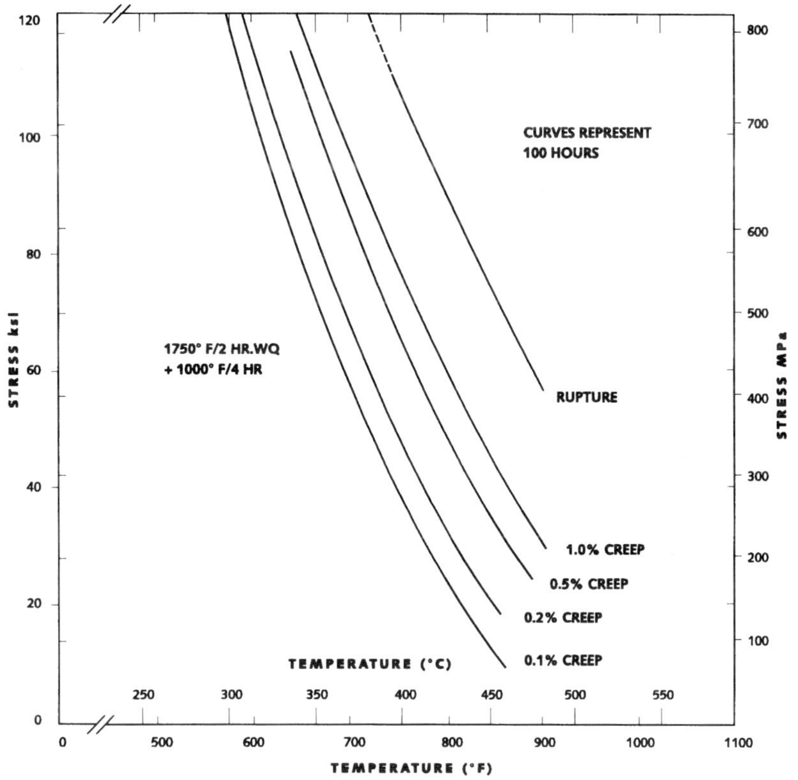

Figure 19 Typical 100 hour creep and rupture stress versus temperature of STA *TIMETAL* 6-4.

Larson–Miller or similar function to extrapolate creep data, it is a good practice to determine the material constants for the product to be actually used by least squares or other acceptable means.

7. Creep Stability

Typical results are given in Table 10. Strength and ductility remain excellent after thermal stress exposure at temperatures up to 850°F (455°C) and times up to 1000 hr.

8. Fatigue Properties

Figure 20 presents typical constant life fatigue diagrams for sheet and bar at room temperature. Fatigue properties are very dependent on surface preparation of the specimen. Moreover, fatigue life often follows a log normal or Weibull statistical distribution. In the absence of pertinent experience and criteria, designers are, therefore, well advised to develop their own fatigue data and criteria for the actual part configuration and surface condition planned for use.

F. General Metallurgy

Titanium-base alloy Ti-6-4 is characterized as an alpha-rich alpha–beta composition. The particular aluminum–vanadium balance provides attractive annealed strength, as

Table 10 Typical Creep Exposed Properties for Ti-6-4

		As Exposed Properties					
		UTS		0.2% YS			
Testing condition	Testing time (hr)	ksi	MPa	ksi	MPa	EL (%)	RA (%)
		Annealed					
Unstressed, 70°F (21°C)	—	134	925	124	855	20	42
Stress-50 ksi (345 MPa)	16	142	980	125	860	18	49
Temperature 650°F (343°C)	100	155	1,070	133	915	15	44
	300	149	1,025	133	915	18	40
	1,000	149	1,025	130	895	13	41
Stress-50 ksi (345 MPa)	16	146	1,005	130	895	16	43
Temperature 750°F (399°C)	100	139	960	128	885	16	43
	300	148	1,020	133	915	20	43
	1,000	147	1,015	130	895	17	45
Stress-50 ksi (345 Mpa)	16	144	995	128	885	17	39
Temperature 850°F (454°C)	100	136	940	123	850	16	48
	300	143	985	133	915	17	34
	1,000	156	1,075	141	970	15	30
		Solution Treated and Aged					
Unstressed	—	166	1,145	153	1,055	18	57
Stress-45 ksi (310 MPa)	150	171	1,180	151	1,040	16	55
Temperature 800°F (427°C)							

well as heat treat response. Aluminum increases the allotropic transformation temperature of titanium. The 6% level is sufficient to markedly strengthen the low temperature alpha phase by solid solution, yet it is not so high that embrittlement results. Vanadium stabilizes the high temperature beta phase, which is manifest by a reduction of the allotropic transformation temperature. The 4% vanadium level exceeds the alpha solubility limit at all the temperatures. This has the effect of stabilizing a small amount of beta to room temperature. Using rapid cooling rates from the solution temperature range permits age hardening of the retained or transformed beta through precipitation.

Although Ti-6-4 is effectively heat treated by the classical solution treat and age procedure, the strengthening mechanism in Ti-6-4 differs markedly in detail from being operative in most hardenable aluminum alloys and precipitation hardening steels. Whereas those materials precipitate submicroscopic compounds coherent with the matrix, Ti-6-4

Figure 20 Typical RT constant-life fatigue diagram for *TIMETAL* 6-4 (a) FTU for unnotched tests was 172 ksi (1186 MPa); For hole type notched tests FTU was 180 ksi (1241 MPa). Gauges: 0.063 and 0.125-IN. (1.6 and 3.2 mm). Surfaces: as rolled. Edges: hand polished through 00 Grit emery paper. Hole (0.0625-IN, 1.59 mm): As drilled and reamed. Test frequency: 25 and 37 Hz (Ref. [19]) (b) FTU for notched plate was 154 ksi (1162 MPa). Gauge: 1.025-IN. (26 mm). Surfaces: As machined. Notches: ground. Test frequency: 100 Hz (Ref. [19]) (c) FTU for unnotched bar was 136 ksi (938 MPA). Bar diameter: 1.25-IN. (31.75 mm) Surfaces: Longitudinal polish through 600 Grit emery belts. Notches: Polished with 600 grit slurry. Test frequency: 29 Hz. (From Ref. 19.)

(a)

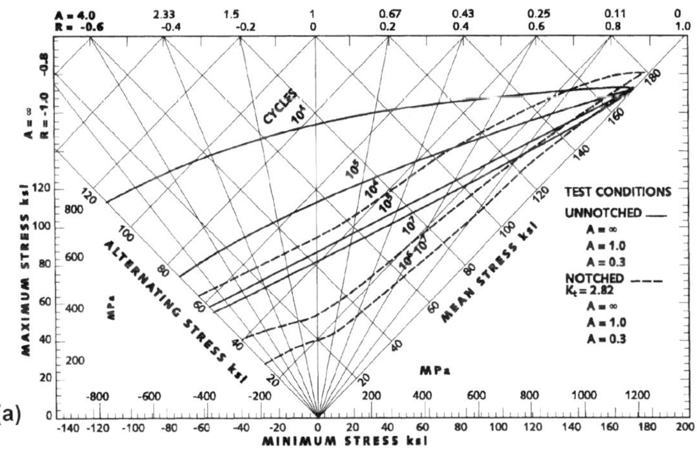
A = 4.0
R = -0.6
CYCLES
MAXIMUM STRESS ksi
ALTERNATING STRESS ksi
MEAN STRESS ksi
MINIMUM STRESS ksi
MPa
TEST CONDITIONS
UNNOTCHED
A = ∞
A = 1.0
A = 0.3
NOTCHED
Kt = 2.82
A = ∞
A = 1.0
A = 0.3

(b)

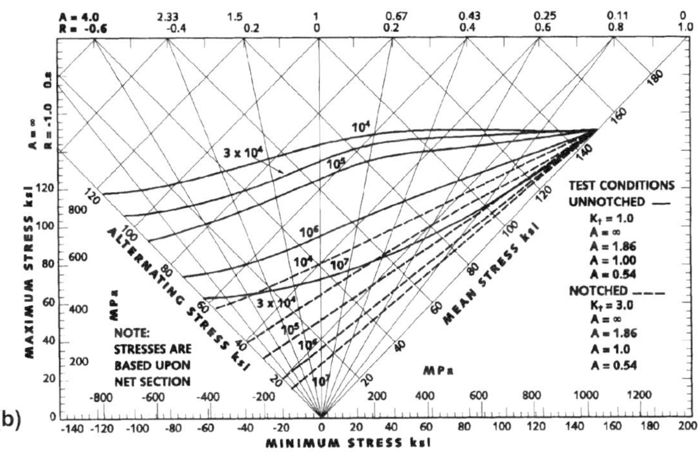
A = 4.0
R = -0.6
MAXIMUM STRESS ksi
ALTERNATING STRESS ksi
MEAN STRESS ksi
MINIMUM STRESS ksi
MPa
NOTE:
STRESSES ARE
BASED UPON
NET SECTION
TEST CONDITIONS
UNNOTCHED
Kt = 1.0
A = ∞
A = 1.86
A = 1.00
A = 0.54
NOTCHED
Kt = 3.0
A = ∞
A = 1.86
A = 1.0
A = 0.54

(c)

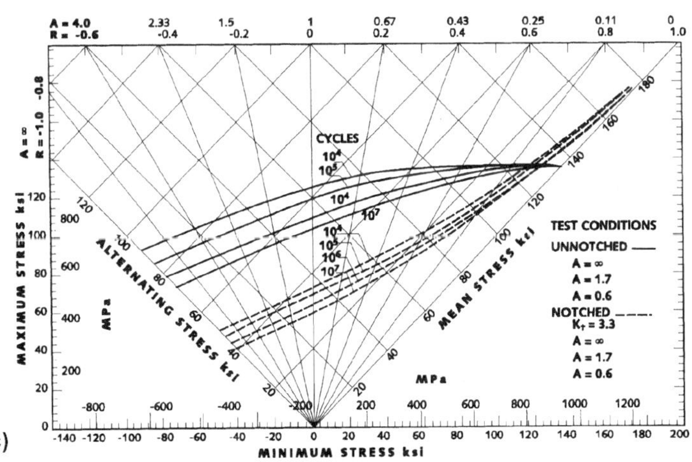
A = 4.0
R = -0.6
CYCLES
MAXIMUM STRESS ksi
ALTERNATING STRESS ksi
MEAN STRESS ksi
MINIMUM STRESS ksi
MPa
TEST CONDITIONS
UNNOTCHED
A = ∞
A = 1.7
A = 0.6
NOTCHED
Kt = 3.3
A = ∞
A = 1.7
A = 0.6

precipitates alpha incoherent with the beta matrix. Ti-6-4 age hardens in a manner much like the 7000 series aluminum alloys given a T7 temper. Ti_3Al, a long-range ordered phase sharing the same basic crystallographic system as alpha, also may precipitate in the alpha phase. Ti_3Al differs basically from alpha in that every other atom in every other row in the basal plane is aluminum. The strengthening effect from Ti_3Al precipitation is of the order of 5 ksi (35 MPa) and is usually accompanied by some loss in fracture toughness and stress corrosion cracking resistance.

A further feature of the aging response in Ti-6-4 is that synergistic effects are evident from the existence of two phases. That is, for identical compositions and solution treatment, there is a microstructural effect apparent in the aging response.

Martensite occurs in Ti-6-4 and other titanium alloys and is quite soft. Its decomposition to alpha plus beta during aging is associated with net strengthening, although the mechanism details are not well established.

The ultimate strength of annealed standard grade Ti-6-4 is above 130 ksi (895 MPa). Most of this strength arises from aluminum and interstitials in solid solution. Vanadium contributes to strength mainly by stabilizing small amounts of beta phase which, when properly dispersed and age hardened, leads to net strengthening. This effect, however, is ordinarily not obtainable in thick sections and in most cases is small relative to what can be achieved through heat treatment. Vanadium also contributes some strengthening of the alpha phase through solid solution. Aluminum strengthens the alpha phase similarly.

The alpha-stabilizing interstitials, oxygen, nitrogen and carbon, as well as the beta-stabilizing interstitial, hydrogen, also play important roles in the metallurgy of Ti-6-4. All of them provide strength increases, but otherwise their effects on properties are largely negative. Oxygen content is varied in commercial practice depending on whether the end use is strength or toughness critical. Hydrogen can be removed by vacuum annealing at temperatures high enough to dissolve residual surface oxide films.

Finally, the flow stress for Ti-6-4 depends on crystallographic texture. The reason for this is that the principal slip directions in alpha lie normal to the prism axis. Deformation parallel to the prism axis is simply more difficult to activate.

Nevertheless, information on Ti-6-4 is so extensive that one can often predict with good accuracy the behavior of the alloy in a new application simply from prior experience.

G. Microstructures

Ti-6-4 may be prepared for metallography by either mechanical polishing or electropolishing. A satisfactory general purpose etchant is an aqueous solution of 1% HF–12% HNO_3. If a more active etchant is desired, the HNO_3 content may be reduced to as little as 3%. Krolls Etch is the term most often used for these etchants.

Figure 21 illustrates the microstructure resulting from a typical solution treat and age heat treatment.

Microstructures resulting from various heat treatments are shown in Fig. 22. Note the dramatic effects of heating temperatures and cooling rates. Water quenching from 1850°F (1010°C) or above produces martensite-like microstructures devoid of any primary alpha. Water quenching from 1800°F (982°C) produces a similar microstructure except that primary alpha (equilibrium alpha that existed at the annealing temperature) appears scattered throughout the transformed matrix. The nominal $\alpha + \beta \rightarrow \beta$ transformation temperature (T_β or beta transus) for the material used to develop Fig. 22 was 1820°F (993°C).

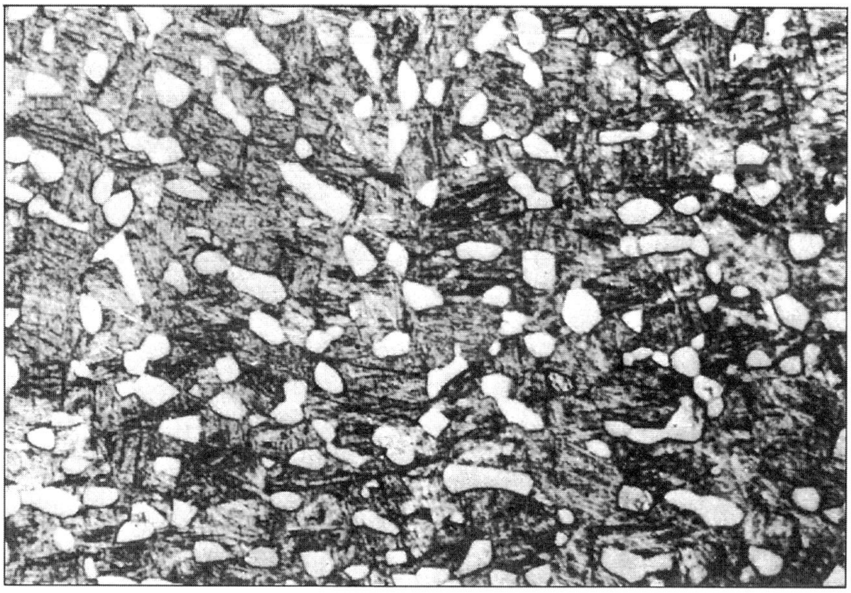

Figure 21 Typical STA microstructure for thin section *TIMETAL* 6-4. 1750°F = 2 hr wQ + 1100°F–2 hr AC (995°C WQ + 595°C–2 hr AC).

As cooling rate decreases, the transformed structure coarsens. After air cooling, the alpha platelets have a finite width and these are coarser still after cooling more slowly. After very slow cooling, the alpha plates are very coarse. For all but perhaps the fastest cooling rates, Ti-6-4 transforms by nucleation and diffusion governed growth processes.

A fast cool must be used to fix the primary alpha content. This is because the primary alpha grains serve as sites for alpha regrowth during cooling. This is most dramatically seen in Fig. 22 series of micrographs showing cooling rate effects from 1800°F (982°C). Therefore, the apparent fraction of primary alpha present in a sample does not necessarily fix the temperature from which cooling began. One must also know the cooling rate.

The micrographs for the encapsulated cool series render the effect of temperature on the microstructure for this particular cooling rate. Cooling rate from 1450°F (788°C) has no obvious effect on the microstructures. At higher temperatures, the amount of primary alpha is observed to decrease with increasing temperature.

Aging does not significantly change these microstructures at the usual optical magnifications.

Omega phase rarely occurs in Ti-6-4. Ti_3Al may be present but is not observable by ordinary optical techniques.

H. Thermal Treatments

Ti-6-4 provides attractive properties in both the annealed- and heat-treated conditions. The various recommended heat treatments are summarized in Table 11.

1. Solution Heat Treating

Solution heat treating is affected by heating between 50°F and 250°F (30–140°C) below the $\alpha + \beta \rightarrow \beta$ transformation temperature and immediately water quenching. The closer the

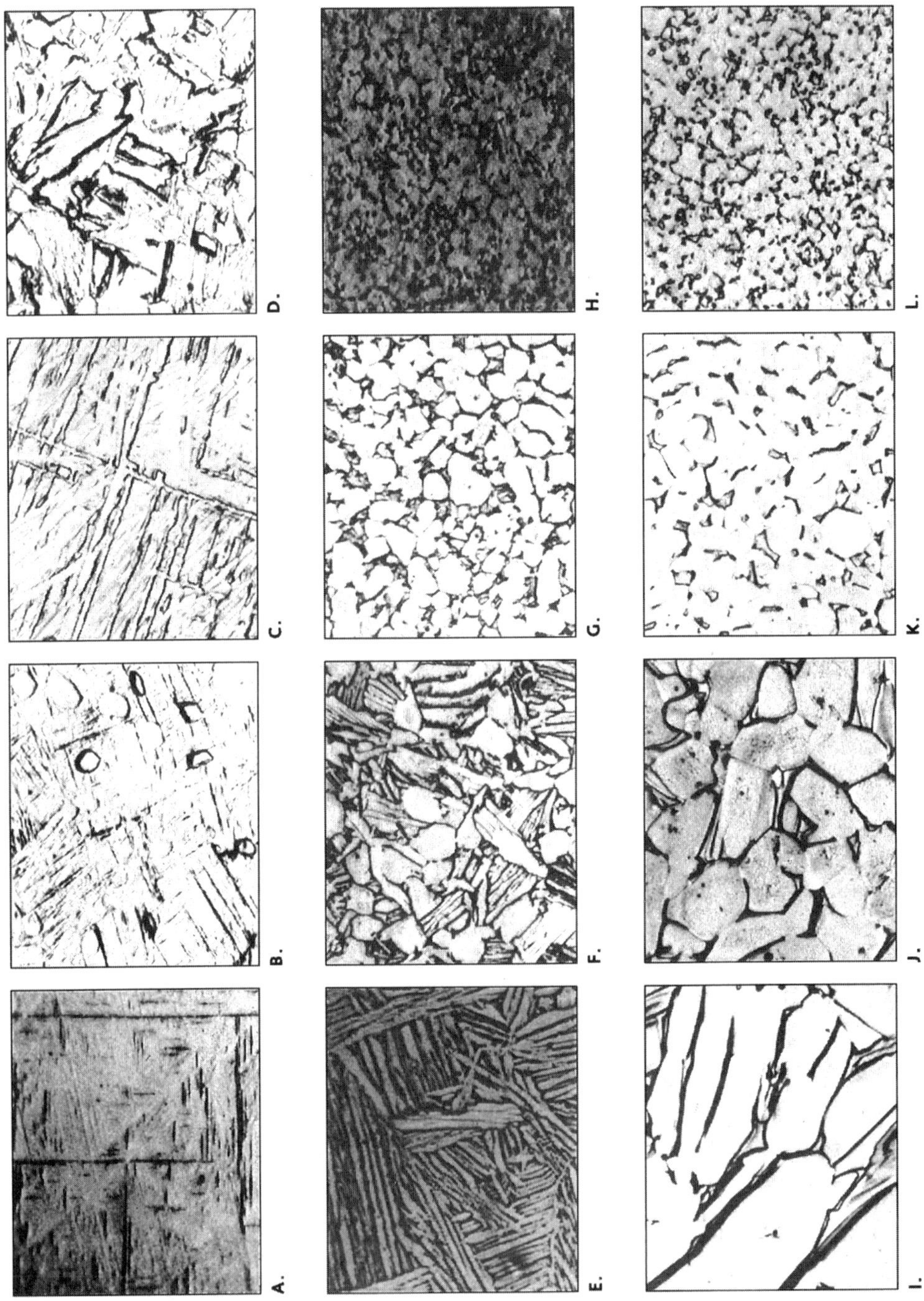
A.
B.
C.
D.
E.
F.
G.
H.
I.
J.
K.
L.

Table 11 Typical Heat Treatments for Ti-6-4

Product Form	Annealing[a]	Solution Treating[b]	Aging[b]
Sheet and light plate	1350 ± 25°F (730 ± 15°C) ½–4 hr AC	1660–1700°F (905 ± 925°C) 5–10 min WQ	1000°F—4 hr AC (540°C)
Plate over ¼ in. (6.4 mm)	1350 ± 25°F (730 ± 15°C) ½–4 hr AC	1700–1750°F (925 ± 955°C) ½ hr WQ	1000°F–4 hr AC (540°C)
Bar, forgings	1350 ± 25°F (730 ± 15°C) ½–4 hr AC	1750 ± 25°F (955 ± 15°C) 2 hr WQ	1000°F–4 hr AC (540°C)

[a]Temperatures up to 1500°F (815°C) may be used provided a protective atmosphere is used. Any contamination resulting from annealing must be removed.
[b]Solution treating and aging cycles may be varied slightly from those listed for specific applications.

solution temperature is to the transformation temperature, the greater is the amount of beta phase present at temperature. At temperatures above approximately 1550°F (845°C), the beta phase is not retained on quenching to room temperature. It is usual, therefore, for the beta phase present at solution temperature above 1550°F (845°C) to transform to martensite if the quench is fast enough or to "Widmanstatten" alpha plus beta if the quench is slower. The metallographic distinction between the two modes of transformation is subtle at critical quench rates. It is good practice to use x-ray or transmission electron microscopy if one needs to establish definitely which transformation mode is operative. The effect of solution temperature on tensile properties and aging response is shown in Figs. 23 and 24.

2. Solution Annealing

Solution heat-treatment temperature and cooling procedures influence toughness. Solution annealing in the beta field provides the highest plane-strain toughness capability. Crack tortuosity as the crack propagates through the transformed microstructure gives

Figure 22 Effect of heating temperatures and Cooling Rates on Microstructures of *TIMETAL* 6-4 Heating temperatures and cooling rates have dramatic effects on the microstructure of TIMETAL 6-4. Cooling from the beta region, 1850°F (1010°C), produces 100% transformed structures. The fraction of primary alpha increases as the heating temperature descends into the alpha + beta phase field, below the transformation temperature (solvus) at 1820°F (993°C). As cooling rates decrease, the transformed structures coarsen and regrowth occurs. Regrowth can increase the apparent primary alpha content. Water quenching thin sections in necessary to fix the actual alpha content at temperature. (A) 1850°F (1010°C) 1 hr. WQ 500×, (B) 1800°F (982°C) 1 hr. WQ 500×, (C) 1850°F (1010°C) 1 hr. AC 500×, (D) 1800°F (982°C) 1 hr. AC 500×, (E) 1850°F (1010°C) 1 hr. encapsulated cool 500×, (F) 1800°F (982°C) 1 hr. encapsulated cool 500×, (G) 1700°F (927°C) 1 hr. encapsulated cool 500×, (H) 1450°F (927°C) 1 hr. encapsulated cool 500×, (I) 1850°F (1010°C) 1 hr. very slow cool 500×, (J) 1800°F (982°C) 1 hr. very slow cool 500×, (K) 1700°F (927°C) 1 hr. very slow cool 500×, and (L) 1450°F (788°C) 1 hr. very slow cool 500×. *Note*: Structures obtained after heating at 1800°F or below are obtained from material which has been processed in the alpha + beta field.

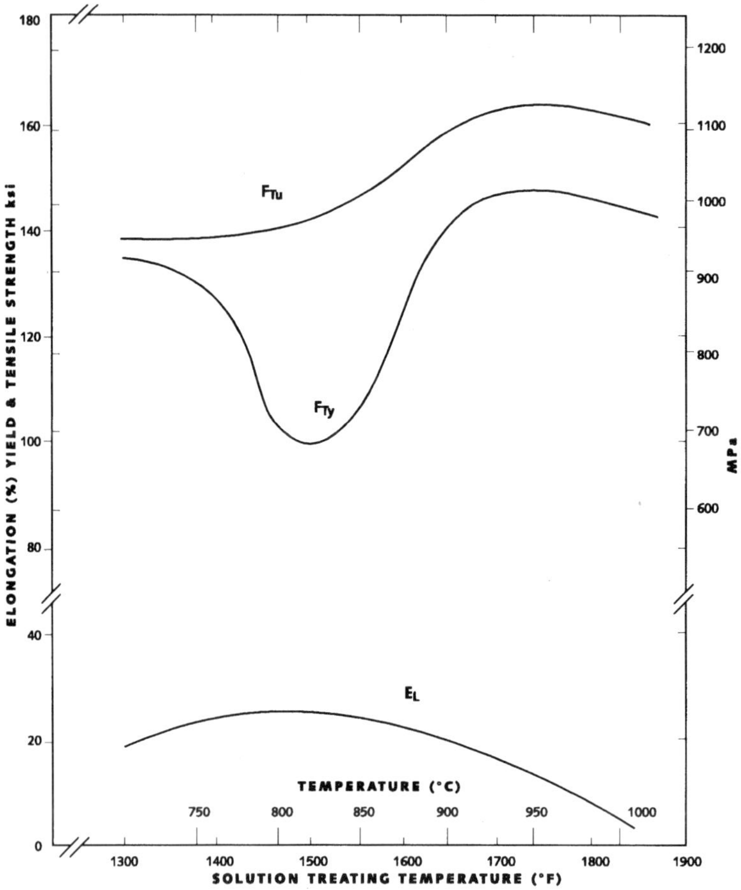

Figure 23 Effect of solution treatment temperature on the solution treated tensile properties of *TIMETAL* 6-4.

rise to this high toughness. By the same token, time to fatigue crack initiation is reduced by beta annealing.

Solution annealing above 1550°F (845°C) but below the $\alpha + \beta \rightarrow \beta$ transformation temperature adds the element of solute partitioning, whereby the primary alpha is somewhat enriched in aluminum and oxygen, and the beta is enriched in vanadium. Upon slow cooling, a toughened background of continuous regrowth alpha isolates the enriched, and rather less tough, primary alpha. Figure 22 illustrates such microstructures. The result is a good combination of strength, ductility, and toughness. Furnace cooling from this temperature range produces a so-called "recrystallized" microstructure. The microstructure is recrystallized in the sense that each phase is essentially dislocation free. If the starting microstructure is equiaxed, the "recrystallized" microstructure will be also. Given sufficient prior $\alpha + \beta$ work, a "recrystallization" anneal will produce equiaxed microstructures.

When high toughness is required and a recrystallization anneal is impractical, a high $\alpha + \beta$ anneal may be used. The element of solute partitioning is still there. The continuous background of transformed beta provides a degree of crack tortuosity and enhances toughness.

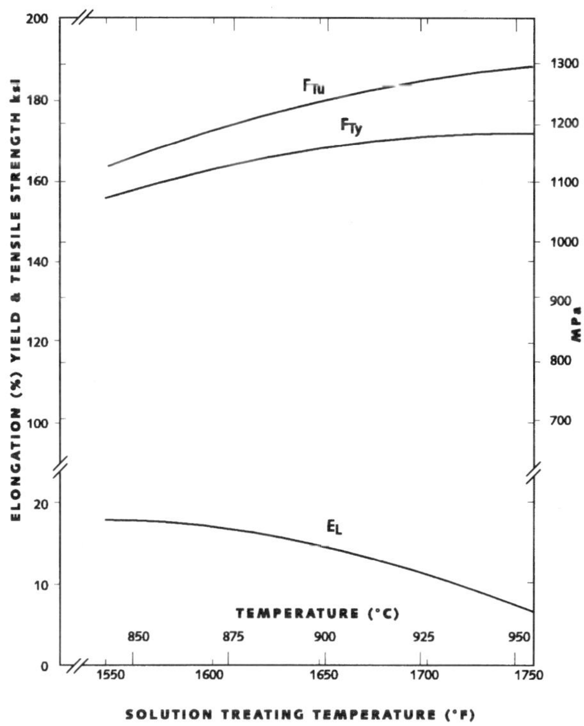

Figure 24 Effect of solution treatment temperature on the STA properties of *TIMETAL* 6-4 aged at 1000°F (517°C), 4 HR, AC.

3. Aging

Aging treatments consist of exposure of previously solution heat-treated material to temperatures from 900°F to 1100°F (480–590°C) from 1 to 24 hr. The lower temperatures provide higher strengths.

Several things happen during aging:

1. Any metastable-beta precipitate alpha.
2. Any martensite will decompose to alpha and beta.
3. The alpha present may precipitate Ti_3Al.

Below 1000°F (540°C), extending the aging times beyond those needed to achieve full strength has little further effect on strength.

4. Stress Relieving

In general, stress relief is accomplished after several hours at 1200°F (650°C). If full stress relief is not required, lower temperatures can be used. Some lower temperature and time effects are illustrated in Fig. 25. Stress relieving in fixtures to remove springback or warpage is common practice.

5. Heat Treat Strategy

The choice of heat treatment depends on many factors. Section size and desired property mix are the main constraints. Sections greater than 1 in. do not fully respond to solution treat and age (STA) types of heat treatment. Time delays in quenching can significantly

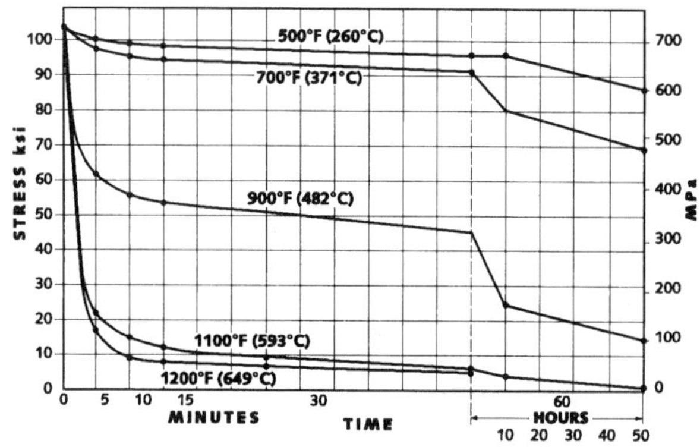

Figure 25 Relief of residual stress versus time for *TIMETAL* 6-4. (From Ref. 19.)

degrade aged strength. This feature is shown in Fig. 26. Figure 27 illustrates the section size effect. Quench-type heat treatments can lead to warpage depending on part configuration.

Because Ti-6-4 is based on the reactive metal titanium, it oxidizes significantly at solution treating temperatures. Solution times and temperatures should, therefore, be minimized consistent with temperature equilibration and response to aging. The thinner the section, the more important oxidation becomes. Solution annealing is best done in vacuum.

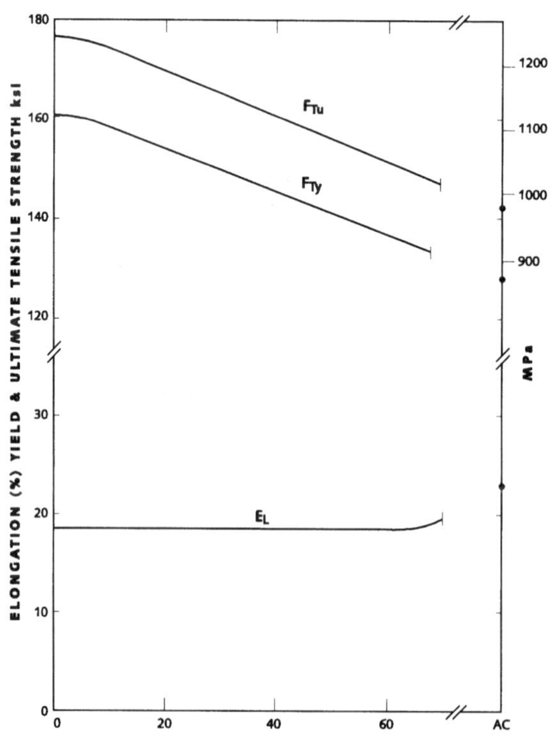

Figure 26 Effect of quench delay on tensile properties of *TIMETAL* 6-4 1/2″ (13 mm) Bar (Solution treated 1759°F (955°C), 1 HR WQ, aged 900°F (480°C), 6 HR, AC).

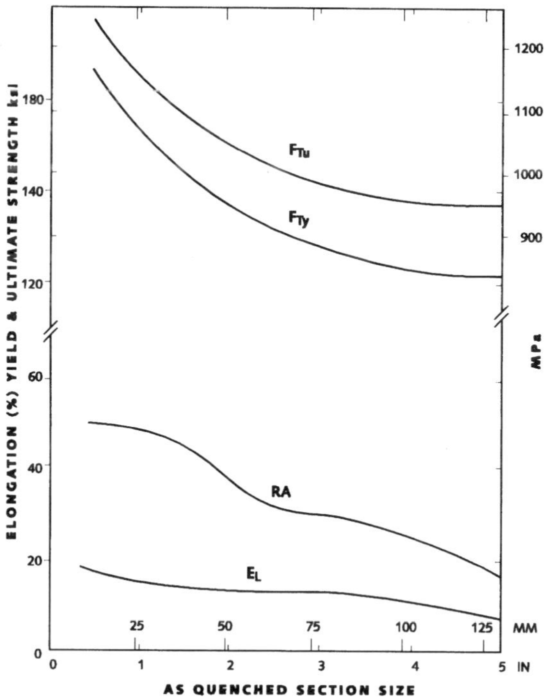

Figure 27 Effect of quenched section size on the tensile properties of STA *TIMETAL* 6-4.

Hydrogen pickup is another reason for limiting solution time and temperature. At solution temperatures, titanium and its alloys readily react with water vapor to form titanium dioxide. Hydrogen is liberated in the process and is largely absorbed in the metal. The less massive the section being treated, the more important this becomes.

Surface conditioning after solution treatment to remove any alpha case is critical. Any procedure must remove enough surface to expose uncontaminated metal. Surface contamination is not always visible macroscopically. Figure 28 shows photomicrographs of contaminated surfaces.

Aging presents less cleanup difficulty; a light pickle suffices. Pickling solutions contain 2–5% HF and 15–35% HNO_3 at an approximate ratio of 1:7. Hydrogen absorption and brightening the metal show a tendency to inhibition by HNO_3.

Microstructures to be avoided in most cases are grain boundary and blocky alpha. These features appear in Fig. 29. Grain boundary and blocky alpha develop on slow cooling through the $\beta \rightarrow \alpha + \beta$ transformation temperature. Such alpha does not spheroidize during heating to the $\alpha + \beta$ field. Grain boundary and blocky alpha result in loss of ductility and toughness. It is evident from Fig. 29 that strain induced porosity is often associated with blocky and grain boundary alpha.

I. Environmental

1. Sea Water Environment

Ti-6-4 is very resistant to general corrosion in seawater at normal ocean temperatures. When coupled with other metals, however, the other metal may become anodic and

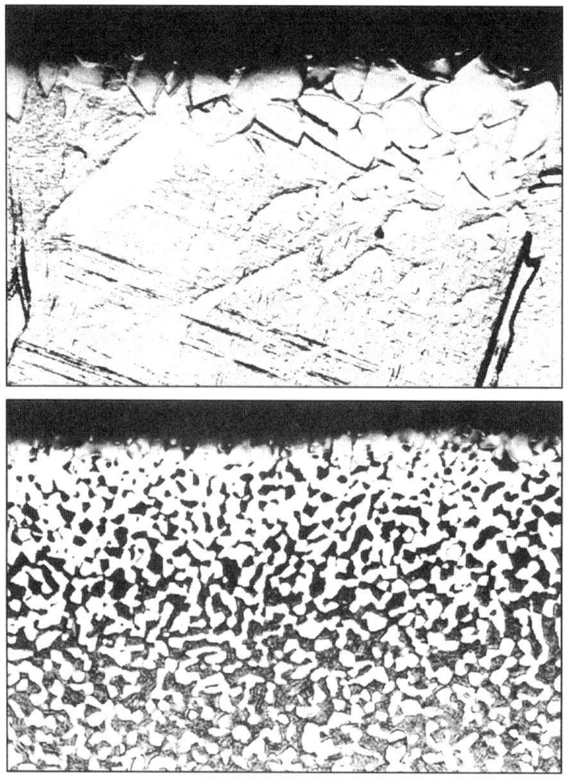

Figure 28 Alpha case arising from oxidation.

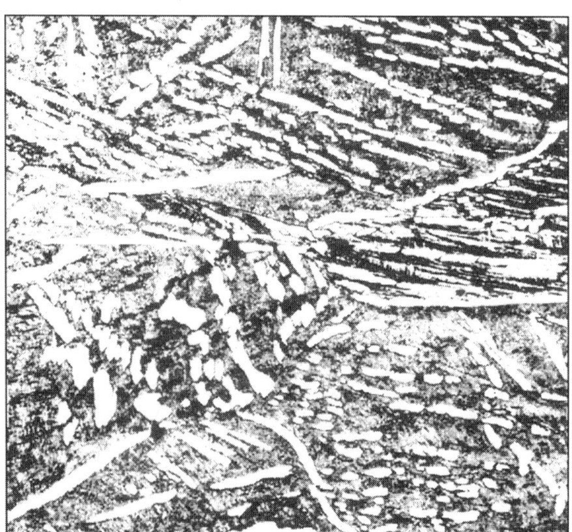

Figure 29 Blocky alpha with grain boundary alpha decorating prior beta grain boundaries. *Note*: strain porosity associated with both types of alpha. Background is transformed beta.

corrode. Ti-6-4 lies near the noble end of the electromotive series and behaves somewhat like austenitic stainless steel in galvanic couples. The resistance of Ti-6-4 toward general corrosion in seawater is due to passivation, which arises from a protective layer of TiO_2. Figure 30 illustrates the cathodic polarization curve for Ti-6-4 in 3.5% ASTM synthetic seawater solution at room temperatures. The polarization characteristics of Ti-6-4 are very similar to those for unalloyed titanium.

The following equations may be used to calculate corrosion rates of the anodic member to be expected when galvanic currents exist:

$$\mathrm{R(mpy)} = 0.13Ie_r$$

or

$$\mathrm{R(mmpy)} = 0033Ie_r$$

Here, I is current density in micro amps per square centimeter, e is the equivalent weight of metal in grams, and r is density in grams per cubic centimeter. Galvanic couples are to be avoided in most situations or accounted for by proper system design.

2. Other Environments

There are certain environments to be avoided. Liquid oxygen, hydrogen under high pressure, red fuming nitric acid, methyl alcohol, nitrogen tetroxide, mercury, solid cadmium,

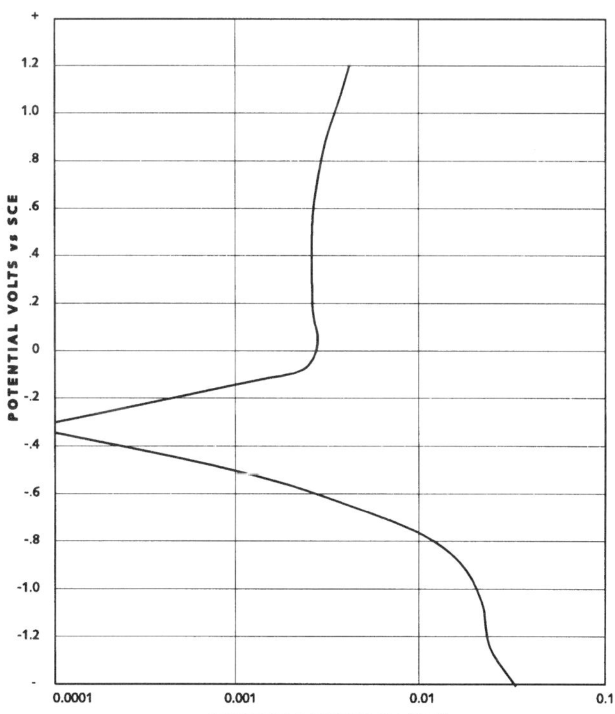

Figure 30 Polarization curve for *TIMETAL* 6-4 in ASTM synthetic seawater PH 8 ambient temperature picked surface scan rate 0.5 mv/sec.

solid silver, and solid gold are environments reported to be destructive to titanium or Ti-6-4 to at least some degree. Embrittlement by solid metal is usually temperature dependent. Cadmium potentially embrittles Ti-6-4 at 300°F (150°C); silver and gold do likewise at 400°F (200°C) or higher. Thermal decomposition of certain organic compounds, such as phosphate–ester base fire resistant hydraulic fluids, can produce acidic by-products, which can result in chemical attack and hydrogen embrittlement. Of course, contact with halides when temperatures exceed 450°F (230°C) is to be avoided under load carrying conditions. In some cases, the effects can be modified or eliminated.

VI. BETA TITANIUM ALLOYS

Beta titanium alloys can be divided into two classes, near-beta alloys and metastable-beta alloys. A near-beta alloy has appreciably higher beta stabilizer content than a conventional alpha–beta alloy such as Ti-6-4, and must be water quenched to retain an all-beta structure even for thin sections. Due to marginal stability of the beta phase in these alloys, they are primarily solution treated below the beta transus to produce primary alpha phase which, in turn, results in an enriched, more stable beta phase as in the microstructure seen in Fig. 31. This enriched beta phase is more suitable for aging. The Ti–10V–2Fe–3Al alloy is an example of a near-beta alloy [1].

The metastable alloys are even more heavily alloyed with beta stabilizers than near-beta alloys and easily retain an all-beta structure upon cooling of thin sections. It is not necessary to heat treat these alloys below the beta transus to enrich the beta phase because of the added stability of the alloy. Therefore, these alloys do not normally contain primary alpha since they are solution treated above the beta transus. These alloys are termed "metastable" because the resultant beta phase is not truly stable—it can be aged to precipitate alpha for strengthening purposes. Examples of such alloys are Ti–15V–3Cr–2Al–3Sn (Ti-15-3) and Beta-21S. Their microstructures can be seen in Figs. 32 and 33[1].

Beta alloys are of interest because they contain a high volume fraction of beta phase, which can be subsequently hardened by alpha precipitation. Thus, these alloys can generate high strength levels (in excess of 200 ksi) with good ductilities and are much more deep hardenable than alpha–beta alloys such as Ti-6-4. Finally, many of the more heavily alloyed beta alloys exhibit excellent cold formability and offer attractive sheet metal-forming characteristics [1].

A. Ti–10V–2Fe–3Al (Ti-10–2-3)

1. Material Properties

Ti–10V–2Fe–3Al is a solute lean beta (near-beta) titanium alloy that was developed primarily as a high strength, deep hardenable forging alloy useful for airframes and engines. It offers the opportunity to use near-net shape forging and provides the best combination of strength and toughness of any of the commercially available titanium alloys [1].

In addition to the high strength condition, the alloy can also be processed to intermediate strength levels for higher fracture toughness. This alloy has also been reported to exhibit a shape memory effect.

2. Product Forms/Manufacturing Considerations

Ti–10V–2Fe–3Al is usually supplied as bar or billet product which has been finish forged (or rolled) in the alpha–beta field. In order to optimize the microstructure for the high

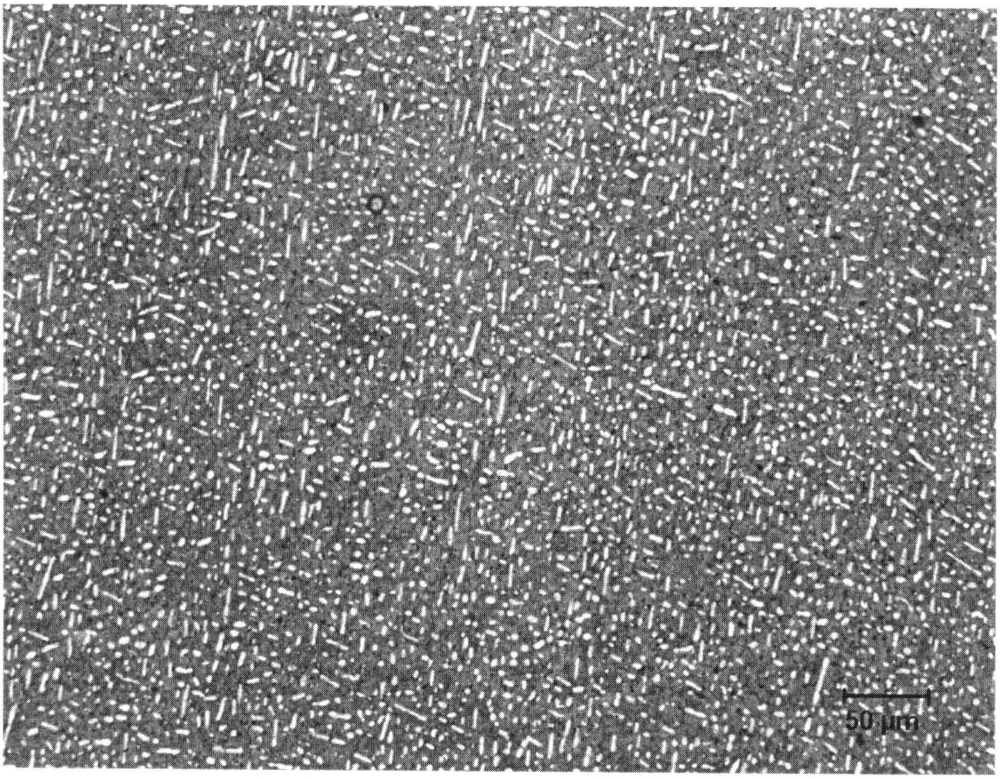

Figure 31 Ti-10V-2Fe-3Al - Lamellar α with a small amount of equiaxed α in an aged β matrix. Heat treatment 1380°F (750°C), 2 hours, water quench + 950°F (510°C), 8 hours, air cooled.

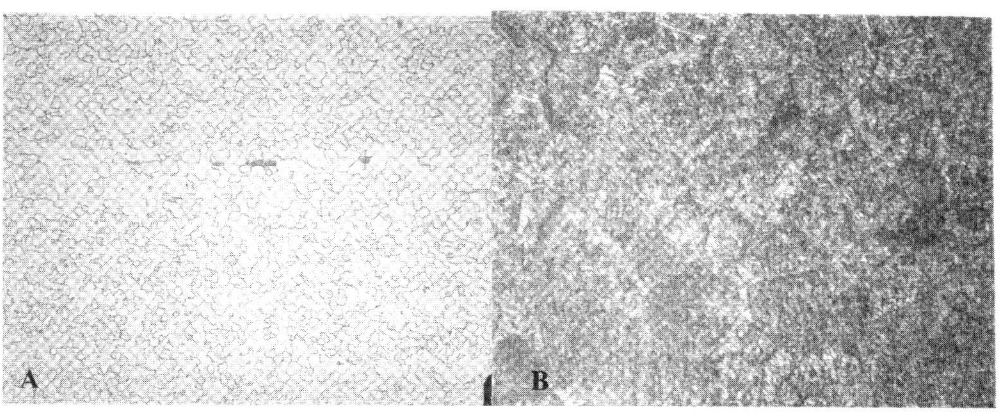

Figure 32 Longitudinal photograph of Ti-15-3. (A) Strip solution annealed and decoration aged at 900°F (482°C) for 30 minutes. (100×). (B) Strip solution annealed and aged for 8 hours. Structure is completely aged (500×).

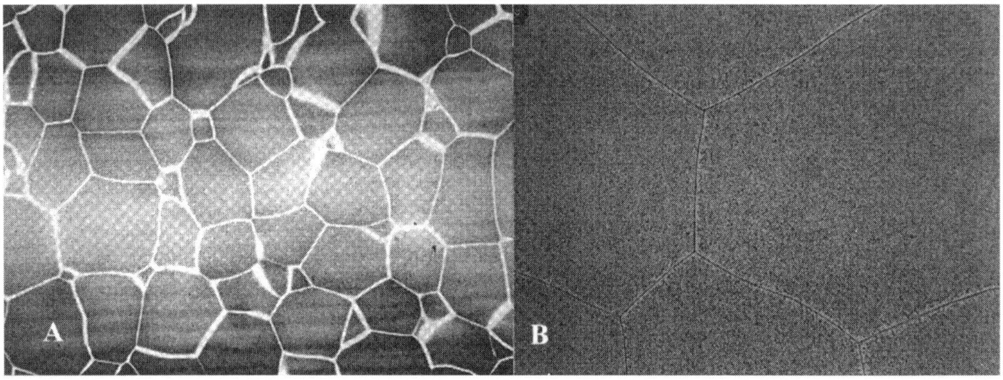

Figure 33 Longitudinal photomicrographs of Beta 21S in supratransus annealed condition. (A) Showing recrystallized beta grains. Decoration aged at 975°F (534°C) to show the microstructure better. (B) Overaged condition showing alpha precipitation in recrystallized beta grains.

strength condition, the forging is usually given a pre-form forge above the beta transus, followed by a 15–25% reduction below the beta transus. The beta forging operation is finished through the beta transus, followed by a quench. The intent of the two-step forging process is to develop a structure without grain boundary alpha, but with elongated primary alpha needles in an aged beta matrix as shown in Fig. 31. The alloy is considered to be deep hardenable, capable of generating high strengths in section thicknesses up to approximately 5 in. and readily weldable by conventional titanium welding techniques [1].

3. Chemical Composition

The chemical composition for Ti–10V–2Fe–3Al is given in Table 12.

4. Physical Properties

Physical properties for Ti–10V–2Fe–3Al are given in Table 13.

Table 12 Chemical Composition for Ti–10V–2Fe–3Al

	Weight%	
Element	Minimum	Maximum
Vanadium	9.0	11.0
Iron	1.6	2.2
Aluminum	2.6	3.4
Oxygen	—	0.13
Carbon	—	0.05
Nitrogen	—	0.05
Hydrogen	—	0.015
Residual elements, each	—	0.10
Residual elements, total	—	0.13
Titanium	Remainder	

Table 13 Physical Properties for Ti–10V–2Fe–3Al

Property	Value
Density	0.168 lb/in.^{3} (4.65 g/cm^3)
Beta transus	1470°F (800°C)
Thermal expansion (75°–800°F)	5.4×10^{-6} in./in./°F
(24°–427°C)	9.7×10^{-6} m/m/°C
Tensile modulus	15.9×10^6 psi (109.6 GPa)
Compressive modulus	16.3×10^6 psi (112.3 GPa)
Shear modulus	6.1×10^6 psi (42.1 GPa)
Poisson's ratio	0.32

5. Mechanical Properties

Minimum mechanical properties for Ti–10V–2Fe–3Al are given in Table 14. Typical elevated temperature mechanical properties are given in Table 15.

6. Heat Treatment

For the high strength conditions, the alloy is generally solution treated approximately from 50°F to 100°F below the beta transus (which is typically 1460–1480°F) for 30 min, followed by a water quench and an 8-hr age at 900–1100°F and then air cooled [1].

7. Beta Fleck

Ti–10V–2Fe–3Al is a segregation prone alloy, which can exhibit a microstructural phenomenon known as "beta-flecks". Certain areas may possess a lower beta transus than the matrix (due primarily to beta stabilizer enrichment) and, as such, can fully transform during heat treatment just below the matrix transus. In severe cases, this condition can lead to lower ductility and a reduction in fatigue strength due to grain boundary alpha formation in the "flecked" region. Care should be exercised to procure only material which has been melted under strict control to prevent severe "fleck" formation [1].

8. Environmental

Ti–10V–2Fe–3Al exhibits excellent resistance to stress corrosion cracking in the solution-treated plus aged condition, typically exhibiting a $K_{Iscc} > 0.8K_{Ic}$. The material in the solution-treated condition should avoid long-term exposure in the 500°F (260°C)–800°F (427°C) range, since such exposure could result in high strength, low ductility conditions. Exposure to cadmium, silver, mercury, or certain other compounds should be avoided [1].

B. Ti–15V–3Cr–2Al–3Sn (Ti-15-3)

1. Material Properties

Ti-15–3 was developed by TIMET during the 1970s in an US Air Force contract and was later scaled up to produce titanium strip. It is a solute-rich beta titanium alloy developed primarily to lower the cost of titanium sheet metal parts by reducing processing cost through the capability of being strip producible and its excellent room-temperature formability characteristics [2]. In the solution-treated condition, it has excellent cold formability; in aged condition, it has high strength. Ti-15–3 is a production alloy that is currently

Table 14 Minimum Mechanical Properties for Ti–10V–2Fe–3Al

Solution treated then aged for 8 hr

AMS specification	Aging temperature °F (°C)	Maximum section thickness in. (cm)	UTS ksi (MPa)	0.2 % TYS ksi (MPa)	Elongation % in. 2 in.	RA (%)	Fracture toughness ksi $\sqrt{in.}$ (MPa $\sqrt{m}$)
4984	900–950 (482–510)	3 (7.6)	173 (1193)	160 (1103)	4	—	40 (44)
4986	950–1000 (510–538)	4 (10.2)	160 (1103)	145 (998)	6	10	55 (60)
4987	1050–1100 (566–593)	4 (10.2)	140 (965)	130 (896)	8	20	80 (88)

Table 15 Typical Elevated Temperature Mechanical Properties for Ti–10V–2Fe–3Al

Solution treated than aged at 950°F (510°C) for 8 hr

Test temperature °F (°C)	UTS ksi (MPa)	0.2 % YS ksi (MPa)	Elongation % in. 2 in.	Reduction of area (%)	Tensile modulus × 10^6 psi (GPa)
75 (24)	176 (1214)	166 (1145)	12	25	16.2 (112)
400 (204)	162 (1117)	152 (1048)	13	33	—
600 (316)	160 (1103)	142 (979)	13	42	13.8 (95)

used for critical applications on air and space vehicles. Ti-15–3 is typically used at temperatures of up to 550°F (288°C).

2. Product Forms/Manufacturing Considerations

Ti-15–3 is primarily supplied as cold rolled strip 0.020–0.090 in. (0.5–2.3 mm) thick, but can also be supplied as ingot, billet, plate, sheet, foil, seamless tube, castings, and welded tube. Ti-15–3 is usually supplied in the solution-annealed condition. In this condition, the alloy has a single phase (metastable beta) structure, and is readily cold formed. To obtain the desired strength of the material, after cold forming, the alloy can be resolution treated in the 1450°F (788°C) to 1550°F (843°C) range and subsequently aged in the 900°F (482°C) to 1100°F (593°C) range. However, if a desired strength is not required, the material can be directly aged after forming and the strength will vary depending upon the amount of cold work in the part. The alloy can also be hot formed. Heating times prior to hot forming should be minimized in order to prevent appreciable aging prior to forming. Ti-15–3 is readily welded by standard titanium welding techniques [1].

3. Chemical Composition

The chemical composition for Ti-15–3 is given in Table 16.

Table 16 Chemical Composition for Ti-15-3

	Weight%	
Element	Minimum	Maximum
Vanadium	14.0	16.0
Chromium	2.5	3.5
Tin	2.5	3.5
Aluminum	2.5	3.5
Oxygen	—	0.13
Nitrogen	—	0.05
Carbon	—	0.05
Hydrogen	—	0.015
Iron	—	0.25
Residual elements, each	—	0.10
Residual elements, total	—	0.40
Titanium	Remainder	

Table 17 Physical Properties for Ti–15–3

Property	Value
Density	0.172 lb/in.^{3} (4.76 g/cm^3)
Beta transus	1375–1425 °F (750–770°C)
Tensile modulus	
Solution treated (annealed)	11.9 Msi (82 GPa)
Solution treated plus aged (1000°F [538°C])	15.5 Msi (107 GPa)
Solution treated plus aged (900°F [482°C])	16.1 Msi (111 GPa)

4. Physical Properties

Physical properties for Ti-15-3 are given in Table 17 and Fig. 34.

5. Mechanical Properties

Minimum mechanical properties for Ti-15-3 strip after aging are given in Table 18. Typical tensile mechanical properties vs. temperature are given in Fig. 35.

6. Heat Treatment

Ti-15–3 should be solution treated for 3–20 min in the 1450°F (788°C) to 1550°F (843°C) range with an air-cooled equivalent. Aging is conducted for 4–16 hr at temperatures between 900°F (482°C) and 1150°F (621°C) followed by air cooling.

7. Environmental

Ti-15–3 appears to be immune to hot-salt stress corrosion cracking below the 440°F (227°C) to 500°F (260°C) range in the aged condition. Crack growth behavior of an aged material does not appear to be affected in the presence of salt water at room temperature. Alloy Ti-15–3 should not be used in the solution-treated condition. Low ductility can result from long time exposure of solution-treated and cold-worked material to service temperatures above approximately 300°F (149°C) or solution-treated material to service temperatures above approximately 400°F (204°C). Under certain conditions, titanium when in contact with cadmium, silver, or mercury can be susceptible to liquid metal embrittlement [1].

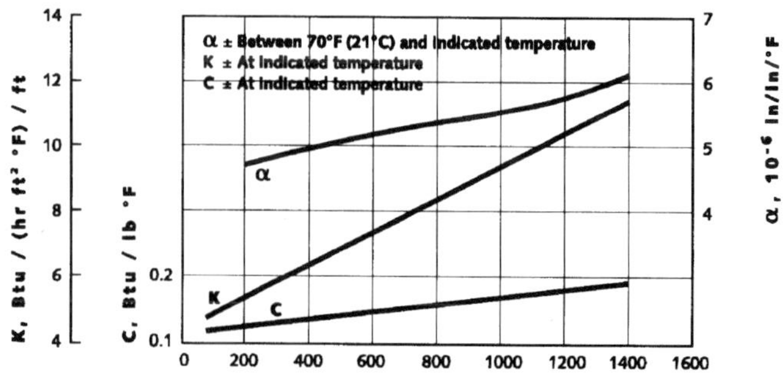

Figure 34 Thermal conductivity (K), specific heat (C), and coefficient of thermal expansion (α)

Table 18 Minimum Mechanical Properties for Ti-15–3

1450°F (790°C) solution heat treatment; air cool				
Aging time (hr)	Aging temperature °F (°C)	UTS ksi (MPa)	0.2 % TYS ksi (MPa)	Elongation % in. 2 in.
8	1000 (538)	145 (1000)	140 (965)	7
8	925 (496)	170 (1172)	160 (1103)	5
16	900 (482)	180 (1241)	170 (1172)	5

C. Beta 21S (Ti–15Mo–3Al–2.7Nb–0.25Si)

Beta 21S is a metastable-beta alloy that offers high specific strength and good cold formability of a metastable-beta alloy, but has been specifically designed for improved oxidation resistance, elevated temperature strength, creep resistance, and thermal stability. Beta 21S is useful for applications from 550°F (290°C) to 1200°F (649°C). Beta 21S is well suited for metal matrix composites because it can be economically rolled to foil, is compatible with most fibers, and is sufficiently stable up to 1500°F (816°C).

1. Product Forms/Manufacturing Considerations

Beta 21S is available as cut sheet, strip, plate, bar, billet, and bloom. It is typically provided in the beta solution-treated condition, which precipitates α to provide strengthening on aging. The morphology and distribution of the α depend on the heat-treatment temperature and the oxygen content. Lower heat-treatment temperatures and higher oxygen contents result in homogeneous spheriodal α; higher aging temperatures and lower oxygen result in lath-type α [2].

Beta 21S is most useful for applications above 550°F (290°C), with thermal stability up to 1200°F (649°C) and creep resistance comparable to Ti-6-4. Special properties include an elastic modulus that is compatible with bone, improved oxidation resistance up to 1200°F (650°C), and resistance to aerospace hydraulic fluid (e.g. Skydrol) [2].

Beta 21S is formed in the solution-treated (as supplied) condition, and then aged to the desired strength level. The forming characteristics of Beta 21S are similar to those of Ti-15-3.

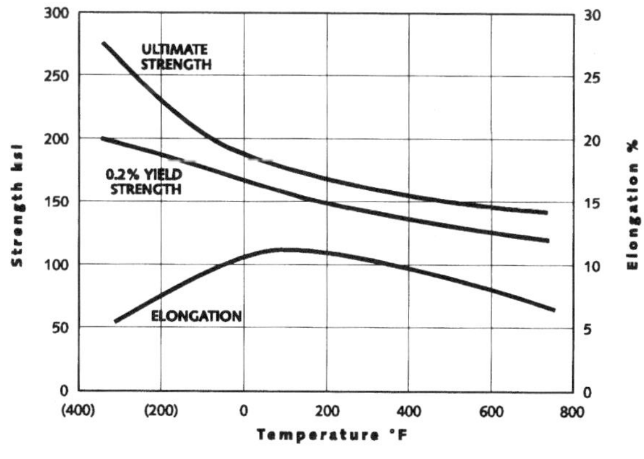

Figure 35 Tensile mechanical properties vs. temperature. 1450°F (790°C) solution, aged 1000°F (538°C) for 8 hrs.

Typical minimum bend radii are 1.0–1.5 times the thickness.

Cold reductions greater than 80% are possible in the compressive operations, including rolling, spinning, and swaging.

Because of the relatively low work hardenablity, maximum tensile deformations are achieved when strains are uniform, such as in hydroforming and bulge forming.

The alloy is strain rate sensitive; therefore, forming should be performed as slowly as practical.

Spring back is relatively severe, but can be compensated for by over-forming or by forming at higher temperatures.

Elevated temperatures [400–1400°F (204–760°C)] increase deformation capability and reduce spring back.

Intermediate anneals can be used between forming operations to restore workability. However, it is essential to choose a combination of cold work and solution heat treatment that produces a high degree of recrystallization with minimal grain growth. Surface contamination (alpha case) must always be removed prior to further forming operations.

Machining should only be performed after aging to avoid a brittle surface that can result from the enhanced aging response of the machining-induced severely cold-worked layer.

Hot forming and hot sizing are best done at the aging temperatures. Exposure of solution heat-treated material to temperatures of 500–800°F (260–427°C) should be kept to less than one to avoid embrittlement. If forming temperatures exceed the beta transus [about 1485°F (807°C)], time at temperature should be minimized to avoid excessive grain growth.

Surface contamination (alpha case), when present, must always be removed prior to forming.

2. Chemical Composition

The chemical composition for Beta 21S is given in Table 19.

3. Physical Properties

Physical properties for Beta 21S are given in Table 20.

Table 19 Chemical Composition for Beta 21S

	Weight%	
Element	Minimum	Maximum
Molybdenum	14.0	16.0
Niobium (Columbium)	2.4	3.2
Aluminum	2.5	3.5
Silicon	0.15	0.25
Iron	—	0.40
Oxygen	0.11	0.17
Carbon	—	0.05
Nitrogen	—	0.05
Hydrogen	—	0.015
Residual elements, each	—	0.10
Residual elements, total	—	0.40
Titanium	Remainder	

Table 20 Physical Properties for Beta 21 S (Aged at 1000°F for 8 hr)

Property	Temperature (°F)	Temperature (°C)	Valuc (English)	Value (SI)
Density	72	22	0.178 lb/in.^{3}	4.93 g/cm^3
Beta transus	1485	807		
Thermal conductivity	91	33	4.4 Btu/[hr ft^2 °F/ft]	7.6 w/m^3 K
	498	259	6.8	11.8
	1024	551	9.8	16.9
	1520	827	12.0	20.8
Specific heat	75	24	0.117 Btu/lb °F	0.117 cal/g °C
	500	260	0.128	0.128
	1000	538	0.142	0.142
	1500	816	0.155	0.155
Electric resistivity	75	24	5.31 (10^{-6}) Ω in.	135.0 (10^{-6}) Ω m
	500	260	5.60	142.3
	1000	538	5.81	147.8
	1500	816	5.84	148.4
Coefficient of thermal expansion	100	38	3.93×10^{-6} in./in./°F	7.07×10^{-6} m/m/K
	200	93	4.41	7.9
	400	204	4.75	8.6
	600	316	4.95	8.9
	800	427	5.11	9.2
	1000	538	5.28	9.5
	1500	816	5.73	10.3
Modulus of elasticity				
Solution treated	75	24	10.5–12×10^6 psi	72–85 GPa
Aged 1000°F (538°C)/8 hr	75	24	15–16×10^6 psi	103–110 GPa
Overaged	75	24	14–15×10^6 psi	96–103 GPa

Table 21 Typical Room Temperature Tensile Properties for Beta 21S (Strip and Sheet to 0.1875 in. Thick)

1550°F (843°C) Solution Heat Treatment, Air Cool

Direction	UTS Strength ksi (MPa)	0.2% TYS ksi (MPa)	Elongation (%)
Aged 1000°F (538°C) for 8 hr			
L	193 (1331)	181 (1248)	5.0
L	189 (1303)	179 (1234)	6.5
T	193 (1331)	179 (1234)	4.5
T	201 (1386)	187 (1289)	4.0
Aged 1100°F (593°C) for 8 hr			
L	164 (1131)	155 (1069)	8.5
L	168 (1158)	159 (1096)	10.0
T	172 (1186)	161 (1110)	10.0
T	172 (1186)	162 (1117)	10.0

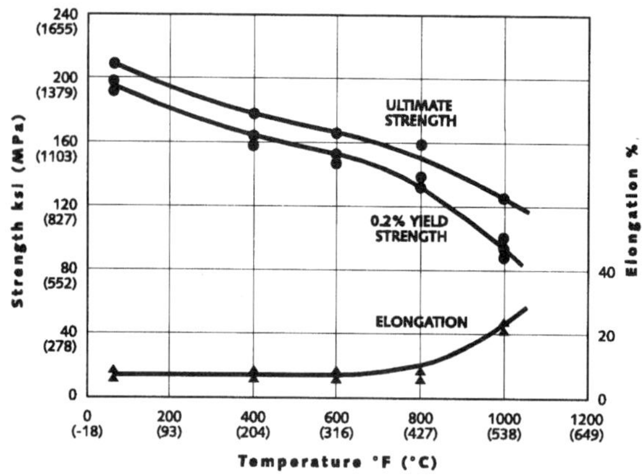

Figure 36 Tensile mechanical properties vs. temperature. (1500°F (816°C) solution treatment, aged 1000°F (538°C) for 8 hrs).

4. Mechanical Properties

Typical room-temperature tensile properties for Beta 21S are given in Table 21. Typical tensile mechanical properties vs. temperature are given in Fig. 36.

5. Heat Treatment

Beta 21S should be solution treated for 3–30 min in the 1500–1550°F (816–843°C) range with an air-cooled equivalent. Aging is conducted for 8–16 hr at temperatures between 950°F and 1250°F (510–691°C) followed by air cooling.

6. Environmental

Beta 21S has outstanding corrosion resistance; in certain instances, it even surpasses the corrosion resistance of CP titanium. Beta 21S was also designed for good oxidation resistance and has the best oxidation resistance of any metastable-beta alloy [20].

REFERENCES

1. *Military Handbook*; MIL-HDBK-5G, November 1, 1994.
2. *Materials Properties Handbook Titanium Alloys*; 1998.
3. Toulouklan, Y.S. *Thermophysical Properties of Matter* In *The TPRC Data Series*, Series Ed. Ed.; Plenum Press 1973; Vol. 12, 1272.
4. Toulouklan, Y.S. *Thermophysical Properties of Matter* In *The TPRC Data Series*, Series Ed. Ed.; Plenum Press: 1973; Vol. 1, 1073.
5. Toulouklan, Y.S. *Thermophysical Properties of Matter* In *The TPRC Data Series*, Series Ed. Ed.; Plenum Press: 1973; Vol. 10, 325.
6. Mote, M.W.; Hooper, R.B.; Frost, P.D. The Engineering Properties of Commercial Titanium Alloys. TML Report No. 92, June 4, 1958.
7. Goldsmith, A.; Waterman, T.E.; Hirschhom, H.H. *Handbook of Thermophysical Properties of Solid Materials*; Amour Research Foundation, McMillan Co.: New York, 1961.
8. Hucek, H.J., Ed. Aerospace Structural Metals Handbook; MCIC, Battelle Columbus Laboratories: 505 King Avenue Columbus, OH 43201.
9. McGee, R.L.; Campbell, J.E.; Carlson, R.L.; Manning, G.K. *The Mechanical Properties of Certain Aircraft Structural Metals at Very Low Temperatures*; WADC-TR-58-386, June, 1958.

10. Weiss, V.; Roy, A. *Further Material Evaluation for Supersonic Transport Aircraft*, Report MET-E 873-6312F NASA Contract No. NASR-43; Syracuse University, August, 1963.
11. Childs, J.K., et al. *Determination of Materials Design Criteria for a Titanium Alloy (TIMETAL 6-4)* at Room and Elevated Temperatures, Progress Report No. P530-6, Southwest Research Institute, WADC Contract AF33 (616)-3348, 1957.
12. Lockheed Georgia Co. *Determination of Design Data for Heat Treated Alloy Sheet*, Vol. 3, Tables of Data Collected, Air Force Contract AF 33(616)-6346, Dec., 1962.
13. MIL-HDBK-5 Committee. *MIL-HDBK5*; Naval Publications and Forms Center: 5801 Tabor Avenue, Philadelphia, PA 19120.
14. Hatch, A.J. Item 65-3, 35th Agenda, MIL-HDBK-5, April, 1968.
15. Mason, W.P.; Wehr, J. Internal friction and ultrasonic yield stress of the alloy 90Ti–A4V. J. Phys. Chem. Solids 1970, 31, 1925–1933, Pergamon Press.
16. *Materials Handbook*; Titanium 6Al–4V, Western Applied Research and Development, NASA Contract No. NAS8-26644, May, 1972.
17. Childs, J.K. op. cit., WADC TR 58-246, August 1958.
18. Hatch, A.J. *Alloy Evaluation Program Summary for 1957–1958*, TIMET Technical Report, October 6, 1958.
19. Wood, R.A.; Favor, R.J., Eds. *Titanium Alloys Handbook*; MCIC-HB-02, Metals and Ceramics Information Center, Battelle, Columbus Laboratories: 505 King Avenue, Columbus, OH 43201, 1972.
20. Eylon, D.; Boyer, R.R.; Koss, D.A., Eds. *Beta Titanium Alloys in the 1990's* TMS 1993, John Fanning, "TIMETAL ® 21S Property Data".

12
Designing with Ni-Base Alloys

Gerhard E. Fuchs
University of Florida, Gainesville, Florida, U.S.A.

David U. Furrer
Ladish Company, Inc., Cudahy, Wisconsin, U.S.A.

Nickel (symbol Ni) is a very important and versatile element for major industrial metals. It is a vital alloying element in a wide variety of cast irons, steels, including austenitic stainless steels, and non-ferrous alloys. In addition, Ni-base alloys are used in numerous demanding applications which require corrosion resistance and high temperature strength (Figs. 1 and 2), and can also be used as coatings which exhibit these environmental resistance. Ni is also used in alloys with controlled thermal expansion or unique magnetic characteristics.

Nickel, whose atomic weight is 58.71 and atomic number is 28, is a lustrous, silvery-white metal, and is one of the transition group of metals in the fourth series in the periodic table, with iron and cobalt. The melting point for Ni is 1453°C and Ni has relatively low thermal and electrical conductivity, high resistance to oxidation and corrosion, excellent strength and toughness at elevated temperatures, as discussed in more detail below. In addition, Ni can be readily alloyed with many other metals.

Nickel occurs in nature primarily as oxides, sulfides and silicates in numerous countries in all the continents. Primary Ni is produced and used in the form of ferro-nickel, nickel oxides and pure nickel metal. Nickel is also readily recycled in many forms and large quantities of secondary or scrap Ni are used to supplement newly mined metal. Typically, over 1 million tonnes of primary Ni is produced and consumed annually in the world, compared with over 800 million tonnes of steel [1].

Ni is widely used in consumer products, and in industrial, military, transportation, aerospace, marine, and architectural applications. Nickel is frequently used in coinage, and as an alloying addition to stainless steels and non-ferrous alloys. About 65% of Ni is used as an alloying addition to stainless steels and about 20% is used in other steels and non-ferrous alloys, including superalloys. About 9% of Ni is used for plating and about 6% is used for coins and nickel chemicals.

I. PHYSICAL PROPERTIES

The normal crystal structure of nickel throughout the range of temperatures up to the melting point is that of a face-centered cube (FCC). The lattice constant for the FCC

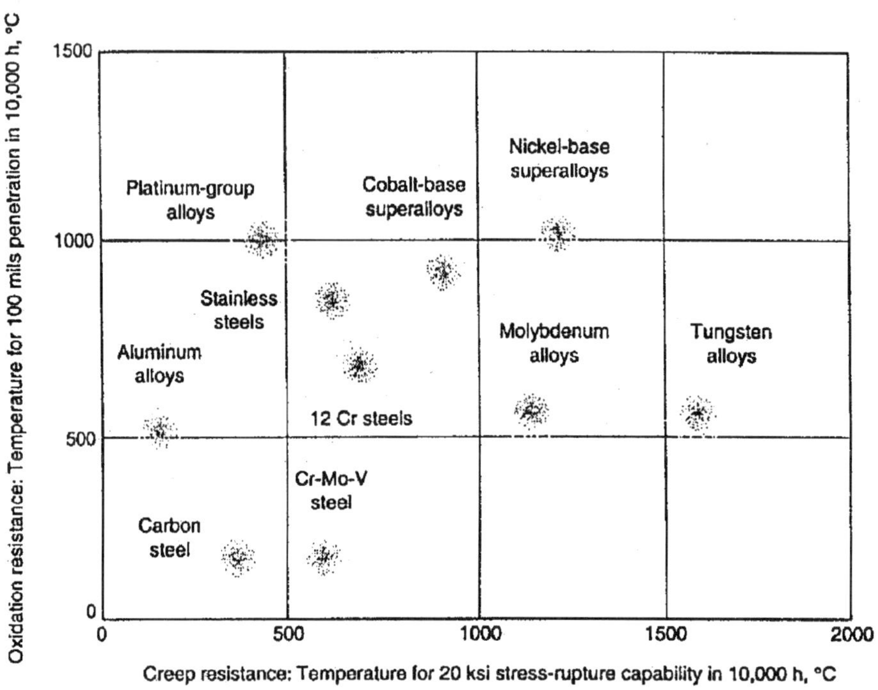

Figure 1 Creep strength and oxidation resistance of several common structural materials. Note that the Ni-base superalloys have a unique blend of creep strength and environmental resistance, compared to other materials [2,5].

form is 0.3167 nm at 20°C. The density of Ni at 25°C is 8.902 g/cm^3. As noted above, the melting temperature for Ni is 1453°C and the boiling point is approximately 2730°C. The coefficient of thermal expansions (CTE) is 13.3 μm/m K at 0–100°C and the thermal conductivity at 100°C is 82.9 W/m K. Ni has a specific heat of 0.471 kJ/kg at 100°C and a recrystallization temperature of 370°C.

II. ELECTRICAL PROPERTIES

The electrical resistivity of Ni is negligible at extremely low temperatures, but increases with increasing temperatures. In addition, the resistivity also increases with increasing levels of impurities or alloying additions.

III. MAGNETIC PROPERTIES

As with the other transition group metals in the fourth series of the periodic table (i.e. Fe and Co), Ni is strongly ferromagnetic at ambient temperatures. Pure Ni can be used in pure form for special applications, such as magnetostriction applications, but is more commonly used as an alloying addition or the base for an alloy for magnetic applications. Most alloys which exhibit high permeability, soft magnetic properties, and permanent magnet alloys are Fe-based with 10–20% Ni or more. The Curie temperature also varies with purity and prior history, but generally lies in the temperature range of 350–360°C.

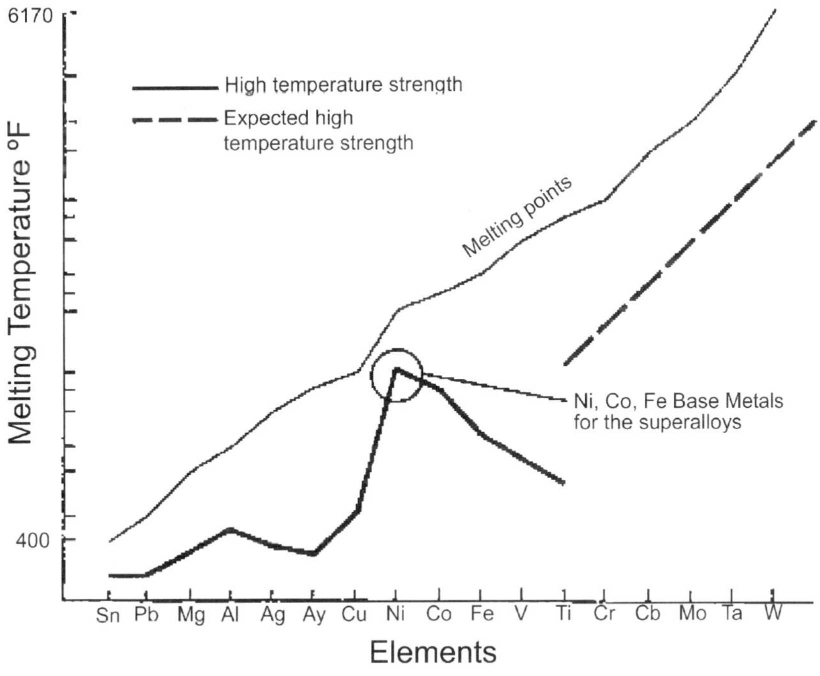

Figure 2 Maximum use temperature and melting temperature of elements in periodic table. Note that although some elements (e.g. W, Ta, etc.) can be used for higher absolute temperatures, Ni-base can be used for higher fractions of the melting point than other elements [3].

IV. MECHANICAL PROPERTIES

Annealed, high purity Ni exhibits moderate levels of yield and tensile strength (approximately 50 and 300 MPa, respectively) and with significant amounts of plasticity (typically greater than 30%). However, significant higher strengths, and decreased ductilities, can be seen in materials that have been thermo-mechanically processed with residual cold work. Typical Vicker's hardness values of 64 HV are reported for high purity Ni. However, similar to the tensile properties, residual cold work and impurities result in significantly higher hardness levels. High purity Ni typically exhibits an elastic (or Young's) modulus of 207 GPa and a shear modulus of 76 GPa. A Poisson's ratio of 0.31 is also observed in Ni.

V. THE USES OF NICKEL

Ni-base alloys are, generally, used for corrosion- and heat-resistant applications, but can also be used for applications including coinage, battery electrodes, filters, and catalysts [2–4]. Nickel and Ni-base alloys are very important to modern industry due to their ability to withstand a wide variety of severe operating conditions involving the combinations of corrosive environments, high temperatures and high stresses in one component or location [5,6]. The reasons for this unique blend of properties stem from a variety of reasons, including the FCC crystal structure which exists in Ni-base alloys and insures high ductility and toughness. In addition, Ni-base alloys are readily fabricated by a variety of conventional processing techniques, including cast, wrought, and powder metallurgy (P/M) methods.

Nickel has good resistance to corrosion in the normal atmosphere, in freshwater and in many acidic and caustic environments. It is an excellent base from which one can develop an alloy with unique properties or specialized applications.

Nickel has extensive solid solubility for many alloying additions, due to the partially filled 3d-orbital. The microstructure of Ni-base alloys consists of an austenitic or FCC solid solution, often called γ (gamma phase). Alloying additions can be done to form dispersoids and/or precipitates that can strengthen the alloy or increase the environmental resistance. Ni forms a complete solid solution with copper and has nearly complete solid solubility with iron. In addition, Ni can dissolve 35% chromium, 20% of molybdenum, and tungsten, and between 5% and 10% of aluminum, titanium, manganese, and vanadium. Therefore, the tough, ductile γ matrix can dissolve extensive amounts of alloying elements in various combinations that can provide solid-solution strengthening, increase corrosion resistance and increase oxidation resistance. In general, the alloying additions with greater atomic size differences result in greater degrees of strengthening. Therefore, additions of tungsten, tantalum, niobium, and molybdenum to the γ matrix result in significantly greater strengthening than iron and cobalt.

Additional strengthening can be obtained in Ni-base alloys by the precipitation of unique intermetallic phases between Ni and aluminum, titanium, niobium, and tantalum. When the appropriate levels of these alloying elements are added to the Ni-base alloy, an age-hardening γ'-Ni_3(Al, Ti, Nb, Ta) (gamma prime) or γ''-Ni_3(Nb, Ti, Al, Ta) [gamma double prime] precipitate is formed, depending on the composition of the resulting alloy. These precipitates result in alloys that exhibit very high strengths to significant fractions of the melting point of the Ni-base alloys. The combination or superposition of the solid-solution hardening of the γ matrix and the precipitation of the γ' or γ'' results in the alloys called superalloys, due to their unique balance of properties and high temperature capabilities.

A. Applications

Ni and Ni-base alloys are extensively used in a wide range of consumer products and in industrial, military, transportation, aerospace, marine, and architectural applications. The unique balance of elevated temperature strength, corrosion resistance, and fabricability makes Ni and Ni-base alloys the material of choice for high-performance applications. For example, Ni-base alloys are widely used in chemical processing, power generation, and aerospace propulsion industries when the need for high elevated temperature strength along with environmental resistance is difficult to meet.

Nickel-base superalloys are used for numerous applications, but none greater than those in the gas turbine engine [7,8]. The high temperature and high stresses encountered in gas turbine engines make superalloys the ideal material for disks in the high pressure compressor and turbine sections, as well as for blades in these same locations within aircraft turbine engines. Superalloys for turbine applications are optimized for creep, tensile strength, crack growth rate, and corrosion considerations in addition to other critical features.

Nickel and Ni-base alloys are used in a wide variety of application, although the majority of applications requires a combination of strength, environmental resistance and/or heat resistance. Some of the more common applications include:

(1) Aircraft gas turbines (Fig. 3): disks, combustors, bolts, casing, shafts, exhaust systems, blades, vanes, afterburners, thrust reversers, etc.
(2) Industrial/power generating gas turbines: disks, combustors, bolts, casing, shafts, exhaust systems, blades, vanes (buckets), etc.
(3) Chemical processing industry: piping, bolts, fans, valves, flanges, reaction vessels, pumps.
(4) Paper mills: tubing, blades, bleaching equipment, scrubbers.
(5) Steam turbines: bolts, blades, tubing.
(6) Metal processing: dies, tolling, fixtures.
(7) Rocket engine: disks, blades, cases, combustors, bolts, shafts.
(8) Heat treating equipment: furnace components, baskets, trays, fans, muffles.
(9) Nuclear power systems: bolts, springs, valve stems, control rod drive mechanisms.
(10) Reciprocating/automotive engines: turbocharger impellers, exhaust valves, glow plugs, catalytic converters.
(11) Medical applications: prosthetic/implant devices, stints.

Ni-base alloys can be processed by a powder metallurgy processing route and both casting and wrought ingot metallurgy processing techniques. Each of the types of processing will be discussed in more detail in subsequent sections. The following sections will discuss the alloying schemes for the different types of alloys.

B. Corrosion-Resistant Ni-Base Alloys

As noted above, the corrosion-resistant alloys are most commonly processed by wrought, ingot metallurgy techniques. However, casting and powder metallurgy techniques can also be used to produce unique corrosion-resistant alloys or conventional alloys

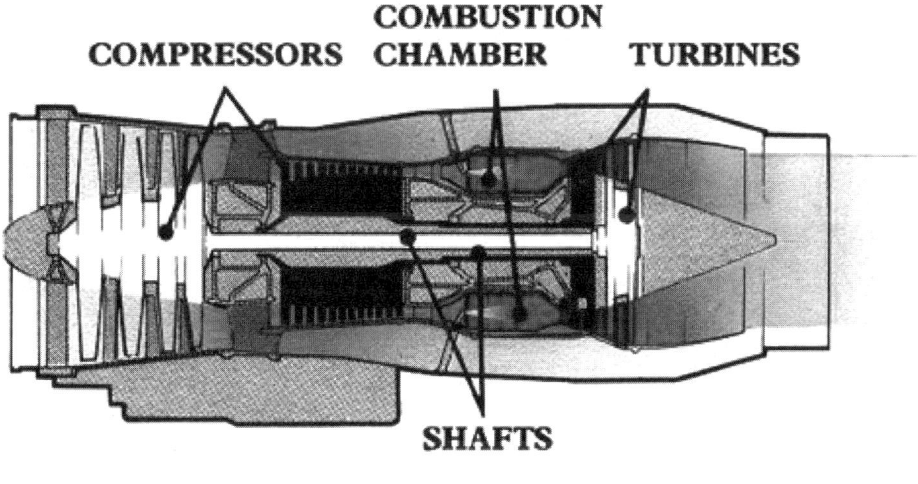

Figure 3 Schematic of gas turbine, showing various components with the engine. Ni-base alloys are typically used in components that are used at elevated temperatures, such as the combustors, turbine blades, turbine vanes, disks, afterburners, nozzles, cases, and the last stages of the compressor.

with unique shapes or properties. The majority of the corrosion-resistant Ni-base alloys has compositions and processing to produce microstructures optimized for corrosion/oxidation resistance. In most cases, the environmental resistance is for aqueous corrosion resistance. However, more aggressive environments (e.g., organic acids, alkalis, salts, sea water) and high temperature oxidation and sulfidation may also be encountered which require Ni-base alloys. In addition, many of the alloys developed for heat resistance, or elevated temperature service, may also exhibit excellent environmental resistance. Therefore, some overlap between environmental and heat-resistant alloys may occur.

Since Ni has extensive solubility for a variety of alloying addition, many different commercial alloys have been developed and are readily available. As noted previously, Ni and Cu exhibit complete solubility, and Co and Fe exhibit very high levels of solubility with Ni. The limit of solubility for Cr is about 35–40% and for Mo is about 20%. A lower limit for solubility for W in Ni is also observed. The γ-matrix in Ni-base alloys can be strengthened by solid-solution strengthening, carbide precipitation or precipitation of an age-hardening phase (e.g. γ' or γ''). Some alloys rely on only one of these strengthening mechanisms, while the others utilize two or more of these methods to produce a corrosion-resistant alloy with increased strength.

The solid-solution strengthening in Ni-base alloys generally comes from the additions of Co, Fe, Cr, Mo, W, V, Ti, and Al. However, the majority of solid-solution strengthening comes from Mo and W. All these elements exhibit size differences, with respect to Ni, that can range from 1% to about 13%, with greater degrees of strengthening observed for greater lattice misfit [9]. It should also be noted that the strengthening resulting from solid-solution alloys is effective at all the temperatures up to the melting point of the alloy. Therefore, the solid-solution strengthened alloys can be used for elevated temperatures service (i.e., $>0.5\,T_m$, where T_m is the melting temperature or solidus of the alloy). At the highest temperatures, Mo and W are the most effective solid solution strengtheners and also result in reduced diffusion rates which further increases the creep strength of the alloy. However, care must be taken to avoid adding excessive levels of elements such as Cr, Mo, and W, since these elements can result in microstructural instabilities that can be detrimental to mechanical properties (e.g., toughness, ductility, creep strength, and fatigue resistance).

Most Ni-base alloys also contain both primary and secondary carbides. Since Ni is not a carbide former, carbon reacts with other elements within the Ni-base alloy, such as Ti, Nb, Ta, and Cr. The presence of these carbides can result in both improved creep strength and corrosion resistance, but may result in reduced fracture toughness and ductility. The primary MC carbides, where M denotes a metallic element, such as Ti, Nb or Ta, form during solidification and are usually relatively coarse in size and exhibit a blocky morphology. These MC carbides generally do not have a significant impact on mechanical properties, although they can act as crack initiation sites in cyclic or fatigue testing. Instead, the MC carbides generally act as a reservoir for the formation of secondary carbides and aid in the control of the γ matrix grain size. The most common secondary carbides have compositions of $M_{23}C_6$ and M_6C, where M denotes the metallic elements, such as Cr, Mo, and W, and are found at grain boundaries. To a large extent Cr typically comprised $M_{23}C_6$ carbides. If the percentage of Mo + W is greater than 6%, M_6C is known to form in alloys. Although the MC carbides form during solidification, the secondary carbides form during subsequent heat treatment and generally improve the strength of grain boundaries. In particular, the secondary carbides are known to increase the creep strength of an alloy by strengthening the grain boundary and preventing or limiting grain boundary sliding. In addition, the secondary carbides are also reported to increase the stress corrosion cracking (SCC) resistance of Ni-base alloys.

The precipitation of γ'-Ni_3(Al, Ti) and γ''-Ni_3(Ti, Nb, Al) in Ni-base alloys results in alloys with significant strength levels at all temperatures [10,11]. γ'' forms in alloys with relatively large amounts of Nb compared to Ti and Al, such as Alloy 718 and Alloy 706. The γ'' precipitate results in significant strengthening due to large lattice mismatch with the γ matrix resulting in high degrees of coherency strengthening. However, the strengthening due to the precipitation of γ'' is only useful at temperatures below approximately 700°C. The high level of lattice mismatch results in instability and coarsening of the γ'' precipitate at elevated temperatures, which results in reduced strength.

The alloys with γ/γ' microstructures are generally referred to as the superalloys (Fig. 4) and can be used when the temperature approaches 0.9 T_m, where T_m is the absolute melting temperature or solidus in an alloy [10,12]. Varying degrees of lattice mis-match are observed between the γ matrix and γ' precipitates depending on the alloy chemistry. Therefore, the γ' precipitates can be engineered to be stable at much higher temperatures with alloy chemistry combinations that provide relatively low mismatch. γ' can be dissolved and precipitated in controlled manners for optimum particle size and distribution. Many heat treatment schemes are based on this capability, with solution temperatures set relative to each respective alloy's γ' solvus temperature. The superalloys typically have between 20 and 70 volume % of γ', with higher strengths being observed in alloys with higher volume fractions of γ'. However, alloys with increased γ' volume fractions typically also exhibit increased difficulty in fabrication and joining.

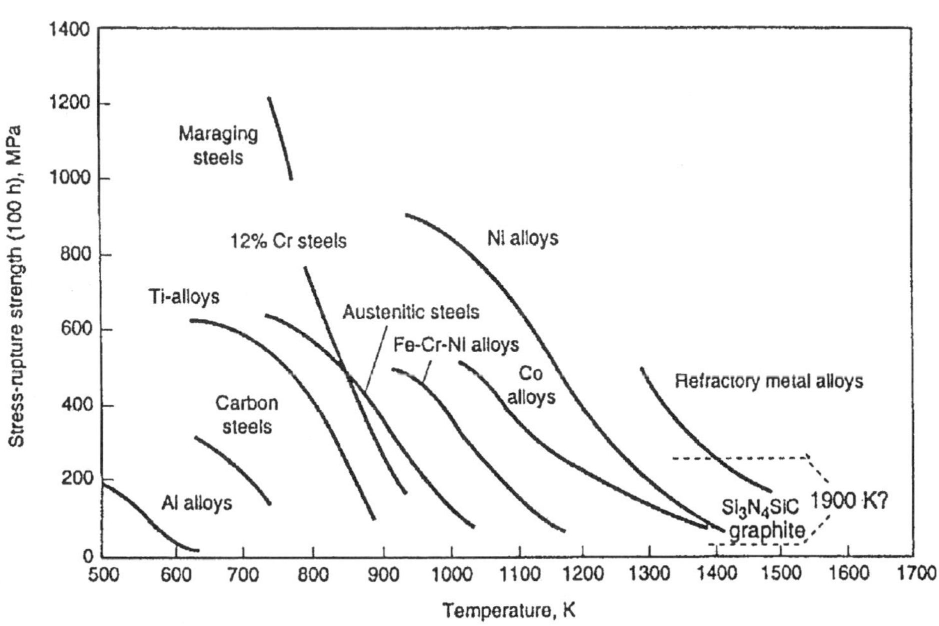

Figure 4 Maximum service temperature, based on 100 hr stress rupture strength, for various creep resistant materials [2,3].

Boride phases are also formed in numerous nickel-base superalloys. These secondary phases (M_3B_2 and MB_2) precipitate upon cooling from high temperature, similar to the manner of secondary carbides. Boride phases act similar to carbides to increase grain boundary properties and increase overall elevated temperature capabilities.

VI. COMMERCIAL NICKEL AND Ni-BASE ALLOYS

An incomplete list of commonly used Ni-base alloys and their properties is presented in Tables 1–5. In addition, a brief description of the various types of alloys is given below.

A. Commercially Pure Ni Grades

Commercial pure Ni grades, such as Nickel 201, typically have minimum Ni contents of 99% and are highly resistant to many corrosive and oxidizing environments. Tight control on impurities and intentional additions is used to maintain the high purity levels. In most alloys, carbon levels of 0.1–0.15 wt% are observed, which can limit the maximum use of temperatures to about 315°C, due to the potential for embrittlement of the alloy. In general, commercial purity alloys exhibit moderate strengths, good ductility and high toughness and are typically used for chemical processing and the electronic industry. Annealed yield strengths of 100–150 MPa and elongations of 40–50% are commonly observed. Cold-worked materials have been reported to exhibit strengths in excess of 600 MPa and ductilities of less than 10%. These alloys are readily fabricated and welded in a variety of applications and are most frequently processed by wrought (ingot) metallurgy techniques.

B. Low-Alloy Nickel Alloys

Alloys which have a minimum Ni content of 94% are considered as low-alloy Ni alloys. Additions of up to 5 wt% Mn (Nickel 211) or up to 5 wt% Al and 0.6 wt% Ti (Duranickel) are used to increase the resistance of the alloy to sulfur embrittlement and corrosion/oxidation, respectively. However, it should also be noted that the high Al + Ti content of Duranickel will result in the precipitation of γ', which will produce significant strengthening. In some alloys (Alloy 360), additions of up to about 2 wt% Be and 0.5 wt% Ti are used for increased strength due to a precipitation-hardening reaction. The precipitation-hardened alloys, including both Duranickel and Alloy 360, exhibit significantly higher strengths and lower ductilities than the commercially pure Ni alloys and the low-alloy Ni alloy, Nickel 211. In most cases, the low-alloy Ni alloys are processed by wrought (I/M) processing techniques, but the Ni–Be alloys (Alloy 360) can be processed by both cast and wrought ingot metallurgy techniques. Joining the low-alloy Ni alloys can be more difficult for the precipitation-hardened alloys due to the potential for of strain-age cracking.

C. Nickel–Copper Alloys

Most Ni–Cu alloys (Monel 400) contain approximately 25–40 wt% Cu for solid-solution strengthening and were developed for corrosion-resistant applications. Some Ni–Cu alloys have higher Cu contents (Monel 401) and some alloys (K-500) contain additions of Al and Ti for precipitation hardening from γ'. Most Ni–Cu alloys exhibit moderate strength, high

Table 1 Compositions (in wt%) of Common Ni-Base Alloys [2–4]

Alloy	Cr	Ni	Co	Mo	W	Nb	Ti	Al	Fe	C	B	Other
Ni 201	—	Bal	—	—	—	—	—	—	—	0.02	—	
Ni 211	—	Bal	—	—	—	—	—	—	—	0.20	—	4.75 Mn
Duranickel	—	Bal	—	—	—	—	0.65	4.5	—	0.3	—	1.0 Si
Alloy 360	—	Bal	—	—	—	—	0.5	—	—	—	—	2.0 Be
Monel 400	—	Bal	—	—	—	—	—	—	2.5	0.3	—	31.0 Cu
Monel 401	—	Bal	—	—	—	—	—	—	0.75	0.1	—	2.25 Mn, 65 Cu
K-500	—	Bal	—	—	—	—	0.65	2.75	2.0	0.25	—	1.5 Mn
Hastelloy B	1.0	Bal	2.5	29.5	—	—	—	—	6.0	0.12	—	
Inconel 600	15.5	Bal	—	—	—	—	—	—	8.0	0.15	—	
Inconel 601	23.0	Bal	—	—	—	—	—	1.35	15.0	0.10	—	
Inconel 690	29.0	Bal	—	—	—	—	—	—	9.0	0.05	—	
Haynes 214	16.0	Bal	2.0	—	—	—	0.5	4.5	3.0	0.05	—	
Incoloy 800	21.5	33.5	—	—	—	—	0.4	0.4	Bal	0.10	—	1.5 Mn
Incoloy 825	21.5	42.0	—	3.0	—	—	0.9	0.2	Bal	0.10	—	1.0 Mn
Hastelloy C	15.5	Bal	2.5	16.0	3.75	—	—		5.5	0.08		0.35 V
Inconel 625	21.5	Bal	—	9.0	—	3.6	0.2	0.2	—	0.05	—	
Inconel 617	22.0	Bal	12.5	9.0	—	—	0.6	1.2	3.0	0.10	0.006	0.5 Cu
Hastelloy S	21.0	Bal	—	15.3	3.7	—	0.2	—	3.0	0.02	0.015	0.35 Cu
Haynes 230	22.0	Bal	—	2.0	14.0	—	—	0.35	3.0	0.10	0.015	0.7 Mn, 0.010 La
Inconel 718	19.0	52.5	1.0	3.1	—	5.2	0.35	0.7	Bal	0.08	0.06	
Inconel 725	21.5	57.0	—	8.2	—	3.2	1.4	0.35	Bal	0.03	—	
625 Plus	21.5	60	—	8	—	3.2	1.4	0.35	Bal	0.03	—	
625 M	21.5	Bal	—	9.0	—	5.2	0.35	0.7	—	0.03	—	
Invar	—	36	—	—	—	—	—	—	Bal	—	—	
Elinvar	4.5	34	—	—	2.0	—	—	—	Bal	1.2	—	1.2 Mn, 1.2 Si
Incoloy 909	—	38	14.0	—	—	4.9	1.6	0.15	Bal	0.06	0.012	0.4 Si

ductility, and high toughness. However, the precipitation-hardened alloys can exhibit much higher strengths, but reduced ductility and toughness. In general, the Ni–Cu alloys exhibit good resistance to corrosion and stress corrosion cracking in a variety of environments. Most Ni–Cu alloys are processed by wrought (I/M) processing techniques, but some alloys can be processed by casting. The fabricability and joining of the Ni–Cu alloys is generally quite good; however, strain-age cracking can occur in the age-hardenable alloys (K-500). The Ni–Cu alloys can be used in the annealed condition or after cold work to increase the strength of the material.

D. Nickel–Molybdenum Alloys

Approximately, 25–30 wt% of Mo is added to Ni–Mo alloys to confer resistance to acid and other reducing environments. The Ni–Mo alloys (Hastelloy B) exhibit excellent resistance to boiling hydrochloric acid and are widely used in the chemical processing industry. Other alloys have been developed with reduced levels of impurities which exhibit even greater resistance to severe environments. However, these alloys exhibit very poor elevated temperature oxidation resistance due to the lack of elements such as Cr and Al. In addition,

Table 2 Compositions (in wt%) of Common Ni-Base Superalloys [2–4]

Alloy	Cr	Ni	Co	Mo	W	Nb	Ti	Al	Fe	C	B	Other
A286	15	26	—	1.25	—	—	2	0.2	55.2	0.04	0.005	0.3 V
AF115	10.7	56	15	2.8	5.9	1.7	3.9	3.8	—	0.05	0.02	0.75 Hf; 0.05 Zr
AF2-1DA	12	59	10	3	6	—	3	4.6	< 0.5	0.35	0.015	1.5 Ta, 0.1 Zr
AF2-1DA6	12	59.5	10	2.75	6.5	—	2.8	4.6	< 0.5	0.04	0.015	1.5 Ta, 0.1 Zr
Alloy 706	16	41.5	—	—	—	—	1.75	0.2	37.5	0.03	—	2.9 (Nb + Ta), 0.15 Cu
Alloy 718	19	52.5	—	3	—	5.1	0.9	0.5	18.5	0.08	—	0.15 Cu
APK12	18	55	15	3	1.25	—	5	2.5	—	0.03	0.035	0.035 Zr
Astroloy	15	56.5	15	5.25	—	—	3.5	4.4	< 0.3	0.06	0.03	0.06 Zr
CMSX-4	6.5	Bal	9.0	0.6	6.0	—	1.0	5.6	—	—	—	6.5 Ta, 0.1Hf
Discaloy	14	26	—	3	—	—	1.7	0.25	55	0.06	—	
EI827	10	67.9	—	7.5	5	—	4.25	4.0	—	—	0.015	0.9 Mn, 0.4 Si
EP199	20.5	55.5	—	5	10	—	1.35	2.35	4.0	0.10	0.008	0.5 Mn, 0.6 Si
EP220	10	56.6	15	5.6	5.5	—	2.4	4.2	—	0.06	—	0.3 V, 0.09 Si
GH586	19	53.9	11	8	3	—	3.4	1.6	—	≤0.08	—	
GH741	15.8	59.2	9	3.9	5.3	—	1.8	5	—	0.04	—	
GH742	14	63	10	5	—	2.6	2.6	2.6	—	0.06	—	0.1 Si
GH4133	20.5	74.3	—	—	—	1.4	2.75	0.95	—	≤0.07	—	
IN100	10	60	15	3	—	—	4.7	5.5	< 0.6	0.15	0.015	0.06 Zr, 1.0 V
IN738	16.0	61.5	8.5	1.75	2.6	2.0	3.4	3.4	—	0.17	0.01	
IN690	29	61	—	—	—	—	—	—	9	0.02	—	0.2 Cu
Incoloy 901	12.5	42.5	—	6	—	—	2.7	—	36.2	0.10	—	
Incoloy 909	—	38	13	—	—	4.7	1.5	0.03	42	0.01	—	
Inconel 600	15.5	76	—	—	—	—	—	—	8	0.08	—	0.2 Cu
Inconel 625	21.5	61	—	9	—	3.6	0.2	0.2	—	0.05	—	
KM-4	12	56	18	4	—	2	4	4	—	0.03	0.03	0.03 Zr
Mar-M 247	8.25	59	10	0.7	10	—	1.0	5.5	—	0.15	—	3 Ta, 1.5 Hf
MERL-76	12.4	54.4	18.6	3.3	—	1.4	4.3	5.1	—	0.02	0.03	0.35 Hf; 0.06 Zr
N18	11.5	57	15.7	6.5	—	—	4.35	4.35	—	0.015	0.015	0.45 Hf; 0.03 Zr
Nimonic 80A	19.5	74.2	1.0	—	—	—	2.25	1.4	1.5	0.05	—	0.01 max Cu
Nimonic 90	19.5	57.1	18	—	—	—	2.4	1.4	1.5	0.06	—	
Nimonic 95	19.5	52.4	18	—	—	—	2.9	2.0	5.0	0.15	—	
Nimonic 105	15	54	20	5	—	—	1.2	4.7	—	0.08	0.005	
Nimonic 115	15	55	15	4	—	—	4	5	1.0	0.20	—	0.04 Zr
PA101	12.5	59	9	2	4	—	4	3.5	—	0.15	0.015	4.0 Ta; 1.0 Hf; 0.1 Zr
PWA 1480	10.0	Bal	5.0	—	4.0	—	1.5	5.0	—	—	—	12.0Ta
Rene' 41	19	55	11	10	—	—	3.1	1.5	< 0.3	0.09	0.01	

(Continued)

Table 2 (Continued)

Alloy	Cr	Ni	Co	Mo	W	Nb	Ti	Al	Fe	C	B	Other
Rene' 88	16	56.4	13.0	4	4	0.7	3.7	2.1	—	0.03	0.015	0.03 Zr
Rene' 95	14	61	8	3.5	3.5	3.5	2.5	3.5	< 0.3	0.16	0.01	0.05 Zr
Udimet 500	19	52	19	4	—	—	3	3	< 4.0	0.08	0.005	
Udimet 520	19	57	12	6	1	—	3	2	—	0.08	0.005	
Udimet 700	15	55	17	5	—	—	3.5	4	< 1.0	0.07	0.02	0.02 Zr
Udimet 710	18	55	14.8	3	1.5	—	5	2.5	—	0.07	0.01	
Udimet 720	18	55	14.8	3	1.25	—	5	2.5	—	0.035	0.033	0.03 Zr
Udimet 720LI	16	57	15.0	3	1.25	—	5	2.5	—	0.025	0.018	0.03 Zr
V57	14.8	27	—	1.25	—	—	3	0.25	48.6	0.08	0.01	0.5 V
Waspaloy	19.5	57	13.5	4.3	—	—	3	1.4	< 2.0	0.07	0.006	0.09 Zr

some of the Ni–Mo alloys can exhibit reduced mechanical properties due to microstructural stability concerns following long-term elevated temperature exposures. The vast majority of Ni–Mo alloys is processed by wrought (I/M) processing techniques and can be readily joined using standard techniques.

E. Nickel–Chromium–Iron Alloys

The addition of 15–30 wt% Cr to Ni–Cr–Fe alloys results in the formation of a very protective Cr_2O_3 surface oxide that offers resistance to both corrosion and oxidation to high temperatures. The Ni–Cr–Fe alloys (Alloys 600 and 690) exhibit moderate strength to elevated temperature, excellent corrosion resistance in a variety of environments, immunity

Table 3 Typical Room Temperature Tensile Properties, Including Yield Strength, Ultimate Tensile Strength, Elongation and Young's Modulus for Select Ni-Base Alloys [2–4]

Alloy	Yield strength (MPa)	Ultimate tensile strength (MPa)	Elongation (%)	Young's modulus (GPa)
Nickel 200	150	450	45	204
Nickel 201	100	400	50	207
Nickel 211	240	530	40	207
Duranickel 301	860	1,150	25	207
Monel 400	250	550	40	180
Monel 401	135	450	50	180
K-500	800	1,100	20	180
Inconel 600	300	650	40	207
Inconel 601	275	600	45	207
Inconel 617	350	750	50	207
Inconel 625	500	900	40	207
Inconel 690	350	725	40	207
Inconel 718	1,000	1,250	15	207
Incoloy 800	300	600	45	193
Incoloy 825	300	700	45	207

Table 4 Typical Room and Elevated Temperature Properties of Selected Cast and Wrought Ni-Base Superalloys [2–4,13]

	Room temperature			540°C			870°C		
Alloy	YS	UTS	Elongation (%)	YS	UTS	Elongation (%)	YS	UTS	Elongation (%)
CMSX-4	1,150	1,225	10	1,275	1,325	17	850	1,100	20
IN 100	850	1,015	9	875	1,100	7	700	885	6
IN 738	895	1,095	7	900	1,090	7	550	770	11
MarM 247	815	965	7	825	1,035	7	690	825	7
Nimonic 80A	620	1,000	39	530	875	37	260	310	30
PWA 1480	895	1,100	4	900	1,130	8	705	995	12
Udimet 700	965	1,400	17	1,170	1,275	16	635	690	27
Waspaloy	800	1,275	23	725	1,170	23	250	275	35

to stress corrosion cracking, and resistance to hot corrosion. Several of the alloys can be used for temperatures approaching 1200°C for long-term applications. Most Ni–Cr–Fe alloys are single-phase austenitic materials with both primary and secondary carbides and can be readily fabricated into a variety of forms, most commonly through wrought (I/M) processing techniques. However, some applications can utilize casting and powder metallurgy processing. Some alloys which are considered Ni–Cr–Fe alloys also contain additions of Al (Alloys 601 and 214) for improved oxidation resistance at the highest temperatures, but may be more difficult to weld and fabricate and may have limited thermal stability due to the formation of γ'. The Ni–Cr–Fe alloys are widely used for elevated temperature service from gas turbines, steam generators, mechanical property testing equipment, chemical processing and nuclear reactors.

F. Iron–Nickel–Chromium Alloys

The Fe–Ni–Cr alloys are similar to the Ni–Cr–Fe alloys, but have increased Fe contents for reduced cost [13]. These alloys are also similar to austenitic stainless steels, but have significantly higher Ni contents. The Fe–Ni–Cr alloys (Alloy 800) exhibit good

Table 5 Typical Elevated Temperature 100 and 1000 hr Stress Rupture Strengths for Selected Cast and Wrought Ni-Base Superalloys [2,3,4,13]

	815°C		870°C		980°C	
Alloy	100 hr	1000 hr	100 hr	1000 hr	100 hr	1000 hr
CMSX-4	800	600	625	450	290	200
IN 100	455	365	360	265	140	90
IN 738	430	315	295	215	130	90
MarM 247	585	415	455	290	195	125
Nimonic 80A	195	85	97	—	15	—
PWA 1480	600	450	475	335	210	145
Udimet 700	400	325	295	200	110	55
Waspaloy	275	200	175	110	45	—

oxidation/corrosion resistance, stress corrosion resistance, and moderate strength to moderate temperatures. These alloys are very fabricable and weldable and are most commonly fabricated by wrought (I/M) processing techniques. The Fe–Ni–Cr alloys are used extensively for chemical processing and power generation. Higher strength Fe–Ni–Cr alloys can be produced by the addition of Mo for solid-solution strengthening and the addition of Al and/or Ti for γ' precipitation (Alloy 825). The Mo addition will also increase the pitting corrosion resistance of the alloy. However, the fabricability and weldability of these higher strength Fe–Ni–Cr alloys will be reduced.

G. Nickel–Chromium–Molybdenum (Tungsten) Alloys

A large number of Ni–Cr–Mo (W) alloys have been developed and are widely used in gas turbines, nuclear applications, chemical processing, pollution control, and waste treatment. These alloys typically have 15–30 wt% Cr and 5–20 wt% Mo and/or W, but other additions of Fe, Co, Al, and Ti are also common. The Ni–Cr–Mo(W) alloys typically are processed by wrought (I/M) processing techniques and can be fabricated in a variety of forms and can be readily joined. Alloys such as the Hastelloy C series and Alloy 625 are widely used in chemical processing, due to their excellent resistance to general corrosion and pitting corrosion. However, care must be taken during welding to prevent the sensitization of the structure to stress corrosion cracking. Most of the alloys, including Alloys 617 and 625, and Hastelloy S, have compositions that have been balanced to produce a stable microstructure. In particular, the levels of Fe, Mo, and W are controlled to prevent the precipitation of deleterious phases. Higher strength can be obtained in these alloys by solid-solution strengthening by increased levels of Mo and W (Alloy 230) and Co (Alloy 617). Additional oxidation resistance can be imparted by the additions of Si, La, and Mn. Generally, higher Fe contents result in reduced creep strength and oxidation resistance. In most applications, these alloys are used in relatively low stress applications that require oxidation/corrosion resistance, weldability, and fabricability (e.g., combustor cans in a gas turbine, heat treatment racks and piping for chemical processing).

H. Precipitation-Hardened Ni-base Alloys

The need for high strength, corrosion-resistant alloys for fastener and drilling equipment applications in sour gas wells led to the development of precipitation-hardened Ni-base alloys. These alloys are related to Alloys 625 and 718 and utilize γ''-Ni_3(Nb, Ti, Al) for increased strength. The alloys (Alloy 725, 625 Plus, and 625 M) typically utilize high Cr and Mo contents for corrosion and stress corrosion cracking resistance (from Alloy 625) and elevated levels of Nb, Ti, and Al for precipitation of γ'' (typical of Alloy 718). These alloys are most frequently processed using wrought (I/M) techniques, but Alloy 625 M was developed for powder metallurgy (P/M) processing. The maximum strength of these alloys relies on some of the residual work from wrought processing. Welding of these materials is possible, since alloys strengthened by γ'' precipitation are less likely to exhibit strain-age cracking (Figs. 5 and 6). However, the strength, and possibly, the corrosion resistance, of the weld region will not be equivalent to the base metal.

I. Ni-base Thermocouple Alloys

Thermocouples are widely used in industry and in research of the measurement of temperatures. The thermocouple relies on two dissimilar materials to produce an electro-motive

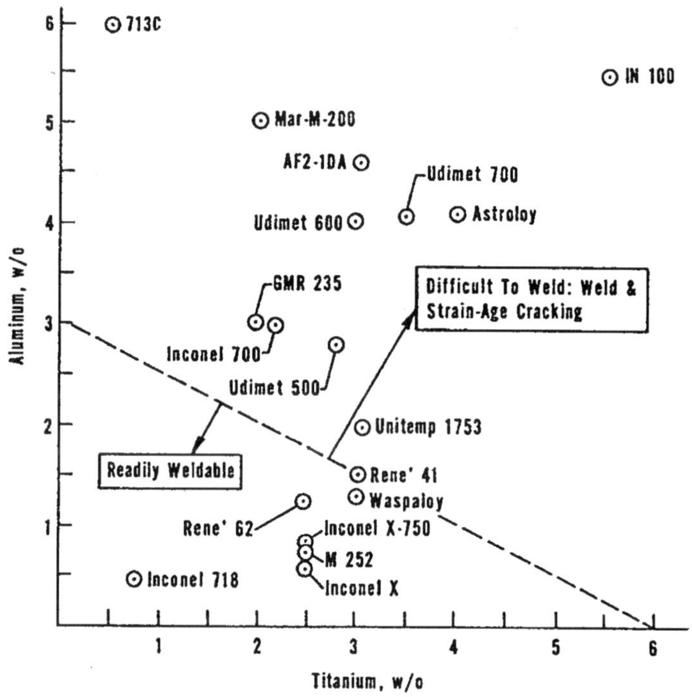

Figure 5 Weldability of Ni-base alloys, as a function of Al and Ti content. Increased Al and Ti contents will result in larger amounts of γ', which forms relatively quickly during cooling. Increased volume fractions of γ' will result in reduced weldability.

force, or emf, that varies with temperature. One of the materials considered is the positive thermoelement (wire) and the other material is the negative thermoelement (wire). A device measures this emf and either reports the emf or the temperature. Any two dissimilar materials can produce an emf signal that is a function of temperature but there are several standard well-defined thermocouples. One of the most commonly used thermocouple pairs is the "type K". This thermocouple comprises of chromel™(Ni–9Cr) which is the positive thermoelement (wire) and the negative thermoelement (wire) is fabricated from alumel™ (Ni–6(Al, Mn, Fe, Si, Co)). The type K thermocouple can be used to temperatures up to 1250°C in air and oxidizing environments. Although type K thermoelements are often called chromel and alumel, these are trade names from Hoskins Corporation, the inventors of these particular wire chemistries. Other manufacturers produced similar elements under other trade names.

J. Nickel–Iron Soft Magnetic Alloys

Soft magnetic Ni–Fe alloys typically contain approximately 30–80 wt% Ni and are used extensively in applications such as electro-magnetic and radio frequency shields, transformers, amplifiers, tape recording heads, ground fault interrupter cores, anti-shoplifting devices, and torque motors. Ni–Fe soft magnetic alloys are characterized as exhibiting high permeability, high saturation magnetostriction, low hysteresis-energy loss, low eddy-current loss in alternating flux, low Curie temperature and constant permeability with

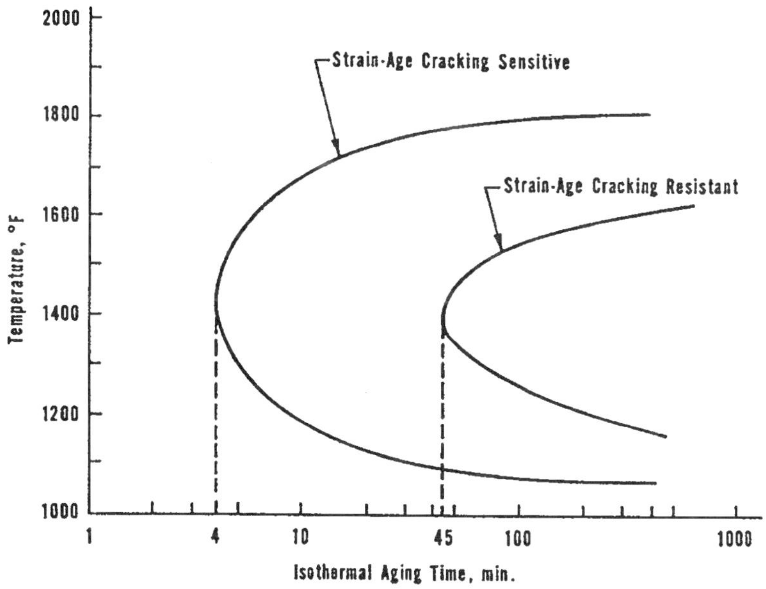

Figure 6 Isothermal, time–temperature transformation curves for the indication of cracking due to strain-age cracking in alloy Rene' 41. The strain-age cracking resistant materials generally form lower volume fractions of γ' and form the γ' at a slower rate [2,3].

changing temperature. The Ni–Fe alloys are readily fabricated using standard wrought (I/M) processing technique for a variety of product forms (e.g., sheet, wire, strip, bar).

Generally, the Ni–Fe soft magnetic alloys are broken down into two (2) categories based on Ni content. The high Ni alloys contain about 75–80 wt% Ni and may contain additions of Mo, Cu, and Cr to accentuate certain magnetic properties. The high Ni alloys generally exhibit high initial and maximum permeabilities and very low hysteresis, but have low saturation induction. However, it should be noted that the magnetic properties of high Ni alloys are very dependent on processing, purity and heat treatment. Significant differences in initial permeability were noted when changes in the cooling rate from heat treatments were made. More rapid cooling resulted in the increased initial permeability. In addition, vacuum induction melting (VIM) is required for processing of the high Ni alloys.

The low Ni alloys typically contain about 40–50 wt% Ni, and may contain additions of up to about 5 wt% Mo. The low Ni alloys exhibit lower initial and maximum permeability than the high Ni alloys, but have a higher saturation induction. The properties of low Ni alloys are also not as dependent on processing, purity, and heat treatment as high Ni alloys are. However, most alloys are heat treated in a protective environment to prevent oxidation.

K. Controlled Expansion Alloys

The coefficients of thermal expansion of most structural alloys range are in the range of $5\text{–}25 \times 10^{-6}$ m/m K. However, for some applications, a reduced coefficient of thermal expansion would be beneficial to reduce stresses due to heating and cooling, to control tolerances between components, and match the thermal expansion of dissimilar materials (e.g., joining metals to ceramics). Typical applications of controlled expansion alloys include pendulums and balance wheels for clocks and watches, glass-to-metal joints, vessels and piping for cyrogenic liquids, superconducting systems in power transmission, integrated circuit lead frames, components for radios and other electronic devices and structural components in precision measurement systems. The most controlled expansion alloys are made up of nickel, cobalt, and iron with small additions of C, Si, Cu, Cr, and Mn.

The controlled expansion alloys contain Ni in addition to Fe since Ni has a profound effect on the thermal expansion of Fe (Fig. 7). Depending on the composition, Ni–Fe alloys exhibit thermal expansions that range from -0.5×10^{-6} m/m*K to 20×10^{-6} m*K. Generally, most controlled expansion alloys contain 30–60 wt% Ni, since virtually any expansion characteristic can be selected from the alloys within this range. The alloys generally contain a total of 1 wt% or less of Mn, Si, and C. Invar (Fe–36 wt% Ni) is one of the most common controlled expansion alloys used in industry and is the Fe–Ni composition which exhibits the minimum in coefficient of thermal expansion. The maximum temperature for minimum thermal expansion is about 275°C, however, it should also be noted that these Fe–Ni alloys would be expected to exhibit poor oxidation/corrosion resistance at high temperatures. The addition of third and fourth elements to the Fe–36 wt% Ni alloys for improved properties will have a significant impact on the thermal expansion characteristics. In addition, processing and heat treatment has a significant impact on the thermal expansion coefficient. Most of the controlled expansion alloys are processed by wrought (I/M) techniques and great care must be taken to obtain reproducible properties.

Small additions of 2–8 wt% Cr to produce Fe–Ni–Cr alloys (Elinvar) with controlled expansion result in an alloy with a limited amount of oxidation/corrosion resistance [2,3]. Further reduction in the thermal expansion coefficient can be achieved by the substitution of a limited amount of Co (about 5 wt%) for Ni, producing Fe–Ni–Co alloys, such as Kovar. These alloys are used for making glass-to-metal seals, but exhibit poor oxidation/corrosion resistance.

High strength, controlled expansion alloys have been developed that contain Nb, Ti, and Al additions to form γ''. These alloys are based on the Fe–Ni–Co system and

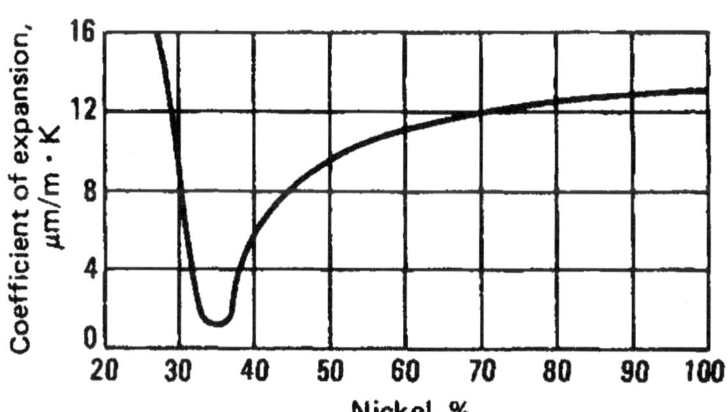

Figure 7 Coefficient of thermal expansion at 20°C for Fe (containing 0.4% Mn and 0.1% C) with the addition of Ni [2,3].

are useful when the temperature approaches 500°C. These alloys (Alloy 909) have been used for applications for improved tolerances, which result in improved efficiencies in gas turbines and rocket engines. Unlike the other controlled thermal expansion alloys, the high strength alloys can be processed by both wrought and cast (I/M) techniques. However, care must be taken in weld repair of large casting due to the potential for strain-age cracking.

L. Nickel–Titanium Shape Memory Alloys

Shape memory alloys can return to their original shape or size when subjected to the appropriate thermal/mechanical procedure. The majority of shape memory alloy is based on the binary, equiatomic (Ni-49–51 wt% Ti) intermetallic compound, NiTi. The intermetallic compound does have a limited amount of solubility for excess Ni and Ti and other metallic elements, too. Some shape memory alloys may also contain limited amounts of Fe, Cr, and Cu. The properties of the shape memory alloys are very dependent on the processing and heat treatment, and most materials require vacuum melting to maintain low levels of impurities. The alloys are readily fabricated using wrought (I/M) processing techniques. In addition to the shape memory effect, these alloys also exhibit superelasticity which enable shape memory alloys to be deformed elastically to very high strains. Typical applications for shape memory alloys include actuators, pipe couplings, orthodontics, eyeglass frames, and bendable surgical tools. Orthodontic arch wires predominantly take advantage of these alloys' extreme superelasticity for ease in shaping, though limited applications have started to utilize the ability to transform reshapen wires in situ through limited thermal cycling. Thermal sensitive switches have taken advantage of shape memory alloys and the ability to change shape through a reversible shear transformation.

M. Precipitation-Hardened Ni-Base Superalloys

The superalloys are generally high strength, corrosion-resistant Ni-base alloys that are used at temperatures in excess of about 540°C. However, there are many subcategories within the field of precipitation-hardened Ni-base superalloys which will be discussed below. There are two major types of precipitation-hardened Ni-base superalloys. The first type achieves high strengths due to the precipitation of γ'' (Alloys 718 and 706) and can be used up to temperatures approaching 750°C. Significantly higher temperature capabilities are observed in the γ'-strengthened superalloys (Mar-M247, IN738, Udimet 720, and CMSX-4). The γ'-strengthened superalloys typically contain 20–70% volume fractions of the precipitate and can be used up to temperatures approaching 90% of the melting temperature (i.e. 1100–1200°C). The primary application for the superalloys includes industrial, power generating, and aerospace propulsion gas turbines. Other applications of the superalloys include nuclear reactors, aircraft structures, spacecraft structures, petrochemical processing, orthopedic and dental prosthesis. Superalloys have been used in cast, wrought, and powder processed forms. Each processing wrought has unique requirements and often unique alloy compositions and will be discussed separately below.

Both types of precipitation-hardened Ni-base superalloys typically contain additions of Co, Cr, Mo, W, Re, Al, Ti, Ta, Nb, V, C, B, Zr, and Hf. Other elemental additions are also utilized for specialized applications. The addition of Co results in increased workability in wrought alloys and improved heat treatment response in high volume fraction cast alloys. Cr additions are typically made for increased oxidation and hot corrosion resistance. However, elevated levels of Cr can result in reduced high temperature strength and reduced microstructural stability. Solid-solution strengthening of the γ matrix phase is accomplished with additions of Mo, W and, in a few cases, V. In high strength single crystal cast superalloys, additional solid-solution strengthening is achieved with the addition of Re. The addition of Al, Ti, Nb, and Ta results in the formation of the precipitation for increased strength. The γ'-hardened alloys typically rely on high levels of Al, with moderate levels of Ti, Nb, and Ta. Excessively high levels of Ti, Nb, and Ta can detrimentally effect the stability of the γ'. The γ'' is formed in alloys with elevated levels of Nb and Ti, and moderate levels of Al. Ta additions are not frequently used in γ''-strengthened alloys. In some γ''-strengthened alloys, both γ' and γ'' are observed and, in some cases, the γ' is reported to be the nucleation site for the γ'' precipitation. It should be noted that γ' quickly precipitates from the supersaturated γ matrix during cooling from elevated temperatures or during aging heat treatments. However, the nucleation kinetics for the formation of γ'' are significantly slow and rely on either specific elements (e.g. Ti) to act as a catalyst, or another phase (e.g. γ') or deformation to act as a nucleation site. The more sluggish formation of γ'' is one of the reasons that γ''-strengthened alloys are much more readily welded than γ'-strengthened alloys.

The grain boundaries of polycrystalline alloys are strengthened by the addition of C, B, Zr, and Hf. The presence of secondary carbides at the grain boundaries prevents grain boundary sliding during creep deformation [2–4,10,12]. In addition, B, Hf, and Zr are believed to increase the strength of the boundaries during elevated temperature service and processing. However, all the grain boundary elements significantly reduce the solidus temperature of Ni-base superalloys. In some alloys, complete solution of the γ' during heat treatments is not possible due to the suppression of the solidus below the γ'-solvus temperature by the addition of C, B, Zr, and Hf. Single crystal alloys, which do not contain high angle grain boundaries, utilize much lower levels of these grain boundary elements, and in some cases, totally eliminate these elements from the alloys. Most of the single crystal alloys, therefore, can be heat treated to allow complete dissolution and re-precipitation of the γ' in a controlled fashion to maximize the effectiveness of the precipitation hardening. Low levels of these grain boundary elements are frequently observed to provide increased castability, defect tolerance and oxidation/hot corrosion resistance.

As noted above, the alloy compositions of the Ni-base superalloys are tailored not just for processing, but also for the application. For example, alloys traditionally intended for industrial gas turbine applications must be resistant to hot corrosion, and, therefore, have high Cr (e.g. 10–18 wt%) contents and increased Ti/Al-ratios. Aircraft engine alloys require greater levels of oxidation resistance and will have higher Al contents (lower Ti/Al-ratios) and lower Cr contents (2–12 wt%). In addition, the aircraft engine alloys are typically utilized at higher temperatures and increased creep strength is required. In order to achieve higher strength, aircraft engine alloys typically have higher Ta contents (increasing γ' volume fraction and strengthening γ') and increased levels of refractory metal additions (W, Mo, Re) for solid-solution strengthening. However, many of these alloy modification for increased strength decreased fabricability and castability.

Wrought alloys typically have somewhat lower volume fractions of γ' (20–50 volume %) and lower C contents than cast alloys. The reduced γ' volume fractions and carbon levels are necessary to increase the hot workability of the alloy. Wrought alloys also typically have higher elevated levels of Co (workability) and Cr (environmental resistance) than many cast alloys. Cast alloys will have elevated levels of all the γ' formers (Al, Ti, Ta, Nb) and carbon, for increased elevated temperature strength. Directionally solidified alloys to produce columnar-grained materials will have elevated levels of Hf (1.0–1.5 wt%) to prevent cracking during casting. As noted above, the single crystal alloys contain much lower levels of the grain boundary strengtheners and even higher levels of the γ' formers (Al, Ta, Ti, Nb) and solid-solution strengtheners (W, Mo, Re, Ru) and decreased levels of Cr. Powder metallurgy alloys are similar to wrought alloys, in that they have reduced levels of the grain boundary elements and increased Cr and Co levels. However, the powder metallurgy alloys also contain higher levels of γ' formers (Al, Ti, Nb, Ta). The powder metallurgy alloys typically have higher γ' volume fractions than the wrought alloys and could not be easily processed by conventional ingot metallurgy techniques. The higher γ' volume fractions and, in some cases, higher solid-solution strengthener levels, would be difficult to cast, without excessive segregation and cracks, and deformation process. Powder metallurgy processing is often the only viable processing method for these alloys (Rene' 95, IN100, AF2-1DA).

VII. MANUFACTURING METHODS

A. Melting and Primary Processing of Nickel-Base Alloys and Superalloys

Ingot metallurgy processing of nickel-based alloys and superalloys has been the traditional method of billet manufacture. Working of cast ingots of iron, iron–nickel and lower alloy content nickel-base materials results in refined billet material. This manufacturing method is often called cast and wrought (C&W) processing.

Casting large ingots inevitably results in segregation and in some cases non-equilibrium phases that are very deleterious to mechanical properties. Considerable attention is paid to melting and casting of ingots in order to control segregation effects. Homogenization cycles are often required to redistribute alloy constituents uniformly throughout the ingot volume. Homogenization cycles are engineered to allow the most effective and cost efficient means of minimizing or eliminating segregation in cast ingot material. Billet macro-segregation must be overcome to provide the best quality materials for critical turbine engine rotating parts; so homogenization and conversion cycles are specifically designed for this reason.

Melting and casting of nickel-base superalloy materials are accomplished by one or more of the following methods: vacuum induction melting (VIM), vacuum arc remelting (VAR) and/or electroslag remelting (ESR)[14]. Figure 8 shows a schematic of the major superalloy ingot melting methods.

Primary melting of virgin melt stock or recycled solid scrap is accomplished by VIM methods. Vacuum induction melting cast ingots are used as electrodes for secondary melting by either VAR or ESR processing for double melt materials, or ESR + VAR processing for triple melt materials. One of the problems with VAR processing is that it can produce excessive macro-segregation, and subsequent uncontrolled microstructures in forging billet stock. Control of the melting process parameters such as melt rate and melt pool depth is a key to obtaining homogeneous, sound, and defect-free ingots.

Segregation, the primary source for chemical inhomogeneity defects in C&W processed superalloys, is a significant problem that results in variations in local transformation temperatures and subsequent variations in grain sizes and grain size banding. Gamma-prime is utilized for many superalloys in conversion operations to control recrystallized grain sizes through grain boundary pinning. Variations in local gamma-prime solvus result in variations in gamma-prime content. If near-solvus processing is used during conversion or secondary forge operations on segregated material, local solvus temperatures can be exceeded and rapid localized grain growth occurs. This results

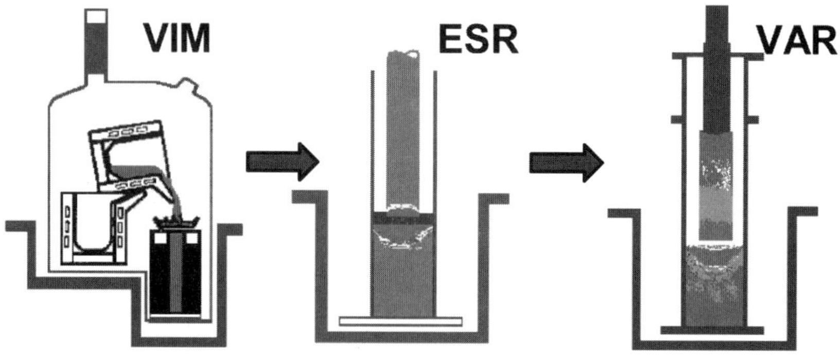

Figure 8 Schematic of vacuum induction melting (VIM), electroslag remelting (ESR), and vacuum arc remelting (VAR) processing methods for superalloy ingots. (Schematic courtesy of Allegheny Technologies.)

in a mixture of grain sizes, typically in bands of fine and coarse grains. In addition to general overall segregation, cast and wrought materials can form deleterious, non-equilibrium phases such as lave phase in extreme cases of segregation and lack of ingot homogenization.

C&W billet materials can have other defects, including oxides, nitrides, carbo-nitrides or slag stringers, which can reduce the component life due to fatigue resistance capability degradation. Another common defect in C&W nickel-base superalloy billet is a feature known as "white spots". A white spot is a "white" macro-etching spot observed on a superalloy billet or forging surface. The white appearance is due to the lack of secondary phases in the nickel matrix in this localized region. White spots can occur in IN718, where it has been studied extensively, but this feature can also appear in other alloys, such as Waspaloy and U720.

There are three types of white spots: (1) Discrete White Spots resulting from solid material dropping into the melt, typically from an unsound electrode, (2) Dendritic White Spots resulting from solid material dropping into the melt, typically from the solidifying

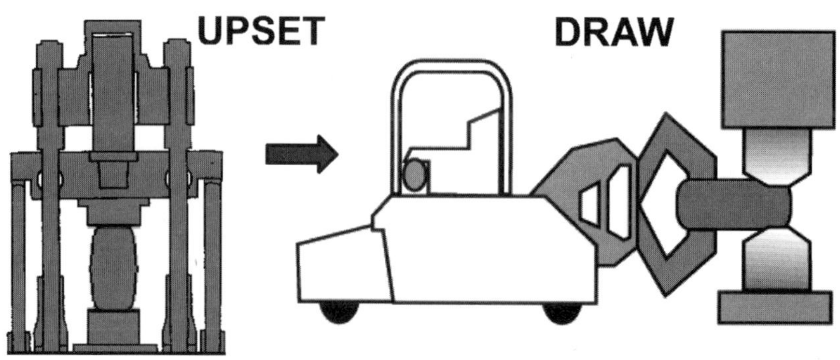

Figure 9 Schematic of upsetting and drawing (cogging) operations in the conversion of cast ingots to wrought billet and bar stock. (Schematic courtesy of Allegheny Technologies.)

shelf or crown, or from the torus formed around the rim of the electrode, and (3) Solidification White Spots resulting from melt fluctuations and local dendrite coarsening.

Freckles are another form of solidification defect that can form in ingot metallurgy superalloys. This defect is formed from strong local convection currents due to local liquid density differences. Freckles have characteristic features such as equiaxed grain columns in ingots, orientations parallel to the solidification direction, contain high concentrations of low density constituents, often have non-equilibrium phases such as Lave phase, exhibit localized variations in gamma-prime solvus and grain size. Cast nickel-base superalloy ingots require conversion into billet, bar, rod, plat or sheet mill products for many applications. Initial deformation steps that provide strain for recrystallization of the coarse as-cast grain size are performed on large open-die forging presses (Fig. 9). Incremental axial and side upsetting is performed until the final product size and microstructure is produced. This primary deformation processing is often termed press cogging, or upsetting and drawing. Ingot conversion steps are critical in developing uniform and repeatable grain sizes, with a maximum possible product yield.

Grain sizes are difficult to control during conversion and subsequently in secondary forging operations as a result of remnant casting segregation. Segregation often limits cast and wrought material to either fine grain or very coarse grain applications, though intermediate grain sizes can be attained through very close control of processing parameters.

C&W billet processing is extensively utilized due to its cost effectiveness. As noted earlier, superalloy technology development and implementation is a careful balance of performance requirements and economics.

Superalloy technology evolved toward powder metallurgy in the late 1960s and early 1970s. High alloy content superalloys are attractive from a performance capability standpoint, but processing by conventional cast and wrought routes were difficult, or near impossible due to segregation and workability issues. The rapid solidification that results for powder production essentially eliminates segregation that plagues C&W materials. Powder metallurgy (P/M) materials have been introduced into modern turbine engines to extend the operating temperatures and to improve reliability. It is thought that the development of advanced land-based gas turbines will lead to increases in P/M nickel-base superalloy use. The costs of P/M nickel-base superalloy materials, however, have restricted their use primarily to turbine engine disks to this point.

Gas atomization, the most common powder manufacturing technique, uses impingement of high velocity argon gas jets on a molten metal stream to produce fine powders, which solidify during free fall in the atomization tower [15]. Gas atomization allows for very fine, uniform, and clean superalloy powders. This method has been tailored by many manufacturers to allow for optimum superalloy powder material.

Superalloy powder must be collected, and consolidated for subsequent forging into component configurations, although certain applications allow direct consolidation of powder into components by hot isostatic pressing (HIP). For all modern billet applications, superalloy powder is canned and consolidated by either HIP or uniaxial hot compaction. The consolidated powder is then extruded at temperatures below the gamma-prime solvus to provide uniform grain refinement and obliteration of any remnant powder particle pattern that exists after the initial consolidation step. Figure 10 schematically shows the steps in the manufacture of P/M superalloy billet material.

The plasma rotating electrode process (PREP) is used to produce very clean powder throughout the world. The PREP process involves rotating an electrode at high speed with a plasma heat source directed at the tip to produce powder. Molten metal droplets are expelled from the rotating electrode by centrifugal force. Plasma rotating electrode process

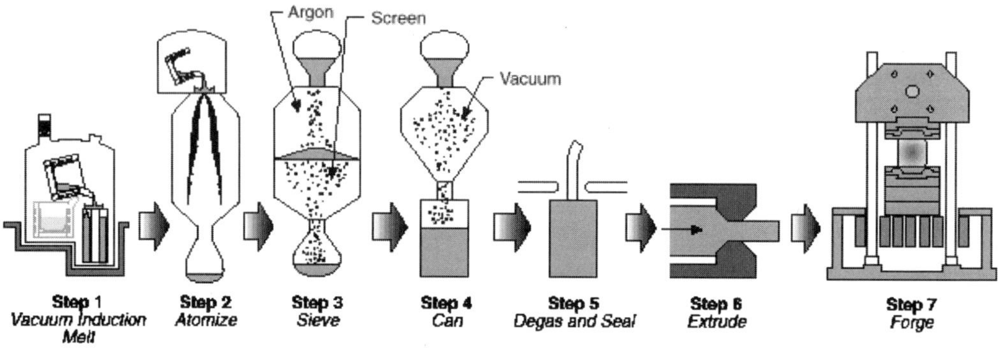

Figure 10 Schematic of P/M processing steps from VIM melting of material heats to final forging of individual disk components. (Schematic courtesy of General Electric.)

does not use large volumes of gas as with gas atomization. Controlled volumes of very clean helium are used to quench the powder produced in this process and are recycled within the processing chamber. The molten metal in this process does not touch ceramic nozzles or other objects during atomization. The low atomization rates, and the cost of the highly sound, super clean electrodes have limited the cost competitiveness of this process.

To a smaller extent, rotating disk atomization and vacuum atomization are used for powder production. In rotating disk atomization, a molten metal stream is poured onto a rotating disk or cup. Centrifugal forces expel droplets, which are gas quenched and collected. In vacuum atomization, the molten metal saturated with soluble gas is drawn into a vacuum chamber through a nozzle, which atomizes the metal stream into powder droplets.

After superalloys are atomized, they are classified by sieving through screens of exact and known mesh sizes. Powder size to a large extent controls ultrasonic inspectability, LCF characteristics, and lifing criteria. Fracture mechanics is used for the design of turbine engine components. The fatigue life of nickel-base superalloys is limited by defects. Defect size and crack growth rate are major turbine disk design criteria. Probability of detection (POD) of a defect is used in lifing and applying safety factors for design lives and in-service inspections. Larger powder sizes allow a greater chance of defects in the final component and hence would have potential application penalties in the form of reduced life capability, increased disk size or reduced allowable service stresses.

B. Casting of Nickel-Base Alloys

Many components are manufactured from nickel and Ni-base alloys by casting. Although casting is thought to be a conventional process, many unique techniques must be utilized to make successful castings. Casting of Ni-based superalloys can result in large amounts of chemical segregation. If this phenomenon is not controlled, deleterious phases can form at the latter stages of solidification, in addition to non-uniform final mechanical properties. Ni-based superalloys also have a large "mushy-zone" or a difference in the liquidus and solidus temperatures. This can cause difficulty in producing sound castings without trapped porosity.

There are several casting processes that can provide economical near-net shape cast components, but all are based to some degree on investment casting. The desired and characteristics physical and mechanical properties can be developed within components that can be of complex shapes in a wide variety of sizes. In addition, the development of preformed ceramic cores has resulted in the capability of casting hollow components, which has been essential for the air-cooling of turbine blades and vanes. Cast components can vary in size from a few centimeters to several meters in major dimensions.

While wrought and powder metallurgy alloy compositions must be restricted to preserve hot workability, defect tolerance, and the desired properties, cast alloys compositions are less constrained in some cases. In fact, cast alloys with much greater strengths can be generated by the casting and heat treatment process (Tables 2–5). The high performance levels in cast alloys are particularly exemplified by the unusually high elevated temperature strength of Ni-base superalloys. In general, cast alloys exhibit superior creep and rupture properties, but higher levels of ductility and fatigue properties are observed in wrought and powder metallurgy alloys.

The investment casting process begins with the production of an exact replica or pattern of the part in wax and/or plastic (Fig. 11). The pattern dimensions must compensate for the wax, mold, and metal shrinkage during processing. If the component contains internal passages and details, a preformed ceramic core is inserted into the die cavity, around which the pattern material is injected. Depending on the size of the component, more than one of the patterns can be assembled into a cluster with gating and runners to control the flow of molten metal into the various die cavities. The component size is inversely related to the number of patterns that can be joined into a cluster. The design of the gating and runners is critical to achieving a sound, metallurgically acceptable casting. Much of the knowledge for the developing of good cluster molds is based on trial and error and previous experience. However, the use of computer modeling tools has allowed for a better understanding of mold design.

Once the clusters have been created, the molds are produced by first immersing the pattern assembly into an aqueous ceramic slurry. A dry, granular ceramic stucco is then applied immediately after dipping to strengthen the shell. These steps are repeated several times to develop a rigid shell. The number of "dips" is dependent on the size of the component. Generally, the larger the size of the component, the greater the mass of molten metal that will be poured into the mold, and the thicker the mold must be to accommodate the increased metal mass. However, the increased thickness of the shell for larger components also results in reduced thermal conduction, which can effect the metallurgical properties of the cast component.

Once the mold has been created, the wax is melted out of the shell in a furnace. Typically, rapid heating rates are used to prevent fracture of the unfired ceramic shell due to the expansion of the wax prior to melting. In some cases, steam flash heating furnaces are used to rapidly remove the wax from the mold. The strength of the shell is then increased by firing, to allow for the handling and the casting operation. An insulation blanket is then tailored to the mold configuration to minimize heat loss during the casting operation and control cooling during solidification.

The metallurgical structure generated in the casting is, to a great extent, controlled by the heat flow during solidification. Casting of components with an equiaxed structure, the mold is preheated to enhance mold filling, control solidification, and develop the proper microstructure. If the component is to be cast in vacuum conditions, the alloy charge is also melted under vacuum in a separate chamber prior to insertion of the preheated mold. During melting and pouring, the vacuum level is maintained a level at or below 1 μm. After pouring the molten metal into the mold, an exothermic material is applied as a hot top to allow for mold filling purposes (i.e. feeding). The mold is then allowed to cool. A different procedure is used if directional microstructures are desired. In some cases, appropriate

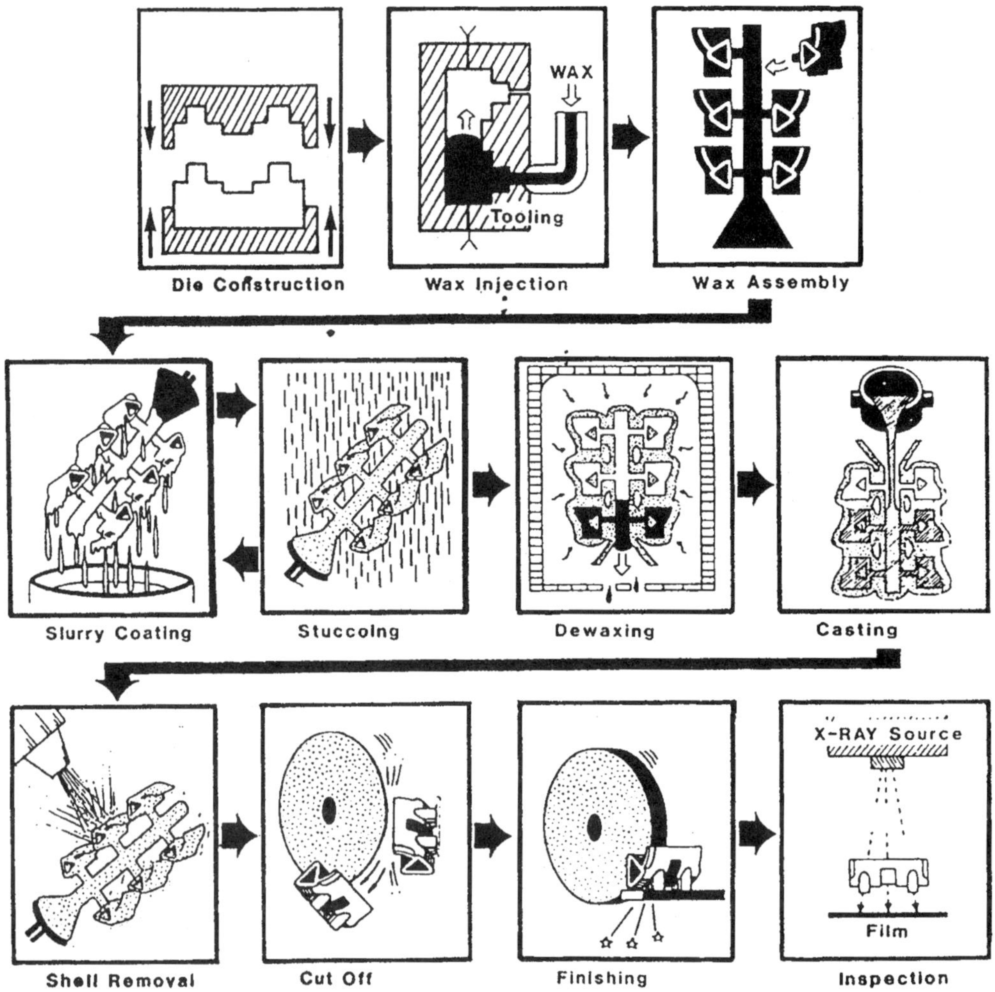

Figure 11 Schematic flow diagram for investment casting. Initial, and most costly step, in the process is the development of an appropriate tool which will be used to make the wax patterns. These wax patterns are then combined with a runner, pour-cup and gating to produce the mold. After multiple dip cycles in a ceramic slurry and stucco, the wax is removed and the mold is ready for casting. The mold is removed from the cast components by a variety of techniques, including grit-blasting and other mechanical means. Only a limited amount of machining is performed after investment cast. The last step is the inspection of the final part [2,3].

preheating and mold wrapping can be used to develop a somewhat directional microstructure due to control of heat flow in the mold. Directional solidification can be used to produce either columnar-grained materials or single crystal casting (Fig. 12). Directional solidification relies on the use of a water-cooled copper chill plate to extract heat from an investment casting mold that is extracted from a preheated mold chamber. In the vast majority of cases, directional solidification is performed in vacuum and cannot utilize convection cooling and must rely solely on radiation cooling and heat conduction through the solidifying part to the chill plate. Since the $\langle 1\ 0\ 0 \rangle$ growth direction is preferred during solidification, the material exhibits an aligned microstructure with the ⟨1 0 0⟩ direction along the growth direction. Single crystals can be produced by the use of selectors that allow only one grain to progress into the mold cavity. Single crystals of virtually any orientation can be produced by using a "seed" that is oriented in the desired direction and placed on the chill plate during the processing. Figure 13 shows examples of three turbine engine blades that were manufactured to produce an equiaxed structure, a columnar or directionally solidified structure, and single crystal microstructure.

Nickel and Ni-base alloys can be melted in air or vacuum [2,3]. Alloys with increased levels of reactive elements, such as Al, Ti, Ta, etc. are particularly sensitive to impurities that are usually melted under vacuum. The more highly alloyed, high performance Ni-base superalloys are generally always melted under vacuum. Some of the simpler alloys are melted in air using induction or indirect arc roll-over furnaces. Most alloys are melted and heated to temperatures approximately 75–175°C above the liquidus temperature. Control of the metal temperature prior to and during casting and the mold temperature is critical for controlling grain size and orientation. For increased consistency, the melting and casting operations are typically automated with programmed closed-loop furnace control. Due to differences in thermal expansions, the shell mold frequently fractures during cooling, which facilitates mold removal. Any remaining mold can be removed by mechanical or hydraulic means. After shell removal, the cluster is checked for composition, casting defects, surface defects, and orientation, if directional microstructures are desired.

The control of the solidification microstructure is critical, since the microstructure determines the mechanical properties and consistency ultimately achieved. To control the microstructure in equiaxed castings, several tools are used, including facecoats to encourage grain nucleation, pour temperature of the metal, mold temperature, shell thickness, part orientation, part spacing, gating locations, insulation to wrap the mold, pouring speed, and shell agitation. The dendrites that forming during solidification are probably the most visible microstructural feature of castings. The spacing of the primary and secondary dendrite arms is controlled by the cooling rate. As the cooling rate is increased, the dendrite arm spacing decreases and the microstructure becomes more homogeneous, ultimately resulting in improved mechanical properties. The cooling rate also impacts the carbide morphology. Similar to the refinement in the dendrite arm spacing, increased cooling rate also results in more refined, blocky primary MC-type carbides. Refined carbides structures are also reported to result in increased fatigue resistance.

Although the composition of cast alloys is usually less constrained than wrought alloys, limits on cast alloys chemistry are used to reduce the amount of eutectic segregation that forms during solidification. The eutectics are usually found in the interdendritic regions of the cast structure and are the last material to solidify. The eutectics typically have a cellular appearance and, in the superalloys, contain excessive amounts of γ'. The eutectic regions can be removed in some alloys by heat treatment to solution the eutectic phases and allow diffusion to reduce the level of segregation.

During solidification, the volume of the metal typically decreases approximately 2–6%, resulting in shrinkage porosity if poor feeding of the molten metal occurs. Proper gating is required to prevent shrinkage porosity. In addition, porosity can occur due to dissolved gas or poor feeding between dendrites. The presence of porosity in the microstructure can result in significant reduction in the fatigue strength since the pores act as fatigue crack nucleation sites. Generally, the occurrence

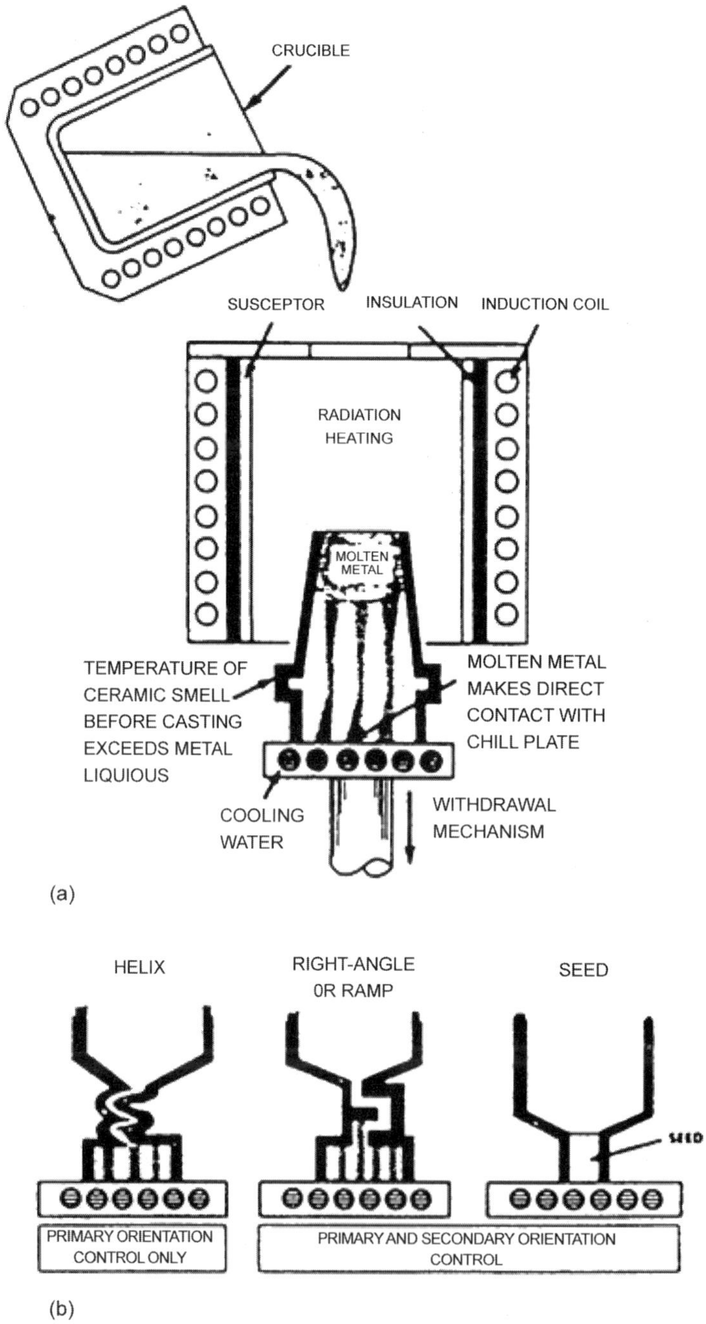

Figure 12 (a) Schematic representation of the directional solidification process to generate columnar grains or single crystal components. The water-cooled copper chill in the withdrawal furnace develops the necessary thermal gradient to produce the columnar grain microstructure. (b) The use of a selector or seed crystal results in the elimination of all but one grain which results in a single crystal [2,3].

Figure 13 The three primary types of investment cast Ni-base superalloys used in gas turbine applications. The equiaxed microstructure is produced by conventional investment casting techniques and may employ the use of agitation to refine the microstructure. The polycrystal and single crystal blades are produced by directional solidification. Photograph provided by Howmet Corporation.

of porosity can be prevented, or at least reduced, by proper gating of the mold and the use of good vacuum melting practices. In addition, hot isostatic pressing or HIP, which is the simultaneous application of pressure and temperature, was applied to casting to eliminate porosity (Fig. 14). Most HIP conditions are in the range of 1000–1200°C and 100–300 MPa. Care must be taken though in highly alloyed superalloys, that temperatures do not exceed the incipient melting temperature during HIP. It has been reported that HIP results in dramatic improvements in fatigue properties and limited improvements in tensile and creep properties.

As noted above, slow cooling can result in the development of a coarse microstructure that has poor mechanical properties. Since cast component cannot utilize subsequent deformation processing to refine the microstructure, other techniques are needed to produce cast components with refined microstructures. Facecoats can be applied to the mold to stimulate nucleation during solidification, as noted above. In addition, with the development of HIP, to reduce porosity in casting, other techniques were developed to refine cast microstructures that result in increased levels of porosity. One technique included the use of mechanical agitation of the mold to break dendrites during solidification which can then act as nucleation sites for additional grains. In addition, the ability to pour the molten metal at lower superheat temperatures was developed which also produced finer microstructures.

Many cast alloys rely on heat treatment to produce optimal properties [2–4]. In particular, the Ni-base superalloys are frequently heat treated for improved properties. These heat treatments can include a solution heat treatment intended to solution all or some portion of the cast γ', including the eutectic regions followed one or more aging heat

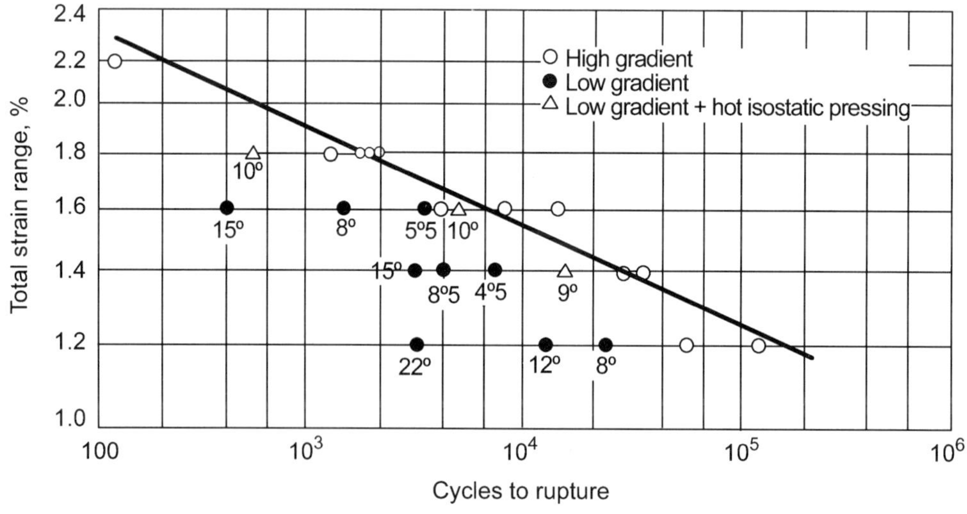

Figure 14 The effect of thermal gradient, orientation and HIP on the strain controlled, fully reversed, low cycle fatigue (LCF) behavior of the single crystal, Ni-base superalloy CMSX-2 at 760°C. The numbers identify for the misorientation of each sample from the desired [0 0 1] orientation. The line represents the average of the data for high gradient samples with limited misorientation ($<5°$). Note that as the misorientation increases, the fatigue life decreases. In addition, the low gradient samples exhibited decreased fatigue life, due to the presence of porosity. However, HIPing of the low gradient samples resulted in a dramatic increase in the fatigue resistance [2,3].

treatments to produce the appropriate size γ'. In some alloys no solution heat treatment can be performed due to incipient melting concerns or no significant improvement in properties is attained by solution heat treatment. Some alloys are given only aging heat treatments and some alloys are used in the as-cast condition. The development of the single crystal alloys allowed for the reduction or removal of the grain boundary elements. These grain boundary elements (e.g. B, C, Hf, Zr) also result in a significant reduction in the solidus temperature, which may be result in the inability to heat treat an alloy [3]. However, the reduction or elimination of these elements in single crystal superalloys allowed for the development of alloys that can be fully solution heat treated. Therefore, the microstructure of the single crystal alloys can be completely tailored for improved properties by solution and re-precipitation of the γ' in a controlled fashion.

The weldability of the alloy is also a concern for casting [2,3]. In many cases, casting defects can be remedied and the useful life of components can be extended by weld repair. The less highly alloyed Ni-base materials can be welded by a variety of techniques including shielded metal arc (SMA), gas–metal arc (GMA) and gas–tungsten arc (GTA) welding techniques. The appropriate preheat techniques are needed to prevent or reduce the incidence of hot-cracking. The solid-solution strengthened Ni-base alloys can also be welded by these techniques, but some additional care must taken to prevent hot-cracking due to the wider melting range and higher strengths. The precipitation-hardened alloys are considerably less weldable. The γ''-strengthened alloys (e.g. Alloys 625, 718) are more weldable than the γ'-strengthened alloys (e.g. IN 100) due to the slower precipitation kinetics of γ''. The γ'-strengthened alloys are prone to strain age cracking and great care must be taken to prevent the build-up of residual stresses during welding. Welding can be performed on these alloys using GTA and GMA techniques and also by electron beam (EB) or plasma-arc (PA) welding techniques. Typical techniques to prevent strain-age cracking include pre- and post-heat control, rapid heating, protective welding environments, limited constraint and post-weld stress relief heat treatments. Generally, the need to weld repair casting increases with increasing component size. However, the constraint on larger components becomes greater, in some cases, and may result in reduced weldability.

Other joining techniques can also be used with Ni-base alloys, including diffusion bonding, friction welding, resistance welding and ultrasonic welding and brazing. In general, the joinability of Ni-base alloys is inversely related to the strength of the alloy. In addition, the casting process is tailored to produce the appropriate microstructure after casting and some subsequent heat treatment. The joining operation may result in reduced properties in the region of the joint, due to the alteration of the local chemistry and microstructure.

C. Deformation Processing of Nickel-Base Alloys

Deformation processing is often required to further transform nickel-base alloy billet stock into useful shapes [16,17]. In addition to developing the required component shapes, the forging process tailors the microstructure of the material to allow for optimum properties.

Forging of components is termed secondary wrought processing, since the conversion of powder cans or cast ingots to billet by deformation methods is called primary wrought processing. Secondary wrought processing methods and sequences are very complicated, and are dictated by many factors, including economics, material processing characteristics, component requirements, and component geometry. For economic reasons, all forging operations are optimized to reduce the amount of metal required to make the final component shape. Nickel-base superalloys are relatively expensive and hard to machine.

Deformation processing can be accomplished at low or high temperatures. Low temperature processing results in cold work and greatly increases strength. This process is utilized primarily for increasing the properties of solid-solution alloys, while hot working, deforming above the recrystallization temperature, is used for all types of nickel-based materials.

Material processing characteristics play a large role in the selection of secondary forging operations. Many nickel-base alloys have relatively low alloy content and consequently limited second phase and elevated temperature solid-solution strengthening. These materials are readily forged by means of hammer, mechanical press, and hydraulic press. These alloys are less sensitive to high strain rates and chilling effects. Materials designed for superior elevated temperature strength and creep resistance are more difficult to deformation process. Highly alloyed cast materials are very temperature sensitive, while the even higher alloy content powder metallurgy materials are both temperature- and strain rate sensitive.

New processing methods have been and continue to be developed to meet the processing requirements of advanced nickel-base superalloys. Hot-die forging has been invented to minimize the chilling effects of forging tools on work pieces during controlled hydraulic press forging. Even more sophisticated forging methods such as isothermal forging have been developed for powder metallurgy superalloys.

Component requirements and geometry play a large role in dictating forging processes and sequences in addition to material response characteristics to deformation. Tight control of heating and forging steps is required for components where controlled grain sizes are needed. For high strength applications, fine grain sizes are often required, mandating subgamma-prime solvus processing. Strain levels need to be engineered into forging processes where large average grain sizes or isolated grains from the mill products are required to be refined in the final component.

Coarsened grain structures are now of great interest for creep and damage tolerance. The achievement of controlled coarse or intermediate grain size material can be done through forging, heat treating or a combination of both. In most cases, combinations of processing steps are required to achieve the controlled microstructure goals. If supersolvus processing is used to coarsen the material after forging, the forging process used to make the component shape is critical in controlling the grain size produced during the heat treatment.

Computer process modeling is an extensive part of modern component and process design. Finite element models are being developed and utilized to predict metal flow and microstructural evolution during forging. This has allowed significant optimization of manufacturing methods with increased component capability and quality.

D. Heat Treatment Processing of Nickel-Base Alloys

The heat treatment of nickel-base alloys is often critical to overall alloy and component performance. The microstructures produced from hot working operations are not optimum for service applications; so the microstructure is tailored to develop the optimum grain size, grain boundary morphology, phases and phase distributions by heat treating.

There are several categories of heat treatments used for modifying the microstructure of nickel-base materials. These include: (1) solution, (2) stabilization, (3) stress relief, and (4) aging. These processes are designed to alter the microstructure within the material and/or reduce internal stresses for improved component performance [17,18].

Solution heat treating is performed on many nickel-base alloys and on nearly all nickel-base superalloys. Solution heat treating is performed to dissolve secondary phases such as γ', γ'' and delta which results from prior forging operation working temperatures or slow cooling rates from forging temperature. Solution heat treating is most often performed just below the solvus temperature for the various precipitate phases or the temperature where significant grain growth occurs. Small amounts of precipitate phases are often left in the matrix to pin the grains from excessive growth. Supersolvus solution cycles are used to fully dissolve precipitate phases into the matrix. This results in maximum precipitation strengthening and added grain growth during the heat treatment process. Cast nickel-base alloys, however, as noted earlier can be very heavily alloyed. Many of these materials have very high γ' contents, which preclude the ability to completely dissolve this phase prior to reaching the solidus temperature for the alloy.

Equally or potentially more important to the high temperature solution cycle portion of the solution heat treatment is the cooling method and part cooling rate profile from the solution cycle. Nickel-based alloys can be oil quenched, gas quenched or still air cooled from solution heat treatment cycles. The imposed cooling rate along with the nucleation and growth kinetics of the precipitating phases produces the morphology and distribution of the secondary phases in the nickel-base component. Often high cooling rates are required for high volume fractions of very fine secondary γ' or γ'', and hence very high strength.

Although mechanical properties are the most significant driving force for selecting rapid cooling methods and rates, residual stress minimization favors slower cooling rates. Large variations in cooling rate associated with rapid quenching methods will give rise to very large thermal stresses, and hence very high residual stresses. High residual stresses can result in problems of part quench cracking, machining distortion or worse yet, unexpected mechanical properties and performance in the final component.

Thermal modeling is increasingly used to guide engineers toward the optimum route to heat and cool nickel-base components. The use of inverse heat transfer coefficient computer programs has allowed engineers to develop excellent predictive tools for various cooling media with surprising accuracy. The use of finite element models and material property/cooling rate databases has allowed a quick and efficient development of effective heat treat cooling methods. Minimum cooling rate criteria can be developed from testing materials cooled at various rates. This minimum cooling rate criteria can be used to iterate a cooling process in a computer model to develop the appropriate shop method, which will in turn develop the required minimum strength level and the lowest possible residual stresses.

Aging cycles for nickel-base alloys are utilized to develop and control the volume fraction, size, and distribution of γ' or γ'' phases and other secondary phases such as carbides and borides. The age cycles may be an intermediate temperature stabilizing cycles where additional particles are precipitated or existing particles are grown, or coalesced to the equilibrium content and morphology at that temperature. Aging, followed by air cooling, is often performed at temperatures slightly above the operating temperature of the component, typically in the range of 649–760°C. Aging is also performed to improve creep and stress rupture properties by carbide precipitation.

Heat treatment of cast nickel-based superalloy components follows the same type of sequence as will cast and wrought precipitation strengthened nickel-base alloys. Generally,

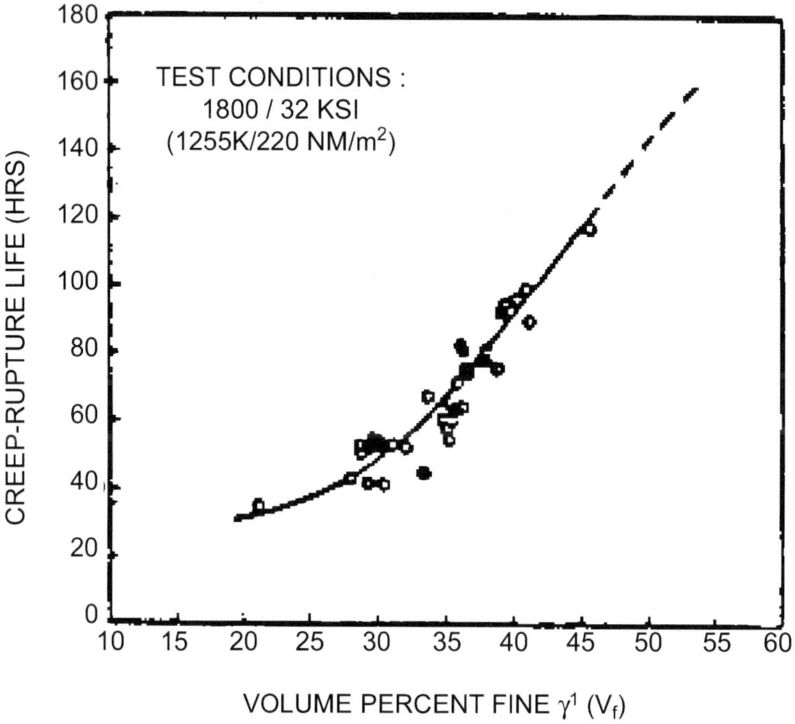

Figure 15 The effect of γ' size on the creep rupture life of single crystal Mar-M200 tested at 982°C/220 MPa. The amount of fine γ' is increased by increasing the solution heat treatment temperature to solution a greater amount of the γ' present after solidification. An increased creep resistance is observed in samples with greater volume fraction γ', and hence a higher solution heat treatment temperature [10].

finer γ' precipitate sizes result in improved properties. Dramatic improvements in properties can be obtained by refinement of the γ' size (Fig.15). In order to achieve the desired microstructures, single crystal superalloys are typically given a solution heat treatment followed by rapid cooling (e.g.inert gas quench) in a vacuum furnace. The protective environment of a vacuum furnace is required to prevent excessive oxidation during the long-term, high temperature solution heat treatments. The material is then given one or more aging steps to precipitate and grow the γ' to the desired size. These γ' precipitates are often called primary γ'. Multiple aging steps are often used to precipitate very fine (i.e. $<0.1\,\mu m$) in the γ channels in between the coarser primary (i.e. 0.3–0.5 μm)γ' precipitates [2–4,10,12]. In some cases, samples are given elevated temperature exposures during the application of the environmental resistance and/or thermal barrier coatings. These additional thermal cycles are included in the heat treatments steps and not done in addition to the previously described heat treatments.

REFERENCES

1. International Nickel Study Group, www.insg.org/nickel, 2001.
2. Davis, J.R., Nickel, Cobalt and Their Alloys, ASM Specialty Handbook, ASM International, Materials Park, OH, 2000. pp. 3–6.

3. Davis, J.R., Heat-Resistant Materials, ASM Specialty Handbook, ASM International, Materials Park, OH 1997, pp. 3–30.
4. Sims, C.T.; Stoloff, N.S.; Hagel, W.C. Eds.; *Superalloys II*; John Wiley & Sons: New York, NY: 1987
5. Rosenberg, S.J., *Nickel and It's Alloys*, National Bureau of Standards Monograph 106, U.S. Department of Commerce, May 1968.
6. Betteridge, W., *Nickel and It's Alloys*, Ellis Horwood Ltd, New York, 1984.
7. Vroman, G., "Nickel expands role in gas turbine engines", In *American Metal Market* Martin-Abott; New York, 1998, October, 10A–11A.
8. Furrer, D., Fecht, H. "Ni-based superalloys for turbine disc applications", JOM 1999, January, 14–18.
9. Morral, F.R., Wrought superalloys, *Superalloys Source Book*; Donachie, M.J., Ed., ASM; Materials Park, OH; 1984, 20–40.
10. Stoloff, N.S., Fundamentals of strengthening. In *Superalloys II*: Sims, C.T., Stoloff, N.S.; Hagel, W.C., Eds.; John Wiley & Sons, New York, NY: 1987, 61.
11. Brown, E.E., Muzyka, D.R. Nickel–iron alloys. In *Superalloys II*, Sims, C.T.; Stoloff, N.S.; Hagel, W.C., Eds.; John Wiley & Sons, New York, NY: 1987; 165.
12. Decker, R.F., Strengthening mechanisms in nickel-base superalloys. Presented at Steel Strengthening Mechanisms Symposium; Zurich, Switzerland. May 5 and 6, 1969.
13. *High Temperature, High Strength, Nickel Base Alloys*; International Nickel Company: Saddle Brook, NJ 1984.
14. Pridgeon, J.W., et al., Principles and Practices of Vacuum Induction Melting and Vacuum Arc Remelting, *Metallurgical Treaitises*; Tien, J.K.; Elliott, J.F., Eds.; TMS: Warrendale, PA; 1981, 261–276.
15. Lenel, F.V.; Ansell, G.S., Powder Metallurgy, *Metallurgical Treatises*, Tien, J.K.; Elliott, J.F., Eds.; TMS: Warrendale, PA; 1981, 345–260.
16. Ruble, H.H., Semiatin, S.L., Forging of Nickel-Base Alloys, *ASM Handbook Volume 14: Forming and Forging*; ASM; Materials Park, OH; 1988, 261–266.
17. Fecht, H.; Furrer, D., Processing of Nickel-Base Superalloys for Turbine Engine Disc Applications, *Advanced Engineering Materials* 2, Issue No. 12, 2000, 777–786.
18. Brooks, C.R., *Heat Treatment, Structure and Properties of Nonferrous Alloys*; ASM, Materials Park, OH; 1982.

13
Designing with Copper Alloys

Morris E. Fine
Northwestern University, Evanston, Illinois, U.S.A.

Junji Miyake
Nippon Mining and Metals Company, Ltd., Koza, Kanagawa, Japan

I. HIGH-STRENGTH, HIGH-CONDUCTIVITY ALLOY DESIGN

A. Introduction

Copper alloys that have both high strength and high conductivity are needed for many electric and electronic applications, such as lead frames, connectors, conducting springs, and sliding contacts. Figure 1 shows the strength–conductivity map of commercially available copper alloys for electronic applications [1]. Since dissolved solutes in a pure metal rapidly reduce electrical conductivity as well as thermal conductivity, solid–solution strengthening is not suitable for designing this class of high-strength, high-conductivity alloys, but they may be designed on the basis of precipitation and/or dispersion hardening [2,3].

The theory of the yield stress of alloys with precipitates or dispersed phases is well in hand and may be used for alloy design. The solubility of the hardening phase in the matrix must be very small, otherwise the conductivity will be degraded too much. Nordheim's rule relates conductivity to dissolved solute in alloys and is also available for alloy design. In a given alloy, decreasing the dissolved solute increases the conductivity and strength due to an increase in volume fraction of precipitates.

Metal–matrix composites (MMCs) can be prepared by dispersing an essentially insoluble phase in a pure metal using powder metallurgy (P/M) techniques, but MMCs usually contain large volume fractions of particles of the dispersed phase that are large compared to precipitates formed by reaction in the solid state. The large size gives inefficient hardening, and the large volume fraction of a phase of relatively low conductivity reduces the conductivity of the composite. There are two avenues for precipitation reactions to occur: changing temperature or changing atmosphere. Both have been used to make high-strength, high-conductivity alloys. Internal oxidation of copper powder containing aluminum followed by consolidating the powder for an aluminum–copper composite is the basis for a series of excellent alloys [4].

Precipitation hardening requires a decrease in solid solubility on lowering the temperature. For the case in hand, the decrease in solid solubility should be very rapid. The heat treatment consists of heating to a high temperature where all of the solute is in solution and then lowering the temperature for the precipitation reaction. Alloys are

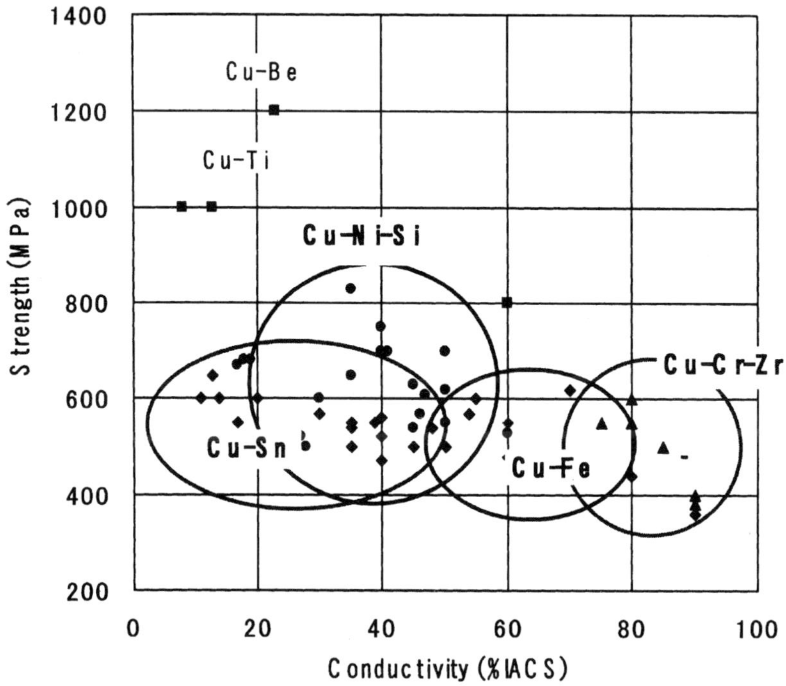

Figure 1 Copper alloys for electronic applications. (From Ref. 1.)

required that have very low solid solubilities at the precipitation temperature. Thermodynamic principles and data can be used very effectively for the integration of this feature into the alloy. Since the solubility decreases with temperature, a low aging temperature is desired, but the precipitation kinetics are slow at low temperatures. A two-temperature aging treatment may be used to increase strength and conductivity.

Solid solubility in an alloy containing precipitates decreases as the precipitate size increases—the Gibbs–Thomson effect. The yield stress first increases as the particle size increases, but after the spacing between particles becomes large enough for dislocation bypass to occur, the yield strength decreases with particle size. In this regime, there is a trade-off between high strength and high conductivity [5].

B. System-Based Design Approach

Smith [6] suggested that materials are hierarchical structures with properties governed by the dynamic evolution that occurs in processing and service. This philosophy has been further developed and formulated into a methodology for the rational design of complex materials. Successful applications have been made for high-toughness, ultrahigh-strength alloy steels [7,8]. Such a design approach utilizes the integration of processing/structure/property/performance relations within a multilevel system structure. The alloy design circle in Fig. 2 shows this for the case of high-strength, high-conductivity alloys. The circle begins with a need, followed by a design concept, the theories needed to achieve the concept, one or more alloy compositions derived from theory, processing of the alloys to give the desired microstructure, and finally back to need, that is, satisfactory conclusion of the alloy design. The steps are in smaller circles connected by arrows. Arrows between

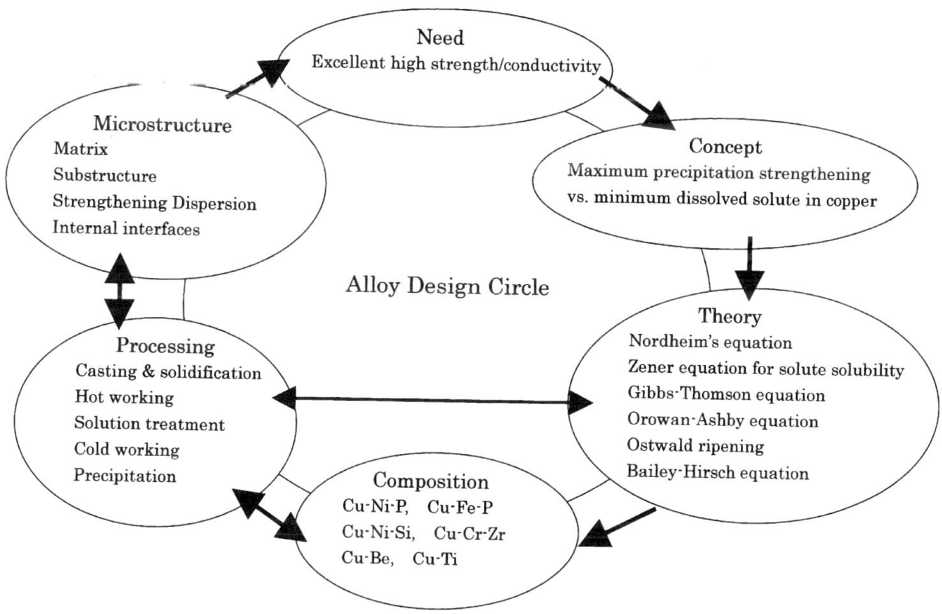

Figure 2 Alloy design circle for high-strength, high-conductivity alloys.

composition and processing and processing and microstructure are double-headed, indicating that there are a number of iterations before the needed properties are obtained. The key composition/structure/property and processing/structure relations identify the critical sciences that are needed. Due to the stringent requirements in the strength–conductivity combination for critical applications, it is unlikely that the best optimizations of strength vs. conductivity can be achieved with binary alloys. The critical science areas needed for the optimization are discussed next.

C. Electrical Resistivity

The theory of resistivity of metals and alloys, even those having complex microstructures, is well established. This has been discussed at great length by Rossiter [10]. In disordered solid solutions, one needs to consider atomic displacements, vacancies, and interstitials, and their static displacements and phonon scattering.

The well-established Nordheim rule [11] relates electrical resistivity, ρ, in a disordered metallic solid solution to its solute content:

$$\rho = \rho_0 + KC(1 - MC) \tag{1}$$

where ρ_0 is the electrical resistivity of the pure solvent, C is the solute concentration, and K and M are two constants. Usually, $K \gg M$, and $\Delta\rho$ $(=\rho-\rho_0)$ is almost proportional to C up to a fairly large concentration ($C < 0.2$). Nordheim's rule has been theoretically established by computing the effective number of conduction electrons and their frequency of scattering. Deviations from Nordheim's rule may occur when the solute and solvent have very different valencies. Figure 3 shows the electrical resistivity of copper solid solutions as a function of solute content for a number of different solute elements in the dilute regime [2,3,12,13].

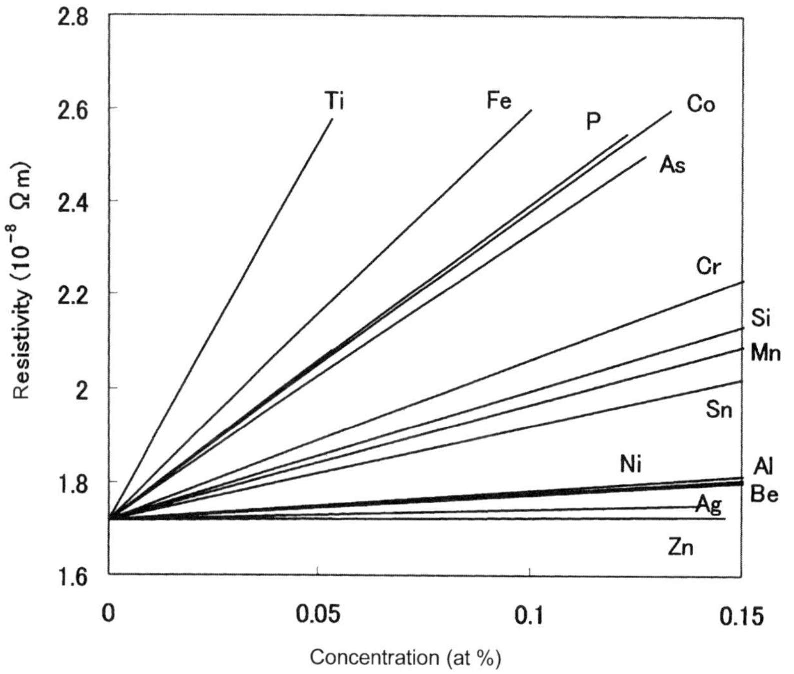

Figure 3 The effect of alloying elements on the resistivity of copper. (From Ref. 2.)

According to Linde's rule [14], when nontransitional substitutional elements are added to copper, the resistivity increases roughly as (ΔZ), where ΔZ is the difference in valency between solute and solvent. This behavior is explained by the difference in potential acting on the conduction electrons in the alloy and in the pure matrix. Thus, the resistivity increase is very small when the solute and solvent have the same valency, because the conduction electrons are slightly attracted or repelled by the solute. In copper, $\Delta Z = 1$, 2, 3, and 4 for Zn, Ga, Ge, and As, respectively. Thus, other things being equal, the valence difference between the solvent and the solute remaining after precipitation should be as small as possible. When transition metals are added to copper, a nonmonotonic increase in resistivity is observed because of resonance with a poorly conducting low-lying d-band. For this reason, iron increases the resistivity of copper rather rapidly. While Cd, Zn, Co, Ni, and Ag seem to be good choices from the viewpoint of resistivity, their solubilities in copper are high.

Engineering materials are polyphase in nature and also contain a variety of defects, such as dislocations, stacking faults, and homophase and heterophase interfaces. The effects of these are small. The electron scattering contributions from interfaces [15,16], dislocations [17,18], and stacking faults [19] have been examined experimentally and theoretically. The contribution of interface scattering becomes appreciable when the mean free path of conduction electrons is comparable to the characteristic microstructural length scale, e.g., in the case of nanocrystalline metals and alloys, thin films, and nanocomposites. If the electrons are specularly reflected at the internal interfaces, the conductivity will not change; however, when the collisions at the internal interfaces are inelastic or partially elastic, the conductivity will decrease due to momentum transfer from the electrons to the lattice [15].

The scattering theory for crystal dislocations and its contribution to electrical resistivity is well established [17,18]. An implicit assumption has been made that the scattering of electrons from dislocations is highly localized and nearly isotropic. This is supported by the experimental results of Basinski and Saimoto [20] in pure copper, where anisotropic distribution of dislocations was introduced by a single-glide deformation but the resistivity increase ($\Delta\rho_d$) was nearly isotropic.

For Cu–Nb in-situ filamentary composites, Verhoeven et al. [21] showed that the experimental resistivity can be decomposed into four scattering components: impurity, phonon, interface, and dislocation. As mentioned earlier, the solute content in the matrix for high-strength, high-conductivity alloys should be very low. Thus, the possibility of any remnant short-range ordering or clustering after the completion of precipitation and its effect on resistivity can be ruled out. In the absence of small precipitates a few nanometers in size that scatter conduction electrons and ordering/clustering effects (and since lattice imperfections have little effect on resistivity at room temperature), the use of Nordheim's relation for design purposes is satisfactory.

D. Solid Solubility

Consider the relation B = B(A), where B is the pure solute and B(A) represents the solute dissolved in solid solvent A. The equilibrium constant for the reaction for a dilute solution is

$$K = X_{B(A)} = \exp(-\Delta G_s/RT) = \exp(\Delta S_s/R)\exp(-\Delta H_s/RT) \tag{2}$$

where $X_{B(A)}$ is the concentration of B in A, and ΔG_s, ΔS_s, and ΔH_s are the standard Gibbs energy, entropy, and enthalpy of dissolving B into A. The natural logarithm of $X_{B(A)}$ may be plotted vs. $1/T$. If ΔH_s is constant, a straight line results. The slope is ΔH_s. ΔS_s may be obtained from the extrapolated intercept at $X_{B(A)} = 1$ based on dilute concentration data. Such a plot is shown in Fig. 4 [2,3,22–25] following Zener [26,27], even though, in many cases, the equilibrium represented is with an intermetallic compound rather than pure B. Such compounds decompose on solid solution. The reaction may be $AB_n = n$B(A) + A, or BC_m = B(A) + mC(A). Obviously, the compounds may be nonstoichiometric, and there may be some solubility of A in the BC_m intermetallic. Nevertheless, Fig. 4 is a useful way to represent the solubilities in copper vs. temperature.

Zirconium appears to be an interesting solute for increasing the strength of copper with minimum degradation of conductivity because its solubility is low at the aging temperature; but the amount of Zr dissolved at the solution-treating temperature is small, so the precipitation hardening is small. This is the basis for a commercial high-conductivity, modest-strength alloy [28].

The solubility of the second component can be reduced by adding a third component. Phosphorus may be added to Cu–Ni alloys that are completely soluble in each other in the solid state in binary alloys to form Ni–P precipitates, thereby reducing the residual solubilities to very low levels. Iron may be added to Cu–P alloys to reduce the solubility of P in copper, and this is the basis for the very successful high-strength, high-conductivity alloy Cu–2.3Fe–0.03P [29].

Equilibrium solute solubility refers to a flat and incoherent precipitate–matrix interface. Both Gibbs–Thomson effect by finite curvature and coherency stress effect will increase the solute content in the matrix. In particular, at the early stage of precipitation, these effects are not negligible. For a given precipitate shape and size, the magnitude of

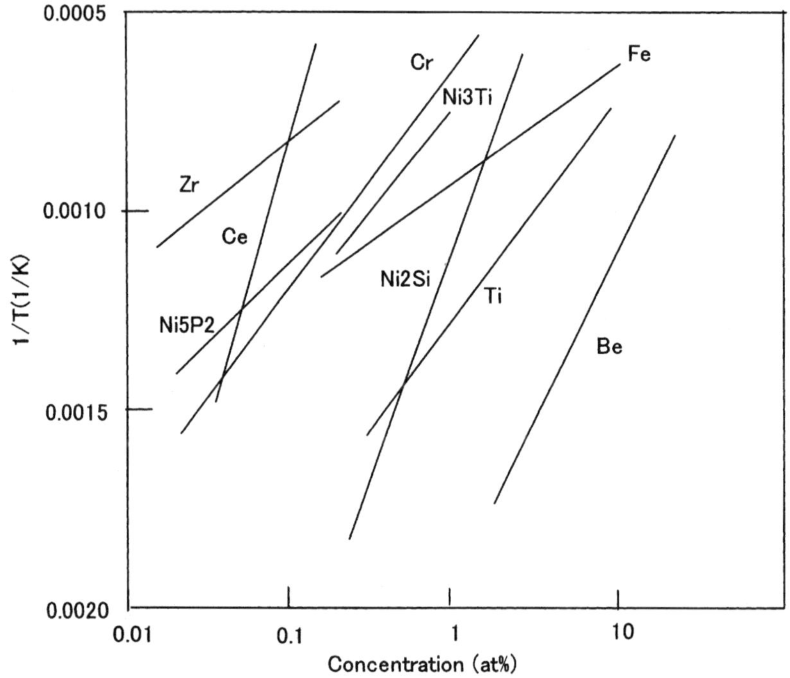

Figure 4 Equilibrium solubility of some elements and compounds in copper vs. $1/T$. (From Ref. 2.)

coherency stress can be approximated within the context of linear elasticity. Fortunately, Thermo-Calc allows the calculation of both constrained equilibria due to curvature, coherency, or any other effects and unconstrained equilibria very conveniently. Thus, Thermo-Calc can simulate the dynamic nature of microstructural evolution and its effect on resistivity.

E. The Gibbs–Thomson Equation

The Gibbs–Thomson equation relates the decrease in solute solubility in the matrix as the precipitate particle size increases. Following Wagner [30], for a spherical particle,

$$C_r = C_{eq} \exp(2\,\sigma\, V_m / r\,\alpha\, RT) \tag{3}$$

where C_r and C_0 are the concentration of solute in the matrix in equilibrium with a particle of radius r and the concentration of a solute in the matrix in equilibrium with a particle of infinite radius, respectively; σ is the interfacial energy; V_m is the precipitate molar volume; α is the stoichiometric factor; R is the gas constant; T is the absolute temperature. Figure 5 represents the solute solubility and Gibbs–Thomson effect, where the solubility curve shifts to the left as the particle radius r increases, and then the solubility becomes the "equilibrium solubility" as shown in the phase diagram when the particle radius r becomes infinite.

The Gibbs–Thomson equation may be combined with the Nordheim relation. Since the solute concentration in the matrix at the precipitation temperature is small in the

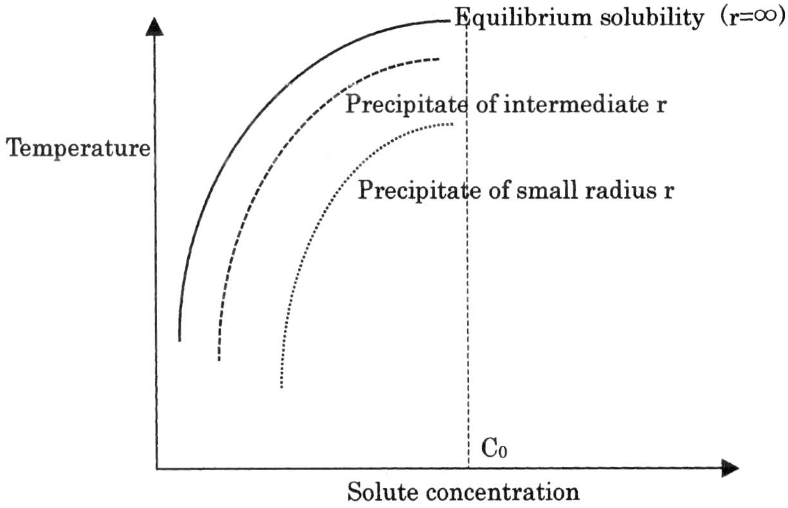

Figure 5 Solute solubility and Gibbs–Thomson effect. Gibbs–Thomson equation: $C_r = C_0 \exp(2\sigma V/rRT)$.

systems of interest because of the required high conductivity and M is near 1, Eq. (1) yields

$$(\rho_r - \rho_0)/(\rho_{eq} - \rho_0) = C_r/C_{eq} \tag{4}$$

where ρ_r and ρ_{eq} are the resistivities of an alloy with particles of radius r and an alloy with a particle of infinite radius, respectively [5]. Scattering of conduction electrons by very small precipitates has been neglected in deriving Eq. (4). Combining Eqs. (3) and (4) gives

$$(\rho_r - \rho_0)/(\rho_{eq} - \rho_0) = \exp(2\,\sigma\, V_m/r\,\alpha\, RT) \tag{5}$$

showing the inverse relationship between resistivity and particle size [5]. In utilizing the Gibbs–Thomson equation, the ρ values are much easier to determine than the matrix compositions. In an alloy containing precipitates, the particles are not all of the same size. Here, r is taken to be an average value.

F. Precipitation Strengthening

Since the solute content in the matrix needs to be low for the present application, the precipitates will be relatively large. The precipitates are expected to be large and strong so that dislocations bypass them, and the Orowan–Ashby equation is assumed to apply [31]:

$$\Delta\tau = \{1/1.8\, Gb/2\,\pi\,(1 - \nu^{0.5})\, \ln(r/b)\}/L \tag{6}$$

where $\Delta\tau$ is the increase in critical resolved shear stress, L is the particle separation, G is the shear modulus of the matrix, b is the matrix Burgers vector, and υ is Poisson's ratio. Taking L to be the mean planar particle separation [32],

$$L = r\left\{(2\,\pi/3f)^{0.5} - 2(2/3)^{0.5}\right\} \tag{7}$$

where f is the volume fraction of precipitates.

Since the solid solubility in the matrix decreases with temperature, a low aging temperature is desired for high conductivity, but the precipitation kinetics will be slow.

G. Ostwald Ripening

In order to obtain the desired low resistivity and full strengthening, it is expected that the precipitation process in the alloy is complete, and the precipitate size is determined only by the state of coarsening (Ostwald ripening). For the later stage of diffusion-controlled Ostwald ripening of spherical particles in the absence of interfacial stresses, the equation based on the LSW theory results [30,33]:

$$r^3 - r_0^3 = kt \tag{8}$$

$$k = 8\phi\sigma D C_{eq} V_m/9RT \tag{9}$$

where ϕ is the volume fraction factor, and D is the diffusivity of the slowest diffusing element. Here, the stoichiometric factor has been assumed to be 1, and r is an average value.

H. Age Hardening Curve and Electrical Conductivity

The age hardening curve and electrical conductivity behavior are depicted in Fig. 6 vs. aging time. As the aging time and particle radius increase, the yield stress increases to a maximum. In the early stages of aging, when the particles are small, they are cut by dislocations. The increase in yield stress is due to increase in precipitate volume fraction from completion of the precipitation reaction and the Gibbs–Thomson effect. The maximum is reached when the Orowan–Ashby and cutting curves vs. particle size cross. In the later stage of aging, precipitation is nearly complete, and dislocations bypass the particles. The yield stress decrease follows the Orowan–Ashby model. On the other hand, electrical conductivity first decreases due to scattering of conduction electrons by the particles. The conductivity then increases after reaching a minimum and continues to increase and nearly saturates in the last stage of aging as the residual solute concentration nearly reaches the equilibrium solubility value. For alloy design of high strength and high conductivity, an optimum aging condition needs to be determined in such a trade-off relationship between yield strength and conductivity. In most cases, the beginning of the overaged regime can be selected for such a design.

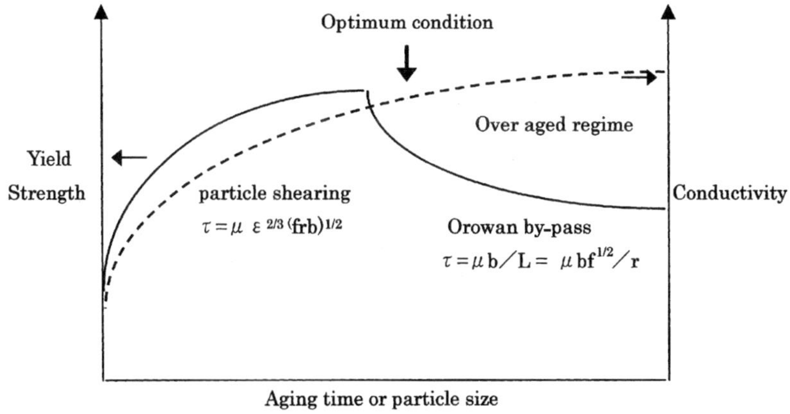

Figure 6 Age hardening curve and electrical conductivity.

I. Dislocation Strengthening

Many critical applications may require strength and conductivity in the vicinity of 1000 MPa and 80% IACS, respectively. Such strength–conductivity combinations cannot be achieved by precipitation or dispersion strengthening alone. Dislocation strengthening improves strength without degrading the conductivity appreciably. The dislocation hardening of polycrystalline fcc metals can be expressed by the so-called Bailey–Hirsch equation (34) as

$$\Delta \tau = \alpha Gb(\rho)^{1/2} \tag{10}$$

where α is a constant (nearly equal to 1), G is the shear modulus, b is the Burgers vector, and ρ is the dislocation density.

The rate of dislocation generation with plastic strain is affected by the presence of solute atoms and precipitates. In the overaged state, multiple glide may operate immediately after yield and dislocation loops form between the precipitates, giving rise to a rapid rate of work hardening [35]. To maximize this effect, the cold working should be done after aging into the Orowan regime. In the Cu–Nb system, where Nb filaments are introduced in situ by cold-rolling or wire drawing, a high density of dislocations contributes significantly to the strength of such alloys [36]. This is also the case for Cu-base Ag alloys. Here, precipitates of high-conductivity Ag solid solution in Cu solid solution increase strength, which is further increased by cold working [37].

Equations (1–10) can be used for alloy design and for selecting the optimum aging time, temperature, and deformation for the desired trade-off between yield strength and resistivity at a temperature where the kinetics are not too slow for practical purposes.

J. Experimental Examination of Theoretical Relations: Cu–Ni–P System

The relationships between precipitate size, resistivity, and yield strength were determined in Cu–0.74 wt% Ni–0.14 wt% P, where the precipitates are cubic Ni_5P_2 [5]. The increase in yield stress, resistivity ratio as defined in Eq. (5), and the precipitate size are shown in Fig. 7 for aging at 450 °C after solution treatment at 900 °C. At this aging temperature, the precipitates are slightly ovoidal in shape, so an equivalent radius is plotted as the size dimension. The solid line is a plot of Eq. (5) assuming σ to be 300 mJ/m^2. The resistivity ratio, with ρ_0 referring to Cu, increases 50% as the particle size decreases from 5 to 2 nm, while the yield stress increases by about 10 MPa. The largest particle size was reached after 500 hr of aging.

Raising the aging temperature to 500°C results in rod-shaped precipitates. The yield stress increases, resistivity ratio, aspect ratio, and equivalent radius particle size relations are given in Fig. 8. Aging for a day at this temperature gives an equivalent particle radius of about 5 nm and an increase in yield stress over the quenched alloy of 30 MPa. The resistivity is higher for the higher aging temperature in keeping with the greater residual solute content in the Cu matrix.

Since the rate of aging decreases as the aging temperature is decreased while the residual solubility decreases, a two-step aging process was investigated in the Cu–Ni–P alloy [38]. The alloy was solution-treated for 60 min at 950°C and then cold-rolled to 60% reduction. The alloy was aged to peak hardness at 450°C (100 min) and then further aged at 375°C. Aging for 400 min increased the hardness from 150 to 160 Vickers numbers, while the conductivity increased from 67% to 71% IACS.

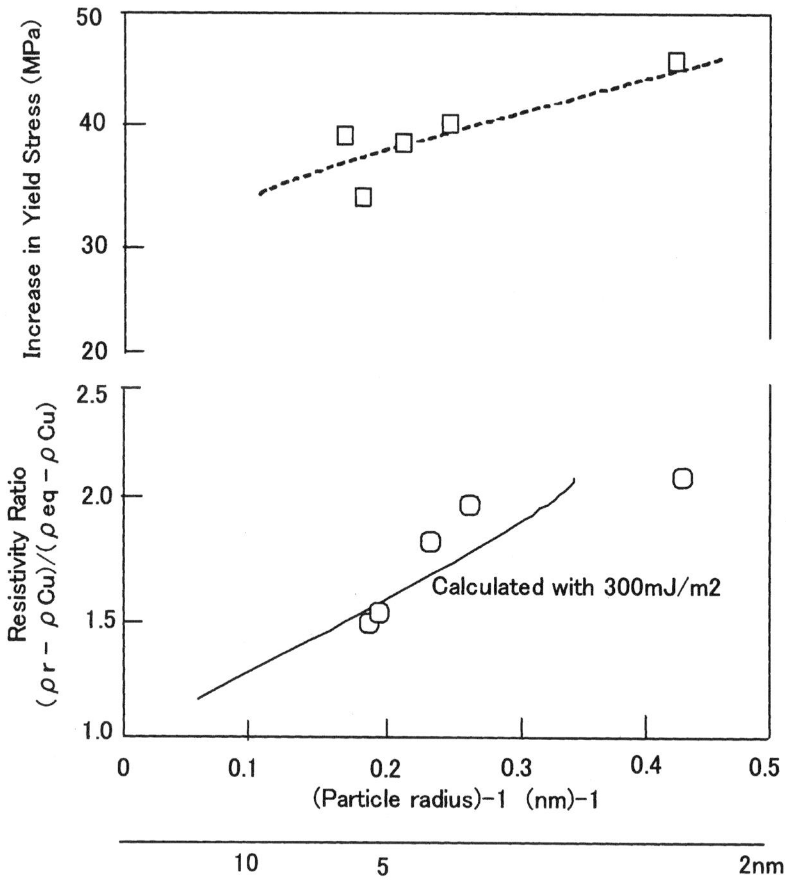

Figure 7 Yield stress increase and resistivity ratio decrease vs. equivalent radius of coherent precipitates for aging at 450°C in Cu–0.74 wt% Ni–0.14 wt% P alloy. (From Ref. 5.)

K. Recent Development of Copper Alloys for Electronic Applications

Figure 9 represents recently developed copper alloys for electronic applications such as lead frames and connectors [39]. The conductivity regimes corresponding to the applications mobile phone, PC, power, and automotive are shown on the x-axis. In the low-conductivity regime, very-high-strength alloys and solid-solution-hardening alloys still dominate. For high strength and high conductivity, the Cu–Ni–Si and Cu–Cr–Zr systems have continued to be the major ones [40,41]. Also, phosphide-containing alloys still dominate in the industry for the medium-conductivity regime. For even further high-strength, high-conductivity applications, new types of precipitates and/or dispersoids need to be explored [42] in combination with new processing routes, including power metallurgy and highly enhanced deformation processes.

II. COPPER FOIL DESIGN FOR HIGH FLEX FATIGUE

A. Introduction

Flexible printed circuits (FPCs) have been widely used for repetitive flex motion devices, such as printers, hard disk drives, and CD-ROM drives, due to their light weight and good

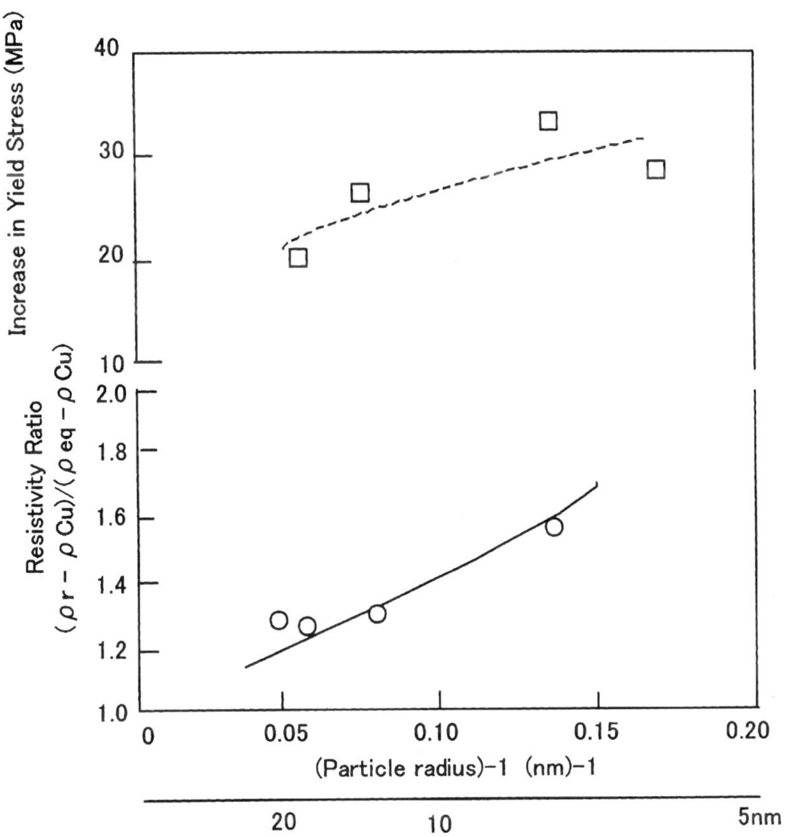

Figure 8 Yield stress increase and resistivity ratio decrease vs. equivalent radius of rod-shaped precipitates for aging at 500°C in Cu–0.74 wt% Ni–0.14 wt% P alloy. (From Ref. 5.)

fatigue performance [43,44]. Two types of copper foils are used for printed circuits and rolled and electrodeposited foils. Since the FPC applications require relatively high fatigue life, rolled and annealed copper foils are generally used. The recent miniaturization of moving components and even higher performance requirements require further improvements in fatigue performance.

During copper foil production, sequential rolling and recrystallization anneals are utilized to attain low final thickness. The final production step that is used in FPCs is a recrystallization anneal. The effects of two factors on the fatigue resistance are important, rolling strain applied during the final reduction and grain size produced by annealing prior to final rolling. Those processing parameters affect microstructure, which in turn impacts the fatigue properties [45].

B. Flex Test and Fatigue Mode

Two testers are used, the MIT folding and the FPC flex tester, as shown in Fig. 10. With the MIT folding endurance tester, the holder rotates repeatedly right and left at an angle of 135°, as shown in Fig. 10(a) [46]. The foil specimen is 12.7 mm wide and 100 mm long. The specimen is held in tension with 6 MPa stress. Both tension and compression are

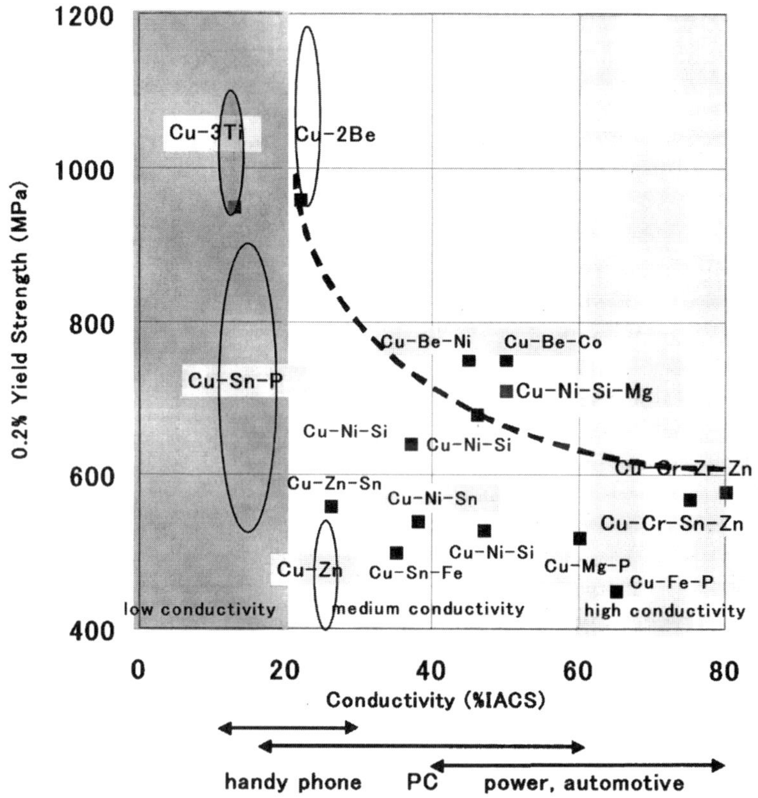

Figure 9 Conductivity vs. strength for some recently developed copper alloys as well as older alloys.

applied alternately on the specimen surface. As the radius of the holder tip is 0.38 mm, the strain on the specimen surface is 4×10^{-2}. The number of cycles to failure of copper foil, N_f, is in the range from 10 to 10^2. The specimen is attached to the FPC tester at the four points indicated by arrows in Fig. 10(b). The foil specimen is 12.7 mm wide and 200 mm long. A 0.1-mm-thick polyester film is attached on both surfaces of the copper foil during testing. The central plate vibrates up and down with a 25 mm stroke at the rate of 25 cycles per second. The strain applied to the specimen varies between zero and compression on the inner surface. Since the bend radius of the specimen is 2.5 mm, the strain on the specimen surface is 6.6×10^{-3}. N_f in this test is in the range of from 10^4 to 10^5. For both tests, the bend line is perpendicular to the rolling direction. Since the cyclic mode is different for the two tests, the cycles-to-failure ranges are much different: the former test is called a "low-cycle test" and the latter a "high-cycle test".

In strain-controlled fatigue tests, there is the well-known relationship between plastic strain amplitude and fatigue life, generally known as the Coffin–Manson relation [47,48], including the elastic strain,

$$\Delta \varepsilon/2 = \Delta \varepsilon_e/2 + \Delta \varepsilon_p/2 = \sigma'_f(2N_f)^b/E + \varepsilon_f(2N_f)^c \quad (11)$$

where $\Delta\varepsilon$, $\Delta\varepsilon_e$, and $\Delta\varepsilon_p$ are total strain, elastic strain, and plastic strain, respectively. σ'_f is fatigue strength coefficient, ε_f is fatigue ductility coefficient, E is Young's modulus, and b

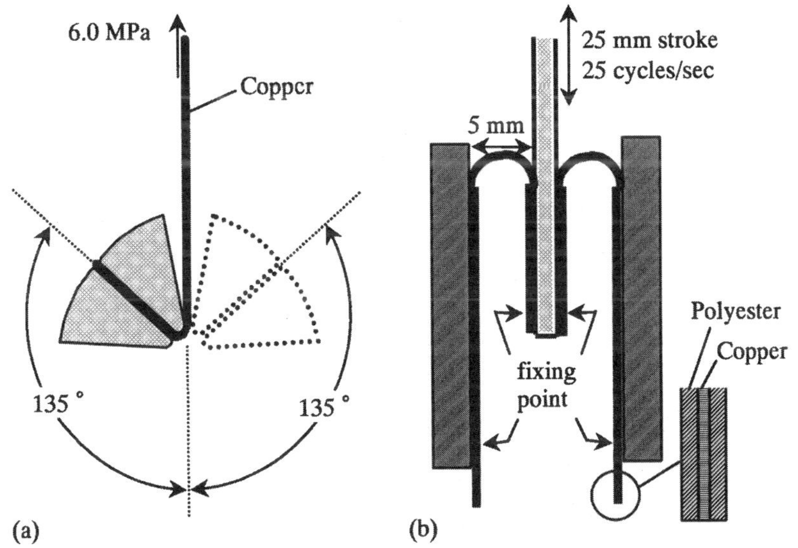

Figure 10 MIT folding and FPC flex tests for copper foil: (a) MIT folding endurance tester (low-cycle test); (b) FPC flex tester (high-cycle test). (From Ref. 45.)

and c are constants. The plastic strain is predominant in the low-cycle region ($\Delta\varepsilon = \Delta\varepsilon_p$), whereas the elastic strain is predominant in the high-cycle region ($\Delta\varepsilon = \Delta\varepsilon_e$). The transition cycle between the two regimes is said to be about 10^3 [52]. With respect to N_f, it is considered that the "low-cycle test" in this experiment is done under the plastic strain condition, and the "high-cycle test" is done under the elastic strain condition. Therefore, the fatigue mode is assumed to be quite different for the two tests.

C. Fatigue and Tensile Tests

Figure 11 shows the relationship between rolling strain, η, and the number of cycles to failure, N_f, for the low-cycle and the high-cycle tests for three levels of grain size, D, before

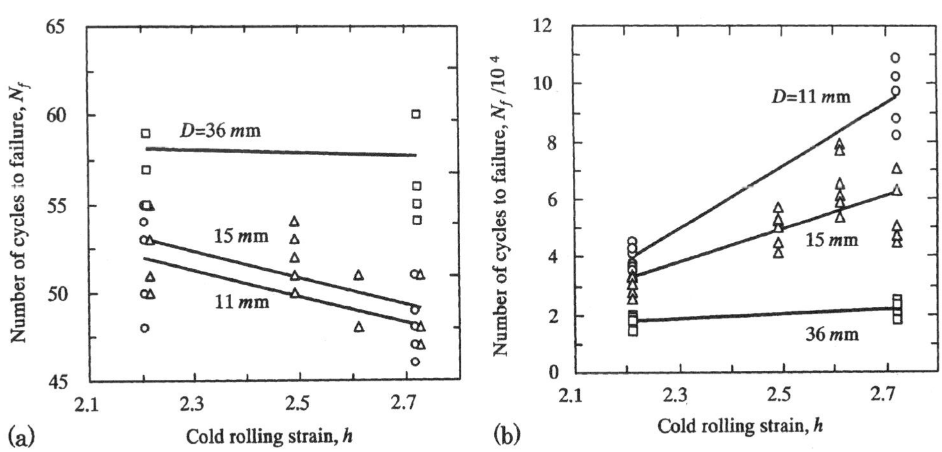

Figure 11 Effects of rolling strain and grain size before rolling on fatigue of copper foil: (a) low-cycle test; (b) high-cycle test. (From Ref. 45.)

final rolling. In the low-cycle test, N_f increases as η decreases and as D becomes larger. On the other hand, in the high-cycle test, N_f increases as η increases and as D becomes smaller. It should be noted that the effects of material processing, such as rolling strain and grain size, on fatigue life are the opposite in the two tests. This also indicates that the fatigue resistance of rolled copper foil for both modes can be enhanced by material processing. Taking the data scatter and the scale of the y-axis into consideration, high-cycle fatigue seems to be more sensitive to the processing variables than does low-cycle fatigue.

D. Fatigue Resistance and Microstructure

In order to clarify the difference in microstructure, two types of specimen were characterized. For specimen (a), which has long life in the low-cycle test and short life in the high-cycle test, grain size before rolling, D, is 36 μm and rolling strain, η, is 2.2. For specimen (b), which has short life in the low-cycle test and long life in the high-cycle test, D is 15 μm and η is 2.6. Figure 12 shows backscattered electron images of rolled surfaces after electropolishing. The microstructure of specimen (a) appears to be a normal recrystallization structure of copper. A number of annealing twins are also observed. On the other hand, the microstructure of specimen (b) is mainly composed of very large grains of over 50 μm diameter. These large grains are elongated along the rolling direction, and the size of annealing twins is very small. This large recrystallized microstructure is considered to be due to abnormal grain growth, that is, secondary recrystallization.

Figure 13 shows N_f as a function of I/I_0, where I is the x-ray diffraction relative intensity for the specimen and I_0 is the intensity for copper powder. It is noted that as I/I_0 increases, in other words, as the cube texture becomes enhanced, the fatigue resistance is enhanced. It is thus noted that the cube texture [49] develops stronger as η is higher and D is smaller.

E. Texture and Fatigue Crack

Figure 14 shows cross sections perpendicular to the bending line after chemical etching at 0.85 times N_f. In specimen (a), cracks propagate in a zigzag manner along the grain boundaries, and typical intergranular cracking occurs. The grain boundary apparently

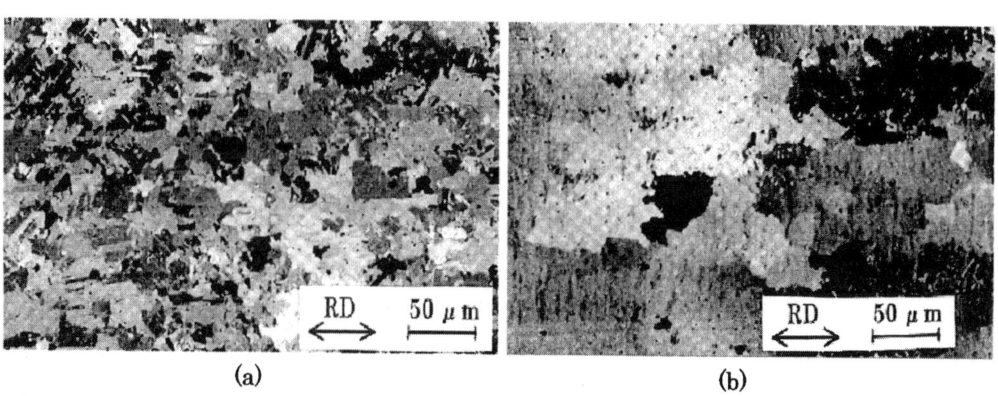

Figure 12 Backscattered electron images of rolled surface after electropolishing. (a) Grain size before rolling: 36 μm; rolling strain: 2.2. (b) Grain size before rolling: 15 μm; rolling strain: 2.6. (From Ref. 45.)

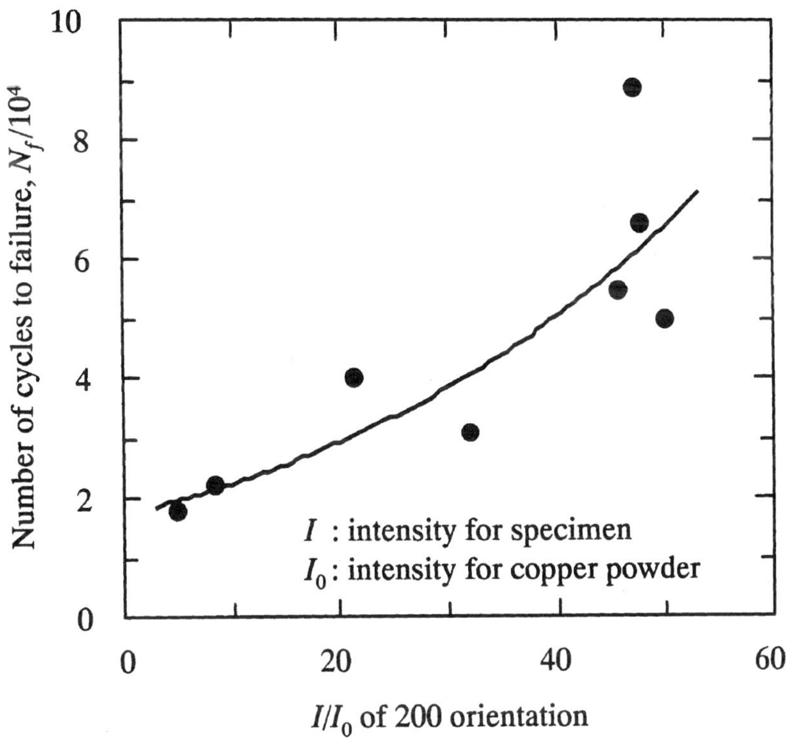

Figure 13 Relationship between x-ray relative intensity of 200 orientation and fatigue life in the high-cycle test. (From Ref. 45.)

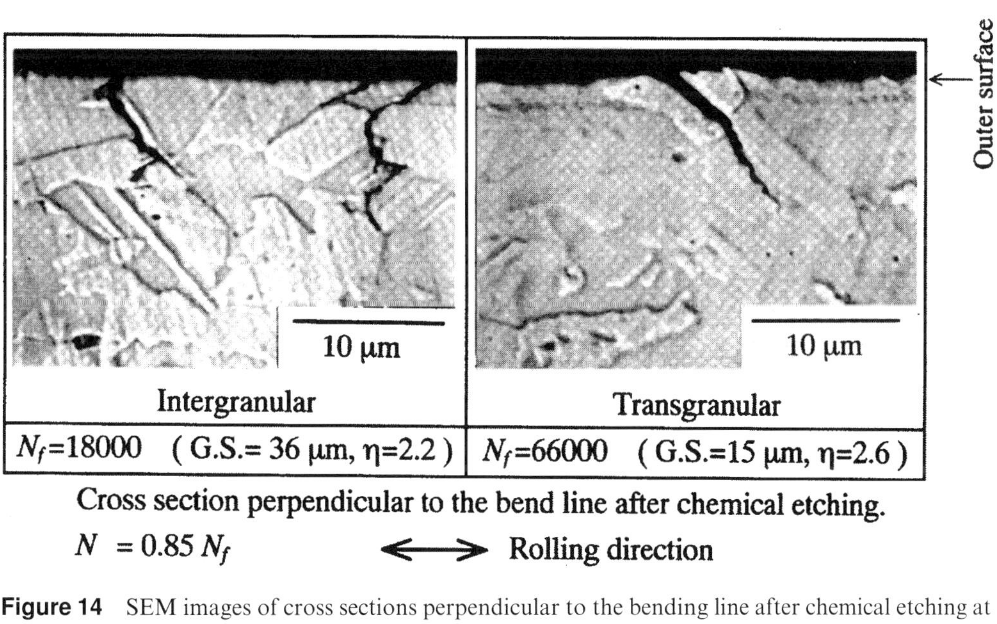

Figure 14 SEM images of cross sections perpendicular to the bending line after chemical etching at the cycle number of $0.85N_f$ in the high-cycle test. (From Ref. 45.)

acts as a crack propagation path, leading to shorter fatigue life. Cracks are randomly distributed over the bend surface. On the other hand, in specimen (b) cracks propagate across the grain boundary, leading to transgranular cracking. Cracks are localized in a particular region. The fatigue life in the high-cycle test is affected by the crack propagation behavior, either intergranular or transgranular [45].

The possible mechanism for two types of cracking is discussed from the viewpoint of cube texture as already shown in Fig. 13. For strong cube texture, misorientation among neighboring grains would be smaller. A weak cube texture is assumed to correlate with crack propagation along grain boundaries. Since Young's modulus varies with crystal orientation, the difference in Young's modulus among neighboring grains becomes large for certain stress directions when the misorientation among neighboring grains is large. In addition, the difference in slip orientation among neighboring grains also becomes large. Assuming that the grains have large differences in Young's modulus and slip orientation, both elastic and plastic deformation behaviors would differ greatly among neighboring grains. This situation causes large stress concentration on grain boundaries of neighboring grains and thus leads to intergranular crack propagation. For the strong cube texture specimen, however, the stress concentration on grain boundary is much smaller in comparison with the weak cube texture specimen. The cube texture microstructure can thus be assumed to characterize the high-cycle fatigue resistance.

F. Flow Diagram for High Flex Fatigue Resistance of Rolled Copper Foil

Through small grain size before rolling and high rolling strain, a highly enhanced cube texture develops and the grain size becomes much larger. High-cycle fatigue resistance is vastly improved, caused by enhanced cube texture. The relationship between material processing, microstructure, and fatigue property of copper foil is well established. Figure 15 represents the flow diagram for high flex fatigue resistance of rolled copper foil. The fine-grained structure and the subsequent high cold-rolling reduction provide the highly developed cube texture after final recrystallization. This contributes to enhanced flex fatigue

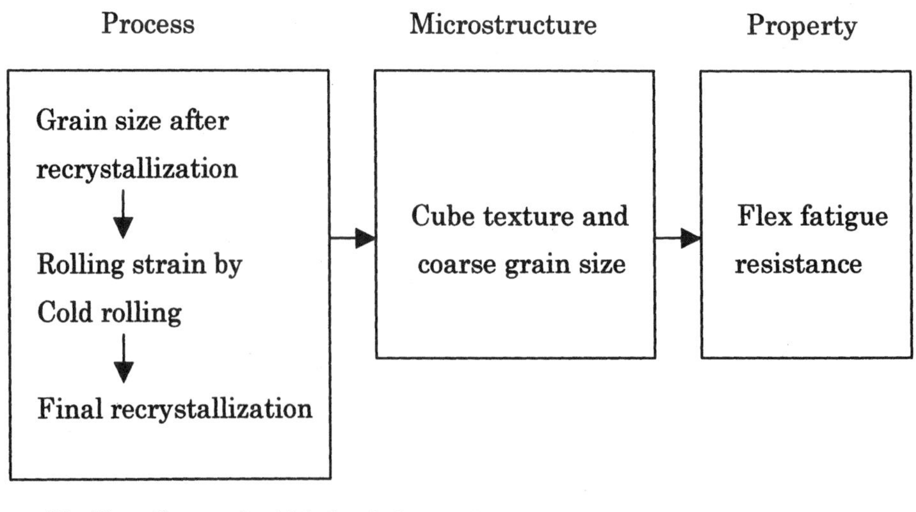

Figure 15 Flow diagram for high flex fatigue resistance of rolled copper foil.

resistance of the foil, and this can be used to further upgrade the fatigue resistance of copper foils.

III. STRESS RELAXATION RESISTANCE OF COPPER ALLOYS

A. Introduction

It is well known that the yield strength of copper alloys is critical for designing electronic connectors, since it determines the initial contact force. The stress relaxation resistance of copper alloys is another key property that determines the contact force of electronic connectors under service conditions such as long period of time and elevated temperatures. Loss of contact force due to stress relaxation at elevated temperatures or/and long periods of service time results in increased electrical contact resistance. Thus, the combination of high yield strength and high stress relaxation resistance needs to be established to ensure high reliability of connectors. Through a uniaxial tensile stress relaxation test for high-strength Cu–Ti alloy [50–53], the stress relaxation of copper alloys is characterized in terms of alloy design principles.

B. Stress Relaxation Measurement

Since electronic connectors operate in bending, a test specimen with straight tapered edges based on ASTM standard E328–86 [54] is made so that the applied stress can be uniform over the bent specimen, and the specimen is bent with an applied stress of 80% of the 0.2% yield stress of each copper alloy and held at 150–200°C for up to 1000 hr. The permanent deflection is measured, and the stress relaxation ratio is calculated using the following equations:

$$Y_0 = l^2 \sigma_0 / Et \tag{12}$$

$$[\text{Stress relaxation ratio}] = Y/Y_0 \times 100\% \tag{13}$$

where σ_0 is the applied stress, E is Young's modulus, t is the specimen thickness, l is the specimen length, Y_0 is the initial deflection, and Y is the permanent deflection.

Figure 16 shows the stress relaxation ratios of Cu–3 wt% Ti alloy and Cu–2 wt% Be alloy tested at 200 °C. The stress relaxation of the Cu–Ti alloy is much smaller than that of the Cu–Be alloy, and the difference between the two alloys is significantly large after a long period of time [52].

C. Grain Size and Rolling Reduction Dependences of Stress Relaxation

The Cu–Ti alloy sample was solution-treated at 850°C to obtain grain sizes of 10, 30, and 50 μm and cold-rolled to 0.3 mm thickness by final cold-rolling reductions of 22% and 47.5%. Then the aging treatment was carried out at 430°C for 7 hr.

In order to conduct the stress relaxation test, an Instron-type tensile machine with a load–strain control unit was used. A strain gauge was attached on both surfaces of the specimen so that the tensile strain detected by the gauge could be kept constant during the test. The grain size, and also rolling reduction dependences of stress relaxation, was measured for the Cu–Ti alloy by a uniaxial tensile stress relaxation test, as shown in Figs. 17 and 18. It is noted that the stress relaxation for 10-μm fine grain is much smaller than that

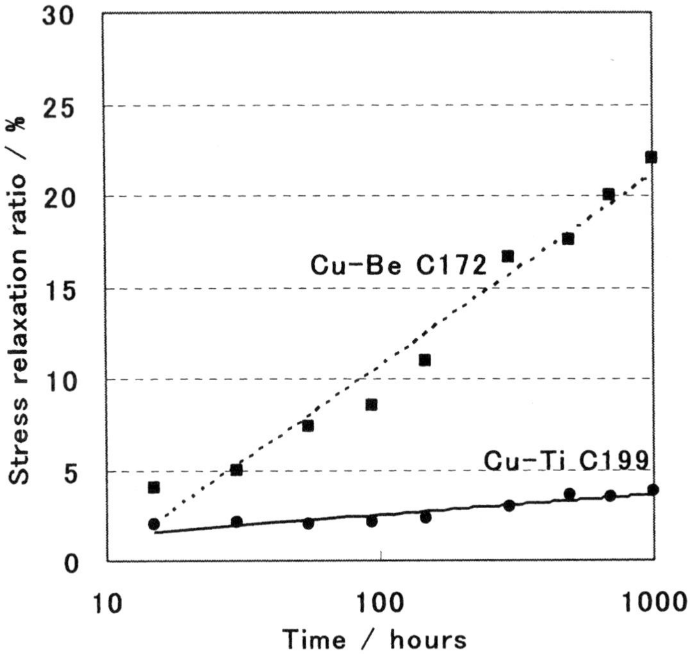

Figure 16 Stress relaxation ratios of Cu–3 wt% Ti C199 and Cu–2 wt% Be C172 at 200°C. (From Ref. 52.)

of 30- and 50-μm specimens with applied stresses of 637 and 686 MPa. The latter two stress relaxation ratios are comparable. For applied stresses of both 637 and 686 MPa, the relaxation for 22% of pre-cold-rolling reduction is smaller than that for 47.5%. This is also noted to be opposite to the yield stress, which is always higher for higher cold-rolling reduction [52].

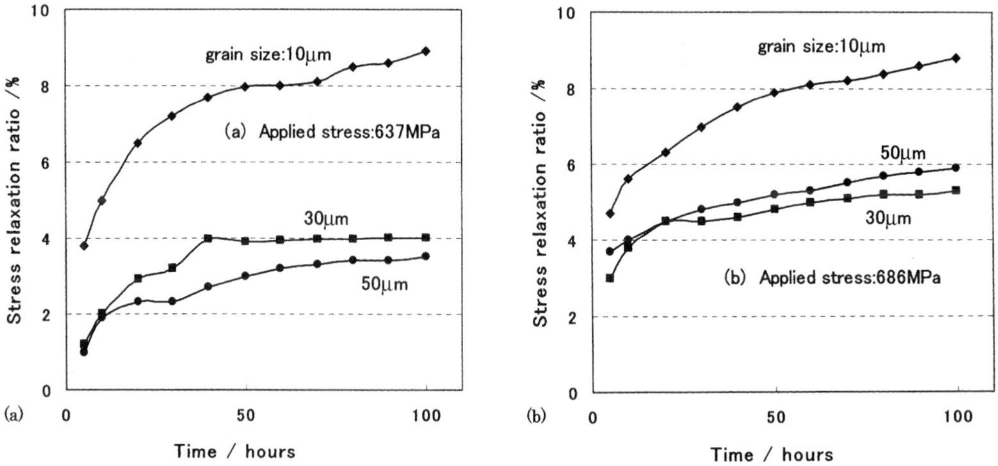

Figure 17 Grain size effect on stress relaxation for Cu–3 wt% Ti alloy rolled by 47.5% and tested with applied stresses of (a) 637 MPa and (b) 686 MPa at 200°C. (From Ref. 52.)

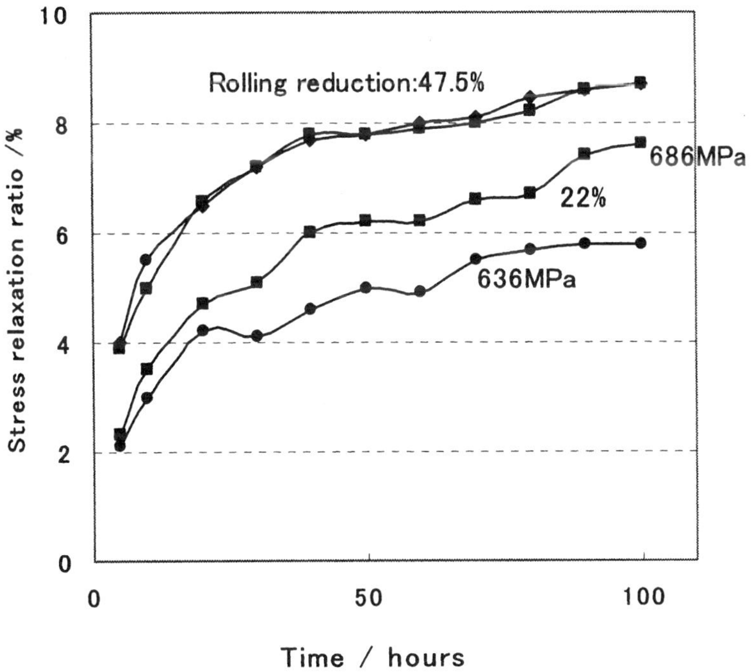

Figure 18 Rolling reduction effect on stress relaxation of 10 μm grain size Cu–3 wt% Ti alloy tested at 200°C. (From Ref. 52.)

D. Strain and Dislocation Displacement During Stress Relaxation

The plastic strain rate, ε_{sr}^{rate}, during stress relaxation is expressed by the following equation:

$$\varepsilon_{sr}^{rate} = -\sigma^{rate}/E \tag{14}$$

where σ^{rate} is the stress relaxation rate, and E is Young's modulus. Integration of the above equation gives the following equation:

$$\Delta\sigma_{sr} = E\varepsilon_{sr} \tag{15}$$

where $\Delta\sigma_{sr}$ and ε_{sr} are the relaxed stress and the plastic strain after time t, respectively. Since $\Delta\sigma_{sr}$ can be measured by experiment, ε_{sr} and ε_{sr}^{rate} after 100 hr were calculated as $\varepsilon_{sr} \fallingdotseq 10^{-4}$ and $\varepsilon_{sr}^{rate} \fallingdotseq 10^{-10}$ sec^{-1}. Those values are extremely low in comparison with the plastic strain, 0.2% at yield strength and the strain rate, 10^{-4} sec ^{1} of conventional tensile test.

If the plastic strain during the stress relaxation test is due to dislocation displacement, it is expressed by the following well-known equation. [55]:

$$\varepsilon_{sr} = \Phi\rho b\Delta s \tag{16}$$

where Φ is an orientation factor, ρ is the mobile dislocation density, b is the Burgers vector, and Δs is the displacement of a moving dislocation. The initial dislocation densities are evaluated by the Bailey–Hirsch equation as already expressed by Eq. (10). The shear modulus G was taken as 4.21×10^4 MPa for pure copper. The dislocation density at 0.2% yield stress (the engineering yield stress) for 10 μm Cu–3 wt% Ti alloy with a pre-cold-rolling of

47.5% is estimated to be 3×10^{15} m^{-2} through Eq. (10). This value leads to nearly 5×10^{-7} m for the displacement of moving dislocations in the tensile test according to Eq. (16).

On the other hand, the dislocation density after 100 hr of testing at 200°C and 637 MPa for the same sample as above is estimated to be 1×10^{14} m^{-2}. The displacement of moving dislocations for the stress relaxation test is then estimated to be nearly 3×10^{-8} m, which is smaller than the former value by one order of magnitude. Hence, the mobile dislocation displacement is less during a stress relaxation test in comparison with a tensile test. Dislocation multiplication plays a major role in plastic deformation for 0.2% yield stress, that is, there is work hardening prior to the 0.2% offset from the long-range stress field of the increase in dislocation density. In the stress relaxation test, the initial stress is less than the engineering yield stress, and this stress becomes lower during the test because of reduction in dislocation density. As dislocation density increases, the yield stress increases, while stress relaxation is enhanced [52].

E. Alloy Design for High Stress Relaxation Resistance Copper Alloy

Based on the effects of grain size after final annealing and final cold-rolling strain on the stress relaxation, the combination of larger grain and small final rolling strain provides higher stress relaxation resistance, but lower yield strength. Electronic connectors need high yield strength as well as high stress relaxation resistance. There is, thus, a trade-off in the alloy process design between stress relaxation resistance and yield strength. The optimum condition of grain size and cold-rolling reduction needs to be established for actual connector applications.

REFERENCES

1. Miyake, J. Overview of connector copper alloys and technical issues. J. Jpn. Copper Brass Res. Assoc. 1997, *36*, 1–7.
2. Miyake, J.; Ghosh, G.; Fine, M.E. Design of high-strength, high-conductivity alloys. MRS Bull. 1996, *June*, 13–18.
3. Ghosh, G.; Miyake, J.; Fine, M.E. The systems-based design of high-strength, high-conductivity alloys. J. Metals 1997, *49*, 56–60.
4. GLIDCOP® dispersion strengthened copper, OM Group, Cleveland, OH.
5. Miyake, J.; Fine, M.E. Electrical conductivity versus strength in a precipitation hardened alloy. Acta Metall. Mater. 1992, *40*, 733–741.
6. Smith, C.S. A Search for Structure; MIT Press: Cambridge, MA, 1981.
7. Olson, G.B. *Innovations in Ultrahigh-Strength Steel Technology*, Olson, G.B.; Azrin, M.; Wright, E.S., Eds.; 1990; 3–66.
8. Olson, G.B. Materials design: an undergraduate course. In *Morris E. Fine Symposium,* Liaw, P.K., et al., Eds.; TMS: Warrendale, PA, 1991; 41–48.
9. Sundman, B.; Jansson, B.; Andersson, J.O. The Thermo-Calc databank system. CALPHAD 1985 , *9*, 153–190.
10. Rossiter, P.L. Electrical resistivity of simple metals and alloys. In *The Electrical Resistivity of Metals and Alloys*; Cambridge University Press: Cambridege, 1987; 138–271.
11. Nordheim, V.L. Zur elecktronen Theorie der Metalle II. Ann. Phys. 1931, *5*, 642–678.
12. Stanley, J.K. *Electrical and Magnetic Properties of Metals*; American Society of Metals: Metals Park, OH, 1963; 53.
13. Taubenblatt, P.W.; Smith, W.E.; Graviano, A.R. Properties and applications of high strength, high conductivity coppers and copper alloys. In *High Conductivity Copper and Aluminum Alloys*, Ling, E; Taubenblatt, P.W., Eds.; TMS: Los Angeles, 1984; 19–29.

14. Mott, N.F.; Jones, H. *The Theory of the Properties of Metals and Alloys*; Dover Publications, New York, 1958; 297 (first published in 1936 by Clarendon Press, Oxford, England).
15. Dingle, R.B. The electrical conductivity of thin wires. Proc. R. Soc. London A 1950, *201*, 545–560.
16. Sondheimer, E.H. The mean-free path of electrons in metals. Adv. Phys. 1952, *1*, 1–42.
17. Brown, R.A. Scattering theory for crystal dislocations. J. Phys. F: Met. Phys. 1977, *7*, 1269–1281.
18. Brown, R.A. Electrical resistivity of dislocations in metals. J. Phys. F: Met. Phys. 1977, *7*, 1283–1295.
19. Broom, T. The effect of temperature of deformation on the electrical resistivity of cold-worked metals. Proc. Phys. Soc. London B 1952, *65*, 871–881.
20. Basinski, Z.S.; Saimoto, S. Resistivity of deformed crystals. Can. J. Phys. 1967, *45*, 1161–1175.
21. Verhoeven, J.D., et al. The resistivity and microstructure of heavily drawn Cu–Nb alloys. J. Appl. Phys. 1989, *65*, 1293–1301.
22. Crampton, D.K.; Berghoff, H.L.; Stacy, J.T. The copper-rich alloys of the copper–nickel–phosphorus system. Trans. AIME 1940, *137*, 354–372.
23. Fedorov, V.N.; Zakhanov, M.V.; Kucherov, V.I.; Osintsev, O.E. Investigation of the copper-rich range of the Cu–Ni–Ti phase diagram. Russ. Metall. 1971, *6*, 128–131.
24. Corson, M.G. Copper alloy systems with variable alpha range and their use in the hardening of copper Proc. Instit. Met. Div. AIME 1927, *77*, 435–450.
25. Subramanian, P.R.; Chakrabarti, D.J.; Laughlin, D.E. *Phase Diagrams of Binary Copper Alloys*; American Society of Metals: Materials Park, OH 1988; 85, 144, 167, 447, 497.
26. Zener, C. Role of statistical mechanisms in physical metallurgy. In *Thermodynamics in Physical Metallurgy*; American Society of Metals: Metals Park, OH, 1950; 16–22.
27. Fink, W.L.; Freche, H.R. Correlation of equilibrium relations in binary aluminum alloys of high purity. Trans. AIME, Institute of Metals Div. 1934, *111*, 304–317.
28. *Copper Alloy Guide*; Olin Brass: Illinois 1992.
29. Olin HSM Copper (Alloy 194) Technical Data; Olin Brass: Illinois.
30. Wagner, C. Theorie der Alterung von Niederschlagen durch Umlossen. Z. Electrochem. 1961, *65*, 581–591.
31. Ashby, M.F. The theory of the critical shear stress and work hardening of dispersion-hardened crystals, oxide dispersion strengthening. In Metallurgical Society Conference; Gordon and Breach: New York, 1968; 143–212.
32. Martin, J.W. Yield and work hardening in absence of recovery. In *Micromechanism in Particle-Hardened Alloys*; Cambridge University Press: Cambridge, 1980; 51–71.
33. Lifshitz, I.F.; Slyozov, V.V. The kinetics of precipitation from supersaturated solid solution. *J. Phys. Chem. Solids* 1961, *19*, 35–50.
34. Bailey, J.E.; Hirsch, P.B. The dislocation distribution, flow stress and stored energy in cold-worked polycrystalline silver. Philos. Mag. 1960, *5*, 485–497.
35. Byrne, J.G.; Fine, M.E.; Kelly, A. Precipitate hardening in an aluminium copper alloy. Philos. Mag. 1961 , *6*, 1119–1145.
36. Bevk, J.; Harbison, J.P.; Bell, J.L. Anomalous increase in strength of in-situ formed Cu–Nb multifilamentary composites. J. Appl. Phys. 1979, *49*, 6031–6038.
37. Sakai, Y.; Inoue, K.; Maeda, H. New high-strength, high-conductivity Cu–Ag alloy sheets. Acta Metall. Mater. 1995, *43*, 1517–1522.
38. Miyake, J.; Fine, M.E. Two-step ageing of copper alloy to optimize combination of strength and electrical conductivity. Scr. Metall. 1991, *25*, 1573–1576.
39. Miyake, J. Copper alloy development and electronic connectors. In Proceedings of the 2001 Conference on Electronic Contact and Connector, Taipei, Taiwan; Electronic Connector Association, 2001.
40. *Olin Alloy C7025, Engineering Guide*; Olin Brass: Illinois 1992.
41. Tomioka, Y.; Miyake, J. A copper alloy development for leadframe. In Proceedings of the 1995 Japan International Electronic Manufacturing Technology Symposium; Omiya, Japan 1995; 433–436.

42. Fine, M.E. Phase transformations in condensed systems revisited: industrial applications. Met. Mater. Trans. A 1996, *27*, 2397–2418.
43. Merchant, H.D.; Minor, Mg.G.; Clouser, S.J.; Leonard, D.T. 18 Micron electrodeposited copper foil for flex fatigue applications. Circuit World 1998, *25*, 38–46.
44. Merchant, H.D.; Minor, M.D.; Rozboril; M.G. Characterization of damage during flex fatigue of 18 micron copper foil. In Proceedings of the IPC National Conference on Flex Circuits, San Diego, March, 1998.
45. Hatano, T.; Kuroswa, Y.; Miyake, J. Effect of material processing on fatigue of FPC rolled copper foil J. Electron. Mater. 2000, *29*, 611–616.
46. Test methods for flexible printed wiring boards. Japanese Industrial Standards (JIS), C5016, 1999.
47. Courtney, T.H. Evaluation of fatigue resistance. In *Mechanical Behavior of Materials*; McGraw-Hill: New York, 1990; 570–580.
48. Dieter, G.E. Fatigue of metals. In *Mechanical Metallurgy*; McGraw-Hill: New York, 1988; 375–431.
49. Ridha, A.A.; Hutchinson, W.B. Recrystallization mechanism and the origin of cube texture in copper. Acta Metall. 1982, *30*, 1929–1939.
50. Harkness, J.C.; Lorenz, C.S. Stress relaxation of beryllium copper in bending. In Proceedings of the 12th Annual Connector and Interconnection Technology Symposium, IICIT, Waretown, New Jersey, 1979; 37–52.
51. Wakamatsu, M.; Miyake, J. High performance titanium copper C199 for electrical contacts. In Proceedings of the 17th International Conference on Electrical Contacts, IICIT, Waretown, New Jersey, 1994; 939–946.
52. Miyake, J.; Kimura, T.; Endo, T. Stress relaxation of Cu–Ti Alloy C199 In *Proceedings, Creep and Stress Relaxation in Miniature Structures and Components*; Merchant, H.D., Ed.; TMS: Warrendale, PA, 1997; 57–74.
53. Laughlin, D.E.; Cahn, J.W. Spinodal decomposition in age hardening copper titanium alloys Acta Metall. 1975, *23*, 329–339.
54. *Annual Book of ASTM Standards, E328-86, Standard Test Methods for Stress Relaxation Tests for Materials and Structures*, ASTM International, West Conschocken, PA, 1995; 353–364.
55. Courtney, T.H. Dislocation density and mechanical strain. In *Mechanical Behavior of Materials*; McGraw-Hill: New York, 1990; 128–131.

14

Designing with Powder Metallurgy Alloys

Joseph W. Newkirk and Ronald A. Kohser
University of Missouri-Rolla, Rolla, Missouri, U.S.A.

I. INTRODUCTION

The objective of any manufacturing operation is the production of products, or components of products, that will adequately perform their intended task. Implicit in this objective is the generation of parts with the required geometrical shape or precision, and design begins with defining the specific needs. Size, shape, complexity, precision, and surface finish all enter into the picture. For ease of manufacture, it is also desirable to include or note axes of rotation, planes of symmetry, and directions of uniform cross-section.

In addition to the geometric requirements, the parts must also possess a desired set of mechanical and physical properties. The mechanical performance of parts opens up a broad area of considerations: how will the part be loaded (static, impact, or cyclic-tension, compression, torsion or bending) and what properties are necessary to withstand the loadings. Strength, stiffness, fatigue resistance, impact resistance, and wear resistance are all considerations to be taken into account. Physical properties include the entire spectrum of electrical, magnetic, thermal, and optical properties, as well as considerations involving weight.

The operating environment must also be defined. Highest, lowest, and normal operating temperatures should be estimated, along with rates of change and the additional properties that must be present at each extreme. Corrosion possibilities should be considered. Other areas of concern include desired product lifetime, product disposal, ease of recycling, and potential liability.

Manufacturing needs form the final set of requirements. Foremost among these might be the number of identical parts to be produced and the desired rate of production or anticipated rate of consumption. Other manufacturing concerns might include the desired level of quality and inspection and quality control measures. Assembly and/or disassembly considerations may impose additional requirements.

Having specified the requirements, the next set of activities generally relates to material selection and process selection. These two are often interrelated and combine to form the proposed manufacturing system. Selection of a process may limit candidate materials, and selection of a material may limit candidate processes. The objective is to arrive at an integrated system of material and process that is the optimum solution for the defined requirements and constraints.

There are a number of ways to produce a shape from an engineering material. Material removal methods begin with an oversize block of material and through processes such as machining remove chips or turnings to produce the shape. This is an ideal approach for one-of-a-kind parts and offers extreme precision, but a portion of the material, often substantial, is converted into some form of scrap. Casting processes utilize the fluidity of liquids and the fact that liquids assume the shape of their container. Shaped molds are created and the material is solidified while in the mold. Process limitations are generally related to the solidification process and the resulting material structures. Deformation processes can be applied to materials with plasticity. Mechanical forces cause the material to flow, altering the shape to that of a desired product. High forces and dedicated tooling is generally required. Consolidation processes form the fourth and final category. These processes consist of "putting pieces together" to build up the desired shape. All the assembly processes, such as welding, brazing, soldering, mechanical fasteners, shrink fits, and others, fall into this group. Since it involves the building up of a shape from smaller units, powder metallurgy can be classified into this group, but in powder metallurgy the number of pieces being joined is of the order of tens to hundreds of thousands.

II. THE POWDER METALLURGY PROCESS

The term "powder metallurgy" is really rather generic and refers to a family of processes that produce parts through the fusing of particulate material. Some specific subgroups include: conventional powder metallurgy (P/M), metal injection molding (MIM), hot isostatic pressing (HIP), cold isostatic pressing (CIP), and powder forging (P/F). Each has its characteristic advantages and limitations, and we will attempt to present their capabilities, beginning with conventional powder metallurgy (Table 1).

In the basic process, high purity, custom-mixed, or prealloyed powders are fed into a die, compacted to a desired shape, and ejected. The pressed powder is then sintered (heated) in a controlled atmosphere furnace to produce metallurgical bonds that link the powder particles. Optional post-sintering operations, such as infiltration, finish machining, or surface treatment, may then be applied to complete the part.

Table 1 Comparison of Four Powder Processing Methods

Characteristic	Conventional press and sinter	Metal injection molding (MIM)	Hot isostatic pressing (HIP)	P/M forging
Size of workpiece	Intermediate <5 lb	Smallest $<\frac{1}{4}$ lb	Largest 1–1,000 lb	Intermediate <5 lb
Shape complexity	Good	Excellent	Very good	Good
Production rate	Excellent	Good	Poor	Excellent
Production quantity	$>5,000$	$>5,000$	1–1,000	$>10,000$
Dimensional precision	Excellent ± 0.001 in./in.	Good ± 0.003 in./in.	Poor ± 0.020 in./in.	Very good ± 0.0015 in./in.
Density	Fair	Very good	Excellent	Excellent
Mechanical properties	80–90% of wrought	90–95% of wrought	Greater than wrought	Equal to wrought
Cost	Low \$0.50–5.00/lb	Intermediate \$1.00–10.00/lb	High $>$\$100.00/lb	Somewhat low \$1.00–5.00/lb

III. A FALSE REPUTATION FOR INFERIORITY

Powder metallurgy is far from being a new process, but for many years, it carried a reputation of inferiority, especially with regard to mechanical properties. Most of the parts being produced had actually been designed for other processes, such as die casting, screw machining, or stamping, and conversion to powder metallurgy tried to retain the same material and geometry. Because of the nature of the process, conventional powder metallurgy products generally retain as much as 10–25% residual porosity. A part containing voids is likely to be inferior to a solid part of the same design and material. As a result, such head-to-head comparisons are destined to cast powder metallurgy in a negative light.

A far more valid comparison, however, is to begin with the set of design requirements, and then give each alternative method of fabrication the freedom to develop an optimized solution based on its particular process. Powder metallurgy could then utilize unique materials that have no casting or forming equivalents. The geometry could be optimized for ease of compaction. Parts could be designed to provide the desired properties with the presence of residual porosity. Since each product would then provide the necessary properties, comparison would be on the basis of net cost, or some other more appropriate criterion. In such comparisons, powder metallurgy is a strong contender, and often the clear winner, for certain types of parts.

IV. ATTRACTIVE FEATURES—PROCESS ADVANTAGES

The products that are commonly produced by powder metallurgy can generally be classified into six groups:

1. *Products of complex shapes that would require considerable machining when made by other processes*. Because of the dimensional accuracy and fine surface finish that is characteristic of the P/M process, many parts require no further processing and others require only a small amount of finish machining. Tolerances can generally be held to within 0.1 mm (0.005 in.). Large numbers of small gears are currently being made by the powder metallurgy process. Other complex shapes, such as pawls, cams, and small activating levers, can be made quite economically.
2. *Porous or permeable products, such as bearings, filters, and pressure or flow regulators*. Oil-impregnated bearings, made from either iron or copper alloys, constitute a large volume of P/M products. They are widely used in home appliance and automotive applications since they require no lubrication or maintenance during their service life. Powder metallurgy filters can be made with pores of almost any size, some as small as 0.0025 mm (0.0001 in.). Unlike many alternative filters, powder metallurgy filters can withstand conditions of elevated temperature, high applied stresses, and corrosive environments.
3. *Products made from materials that are difficult to machine or with high melting points*. Some of the first modern uses of powder metallurgy were the production of tungsten lamp filaments and tungsten carbide cutting tools.
4. *Products where the combined properties of two or more metals (or metals and nonmetals) are desired*. This unique capability of the powder metallurgy process is applied to a number of products. In the electrical industry, copper and graphite are frequently combined in applications like motor or generator brushes where

copper provides the current-carrying capacity and graphite provides lubrication. Bearings have been made of graphite combined with iron or copper, or from mixtures of two metals, such as tin and copper, where the harder material provides wear resistance and the softer material deforms in a way that better distributes the load.

5. *Products where the powder metallurgy process produces clearly superior properties.* The development of processes that produce full density has resulted in P/M products that are clearly superior to those produced by competing techniques. In areas of critical importance such as aerospace applications, the additional cost of the processing may be justified by the enhancement of properties.
6. *Products where the powder metallurgy process offers definite economic advantage.* Consideration of the process advantages that follow in the next section reveals features that may make powder metallurgy the most economical among two or more alternative ways to produce an equivalent part.

V. THE STARTING MATERIAL—METAL POWDER

The starting material for a casting is some form of melt shock. Since the structure and properties will be set by the solidification process, the only properties of concern will be chemistry and purity, and some properties relevant to the processing of liquids, such as melting temperature and fluidity.

In contrast, the starting material of powder metallurgy plays a far more significant role in the success of the process. While chemistry and purity continue to be important variables, they are supplemented with concerns for particle size, size distribution, particle shape, and the surface texture of the particles. A variety of processes can be used to produce powdered material, and each process imparts its unique characteristics.

Over 80% of a commercial powder is produced by some form of melt atomization. A stream of liquid is fragmented into small molten droplets which then solidify into the powder particles. The fragmentation is usually induced by a stream of pressurized gas or liquid, and the fragmentation and solidification is usually conducted under the protection of an inert gas shroud. Surface tension tries to pull the liquid fragments into a spherical shape, while cooling brings about solidification. The relative influence of these effects dictates the final geometry of the powder. Since each particle is a fragment of the melt, the individual particles have the same chemistry as starting material.

Powders can also be produced by the chemical reduction of particulate compounds, electrolytic deposition from solutions or fused salts, pulverization or grinding of brittle materials, thermal decomposition of carbonyls and hydrides, precipitation from solution, and condensation of metal vapors. Each method imparts its unique features to the resulting powder, and these features influence the response to subsequent processing.

VI. POWDER CHARACTERIZATION

Powdered material is generally characterized by a number of properties that include bulk chemistry, surface chemistry, particle size and size distribution, particle shape, surface texture, and internal structure. Still other tests assess the suitability of a powder for further processing. Flow rate determines the ease of filling and distributing powder into a die. Apparent density measures the ability of a powder to fill space without the application of external pressure. Combined with tap density, it is also important to the design of the compaction and ejections steps. Compressibility assesses the response of the powder to applied pressure, and green strength describes the strength of the powder after pressing.

VII. MIXING AND BLENDING

It is rare that a single powder will possess all the desired characteristics. The starting material is most likely a mixture of various types or sizes of powder, powders of different chemical compositions (which may diffuse to form an alloy or remain separate to form a composite), and additions of lubricants or binders. Powdered graphite can play a dual role, serving as a lubricant during compaction and a source of carbon as it alloys with iron during sintering. Particulate wax is a common lubricant, improving flow and compaction at the expense of green strength (the strength of a powder metallurgy part before sintering). Binders produce the reverse effect.

Alloys are created by mixing powders of different compositions. As opposed to prealloyed powders produced by melt atomization, mixed powders offers easier compaction. This is an important consideration for the fabrication of the part and leads to high production volumes. The individual elements or master alloys usually constitute a small percentage of the volume of the mixed powder, and subsequently are quite far apart in the mix. This leads to the need to homogenize the resultant sintered compact, or leave the part with significant inhomogeneity. Sintering is usually insufficient to fully homogenize the resulting part, leaving the properties strongly effected.

Another choice is diffusion alloyed powders. These are mixed powders where some prealloying is done on the surfaces of the particles only. Compaction is still good, but the homogenization is now greater, since the islands of alloying elements are smaller and closer together. This leads to improved properties.

VIII. COMPACTION

Compaction is probably the most critical step in the P/M process since it sets both the density of the powder and the uniformity of that density throughout the product. Figure 1 shows a typical compaction sequence using a mechanical press. With the bottom punch in its fully raised position, the powder feed moves into position over the die. The bottom punch then descends to a preset fill depth, and the powder feed retracts, leveling the powder. An upper punch then descends into the die, compacting the powder. The upper punch then retracts, and the bottom punch raises to eject the powder compact. As the powder feed advances the next cycle, its forward edge clears the compacted part from the press, and the cycle repeats (Figs. 2 and 3).

The objective of compaction is to produce a high-density compact with sufficient green strength for in-process handling and transport to the sintering furnace. Since final properties have a strong dependence on density, uniform properties require uniform density. Most compaction is performed with mechanical or hydraulic presses and rigid tooling. Compacting pressures range from 3 to 120 tons/in.^{2}, with 10 to 50 tons/in.^{2} being the most common. Because of press capacity limitations, most powder metallurgy products have pressing areas less than 10 in.^{2}.

During compaction, the powder particles move primarily in the direction of punch travel. Since the loose-fill height is generally two or more times the pressed height, the

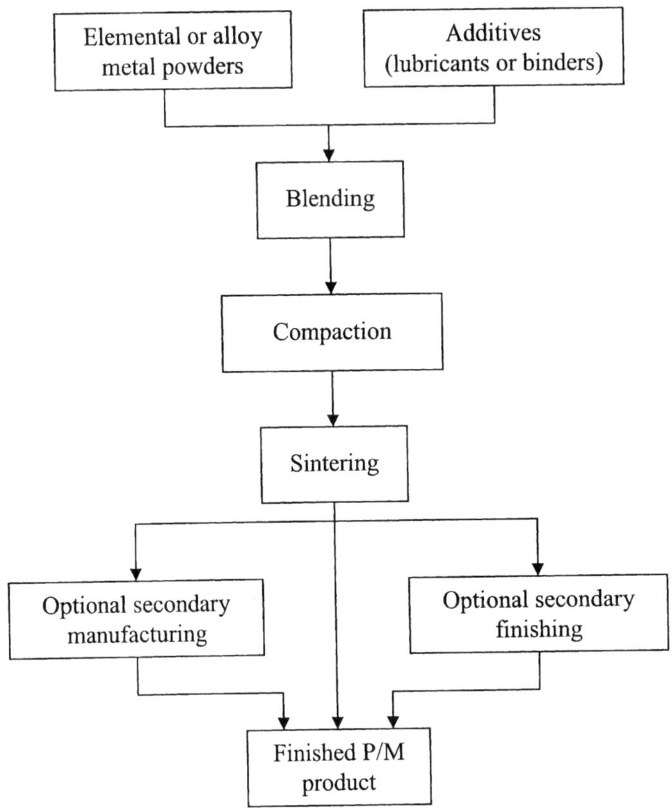

Figure 1 Simplified flowchart of the press and sinter process.

amount of particle travel in the pressing direction can be rather substantial. Lateral flow, however, is rather limited, such that particles rarely move more than two to three particle diameters from the original axis of pressing. Axial movement continues until the powder produces an equal and opposing force against the moving punch. This opposition is usually a combination of the reaction force on the bottom punch and frictional forces generated between the powder particles and the side walls of the die cavity. Since the side walls can carry load, the powder density is greatest beneath the punch and decreases as one moves down the column. Since we are compacting powder and not pressurizing a liquid, the transmitted pressure is not uniform and the resulting density is not uniform throughout the compact. Because the voids play a critical role in establishing final properties, a non-uniform density will certainly be accompanied by non-uniform properties.

Another important consideration in the compaction of P/M parts is the ability of the green part to survive the ejection process and subsequent handling prior to sintering. Since the greatest stress is usually seen during ejection from the die, and die wall friction also leads to greater bending moments on the part, surviving ejection typically means the part will survive until sintering. The green strength of the part is critical to this step and impacts design considerations for compaction, even at the expense of final properties.

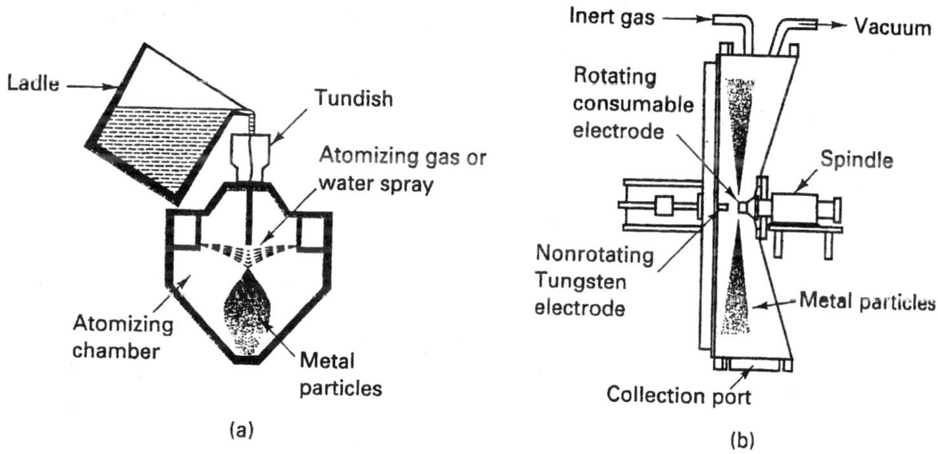

Figure 2 Two of many methods available for producing metal powders: (a) melt atomixation; (b) rotating electrode atomization. (From Ref. 7.)

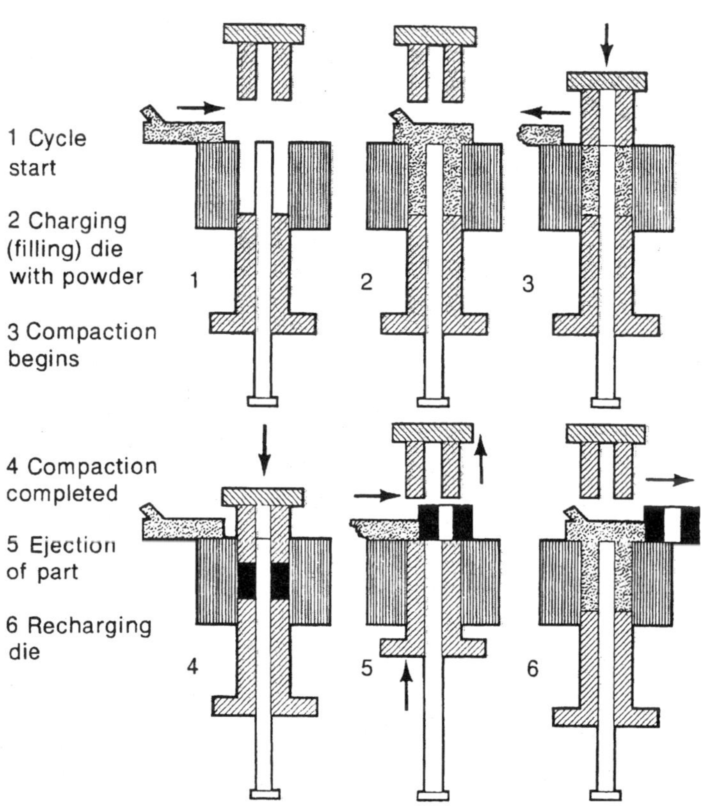

Figure 3 Typical compaction sequence for a single level P/M part, showing the various steps. The sequence is cyclic, where step 3 would follow after step 6 once the process is started. (From Ref. 8.)

IX. THE IDEAL POWDER METALLURGY PRODUCT

The ideal powder metallurgy product has uniform high density and the attractive properties that accompany a low volume fraction of residual voids. The design features that enable such results include:

1. *A uniform thickness that is relatively thin.* Single level parts can be compacted by a single pressing motion, and thin parts have the least side wall area to carry load and create non-uniform pressures and density.
2. *A uniform cross-section in the direction of pressing.* Since the loose powder occupies a volume that is at least double that of the final product, shape and dimensions must be maintained as the powder is compacted.
3. *A small cross-sectional area.* Compacting pressures can be quite high, typically between 10 and 50 tons/in.^{2} (Table 2). The required capacity of the press is the compacting pressure times the pressing area. Large areas require large, and therefore expensive, presses.
4. *Holes aligned with the direction of pressing.* Core rods can be inserted through holes in the punches to produce aligned holds in the product. The powder moves down around the rods during the compaction process, and the compacted part is removed from the rods during the ejection process. Core rods at any angle other than the direction of pressing would interrupt the flow of powder and result in gross variations in powder density. Therefore, angled holes will have to be machined after pressing.

Because of their simplicity, thin, single thickness parts are termed as Class 1 powder metallurgy parts, and are clearly the simplest to produce. They are generally compacted by a single moving punch. Thickness is generally less than $\frac{1}{4}$in (Fig. 4).

X. VARIATIONS FROM IDEAL

When thicker parts are to be compacted, the resulting density can be made more uniform through the use of a double-action press. During compaction, both the upper and lower punches move with respect to the cavity walls. The descending upper punch compacts the powders near the top of the part, while the upward moving bottom punch compacts the bottom region. The result is a more uniform density and the ability to compact thicker

Table 2 Typical Compacting Pressures for Various Applications

Application	Compaction pressures tons/in.^{2}	Compaction pressures MPa
Porous metals and filters	3–5	40–70
Refractory metals and carbides	5–15	70–200
Porous bearings	10–25	146–350
Machine parts (medium density iron and steel)	20–50	275–690
High-density copper and aluminum parts	18–20	250–275
High-density iron and steel parts	50–120	690–1,650

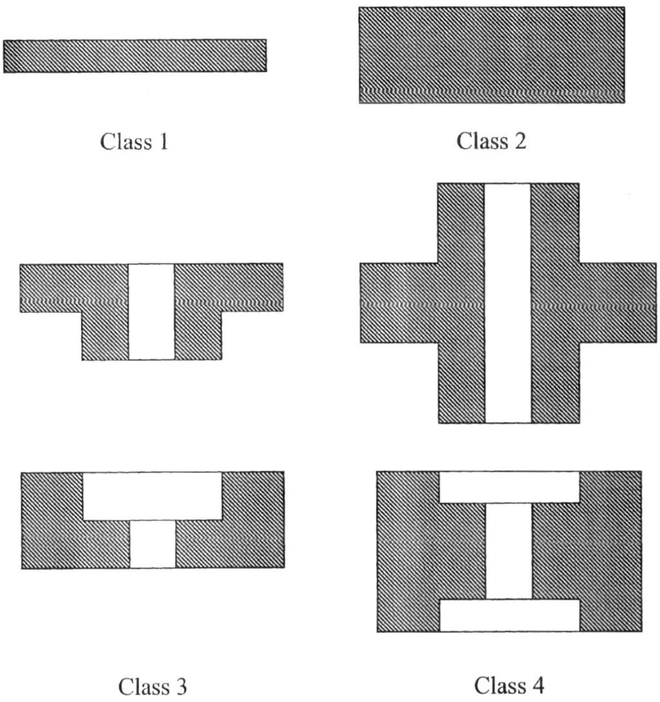

Figure 4 Examples of the four classes of press and sintered P/M parts.

parts. Class 2 parts are one-level parts of any thickness that requires double-action pressing (Fig. 5).

Since side wall friction significantly affects the uniformity of density, the ideal part has a low thickness/width ratio (ideally below 2) and a large ratio of pressing area to perimeter. Parts failing to meet these guidelines tend to exhibit large variations in density.

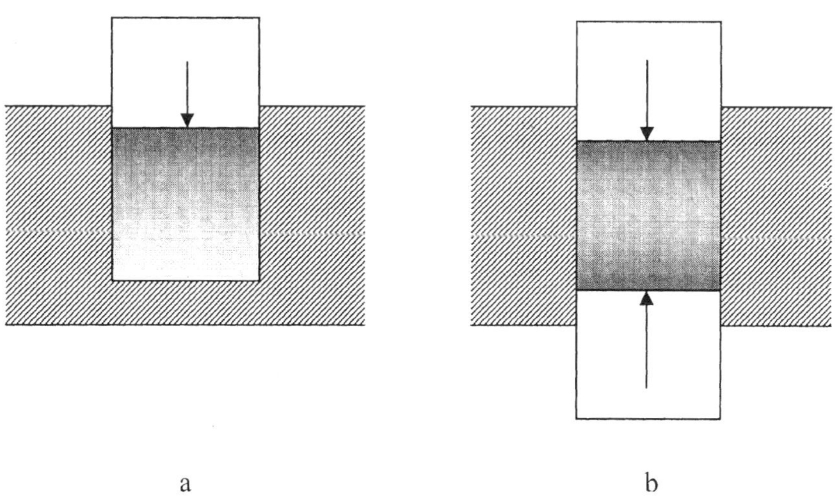

Figure 5 A comparison of the uniformity of the density between single and double action pressing: (a) single action; (b) double action.

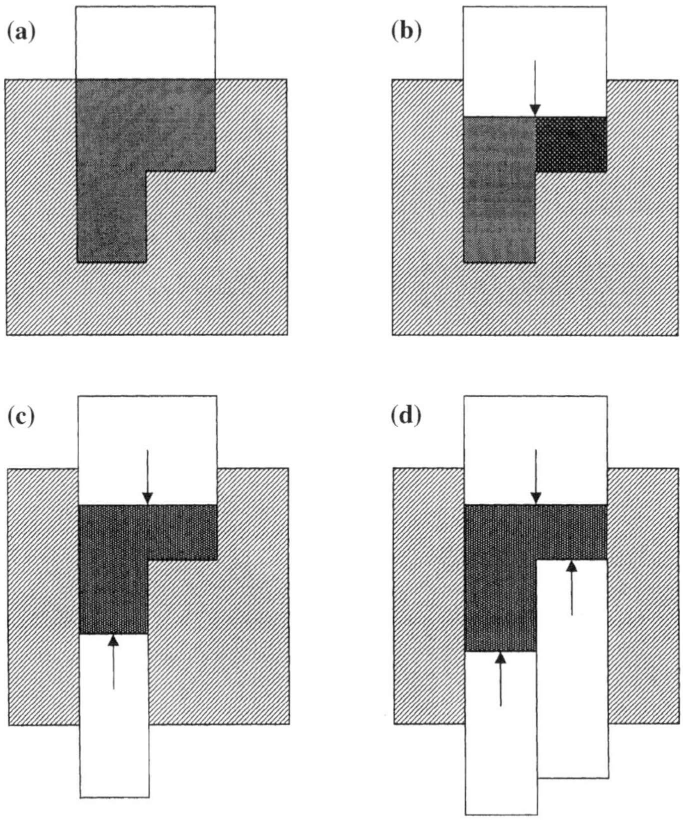

Figure 6 Multi-level parts require multiple punches: (a) a single action pressing of a multi-level part before pressing; (b) after pressing the density is non-uniform; (c) and (d) uniform density can be achieved by several means.

If two or more thicknesses are present in a single part, a single displacement, such as the movement of a single punch (or upper and lower punches that span the entire pressing area), will result in different degrees of compaction. Therefore, if a product incorporates multiple thicknesses, it will require more complicated presses or compaction methods. In general, the uniform compaction of multiple thickness parts will require the synchronous movement of multiple punches, each traveling different distances. Class 3 parts are two-level parts that are pressed from the top and with two punches on the bottom. Class 4 parts are the most complex of the parts produced by rigid-die compaction, and are multi-level parts that require two or more pressing motions (Fig. 6).

Consider the compaction of a simple two-level part. If both levels are filled with loose powder, and a single punch descends a given distance, the thinner segment will be reduced to a smaller fraction of its original volume and will have a greater density. If both regions are to be compacted the same amount, say to one-half of the original thickness, the punch in the thinner segment will move a smaller distance compared to the punch in the thicker segment. In addition, the movements must be synchronized. If the punch in one region moves first, the subsequent movement of the second punch will shear the particles beneath it over the already compacted material in the first region. Instead of producing a product with interlocked particles throughout, there will be a distinct plane of weakness at the interface between the two regions. Sintering will not heal this defect and, as is usually the case, will not heal any defects produced during compaction. In order to prevent the formation of this defect, the punch in the thicker region must complete its longer stroke in the same time that the punch in the thinner region completes its shorter one. Thus, not only must the movement of the two punches be synchronized, but the two punches must have different strokes and different velocities! The more levels, the more punches that must be similarly synchronized.

When part complexity becomes high, rigid tooling is abandoned, and compaction is performed with flexible molds that are acted upon by pressurized gas or liquid. This process is known as isostatic compaction when the pressure is applied to all sides simultaneously. If performed at ambient temperatures, this is called cold isostatic pressing (CIP). While production rates are extremely slow, multiple molds can be compacted in a single pressurization. Part size can be as great as several hundred pounds. When high density is desired, heating and pressing can be combined in a process known as hot isostatic pressing (HIP). Small parts of high complexity can be produced by a process known as metal injection molding that will be discussed later in this chapter.

Regardless of the compaction process, powder metallurgy products are usually rather small. Compared to the starting material for other processes, such as casting, forming, or machining, the cost of a powdered material is usually quite a bit greater. The elimination or reduction in subsequent machining and scrap generation can often make powder metallurgy the favored process, but this is difficult if the size of the piece makes the material cost differential too great.

XI. EFFECTS OF COMPACTION

The properties of powder metallurgy parts are strongly dependent upon the density of the part, and part density in turn depends upon the amount of pressure that is applied and the characteristics of the powder (including size, shape, surface texture, and mechanical properties). When the powder particles are pressurized, density can increase for a variety of reasons, including the bulk movement and rearranging of particles, the deformation of individual particles, and particle fracture or fragmentation. The characteristics of the powdered material control which of these mechanisms dominates. Elemental metal powders are usually high deformable and a great deal of deformation can occur during compaction. Ceramic powders are usually brittle and will fracture. Metal powders that are cold worked or are highly allowed will be between the two extremes.

Densities are usually reported as a pure value (such as grams per cubic centimeter) or as a percentage of theoretical, where the difference between this number and 100% is the amount of void space left within the compact. Adjectives before the word density define the conditions of the part when the density was determined. Apparent density refers to the density of the loose, uncompacted powder. Green density provides the value after compaction, but before sintering. Final density is determined after sintering (Fig. 7).

Since uniform, high density is a common objective of compaction, various means have been developed to assist the densification. As mentioned above, lubricants are frequently added to reduce die wall friction and promote transmittal of the compacting pressure. Unfortunately, lubricants also reduce the green strength, and too much lubricant can actually reduce mechanical properties because of the voids produced when the lubricant is removed from the compact.

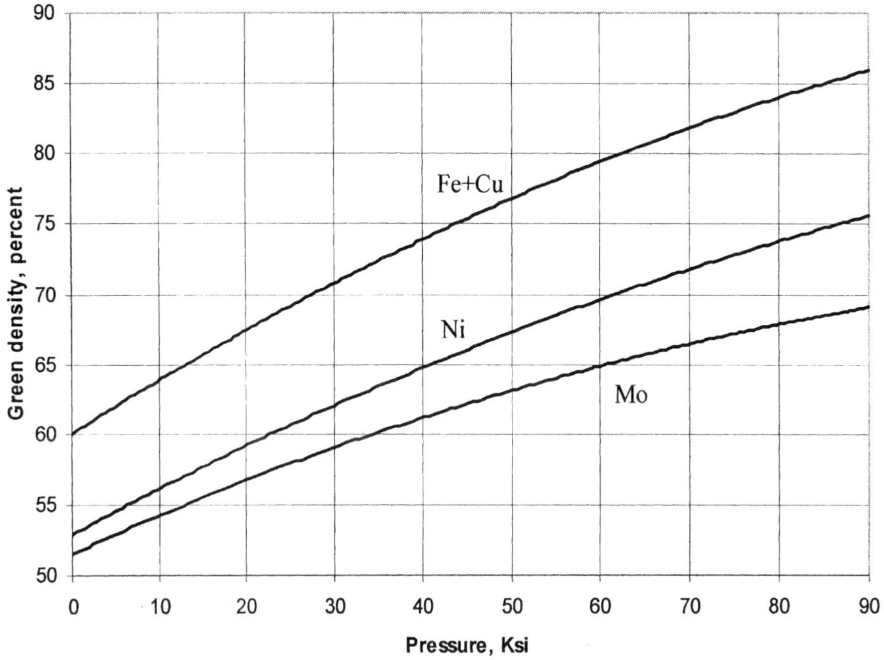

Figure 7 The relationship between compaction pressure and green density of different metal powders.

Another technique that is seeing expanding use, particularly in automotive applications, is warm compaction. The powder is preheated prior to pressing, softening the metal for better response to the applied pressure. Heated dies are typically used to prevent cooling of the powder during application of the pressure. Advantages of warm compaction are higher green and sintered densities, higher green and final mechanical properties, and greater uniformity of density in the sintered part (Fig. 8).

With good mechanical compaction practice, the density of loose powder can be raised to about 80% of an equivalent cast or forged material. At this stage, the parts generally have sufficient strength (green strength) to hold the shape and permit a reasonable amount of handling. In addition, the compaction operation determines both the nature and distribution of the remaining porosity.

XII. COMPACTION TOOLING

During the compaction operation, powder particles flow across die surfaces, often under considerable pressure. As a result, wear is a major concern, and compaction tools are generally made from wear resistant tool steels. For particularly abrasive powders or high-volume production runs, cemented carbides may be specified, with the carbide frequently being an insert in a steel or tool steel support. Core rods have similar requirements to the dies, but the punches tend to experience more impact or shock loadings as they rapidly descend against the powder fill. As a result, the punch material is generally selected more

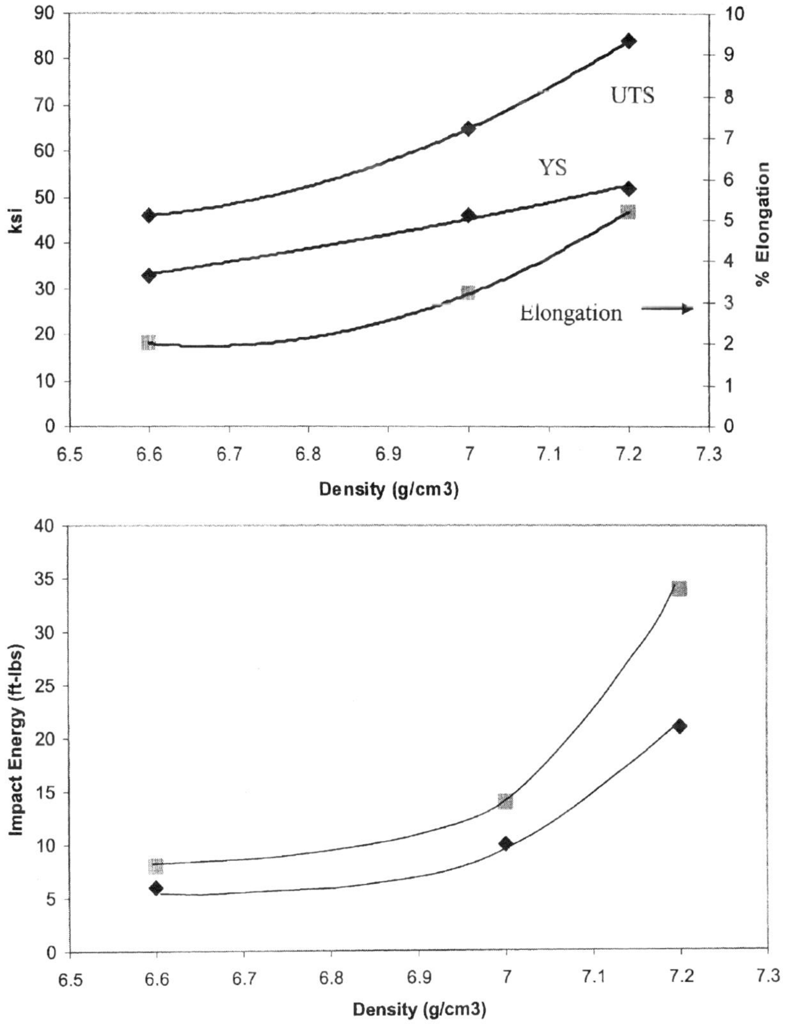

Figure 8 Sintered density vs. mechanical properties of FN-0205 (squares) and FN-0205 (diamonds).

for a balance of shock and wear resistance, while dies and core rods have more emphasis on wear.

Still another concern when selecting tool material is the geometry of the piece. Tool steels undergo a heat treatment to establish the high hardness desired for resistance to wear. During the heat-treatment cycle, the various heatings and coolings often produce expansions, contractions, warpage and distortion. After heat treatment, however, the material is extremely hard, and not able to be easily machined. Any adjustments to tool geometry must then be made by some form of abrasive grinding. The external surface of the male-geometry punches and core rods are generally more accessible than the interior surfaces of the die cavities. As a result, the more distortion resistant air-hardenable tool steels are often preferred for dies, while the less expensive oil- or water-hardenable tool steels may be preferred for punches and core rods, especially ones of simple geometry.

XIII. SINTERING

The strength of pressed-powder compacts is attributed primarily to the mechanical interlocking of the particles. It is clearly insufficient for use, but permits gentle handling with retention of shape and detail.

In the sintering operation, the pressed compacts are placed in a controlled atmosphere environment and heated. A burn-off or purge is first conducted to combust any air, volatilize and remove lubricants or binders, and raise the temperature of the compacts in a controlled manner. The temperature is then increased to a level where solid state diffusion occurs (typically 70–80% of the melting point for metals and near 90% for refractory materials), and sufficient time is provided to form bridging bonds between the particles. The mechanical bonds of compaction become metallurgical bonds.

With additional time, diffusion acts to decrease the size of remaining pores, with a companion improvement in properties. However, pore reduction is strongly effected by the driving force for densification, the reduction in surface area, and the ability to transport atoms to pores. As densification reduces the driving force, the process slows down. In addition, at densities exceeding 90%, grain growth leads to the separation of grain boundaries from the pores and the transport of atoms slow significantly. The closer it is to full density, the harder it is to fully densify by sintering alone. So, the additional cost of maintaining the temperature and protective atmosphere is weighed against the ever-increasing level of properties, and an acceptable compromise is determined. The protective atmosphere is then maintained while the product is cooled to room temperature.

Atmosphere control is critical to successful sintering. Compacted powder typically contains between 10% and 25% residual porosity, and some of these voids are interconnected to exposed surfaces. If the material were heated in air, rapid oxidation would occur, and this would severely impair the bond formation and resultant properties. Reducing atmospheres break down any oxide that may be present on particle surfaces, and combust any harmful gases that may enter the furnace or be generated during the sintering. Inert gases cannot perform these tasks, but serve to prevent the formation of any additional contaminants. Vacuum sintering is often performed, and nitrogen atmospheres are also common.

Ferrous alloys are often produced by mixing graphite with iron powder to form a steel or stainless steel. While the high mobility of carbon allows for good homogeneity of the interstitial element, this mobility also means that carbon is free to interact with the atmosphere. This situation is similar to that found in heat-treating wrought or cast steel parts and has similar answers. Control of the carbon potential of the atmosphere may be necessary to maintain correct carbon levels.

During the sintering operation, metallurgical bonds form between the powder particles. With the increase in density, strength, ductility, toughness and electrical and thermal conductivities also increase. If different materials were blended together, interdiffusion may promote the formation of alloys or intermetallic phases. As a consequence of the density increase, size decreases. To meet desired tolerances, the compacted shape must be appropriately oversize. Not all the porosity is removed, however, and final pressed-and-sintered powder metallurgy products generally retain between 5% and 25% residual porosity.

XIV. HOT PRESSING AND ALTERNATIVE METHODS OF COMPACTION

Compaction is generally performed under conditions of high pressure at room temperature. Sintering involves elevated temperature at atmospheric pressure. It would appear that process efficiency could be gained by simultaneously exposing loose powder to elevated temperatures and elevated pressures to concurrently compact and sinter. While there are attractive features to this approach, the dominant problem is the need to keep the particle surfaces free from oxidation. Since elevated temperatures accelerate reactions, and we want the applied pressure to densify the material, hot-pressing operations generally begin by encapsulating the loose powder in an evacuated container. Another alternative is to use a press that is built with enclosed dies that can be evacuated. Since each piece must be separately encapsulated, hot-pressing operations are generally limited to high-cost products where the unique or superior properties justify the expense. Hot-pressed parts are usually full density, with uniform, isotropic properties. The vacuum enables the processing of reactive materials. The absence of die compaction enables the production of large parts and shapes that would be difficult to eject from rigid dies.

Conventional forming processes have also been adapted to perform both room temperature and elevated temperature powder compaction. Sheet material can be produced by feeding powder through the rolls of a rolling mill. Powders can also be rolled into a continuously fed wire mesh screen, which provides support for handling prior to sintering. Rods, wires, and small billets can be produced by the hot or cold extrusion of encapsulated powder.

Forging can be used to form complex shapes from canned powder or simple shape sintered preforms. One approach to making a complex-shaped product that would be difficult to compact is to first produce a simple shape preform by conventional press-and-sinter manufacture. Subsequent hot forging converts this material to the desired final shape, while adding the benefits of both metal flow and further densification as a result of the shear under compression that occurs during forging. By starting with powder metallurgy, forged products can be produced that are free of segregation, have uniform fine grain size throughout, and utilize the novel alloys and unique combinations that are possible through the blending of powders. Because of the initial porosity and permeability, protective atmospheres may be required during the heating and forging operation. Powder forged properties can be exceptionally good.

XV. SECONDARY PROCESSING

While many powder metallurgy parts are ready to use following the sintering operation, others utilize one or more secondary operations to provide enhanced precision, improved properties, or special characteristics.

In a conventional press-and-sinter operation, the compaction operation sets dimensions, but these dimensions then change as the part shrinks during the densification of sintering. In addition, warping or distortion may occur as a result of non-uniform heating and cooling. Repressing, coining or sizing may be employed to restore or improve dimensional precision. The accompanying cold working and further increase in density can combine to provide a 25–50% increase in product strength.

Conventional finishing operations, such as heat treatment, machining, and surface treatment, can also be performed. High-density parts, those with less than 10% porosity, can be treated as conventional parts. Parts with lower density usually require some form of special precautions, such as protective atmospheres, special coolants or quenchants, or adjusted processing conditions. This is due to the fact that the porosity will most likely be connected throughout the part when the porosity exceeds 10%, allowing the atmosphere to penetrate to the internal surfaces on the part.

Powder metallurgy steel parts can be quenched and tempered to achieve high strengths or hardnesses as required. However, the hardenability of steels depends in part on the density of the parts since the thermal conductivity of the metals is effected by the density. In addition, the homogeneity of the parts made from mixed powders will be effected by sintering conditions and can result in changes in hardenability due to composition variations. In addition to these problems, hardness measurements on a powder metallurgy sample will result in a lower hardness value due to residual porosity. If a certain heat-treated hardness is expected and used to measure the formation of martensite, the resulting lower measured hardness, called the effective hardness, due to porosity will make the steel appear to not be fully hardened. Microhardness tests, which do not interact with the pores in the material, do give a true reading of the state of the part. These factors must be taken into account when specifying the heat treatment of sintered steel parts.

Sinter hardening is a process where the austenization treatment of steel compacts is replaced by the high temperature sintering. Quenching then must be done in the furnace before removal of the part and the subsequent tempering. This process provides a significant economic benefit by the elimination of the reheating to the austenite temperature. However, materials with very good hardenability are required to allow for sufficient depth of martensite formation at the relatively slow cooling rates possible in the sintering furnace.

The presence of porosity and permeability also opens up the possibility of unique secondary processing, such as impregnation and infiltration. During impregnation, oil or other liquid, such as polymeric resin, is introduced into the porous network. The most common application is the oil-impregnated bearing, which contains between 10% and 40% oil by volume. Parts impregnated with Teflon offer the combined properties of high strength and low friction. If the pores are filled with a solid material, such as a lower melting point metal, the process is called infiltration. Strength and toughness improve, due to the absence of pores. Components can be made gas- or liquid-tight, machinability and corrosion resistance improve, and smooth-surface platings are now possible.

XVI. PROPERTIES AND CAPABILITIES

The properties of powder metallurgy products are dependent upon a number of variables, including the type and size of powder, the amount and type of lubricant, the type of pressing and the pressing pressure, the sintering time and temperature, and the details of secondary processing. As a result, products can range from low-density, highly porous parts with tensile strengths as low as 10 ksi. up to high-density pieces with strengths in excess of 180 ksi.

Most mechanical properties show a strong dependence on final density. Since the voids in powder metallurgy parts act as stress raisers and assist in starting and propagating fractures, the fracture-limited properties (toughness, ductility, and fatigue life) tend to be more sensitive to changes in density than strength or hardness. The higher the strength of the material, the more the properties of porous powder metallurgy parts fall below those of full-density wrought or cast equivalents. If special processing is used to increase the density of the powder metallurgy part, the properties approach those of a wrought or cast material. If full density is achieved with accompanying fine grain size, the properties of the powder metallurgy product can actually exceed those of wrought or cast equivalents. Because of the extremely significant role played by residual porosity, it is vitally important that powder metallurgy parts be designed and materials selected so that the desired properties are obtained with the level of residual porosity that is likely to result.

Physical properties are also affected by porosity. Electrical, thermal, and magnetic properties all vary with density, with performance usually decreasing in the presence of voids. Corrosion resistance decreases as a result of pockets, which entrap corrosive media. In the areas of sound and vibration damping, the voids can make a positive contribution, and powder may be preferred as a means of enhancing these characteristics.

Many powder metallurgy applications utilize the unique capabilities of powder processing. Some of the chemical means of producing powdered material can maintain extremely high levels of purity—making powder metallurgy an attractive means of producing high purity products. Segregation, or non-uniformity of chemistry within a part, is characteristic of a number of manufacturing processes, especially casting. By producing parts from well-blended powders, the chemistry of the product is assured to be uniform throughout.

Product chemistries are not limited by solubility limits or miscibilities, since the materials are combined in the form of particulate solids, and remain solid throughout all subsequent processing. As a result, compositions can be used that cannot be produced by casting or metal forming (since the material being formed is usually cast at some point in its prior history). Unique combinations or blends are possible, such as that of iron–copper.

Particulate composites can be made in which the components come from diverse families of materials, such as metals and ceramics, or metals and polymers.

High levels of porosity or controlled amounts of permeability open the way to another group of unique products. Porous products have a high level of exposed surface for a given amount of material, making them ideal for applications such as electrodes where performance is related to surface area. Permeable parts can act as filters, where the component materials provide the required levels of strength or corrosion resistance, as well as the ability to operate at elevated temperature.

XVII. DESIGN CONSIDERATIONS IN CONVENTIONAL POWDER METALLURGY

The ultimate objective of any manufacturing system is to economically produce products for specific engineering applications. Success begins with good design and follows with good material and proper processing. In designing parts that are to be made by conventional powder metallurgy, it must be remembered that P/M is a special manufacturing process and provision should be made for a number of unique factors. Products that are converted from other manufacturing process without modification in design rarely perform as well as parts designed specifically for manufacture by powder metallurgy. Some basic rules for the design of P/M parts are:

1. The shape of the part must permit ejection from the die. Sidewall surfaces should be parallel to the direction of pressing. Holes or recesses should also have

uniform cross-section with axes and sidewalls parallel to the direction of punch travel.

2. The shape of the part should be such that powder is not required to flow into small cavities such as thin walls, narrow splines, or sharp corners.
3. The shape of the part should permit the construction of strong tooling.
4. The shape of the part should be within the thickness range for which P/M parts can be adequately compacted and handled.
5. The part should be designed with as few variations in section thickness as possible.
6. Parts can be designed to take advantage of the fact that certain forms and properties can be produced by P/M which are impossible, impractical, or uneconomical to obtain by any other method.
7. If necessary, the design should be able to be produced with available equipment. Pressing areas should match press capability, and the number of thicknesses should be consistent with the number of available press actions.
8. Consideration should also be made for product tolerances. Higher precision and repeatability is observed for dimensions in the radial direction (set by the die) than for those in the axial or pressing direction (set by punch movement).
9. Finally, design should consider and compensate for the dimensional changes that will occur after pressing, such as the shrinkage that occurs during sintering.

XVIII. ADVANTAGES AND DISADVANTAGES OF CONVENTIONAL POWDER METALLURGY

All manufacturing processes have their characteristic advantages and disadvantages that should be considered if the process is to be employed economically and successfully. For powder metallurgy, the most common advantages are:

1. *Elimination or reduction of machining.* The dimensional accuracy and surface finish of P/M products are such that subsequent machining operations can often be eliminated.
2. *Scrap is eliminated or reduced.* Powder metallurgy is the only common manufacturing process in which no material is wasted. In casting, machining, and press forming the scrap can often exceed 50% of the starting material. This is particularly important where expensive materials are involved and may make it possible to use more costly materials without increasing the overall cost of the product.
3. *High production rates.* All the steps in the P/M process can be readily automated. Labor requirements are low, and product uniformity and reproducibility are among the highest in manufacturing.
4. *Complex shapes can be produced.* Subject to the limitations discussed previously, complex shapes can be produced, such as combination gears, cams, and internal keys.
5. *Wide variations in composition are possible.* Parts of very high purity can be produced. Immiscible materials can be combined, and solubility limits can be exceeded. Metals and ceramics can be intimately mixed. In most cases, the chemical homogeneity of the product exceeds that of all competing techniques.

6. *Wide variations in properties are possible.* Products can range from low-density parts with controlled permeability to high-density parts with properties that equal or exceed those of equivalent wrought counterparts. Damping of noise and vibration can be tailored into a P/M product. Other properties can often be designed to meet the specific needs of an application.

Commonly cited disadvantages include:

1. *Inferior strength properties.* As a result of the residual porosity, the mechanical properties of powder metallurgy parts are generally inferior to those of wrought or cast products of the same material. The use of P/M parts may be limited when high stresses are involved. However, the required strength and fracture resistance can often be obtained by using different materials or by employing alternate or secondary processing techniques that are unique to powder metallurgy.
2. *Density variations produce property variations.* The variation in density that is frequently observed in compacted products generally results in property variations throughout the part. For some products, these variations may be large and unacceptable.
3. *High material cost.* On a per pound basis, powdered metals are considerably more expensive than wrought or cast material. The absence of scrap and the elimination of machining, however, can often offset the higher cost of the starting material. In addition, powder metallurgy is usually employed for rather small parts where the material cost per part is not very great.
4. *High cost of tooling.* Because of the high pressures and severe abrasion involved in the compaction process, the P/M dies must be made of expensive materials and be relatively massive. Because the tooling is generally unique to a part, production quantities of less than 10,000 identical parts are normally not practical.
5. *Design limitations.* The powder metallurgy process may not be feasible for many shapes. Parts must be able to be ejected from the die. Good compaction places limitations on the thickness/diameter (or thickness/width) ratio. Thin vertical sections are difficult, and the pressing area must be within the capacity of available presses. Few parts exceed 25 in.^{2} in pressing area.
6. *Health and safety hazards.* Many metals, such as aluminum, titanium, magnesium, and iron, can ignite or explode when in particle form (i.e. with large surface/volume ratios). Fine particles of metal can also remain airborne for long times and can be inhaled by workers. The handling of metal powders, therefore, frequently requires the use of inert atmospheres, dry boxes, and hoods, as well as special cleanliness of the working environment.

XIX. EXTENDING THE LIMITS

For many years, powder metallurgy products carried the stigma of "low strength" or "inferior mechanical properties". This label was largely the result of comparisons where different processes were used to produce parts of the same size, shape, and *material.* In the more valid comparisons, where the size, shape, and *desired mechanical properties* are specified, and both process and material can be varied, powder metallurgy frequently emerges as a valid competitor. Unique materials can be used, such as iron–copper blends, for which there are no cast or wrought equivalents, and parts are designed that will provide the desired properties while containing the normal amount of residual porosity.

Processing by conventional press-and-sinter methods still imposes significant constraints with respect to size, shape, production quantity, and possible mechanical properties. A variety of special processes, such as metal injection molding, hot isostatic pressing, and P/M forging, have been developed to overcome many of these limitations and significantly expand the capabilities of the powder metallurgy process. These processes and adaptations will now be discussed.

XX. METAL INJECTION MOLDING

Metal injection molding is similar to conventional powder metallurgy process in several ways and different in other important areas. This leads to new considerations for the design of MIM parts when compared to conventional P/M. It also offers many new advantages that can be utilized to produce cost-effective, high performance parts.

Like conventional P/M, MIM uses powdered metals as the starting material, and properties are controlled by the density produced during sintering. Every other aspect of the process is different. An overview of the process is shown in Fig. 1. The metal powder is mixed with a specially designed plastic binder, often with thermoplastic properties. This mixture is often produced in the form of pellets, which makes for ease of handling and uniformity of composition. The mixture is called the feedstock and can contain up to 50 vol% binder. The feedstock is heated to the plastic range and injected into a mold under pressure. The mold temperature is such that the part becomes harder and can then be ejected from the mold with excellent green strength. The part is now treated to remove the plastic binder. Techniques such as solvent extraction, controlled heating to volatilize the binder, or catalytic debinding using acids are used to remove the binder. The resulting part is very fragile and must be treated carefully before sintering. The part is sintered similar to conventional P/M parts, except that the lower density of the part before sintering leads to large amounts of shrinkage, often up to 25% or more.

In conventional P/M, the powder is often chosen to flow well during the filling and compaction, as well as to result in low shrinkages so that dimensions are closely held. Since the powder is carried by the plastic binder, flow characteristics are not the major attribute of the powder chosen, and instead the powders are often chosen on the basis of sinterability instead. The powders are very fine, usually in the range of 2–20 μm in diameter. The fine powder sinters to high densities. Densities of 99% of theoretical are common. This has a significant effect on the properties of MIM parts usually making them superior to other fabrication techniques (Fig. 9).

While the size of conventional P/M products is generally limited by press capacity, the size of the MIM is more limited by two things, the cost of the fine powders and the ability to completely remove the binder from all areas of the part. Typically MIM parts are complex-shape metal parts with thicknesses as large as 6.3 mm ($\frac{1}{4}$in.) and weights under 60 g (2oz). Section thicknesses as small as 0.25 mm (0.010 in.) are possible because of the fineness of the powder. Special binders are now available which allows much larger parts to be successfully debound [1].

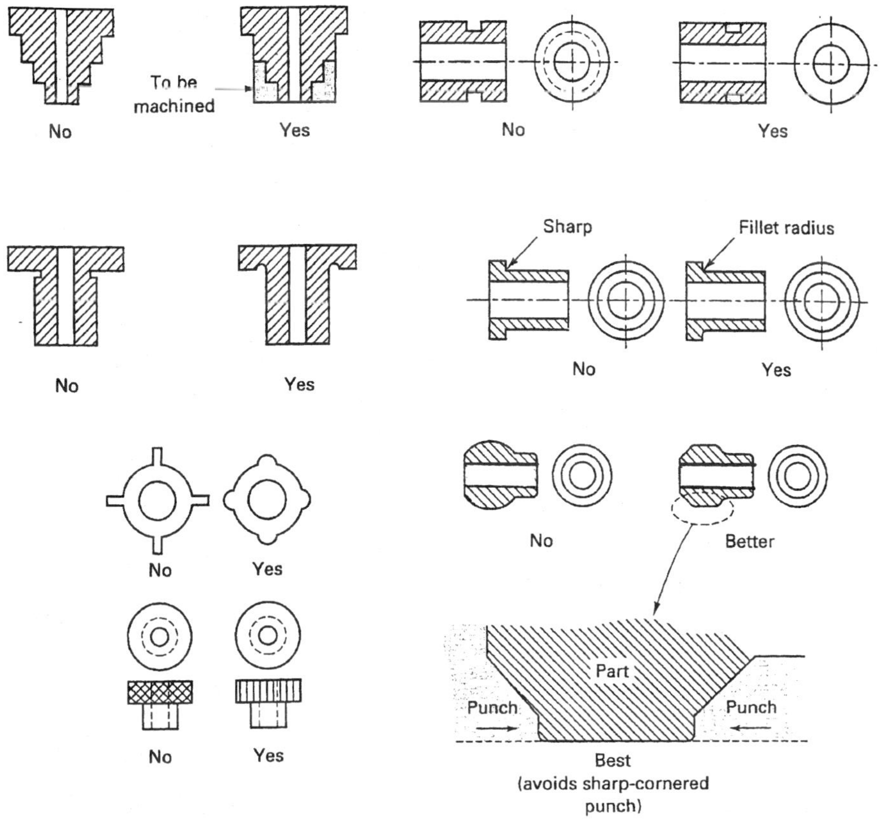

Figure 9 Examples of part designs poorly suited to P/M fabrication and design changes which makes them well suited. (From Ref. 9).

XXI. ATTRACTIVE FEATURES OF MIM PROCESS

The process of metal injection molding leads to many attractive features that can be used by designers (Fig. 10). Some of these are:

1. Highly complex shapes can be molded, limited only by the ability to be ejected, sımılar to die casting.
2. Good dimensional tolerances and good surface finishes are possible, despite the large amount of shrinkage common to the process.
3. Excellent mechanical properties can be achieved with a variety of material choices.
4. Complex shapes often allow the elimination of most to all subsequent machining and assembly steps, lowering costs.
5. High density allows the parts to be treated with the same secondary processes as machined parts.
6. Medium to large part runs can be made quickly.
7. Often feedstock can be reground and reused, which lowers material costs.

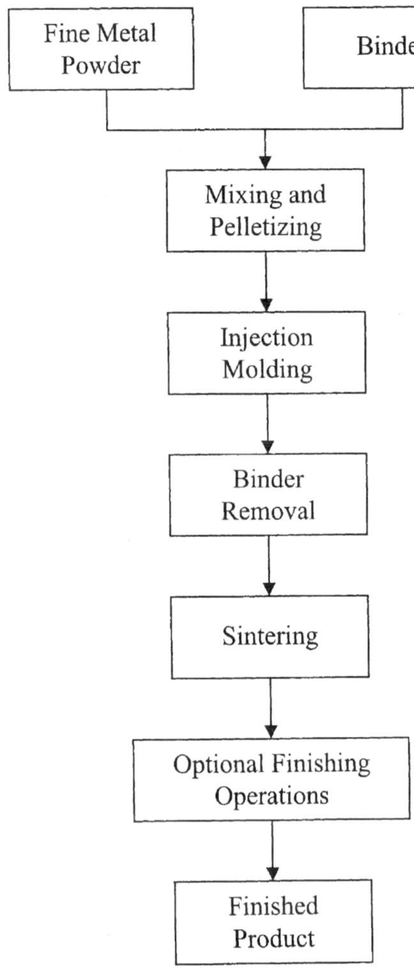

Figure 10 Simplified flowchart of the metal injection molding process.

XXII. DESIGN CONSIDERATIONS OF MIM

The cost of MIM parts is driven by several factors. In many ways, they are similar to the factors controlling costs of die casting and plastic injection molding, but differ in the amount that each factor affects the cost. Two major factors are the hourly molding machine costs and the cycle time. Lower machine costs and short cycle times reduce costs. Larger parts or parts with high complexity may increase the size of the machine required and increase cost. Tooling for MIM requires complex dies able to withstand the erosive feedstock and hence tend to be expensive when compared to other techniques. A larger volume of parts allows this cost to be averaged over the entire production run.

The MIM process usually requires the extraction of heat from the mold in order for the binder to solidify. Design features, such as thin sections, which accommodate higher cooling rates, can reduce cycle time and decrease costs. Both for molding time and debinding times, thinner sections are helpful. Sintering costs may be slightly higher than conventional P/M if longer cycle times are required to reach the required densities.

Metal injection molding material costs are higher than alloy billets or ingots as it is for conventional P/M, and the formulated feedstocks with finer powder sizes further increases material cost. However, the process is very material efficient. A large percentage of the feedstock ends up in finished parts as there is little flash or and small runners. Green parts, sprues, runners and left over feedstock can usually be reground and reused with a significant loss of properties. Since MIM often replaces many machining steps if fabricating the final part, there is little to no turning generated. Finally, the good properties developed by sintered MIM parts often mean that less material needs to be used to meet the required performance goals.

When considering how to design the shape of an MIM part, it is necessary to remem ber several artifacts of the process. Since the mold is usually in two halves, there is a parting line that is created. In addition, the gate(s) will leave a visible mark on the part. Finally, the ejection of the part is carried out with pins, which will once again make a noticeable mark on the green part. It is important to design the part such that these marks are in places where they do not effect the performance or appearance of the part.

The parting line plays an important role in designing the mold as the orientation of the part should be such that the part is easy to eject from the mold. This is accomplished through orienting the largest plane of the part to coincide with the parting line and most of the vertical features should be perpendicular to it. Additional mold members can be used not only to create more complex shapes, but also produce a parting line and increase cost.

The gate locations should be placed at locations on the part where they will not require additional steps to remove. This should be in accordance with locations to produce proper mold filling. Ejector pin locations should also be unobtrusive and located where the green part is strong enough to withstand the stress. A minimum of ejector pins should be used consistent with safe ejection of the part.

During debinding and subsequent sintering, the parts are very fragile and subject to cracking or distortion. It is important to also consider how to support the part during these operations. Special fixtures or setters can be used but significantly increase the cost. If possible, a flat surface that can lie on a standard refractory should be incorporated to decrease the cost of creating specialized support structures.

Wall thicknesses should be kept as uniform as possible to prevent distortion, internal stresses, cracks, and other defects. Additionally, varying thickness can lead to non-uniform shrinkage and loss of dimensional control. Ribs and attachment points can lead to a sink line at the point opposite the point of attachment and attention needs to be paid to proper use of coring to reduce wall thicknesses and reduce sinking.

Finally, holes can usually be produced easily with core rods to eliminate machining. Additional holes may be used to reduce material use and help control wall thickness for the reasons mentioned above.

XXIII. PROPERTIES OF MIM PARTS

The properties of metal injection molded parts are significantly better than the typical press-and-sinter P/M part. This is due to the lower porosity of the typical MIM part and the smaller powder used in the feedstock. Metal injection molding parts can easily achieve 5% or less porosity, allowing for mechanical properties closer to wrought. A wide selection of alloys is available, and many more are possible with custom blending of feedstock. Since compaction is not a part of the MIM process, prealloyed powders can be used without any significant penalty. Any wrought composition can be molded and sintered.

Table 3 Comparison of Properties of Powder Metallurgy Materials and Equivalent Wrought Metals

Material[a]	Form and composition	Condition[b]	Theoretical density (%)	Tensile strength		Elongation in 2 in. (%)
				10^3 psi	MPa	
Iron	Wrought	HR	–	48	331	30
	P/M—49% Fe min	As sintered	89	30	207	9
	P/M—99% Fe min	As sintered	94	40	276	15
Steel	Wrought AISI 1025	HR	–	85	586	25
	P/M—0.25% C, 99.75% Fe	As sintered	84	34	234	2
Stainless steel	Wrought Type 303	Annealed	–	90	621	50
	P/M Type 303	As sintered	82	52	358	2
Aluminum	Wrought 2014	T6	–	70	483	20
	P/M 201 AB	T6	94	48	331	2
	Wrought 6061	T6	–	45	310	15
	P/M 601 AB	T6	94	36.5	252	2
Copper	Wrought OFHC	Annealed	–	34	234	50
	P/M copper	As sintered	89	23	159	8
		Repressed	96	35	241	18
Brass	Wrought 260	Annealed	–	44	303	65
	P/M 70% Cu–30% Zn	As sintered	89	37	255	26

[a]Equivalent wrought metal shown for comparison.
[b]HR, hot rolled; T6, age hardened.

Using the same composition as wrought alloys, there is a slight degradation of properties. For example, 316L in the annealed condition has a yield strength of 30 ksi and an ultimate strength of 75 ksi. Typical MIM-316L properties are 25 ksi yield and 75 ksi ultimate. The ductilities are similar. However, since it is usually just as easy to fabricate the part from another alloy, the substitution of a higher performance alloy can be made with an increase of properties (Table 3).

The heat-treatment and other secondary operations typically require little to no adjustment compared to the treatment of wrought alloys. Steels, stainless steels, copper alloys, and nickel-base superalloys are routinely molded, while titanium is increasingly being molded.

XXIV. DESIGN CONSIDERATIONS OF P/M PROPERTIES

When designing P/M parts, one value often used is the minimum value. The concept of a minimum value, such as minimum strength, is often used to quote P/M properties [2,3].

Table 4 Values for Common Ferrous P/M Alloys

	Minimum strength		Typical properties				
	Yield (ksi)	UTS (ksi)	UTS (ksi)	Yield (ksi)	Elongation % (1″ gage)	Hardness	Density
FL-4205-35	35		52	42	1.0	60HRB	6.80
-40	40		58	47	1.0	66	6.95
-45	45		66	52	1.5	70	7.10
-50	50		73	58	2.0	75	7.30
FL-4205-80HT		80	90		<0.5	28HRC	6.60
100HT		100	110		<0.5	32	680
120HT		120	130		<0.5	36	7.00
140HT		140	150		<0.5	39	7.20

For P/M materials in the as-sintered condition, the yield strength is used as the minimum strength value, while for the heat-treated condition (quench and tempered ferrous), the ultimate is used for the minimum strength value. Some published values for common ferrous P/M alloys are shown in Table 4.

The minimum value is not determined statistically. It is instead a result of round-robin testing from various manufacturers. The minimum value that is listed actually represents the lowest value that every contributing parts producer can meet consistently. It does not accurately represent the properties that can consistently be achieved by a parts producer using procedures to optimize the properties. The performance of parts in service cannot be predicted by the use of the minimum values listed. In addition, the typical values that are given alongside the minimum values represent the average of the various producers and not the range of properties that can be achieved with a little effort.

When working with a P/M fabricator, it is often best to set up a proof test that allows the parts to be tested as a whole, rather than cutting samples from the parts. This takes into account the variations in density that sometimes are created within the part by the need to meet dimensional tolerances. If a proof test cannot be decided upon, then the location of test specimens must be specified, so that reliable comparisons between batches of parts can be made.

In statistics, when working with reliability or lifetime prediction data, the standard frequency distribution used to describe the data is a Weibull distribution [4]. While the Weibull distribution is often used to describe the mechanical properties of ceramic materials, it is rarely used to describe metallic materials. One exception is its occasional use to describe high cycle fatigue data of alloys. Most often, mechanical properties data for metals and alloys are fit to a normal, or Gaussian, distribution, and subsequently represented by a mean and standard deviation.

The treatment of mechanical properties of P/M materials is no exception to the practice of using normal statistics to describe their distribution. German points out that the transverse rupture strengths of P/M materials typically show scatter that are assumed to be caused by flaws and a Weibull distribution can be used to model the variation [5]. Dieter makes a stronger statement and says [6], "While the Gaussian or normal frequency distribution is often taken as the accepted statistical distribution for failure strengths there is no theoretical or experimental justification for this situation. In fact, since the normal frequency distribution has long tails on each side of the mean it can be argued that this is unrealistic and does not represent the observed facts. Therefore, the most generally applicable frequency distribution in fracture problems is the Weibull distribution." If the Weibull is the only distribution that should be applied to fracture properties, the question must be raised as to why the normal distribution is so widely used instead? The best answer available seems to be that it is simply easier to use the normal distribution, although it is not accurate to do so.

The three-parameter Weibull distribution is represented in the following equation for the cumulative probability:

$$P(x) = 1 - \exp\left[-\left(\frac{x - x_0}{\Theta - x_0}\right)^m\right] \tag{1}$$

In this equation, x is the mechanical property of interest, Θ is the scale parameter, m is the shape parameter, also called the Weibull modulus, and $x_0$is the minimum allowable value of x, or the threshold value of x. For flaw limited materials, such as ceramics, the threshold value, x_0, is 0 and the Weibull can be described by just two parameters. Metals typically have a threshold value for most mechanical properties due to the ability of metals to plastically deform. Many P/M materials are low in ductility and can behave either way.

If there is no threshold value, then x_0=0, and the distribution simplifies to the two-parameter Weibull distribution:

$$P(x) = 1 - \exp\left[-\left(\frac{x}{\Theta}\right)^m\right] \tag{2}$$

A material whose properties fit this simpler distribution would have a finite probability of failing at any property level, although it may be small. Materials which have a threshold value are more suited to critical applications where unexpected failures cannot be tolerated.

REFERENCES

1. Barton, K.; Das, S.; Lu, J.; Goldenberg, M; Olsen, D.; LaSalle, J. *Advances in Powder Metallurgy and Particulate Materials—2001*; Metals Powder Industries Federation: 2001, 4-159–4-165.
2. *Materials Standards for P/M Structural Parts*, 2000 Ed.; Metals Powder Industries Federation: 2000; 2–3.
3. *Materials Standards for Metal Injection Molded Parts*, 2000 Ed.; Metals Powder Industries Federation: 2000; 2–9.
4. Weibull, W. J. Appl. Mech. 1951, *18*, 293–297.
5. German, R.M. *Powder Metallurgy Science*, 2nd Ed.; Metal Powder Industries Federation: 1994; 378.
6. Dieter, G.E. *Mechanical Metallurgy*, 3rd Ed.; McGraw-Hill: 1986; 370.
7. Deharmo et al. *Materials and Processes in Manufacturing*, 9th Ed.; John Wiley, 2003.
8. *Powder Metallurgy Design Solution*, Metal Powder Industries Federation, 1993.
9. Deharmo et al. *Materials and Processes in Manufacturing*, 9th Ed.; John Wiley, 2003.

15

Designing with Metal-Matrix Composites

Veikko K. Lindroos, J. T. Hellman, D. Lou, R. Nowak, E. Pagounis, X. W. Liu, and I. M. Penttinen
Helsinki University of Technology, Espoo, Finland

I. INTRODUCTION

The idea of the composite material concept is that two or more different materials and sets of properties are combined in the composite and the resulting new material possesses a unique combination of properties, which is not available in conventional engineering materials. In a metal-matrix composite (MMC), the continuous phase—matrix—is a monolithic alloy and the separate discontinuous phase (or phases)—reinforcement—consists of metallic, carbon or ceramic additions (Fig. 1). The motives for design of MMCs most often arise from such objectives as enhancement of material stiffness, specific stiffness, strength, creep resistance, or wear resistance, which are common objectives for traditional engineering alloy design as well. Consequently, analogies with other approaches of alloy design, such as precipitation hardening, dispersion hardening, and dual-phase steels, are obvious.

Since MMCs consist of at least two different components—most often metallic matrix and ceramic reinforcement—the chemical and physical compatibility of the components is of major importance in the MMC design (Fig. 1). The compatibility of matrix and reinforcement determines, e.g., how well the components are bonded together, do any chemical interactions take place at the interface, which interphases, if any, are formed and what are their properties. Furthermore, it governs the stress-state of the composite during and after manufacturing process and in later thermal cycling and how the microstructure, corrosion resistance and mechanical failure mechanisms of the matrix are affected by the incorporation of the reinforcement. It is obvious that the differences in the properties of the components of the composites enable revolutionary innovations in tailoring new materials, but at the same time, they are problematic in terms of compatibility and consequently material processing as well as operation in the conditions of practical applications.

The history of modern composite material development stretches back to the first few decades of the 20th century. Those first attempts to consolidate composites from aluminum and alumina powders were followed by an incubation period until the late 1940s and 1950s, when the needs of military and civil aviation technology and the starting space programs initiated massive human and capital investments into the development of new materials, such as light alloys and light alloy matrix composites. The development of fiber-reinforced composites continued to flourish during the 1960s but gradually the

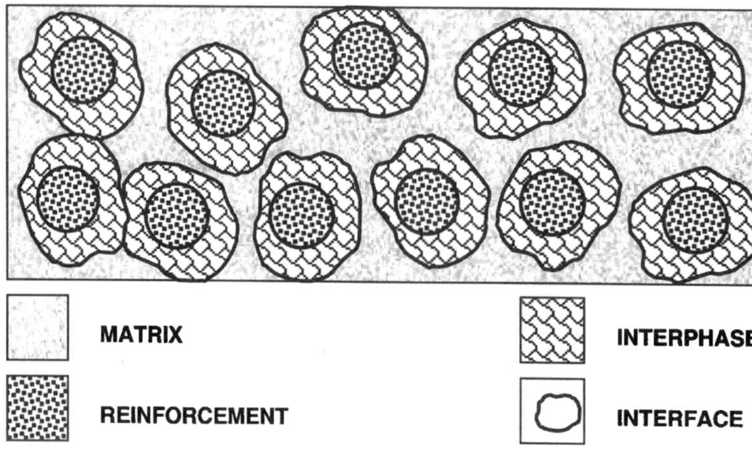

Figure 1 The nomenclature in the structural analysis of metal-matrix composites.

focus was shifted to discontinuously reinforced composites. This widened the possible application areas of composites and significant engineering innovations, which exploit the improved material properties. One important step in the recent (from 1980s to the present) evolution of MMCs has been the adoption of new matrix materials, i.e., steels, superalloys, and intermetallic alloys which again extends the variety of engineering applications of MMCs into wear, corrosion, and high temperature applications. The improvements in the material manufacturing technologies (such as modern powder metallurgy) have enabled the property optimization and commercial utilization of these new materials. What is more, emergence of powerful computer technology and the successive development of modeling tools has enhanced the design and optimization of material properties of the MMCs.

The above-presented brief insight into the history of MMCs indicates that the focus of MMC development has been in light alloy matrix composites, which indeed have been widely discussed in the current existing literature. Against this background, the intention of the following discussion is to broaden the public awareness of most recently developed MMC groups by concentrating on the introduction of steel matrix composites and their powder metallurgical (P/M) manufacturing technologies and applications. It is believed that the approach will open new views in the work of material designers and engineers.

A. Types of Metal-Matrix Composites

The classification of metal-matrix composites is made according to reinforcement type or matrix alloy. There are continuous and discontinuous reinforcements, and the latter ones may be further divided into several categories such as particulates, whiskers, staple fibers, flakes, and platelets as illustrated in Fig. 2. The most common classes according to matrix material are light alloy matrix composites (Al, Ti, and Mg alloys), copper matrix composites, steel matrix composites, superalloy matrix composites, and intermetallic matrix composites.

The most important group of MMCs, especially if production volume or market value is considered, is the aluminum matrix composites. The matrix alloys of Al-based MMCs are most often from alloy series 2xxx and 6xxx, while series 5xxx, 7xxx, and 8xxx are less common. The light weight, reasonable mechanical properties and environmental

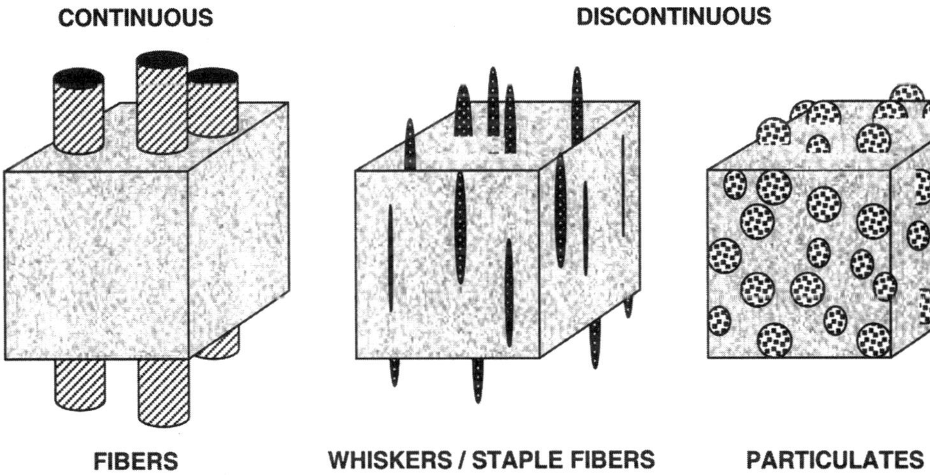

Figure 2 Continuous and discontinuous reinforcement of metal matrix composites.

resistance explain the applicability of Al alloys as matrix alloys. Suitable reinforcements remarkably improve the mechanical and wear properties of Al alloys but, as discussed later in this chapter, the corrosion resistance may be decreased and the interaction of matrix and reinforcement affects the aging behavior of the matrix alloys, which has to be taken into consideration in the heat treatment and elevated temperature operation of the Al-based MMCs. The reinforcements used are fibers, whiskers, particulates, or platelets. Silicon carbide (particulates, whiskers, or fibers/staple fibers) is the dominating reinforcement material in Al-based MMCs, although graphite and boron fibers and Al_2O_3 (fibers, short fibers, or particulates) are widely used as well.

Magnesium matrix composites offer essentially same kind of properties as Al-based MMCs, i.e. light weight, high stiffness, and low coefficient of thermal expansion. Pure Mg or Mg–Al alloys with Mn and Zn addition are the most common matrix alloys and the selection of fiber reinforcement turns to graphite (fibers), Al_2O_3 (staple fibers) or discontinuous B_4C or SiC in most of the cases. Actually, the reinforcements and respective manufacturing methods are more or less the same as in case of Al-based MMCs, and consequently, the selection between Al-based MMCs and Mg-based MMCs is usually made according to weight and/or corrosion resistance requirements.

High specific strength even at moderately elevated temperatures and excellent corrosion resistance are the reasons for the use of titanium alloys as MMC matrix alloys. In practice, the selection of matrix turns to Ti alloys if the performance of Al alloys, especially at elevated temperatures, is not sufficient for the applications. The matrices of Ti-based MMCs are typically $\alpha+\beta$-alloys (e.g. Ti–6Al–4V) or metastable β-alloys and the reinforcement, usually in fiber form, is most often either B_4C/B, SiC, SCS-6 or BORSIC.

If the MMCs are designed for applications requiring high thermal conductivity and moderate high temperature strength, copper becomes an interesting matrix candidate. Elevated temperature strength of copper is effectively increased by tungsten fiber reinforcement as well as by graphite fibers, which, in addition to that, enable reduction of weight and increase of conductivity, and provide the possibility for tailoring the coefficient of thermal expansion (CTE), e.g. in materials for advanced aircraft structures or satellite radiator panels.

Steels are available as thousands of different grades, some of which are inexpensive low alloyed general-purpose materials while others may have high strength, hardness, corrosion resistance, or some other special properties for special purposes. However, by ceramic particulate addition, which is the most common way to design steel matrix composites, the hardness and wear resistance of steels can be significantly improved. Although the reinforcement additions decrease ductility and corrosion resistance, it is possible to achieve a very attractive overall performance of MMCs and replace more expensive, less practical (manufacturing aspects) and less corrosion resistant materials such as cemented carbides. Provided the corrosion resistance, heat resistance, or strength of the steel matrix composites is below the desired level, the steel matrix may be replaced with nickel or cobalt-based superalloy. Both in steel and superalloy matrix composites, the reinforcement particulates are typically carbides such as Cr_3C_2, TiC, TaC, SiC, VC, WC or NbC, oxides such as Al_2O_3, Zr_2O_3, Y_2O_3, nitrides such as TiN or borides such as TiB_2.

Intermetallic compounds, such as aluminides of nickel, titanium, and iron, do not have such fixed compositional short-range order as alloys, but a long-range order of the arrangement of atoms within the lattice and a more limited number of slip systems available for plastic deformation. They have relatively low densities and high melting points, which increase their potential as matrix alloys of MMCs. The optimization of chemical compatibility and CTE mismatch are the key issues in the selection of suitable reinforcement for each intermetallic compound. In practice, Al_2O_3, SiC, and refractory metal fibers and TiC or TiB_2 particulates appear to be applicable as reinforcements.

B. Particulate-Reinforced Powder Metallurgical Composites

The main objectives of the primary processing stages of MMCs comprise introduction of reinforcement into matrix in the desired amount and location as well as proper bonding of the constituents without excessive, possibly deleterious, reactions. In addition to controlled bonding, densification and microstructure of the constituents in the MMCs, the use of modern P/M technologies, such as inert gas atomization in powder production and hot isostatic pressing (HIP) in powder densification, offers great freedom in tailoring the composition and properties of the MMCs. For instance, the matrix alloy may be selected according to the requirements for mechanical properties, corrosion resistance, and thermal conductivity, etc., in each application and the ceramic reinforcement according to specifications for hardness, toughness, and wear resistance. Additionally, it is possible to create gradients of physical and thermal properties into components by gradual variation of the relative proportion of the reinforcement in the matrix. The P/M MMC materials are fully dense and the possibilities for controlling the distribution of the ceramic particulates are far better than, e.g. in all spray coating and casting technologies. Such a defect-free microstructure of material and the uniformity of material properties over the whole component are essential in critical process components with high reliability requirements. What is more, the P/M manufacturing processes allow near-net-shape (NNS) production of the components, which significantly reduces the machining costs and delivery times of the products.

Liquid or semi-solid processing techniques are commonly used in the production of Al-based MMCs and Mg-based MMCs, but especially in case of discontinuous SiC and Al_2O_3 reinforcements and continuous graphite fiber reinforcement, the use of P/M technologies increases because of better process control and resulting material properties. The high reactivity of Ti-alloys with many common reinforcement materials has hampered the development of Ti-based MMCs, and besides rapid low-temperature roll bonding, only

P/M methods, such as HIP and vacuum pressing, have proven their applicability in the manufacturing of Ti-based MMCs. Because of the limited machinability of the Ti-based MMCs, the possibilities of NNS technology have inevitable economical importance in their case as well.

Powder metallurgically manufactured ceramic particulate-reinforced MMCs with steel, superalloy, or intermetallic matrix have excellent performance potential for wear and corrosion resistant parts of the chemical, petrochemical, process, metallurgical, and offshore industry as well as mining and rock engineering, energy conversion, and engine technology. Their potential is based on excellent abrasion and adhesion wear resistance, which is superior to alloy steels, tool steels, high-speed steels and superalloys including stellites and in some cases WC cermets. Owing to their better fracture properties and more flexible manufacturing technologies, these MMCs tend to replace the conventional materials, such as cemented tungsten carbides (e.g. WC/Co), in applications requiring exceptional wear resistance. Certain sintered hard metals and sprayed coatings have almost equal wear properties, but the former ones are applicable only as small components or joined claddings and the latter ones suffer from thickness and size limitations, porosity, and poor adhesion to substrate. What is more, by means of optimal matrix metal alloying and ceramic particulate selection, the oxidation resistance and the tribological properties of the P/M MMCs are clearly better than those of conventional engineering materials.

C. Technical and Economical Applicability of Metal-Matrix Composites

Light alloy matrix composites are ideal candidates for moving, especially rotating, components of machines and equipment in which the high specific strength and stiffness at moderately elevated temperatures, possibly combined with fatigue resistance, dimensional stability and/or tribological properties, determine the material selection. In these applications, the relatively high unit price of MMCs is more or less compensated by the resulting savings in material consumption and reduced costs of operation and maintenance. On the other hand, superior performance and reliability are sometimes necessary independently from the costs. In aerospace structures and rocket engine components, e.g. the material costs are secondary issues, but in the mass production of every-day consumer goods, they usually have decisive role. Such examples as the Toyota diesel engine piston selectively reinforced with an aluminosilicate ceramic compact or certain Mg-based MMCs for electronic packaging are good examples of technically successful and cost-efficient application of MMCs in mass products.

Both from technical and economical points of view, the most appealing way to use wear and corrosion resistant steel and superalloy MMCs is to construct MMC-cladded compound structures, such as those presented in the introduction of industrial applications later in this chapter. In such cases, the MMC-cladding provides the desired surface properties of the component while the metallic part reduces the total cost of the component, improves the toughness of the structure and enables the joining of the component by means of conventional methods such as welding. Compound structures may be the only solution, for instance, when high dimensional stability and strength in wide temperature region must be combined with excellent wear resistance in one single component. The P/M processes, which best suit the production of MMCs, are the most applicable techniques for constructing compound structures as well. Due to the CTE mismatch between different metal alloys and especially metals and ceramics, some crack initiation promoting stress concentrations may appear at the interface of the multimaterial joint areas especially during thermal cycling. Even these problems may be solved by means of P/M gradient structures,

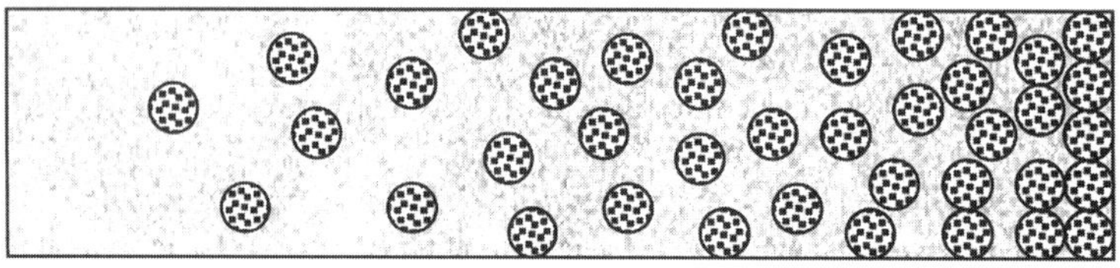

Figure 3 The reinforcement concentration (dark phase) increases gradually from left to right in the schematic illustration of the cross-section of a functionally graded metal matrix composite (FG MMC).

i.e. functionally graded (FG) materials. In the P/M FG materials, the powders are mixed in such a way that instead of sharp borderline between the two different materials a gradually changing composition of powder mixture is formed at the joint area (Fig. 3).

II. MATRIX ALLOY DESIGN

The great advantage of MMC materials is a superior combination of different properties of metallic matrix and hard reinforcement, but these properties can only be attained with the proper selection of matrix alloy, reinforcement, fabrication method, and the adjustment of matrix microstructure in the secondary processing.

A. Design and Selection of Matrix Alloy

The matrix of the composites has to be designed for two tasks: firstly, to bind and support the reinforcing phase [1] and, secondly, to satisfy special properties based on the requirements in service [2–5]. Binding strength between the reinforcement and the matrix, which is supposed to transfer and distribute the load to the reinforcement, depends on the type of matrix and reinforcement and the fabrication technique [5]. The matrix hardness is a key factor in supporting the reinforcing phase [1]. In addition, other considerations, such as cost, weight, fabricability, and availability of the matrix, should be made [6].

For design of non-ferrous alloy (e.g. Al, Ti, Mg) matrix composite, useful information about the selection of matrix is widely available [5,7,8], while in case of ferrous matrix composites, information about the selection of matrix alloys is very limited. Since the objectives of design of iron and steel matrix composites most often include extreme wear resistance and a combination of different specific properties, such as resistance to corrosion and oxidation [1,9–12], the design parameters deviate from those of light alloy matrix composites to certain extent. Against this background, this part is aimed to provide information, which serves the design of steel and superalloy matrices of particulate-reinforced composites for some of the most common service conditions.

1. Aqueous Corrosive Plus Wear Environments

Metal-matrix composites generally consist of a non-metallic reinforcement and a metallic matrix. The corrosion of non-ferrous alloy, such as aluminum-, copper-, magnesium-, titanium-based MMCs has been extensively studied and summarized in Refs. [6,13,14]. As the research results showed that pitting corrosion is the major mechanism of this kind of composites in seawater [6,15], coating is believed to be the best way to protect them [15]. In ferrous alloy matrix composites, however, coating or self-protective film seems less effective because they are usually under a combination environment of corrosive and wear. Very limited research has been conducted on corrosion of super-alloy and ferrous alloy matrix composites [9], but it is quite reasonable to conclude that electrochemical reaction such as pitting corrosion will also be a key problem for these composites in aqueous environment. This is because of the potential differences between reinforcement and matrix. For instance, if reinforcement has higher potential than matrix alloy, the matrix will be corroded in the electrochemical cell system where the matrix alloy acts as anode and the reinforcement as cathode. In order to reduce the reaction rate, the potential difference (ΔE_{therm}) between the reinforcement and matrix should be as low as possible, as shown in the following formula:

$$\Delta E_{\text{therm}} = (\text{Ee})\text{c} - (\text{Ee})\text{a} \tag{1}$$

where (Ee)c and (Ee)a are equilibrium potential of reinforcement (as cathode) and alloy matrix (as anode), respectively.

In other words, for corrosive plus wear environments, the design and selection of matrix alloys are based on: (i) their common mechanical and chemical properties which was well documented [6,16], and (ii) with less potential difference with its reinforcement. Austenitic (e.g. 316L) and ferritic stainless steel (SS), maraging steel and martensitic SS were usually used as the matrices under corrosion environments [1,11,12,17–19], such as corrosion–erosion in wet steam flow or gases owing to the good wear and corrosion properties of the stainless steel matrix composites.

2. Oxidizing Plus Wear Conditions

In many industrial cases, the need for wear resistant components which are capable of operating at high temperature (e.g., exceeding 500–600°C) without severe oxidizing has led to the development of super alloy matrix composites, such as those with 71%Ni, 18%Cr, 8%Fe, and Ti+Al superalloy matrix [2]. The alloying controls matrix oxidation at elevated temperatures and the excellent wear resistance is based on the addition of hard and stable reinforcements, which offer 20–30 HRC higher hardness than the commercial superalloys with similar chemical properties. Various superalloy matrix MMCs having matrices, such as Nimonic 81, Inconel 671, and reinforcements, such as Al_2O_3, TiC, or Cr_3C_2 have been designed to provide erosion and corrosion protection in oxidizing environments at 800–850°C.

3. Severe Wear Conditions

Among wear mechanisms, abrasive wear and adhesive wear (either alone or in combination) are the most destructive wear mechanism, which account for the majority of industrial wear problems [20,21]. The present materials used as a protection against abrasion can be divided into four main categories: heat-treated steels, austenitic manganese steels, high chromium cast irons, and cemented carbides. Among them, cemented carbides have the highest abrasion resistance, but, they are too brittle to withstand impact. On the contrary, heat-treated steels such as tool steels have the highest toughness and the lowest abrasion resistance. New materials falling in the gap between these two extreme cases would be the most promising as abrasion resistant materials [21].

For this purpose, over the last decade, attempts to produce ferrous matrix composites with different carbides have intensified [2,6,22–24] because of their lower costs, possibility of heat treatment, good machinability (in the annealed condition), and weldability [24]. The matrix alloy selection for these composites is focused on cold and hot work tool steels (such as PM10V, H11, and H13), high-speed steels (M3/2, M2, T1, T2, T15, T42) or white cast iron [9,10,25–29], one example being the tool steel matrices in the family of ferro-TiC composites [4]. A variety of MMCs with specific properties can be obtained by incorporating the different types and volume fractions of carbides or ceramic particles into various matrix alloys. Wear resistance of such composites is three to ten times higher than those of unreinforced alloys, which makes them ideal candidates for tooling (dies, rolls, cutting tools) and wear application in process (metallurgical) industries.

B. Selection of Size and Volume Fraction of Matrix Powder

The selection of size and volume fraction of matrix powder in relation to the size and volume fraction of reinforcement is very important for the final microstructure of a composite. A homogeneous and fine dispersion of hard particles (HP) in the ductile metal matrix (MM) will be inevitably beneficial for both fracture strength and toughness of the composite and contacts of HPs as in a duplex or net-like distribution lower the fracture strength and toughness [30]. Although there is no mathematical solution to this topological problem, Fig. 4 may give some support to the design and selection of powder size and volume fraction of matrix and reinforcement. In powder metallurgically manufactured composites, the toughness has been found to be strongly dependent on the ratio of ceramic powder size to matrix powder size, as long as the ratio increases the fracture toughness increases [7]. Microstructural observations reveal that densification occurs mainly to the

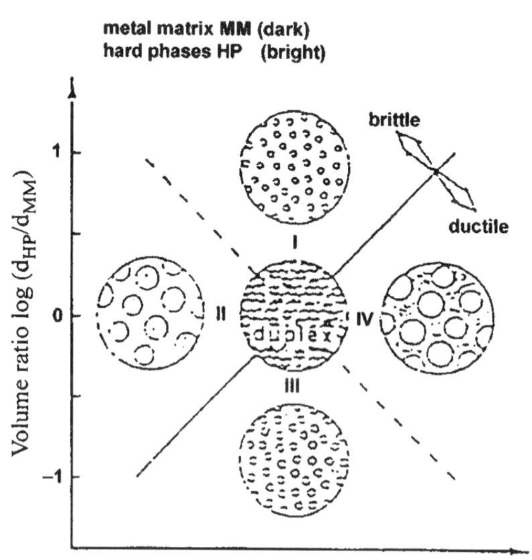

Figure 4 The ratio of particle sizes and volume fractions of matrix and reinforcement determine the microstructure of MMC after consolidation. Such dispersion of HP in MM as in Quadrants III and IV result in ductile material.

smaller matrix particles, while the coarser remain almost unchanged [24]. Nevertheless, the optimal design of matrix powder size and reinforcement volume fraction is, of course, very much based on the requirements for materials in actual applications [30]. The experiments have, e.g. showed that the net-like distribution of HP in MM may result in better abrasion resistance than the duplex distribution of HP regardless of the lower transverse rupture strength and impact toughness.

Interparticle spacing is an important factor in the design of wear resistant composite materials. Usually, it is beneficial for the abrasion resistance to have small interparticle distance, which leaves less matrix exposed to wear phenomena. Assuming that a spherical reinforcement (diameter d_r) is ideally distributed in the matrix, the inter-particle spacing λ and the volume fraction V_f can be given as [31]

$$\lambda = \left[\left(\frac{\pi}{6V_f}\right)^{1/3} - 1\right] d_r \tag{2}$$

In case secondary processing of the composite includes forming processes, such as extrusion, the uniform distribution of reinforcement can be obtained provided the inter-particle spacing λ is greater than the final transverse length of matrix powder d_f, i.e.

$$\lambda \geq d_f \tag{3}$$

In fact, d_f can be obtained by measuring the matrix powder boundaries on the cross-section of the as-processed composite.

C. Development of Ferrous Alloy Matrix Composites

A variety of iron and steel matrix composites have been developed to offer wear resistance combined with certain special properties, such as maximum operating temperature, hardness, toughness, strength, thermal shock, corrosion and oxidation resistance, just to mention a few. Some of these commercial and experimental materials are introduced below.

Since 1960s, P/M technology has been applied for production of composite materials for high performance tooling and engineering applications. The early P/M composites, e.g. ferro-TiC materials, have a combination of good wear, heat and/or corrosion resistance as well as additional properties, such as high elastic modulus and good edge retention [2]. Some properties of these early MMCs are presented in Table 1.

The more recently developed (late 1990s) ferrous matrix composites extend the variety of performance profiles and respective applications more into the direction of corrosion exposing wear applications, which is reflected in the matrix alloy selections. The corrosion resistance of the superaustenitic stainless steels and superduplex stainless steels is very good even in the most corroding operating conditions of pulp and paper as well as chemical industry. Furthermore, some of these second generation MMCs can be used in as-HIPed condition without secondary processing steps, such as heat treatment, which is very beneficial for the cost formation and duration of the production. Some most recently commercialized steel matrix composites are introduced in Table 2 and Fig. 5; furthermore, their abrasion wear resistance compared with some other engineering materials is presented in Fig. 6

The composite materials introduced above have already proven their potential in actual applications. Additionally, there are a large number of experimental or

Table 1 Selected Properties of Ferro-TiC® Composites [2,6]

Grade	Matrix type	TiC content, vol%	Hardness, HRC	Temperature limit, °C
C	Medium alloy steel	45	44–70	190
CM	High Cr tool steel	45	48–69	525
CH-25	H13 hot work tool steel	25	32–66	540
CHW-45	H13 hot work tool steel	45	45–64	540
CHW-25	H13 hot work tool steel	25	30–61	540
SK	Impact resistant tool steel	35	30–62	540
CS-40	Martensitic stainless steel	45	50–68	370
PK	Maraging steel	42	50–62	450
MS-5A	Age hardenable martensitic SS	41	48–61	450
HT-6A	Age hardenable Ni-base alloy	40	46–52	980
HT-2A	Age hardened Ni–Fe alloy	40	46–53	760

Table 2 Selected Properties of Ralloy® Composites [32,33]

Grade	Matrix type	Reinforcement	Hardness, HRC	Temperature limit, °C
AMMC 30	Austenitic stainless steel	Al_2O_3, 30 vol %	35	700
DMMC 20	Super-duplex stainless steel	Cr_3C_2, 20 vol %	38	700
DMMC 30	Super-duplex stainless steel	Cr_3C_2, 30 vol %	50	700
MANU 30	Hot work tool steel	Cr_3C_2, 30 vol %	66–69	500

semi-commercial materials, which may appear in the market during the next few years. They include such materials as tool steels reinforced with WC, NbC, TaC or Al_2O_3 [25–29], superaustenitic stainless steels reinforced with TiC or TiN [34] and nickel-based superalloys reinforced with Al_2O_3, TiC or Cr_3C_2

D. Development of Matrix Microstructure

The studies on the influence of reinforcement on the matrix microstructural development during fabrication, heat treatment, and deformation are numerous for light alloy matrix composites, while in the case of ferrous and superalloy matrix composites, many aspects have not been thoroughly investigated, and, therefore, only limited information is available.

1. Internal Stresses in Matrix

Internal stresses are generated in MMCs due to different coefficients of thermal expansion (CTE) and elastic properties between the ceramic particles and the metal matrix. In the calculation of internal stress fields in MMCs, such methods as Eshelby's inclusion method modified for the case of finite volume fraction of reinforcement, are used. Provided the particles are close to a spherical shape and assuming no plastic relaxation of stresses, the residual stress σ_p in the particle during cooling from the high temperature can be presented by the following equation [24]:

$$\sigma_p = \frac{-(\alpha_m - \alpha_p)}{(0.5(1+\nu_m) + (1-2\nu_m)f)/E_m(1-f) + (1-2\nu_p)/E_p} \Delta T \quad (4)$$

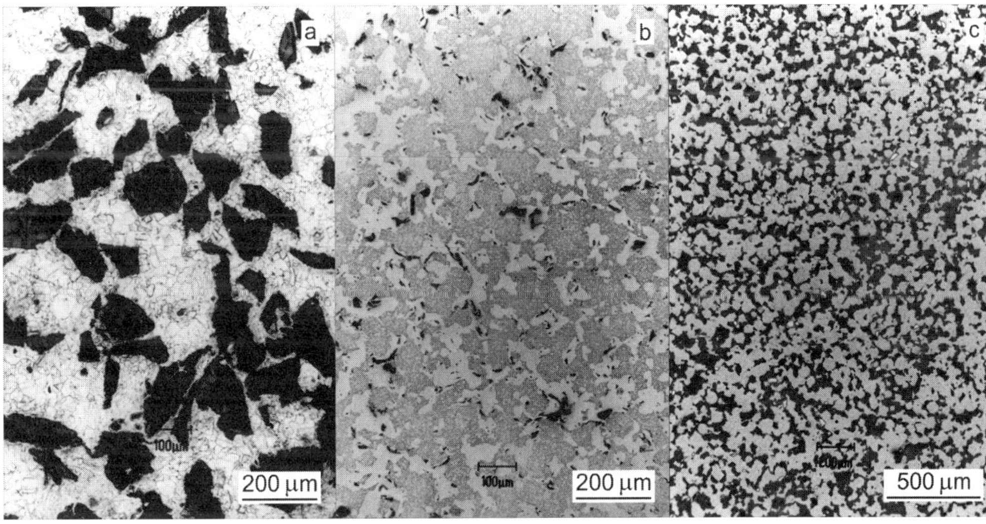

Figure 5 The microstructures of recently developed steel matrix composites: (a) AISI 316+30 vol% Al_2O_3, (b) Duplok 27+20 vol% Cr_3C_2, and (c) H11+30 vol% Cr_3C_2.

The stress-state in the metal matrix is obtained according to the following formulae: Radial:

$$\sigma_{\rm m}^{r} = \frac{\sigma_{\rm p}}{1-f}\left[\frac{r^3}{R^3} - f\right] \tag{5}$$

Tangential (θ):

$$\sigma_{\rm m}^{\theta} = -\frac{\sigma_{\rm p}}{1-f}\left[\frac{r^3}{2R^3} + f\right] \tag{6}$$

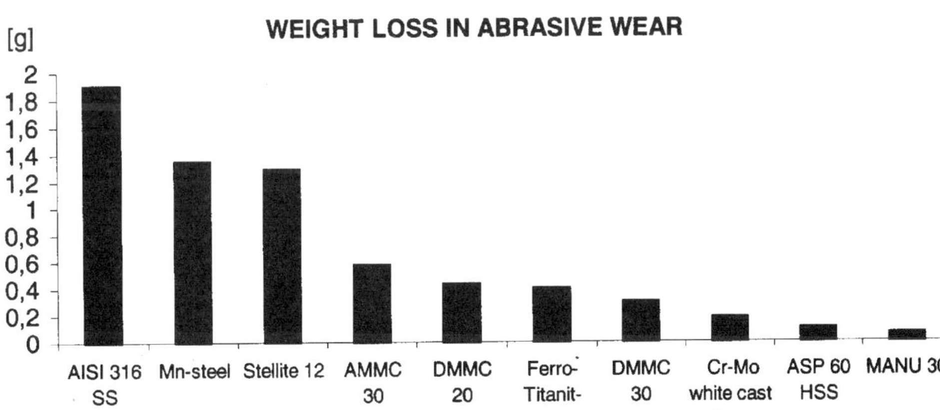

Figure 6 The abrasion wear resistance of second generation MMCs compared with some other engineering materials [33].

In the equations above, α is the CTE, ν Poisson's ratio, E Young's modulus, f reinforcement volume fraction and ΔT is the temperature change during quenching. The subscripts m and p refer to the matrix and reinforcement, respectively, r the particle radius and R the distance from the center of the particle ($R \geq r$).

This model assumes a purely elastic behavior for the ceramic and the metal; however, provided the yielding criterion ($|\sigma_m{}^r - \sigma_m{}^\theta| > \sigma_y$; σ_y = yield strength of the matrix) is satisfied, plastic deformation of the matrix will occur, which results in stress relaxation [35].

Additionally, in the case where $R = r$, i.e. at the interface,

$$\sigma^r_{\mathrm{m}} = -\sigma_{\mathrm{p}} \tag{7(a)}$$

the stress in the matrix reaches maximum, whereas in cases where $R \gg r$, the stress reaches the minimum:

$$\sigma^r_{\mathrm{m}} = \frac{-\sigma_{\mathrm{p}} f}{1-f} \tag{7(b)}$$

The tensile stress in the matrix and compressive stress in the particle are hydrostatic in nature, in the cases where the particles are relatively uniformly distributed [24]. The inherent stresses may cause decrease in strength values, such as transverse rupture strength and toughness, as well as affect the martensitic phase transformation during heat treatment, as discussed in more detail later.

2. Microstructural Development in Sintering and Hot Isostatic Pressing

The P/M techniques, such as sintering and hot isostatic pressing, are successfully used to produce metal-matrix composites and other special property materials, which contain a high volume fraction of carbides or ceramic particles for improved wear resistance [36]. However, the interaction between the reinforcement and matrix makes the processes more complicated than conventional P/M processing. Some reinforcements interact with matrix powder and change the chemical composition of matrix (Table 3) and consequently modify the composition of the natural carbides M_6C and MC during sintering or HIPing [37]. What is more, the reinforcement also inhibits the grain growth of matrix at high processing and operating temperatures. However, the desired microstructure of the MMC is a result of careful adjustment of processing parameters (i.e. thermodynamical and kinetic prerequisites for interaction between matrix and reinforcement) in sintering in vacuum or HIPing.

3. Development of Microstructures by Heat Treatment

Additions of carbides and other particles into ferrous alloys matrices have shown a considerable influence on the matrix microstructure during heat treatment [25]. This is due to compositional changes in matrices of composites as a result of dissolution of some reinforcement carbides during heat treatment procedures. An example of this is seen in Table 3. The grain sizes of the composites containing different reinforcement are plotted as a function of austenitizing temperature in Fig. 7, which suggests that the reinforcement could retard grain growth by boundary pinning when austenitizing temperature exceeds a certain value. For instance, in the case of TiC-reinforced HSS matrix composite, the M_6C carbides were replaced by the less soluble MC carbides in matrix microstructure as result of reaction between the TiC and the steel [25].

Table 3 Matrix Compositions for Composite in as Sintered and Quenched States Determined by EDS

Material	Condition	Elements, wt.% (Fe bal.)						
		Cr	Mo	Nb	Ti	V	W	$\sum$(Mo + W)
M3/2	As sintered	4.41	2.08			1.22	0.83	2.91
	Quenched (1140 °C)	4.50	1.86			0.85	0.57	2.43
	Quenched (1180 °C)	4.43	1.69			0.88	0.59	2.28
	Quenched (1200 °C)	4.60	2.33			1.16	0.90	3.23
	Quenched (1220 °C)	4.60	2.33			1.27	0.90	3.23
M3/2 +7.74% NbC	As sintered	4.0	1.0			1.0	1.0	2.0
	Quenched (1150 °C)	4.70	1.37	0.49		1.49	1.21	2.58
	Quenched (1180 °C)	4.59	1.36	0.26		1.12	1.29	2.65
	Quenched (1200 °C)	4.72	1.52	0.45		1.29	1.38	2.90
M3/2 + 5% TIC	As sintered	4.33	2.62		0.08	1.15	1.23	3.85
	Quenched (1100 °C)	4.51	2.74		0.14	0.85	1.0	3.74
	Quenched (1180 °C)	4.36	2.79		0.08	1.11	1.26	4.05
	Quenched (1200 °C)	4.69	2.89		0.12	0.95	0.95	3.84

Compared with unreinforced alloys, one of the most substantial effects of reinforcement on matrix phase transformation is the decrease in the amount of retained austenite (RA) in the composite when heat treated in the same way as the unreinforced matrix alloy [25,25]. This is attributed to the effect of internal tensile stresses of MMC on the martensitic transformation. The stresses raise the M_s temperature by decreasing the strain energy $\Delta G_i^{\gamma-\alpha}$ needed.

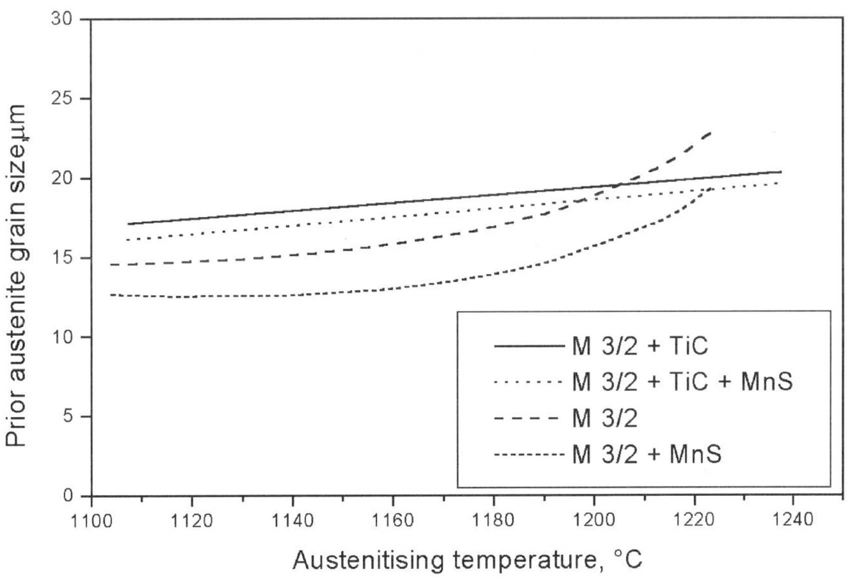

Figure 7 The effect of reinforcement on austenite grain size of composites as a function of austenitizing temperature.

Additionally, the tempering response of the composite, e.g., carbide precipitation and secondary hardening, differs from the one of unreinforced matrix because of different solubility of reinforcements.

4. Matrix Microstructure and Hardness Design

Matrix microstructure has, of course, a strong impact on the properties of the composites. For example, austenite and bainite are more abrasion resistant than ferrite, pearlite or martensite with the same hardness [38]. This is due to the higher strain-hardening capacity and ductility of the austenite. Furthermore, matrix with cubic crystal structure has been found to be more resistant to abrasive wear than hexagonal crystal structure because of higher work-hardening rate of cubic structure [38].

Considering both the mechanical and wear properties of MMCs, the matrix hardness is one of the key factors affecting it. Matrix should have sufficient hardness in order to be able to embed and support the reinforcing particles, as well as to protect the matrix area between reinforcement particles from wear. However, the optimal matrix hardness is usually determined according to the service environment or based on the reinforcement size used. For instance, in abrasive wear tests, some results exhibit that composites with fine reinforcement (e.g. 3 μm) benefit from high hardness of matrix, while the composites containing large reinforcement particles (e.g. 30 μm) benefit from lower hardness of matrix [1]. Similar conclusions can be made based on high-stress abrasion tests [5]. Most wear test results of MMCs demonstrate a strong linear relationship between wear resistance and hardness [1,5], which is also suggested in Fig. 4. However, poor correlation between hardness and wear loss has also been suggested in some reports [22]. The controversy in results led to the conclusion that many other factors influence the wear properties of composites, and matrix hardness value is only one rough criterion in the prediction of wear resistance of a composite.

III. INTERFACES IN METAL-MATRIX COMPOSITES AND SELECTION OF REINFORCEMENT

The present section introduces short review of interfacial phenomena in metal-matrix composites, which in a large part, are responsible for mechanical performance of the composites, since microstructural integrity of fiber matrix interfaces influences the mechanical behavior of MMCs. Such a topic involves inevitably conclusions concerning selection of the reinforcement, which would secure the desired properties of composite.

A. The Interfacial Phenomena in Metal-Matrix Composites

As it is mentioned in Section 1, the essence of the MMC-concept lies in the interfacial phenomena, which allow one to explore the benefits offered by the combination of elasto-plastic as well as fracture properties of normally very different materials. It is no doubt that the junction of the two components, the matrix and the applied reinforcement, would considerably affect the properties of the composite, despite the final outcome being difficult to predict exclusively from the interface characteristics. The ultimate assets of composite, and consequently its performance, would to a great degree depend on such factors as the size, distribution, volume percentage, and orientation of the reinforcement, as pointed out in several reviews of the subject (see e.g. the overview of structure/property relationship in

Ref. [39]). Although structural factors are of the primary concern when one designs a new composite, the significance of interfacial bond strength and related mechanical conduct has long been appreciated, namely, since the origin of bonding at metal–ceramic interfaces has been elucidated a decade ago [40]. This radical step in our knowledge came as a result of first principle calculations of bond structure, charge transfer, and point defects, coupled with high-resolution microscopic observations by means of either AFM or TEM, which support ab initio calculations.

Already 10 years ago, Clyne and Withers [39] indicated that the interfacial bond strength is of main interest for polymer-based composites, while it has a lesser impact on those ceramic embedded in metal matrix, since their fracture toughness is detrimental for their applications, i.e., their design require weaker interface. This contrasts with MMCs for which strong bond is highly desirable (e.g. [40]), since the presence of metal matrix usually assures high toughness characteristic for this kind of composites.

The earlier works which review properties and application of MMCs [41–43] normally address the interfacial effects in separate sections devoted to: (i) origin and estimation of interfacial stresses including evaluation of critical stress level, (ii) initial debonding process (interfacial fracture) and interfacial frictional sliding, (iii) thermodynamical stability of MMCs, interfacial reactions and their impact on mechanical behavior of composites, as well as (iv) the possible modification of interface by means of fiber coating with various film deposition techniques (PVD, e.g. [43]; CVD, e.g. [44]), wetting or even ion-beam modification [45]. The present authors, however, decided to refrain from similar arrangement of the subject. Instead, the topic is discussed in one body according to chronologically emerging works to grasp the changes in approach to the MMCs design which occur in past 10 years, and to extract the pertinent information which become available for design of these kinds of materials.

1. Early Works on Metal/Ceramic Interfaces

Considerable interest in the interfacial phenomena was expressed as early as in 1967, by Grant et al. [46], who indicated unsolved problems concerning cohesion, and structural stability in fiber- or dispersion-strengthened systems with metallic matrices. Furthermore, thorough studies of the structural stability of composites affected by changing in surface free energy, solubility, and diffusivity were carried out by Fischmeister et al. [47], who aimed to control the listed parameters by alloying additions. Their experiments and calculations for alumina particles embedded in nickel and iron matrix with minor alloying additions proved that these additions can strongly affect the work of adhesion, since they change the contact angle between alumina and liquid metal [47]. Further, the particle/matrix bonding strongly depends on interfacial energy, and in consequence, alloying effects. The solute additions appeared to be often detrimental, while sometimes beneficial for structural stability understood by Fischmeister et al. [47] as resistance to particle coarsening. Hence, the presented results clearly indicate that structural stability of a composite can be controlled by a proper design (alloying) of its matrix. This conclusion was confirmed by the further study of the same authors [48] which focused on particle-to-matrix bonding in austenitic and ferritic iron alloys, and which most prominent outcome is the proposed criterion for brittle fracture of the matrix:

$$\sigma_a > \frac{E}{30\left(1 + 3.36/\sqrt{\theta}\right)} \tag{8}$$

where σ_a denotes maximum applied tensile stress, E stands for Young's modulus of the matrix, while θ defines contact angle between the matrix and reinforcement (the measure of bonding strength W).

Furthermore, Easterling et al. [48] presented successful prediction of the strain to cavitation using Ashby's model based on the assumption that decohesion occurs when the back stress from the pile-up of the dislocation loops exceeds the bonding strength. This approach combined with energy criterion of Brown and Stobbs [49] allowed the authors to explain a sudden fall in observed rates of work hardening in terms of critical strain at the onset of cavitation, and finally to assess the relationship between adhesion on fracture mode.

The predictive methodology for interfacial properties was introduced by Hondros [50] who used surface analysis spectroscopy together with Gibbs Adsorption Theorem in order to measure interfacial energies for variety of solid/solid and solid/liquid interfaces. Hondros [50] argued that an accurate assessment of interfacial surface energies should be based on complex interactive contributions from, composition, crystalline orientation, and temperature. He systematized much of available data for surface energy of solid metals (Fig. 8), liquid metals (Fig. 9) and interfacial energy between metal and non-metal (Fig. 10), deriving from a variety of experimental techniques such as conventional calorimetric measurements, differential scanning calorimetry (DSC), particle-to-decohesion technique as well as their derivatives. The final conclusions of Hondros [50] are that in metal–ceramic systems, good adhesion is normally associated with low value of interfacial energy, while high work of adhesion reflects that good practical bonding had considerable impact on further research in this field. Moreover, his idea that the presence of active elements at the interface may stabilize the particles against coarsening, and

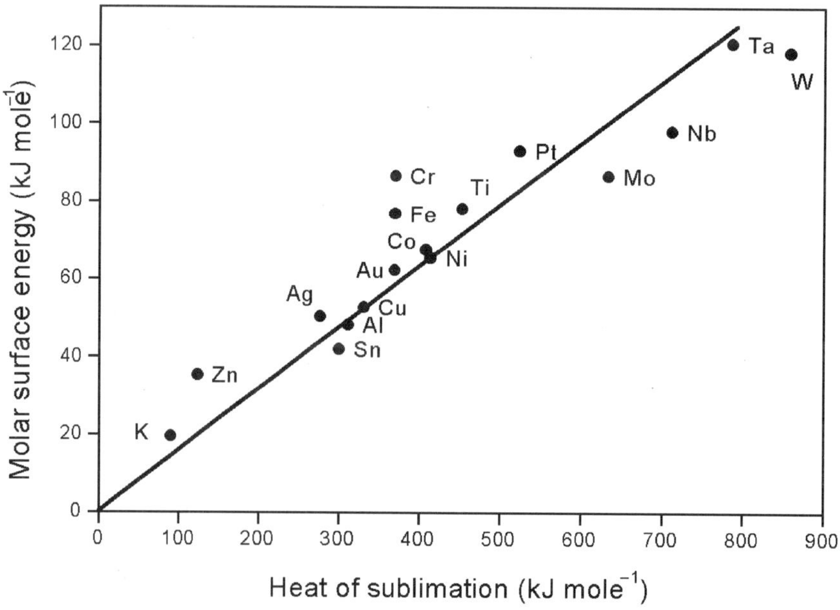

Figure 8 Relationship between the surface energy of solid metals at their melting point and the heat of sublimation. (From Ref. 50.)

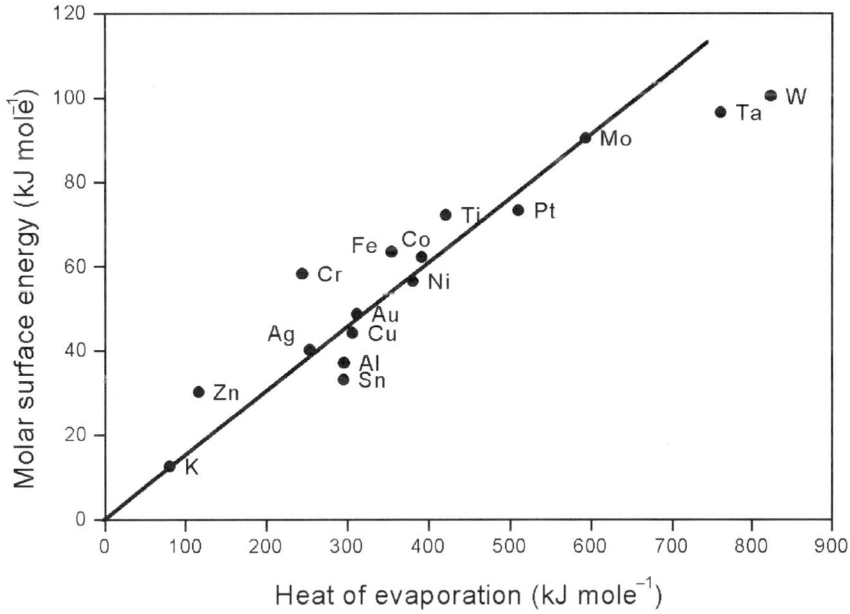

Figure 9 Relationship between the surface energy of liquid metals and the heat of evaporation. (From Ref. 50.)

simultaneously, reduce the interfacial energy, has had important implication on the design of dispersion strengthening alloys and composite materials.

In summary, the early works on metallic surfaces and interfaces introduced and categorized the basic concepts appear useful for metal-matrix composites. They indicated that

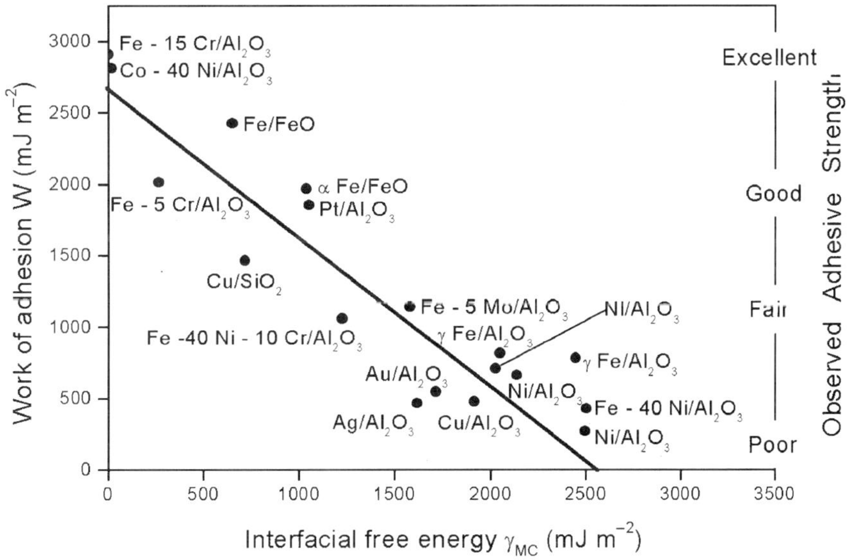

Figure 10 Correlation between the work of adhesion and the interfacial energy for metal/nonmetal junctions. (From Ref. 50.)

when ceramic surface is brought into contact at high temperature with metal matrix, bonds are established through either physical or chemical interactions (for review refer to Pagounis [51]). The former bonding occurs when the surfaces approach interatomic distance, which enables charge transfer through the interface, while the latter concerns chemical interactions, enabling formation of extended interdiffusion zones or new phases. It was also recognized that interface bonding in MMCs should be strong enough to transfer load from matrix to reinforcement.

Furthermore, it was recognized that a convenient description of the metal–ceramic physical bonding includes the work of adhesion W_a (details in Ref. [51]), and alloying conditions affect the contact angle, and consequently, the value of W_a being dependent on θ [47]:

$$W_a = \gamma_c + \gamma_m + \gamma_{mc} = \gamma_m(1 - \cos\theta) \tag{9}$$

where γ_c, γ_m and γ_{mc} define surface energies of ceramic, metal matrix and metal–ceramic interface, respectively. Indeed, based on Eq. (8) they derived, Easterling et al. [48] found that their MMC samples with W_a below 100 mJ/m^2 are susceptible to brittle fracture (see Fig. 11). It is worth noting that the indicating equation [Eq. (8)] constitutes the first

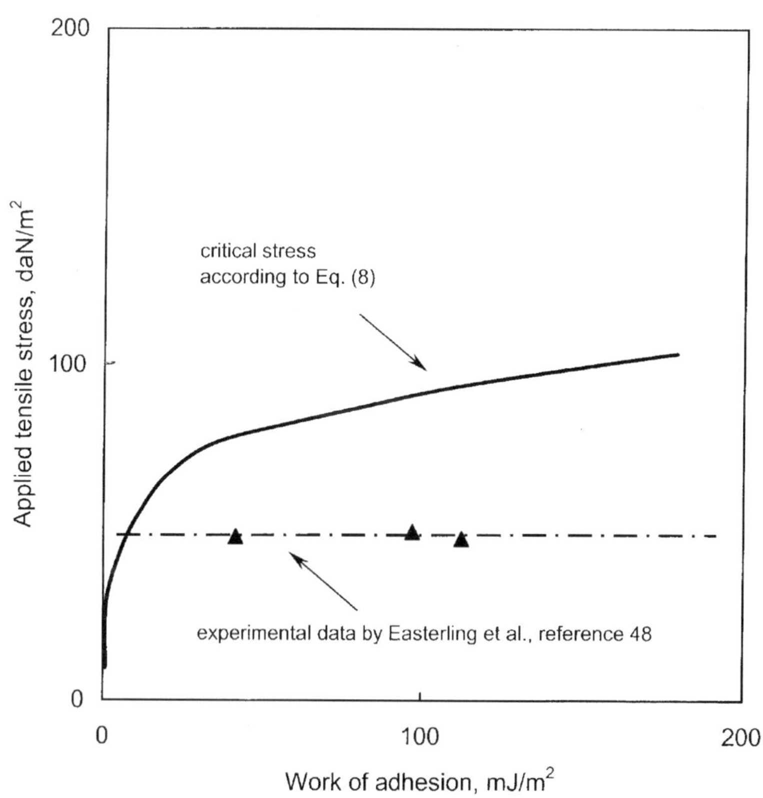

Figure 11 Applied tensile stress sa versus work of adhesion for Fe-Ni alloys with additions of Cr and Mn and the critical stress according to criterium proposed by Easterling, Fischmeister and Navara. (From Ref. 48.)

possible criterion concerning selection of the matrix and reinforcement in order to obtain specific properties of the composite.

Finally, Eq. (9) allows one to conclude on the level of the work of adhesion from the measurements of the contact angle θ between ceramic and metal. Hence, this measurement constitutes a crucial point of the reviewed pioneering research concerning metal–ceramic interfaces [47,48,50–52]. However, measurement of contact angle, and consequently work of adhesion, is not an easy task, since it requires complex procedure which includes:

- deformation of the prepared alloy with fine ceramic particles until initial cavitation appears,
- subsequent annealing of the specimen to achieve equilibrium conditions, and
- the measurement of the θ-angle in the TEM micrographs of the samples.

2. Dynamic Development of Research Concerning Metal/Ceramic Interface Phenomena

The results of early works on metal–ceramic interfaces stimulated an avalanche of research on interface phenomena, since already in early 1980s it became obvious that they would be essential for design of new MMCs with entirely new properties, and hence offering a variety of applications. A considerable number of studies targeted such topics as wetting [53–59], interface reactivity [15,60–70], problems with internal stress in the interface and associated debonding [35,71–84], formation of new phases in the interface region [89,90], as well as interfacial shear and fiber pushout (e.g., [91]).

Moreover, the structure of both matrix and interface was modeled on atomic level using extensive computer simulation [92,93]. The latter gave us a new insight into the interface phenomena tackled previously at the stage available for experimental examination. Moreover, while the results aimed originally for MMCs applications appeared to be of great use for deposition of metallic thin films on the ceramic substrates [94]. Nowadays, the knowledge gathered due to such an extensive research enables us to determine the suitable bonding conditions for MMCs materials (see Ref. [51]).

a. Wetting. The estimation of the value of contact angle θ (refer to precedent section) allows one to conclude on the wetting of ceramic by metal (refer to Fig. 12). Hence, good wetting occurs usually for small θ-angle, while high θ values, especially those exceeding 90° indicate poor wetting [51], which confirms that the contact angle can be regarded as a parameter characteristic for wetting ability—as it was used by Fischmeister et al. [47,48].

The techniques which allow one to obtain desired distribution of dispersed particles in cast alloy matrices by controlling solidification parameters and introducing melt particle wettability were reviewed as early as in 1986 by Rohatgi et al. [53]. The authors indicated the methods, which are particularly useful to promote wetting between ceramics, and metallic melts, including metal coatings on refractory particles, addition of reactive elements to the melt, heat treatment of particles and ultrasonic irradiation of the melt. Furthermore, they thoroughly discussed the correlation between wetting and interface bonding for a number of metal matrices and different ceramic particles, which led to conclusions of practical importance concerning fabrication of several metal–ceramic particle composites designed for a variety of automotive and electromechanical applications. Rohatgi et al. [53] emphasized that thermodynamic analysis points towards wetting as an essential process for the transfer of particles into the melt.

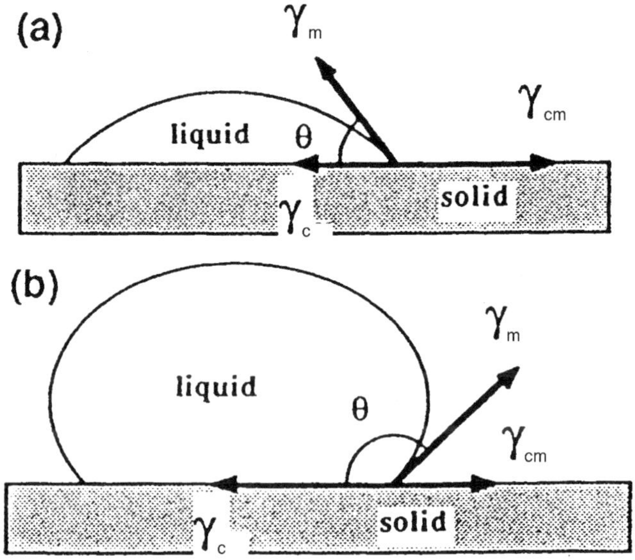

Figure 12 Schematic of the contact between a liquid drop and a solid, where γ_c, γ_m and γ_{cm} define surface energies of ceramic, metal-matrix and ceramic-metal interface, respectively.

Wetting of various types of solids, namely metals, oxides, carbon, and carbides was addressed by Delannay et al. [54], who offered critical evaluation of all the reports published until 1987, which concern preparation of MMCs by liquid metal infiltration method. The work of Delannay et al. [54] provides a kind of bridge between the fundamentals of ceramic and metals coating by liquid metals and the production of metal-matrix composites. An important advantage of this study is consideration of wetting under dynamic condition, in contrast to the commonly used contact angle measurement by the sessile-drop method that constitutes static approach with assumed equilibrium. The work focused in a large part on multifilament reinforcement of MMCs.

Thorough theoretical and experimental study on ceramic–metal wetting allowed Russel et al. [55] to compare the theoretically estimated work of adhesion between alumina and several metals, with the measured values (see Fig. 13). They found that wettability improves in the presence of magnesium, when holding time and temperature increase.

Wettability of silicon nitride-based materials and Al_2O_3–TiC/N fibers immersed in liquid iron nickel alloys was addressed by Yeomans and Page [56], together with the possible ceramic–liquid metal reaction interfaces. The authors demonstrated that Al–O bonds in alumina are resistant to attack from iron and nickel, while the wettability of titanium carbonitride is higher, which makes this interface more reactive.

To the best of our knowledge, the studies of Estathopoulos et al. (see e.g., Ref. [57]) are among the most important contributions to this particular research area. The authors presented numerous examples of different wetting behavior of reactive and non-reactive pure metal–ceramic systems (see e.g., [57]), affected by presence of oxygen as well as alloying elements. The results were discussed in terms of the thermodynamic model by Estathopoulos et al. [57] that gained common acceptance and allowed the authors to draw general conclusions on metal–ceramic wetting which hold as well for other ceramics (refer to discussion in Ref. [55]).

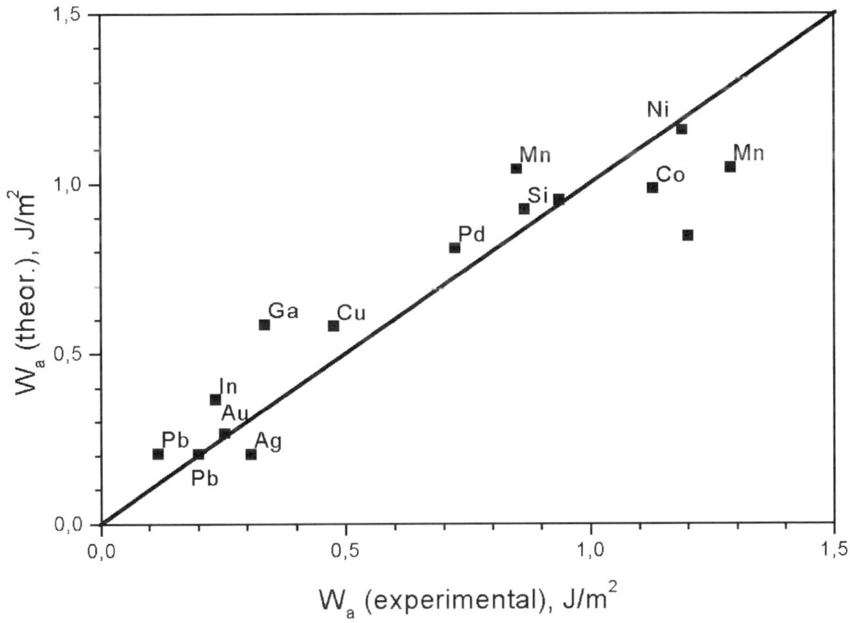

Figure 13 Comparison of theoretically and experimentally determined work of adhesion between alumina and indicated metals. (From Ref. 55.)

The interfacial phenomena, namely the capillary effects that occur during solidification processing of MMCs, were discussed by Mortensen [58], who focused on non-chemical aspects of wetting of reinforcement by liquid metal, including infiltration process. In a due course Nakashima et al. [59] confirmed the effect of oxygen on surface tension between Fe–O alloy and Al_2O_3. The similar research based on sessile drop technique for the measurement of contact angles was carried out for ZrO_2–liquid metal systems by Sotiropoulou and Nikolopoulos [95]. The later proposed the new model for evaluation of adhesion work, based on the assumption that the interactions in the interface are of either of Van der Waals type, or they establish chemical bonding.

Finally, the comprehensive review on the wettability concerning MMCs fabrication was given by Pagounis [51], who emphasized that the interfacial energy largely depends on bonding type in the ceramic phase. He argued that ionic ceramics (most of oxides) possess low wettability due to strong binding of electrons and charge discontinuity, while covalent solids, e.g., carbides such as SiC and B_4C, are prone to be wetted, and in consequence react with metals. The same applies to metallic-like carbides (Cr_3C_2, Mo_2C) or sub-oxide ceramics formed by ion-bombardment, which exhibit high wettability [51].

b. Interface Reactivity. The physical interactions described in the precedent section might not be sufficient in order to produce metal–ceramic interface of high strength, and the chemical reactions are used to achieve a suitably strong binding. Pagounis [51] emphasized, however, that good physical contact between ceramic and metal acts in favor of bonding and prompts chemical reactions, which result in mass transfer through the interface. Indeed, such reactions would result in formation of transition interlayer with different physical properties, similarly to that formed by ion-beam mixing when the multilayer system is bombarded with energetic ions [68,69]. However, the reaction products in

MMCs' interfaces are of greater variety than in a case of ion-mixing, since they include solutions, intermetallics, spinels, amorphous solids, recrystallized zones, to name only a few.

Systematic studies of general rules that govern interfacial reactions, crystallographic relationship, and atomistic structure of interfaces were addressed by Rühle and Evans [15], who emphasized the role of chemistry of interfaces in adequate mechanical integrity MMCs. Their review is followed by a legion of reports concerning particular metal–ceramic combinations. Hence, the reactions that occur at the interface of oxidized SiC particles used as reinforcement in Al–Mg alloys, discussed by Legoux et al. [60], serve as an example of such studies.

The similar category of the research comprises such subject as: the thermodynamic analysis of aluminate formation at the Fe/Al_2O_3 and Cu/Al_2O_3 interfaces by Trumble [62], which required correction of ternary phase relationships to provide basis for thermodynamic considerations; characterization of interfacial reactions between Ti matrix and outermost coating layer using energy dispersive spectroscopy observations [63]; determination of reaction zone thickness in which (Cr, Co) carbosilicates are formed in cobalt-based alloys [64]; and the study of kinetics of the diffusion controlled reaction zone formation at the interface of the Fe–SiC metal-matrix composite systems [66].

For the sake of completeness of the present report, it is worth to mention the extended review of chemical reactions which occur in metal–ceramic interfaces by Howe [61], in which the author placed emphasis on Al and Ti alloys bonded to Al_2O_3 and SiC ceramics. Interesting experiments were carried out on Si-thin film/Ti-substrate, and C-thin film/Ti-substrate systems used as a model of interfaces present in silicon carbide-reinforced MMCs [65]. Consequently, Derby and Qin [67] provided an original concept of closure of interface voids during metal–ceramic diffusion bonding of metal foils in contact with TZP zirconia (see Fig. 14).

Following Pagounis [51], one should notice that the redox reactions (see also Ref. 70) play an important role in MMCs, since they are common for the systems with low wettability, and result in diffusion of oxide ceramic elements into the metal. Pagounis [51] argued that in the case of highly reactive materials, two ceramic components are dissolved in the metal phase where they would form new phases. The author emphasized that reaction layers which are of considerable thickness, i.e., exceeding 1 μm, deteriorate mechanical properties of MMC, while the layers thinner than 0.1 μm might be beneficial, or even in particular cases, essential for strong bonding.

B. Selection of a Proper Combination of the Metal-Matrix Composite Components

The selection of a proper combination of matrix and reinforcement was from the beginning the main goal of design and production of metal-matrix composites. At the beginning of the present section, we indicated the relationship (8) derived by Fischmeister et al. [48,48] as a first rule of reinforcement ever proposed. It is clear, however, that proper design of MMCs requires far more complete and complex approach. Although there were many efforts which resulted in numerous proposals of essential rules concerning selection of reinforcement, frequently based on advanced solid-mechanics considerations (see e.g., Ref. [60]), our present knowledge in this area seems mainly from the works by Ashby [96–99].

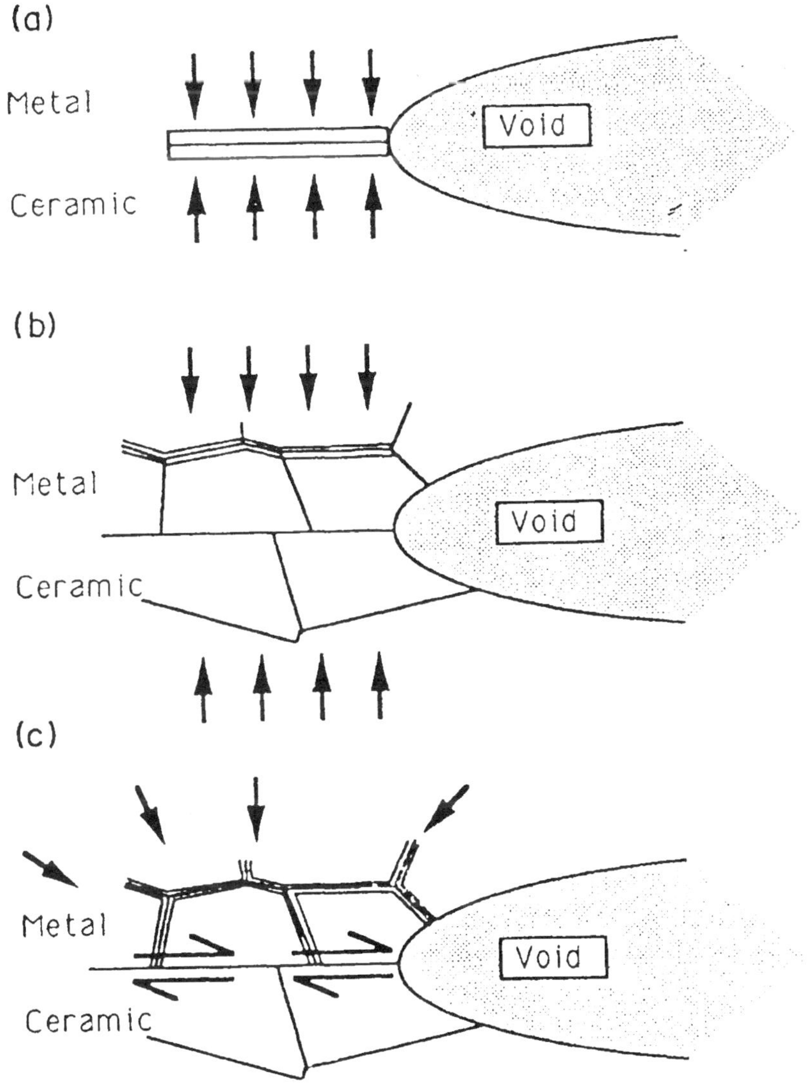

Figure 14 The mechanism of void closure by vacancy transport exclusively in the metal foil (a), when metal ceramic interface acts as sink (b), and when the metal grain boundaries act as sink (c), proposed by Derby and Qin [67].

It is worth noting that despite of foundation of his criteria of reinforcement-matrix selection already in early 1990s [96,97], the concept by Ashby undergoes evolution according to emerging novel results and development of common knowledge concerning the composites.

Hence, in his early overview of the problem Ashby [97] elaborated the details of conceptual tools which guide the selection of composite components in such a way as to obtain its best thermo-mechanical performance. He emphasized the use of so called "merit indices" which describe material performance (most basically—combination of Young's modulus E, thermal expansion coefficient α, density r, fracture stress σ_f, fracture toughness J_{IC}, thermal conductivity λ, thermal diffusivity, etc.), together with construction of material–property charts (see example in Fig. 15), where these indices are visualized and enveloped by "bounds for composite properties". Ashby [97] argued that the method he used is helpful to have an idea of what strength and weakness of a new composite might be and where its application might lie, before starting developing composite. The conceptual design of new MMC systems has been elaborated in detail by Shercliff and Ashby [98]. Recently, Ashby [99] gave a new insight into this subject by focusing on multiobjective optimization methods which can be adapted for MMCs design.

IV. POWDER METALLURGICAL FABRICATION

There are currently three major production routes used to manufacture particle-reinforced metal-matrix composites. These include:

- Molten metal mixing (or stir casting).
- Spray deposition.
- Powder metallurgical techniques.

The selection of one particular process over the others depends on a number of factors, such as:

1. Type of materials and their properties.
2. Final desired properties with respect to part performance.
3. Size, shape, and complexity of the part.
4. Subsequent processing.
5. Availability of equipments and operating experience in the field.
6. Overall cost of processing.

Powder metallurgy (P/M) is an attractive processing route for particulate metal-matrix composites because it offers the possibility to use a wide range of reinforcement volume fraction and size, and it achieves a better distribution into the matrix material. Additional advantages of the P/M process are that a high dislocation density can be introduced into the matrix, the produced subgrain size is small, and a reaction between the matrix and the reinforcement can be minimized by using solid state processing [51]. This technique is most widely used in the case of composites with steel or other high performance materials (tungsten, molybdenum, niobium, and tantalum) as matrix, which are difficult to fabricate by the conventional liquid metallurgy processes, due to the high processing temperatures involved. The basic manufacturing in P/M includes mixing (or blending) of the metal and ceramic powders, degassing, compaction, and consolidation. The initial process of mixing the raw materials is an important first step, since it controls the distribution of particles and porosity in the composites. Compaction is the process where the powder is given its desired shape. The compacted specimens are called *green specimens*. After compaction the powders are consolidated by various methods, which include:

- sintering,
- hot isostatic pressing (HIP),
- extrusion,

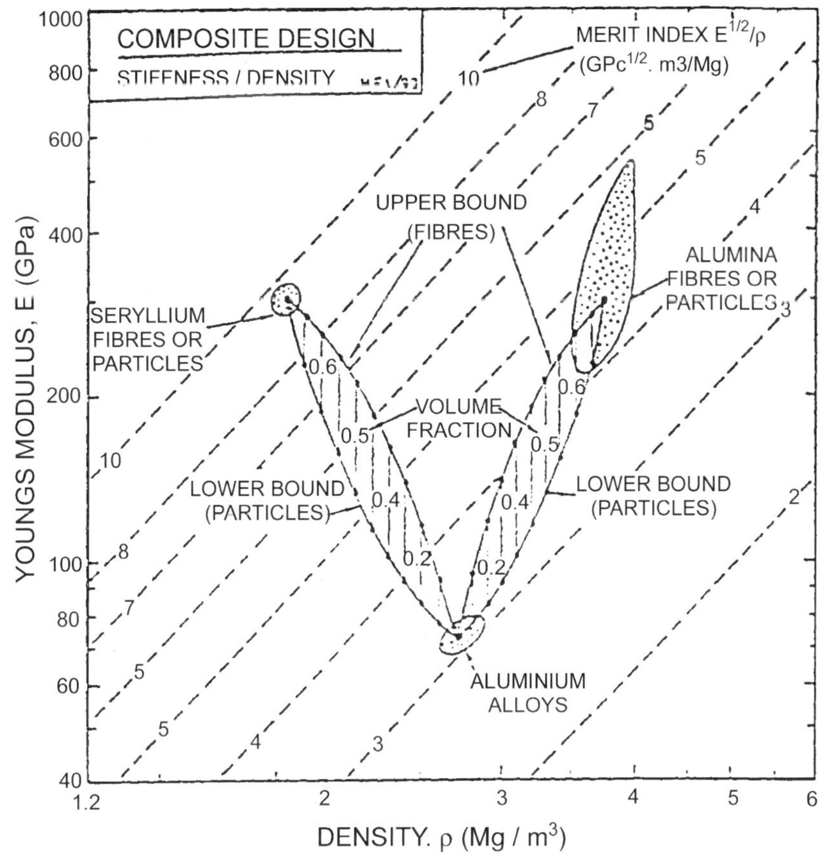

Figure 15 An example of modulus-density space used by Ashby [61] to design new Al-Be and Al-Al_2O_3 composites.

- high energy high rate processing (HEHR), and
- resistance sintering (RS)

Among these, sintering and hot isostatic pressing are the most widely used in composite volume production.

A. Preparation of Powder Mixtures

The first operation in P/M is the mixing of the powders, which means the preparation of a uniform mixture of two or more components. There are three major mixing procedures: dry mixing, wet mixing, and ball milling (blending). Which of them is the best depends on the materials to be mixed, their size, and properties. The most common mixing process in P/M is dry mixing, e.g. in a twin shell or cone mixer. Most types of mixing equipment achieve their highest degree of particle distribution rather quickly when dealing with dry, free flowing materials. Any extension of the mixing cycle usually results in a variation of some acceptable mixing norm. Therefore, it is prudent to determine the minimum processing time needed to achieve the desired result and to terminate processing at the given time.

The two most common problems in mixing of the powders are segregation and agglomeration. Segregation occurs when particles possessing some particular property show a preference for being at some part of the system. There is a general agreement that the primary reason for segregation is the different flow rates caused by powder properties, or velocity gradients created by mixing, handling, transporting, or movement of materials. Some of the variables affecting flow, and therefore segregation are:

- particle shape, size, and size distribution,
- particle density,
- ratio of components, and
- electrostatic phenomena.

As a general rule, the greater the uniformity of particle size of the materials to be mixed, the greater the possibility for consistent and uniform results. The shape of particles also influences the mixing result; it is easier to mix powders with spherical particles than with irregular or flake-like particles. Segregation caused by differences in size or shape can only take place when the particles are free flowing.

When dealing with composite materials the most important reason for segregation is the difference in densities between the components. Low-density ceramic particles tend to "float" and collect at the top of the mass, while high-density metal matrix powder tends to "sink" to the bottom of the mix. The situation is aggravated, in case where the heavy particles are coarse and the light particles are fine. Size and density differences can balance out, on the other hand, if the fine are denser than the coarse components. For mixtures where the small particles tend to coat the larger ones a good mix usually depends on the uniform size of the small particles and a narrow ratio of the two components.

Agglomeration is the building of clusters from the reinforcing powder into the matrix powder. Whereas in one component metallic powder the production of agglomerates of small particles can reduce segregation, in composite materials this phenomenon reduces the mechanical properties. Segregation and agglomeration can be avoided provided:

1. Mixing in a vacuum.
2. Slowing down the motion of the mixer.
3. Introduce the mixture in alternate layers instead of completely separated.
4. Introduce surfactants (wetting agents) to the mix if electrostatic forces built up. This is the case when the shell is covered with small particles.
5. Avoid overmixing.
6. Mixing in steps.
7. Other mixing methods, e.g. wet mixing or ball milling, are used.

1. Wet Mixing

One of the suggestions for eliminating segregation is wet mixing, which includes the formation of a slurry. The slurry can be used to produce a uniform mixture, which is subsequently dried. The sole mixing of the powders in a solvent (water or organic) often proves unsuccessful because of settling of the less dense materials; this can be inhibited by the addition of *suspending agents* which increase the viscosity. *Binders* are added in order to obtain the appropriate viscosity in the slurry and a sufficient green strength. *Plasticizers* soften the binder in the dry or near dry state. They prevent cracking during drying and provide the green bodies with flexibility. In the case of pressing of the powders they cause easy flowing of the powders. *Deflocculating* or dispersing agents prevent agglomeration. Table 4 summarizes a number of binders, plasticizers, suspending and deflocculating agents commonly used in P/M processing.

2. Ball Milling

When very thorough mixing is required, ball milling is applied. This can be done either dry or wet. Dry milling is usually considered with mechanical alloying. Ball milling is connected with comminution but several other processes can take place. The stages during ball milling are microforging, fracture, and agglomeration.

Impact forces are proportional to the mass of the milling medium. Fine grinding is achieved with a small diameter of the balls whereas large balls should be used for large strong particles. Rough balls result in high friction and abrasion, i.e. wear of the particles. Chemical additives can be added in order to prevent cold welding and agglomeration. Surface active agents lower the surface tension of the materials to be milled, resulting in shorter milling times and finer powders. Examples of ball milling and solvents used in composite production are shown in Table 5.

Ball milling for the production of a uniformly dispersed mixture of fine ceramic powder and pure metal powders in the proper proportion required for the composition of the binder alloy has been widely used in the manufacturing of cermets. The same method has also been utilized in many cases [100,101] in the production of iron-based composites. The preparation of powder mixtures is made by ball milling the ceramic phase with iron, carbon, and elemental alloying metals in powder form in the proportions needed for obtaining a specific iron alloy matrix. In many cases, the properties of iron-based composites in which the iron alloy matrix is synthesized from ingredients are superior to composites produced by applying prealloyed powders [101]. This happens because the flow stress of fine elemental ductile metal powders is low compared to the flow stress of prealloyed atomized powders, and because the diffusion distances are shorter in fine powders.

Table 4 Mixing Additives to Slurries

Binders	polyvinyl alcohol, gum arabic, polyvinyl acetate, ethylene oxide polymers, acrylic resin emulsion, borax, sodium silicate
Plasticizers	glycerine, ethylene glycol, ethanolamines
Suspending agents	methyl cellulose, ethylene oxide polymers
Deflocculating agents	monoethylamine, ammonium tartrate

Table 5 Examples of Ball Milling in Steel Matrix Composite Production

Materials	Solvent	Time	Ball material
$316L + Y_2O_3$	Acetone	6–36 hr	Cemented carbide
WC + Fe + Ni + C	Benzene	72 hr	WC
Fe + VC, WC, TiC	—	40–48 hr	Carbide
Steel, TiC, Mo	Alcohol	72 hr	
Fe + MC	N-heptene	1 hr, attritioning	Steel
$434L + Al_2O_3$	Acetone	7–62 hr	Steel (diameter 2 cm)

Several types of mills are available, e.g. tumbler ball mills, vibratory mills, and attrition mills. Attritioning mills are much more efficient than conventional mills, resulting in a much shorter milling time. They are also usually applied in mechanical alloying of metal and ceramic powders, where considerable energy is needed.

Mechanical alloying (MA) is a dry, high energy ball milling process for producing composite metal powders with a controlled, fine microstructure. It is carried out in a highly agitated ball charge by the repeated cold welding and fracturing of a mixture of metal powders to which some non-metal powders can be added. In addition to the strengthening caused by uniformly dispersed fine ceramic particles, the fine grain size and the high dislocation density of the metal matrix, as a result of work hardening of metal powders, also contribute to the strengthening of mechanically alloyed MMCs. The widest use of this technique has been in the production of oxide dispersion-strengthened (ODS) nickel and iron-based superalloys for service at elevated temperatures [102].

Steel powders are mechanically alloyed with hard ceramic powders, by a step by step implantation of the hard ceramic particles into the softer and coarser steel powder. The continuous multiple fracture and re-welding results in the even better distribution of the hard phase in the steel matrix. Thummler and Gutsfeld [10] have reported that hardening effects by cold deformation as well as the non-metallic constituents and the flake type shape of the mechanically alloyed composite powders lead to some decrease of the compactibility. However, annealing of the powder mixture at 600–800°C improves the pressing behavior to a considerable extent and the compactibility is then acceptable for further processing.

Mechanical alloying increases also the sinterability of the composites since it drastically improves the number of metal/metal contacts in the compact on account of the dominating metal/non-metal contacts in the mixed powders. Preferably, the metal/metal contacts are more active in sintering than the metal/ceramic contacts. After consolidation of MA powders by sintering or HIPing, an annealing treatment is suitable to recrystallize the flake type grains and to produce a material with equiaxed microstructure and uniform reinforcement distribution. Koss et al. [91] have demonstrated that Fe/Y_2O_3 mechanically alloyed composite materials exhibit some degree of grain size strengthening. Therefore, mechanical alloying before sintering has been found to increase mechanical properties in steel matrix composites.

B. Compaction

The most common way to obtain the desired shape of a powder mixture is by compaction. Other methods are extrusion, rolling or slip casting. The mixture is pressed in a die at pressures which make the powders adhere to each other to form a green compact with an appropriate density. During compaction different processes take place depending on the pressure. At low pressures, the particles rearrange, then they become deformed (plastically or elastically) and if they are brittle they will fragment at higher pressures. The plastic deformation can result in work hardening, which explains why the theoretical density is hard to achieve by simple compaction. When the material gets harder it becomes more difficult to compress. The density of the produced specimen is dependent on the compaction pressure. The density also influences the green strength of the specimen. The strength is mainly due to mechanical interlocking, which is favored by irregular (i.e. water atomized) particles. Cold welding of the particles is not verified. Binders, which serve as lubricants during compaction, are added to increase the green strength of ceramic specimens, but they may decrease the green strength of metallic products [103].

In the conventional method of compaction, the pressure is usually applied in one direction, resulting in an uneven distribution of consolidation, and sometimes, insufficient densities. This affects the subsequent processing of the green compact such as sintering and secondary treatments. To control better the quality of green compacts, the cold isostatic pressing technique has been developed. Cold isostatic pressing (CIP) is mainly used to press powders under a high pressure. The advantages of CIP over the other compaction methods include the uniformity of the achieved density, regardless of the size and shape of the powders. Furthermore, it controls shrinkage and limits residual stresses resulting from the wall friction in one-dimensional pressing.

C. Powder Densification

Densification of green compacts is achieved when heat is applied to the powder body. The densification process is the most critical step in powder metallurgical production of composites as it affects the metal–ceramic interface bond, the distribution of the reinforcements and, thus, the composite mechanical and physical properties.

1. Sintering

Sintering occurs when heat is applied to a powder compact containing pores. The driving force of sintering is the reduction in the surface area associated with pores. This process is energetically favorable because it decreases the surface energy (and therefore the overall system's energy) by decreasing the surface area. During sintering, the compacted specimen attains its desired strength. Sintering is governed by diffusion, and the atoms must have a certain energy, an activation energy, before transport can take place. This energy is given from the high temperature.

There are two important parameters affecting the sintering process, namely sintering atmosphere, and sintering temperature. Ideally, the sintering atmosphere should contain three elements:

1. Hydrogen for reduction of oxides in the metal powder.
2. Nitrogen inert atmosphere, capable of preventing oxygen and moisture from entering the furnace.
3. Carbon controlling gas such as CO or CH_4 (but not CO_2).

For most scientific use, vacuum or pure hydrogen is applied.

The sintering temperature depends on the material which will be sintered. With composites, the sintering temperature also depends on the reinforcement volume percent. The usual sintering temperature for iron is 1120°C. However, vast research indicates that the use of a higher temperature increases density and mechanical properties of the product. This has been attributed either just to the higher temperature or to liquid phase sintering. Liquid phase formation at lower temperatures during sintering can be achieved by using sintering additives. These are, for iron alloys, phosphorous (P), copper (Cu), and boron (B). Carbon works in the same way, but only if it exceeds 2 wt% [103]. The use of such additives should be carried out carefully because they may limit the temperature at which the component is used after sintering.

In composite materials, problems occurring commonly during sintering are the presence of oxide films, the imperfect distribution of particles, and the sweating during liquid phase sintering or the poor bonding strength in solid phase sintering. The tendency

Table 6 Wetting Angles Between Melts and Hard Compounds [5]

1280°C		1450°C		1550°C	
Fe-P-C/NbC	21°	Fe-P-C/NbC	28°	Fe/NbC	25°
Fe-P-C/TiC	32°	Fe-P-C/TiC	45°	Fe/TiC	39°
Fe-P-C/TiN	33°	—	—	—	—
Fe-P-C/Al_2O_3	92°	—	—	Fe/Al_2O_3	141°

of liquid metals to sweat out during liquid phase sintering is due to the poor wettability of the ceramic particles by the liquid metal. On the other hand, liquid phase sintering can in some cases enhance the poor bonding obtained during solid state sintering.

In iron-based composite materials, the formation of a limited liquid phase during sintering is particularly important because it may improve the interfacial bond between matrix and reinforcement [104]. Elements such as C, P, and B, introduced into the powder mixture in the form of graphite [10], Cu_3P [105], Fe_2P, and CrB [104], have been reported to favor the creation of a liquid phase in lower temperatures during sintering, which increases the density and strength of the composite. These sintering additives may be applied in the case of ceramic particles with high liquid phase formation temperature in contact with iron and/or low solubility in it, like Al_2O_3, NbC, TiC, and TiN (see Table 6). Particularly, the addition of carbon and phosphorous, introduced into the powder by dry mixing, increases the wettability of the hard compounds by the Fe–C–P eutectic melt and lowers the sintering temperature. This behavior is represented in Table 6.

However, although liquid phase sintering produces dense composite bodies, it should be carried out carefully since the presence of cast grain boundary regions may also deteriorate the mechanical properties of the composite, e.g. its hardness and wear resistance. Atomistic scale of sintering phenomenon, which is not meant to be discussed in detailed in this review, is given elsewhere by, e.g., Lindroos [106].

2. Hot Isostatic Pressing

Hot isostatic pressing is a material processing technique in which high isostatic pressure is applied to a powder part or compact at elevated temperatures to produce particle bonding. The HIP process is suitable for the fabrication of metal-matrix composites, because in a commercial P/M production route, the only additional step is the mixing between the metal and the ceramic powder. This process usually results in a fully dense body, although partially dense bodies can also be intentionally produced. During processing, the compact is subjected to equal pressure from every side. On an atomic scale, the isostatic pressure arises from molecules or atoms of gas colliding with the surface of the object. Each gas atom is acting as an individual "hot forge". These tiny atomic forges have the ability to reach all surfaces of the component, including re-entrant angles, and to act reliably and consistently independent of the shape. On average, the numbers of gas atoms moving through unit area and their velocities are the same in all directions. Therefore, the pressure on every surface of the component is the same and acts in a direction normal to the surface.

The aim of the HIPing process is the removal of pores originated from the packing of powder particles, to produce a body with optimum density. Plastic flow, shear and creep processes, and diffusion may have key roles during HIPing [22]. The combination of pressure and temperature can be used to achieve a particular density at a lower temperature than would be required for sintering alone. The effect of the lower temperature is that unacceptable grain growth can be avoided. Furthermore, the methods for enhancing densification of powders by introducing additives such as low melting point constituents (which may have deleterious effects on mechanical properties) are not needed. When pressure is applied in addition to heat, the densification mechanisms are modified. The pressure may, e.g. cause rearrangement of the particles, it may induce plasticity and creep in the compact, and it may augment the effects of surface tension as a driving force for diffusion. Densification mechanisms are then dependent on the pressure, the temperature, the pore size and the location of the pores.

Gas atomized powders are suitable in HIP and their production route is well described. Moreover, problems encountered frequently in conventional sintering such as wettability, porosity, sintering activity or evaporation of components are almost totally eliminated. The advantages of the HIP process in composite production can be easily deduced.

The HIP technology provides a quick, precise, and economic method for the manufacture of near-net-shapes. It significantly reduces energy and material consumption as well as final machining. As an inherent feature, the HIPed materials are characterized by a homogeneous microstructure resulting in improved mechanical properties (e.g. strength, toughness). The HIP technology allows for the development of new alloys (e.g. carbide rich materials), which cannot be produced by conventional means.

Since the HIP process is a gas pressure diffusion bonding process, it is extremely well suited for the fabrication of composite structures. Due to the process conditions, it is possible to combine metals with identical metals, metals with reactive metals, metals with refractory metals and metals with ceramic materials. Uniform bonding is achieved, almost totally independent of the geometry of the contact area.

Although densification of monolithic materials has been studied extensively, in composites little work has been done on the deformation processing [107]. This is quite surprising, considering that HIP is an attractive and widely used production route for composite materials. Experimental studies confirm that the presence of relatively undeformable particles greatly reduces the densification rate. Creep plays a minimal role in composite densification when hard inclusions were introduced in a much softer matrix. Accordingly, yielding and diffusion alloying are assumed to be the dominant mechanisms during the densification of composite mixtures. On the contrary, Suryanarayanan et al. [108] have demonstrated through metallographic observations that power law creep dominates densification in $MoSi_2$–Nb and $MoSi_2$–SiC composites. Finally, Kaysser [109] has shown in a W–Ni composite hard material, that during the final stage of HIPing, densification by power low creep is small compared to pore shrinkage and the subsequent elimination of pores by vacancy transport along the grain boundaries. Similar phenomena have been described by Derby and Qin [67] for diffusion bonding.

An important conclusion involved with the HIPing of composite materials is that densification can be retarded by the presence of contacts between undeformable reinforcing particles which support a significant portion of the applied pressure. In addition, partitioning of deformation between soft matrix and hard inclusion powder will result in the increased deformation of the softer material. Hence, the need for the softer particle to absorb the deformation at hard–soft particle contacts has a dominant effect. The softer particles are not only more heavily deformed within the composite mixture than the hard particles, but they are even more heavily deformed in the composite than in the monolithic powder [110]. Pagounis et al. [110] have studied the influence of the hard reinforcements (inclusions) on densification. Their results indicate that random mixing, high volume fraction, small relative size, and high aspect ratios of the reinforcing phase inhibit densification.

D. Process Equipment and Procedure

Typical hot isostatic pressing units consist of a pressure vessel (usually made from a ductile low alloy steel), gas storage and handling system, furnace, tooling, power supply, controls, and instrumentation. The basic process flowchart for the production of various sized encapsulated HIP parts includes filling of the cans (encapsulation), welding, leak testing, outgassing, and finally the HIPing cycle.

Encapsulation. Encapsulation is the filling of the powders (without using lubricants) in a capsule, which will be submitted in the HIPing procedure. The basic requirements for the capsule are that it should be relatively strong, gas tight, inert and plastic under the applied temperature and pressure conditions, and compatible with the material to be pressed so as to minimize diffusion reactions. Powder encapsulation is used for spun cylindrical, simple cylindrical, or rectangular shapes. However, combinations of metal forming and welding can produce additional shapes. Sheet metals are by far the most widely used capsule material and they are generally of mild steel, stainless steel or nickel alloys. For high temperature service, capsules are made out of molybdenum. In each case, care should be given to avoid any reactions (interdiffusion alloying, carbide formation, etc.) between powders and capsule. Correct capsule design is essential to ensure that the advantages of the HIP technology for near-net-shape production are used to the maximum extent possible. Information about capsule design is given in the literature.

Welding, Leak Testing, Outgassing. After encapsulation, the capsules are closed by gas tungsten arc welding or electron beam welding (for molybdenum capsules). Leak testing is important since HIPing can be achieved successfully only with leak-free containers. Finally, outgassing is performed to remove air, water vapor, and liquid mixing agents from the packed powder blend. Therefore, formation of particle surface oxide and nitride films during HIPing, which reduce workability and mechanical properties is avoided.

HIPing Cycle. Once the capsule is filled, closed and outgassed, it is inserted into the HIP equipment, where pressure and temperature are applied. The main parameters affecting the HIPing procedure and the produced material are pressure, temperature, holding time, and also heating and cooling rate. No general statements on the quantities for these parameters can be given, since they depend on the material to be HIPed. As a general rule, the processing temperature is greater than $0.7T_m$ where T_m is the material's solidus temperature. In Table 7, HIPing conditions for some materials are given. With the time taken for evacuation, pressurization, depressurization, and charge handling, the total cycle time for a production plant is usually between 7 and 12 hr.

Table 7 HIPing Parameters for Selected Materials

Material	HIP temperature (°C)	HIP pressure (MPa)	Yield stress (MPa)
SiC	1850	200	10000
Al_2O_3	1500	100	5000
Al and its alloys	500	100	100–627
Superalloys	1100–1280	100	200–1600
Steels	950–1200	100	500–1980
Ti and its alloys	920	100	180–1320
Al/SiC composites	520	100	412–510
WC/Co hardmetals	1350	100	6000
Ferro-Titanit K 36/5	1330	100	1250

E. Secondary Treatments

After sintering or HIPing of the composites, the samples are heat treated to optimize the properties. Other secondary treatments include forging, extrusion or post-HIPing, which increase the density of the materials.

In iron-based composites, preliminary heat treatment (annealing) is usually performed to receive spheroidized pearlite and, therefore, more machinable composites. The final heat treatment consists of hardening and tempering. Kubarsepp [100] has reported for steel bonded hardmetals that the heating time should be at least twice as long as that used for steels. On the other hand, typical heat treatments for *Ferro-TiC* alloys are done by fast immersion of the samples into isothermal baths at the austenitizing and tempering temperatures [2]. Accordingly, a coarsening of the austenitic grains during slow heating is avoided. In most cases, oil is used as the hardening agent, but if the reinforcement content increases, air, inert gases or vacuum can be used.

A specific feature of these composites is a very slight change in dimensions during hardening. Because of this, frequently no additional grinding or other finishing operations are required [2]. This extremely small size change is caused by the restraining effect of the thermodynamically stable TiC particles, which avoid extensive changes in the lattice structure and size of the composite materials.

Iron-based composites containing up to 50 vol% reinforcements are machinable in the annealed or solutionizing condition by using conventional tool equipment. It is important to notice that in contrast to cemented tungsten carbides, which need diamond wheels for cutting and finishing operations, these composites can be ground both in the annealed and hardened condition with common Al_2O_3 wheels.

V. MECHANICAL AND PHYSICAL PROPERTIES OF METAL-MATRIX COMPOSITE

In design of an MMC system, it is essential to predict mechanical properties with accuracy. This requires a thorough database on matrix and reinforcement properties, as well as the equations describing the correlation between the composite properties and those of the components. To assess the mathematical models in order to select suitable ones is often difficult because of the limited databases available. Moreover, the existing models are found unsatisfactory in predicting certain composite properties.

A. Mechanical Performance of Metal-Matrix Composite

In the following parts, the mechanical properties of MMC will be introduced in brief. Rather than presenting a set of collected figures on each property from different composite systems, the idea here is to highlight the dominant influence parameters that govern the final composite properties.

1. Deformation and Failure of Metal-Matrix Composite

The deformation behavior of an MMC is the result of the interaction between two inherently different classes of materials that are brought together. The matrix material in an MMC usually shows an elastic–plastic behavior when undergoing deformation, while the plasticity of reinforcement is very low. A comprehensive coverage on MMC materials [39] has demonstrated that the mechanics of elastic deformation of MMCs is relatively easy to reveal, as one can treat it through an evaluation of the average stresses in each phase. This evaluation is still valid when dealing with the plastic deformation, since the global behavior of MMCs during plastic deformation is still governed by the state of average stresses, although in the matrix and reinforcement interfacial area, the stress-states can be greatly different to the global ones. However, average stress analysis is obviously not an adequate basis for the treatment of fracture and failure of MMCs, which depend on the local processes that control the initiation and propagation of a crack.

The characteristics of deformation and failure reflect the situation in the mathematical modeling of mechanical properties of MMCs, in which model predictions for composite strength, fracture toughness, and fatigue life are still under development [111].

2. Mechanical Properties of Metal-Matrix Composite

The observed mechanical behavior of a given MMC system is the consequence of the joined effects from all the influence factors such as the properties of the matrix and reinforcement, the reinforcement volume fraction, geometry and geometric arrangement, the characteristics of the interfacial area, and the fabrication method. However, the major contributing factors for each specific property are different. Based on the information from a comprehensive work in developing an expert system for design of metal-matrix composite [111], the major mechanical properties of MMCs are characterized below.

Modulus of Elasticity. This property is one of the least sensitive properties to the composite microstructure, though variations in the elastic moduli among the particulate and whisker-reinforced MMCs are common. This is mainly attributed to the difficulties in obtaining accurate measurements due to the short proportional regime on the engineering stress–strain curves of MMCs. The major controlling factors on modulus of elasticity are matrix and reinforcement modulus, reinforcement aspect ratio, volume fraction, and alignment.

Examples of mathematical model that are used in dealing with the elastic moduli of particulate-reinforced MMCs are Hashin–Shtrikman Bounds [112–115], Paul's Model [114,116] and Nomura–Chou Bounds [39,117]. For aligned continuous fiber-reinforced MMCs along longitudinal direction, the rule of mixtures [Eq. (10)] can be simply applied.

$$E_c = (1 - f)E_m + fE_f \tag{10}$$

In the equation, f is the volume fraction of reinforcement, while E_c, E_m, and E_f are the elastic moduli of the composite, matrix and reinforcement, respectively.

Shear Modulus and Poisson's Ratio. The shear modulus and Poisson's ratio of MMCs are the two other properties along with elasticity modulus that are least sensitive to composite microstructure. The major controlling factors on composite shear modulus and Poisson's ratio are similar to those on modulus of elasticity.

Mathematical models that deal with the shear modulus and Poisson's ratio of particulate-reinforced MMCs include Hashin–Shtrikman Bounds [112–115], and Paul's Model [114,116]. Again, for aligned continuous fiber-reinforced MMCs along longitudinal direction, the rule of mixtures [Eq. (10)] can be applied to predict composite shear modulus and Poisson's ratio.

Strength. Despite numerous models having been developed, there is still a lack of reliable predictive capability in dealing with the strength of MMC materials. Strength depends in a complex manner on the composite microstructure [118]. The effects of residual stresses, non-linear and temperature-dependent work hardening of the matrix, interface strength and load transfer between reinforcement and matrix, statistical parameters associated with the intrinsic strength of the reinforcement, as well as the variations in microstructural parameters need all to be considered.

The majority of the models that predict composite strength can be classified into four types: law of mixtures [119], shear leg [119–123], dislocation [119,124,125] and Eshelby [123,127]. The dislocation type of models can be further classified into Orowan strengthening, grain and substructure strengthening, quench hardening and work hardening [119,126,128]. Since the matrix properties and the nature of the matrix/reinforcement interface have strong influence on the composite strength, models that incorporate both load transfer and matrix strengthening should give better prediction on composite strength [125,129].

The effects of secondary factors on the composite strength can be significant. For example, the primary factors in determining the *yield strength* of MMC are, theoretically, matrix yield strength, reinforcement volume fraction, and alignment. However, the secondary factors including matrix/reinforcement interface and reinforcement shape often have a strong influence that they can render a well-developed model incapable in predicting composite yield strength with accuracy.

Fracture Toughness. Model prediction of fracture toughness of MMCs is far from satisfactory. The knowledge of fracture mechanics (particularly linear elastic fracture mechanics) has been shown to be inappropriate to characterize the fracture resistance of many MMC systems [130,131]. This is due to the great sensitivity of composite fracture toughness to local parameters that can vary considerably, often in a poorly controlled manner.

The existing models in predicting the fracture toughness of MMCs can be classified into three groups: crack path models [132], fractography-based models [133], and energy-based models [134]. In summary [39], the crack path models do not show the correct variation with parameters such as volume fraction of the reinforcement and composite strength and, therefore, it is unsatisfactory. The fractography-based models relay on post-mortem examination of the fracture surface so that they are interpretive rather predictive. While, in developing the energy-based models, attempts to predict all the contributions to the fracture energy are still in the embryonic stages.

Fatigue Resistance. The fatigue resistance of MMC materials cannot be well described also by linear elastic fracture mechanics. Fatigue fracture of MMC materials is governed by the growth of very small microcracks. One of the important parameters concerning fatigue is ΔK, the difference in stress intensity between the maximum and minimum loading. The performance of particulate-reinforced MMCs in resisting fatigue cracking depends largely on the magnitude of ΔK, the matrix fracture toughness, the intrinsic strength of the reinforcing particles and their alignment. Models predicting the fatigue life of particulate-reinforced MMCs with general satisfactory are not available, but explanations on the fatigue behavior of composites under low ΔK and high ΔK loading conditions have been proposed [135–137].

Thermal Conductivity. The thermal conductivity of an MMC is a function of matrix and reinforcement thermal conductivity, reinforcement aspect ratio, orientation, and volume fraction. However, unlike elastic moduli, the sensitivity of this composite property to the reinforcement aspect ratio and volume fraction is less profound since the thermal conductivity values of matrix and reinforcement are usually similar. For the same reason, the inhomogeneous distribution, or clustering of reinforcement, has little affect on the global thermal conductivity of MMC materials [138]. However, thermal resistance at the matrix/reinforcement interface due to presence of reaction layer or porosity can affect the thermal conductivity of MMCs. This is important when designing a microelectronic device using MMC materials.

For particulate-reinforced MMCs, Paul's Model [114,116], Hashin–Shtrikman Bounds [112–115], and Hasselman–Johnson's model [139] are the examples that can be used to predict the thermal conductivity of MMCs.

Coefficient of Thermal Expansion. The coefficient of thermal expansion of MMC materials (CTE) is controlled by the volume fraction, aspect ratio, the CTE value and the alignment of the reinforcement. In addition, the composite CTE can also be influenced by the presence of an interfacial reaction layer. Depending upon the interfacial reaction products, either an increase or decrease in composite CTE can be expected.

Modeling of composite CTE is based on the understanding of the internal stresses arising upon changing the temperature. For aligned continuous fiber-reinforced MMCs along longitudinal direction, the Shapery equation [140] gives good prediction of composite CTE. In predicting the CTE of particulate-reinforced MMCs, the Hashin–Shtrikman Bounds [112–115] should be considered.

In summary, good prediction of MMC properties can be achieved when the property values of the matrix and reinforcement are available, the reinforcement shape, volume fraction, and orientation are well characterized and, in addition, the interfacial reaction layer is negligible. Examples in this case are the properties of aligned continuous fiber-reinforced MMCs along longitudinal direction, for which analytical models can give good prediction. On the other hand, in particulate-reinforced MMCs with the reinforcements being randomly distributed, it is common that the phase geometry is unknown, the volume fraction of the reinforcement is inaccurate and the reinforcement shape is difficult to characterize. In this case, selected bounds are usually employed and a rough estimation of composite properties can only be expected.

3. Mechanical Property Testing in Metal-Matrix Composites

In the development of MMC materials, lack of standard testing methods has long been one of the main reasons that inhibit the commercialization of this material. Progress in understanding of the behavior of MMCs is heavily reliant on the range of available experimental techniques and the correct selection of the most suitable method for specific tasks. In design of an MMC material for a proposed application, there is a requirement for a comprehensive and widely agreed testing method to ensure that the obtained engineering data are reliable and relevant.

The difficulties in standardizing the test procedures for MMCs arise from the deformation nature of composite materials that is highly sensitive to local variations of microstructure. Except the ASTM D3552 that is specially designed for tensile test of fiber-reinforced MMCs [141], all the methods used on conventional materials are adopted on MMCs with special care.

For example, special care must be taken when measuring the elastic modulus of a particulate-reinforced MMC material. It is known that truly reversible stress–strain behavior of a metallic material relies upon the absence of plastic deformation. In MMCs with particulate reinforcement, large point-to-point variations in stress, arising from mismatches in stiffness and thermal expansion coefficient between the components, often give rise to plastic deformation at very low applied load. The consequence is that the straight-line portion on the stress–strain curve may be very short, in two parts or, it may not exist at all [142,143]. Therefore, when attempting to determine the elastic modulus by measuring the slop of the straight-line portion, techniques like prestraining that extends the proportional regime or those that use very low stress and strain amplitude such as ultrasonic wave and resonant vibration should be considered.

Accurate measurement of displacement and strain of MMCs is another example in which special care must be taken. Because of the high strength and low ductility of composite materials, MMCs are prone to the effect of off-axis stresses, stress concentration, and gripping configuration for the specimen. It is demonstrated [143] that two strain gauges mounted on either side of a longitudinal rectangular specimen should be used in order to have accurate measurements on proportional limit of particulate-reinforced MMCs. If only one strain measurement system is used for each specimen, it is often that the stress–strain curve does not have an initial linear portion and shows curvature almost from the origin.

The conventional techniques in hardness test are used for MMC materials, especially for particulate-reinforced MMCs. However, care must be taken since in composite materials hardness is not independent of load. At low loads, the indentation might not be large enough to sample a representative number of reinforcing particles, resulting in lower hardness readings. The influence of surface preparation and residual stresses are also important and as a result large data scatter in composite hardness is observed [143]. In testing the impact strength of MMCs, the standard Charpy type test can be used. Careful instrumentation is required since the impact strength of MMCs can be very low, particularly for those particulate-reinforced MMCs that are tested with notched specimens.

The effect of sample preparation method on the mechanical behavior of MMCs must not be overlooked. For instance, the tensile strength of a particulate-reinforced MMC can be markedly different if the specimens are sectioned using different technique [143]. Specimens prepared by diamond wheel or diamond end mill have clearly higher tensile strength than those prepared with spark machining. Avoiding surface relief is essential in order to see particular microstructural features and this can be especially difficult with MMCs as a result of the large difference in hardness between matrix and reinforcement. Therefore, an automated polishing device that controls the contact pressures, lubricant dispensing rates, and wheel speeds is necessary in order to achieve good flatness of specimen surface.

B. Wear Properties

Wear can be defined as a damage (mechanical and/or chemical) to a solid surface, generally involving progressive loss of material, due to the relative motion between that surface and a contacting substance or substances. The types of wear are adhesive, abrasive, erosive, corrosive, and fatigue. The two first types are the most common but the interest in the present work will concentrate on abrasion wear. The wear of materials by the action of solid particles has been largely studied in terms of two distinct regimes: abrasion and erosion. Abrasive wear, as defined by ASTM, is due to hard particles or hard protuberances that are forced against and move along a solid surface. A distinction is made between two-body abrasion, in which wear is caused by hard protuberances on one surface which can only slide over the other, and three-body abrasion, in which the particles are trapped between two solid surfaces but are free to roll as well as slide. Although laboratory tests are usually designed to simulate two-body abrasion (pin-on-disc test), three-body wear (rubber wheel test) is more common in practice.

The hardness of the particles involved in abrasion has an important influence on the rate of wear. Particles with lower hardness than that of the wearing surface cause much less wear than harder particles. Experimental results have proved [144] that abrasive grits of any shape will cause plastic scratching only if $H_a/H_s > 1.2$, which is the ratio of abrasive hardness H_a to surface hardness H_s. Under this condition, the processes which are possible when a single abrasive tip traverses a surface include ploughing, wedge formation, cutting, microfatigue, and microfracture. Mechanism or mechanisms that will dominate depends on a number of parameters such as the type of wearing material (ductile or brittle), degree of penetration, Young's modulus, surface hardness, etc.

1. Ploughing

Ploughing is the process of displacing material from a groove to the sides. It occurs under light loads and does not result in any real material loss. However, the build up of dislocations through cold working can result in later scratches if additional work occurs; this is caused by *microfatigue*. *Wedge formation* is another form of mild abrasive wear and takes place when the ratio of the shear strength of the contact interface relative to the shear strength of the bulk rises to a level high enough (from 0.5 to 1.0). A wedge can then develop on the front of an abrasive tip, and in that case, the total amount of material displaced from the groove is greater than the material displaced to the sides. The most severe form of abrasion wear for ductile materials is *cutting*. During the cutting process, the abrasive tip removes a chip and the result is a loss of material.

For ductile materials, the mechanisms of ploughing, wedge formation, and cutting have been observed. It is found that wedge formation and cutting lead to wear loss. A simple model has been developed for the abrasive wear of ductile materials, involving removal of material by plastic deformation [145]. The model predicts that the total volume removed per unit sliding distance W is proportional to the applied load σ and inversely proportional to the surface hardness H (like Archard's equation for sliding wear):

$$W = \frac{\sigma}{3H}\tan\omega \tag{11}$$

where ω is the angle between the abrasive and material surface. Brittle materials have an additional mode of abrasive wear, namely *microfracture*. When a hard angular particle is pressed against the surface of a brittle material, local plastic deformation can occur at the point of contact, followed by the formation of cracks, which can lead to the detachment of material and, hence, to wear. For the microfracture mechanism, the dependence of wear volume loss is much more complicated [146]

$$W = \alpha\frac{\sigma^{5/4}D^{1/2}}{A^{1/4}K_c^{3/4}H^{1/2}} \tag{12}$$

where D is the diameter of abrasive grit (mm); A the contact area (mm^2); K_c the fracture toughness ($MN/m^{3/2}$); and α is a constant.

As can be seen, some important differences between this mechanism and that involving plastic deformation [Eq. (11)] occur. This model predicts an inverse correlation between the wear rate and some power of the material's fracture toughness; wear often depends more strongly on toughness than on hardness [147]. It also suggests that the wear rate should increase more rapidly than linearly with the applied load, while it takes into account the influence of the size D of the abrasive particles. Finally, it suggests that wear by fracture will occur only when a critical load on each abrasive particle is exceeded. According to this idea, Hutchings [148] has found a transition in the wear mechanism of brittle materials with an increasing load or abrasive particle size: at low loads or with small abrasive particles, fracture may be suppressed and abrasive wear may occur by plastic processes. For higher loads, or larger particles, brittle fracture occurs, leading to quickly increased wear rates.

The effect of material properties on abrasive wear resistance has received great attention, since a variety of material characteristics have been shown to either form a correlation with abrasive wear or to have some effect on it. These properties include hardness, fracture toughness, elastic modulus, yield strength, microstructure, crystal structure, and composition.

The hardness of the material has been shown to correlate with its abrasion rate because it primarily determines the depth of penetration of abrasive particles. Khruschov [149] has found an inverse relationship between abrasion rate and hardness for pure metals. However, increasing the hardness of wearing materials to reduce wear may also be limited owing to the risk of brittle failure of the components. On the other hand, Zum-Gahr [150] has demonstrated that hardness is not a suitable parameter to describe reliable wear resistance of inhomogeneous or multiphase materials. Fracture toughness is important in determining abrasive wear for ceramics and other brittle materials. Fischer et al. [151] have studied the wear rate of zirconia samples with equal hardness but different toughness and found that the wear decreased with the fourth power of the toughness.

The microstructure of the abraded material is also important. Austenite and bainite of equal hardness are more abrasion resistant than ferrite, pearlite, or martensite. This is because of the higher strain-hardening capacity and ductility of the austenite. Also, crystal structure has been found to influence abrasive wear since cubic metals wear at about twice the rate of hexagonal metals [152]. This is attributed to the lower work hardening rate of the hexagonal metals. Finally, heterogeneous materials with embedded ceramic particles exhibit greater resistance to softer abrasive particles (i.e. softer than the ceramic reinforcements) than homogeneous materials of equal hardness. Homogeneous materials are superior in the case where wear occurs on the high level, i.e. due to harder abrasive particles [153].

The use of relatively large, hard, incoherent particles such as ceramics in a soft metal matrix to produce a composite material can be useful in decreasing abrasive wear. Composites may offer the answer to achieve high hardness and sufficient fracture toughness to avoid brittle fracture. The abrasive wear resistance of these materials depends on different microstructural parameters such as the hardness, shape, size, volume fraction, and distribution of the embedded particles, the properties of the matrix and the interfacial bonding between the two phases. Zum-Gahr [153] has proposed a model, which shows the effects of varying the particle parameters on the abrasion wear resistance. The particle characteristics that work best for wear protection are hard, tough, and blocky. A high hardness value makes them harder to cut. Toughness makes them resistant to breakage, while blocky particles reduce crack propagation and breakage. Different interactions between abrasive particles and embedded phases can occur, depending on their hardness ratio and the size ratio of the wear grooves produced to the second phase particles. Digging out, cutting, cracking, and pulling out of the second phase and blunting or cracking of the abrasively acting particles have to be considered. The most important parameters affecting wear resistance in particulate-reinforced metal-matrix composites will be discussed individually.

2. Effect of Reinforcement Volume Fraction

Most studies indicate that the wear resistance of PRCs, regardless of the processing route, increases with the increasing volume fraction of the reinforcement [154–164]. This is because extensive plastic ploughing and cutting of the unreinforced matrix is reduced by the presence of hard, undeformable ceramic particles. However, the increase in volume fraction of the reinforcing phase can continuously reduce the wear loss provided spalling of this phase is avoided. Otherwise, the wear loss of the composite is increased, or a minimum occurs at about 30–40 vol% with an increasing volume fraction of the reinforcement [153,165].

3. Effect of Reinforcement Particle Size

The literature on the effects of the reinforcement particle size on the abrasion wear resistance of PRCs is limited. On the other hand, various research results disagree about the influence of this critical microstructural factor on the wear resistance of these materials. Lee et al. [154] have demonstrated in an Al/SiC_p composite that, for equal volume fractions, composites with smaller particle size exhibit lower wear resistance. They have attributed this behavior to the ability of the larger particles to remain embedded a longer time in the matrix, until it can no longer support them or until they are broken down into smaller particles. On the contrary, most authors have reported increased wear resistance with a decreased reinforcement particle size [166–169]. They have correlated this behavior with the decrease in the interparticle spacing (or mean free path) λ of the composite when the particle diameter decreases in a constant volume fraction of reinforcement. This presupposes that the reinforcing phase is strongly embedded and mechanically supported by a sufficiently hard matrix.

At a first approximation, both opinions seem plausible. On the other hand, when studying the influence of the reinforcement particle size on the abrasion wear resistance of PRCs, some other parameters (abrasive grit size and hardness, matrix microstructure) affecting wear should also be taken into account.

A very interesting work about the influence of the reinforcement particle size on the abrasion wear resistance of a steel/TiC particulate composite has been carried out by Axen and Zum-Gahr [1]. They have tried to correlate the dependence of wear on the reinforcement particle size with other parameters affecting wear. According to this work, different interactions can occur between abrasive grits and reinforcing particles. A distinction between hard and soft abrasive grits, i.e. harder or softer than the reinforcing particles, and also between small and large grit sizes, was made. Hard abrasive grits can easily dig out small phases, while soft abrasive grits are able to produce large pits and rub the reinforcing particles. Large particles deficiently bonded to the matrix can be pulled out by soft abrasive grits. However, large particles strongly bonded to the matrix can blunt or fracture soft abrasive grits. Their experimental results confirmed that at a given volume fraction, high protection of the matrix is obtained at smaller abrasive grit sizes by small reinforcing particles owing to the smaller mean free paths in the matrix. Furthermore, with an increasing abrasive grit size, the wear intensity was lower for smaller reinforcement particle sizes provided they were supported by a sufficiently hard matrix. Therefore, the matrix microstructure plays also a significant role; at a given volume fraction, the smaller particles were of advantage within a hard matrix (martensite) but the larger particles within a softer matrix (austenite).

From the above, it is clear that the effect of the reinforcement particle size on the abrasion wear resistance of PRCs cannot be studied individually but in close correlation with the abrasive grit size and hardness, and the matrix microstructure and hardness. Hence, the influence of the reinforcement particle size depends on the abrasive system to be studied (particle size and hardness in correlation with abrasive grit size and hardness, and matrix microstructure and hardness) and the results differ in each case.

4. Effect of Penetration Depth

Modeling of the abrasive wear resistance of PRCs must include, in addition to the volume fraction, the mean free path and the reinforcement particle size; the penetration depth h of the abrasive grits, which is closely related to their size D and the hardness of the wearing surface H [167]:

$$h = \left(\frac{2}{\pi K}\right)^{1/2}\left(\frac{\sigma}{H}\right)^{1/4} D \tag{13}$$

where σ is the applied load and K is a constant.

Zum-Gahr [38] has indicated that the depth of subsurface deformation in abrasive wear should depend upon the depth of penetration by the abrasive particles. A better approach for the study of abrasion wear of PRCs, which also includes the ratio grit size-to-reinforcement particle size D/d (which is very important as indicated in the previous paragraph), is the use of the relative penetration depth h_r, i.e., the ratio of the penetration depth to the reinforcement particle diameter d:

$$h_r = \frac{h}{d} = \left(\frac{2}{\pi K}\right)^{1/2}\left(\frac{\sigma}{H}\right)^{1/4}\frac{D}{d} \tag{14}$$

Wang and Rack [167] have shown that the wear resistance increases substantially in case where the relative penetration depth decreases below unity, i.e. $h < d$.

5. Effect of Matrix Microstructure

The matrix microstructure has a significant influence on the abrasion wear resistance of PRCs. The matrix shall be designed so that only limited deformation shall occur during the attack of abrasive grits to avoid spalling of the reinforcing particles. At larger mean free paths in the composite, i.e., large particle size or small reinforcement volume fraction, wear may primarily be determined by mass loss of the matrix; therefore it should be hard enough.

Experiments in aluminum matrix composites have shown that peak aged matrices show better wear resistance [166]. In steel/TiC composites, the matrix hardness influences the wear rate if small particles are used [1]. Axen and Zum-Gahr [1] have found that for relatively large particle sizes (30 μm), the wear resistance against larger abrasive grits (>125 μm) increases substantially with increasing the amount of retained austenite in the steel matrix. The favorable effect of retained austenite is caused by strain induced transformation of austenite to martensite during abrasive wear. This happens because both large grit size and mean free path enhance the plastic deformation of the matrix and, hence, promote strain induced $\gamma \rightarrow \alpha$ transformation. Finally, the hardness of matrix microstructures may also be limited, owing to the risk of brittle failure of components.

It is therefore clear that the abrasive wear performance of a PRC material depends on many factors that are varied under different conditions. This can be seen, for example, when studying the effect of the reinforcement particle size on the wear resistance of a composite. The observation that decrease in reinforcement particle size, as of the interparticle spacing, will increase the wear resistance of a composite may not be proven in practice, if one does not take into account the influence of matrix microstructure and reinforcement particle size-to-abrasive grit size ratio.

In summary, the first requirement for obtaining better abrasion resistance in PRCs is to use a reinforcing phase with higher hardness than the abrasive grit. In addition, the matrix hardness should be as high as possible and the inequality $\lambda < D_g < d$ is satisfied, here λ is the interparticle spacing, D_g and d are the groove size and reinforcement particle size, respectively.

C. Corrosion of Metal-Matrix Composites

Studies on the corrosion behavior of MMCs are sparse comparing to those devoted to the mechanical behavior and processing methods. Metal-matrix composites are typically designed for superior mechanical and physical properties, not for resistance to environmental degradation. Consequently, some MMCs have very low resistance to corrosion and undergo severe degradation in even mild environments. However, in many cases, for example in marine environments, an adequate corrosion resistance is essential for successful application of MMCs. It has been found that severe deficiencies in resisting environmental degradation are undesirable even for MMC components that will be employed in the dry atmosphere in outer space, since the components must retain adequate chemical stability on the assembly line, during storage and transport [170].

The corrosion resistance of MMCs is often overlooked during the stage of design. Corrosion scientists become involved only in the testing and evaluation stage of MMC development. In the study of corrosion behavior of MMCs, the majority of work has been concentrated on marine corrosion of aluminum-based MMCs [171–177], which are the only type that are widely available. Information on the corrosion behavior of less common MMCs can also be found, but the volume of the literature is small when comparing to that related to aluminum-based composites. A good knowledge about the fundamentals in the corrosion of MMCs is essential toward successful design of MMCs for intended applications.

1. Fundamentals in Corrosion of Metal-Matrix Composites

There are no special mechanisms in point of corrosion science that are behind the corrosion behavior of MMCs. Corrosion of an MMC material in aqueous environments involves an electrochemical process consisting of anodic (oxidation) and cathodic (reduction) reactions. The oxidation reaction at the anode results in metal dissolution and can be represented by

$$M \rightarrow M^{n+} + ne^- \tag{15}$$

The electrons of the anodic reaction must be consumed by the cathodic reaction for the corrosion to proceed. In aqueous solutions, oxygen reduction and proton reduction are the predominant cathodic reactions

$$O_2 + 4e^- + 2H_2O \rightarrow 4OH^- \tag{16}$$

$$2H^+ + 2e^- \rightarrow H_2^- \tag{17}$$

In practice, a corroding surface of a metal exposing to an aqueous solution distributes enormous amount of microgalvanic cells. The cathodic sites on the surface at which the reduction reactions occur are determined by many factors. The sites may distribute evenly and/or they may shift over the entire surface from time to time resulting in general corrosion or, the sites may rather concentrate in certain places causing localized corrosion. It is apparent that the reinforcing phase in an MMC will act as cathode to promote dissolution of the matrix alloy which becomes the anode, as the reinforcements are usually more noble than the matrix materials. However, the actual roles of reinforcement in the corrosion of MMC are more complicated. Under special conditions reinforcement can even enhance the corrosion resistance of MMC in against certain types of corrosion attack.

2. The Role of Reinforcement

The role of reinforcement in the corrosion of MMC depends on many factors. One of the important factors is the electrical conductivity, as it determines the way that the reinforcing phase contributes to the corrosion of MMC. An adequate conductivity is necessary for a reinforcement to have a direct contribution to the corrosion of an MMC. In this case, the reinforcing phase is a noble conductor that severs as inert electrodes for proton and oxygen reduction. As a result, the metal matrix that is galvanically coupled to the noble conductor is corroded at an accelerated rate. A good example is the corrosion attack in graphite-reinforced aluminum MMCs in chloride-containing environments [171,178,179]. As a good conductor, the graphite serves as cathodic sites for oxygen or proton reduction causing severe corrosion of the matrix as the anode. Another example is the corrosion of SiC-reinforced aluminum composites, one of the most popular MMCs that is widely available. Although SiC is a semiconductor, its conductivity can be rather high as the impurity level is increased. Consequently, significant galvanic corrosion between aluminum and SiC in aerated chloride-containing environments has been found [171,180].

The indirect contribution of reinforcing phase to the corrosion of MMC is also important. Since it is the result of the interaction between the reinforcement and the matrix, this contribution is often inevitable. The reinforcement can have an indirect effect on the corrosion of MMC in three ways [181]. First, the presence of reinforcement may accelerate the aging behavior with aging precipitates nucleating at interfaces, grain boundaries and matrix [53,182,183]. The precipitates enhance the dissolution of the matrix alloy as they are usually noble and then act as extra sites for cathodic reactions. A good example is the preferential attack at the SiC/matrix interface in SiC-reinforced aluminum composites due to the presence of the intermetallic precipitate $CuAl_2$ in the vicinity of the SiC phase [172,176,177,181]. Clear evidence has suggested that it is the $CuAl_2$ phase rather than the SiC reinforcement that plays a dominant role in the corrosion process. When comparing to SiC, $CuAl_2$ is a good conductor. Second, the formation of reaction products and segregation of alloying elements at or along the reinforcement/matrix interface render the corrosion resistance of MMC lower. It has been reported that the occurrence of Al_4C_3 [173] at the interface of carbon fiber/aluminum and the formation of $MgAl_2O_4$ and Mg_2Al_3 at the Al_2O_3/Mg-added aluminum interface [170,183] has a detrimental effect on the corrosion of MMC. Third, dislocation pile-ups near hard ceramic particle/matrix matrix interfaces due to thermal mismatch and/or during fabrication processes may change the localized corrosion characteristics. It has been found that in an Al_2O_3/Al composite the high dislocation density in composites enhances the solute diffusivity, one of the consequences is the possibility of Mg diffusing to the reinforcement/matrix interface is increased leading to higher growth rate of detrimental intermetallic compounds [183].

The influence of geometry and geometrical arrangement of reinforcement should also be of concern when designing an MMC having good corrosion resistance. For example, the reinforcement type can have an effect on the propagation of corrosion pits. It has been observed [173] that in MMC reinforced with continuous fibers, corrosion is generally channeled along the fibers to form pits deep within the material. The pit depths are similar to those in the respective unreinforced alloys when same material but in the form of short fiber or particulate is used as the reinforcement [172]. In another study [177], higher corrosion resistance is achieved in a SiC/Al 6013 composite when the homogeneity of the distribution of the reinforcement is improved.

3. Corrosion Types in Metal-Matrix Composites

Galvanic corrosion is the primary concern regarding to the corrosion behavior of MMCs. This is particularly true when an active metal such as aluminum or magnesium is galvanically coupled to a noble conductor such as graphite fiber. However, it should be kept in mind that passivity influences the galvanic corrosion behavior of metals and alloys. For example, the position of stainless steels in the galvanic series depends on their state, in which more noble position is assumed by the stainless steels in passive state as compared with the lower position of these materials when in the active condition [184]. This explains why in same cases MMC exhibits a pitting-like corrosion phenomenon while theoretically galvanic driving force, namely the potential difference between reinforcement and matrix, is supposed to have a dominant role.

Pitting does occur in MMCs, especially in those reinforced with particulate reinforcements having limited conductivity [176,177,185]. The presence of reinforcement particles renders the passive film on the composite surface more vulnerable to breakdown in chloride-containing environments, and pits are found preferentially at the reinforcement/matrix interface regions. In one study [185], it is found that in the SiC/2024 Al composites, the breakdown of passive film and initiation of micropitting corrosion can take place even at an open-circuit potential that is more negative than the pitting potential. It states that the SiC reinforcement not only leads to imperfection of the oxide film, but also changes the kinetics of the precipitation of the second phase in the matrix. This change causes a variation in the composite microstructure and, as a result, changes the mechanics of pit initiation [177,185]. It is also found that in MMCs the oxide film can hardly be repassivated once the localized attack is initiated, as numerous microcrevice sites exist in the interface area.

The stress corrosion cracking (SCC) behavior of particulate-reinforced aluminum composites has also been demonstrated [181,186,187]. Though the general trend of MMCs in against SCC shows a higher susceptibility comparing to the equivalent monolithic alloys, some composites do exhibit an improved resistance to SCC. This improvement has been attributed to a reduced crack tip strain rate due to the presence of reinforcing phase [188].

4. Corrosion Protection in Metal-Matrix Composites

Methods used in the corrosion prevention of metals and alloys can also be used in composite materials. Accordingly, three kinds of method are usually concerned in corrosion prevention of MMCs: protective coating, anodization, and decoupling reinforcement from matrix. When selecting protective coating for an MMC system, one should be aware that a proven coating system for an alloy might not be suitable for an MMC based on that alloy. Poor adhesion and wettability between the coating and the reinforcement or differences in the electrochemical properties of the alloy and the MMC can render a good coating system for the alloy ineffective for the MMC [173,189]. The thickness and evenness of the anodic film appear to be the critical factors in anodization of MMCs. Sulfuric acid has been found to be a successful electrolyte in anodizing aluminum-based composites reinforced with SiC particles [173,190]. Decoupling the reinforcement from matrix can be achieved by coating the reinforcement with an electric insulator. The insulator restricts the flow of galvanic current and increases the ohmic (*IR*) drop between the reinforcement and matrix, as shown in Fig. 16. As a result, the corrosion rate (I_{GALV}) is significantly reduced.

5. Design of Metal-Matrix Composite with Corrosion Resistance

Prediction of the corrosion behavior is the first step in design of MMC with corrosion resistance. This can be done by using the mixed potential theory. With the knowledge of the polarization behavior of the matrix and the reinforcing phases, the mixed electrode diagram of the composite can be calculated. However, the corrosion behavior of an actual MMC can be different from the predicted one, and this is more likely when the major constituents (matrix and reinforcement) are the only components considered in the mixed potential theory [170,189]. Figure 17 shows an example of dissimilarity between the theoretical polarization curve and the experimental one.

Microstructural changes and contaminants in the MMC are the main reasons to cause the dissimilarity in the polarization behavior. The contamination sources are, e.g., segregation of alloying elements, formation of interphases and reaction products between the matrix and the reinforcement, and the processing related impurities. Although microstructural changes in the matrix of an MMC can be positive in terms of mechanical strengthening, they impart difficulties in predicting the corrosion behavior of composites. Choosing suitable processing method is often a critical step in limiting the contamination level and controlling the microstructural changes in the matrix. This requires, therefore,

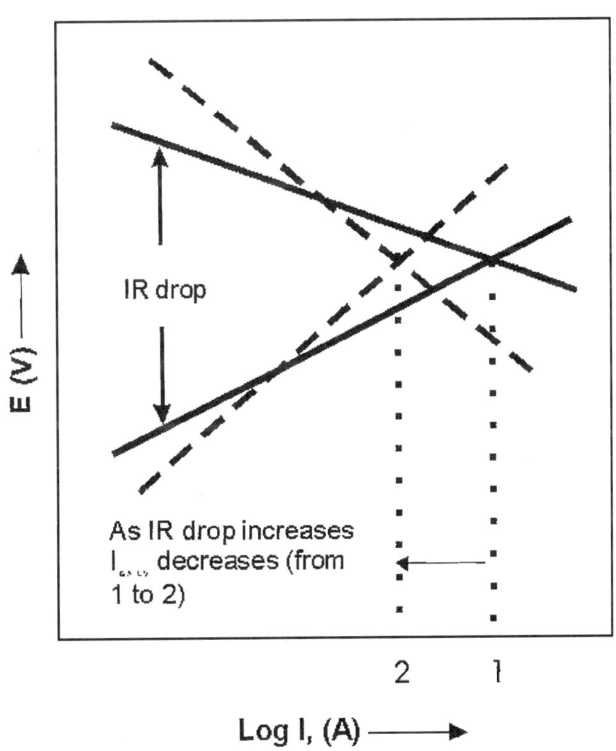

Figure 16 As *IR* drop between cathode and anode increases, the corrosion rate (I_{GALV}) decreases [1].

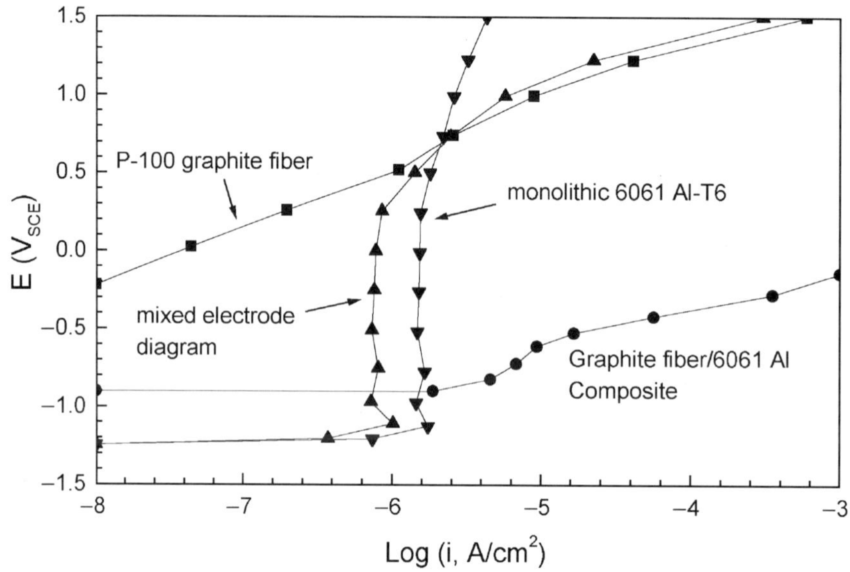

Figure 17 Anodic polarization curves of graphite fiber, monolithic aluminum, and 50 vol. % graphite-fiber/aluminum composite. In comparing the mixed electrode diagram curve with the graphite-fiber/6061 aluminum composite curve, the dissimilarity between the calculated composition curve and the measured one is dramatic. (From Ref. 170.)

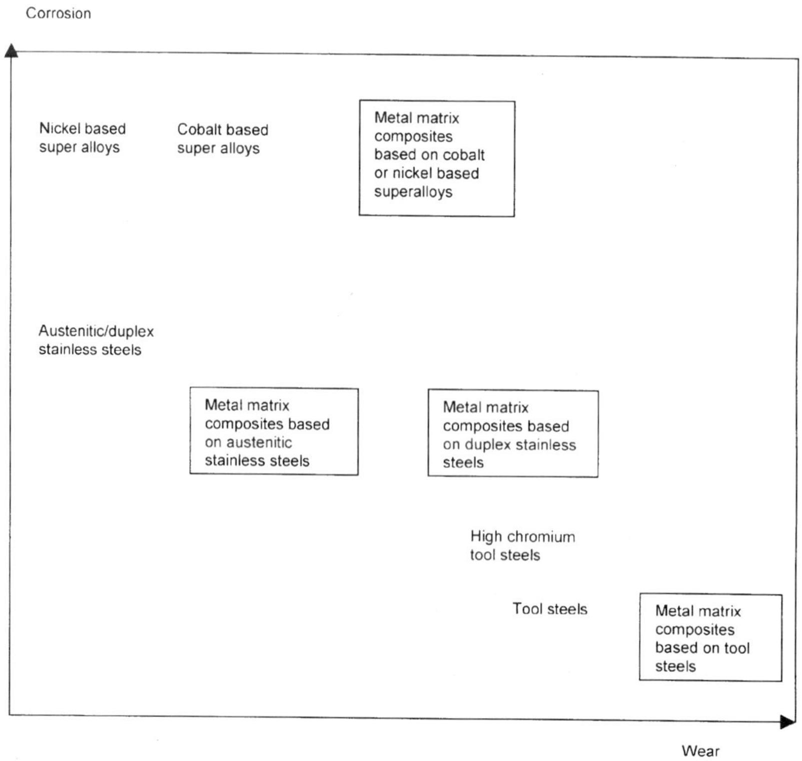

Figure 18 The performance of steel and superalloy based metal matrix composites in combined wear and corrosion protection.

a more interactive approach where corrosion scientists become involved in the design stage of MMC development to produce composites of adequate resistance to environmental degradation as well as superb mechanical and physical properties (Fig. 18).

VI. APPLICATIONS

When the performance profile of metal-matrix composites (MMC) is compared to those of more traditional engineering materials, the advantages of MMCs are apparent; better specific tensile/compressive properties and exquisite specific stiffness, excellent wear properties, advanced creep behavior, superior fatigue resistance, and good thermal and electrical conductivity together with moderate coefficient of thermal expansion (CTE). In addition, both physical and mechanical properties as well as the structure of MMCs can be modified broadly, which renders possible tailor-made performance. This kind of superior performance profile makes MMCs suitable for a wide range of exacting applications.

The concept of MMC has been already utilized for several decades. Materials such as CerMet (TiC-particles in steel matrix) for machine parts (e.g. rolling guides) and superconductors consisting of NbTi in a copper matrix are early commercial applications. Advanced properties such as high specific stiffness and readiness for use in thermal management applications make MMCs suitable materials in aerospace field. During the 1980s, it was already predicted that defense/aerospace applications would cover about 85% of the MMC market in 2000. Environmental concern has however forced car manufacturers to seek for better fuel economy and lighter vehicles. These requirements have increased the usage of MMCs in automobiles and it has been estimated that more than 60% of the MMC material produced in 1999 was used in ground transportation applications. The total MMC world market in 1999 was approximately 2.5 million kg and it has been estimated to grow 14% on the average annually from 1999 to 2004 [191].

In spite of the above-mentioned advanced properties and usage, most MMCs are still niche materials as compared to monolithic materials. The reasons for this situation are many. Provided the investment costs alone are considered, the cost of MMC is typically high. These usually quite large costs include e.g. materials and process qualification and engineering design and they fall before any benefit of the MMC material has been realized [192–194]. New low cost MMC manufacturing processes are under development and achievements like castable 39/SiC/20p aluminum MMC product for close to $2.20 per kg have already been published [195]. The redesign and application of NNS processing methods are key issues in full utilization of the improved material properties of MMCs.

One problem in hampering the wider use of MMCs is the lack of design data. Metal-matrix composite materials can typically be produced with wide range of properties using a variety of processes. This feature gives the possibility for tailor-made properties/structure but at the same time makes the writing of usable material data sheet complicated for design engineer. Another drawback of MMCs, the lack of theoretically predicted properties, is strongly connected with the fact that fundamental means of characterizing particle distributions and qualifying microstructures have yet to be developed [196]. Consequently, demonstration projects are often necessary prior to large-scale commercial exploitation of the materials innovations. Some commercially successful applications of these MMC materials will be contemplated shortly below.

A. Classification of Applications

Metal-matrix composites have been used in several areas such as in ground transportation, electrical/thermal management, commercial and military aerospace applications, industrial and defense applications as well as recreational applications.

1. Ground Transportation Applications

In automotive industry, the main reason to apply new MMC components is weight reduction. This goal prefers aluminum-based MMCs containing coarse filler particulates. The increment of yield strength and ultimate tensile strength of MMC parts at reasonable cost level is typically between 30% and 50% compared to monolithic aluminum. Here, the quality/cost ratio favors more traditional materials and consequently structural applications of MMCs are sparse. However, the situation is different when increased wear or thermal fatigue resistance is required. Metal-matrix composite car parts containing even coarse (>800 μm), low cost particles show 300% increased wear resistance over the counterparts [192,197].

Among the most successful MMC applications in auto manufacturing have been top ring groove reinforced pistons in diesel engines (Toyota), connection rods like stabilizer bars (Honda) and gas-engine cylinder blocks (Honda), crank dampers (Toyota), cylinder liners in high-rpm motorcycle engines and brake components. Low or moderate volume fraction products such as extruded driveshafts (General Motors, Chevrolet, Ford), brake rotors (Plymouth) and studs have also been significant MMC applications [192,197,198].

2. Commercial and Military Aerospace Applications

The aerospace applications introduce the most innovative and demanding area of materials science. In addition to already mentioned properties of MMC materials (high specific strength and stiffness, low CTE coupled with good thermal conductivity, low density), they offer properties such as good fatigue resistance, atomic oxygen resistance, and compatibility. All kinds (Al-, Ti-, Be-, Ni-, Cu- and steel-based) of MMC materials have been utilized in aerostructure applications (aircraft landing gears, ventral fins), in aeroengine applications (actuator pistons, guide vanes, bladed rings, links and frames and shafts) and in spacecraft and rocket propulsion applications [199–202].

3. Defense Applications

Only few military MMC applications outside aerospace region are publicly known. Inertia guidance spheres for Trident missiles are made of high volume fraction particle-reinforced aluminum. Here, Al-based MMC material has replaced beryllium giving cheaper and nontoxic alternative. Other applications where MMC materials have been at least tested are seeker supports for AMRAAM missile, some tank tread components and lightweight bridging applications [197,198,203].

4. Industrial Applications

Metal-matrix composites have several attractive performance features, which can be utilized in development of material solutions for industrial applications. The main benefit from material development point of view is the possibility to design the microstructure and composition of the metal-matrix composite to fit exactly to the intended application. Perhaps the largest potential of metal-matrix composites is found in applications where material degradation occurs due to combined environmental load (corrosion, oxidation) and wear in various modes. The existing material candidates for these applications are either various types of tools steels, cobalt and nickel-based superalloys. However, the properties of these material groups are in most cases inferior as compared to tailored metal-matrix composites. Figure 1 illustrates the position of various material candidates based on qualitative material selection.

The existing metal-matrix composite options for energy, oil-cracking process, metallurgical, mining, and pulp/paper industries are presented in Table 8.

5. Electrical/Thermal Management

Here, the main applications are microprocessor lids and heatspreaders, wireless RF/microwave communications and base plates for other high-powder devices and modules (e.g. matched expansion substrates). Metal-matrix composite materials like PM-DRA, aluminum–beryllium, and beryllium-based MMCs have been utilized for these parts mainly in space applications. The use of MMC materials is based on tailorable physical properties such as thermal expansivity together with electrical and thermal conductivities [194,198,203,204].

6. Recreational Applications

For MMC materials, recreational applications supply a low volume area with moderate importance. It is common in this market sector to apply advanced materials more for their high-tech image than their technical performance. The market's relative insensitivity to materials cost makes this sector a prime target for MMC producer and help them to bring down the material cost in other new applications. Typical applications are golf club faces, horse shoes, tennis racquets, and bicycle components (frames, wheel rims, etc.) made of Al-based MMC, castable DRA. Probably the largest single piece of MMC made to date can be found inside this sector: cross booms for catamaran competing in the America's Cup race [197,198,203].

Table 8 Existing MMC Applications in Industry

Application field		Material degradation forms	Metal matrix composites
Energy	Erosion shields Valves Pumps	Combined corrosion, oxidation and erosion	Stainless steel and nickel based superalloy based metal matrix composites
Oil cracking process	Riser pipes Valves Thermocouple shields	Combined oxidation and erosion	Stainless steel based metal matrix composites
Metallurgy	Hot working rolls Extrusion tools	Combined oxidation, abrasion and thermal fatigue	Tool steel, cobalt or nickel based superalloy metal matrix composites
Mining	Stone crusher wear parts Loader bucket wear parts	Combined gouging abrasion, impacts and mild corrosion	Tool steel based metal matrix composites

B. Industrial Case Studies

Two advanced MMC applications have been selected for particular perusal; reinforced piston for diesel engines and industrial components for process industry.

1. Reinforced Piston for Diesel Engines

The most successful and widespread application of MMCs in automotive industry is selectively reinforced piston in Toyota diesel engines. The MMC piston was developed in 1980 and since 1981 Toyota has applied these FRM pistons for all types of their diesel engines. The total amount of these car parts seems to be more than 10 million.

In the pilot FRM product, top ring groove of the piston was reinforced by alumina–silica short fibers. The piston was produced using infiltration squeeze casting. The advantages of this construction included reduced weight of piston, improved wear properties around the piston ring grove, possibility to use advanced engine design with increased cylinder pressure and the fact that cylinder bores can be designed with a closer fit by exploiting the lower thermal expansion coefficient of MMC. The piston replaced Ni-resist insert piston with no add-on costs. It did not also require any significant changes in production processes and investments for new implementation were small [192,197,205].

Since 1981, various improvements in fiber composition, fiber size and hardness control by crystallization of fibers have been applied to production. At the first stage of improvements, the original reinforcement was replaced with the combination of alumina–silica short fibers and $NiAl_2$ particulates. At the end of last millennium, presintered composite steel replaced Al-based MMC at top ring groove. At the same time, a new improvement, MMC bowl lip made of alumina–boria whisker with surface modification, was introduced.

2. Maintenance-Free Components for Oil Catalytic Cracking Process

Catalyst material used in the catalytic cracking refinery process converting heavy oils to more valuable products causes intense wear of some critical components of the process. Short operation life of such components may dramatically reduce the cost effectiveness of the operation.

One of the most critical components in thermal catalytic cracking plant is the raiser tube. The tube returns catalyst material from the bottom of the column back to the process. The total length of the raiser tube is 78 m. Especially the lower part of a riser tube is subjected to very heavy erosive attack due to turbulence in high velocity (about 40 m/sec) catalyst flow.

The 3.8 m long and 600 mm diameter bottom part of the raiser tube for Fortum oil refinery in Naantali, Finland, was manufactured by Metso Powdermet Oy as a PM–HIP compound structure. The outer shell of the tube is 10CrMo 910 pressure vessel steel cladded internally with 10-mm layer of metal-matrix composite. The tube was HIPed in three parts that were joined together by welding. Due to high operation temperature (660–670°C), the matrix of the MMC is stainless steel. The addition of 20 vol% chromium carbides with optimized particle size and size distribution ensures the good erosion resistance combined to adequate toughness and thermal shock resistance (Fig. 19).

The 6-mm welded overlay used previously also in most critical part of the riser tube was locally penetrated by catalyst flow after quite short operating periods. The reasons for unexpected local failures are connected to inhomogeneous welding structure, high crack density and high internal stresses after welding and dilution with base material. All these detrimental phenomena could be avoided by the use of HIPed compound structure with 10-mm layer of MMC-internal cladding.

Metal-matrix composites are with great success also utilized in other wearing components of oil-cracking process, i.e., bottom plate of riser, erosion shields of thermocouples and valve gates.

Another representative example of utilizing good erosion resistance of MMCs is erosion shields supplied to metallurgical industry for use in high temperature erosion applications (Metso Powdermet Oy, Finland). The material in the wearing side is oxidation resistant metal-matrix composite combined with weldable stainless steel for joining the wear parts in the construction (Fig. 20). The thickness of wearing surface is 120 mm,

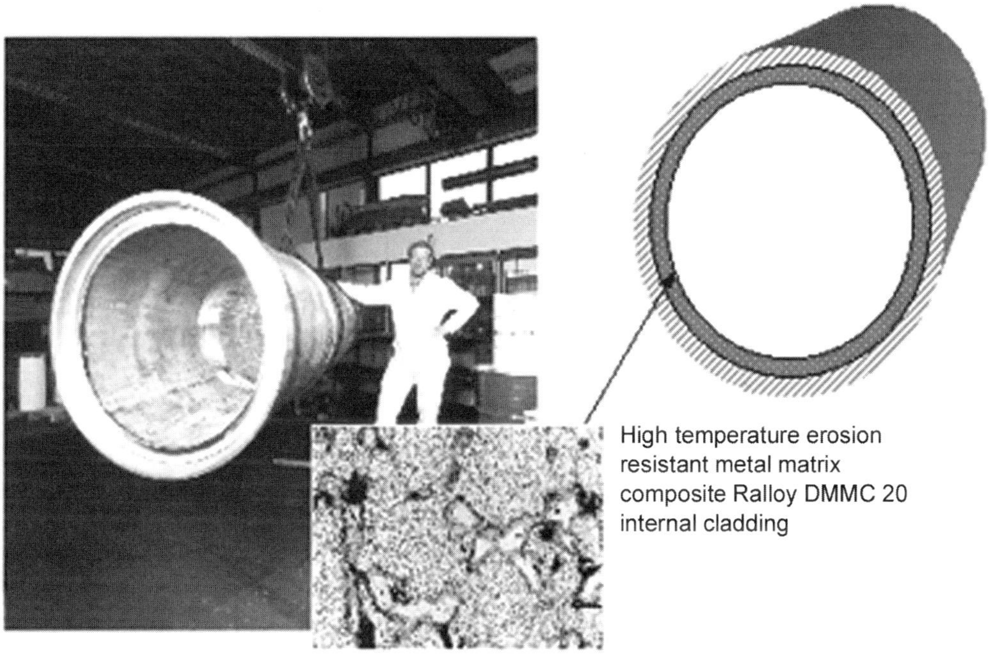

Figure 19 The 3.8 meter long compound structured riser tube for oil catalytic cracking process manufactured by PM-HIP-technology (courtesy of Metso Powdermet Oy).

Figure 20 The weldable construction steel cladded with MMC in the HIPed high temperature erosion shield blocks (courtesy of Metso Powdermet Oy).

which means substantial improvement as compared to welded hardfacing overlays with typical maximum thickness of 5–15 mm or thermally sprayed coatings with maximum thickness less than 1 mm. Similar erosion shields have also been used to protect heat exchange tubes in fluidized bed boilers.

VII. SUMMARY

The sophisticated composite materials offer such property combinations and performance profiles, which are not available in any conventional engineering materials. One might say that they move the objectives and means of traditional alloy design several steps forward and open brand new views in innovative material design.

The advanced material manufacturing technologies, such as the latest methods of powder metallurgy, have had a key role in overcoming the problems related to combining of dissimilar constituents of MMCs—metals and ceramics—which in many cases are inherently incompatible due to remarkable differences in physical and chemical properties. Besides control of interfacial misfit, the ability to gradually variate the relative proportion of reinforcement in the matrix allows the design of functional gradients of physical and thermal properties into components made of MMCs. Additionally, the emergence of powerful computer technology and the development of modeling tools has enabled the simulation of interactions between matrix and reinforcement during both manufacturing and operation, and has thus facilitated the design and optimization of material properties of the MMCs. What is more, the modern P/M manufacturing allows near-net-shape production of the components, which significantly reduces the cost and duration of the production processes.

According to the current market volumes of MMCs, the most important group is the aluminum matrix composites. Magnesium matrix composites offer essentially same kind of properties as Al-based MMCs, i.e. light weight, high stiffness and low coefficient of thermal expansion, and consequently, the selection between Al-based MMCs and Mg-based MMCs is usually made according to weight and/or corrosion resistance. The selection of matrix often turns to Ti alloys or intermetallic compounds, provided the performance of Al alloys, especially at elevated temperatures, is not considered to be sufficient.

Wear and corrosion are the factors number one and number two, which cause the destruction of engineering materials. Among very few materials, steel matrix composites have tolerance for both. In cases where the importance of corrosion resistance or heat resistance are most pronounced in the material optimization, the steel matrix may be replaced with nickel or even cobalt-based superalloy. As illustrated in the introduction of industrial applications above, the MMCs with steel or superalloy matrix have remarkable performance potential for the most critical wear and corrosion resistant components of chemical, petrochemical, process, metallurgical and offshore industry as well as mining and rock engineering, energy conversion and engine technology.

Highest quality and performance are never available at lowest price. In case of the new top performance metal-matrix composites, the investment costs may not be directly competitive, but owing to extended durability and reliability in operation, the total life cycle costs can be significantly lower than those of conventional material selections. Considering the indirect savings due to reduced maintenance and unscheduled down-time in industrial applications, such as tooling, multiple investment costs may be acceptable. Still, the pay-back period of the additional investment in high-performance composites is often surprisingly short in practice. Other aspects to be considered in the calculations of design engineers are how to discount the process development that the new materials enable, how their use is reflected in the quality of final products and, furthermore, what is the total environmental effect of the entire production chain.

ACKNOWLEDGMENTS

The authors would like to express their appreciation for the extensive funding provided to the long-term metal-matrix composite research in the Laboratory by the Academy of Finland and National Technology Agency of Finland. The mutually fruitful co-operation, together with the broad spectrum of the Finnish industry, in particular the long-term co-operation together with Metso Powdermet Oy, as well as the technical assistance of the Laboratory personnel is gratefully acknowledged.

Finally, the authors would like to express their appreciation to the handbook editors Dr. George E. Totten, Dr. Kiyoshi Funatani and Dr. Lin Xie for their co-operation and advice during the course of the handbook project.

REFERENCES

1. Axen, N.; Zum-Gahr, K.-H. Abrasive wear of TiC–steel composite clad layers on tool steel. Wear 1992, *157* (1), 189–201.
2. Panchal, J.M.; Vela, T.; Robisch, T. Ferro-TiC metal matrix composites for high performance tooling and engineering applications. In *Fabrication of Particulates Reinforced Metal Composites*, Montreal, Quebec, Canada, Sept 17–29, 1990; ASM International: Materials Park, OH 44073, USA.
3. Sadagopan, D. A combinatorial optimization approach to composite materials tailoring. Trans. ASME 1997, *119*, 494–503.
4. Sadagopan, D.; Pitchumani, R. Property-based optimal design of composite materials and their internal architectures. J. Composite Mat. 1998, *32* (19), 1714–1752.
5. Huda, D.; Baradie, M.A.E.; Hashmi, M.S.J. Metal-matrix composites. Part II. Materials aspects. J. Mat. Proc. Tech. 1993, *37* (1–4), 529–541.
6. Dogan, O.N.; Hawk, J.A.; Tylczak, J.H.; Wilson, R.D.; Govier, R.D. Wear of titanium carbide reinforcement metal matrix composites. Wear 1999, *225–229*, 758–769.
7. Ibrahim, I.A.; Mohamed, F.A.; Lavernia, E.J. Particulate reinforced metal matrix composite—review. J. Mater. Sci. 1991, *26* (5), 1137–1156.
8. Eliasson, J.; Sandstrom, R. Materials selection and grade optimization applied to aluminium matrix composites. J. Mat. Eng. Performance 1995, *4* (3), 358–367.
9. Stevenson, A.N.J. Development of the dry sand/rubber wheel abrasion test. Wear 1996, *195*, 232–240.
10. Thummler, F.; Gutsfeld, C.H. Sintered steel with high content of hard phase: a new class of wear resistant materials. Powder Met. Int. 1991, *23* (5), 285–290.
11. Lal, S.; Upadhyaya, G.S. Effect of phosphorous and silicon addition on the sintered properties of 316L austenitic stainless steel and its composites containing 4 vol% yttria. J. Mater. Sci. 1989, *24* (9), 3069–3075.
12. Vardavoulias, M. Dry sliding wear mechanism for P/M austenitic stainless steel and their composites containing Al_2O_3 and Y_2O_3 particles. Tribology Int. 1996, *29* (6), 499–506.
13. McKimpson, M.G.; Scott, T.E. Processing and properties of metal matrix composites containing discontinuous reinforcement. Mat. Sci. Eng. A 1989, *A107* (1–2), 93–10.
14. Lloyd, D.J.; Lagace, H.; McLeod, A.; Morris, P.L. Microstructure aspects of aluminium-silicon carbide particulate composites produced by a casting method. Mat. Sci. Eng. A 1989, *107* (1–2), 73–80.

15. Rühle, M.; Evans, A.G. Structure and chemistry of metal/ceramic interfaces. Mat. Sci. Eng. A 1989, *107* (1–2), 187–197.
16. Begg, A.R. Metal matrix composites by powder metallurgy. Powder Metall. 1993, *36*, 107–110.
17. Mukherjee, S.K.; Upadhyaya, G.S. Mechanical behaviour of sintered ferritic stainless steel–Al_2O_3 particulate composites containing ternary additions. Mat. Sci. Eng. 1985, *75* (1–2), 67–78.
18. Pantankar, S.N.; Tan, M.J. Role of reinforcement in sintering of SiC/316L stainless steel composite. Powder Metall. 2000, *43* (4), 350–352.
19. Liu, X.; Pagounis, E.; Hellman, J.; Lindroos, V.K. The influence of reinforcement particle size distribution on the mechanical behaviour of a stainless steel/TiN composition. Metall. Trans. A 2000, *31A* (Jan.), 309–318.
20. Stasko, W.; Pinnow, K.E.; Eisen, W.B. *Development of Ultra-High Vanadium Wear Resistant Cold Work Tool Steels*. In *Advances in Powder Metallurgy and Particulate Materials*, Washington, DC, USA; Metal Powder Industries Federation, 1996, 179–188.
21. Champane, B.; Anders, R.; Fiset, M. Properties of WC–Co/steel composites. Int. J. Refract. Hard Met. 1987, *6* (3), 155–160.
22. Pagounis, E.; Talvitie, M.; Lindroos, V.K. Influence of reinforcement volume fraction and size on the microstructure and abrasion wear resistance of hot isostatic pressed white iron matrix composites. Met. Mat. Trans. A 1996, *27* (12), 4171–4181.
23. Chawla, K.K. Metal matrix composites. In *Materials Science and Technology*, Chan, R.W.; Hasen, P.; Kramer, E.J., Eds.; VCH: Berlin, 1993; 143.
24. Pagounis, E.; Talvitie, M.; Lindroos, V.K. Effect of thermal expansion coefficients on the martensitic transformation in a steel matrix composite. Scripta Mater. 1996, *34* (3), 4070–4113.
25. Bolton, J.D.; Gant, A.J. Heat treatment response of sintered M3/2 high speed steel composites containing additions of manganese sulphide, niobium carbide, and titanium carbide. Powder Metall. 1996, *39* (1), 27–35.
26. Zapata, W.C.; Costa, C.E. Wear and thermal behaviour of M2 high-speed steel reinforced with NbC composite. J. Mater. Sci. 1998, *33*, 3219–3225.
27. Oliveira, M.; Bolton, J.D. Effect of ceramic particles on the mechanical properties of M3/2 high speed steel. Int. J. Powder Metall. 1996, *32* (1), 37–49.
28. Jouanny-Tresy, C.; Vardavoulias, M.; Jeandin, M. Microstructural evolution during liquid-phase sintering of high-speed steel-based composites containing TiN-coating Al_2O_3, TiC, or Al_2O_3 particles. J. Mater. Sci. 1993, *28* (22), 6147–6154.
29. Zapata, W.C.; Costa, C.E.; Torralba, J.M. Sinterability and wear behaviour of P/M M2 high speed steel reinforced with NbC composite. J. Mat. Proc. Tech. 1995, *53* (1–2), 483–490.
30. Berns, H.; Franco, S.D. Effect of coarse hard particles on high-temperature sliding abrasion of new metal matrix composites. Wear 1997, *203–204*, 608–614.
31. Tan, M.J.; Zhang, X. Powder metal matrix composites: selection and processing. Mat. Sci. Eng. A 1998, *A244* (1), 80–85.
32. Pagounis, E.; Lindroos, V.K. Processing and properties of particulate reinforced steel matrix composites. Mat. Sci. Eng. A 1998, *246*, 221–234.
33. Pagounis, E.; Lindroos, V.K. Development and performance of new hard and wear resistant engineering materials. J. Mat. Eng. Performance 1997, *6*, 749–756.
34. Skolianos, S.; Kattamis, T.Z.; Chen, M. Cast microstructure and tribological properties of particulate TiC-reinforced Ni-based or stainless steel matrix composites. Mat. Sci. Eng. A 1994, *183* (1–2), 195–204.
35. Sun, Z.M.; Li, J.B.; Wang, Z.G.; Li, W.J. Residual stresses in silicon carbide particulate reinforced aluminium composites. Acta Metall. Mater. 1992, *40* (11), 2961–2966.
36. Torralba, M.J.; Cambronero, L.E.G.; Ruiz-Prieto, J.M.; das Neves, M.M. Sinterability study of PMM2 and T15 high speed steel reinforced with tungsten and titanium carbides. Powder Metall. 1993, *36* (1), 55–66.

37. Wronski, A.S.; Martin, B.; Vincent, L.; Wright, C.S.; Eagles, A.M. Fretting wear and cracking in sintered metal matrix composites. Wear 2001, *248* (1–2), 65–74.
38. Zum-Gahr, K.-H. How microstructure affects abrasive wear resistance. Metal Progress 1979, (Sept.), 46–52.
39. Clyne, T.; Withers, P. *An Introduction to Metal Matrix Composites*; Cambridge University Press: Cambridge, UK, 1993; 13–270.
40. Ruhle, M.; Heuer, A.; Evans, A.G.; Ashby, M.F. Bonding, structure and mechanical properties of metal/ceramic interfaces. Acta Metall. Mater. 1992, *40*, vii–viii.
41. Krieder, K.G. *Interfaces in Metal Matrix Composites*; Academic Press: New York, 1974.
42. Vinson, J.R.; Taya, N. *Recent Advances in Composites in the United States and Japan*; ASTM STP 864: Philadelphia, 1985.
43. Everett, R.K.; Arsenault, R. *Metal Matrix Composites: Processing and Interfaces*; Academic Press: Boston, MA, 1991.
44. Alam, M.K.; Jain, S.C. JOM 1990, *42*, 56–58.
45. Takagi, T. Ion beam modification of solids, towards intelligent materials in modification of ceramics and semiconductors by ion bombardment. Mater Sci. Eng.; 1998; *253*, 30–41.
46. Grant, N.J.; Siegel, H.J.; Hall, R.W. NASA Report (SP-143), 1967.
47. Fischmeister, H.F.; Navara, E.; Easterling, K.E. Effects of alloying on structural stability and cohesion between phases in oxide/metal composites. Metal Sci. J. 1972, *6*, 211–215.
48. Easterling, K.E.; Fischmeister, H.F.; Navara, E. The particle-to-matrix bond in dispersion-hardened austenitic and ferritic iron alloys. Powder Metall. 1973, *16*, 128–145.
49. Brown, L.M.; Stobbs, W.M. Word-hardening of Cu-silica. Pt 1. Model based on internal stresses with no plastic relaxation. Phil. Mag. 1971, *23*, 1185–1195.
50. Hondros, E.D. Interfacial energies and composition in solids. In *Precipitation Processes in Solids*; AIME: Warrendale, 1978; 1–30.
51. Pagounis, E. Production, microstructure, and mechanical properties of iron-based composites. Thesis; Laboratory of Physical Metallurgy and Materials Science, Helsinki University of Technology: Helsinki, 1996; 53.
52. Hancock, J.W.; Dillamore, I.L.; Smallman , R.E. Effect of matrix stacking fault energy on the creep of recrystallized Ni-Co-Al_2O_3 alloys. *Second International Conference on the Strength of Metals and Alloys*; ASM: Metals Park, 1970; Vol. 3, 1118–1123.
53. Rohatgi, P.K.; Asthana, R.; Das, S. Solidification, structures, and properties of cast metal-ceramic particle composites. Int. Mat. Rev. 1986, *31* (3), 115–139.
54. Delannay, F.; Froyen, L.; Deruyttere, A. The wetting of solids by molten metals and its relation to the preparation of metal-matrix composites. J. Mater. Sci. 1987, *22*, 1–16.
55. Russel, K.C.; Oh, S.Y.; Figueredo, A. Theoretical and experimental studies of ceramic:metal wetting. MRS Bulletin 1991, (Apr.), 46–52.
56. Yeomans, J.A.; Page, T.F. Studies of ceramic-liquid metal reaction interfaces. J. Mater. Sci. 1990, *25*, 2312–2320.
57. Estathopoulos, N.; Chatain, D.; Coudurier, L. Wetting and interfacial chemistry in liquid metal–ceramic systems. Mat. Sci. Eng. A 1991, *135*, 83–88.
58. Mortensen, A. Interfacial phenomena in the solidification processing of metal matrix composites. Mat. Sci. Eng. A 1991, *A135*, 1–11.
59. Nakashima, K.; Takihira, K.; Mori, K.; Shinozaki, N. Wettability of Al_2O_3 substrate by liquid iron—effects of oxygen in liquid Iron and purity of Al_2O_3 substrate. Mat. Trans. JIM 1992, *33* (10), 918–926.
60. Legoux, J.G.; L'Esperance, G.; Salvo, L.; Suéry, M. Influence of particle oxidation on the interfacial structure of SiC reinforced Al-1% Mg alloys. *Fabrication of Particulate Reinforced Metal Composites*; ASM International: Montreal, Canada, 1990.
61. Howe, J.M. Bonding structure and properties of metal/ceramic interfaces: Part 1 Chemical bonding, chemical reaction, and interface structure. Int. Mat. Rev. 1993, *38* (5), 233–256.

62. Trumble, K.P. Thermodynamic analysis of aluminate formation at Fe/Al_2O_3 and Cu/Al_2O_3 interfaces. Acta Metall. Mater. 1992, *40*, S105–S110.
63. Guo, S.Q.; Kagawa, Y.; Saito, H.; Masuda, C. Microstructural characterization of interface in SiC fiber-reinforced Ti–15V–3Cr–3Al–3Sn matrix composite. Mat. Sci. Eng. A 1988, *246*, 25–35.
64. Schneibel, J.H.; Sabol, S.M.; Joslin, D.L. On the high-temperature reactions between advanced ceramics and a cobalt-base alloy. Mat. Sci. Eng. A 1998, *246*, 124–132.
65. Vahlas, C.; Hall, I.W.; Haurie, I. Investigation of interfacial reactivity in composite materials. Mat. Sci. Eng. A 1999, *259*, 269–278.
66. Pelleg, J. Reactions in the matrix and interface of the Fe–SiC metal matrix composite system. Mat. Sci. Eng. A 1999, *A269*, 225–241.
67. Derby, B.; Qin, C.-D. Metal–ceramic interfaces: sources and sinks for mass transfer. Acta Metall. Mater. 1992, *40*, S53–S58.
68. Bolse, W. Ion-beam induced atomic transport through bi-layer interfaces of low- and medium-z metals and their nitrides. Mat. Sci. Eng. R 1994, *12*, 53–120.
69. Nowak, R.; Yoshida, F.; Morgiel, J.; Major, B. Post deposition relaxation if internal stress in sputter-grown thin films caused by ion bombardment. Brit. J. App. Phys. 1999, *85*, 841.
70. Ustundag, E.; Subramanian, R.; Vaia, R.; Dieckmann, R.; Sass, S.L. In situ formation of metal–ceramic microstructures, including metal–ceramic composites, using reduction reactions. Acta Metall. Mater. 1993, *41* (7), 2153–2161.
71. Takao, Y.; Taya, M. Thermal expansion coefficients and thermal stresses in an aligned short fiber composite with application to short carbon fiber/aluminium Trans. ASME 1985, *52*, 806–810.
72. Vogelsang, M.; Arsenault, R.J.; Fischer, R.M. An in situ HVEM study of dislocation generation at Al/SiC interfaces in metal matrix composites. Metall. Trans. A 1986, *17A*, 379–389.
73. Arsenault, R.J.; Shi, N. Dislocation generation due to differences between the coefficients of thermal expansion. Mat. Sci. Eng. 1986, *81*, 175–187.
74. Taya, M.; Mori, T. Dislocations punched-out around a short fiber metal matrix composite subjected to uniform temperature change. Acta Met. 1987, *35* (1), 155–162.
75. Ledbetter, H.M.; Austin, M.W. Internal strain (stress) in an SiC–Al particle-reinforced composite. Mat. Sci. Eng. A 1987, *89*, 53–61.
76. Kim, C.T.; Lee, J.K.; Plichta, M.R. Plastic relaxation of thermoelastic stress in aluminum/ceramic composites. Metall. Trans. A 1990, *21A*, 673–682.
77. Shi, N.; Arsenault, R.J. Analytical evaluation of the thermal residual stresses in SiC/Al composites. JSME Int. J. Series I 1991, *34* (2), 143–155.
78. Dunand, D.C.; Mortensen, A. On the plastic relaxation of thermal stresses in reinforced metals. Acta Metall. Mater. 1991, *39* (2), 127–139.
79. Pickard, S.M.; Schmauder, S.; Zahl, D.B.; Evans, A.G. Effect of misfit strain and reverse loading on the flow strength of particulate-reinforced Al matrix composites. Acta Metall. Mater. 1992, *40* (11), 3113–3119.
80. Shibata, S.; Taya, M.; Mori, T.; Mura, T. Dislocation punching from spherical inclusions in a metal matrix composite. Acta Metall. Mater. 1992, *40* (11), 3141–3148.
81. Smith, L.F.; Krawitz, A.D.; Clarke, P.; Saimoto, S.; Shi, N.; Arsenault, R.J. Residual stresses in discontinuous metal matrix composites. Mat. Sci. Eng. A 1992, *19*, L13–L15.
82. Arsenault, R.J.; Zhou, X. Effect of thermal history on the strengthening of SiC/Al composites. Scripta Met. et Mater. 1993, *28*, 875–878.
83. Lewis, C.A.; Stobbs, W.M.; Withers, P.J. Internal stress induced debonding in a zirconia-reinforced 6061 aluminium alloy composite. Mat. Sci. Eng. A 1993, *171*, 1–11.
84. Suéry, M.; Teodosiu, C.; Menezes, L.F. Thermal residual stresses in particle-reinforced viscoplastic metal matrix composites. Mat. Sci. Eng. A 1993, *167*, 97–105.
85. Bullough, R.; Davis, L.C. The residual deformation fields in particle reinforced metal-matrix composites. Acta Metall. Mater. 1995, *43*, 2737–2742.

86. Kumar, S.; Singh, R.N. Three-dimensional finite element modeling of residual thermal stresses in graphite/aluminum composites. Acta Metall. Mater. 1995, *43*, 2417–2428.
87. Kolhe, R.; Hui, C.Y.; Ustundag, E.; Sass, S.L. Residual thermal stresses and calculation of the critical material particle size for interfacial crack extension in metal–ceramic matrix composites. Acta Metall. Mater. 1996, *44* (1), 279–287.
88. Pyzalla, A.; Berns, H. Theoretical and experimental methods for the determination of thermal residual microstresses in particle reinforced metal matrix composites. Mat-Wiss U Werkstofftech 1997, *28*, 180–197.
89. Backhaus-Ricoult, M. Solid state reactions between silicon carbide and (Fe, Ni, Cr)-alloys: reaction paths, kinetics and morphology. Acta Metall. Mater. 1992, *40* (Suppl.), S95–S103.
90. Soni, K.K.; Chabala, J.M.; Mogilevsky, R.; Levi-Setti, R.; Tseng, M.W.; Williams, D.B. Ion microprobe investigation of interfacial phenomena in composites. JOM 1993, (Mar.), 29–33.
91. Koss, D.A.; Hellmann, J.R.; Kallas, M.N. Fiber pushout and interfacial shear in metal-matrix composites. JOM 1993, (Mar.), 34–37.
92. Duffy, D.M.; Harding, J.H.; Stoneham, A.M. Atomistic modelling of the metal/oxide interface with image interactions. Acta Metall. Mater. 1992, *40*, S11-S16.
93. Stoneham, A.M.; Harding, J.H. Computer simulation of interfaces: what do we need to know? Acta Metall. Mater. 1998, *46*, 2255–2261.
94. Takagi, T. Role of ions in ion-based film formation. Thin Solid Films 1982, *92*, 1–34.
95. Sotiropoulou, D.; Nikolopoulos, P. Work of adhesion in ZrO_2-liquid metal systems. J. Mater. Sci. 1993, *28*, 356–360.
96. Ashby, M.F. On the engineering properties of materials. Acta Met. 1989, *37*, 1273–1293.
97. Ashby, M.F. Criteria for selecting the components of composites. Acta Metall. Mater. 1993, *41* (5), 1313–1335.
98. Shercliff, H.R.; Ashby, M.F. Design with metal matrix composites. Mater. Sci. Tech. 1994, *10*, 443–451.
99. Ashby, M.F. Multi-objective optimization in material design and selection. Acta Mater. 2000, *48*, 359–369.
100. Kubarsepp, J. Steel-bonded hardmetals. Thesis, Tallin Technical University: Tallin, Estonia, 1992.
101. Gutmanas, E.Y. Design of iron-based wear resistant materials by advanced powder metallurgy International Conference of Powder Metallurgy; Institute of Metals: London, 1990.
102. Arzt, E.; Schultz, L. Eds.; *New Materials by Mechanical alloying*, Deutsche Gemeinschaft für Metallkunde: Oberusel, 1989.
103. Petersen, N. Wear resistance MMC. Thesis, Technical University Denmark: Kopenhagen, Denmark, 1990.
104. Petersen, N. *Properties of Liquid Phase Sintered Stainless Steel—Particulate Composites, Metal Matrix Composites—Processing, Microstructure and Properties*; Riso National Laboratory: Denmark, 1991; 581–586.
105. Vardavoulias, M.; Jouanny-Tresy, C.; Jeandin, M. Sliding-wear behaviour of ceramic particle-reinforced high-speed steel obtained by powder metallurgy. Wear 1993, *165* (2), 141–149.
106. Lindroos, V.K. Annihilation of vacancies by small angle boundaries during sintering. Metall. Trans. A 1971, *2*, 3231–3233.
107. Pagounis, E.; Talvitie, M.; Lindroos, V.K. Influence of metal/ceramic interface on the microstructure and mechanical properties of HIPed iron-based composites. Composite Sci. Techol. 1996, *56*, 1329–1337.
108. Suryanarayanan, R.; Sastry, S.M.L.; Jerina, K.L. Consolidation of molybdenum disilicide based materials by hot isostatic pressing (HIP) comparison with models. Acta Metallurgica Materialia 1994, *42*, 3741–3750.
109. Kaysser, W. *Hot Isostatic Pressing: Theory and Applications*; ASM International: Materials Park, OH, 1990; 3741, 1–13.

110. Pagounis, E.; Talvitie, M.; Lindroos, V.K. Consolidation behavior of a particle reinforced metal matrix composite during HIPing. Mat. Res. Bull. 1996, *31* (10), 1277–1285.
111. Legzdins, C.; Samarasekera, I.; Meech, J. MMCX—an expert system for metal matrix composite selection and design. Can. Met. Quarterly 1997, *36* (3), 177–201.
112. Hashin, Z.; Shtrikman, S. A variational approach to the effective magnetic permeability of multiphase materials. J. Appl. Phys. 1962, *33*, 3125–3131.
113. Hashin, Z.; Shtrikman, S. A variational approach in the theory of the elastic behavior of multiphase materials. J. Mech. Phys. Solids 1963, *11*, 127–140.
114. Christensen, R.M. A critical evaluation for a class of micromechanics models. J. Mech. Phys. Solids 1990, *38* (3), 379–404.
115. Hale, D. Review: the physical properties of composite materials. J. Mater. Sci. 1976, *11*, 2105–2141.
116. Paul, B. Prediction of elastic constants for multiphase materials. Trans. Metall. Soc. 1960, *218* (1), 36–41.
117. Nomura, S.; Chou, T. Bounds of effective thermal conductivity of short-fiber composites. J. Composite Mat. 1980, *14* ((4), 120–129.
118. Mortensen, A. Metal matrix composites: an overview In *Encyclopedia of Advanced Materials 1*, Bloor, D.; Flemings, R.; Mahajan, S., Eds.; Pergamon Press: Cambridge, 1993; 1497–1502.
119. Taya, M. Strengthening mechanisms of metal matrix composites. Mat. Trans. JIM 1991, *32* (1), 1–19.
120. Rangaraj, S.; Bhaduri, S. A modified rule of mixtures for prediction of tensile strengths of unidirectional fiber-reinforced composite materials. J. Mater. Sci. 1994, *29*, 2795–2800.
121. Nardone, V.; Prewo, K. On the strength of discontinuous silicon carbide reinforced aluminum composites. Scripta Met. et Mater. 1986, *20*, 43–48.
122. Clyne, T. A simple development of the shear lag theory appropriate for composites with a relatively small modulus mismatch. Mat. Sci. Eng. A 1989, *122*, 183–192.
123. Taya, M.; Arsenault, R. A comparison between a shear lag type model and an Eshelby type model in predicting the mechanical properties of a short fiber composite. Scripta Met. et Mater. 1987, *21*, 349–354.
124. Kelly, P.M. The quantitative relationship between microstructure and properties in two-phase alloys. Int. Mat. Rev. 1973, *18*, 31–36.
125. Ramakrishnan, N. Analytical study on strengthening of particulate reinforced metal matrix composites. Acta Mater. 1996, *44* ((1), 69–77.
126. Arsenault, R.; Wang, L.; Feng, C. Strengthening of composites due to microstructural changes in the matrix. Acta Metall. Mater. 1991, *39* (1), 47–57.
127. Eshelby, J.D. The determination of the elastic field of an ellipsoidal inclusion, and related problems. Proc. Roy. Soc. A 1957, *241*, 376–396.
128. Wu, Y.; Lavernia, E. Strengthening behavior of particulate reinforced MMCs. Scripta Met. et Mater. 1992, *27*, 173–178.
129. Miller, W.; Humphreys, F. Strengthening mechanisms in particulate metal matrix composites. Scripta Met. et Mater. 1991, *25*, 33–38.
130. Rack, H.; Ratnaparkhi, P. Damage tolerance in discontinuously reinforced metal-matrix composites. JOM 1988, *(Nov.)*, 55–57.
131. Friend, C. Toughness in metal matrix composites. Mater. Sci. Tech. 1989, *5* (1), 1–7.
132. Hahn, G.; Rosenfield, A. Metallurgical factors affecting fracture toughness of Al alloys. Metall. Trans. A 1989, *37*, 653–670.
133. Crowe, C.; Gray, R.; Hasson, D. Microstructure controlled fracture toughness of SiC/Al metal matrix composites. *Fifth International Conference on Composite Materials (ICCM V)*; TMS-AIME: San Diego, CA, 1985.
134. Davidson, D. Fracture characteristics of Al-400 Mg mechanically alloyed with SiC. Metall. Trans. A 1987, *18A*, 2115–2128.
135. Padkin, A.; Boreton, M.; Plumbridge, W. Fatigue crack growth in two-phase alloys. Mater. Sci. Tech. 1987, *3*, 217–223.

136. Williams, D.; Fine, M. Quantitative determination of fatigue microcrack growth in SiC_w reinforced 2124 Al alloy composites. *Fifth International Conference On Composite Materials (ICCM V)*; TMS-AIME: San Diego, CA, 1985.
137. Christman, T.; Suresh, S. Effects of SiC reinforcement and aging treatment in fatigue crack growth in an Al–SiC composite. Mat. Sci. Eng. A 1988, *102*, 211–216.
138. Aradhya, K.; Surappa, M. Estimation of mechanical properties of 6061 Al–SiC_p composites using finite element method. Scripta Met. et Mater. 1991, *25*, 817–822.
139. Hasselman, D.; Johnson, L. Effective thermal conductivity of composites with interfacial thermal barrier resistance. J. Composite Mat. 1987, *21* (6), 508–515.
140. Shapery, R. Thermal expansion coefficient of composite materials based on energy principles. J. Composite Mat. 1968, *2* (3), 380–404.
141. ASTM D3352. Test method for tensile properties of fiber-reinforced metal matrix composites, 1977/1982.
142. Nieh, T.; Chellman, D. Modulus measurements in discontinuous reinforced Al composites. Scripta Met. 1984, *18*, 925–928.
143. Roebuck, B.; Gorley, T.; McCartney, L. Mechanical property test procedures for metal matrix composites. Mat. Sci. Eng. 1989, *5*, 105–117.
144. Chaudhri, M.M.; Hutchings, I.M.; Makin, P.L. Plastic compression of spheres. Phil. Mag. A 1984, *49*, 493–503.
145. Rabinowicz, E. *Friction and Wear of Materials*; Wiley: London, 1966.
146. Khuri-Yakub, B.T., Evans, A.G., Kino, G.S. Acoustic surface wave measurements of surface cracks in ceramics. In *The Science of Ceramic Machining and Surface Finishing II*, National Bureau of Standards, Washington, D.C., 1979, 379–393.
147. Hornbogen, E. The role of fracture toughness in the wear of metals. Wear 1975, *33*, 251–259.
148. Hutchings, I.M. Mechanisms of wear in powder technology: a review. Powder Techn. 1993, *76*, 3–13.
149. Khruschov, M.M. Principles of abrasive wear. Wear 1974, *28*, 69–88.
150. Zum-Gahr, K.-H. Abrasive Wear. Fifth International Congress on Tribology (Eurotrib '89); Eurotrib: Helsinki, 1989.
151. Fischer, T.E.; Anderson, M.P.; Jahanmir, S. Influence of fracture toughness on wear resistance of yttria-doped zirconium oxide. J. Am. Ceram. Soc. 1989, *72*, 252-257.
152. Alison, P.J.; Wilman, H. The different behavior of hexagonal and cubic metals in their friction, wear and work hardening during abrasion. Brit. J. App. Phys. 1964, *15*, 281–289.
153. Zum-Gahr, K.-H. *Microstructure and Wear of Materials*; Tribology Series; Elsevier Publishers: Amsterdam, 1987; Vol. 10.
154. Lee, H.L.; Lu, W.H.; Chan, S.L.-I. Abrasive wear of powder metallurgy aluminum alloy 6061—SiC particle composites. Wear 1992, *159* (2), 223–231.
155. Lee, H.L.; Lu, W.H.; Chan, S.L. Abrasive wear of sintered aluminium alloy and its composite containing SiC particles. Chi. J. Mat. Sci. 1992, *24* (1), 40–52.
156. Erich, D.L. Metal-matrix composites: problems, applications and potential in P/M industry. Int. J. Powder Met. 1987, *23*, 45–54.
157. Zamzam, M.A. Abrasive wear of aluminium-matrix composites. Metall 1991, *45*, 250–254.
158. Skolianos, S.; Kattamis, T.Z. Tribological properties of SiC_p—reinforced Al–4.5Cu–1.5Mg alloy composites. Mat. Sci. Eng. A 1993, *163* (1), 107–113.
159. Prasad, S.V.; Rohatgi, P.K. Tribological properties of Al alloy particle composites. JOM 1987, *39* (11), 22–26.
160. Lin, S.J.; Liu, K.S. Effect of aging on abrasion rate in an Al–Zn–Mg–SiC composite. Wear 1988, *121*, 1–14.
161. Hoskins, F.M.; Portillo, F.F.; Wunderlin, R.; Mehrabian, R. Composites of aluminium alloys: fabrications and wear behavior. J. Mater. Sci. 1982, *17*, 477–498.
162. Sato, A.; Mehrabian, R. Aluminum matrix composites—fabrication and properties. Metall. Trans. B 1976, *7B*, 443–441.

163. Bhansali, K.J.; Mehrabian, R. Abrasive wear of aluminium-matrix composites. JOM 1982, *34* (9), 30–34.
164. Alpas, A.T.; Embury, J.D. Sliding and abrasive wear behavior of an aluminium (2014)-SiC particle reinforced composite. Scripta Met. 1990, *24*, 931–935.
165. Zum-Gahr, K.-H. Abrasive wear of two phase metallic materials with a coarse microstructure In *Wear of Materials '85*, Ludema, K.C., Ed.; ASME: New York, 1985; 45.
166. Zongyi, M.; Jing, B.; Yuxiong, L.; Hongwei, S.; Yinxuan, G. Abrasive wear of discontinuous SiC reinforced aluminium alloy composites. Wear 1991, *148*, 287–293.
167. Wang, A.; Rack, H.J. Abrasive wear of silicon carbide particulate- and whisker-reinforced 7091 aluminium matrix composites. Wear 1991, *146*, 337–348.
168. Larsen-Basse, J.; Tanouye, P.A. *Abrasion of WC–Co Alloys by Loose Hard Abrasives*, Komanduri, R., Ed.; Hard Material Tool Technology: Pittsburgh, 1976; 188.
169. Larsen-Basse, J. Wear of hard-metals in rock drilling: a survey of the literature. Powder Metall. 1973, *16*, 1–32.
170. Hihara, L.H.; Latanision, R.M. Corrosion of metal matrix composite. Int. Mat. Rev. 1994, *39* (6), 245–246.
171. Hihara, L.H.; Latanision, R.M. Galvanic corrosion of aluminium-matrix composites. Corrosion 1992, *48* (7), 546–552.
172. Coleman, S.L.; Scott, V.D.; McEnaney, B. Corrosion behaviour of aluminium-based metal matrix composites. J. Mater. Sci. 1994, *29*, 2826–2834.
173. Chen, C.; Mansfeld, F. Corrosion protection on an Al 6092/SiC_p metal matrix composite. Corr. Sci. 1997, *39* ((6), 1075–1082.
174. Cooper, K.P.; Slebodnick, P.L. Microstructural inhomogeinities and sea water corrosion in laser-deposited Ti–6Al–4V alloy matrix/carbide particulate composite surface. J. Mater. Sci. 1998, *33*, 3805–3816.
175. Bache, M.R.; Evans, W.J.; Shakesheff, A.L.; Shield, J.; Uygur, I. Influence of aqueous salt environment on fatigue response of SiC_p reinforced aluminium alloys. Mater. Sci. Tech. 2000, *16*, 825–829.
176. Yue, T.M.; Wu, Y.X.; Man, H.C. On the role $CuAl_2$ precipitates on pitting corrosion of aluminium 2009/SiC_w metal matrix composites. J. Mater. Sci. Lett. 2000, *19*, 1003–1006.
177. Ahmad, Z.; Paulette, P.T.; Aleem, B.J.A. Mechanism of localized corrosion of aluminium–silicon carbide composites in a chloride containing environment. J. Mater. Sci. 2000, *35*, 2573–2579.
178. Evans, J.M.; Braddick, D.M. Corrosion behavior of fiber-reinforced Al composites. Corr. Sci. 1971, *11*, 611–614.
179. Saxena, M.; Modi, O.P.; Yegenswaran, A.H.; Rohatgi, P.K. Corrosion characteristics of cast aluminum Alloy—3 wt.% graphite particulate composites in different environments. Corr. Sci. 1987, *27*, 249–256.
180. Aylor, D.M.; Moran, P.J. Effect of reinforcement on the pitting behavior of aluminium-base metal matrix composites. J. Electrochem. Soc. 1985, *132*, 1277–1281.
181. Singh, P.M.; Lewandowski, J.J. Stress corrosion cracking of discontinuously reinforced aluminium (DRA) alloy 2014 during slow strain rate testing. J. Mater. Sci. Lett. 1996, *15*, 490-493.
182. Suresh, S.; Christman, T.; Sugimura, T. Accelerated aging in cast Al alloy-SiC particulate composites. Scripta Met. 1989, *23*, 1599–1602.
183. Dutta, I.; Allen, S.M.; Hafley, J.L. Effect of reinforcement on the aging response of cast 6061 Al–Al_2O_3 particulate composites. Metall. Trans. A 1991, *22A*, 2553–2563.
184. Fontana, M.G.; Greene, N.D. *Eight Form of Corrosion Engineering*; McGraw-Hill Book Company: New York, 1987; 41–45.
185. Feng, Z.; Lin, C. Pitting behaviour of SiC_p/2024 Al metal matrix composites. J. Mater. Sci. 1998, *33*, 5637–5642.
186. Trazskoma, P.P. Pit morphology of aluminium alloy and silicon carbide/aluminium alloy metal matrix composites. Corrosion 1990, *46* (5), 402–409.

187. Mansfeld, F.; Liu, S.; Kim, S.; Shih, H. Corrosion protection of Al alloy and Al-based metal matrix composites by chemical passivation. Corrosion 1989, *45*, 615–630.
188. Jones, R.H. Stress corrosion cracking of metal matrix composites: modeling and experiment. In *Environmental Effect on the Advanced Materials*, Jones, R.H.; Ricker, R.E., Eds.; TMS: Warrendale, 1991; 283–296.
189. Hihara, L.H.; Latanision, R.M. Localized corrosion induced in graphite/aluminium metal-matrix composites by residual microstructural chloride. Corrosion 1991, *47*, 335–341.
190. Shahid, M. Mechanism of film growth during anodizing of Al-alloy-8090/SiC metal matrix composite in sulphuric acid electrolyte. J. Mater. Sci. 1997, *32*, 3775–3781.
191. Rittner, M.N. Expanding world markets for MMCs. JOM 2000, (Nov.), 43–45.
192. Kevorkijan, V.M. Aluminium composites for automotive applications: a global perspective. JOM 1999, (Nov.), 54–58.
193. Kunze, J.M.; Bampton, C.C. Challenges to developing and producing MMCs for space applications. JOM 2001, (April), 22–25.
194. Zweben, C. Metal matrix composites. Ad Mat. Proc. 1994, (1), 28–30.
195. Herling, D.R.; Grant, G.J.; Hunt, J.W., Jr Low-cost aluminium metal matrix composites. Ad Mat Proc 2001, (July), 37–40.
196. Hunt W., Jr.; Maruyama, B. The World still won't beat a path to your door: transitioning DRA to the marketplace. JOM 1999, (Nov.), 62–64.
197. Feest, E.A. Exploitation of the metal matrix composites concept. Met. Mat. 1988, (May), 273–278.
198. Maruyama, B.; Hunt W.H., Jr. Discontinuously reinforced aluminium: current status and future direction. JOM 1999, (Nov.), 59–61.
199. Shelley, J.S.; LeClaire, R.; Nichols, J. Metal-matrix composites for liquid rocket engines. JOM 2001, (April), 18–21.
200. Hooker, J.A.; Doorbar, P.J. Metal matrix composites for aeroengines. Mater. Sci. Tech. 2000, *16* (July–August), 725–731.
201. Doychak, J. Metal- and intermetallic-matrix composites for aerospace propulsion and power systems. JOM 1992, (June), 46–51.
202. Rawal, S. Metal-matrix composites for space applications. JOM 2001, (April), 14–17.
203. Begg, A.R. Application for metal matrix composites. MPR 1991, (Oct.), 42–45.
204. Parsonage, T. Beryllium metal matrix composites for aerospace and commercial applications. Mater. Sci. Tech. 2000, *16* (July–August), 732–738.
205. Funatani, K. Personal communication, 2001.

Index